Foundations of Technical Mathematics
Canadian Edition

Paul Calter
Vermont Technical College

Carol Felsinger Rogers
Vermont Technical College

Canadian Edition prepared by
John Giguere
Sault College of Applied Arts and Technology

Prentice Hall Canada Career & Technology
Scarborough, Ontario

Canadian Cataloguing in Publication Data

Calter, Paul
    Foundations of technical mathematics

Canadian ed.
Includes index.
ISBN 0-13-999210-3

1. Engineering mathematics. I. Rogers, Carol Felsinger. II. Giguere, John F. 1943- .
III. Title.

TA330.C34 2000   510   C99-930528-X

© 2000 Prentice-Hall Canada Inc., Scarborough, Ontario
Pearson Education

ALL RIGHTS RESERVED

No part of this book may be reproduced in any form without permission in writing from the publisher.

Allyn and Bacon, Inc., Needham Heights, MA
Prentice-Hall, Inc., Upper Saddle River, New Jersey
Prentice-Hall International (UK) Limited, London
Prentice-Hall of Australia, Pty. Limited, Sydney
Prentice-Hall Hispanoamericana, S.A., Mexico City
Prentice-Hall of India Private Limited, New Delhi
Prentice-Hall of Japan, Inc., Tokyo
Simon & Schuster Southeast Asia Private Limited, Singapore
Editora Prentice-Hall do Brasil, Ltda., Rio de Janeiro

ISBN 0-13-999210-3

Vice President, Editorial Director: Laura Pearson
Acquisitions Editor: David Stover
Signing Representative: Jaycee Jovic
Developmental Editor: Marta Tomins
Production Editor: Mary Ann McCutcheon
Copy Editor: Catharine Haggert
Production Coordinator: Wendy Moran
Cover Design: Sarah Battersby
Cover Image: Kevin Ghiglione
Page Layout: Bi-Comp, Inc.

Original English Language edition *Introductory Algebra and Trigonometry with Applications,* published by Prentice-Hall, Inc.
Upper Saddle River, New Jersey, 07458
Copyright © 1997

13 14 15 DPC 08 07 06

Printed and bound in Canada.

Visit the Prentice Hall Canada web site! Send us your comments, browse our catalogues, and more at **www.phcanada.com**. Or reach us through e-mail at **phabinfo_pubcanada@prenhall.com**.

Every reasonable effort has been made to obtain permissions for all articles and data used in this edition. If errors or omissions have occurred, they will be corrected in future editions provided written notification has been received by the publisher.

Contents

Preface vi

1 Arithmetic of Whole Numbers 1

- 1-1 The Real Numbers 1
- 1-2 Adding and Subtracting Whole Numbers 7
- 1-3 Multiplying Whole Numbers 12
- 1-4 Dividing Whole Numbers 15
- 1-5 Powers and Roots of Whole Numbers 17
- 1-6 Scientific Notation 20
- 1-7 Combined Operations 22
- Review Exercises 25

2 Common Fractions 27

- 2-1 Terminology 27
- 2-2 Changing the Form of a Fraction 31
- 2-3 Multiplying Fractions 34
- 2-4 Dividing Fractions 37
- 2-5 Adding and Subtracting Fractions 41
- 2-6 Complex Fractions 48
- Review Exercises 50

3 Arithmetic with Decimal Numbers 54

- 3-1 Decimal Numbers 54
- 3-2 Adding and Subtracting Decimal Numbers 58
- 3-3 Multiplying Decimal Numbers 60
- 3-4 Dividing Decimal Numbers 62
- 3-5 Powers and Roots of Decimal Numbers 64
- 3-6 Scientific Notation 67
- 3-7 Combined Operations 72
- Review Exercises 74

4 Units of Measurement 77

- 4-1 The SI (Metric) System of Units 77
- 4-2 British/American Customary System of Weights and Measure 81
- 4-3 Converting Between British/American and Metric Units 84
- 4-4 Arithmetic with Numbers Having Units of Measure 85
- 4-5 Converting Areas and Volumes 88
- 4-6 Converting Rates 90
- Review Exercises 92

5 Percentage 94

- 5-1 Converting to and from Percent 94
- 5-2 Solving Percentage Problems 97
- 5-3 Percent Change 101
- 5-4 Percent Error, Percent Concentration, and Percent Efficiency 104
- Review Exercises 106

6 Arithmetic with Signed Numbers 108

- 6-1 Signed Numbers 108
- 6-2 Adding and Subtracting Signed Numbers 110
- 6-3 Multiplying Signed Numbers 114
- 6-4 Dividing Signed Numbers 117
- 6-5 Converting Within SI 119
- 6-6 Powers and Roots of Signed Numbers 121
- 6-7 Scientific and Engineering Notation 124
- 6-8 Combined Operations 127
- Review Exercises 129

7 Introduction to Algebra 131

- 7-1 Algebraic Expressions 131
- 7-2 Adding and Subtracting Polynomials 136
- 7-3 Laws of Exponents 140
- Review Exercises 146

8 Multiplication of Algebraic Expressions 148

- 8-1 Product of Two or More Monomials 148
- 8-2 Product of a Multinomial and a Monomial 151
- 8-3 Product of Two Binomials 152
- 8-4 Product of Two Multinomials 155
- 8-5 Powers of Multinomials 156
- 8-6 Removing Symbols of Grouping 158
- Review Exercises 160

9 Division of Algebraic Expressions 162

- 9-1 Quotient of Two Monomials 162
- 9-2 Dividing a Polynomial by a Monomial 166
- 9-3 Quotient of Two Polynomials 167
- Review Exercises 170

10 Simple Equations 172

- 10-1 Substituting into Equations and Formulas 172
- 10-2 Solving Simple Equations 177
- 10-3 Simple Equations with Approximate Coefficients and Literals 183
- Review Exercises 184

11 Solving Word Problems 187

- 11-1 Solving Word Problems 187
- 11-2 Uniform Motion Problems 192
- 11-3 Financial Problems 196
- 11-4 Mixture Problems 198
- Review Exercises 202

12 Functions and Graphs 204

- 12-1 Functions 204
- 12-2 Functional Notation 209
- 12-3 Rectangular Coordinates 213
- 12-4 Linear Functions 218
- 12-5 Quadratic and Other Functions 226
- 12-6 Reading Graphs 231
- 12-7 Graphing Calculators 236
- Review Exercises 241

13 Plane Geometry 244

- 13-1 Lines and Angles 244
- 13-2 Triangles 254
- 13-3 Right Triangles 261
- 13-4 Quadrilaterals 267
- 13-5 Circles 273
- Review Exercises 278

14 Solid Geometry 280

- 14-1 Prisms 280
- 14-2 Cylinders 287
- 14-3 Pyramids and Cones 290
- 14-4 Spheres 299
- Review Exercises 301

15 Right Triangle Trigonometry 304

- 15-1 Sine, Cosine, and Tangent Functions 304
- 15-2 Inverse Trigonometric Functions 308
- 15-3 Solution of Right Triangles 311
- 15-4 Applications 318
- 15-5 Cosecant, Secant, and Cotangent Functions 325
- Review Exercises 329

16 Factoring 332

- 16-1 Common Factors 332
- 16-2 Difference of Two Squares 339
- 16-3 Simple Trinomials 343
- 16-4 General Trinomials 347
- 16-5 Perfect Square Trinomials 353
- 16-6 Sum or Difference of Two Cubes 355
- Review Exercises 358

17 Algebraic Fractions 360

- 17-1 Equivalent Fractions 360
- 17-2 Multiplying and Dividing Algebraic Fractions 368
- 17-3 Adding and Subtracting Algebraic Fractions 372
- 17-4 Complex Fractions 379
- Review Exercises 383

18 Fractional Equations 385

- 18-1 Solution of Fractional Equations 385
- 18-2 Literal Equations and Formulas 390
- 18-3 Applications 394
- Review Exercises 398

19 Systems of Linear Equations 401

- 19-1 Graphical Method 401
- 19-2 Addition Method 408
- 19-3 Substitution Method 413
- 19-4 Applications of Systems of Two Equations 416
- 19-5 Systems of Three Equations 422
- 19-6 Solution of Systems of Two Linear Equations by Determinants 427
- 19-7 Solution of Systems of Three Linear Equations by Determinants 432
- Review Exercises 436

20 Exponents and Radicals 439

- 20-1 Fractional Exponents 439
- 20-2 Simplification of Radicals 445
- 20-3 Adding and Subtracting Radicals 450
- 20-4 Multiplying Radicals 454
- 20-5 Dividing Radicals 457
- 20-6 Radical Equations 460
- Review Exercises 465

21 Quadratic Equations 468

- 21-1 Introduction to Quadratic Equations 468
- 21-2 Solution by Factoring 475
- 21-3 Quadratic Formula 480
- 21-4 Applications of Quadratic Equations 487
- 21-5 Applications Involving Rates 492
- 21-6 Quadratic Functions 496
- Review Exercises 500

22 Oblique Triangles 504

- 22-1 Trigonometric Functions of Any Angle 504
- 22-2 Finding Angles from Trigonometric Functions 511
- 22-3 Law of Sines 517
- 22-4 Ambiguous Case 523
- 22-5 Law of Cosines 527
- Review Exercises 532

23 Vectors 534

- 23-1 Vectors and Components 534
- 23-2 Applications of Vector Components 539
- 23-3 Graphical Addition of Vectors 542
- 23-4 Addition of Perpendicular Vectors 546
- 23-5 Vector Addition by Components 549
- 23-6 Vector Addition Using the Laws of Sines and Cosines 554
- Review Exercises 557

24 Applications of Radian Measure 560

- 24-1 Radian Measure 560
- 24-2 Arc Length and Area of a Sector 564
- 24-3 Uniform Circular Motion 568
- Review Exercises 571

25 Graphs of Trigonometric Functions 573

- 25-1 Graph of $y = a \sin x$ 573
- 25-2 Graph of $y = a \sin bx$ 581
- 25-3 Graph of $y = a \sin (bx + c)$ 587
- 25-4 Graph of $y = a \cos (bx + c)$ 590
- 25-5 Graph of $y = a \tan (bx + c)$ 594
- Review Exercises 599

26 Ratio, Proportion, Variation 602

- 26-1 Ratio 602
- 26-2 Proportion 606
- 26-3 Direct Variation 613
- 26-4 Joint, Inverse, and Combined Variation 619
- Review Exercises 626

27 Exponential Functions 629

- 27-1 Introduction to Exponential Functions 629
- 27-2 Graph of the Exponential Function 631
- 27-3 Exponential Growth 636
- 27-4 Exponential Decay 640
- Review Exercises 644

28 Logarithms 646

- 28-1 Introduction to Logarithms 646
- 28-2 Common and Natural Logarithms 650
- 28-3 Properties of Logarithms 655
- 28-4 Solution of Exponential Equations Using Logarithms 661
- 28-5 Applications of Logarithmic Equations 665
- Review Exercises 667

29 Complex Numbers 670

- 29-1 Imaginary Numbers 670
- 29-2 Complex Numbers 676
- 29-3 Operations with Complex Numbers 679
- 29-4 Graph of a Complex Number 684
- 29-5 Polar and Trigonometric Forms 686
- Review Exercises 691

Appendix A, Summary of Facts and Formulas 693
Appendix B, Conversion Factors 702
Answers to Selected Odd-Numbered Exercises 707
Index to Applications viii
Index to Writing Questions (Exercises) x
Index to Team Projects (Exercises) x
General Index 756

Preface

This book is intended for students at community colleges, technical schools, and technical colleges. It has been designed to meet the needs of technical students, as well as students who will benefit from an applications approach to learning algebra and trigonometry. Applications have been drawn from many different technical and nontechnical areas, such as the electrical, financial, navigation, business, automation, computer, and allied health fields. The text can also be used for courses in intermediate or college algebra. There is sufficient material provided here for either a one-semester or a two-semester course, depending upon the pace and depth of presentation.

Given the wide adoption of the graphing calculator in the mathematics classroom, we have thoroughly integrated this powerful tool into the book, both in the text and in the exercises. However, we have still retained most of the manual methods, such as graphing by computing and plotting a table of point pairs, for those classes not using the graphing calculator.

Features of the Book

- **Examples.** Many fully worked examples form the backbone of the textbook.
- **Exercises.** Practice is essential for learning mathematics, so we include a large number of exercises. Those given after each section are arranged by difficulty and grouped by type to allow pratice on a particular area.
- **Chapter Review Exercises.** These are scrambled as to type and difficulty, requiring the student to be able to identify type.
- **Answers.** Answers to selected exercises are provided, allowing for checking of work.
- **Summary of Facts and Formulas.** All important formulas are boxed and numbered in the text. We also list these formulas in Appendix A, Summary of Facts and Formulas. This listing can provide a common thread between chapters, and help students to see interconnections that might otherwise be overlooked. The formulas are grouped logically in the Summary and are numbered sequentially there. Therefore, the formulas do not necessarily appear in the text in numerical order.
- **Common Error Boxes.** There are many mistakes that students make over and over, from one class to the next.
- **Graphing Calculator.** Calculator instructions are given in the text, and calculator problems are given in the exercises. The book is not designed for use with any particular brand of calculator. Actual keystrokes are shown in the earliest chapters, for operations that are essentially similar from one calculator to another. Thereafter, for operations that vary widely between calculators, written descriptions are given.
- **Applications.** Applications are provided wherever appropriate as motivation for the students. An Index to Applications is provided.
- **Team Projects.** A Team Project is given at the end of most chapters for the instructor who would like to try a cooperative learning approach to some units.
- **Instructor's Solution Manual.** An Instructor's Solution Manual contains worked-out solutions to even-numbered problems in the text.

Despite our best efforts, errors will creep in. If you find any, we hope you will contact the authors in care of the publisher.

Acknowledgments for the U.S. Edition

We are grateful to the following colleagues for their valuable reviews of the manuscript: Dr. Tony Banach, ECPI Computer Institute; Aaron Loggins, Albuquerque Technical-Vocational Institute; Lois McBride, Stark Technical College; Larry Moore, Midlands Technical College; Len Mrachek, Hennepin Technical Center; James Newsom, Tidewater Community College; Charles Oster, Sauk Valley Community College; Dale Thielker, Ranken Technical College; Roy Wilson, Ceritos College.

We would like to thank especially Floyd McPhetres for reviewing most of the manuscript and for providing invaluable suggestions, and Michael Calter for his extensive work on the early chapters. To check for accuracy, all the examples and exercises were solved by Jennifer Rogers, with help from Emily Davis and Marlys Eddy. Writing Exercises and Team Projects were reviewed by Linda Davis and Ginny Richburg. The *Instructor's Solution Manual* was prepared by Susan Porter.

Paul Calter
Carol Felsinger Rogers

Preface to the Canadian Edition

Foundations of Technical Mathematics, Canadian Edition, is an adaptation of *Introductory Algebra and Trigonometry with Applications* by Paul Calter and Carol Felsinger Rogers. The main features of the text were retained throughout. The adaptation adjusts certain elements of the text to better fit the curriculum of Canadian educational institutions. As well, the text was revised to follow the International System of Units (SI) for the most part, with lesser emphasis on the British/American system.

- **Units.** Most examples adhere to SI standards, with only limited use of British/American customary units. The use of SI represents a change from the old CGS metric standards in the original text, which are no longer generally used by the international scientific and business communities. In most exercise sets the first group of exercises uses SI units. These are followed by a few exercises using British/American units. The chapter on units of measure emphasizes SI units and includes an expansion of the metric prefix table from twelve prefixes to the full twenty prefixes currently approved by the International Committee of Weights and Measures (CGPM).
- **Outcomes.** Found at the beginning of each chapter, they help the student focus on desired learning objectives.
- **Concepts.** Some concepts have been added, such as prime numbers in simplifying fractions, engineering notation and its relationship to SI, the multiplicative identity, and a simpler method of converting within the metric system.
- **Frustum.** In the section on solid geometry, the applications of the frustum of a cone or pyramid have been expanded as they are useful in many fields.

Acknowledgments

Several reviewers provided excellent suggestions for this Canadian adaptation. Special thanks are extended to Waldo Dyck, Red River College, for his detailed review during the writing phase and to Allan MacGowan, University of Guelph, who carefully reviewed this project at a later stage. Many suggestions were made by Bob MacMillan, Georgian College; David Sabo, British Columbia Institute of Technology; and Ngok Yeung Lai, Nova Scotia Institute of Technology.

I would like to thank my colleagues at Sault College for the discussions and reviews of various sections, especially Brad Kirk, Bob Hamel, John McGauley, Geoffrey Disano and Wally MacQuarrie (retired).

A special thank you to my wife Elaine, whose encouragement and assistance was vital to the successful completion of this project.

John Giguere

Index to Applications

ASTRONOMY, SPACE
Aircraft speed, position, 196, 202, 322, 396, 399, 421, 495, 502, 522, 541, 554, 557, 558
Atmospheric pressure at various altitudes, 666
Earth's orbit, 277
Gravitation, 625
Gravity on moon, 492
Head wind, 495
Propeller speed, 577
Rockets, 75, 234
Speed, position of a satellite or spacecraft, 62, 205, 203, 571

AUTOMOTIVE
Antifreeze mixture, 104, 202, 611
Electric car, 100
Engine displacement, 290, 619
Engine power, 619
Fuel mixtures, 105, 202,421, 605
Gas turbine, 103
Stopping distance, 500
Wheels, tires, 277, 571

DYNAMICS
Centrifugal force, 242, 475
Freely falling body, 19, 26, 67,211, 242, 359, 491, 492, 502
Kinetic energy, 342,390, 472, 497, 577, 619
Machine efficiency, 605
Newton's second law, 625
Object thrown upwards or downwards, 224, 232, 234, 235, 339, 347, 353, 489, 490, 492, 502
Pendulum, 19, 67, 577, 619
Power to propel a ship, 628
Resultants and components of velocity vectors, 539
Rotation, 568
Simple harmonic motion, 464, 467
Trajectories of projectiles, 115, 235, 492, 500, 627
Uniform circular motion, 61,236, 568
Uniform linear motion, 16, 25, 47, 51, 61, 62, 192–195, 203, 213, 214, 419, 420, 492, 625
Uniformly accelerated motion, 177, 185, 421, 488
Vibrations of a stretched wire, 625
Work, 208, 213, 225, 242

ELECTRICAL
Alternating current, 590, 594
Battery voltage, 338
Capacitors, 105, 383, 394, 431, 643, 664, 668
Charge, 501
Current in a circuit, 393, 500, 625, 664
Current in a resistor or wire, 16, 25, 36, 63, 72, 117, 130, 625, 626
Decibels gained or lost, 665, 666, 669
Electric car, 99
Impedance, reactance, and phase angle, 67
Magnetic data storage, 577
Magnetic tape head, 577
Motor efficiency, horsepower, torque, 61, 104, 105, 177, 178, 185
Ohm's law, 234, 372
Power dissipated in a resistor, 18, 19, 40, 51, 67, 72, 127, 130, 214, 372, 473, 474, 475, 500, 501, 502, 619
Power supply ripple, 100
Resistance change with temperature, 177, 185
Resistance of conductors, 243, 626
Resistor networks, 467
Resistors in parallel, 64, 72, 382, 391, 394
Resistors in series, 60, 72, 81, 426, 431, 435, 496,
Resonance, 465
Standard cell, 100
Tolerances for resistors and capacitors, 106
Transformer, 626
Voltages in circuits, 113, 130, 225, 426, 485

ENERGY
Demand for electric power, 640
Electric car, 100
Fuel consumption, 51, 103
Fuel mixtures, 105, 202, 421, 605
Fuel tanks, 280, 290, 301
Insulation, 103, 626
Kinetic energy, 342, 390, 472, 497, 577, 619
Monthly electric bill, 242
Oil consumption and imports, 100, 103
Oil deposits, 100
Oil pipeline, 301
Oil slicks and spills, 195, 277
Population growth, 635, 640, 664
Powdered coal pipeline, 106
Radioactivity, 635, 641, 642, 643, 645, 664, 668
Solar energy, 89, 99, 100
Solar panel, 531
Water turbine, 571
Wind generator, 422, 569, 571

FINANCIAL
Bank balance, 11, 100, 130
Bonus, 494
Financial problems, 196–198
Inflation, 640, 664
Interest, compound, 67, 177, 185, 231, 242, 636, 637, 640, 664, 666, 668
Interest, simple, 177, 185, 197, 202, 225, 338, 421, 626
Investments, 426
Labor costs, 203
Land sale 120
Land purchase, 437
Manufacturing cost, 500
Markup 47
Profit and loss, 59, 99, 100, 103
Revenue equation, 230
Stocks and bonds, 47, 63, 103, 117, 130, 198, 383, 421, 611
Taxes, 196, 198, 203, 612
Unit cost, 62, 208, 226

FLUIDS
Atmospheric pressure at various altitudes, 644
Boyle's law, 623
Charles' law, 619
Chemical reactions, 643
Concentration, 104, 199
Drying rate of wet fabrics, 643
Flow over a dam, 176
Flow in pipes and gutters, 177, 185, 617
Flow to or from tanks, 397, 619, 625, 626
Gas laws, 394, 459, 625
Mixtures problems, 198–202, 418, 421, 424, 426, 435, 437, 626
pH, 665, 666, 669
Pressure head, 291
Water main, 291

GEODESY
Earthquake intensity, 667
Latitude, 567
Speed or a point on the equator, 571

HEALTH
Diets, nutrition, 418, 431, 437
Growth of bacteria, yeast, 44, 465, 633, 635, 640, 643, 644, 664, 668
Medicine dosage, 393
Population growth, 635, 640, 664

LIGHT, OPTICS
Illumination, 625
Lenses, 64, 378, 382, 394
Photographic exposures, 619
Prism, spectrum, 286
Refraction of light, 244
Windows, 572

MACHINE SHOP
Bolt circles, 325
Cutting speed, 195
Depth of cut, 325
Dovetail, 325
Hexagonal stock, 267, 325
Machine production times, 47, 397
Rivet head, 302, 325
Scale drawings, 604, 612
Screw head, 325
Sheet metal pattern, 346, 491, 567
Shop trigonometry, 321
Washers, 342

MATERIALS
Alloys, 200, 202, 203, 421, 437, 438, 611
Concrete Mixtures, 60, 106
Density and specific gravity, 16, 25, 37, 39, 51, 63, 226, 342, 372
Mineral growth, 644
Radioactivity, 635, 641, 642, 643, 645, 664, 668

MECHANISMS
Flywheel, 61, 569, 643, 664
Pumps, 106
Slider mechanism, 37
Speed reducer, 106, 107
Springs, 232, 235

NATURAL RESOURCES/ECOLOGY
Anti-rabies vaccine, 51
Azimuth compass, 523, 538
Bacterial growth, 444, 633, 635, 664, 669
Difference in elevation, 342
Fire tower sighting, 331
Height of a tree, 322, 612
Map distance using scale, 612
Municipal landfill, 37
Oil consumption, 103
Radioactive decay, 645
Solar collector, 89
Surface area of a lake, 89
Volume of a lake, 120, 296
Volume of a log, 296

NAVIGATION
Aircraft air speed and ground speed, 396
Azimuth circle, 523, 538
Compass directions, 319
Latitude, 567
Position or speed of an aircraft or ship, 324, 398

STATICS
Force vectors, 540, 541, 542, 546, 552
Friction, 235, 242
Gravitational attraction, 625
Mechanical advantage, 575
Resultants and components of force vectors, 540
Springs, 40, 51, 232, 235, 615

STRENGTH OF MATERIALS
Beams, 117, 339, 487, 625
Modulus of elasticity, 71, 72, 127, 130, 172, 185
Stress and strain, 40, 51, 72, 127, 130
Thermal expansion, 100
Wire strength versus temperature, 444

STRUCTURES
Beams, 117, 339, 487, 625
Building foundation, 301
Escalator, 541, 557
Hemispherical dome, 300
I-beam, 286
Poles, towers antennas, 260, 324, 520, 507, 530, 531
Radome, 302
Ramps, 530
Retaining wall, 286
Roofs, rafters, 267, 278, 324, 530, 606
Silo, 290, 301
Staircase, 324
Steeple, 290
Wind pressure on building, 75

SURVEYING
Circular highway curves, 486, 577
Elevations, 112, 341
Elevations found by barometer readings, 666
Land areas, 81, 83, 92, 261, 566
Latitude and longitude, 577
Maps and scale drawings, 604, 612
Measuring inaccessible distances, 260, 267, 278, 319, 323, 522, 523, 529, 531, 530, 612

THERMAL
Exponential rise or fall in temperature, 642, 643, 644, 664, 665, 669
Heat flow by conduction, 619, 625, 626
Heat needed to melt ice, 619
Steam boiler, 301
Temperature change, gradient, 102, 103, 113, 130
Temperature conversion, 177, 121, 185
Thermal expansion, 100
Velocity of sound at different temperatures, 226,
Wire strength versus temperature, 444

Index to Writing Questions

1. Give your thoughts on the distinction between exact and approximate numbers, 24
2. Write a letter to the editor giving your opinion on the value of studying fractions, 52
3. Write a memo to your boss defending your rounding practices, 75
4. Write a few lines about calculator display formats, 75
5. Write a letter to your company president stating why or why not your company should "go metric," 92
6. State your opinions on the importance of percentage in everyday life, 107
7. Explain the difference between the two keys on a calculator marked with a negative sign, 130
8. Write a note to a friend explaining what algebra is, 147
9. Give your opinion on the value of learning rules like the FOIL rule, 160
10. Explain whether you think its valuable to simplify fractions or not, 171
11. Write your thoughts on how to solve an equation, 171
12. Find and explain an equation from the field you expect to enter, 186
13. Make up a word problem, 203
14. Describe what is meant by a function, 243
15. Write an essay on geometry in either art, nature, or architecture, 279
16. Describe the various geometric solids, 302
17. Write an essay on how you might use trigonometry, 331
18. State how to recognize different types of factorable expressions, 359
19. Give examples of algebraic fractions in science or business, 384
20. Make up and solve a fractional equation with an extraneous solution, 399
21. Describe the methods of solving systems of equation, 438
22. Explain how exponents and radicals are different ways of writing the same thing, 467
23. Describe different ways of solving a quadratic equation, 502
24. Write a letter to a client defending your use of the law of cosines, 533
25. Describe the three methods for adding vectors, 533
26. Explain why we need to study radians, 572
27. Explain how changing the constants in the equation for the sine wave affects its graph, 600
28. Write a presentation to your company president pointing out why he is wrong about the dimensions of plastic trays, 628
29. Describe how exponential functions are used in business and biotechnology, 645
30. Explain what a logarithm is, 669
31. Explain the relationship among real, imaginary, and complex numbers, 691

Index to Team Projects

General suggestions for team projects, 52
Study the ancient Egyptians method of writing fractions, 53
Find the new dimensions of a template after changes are made, 76
Study the Chinese system of weights, 93
Write an hour exam on the topic of percentage, 107
Write an hour exam on division of algebraic expressions, 171
Use a graphing calculator to graph a given function, 279
Use Sherlock Holmes' method to find the height of a tree, 279
Compare the price of brick of gold to that of a brick of platinum, 303
Make a simple transit and use it to measure some distance, 331
Factor trinomials by representation as areas of rectangles, 359
Determine the positions of employees in a company, 384
Solve an alphametric puzzle, 399
Solve a system of three equations to find the amounts of elements in an alloy, 438
String model of a right triangle, 467
Analyze the flight of a baseball, 503
Analyze four mutually tangent circles, 533
Create displacement vectors on your campus, 559
Compute the kinetic energy of rotation of various balls, 572
Analyze the equation for a sound wave, 600
Prove or disprove whether the weights of people are proportional to the cube of their heights, 628
Graph the temperature of a kettle of boiling water as it cools, 645
Graph and analyze grades versus number of hours studied, 669
Find the three complex cube roots of one, 692

Arithmetic with Whole Numbers

Outcomes

When you have completed this chapter, you will be able to

1. Identify the different classifications of numbers such as natural number, whole number, and integer.
2. Add and subtract positive whole numbers.
3. Multiply and divide positive whole numbers.
4. Evaluate powers and roots of positive whole numbers.
5. Perform the basic operations of scientific notation using whole numbers.
6. Correctly perform calculations involving addition, subtraction, multiplication, division, powers, and roots.

We start our study of mathematics with a review of arithmetic. First we identify the different types of numbers that we will work with in technical mathematics, and then go on to do arithmetic with the simplest kind, the positive whole numbers.
  We do addition and subtraction, then multiplication and division, entirely by calculator. We then calculate powers of whole numbers and introduce scientific notation in a simple way. We finally calculate the roots of some whole numbers.
  This practice with whole numbers will prepare us for arithmetic with fractions in Chapter 2, and arithmetic with decimal numbers in Chapter 3.

1-1   THE REAL NUMBERS

Integers

The numbers
$$\ldots, -4, -3, -2, -1, 0, 1, 2, 3, 4, \ldots$$
are called *integers*. They include zero and negative values as well.

FIGURE 1-1
The real numbers.

Example 1

Some integers are
$$8, \quad 362, \quad 3, \quad -372, \quad 0, \quad \text{and} \quad 1\,000\,000$$

Whole Numbers

The nonnegative integers including zero
$$0, \quad 1, \quad 2, \quad 3, \ldots$$
are called *whole numbers*.

> **Example 2**
> Some whole numbers are
> $$82, \quad 2193, \quad 0, \quad 5, \quad \text{and} \quad 1\,000\,000$$

Sometimes zero is not included in the definition of the whole numbers, but we will include it in our definition.

Natural Numbers

The whole numbers

$$1, \quad 2, \quad 3, \quad \ldots$$

are called the *natural numbers*. They do not include zero.

Digits

The ten whole number symbols

$$0, \quad 1, \quad 2, \quad 3, \quad 4, \quad 5, \quad 6, \quad 7, \quad 8, \quad 9$$

are called *digits*.

In this chapter we will do arithmetic with whole numbers only, and we will cover negative numbers in Chapter 6. However, we will show other kinds of numbers now, and tell where in the text we will cover them.

Rational Numbers

The *rational numbers* are all those numbers that can be expressed as the quotient of two integers (except that the denominator may not be zero). The rational numbers include the integers.

> **Example 3**
> Some rational numbers are
> $$\frac{1}{5}, \quad -\frac{3}{8}, \quad -\frac{27}{13}, \quad -\frac{3}{5}, \quad \frac{7}{3} \quad \text{and} \quad -4$$

We cover rational numbers in Chapter 2.

Irrational Numbers

Numbers that cannot be expressed as the quotient of two integers are called *irrational*.

> **Example 4**
> Some irrational numbers are
> $$\sqrt{5}, \quad \sqrt[3]{7}, \quad \text{and} \quad -\sqrt{8}$$

We cover irrational numbers in Chapters 3 and 21.

Real Numbers

The rational and irrational numbers together make up the *real numbers*.

---- **Example 5** ----
Some real numbers are

$$88, \quad -6, \quad \sqrt{72}, \quad 0, \quad -\sqrt{23}, \quad \frac{3}{4}, \quad \text{and} \quad -\frac{55}{57}$$

Imaginary Numbers

Numbers such as $\sqrt{-4}$ do not belong to the real number system. They are called *imaginary numbers*.

---- **Example 6** ----
Some imaginary numbers are

$$\sqrt{-3}, \quad \sqrt{-1}, \quad -\sqrt{-4}, \quad \text{and} \quad \sqrt{-81}$$

Imaginary numbers are discussed in Chapter 29, where we will also cover *complex* numbers. Except when otherwise noted, all the numbers we work with are real numbers.

Decimal Numbers

Most of our computations are with numbers written in the familiar decimal system. We say that the decimal system uses a *base of 10* because it takes 10 units in any place to equal 1 unit in the next higher place.

---- **Example 7** ----
Some decimal numbers are

$$93.2, \quad -29.5, \quad 26, \quad -123, \quad 0.0034, \quad \text{and} \quad -23\,937.5$$

We will do arithmetic with decimal numbers in Chapter 3.

Place Value

The names of the places relative to the decimal point are shown in Fig. 1-2.

■ **FIGURE 1-2**
Values of the positions in a decimal number.

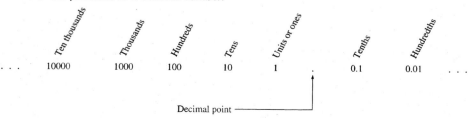

Example 8
What is the place name of the 5 and the 7 in the number 25 471?

Solution: The 5 is in the thousands place, and 7 is in the tens place.

The place names of digits to the right of the decimal point are covered in Chapter 3.

Approximate Numbers

Most of the numbers we deal with in technology are approximate. All numbers that represent measured quantities are approximate.

Example 9
A certain shaft is approximately 2.75 in. in diameter.

Some fractions can be expressed approximately in decimal form.

Example 10
$\frac{2}{3}$ is approximately equal to 0.6667.

Some roots such as $\sqrt{2}$ when left in root form are considered to be exact numbers. These values can also be expressed in approximate form.

Example 11
$\sqrt{2}$ is approximately equal to 1.4142.

We will do arithmetic with approximate numbers in Chapter 3.

Exact Numbers

Exact numbers are those that have no uncertainty. Typical exact numbers are those used for counting.

Example 12
A certain classroom contains exactly 32 chairs.

Exact numbers often occur in definitions.

Example 13
There are exactly 24 h in a day, no more, no less.

Exact numbers are usually, but not always, integers.

FIGURE 1-3
The number line.

--- **Example 14** ---
There are exactly 2.54 cm in an inch, by definition.

Not all whole numbers or integers are exact.

--- **Example 15** ---
A certain town has a population of about 23 400 people.

You may assume that all numbers in this chapter are exact.

Number Line

We can graphically represent every real number as a point on a line called the *number line* (Fig. 1-3). It is customary to show the positive numbers to the right of the zero and the negative numbers to the left.

--- **Example 16** ---
Place the following numbers on the number line.

(a) 5   **(b)** -3 **(c)** 2.5   **(d)** $-3\frac{1}{2}$

Solution: The given numbers are shown on the number line of Fig. 1-4.

FIGURE 1-4

Signs of Equality and Inequality

Several signs are used to show the relative position of two numbers on the number line.

$a = b$ means that *a equals b*, and that a and b occupy the same position on the number line.

$a \neq b$ means that a and b are *not equal* and have different locations on the number line.

$a > b$ means that a is *greater than b*, and a lies to the right of b on the number line.

$a < b$ means that a is *less than* b, and a lies to the left of b on the number line.
$a \geq b$ means that a is *equal to or greater than* b, and a occupies the same position or lies to the right of b on the number line.
$a \leq b$ means that a is *equal to or less than* b, and a occupies the same position or lies to the left of b on the number line.
$a \approx b$ means that a is *approximately equal to* b and that a and b are near each other on the number line.

Example 17

Place the proper equality or inequality sign between $\frac{1}{3}$ and 0.3.

Solution: Since $\frac{1}{3}$ is equal to 0.3333. . . , it is greater than 0.3, so we write

$$\frac{1}{3} > 0.3$$

SECTION 1-1 Exercises The Real Numbers

Kinds of Numbers

State whether each number is an integer, whole, natural, a digit, rational, irrational, real, imaginary, decimal, positive, or negative. Each number may have more than one name.

1. 7
2. −3
3. $\frac{4}{5}$
4. $\sqrt{7}$
5. −3.72
6. $\sqrt{-6}$
7. $-\frac{7}{8}$
8. 88.2
9. $-\sqrt{82}$
10. $2\frac{4}{5}$

Place Value

11. State the place value of the 3 and the 4 in the number 35 841.

12. State the place value of the 1 and the 5 in the number 35 841.

Exact and Approximate Numbers

State whether the number given in each statement is exact or approximate.

13. A person has 10 fingers.

14. A certain table is 2 metres long.

15. The temperature outside is 10 degrees Celsius.

16. A week has seven days.

Number Line

Draw a number line, using a scale to mark divisions from −10 to +10. Place each of the following numbers on your number line.

17. 8
18. −3
19. −6
20. 5
21. $5\frac{1}{2}$
22. $-\frac{3}{4}$
23. 5.4
24. 9.7
25. −4.6
26. −3.2
27. 0
28. $-3\frac{1}{4}$

Equality and Inequality Signs

Insert the proper sign of equality or inequality (=, ≈, <, >) between each pair of numbers.

29. 8 and 11   **30.** 5 and −3   **31.** −2 and 6

32. −10 and −12   **33.** $\frac{1}{4}$ and 0.25   **34.** $\frac{1}{3}$ and 0.333

1-2   ADDING AND SUBTRACTING WHOLE NUMBERS

Adding Whole Numbers

To add two numbers by calculator, simply enter the first number, press $\boxed{+}$, enter the second number, and press $\boxed{\text{ENTER}}$. (Some calculators use the $\boxed{=}$ key, or $\boxed{\text{EXE}}$, for *execute,* instead of an $\boxed{\text{ENTER}}$ key.) The number we get is called the *sum.*

Example 18

Evaluate 2845 + 3273 by calculator.

Solution: There are many types of calculators in use, so the names of the keys and the appearance of the display may be slightly different than shown here. You should, however, be able to use the keystrokes shown here as a guide in operating your own calculator, but be sure to check your operating manual. The keystrokes are

Horizontal and Vertical Addition

Numbers to be added may be arranged horizontally or vertically. Of course, when using the calculator it does not make any difference how the numbers are arranged, as they must be keyed in one at a time anyway.

Example 19

The addition of 335, 103, and 224 may be represented horizontally,

$$335 + 103 + 224 = 662$$

or vertically

$$\begin{array}{r} 335 \\ 103 \\ \underline{224} \\ 662 \end{array}$$

Commutative Law

If *a* and *b* are two numbers, the commutative law simply states that you can add these numbers in any order.

| COMMUTATIVE LAW FOR ADDITION | $a + b = b + a$ | 1 |

*

Example 20

It is no surprise that

$$5 + 2 = 2 + 5$$

Symbols of Grouping

We will often need *symbols of grouping*. They are

      ( ) parentheses   [ ] brackets   { } braces

We will show the use of these symbols many times as we go along. For now, just remember that the operations *inside* the symbols of grouping must be done before any other operations.

The division line of a fraction, called a fraction line (also called a vinculum), acts like a grouping symbol. Operations above or below a fraction line must be performed before any other operations. The computations $(4 + 5) \div (5 - 2)$ can also be written $\frac{4 + 5}{5 - 2}$. The addition and subtraction are performed before the division. A vinculum also makes up part of the radical sign and operations under the radical sign must also be done first. A vinculum (or overbar) can also be used as a grouping symbol.

Example 21

In the expression

$$5 - (1 + 3)$$
$$\text{or} \quad 5 - [1 + 3]$$
$$\text{or} \quad 5 - \{1 + 3\}$$
$$\text{or} \quad 5 - \overline{1 + 3}$$

the 1 and the 3 must be added *before* subtracting from 5.

Associative Law

If a, b, and c are any numbers, the associative law says that you can group numbers to be added in different ways.

| ASSOCIATIVE LAW FOR ADDITION | $a + (b + c) = (a + b) + c$ | 3 |

---

* All boxed and numbered formulas are listed in Appendix A, "Summary of Facts and Formulas."

Example 22

Add 1, 3, and 5.

Solution: We might choose to add the 1 and the 3 first,

$$1 + 3 + 5 = (1 + 3) + 5$$
$$= 4 + 5$$
$$= 9$$

Or we might want to add the 3 and the 5 first.

$$1 + 3 + 5 = 1 + (3 + 5)$$
$$= 1 + 8$$
$$= 9$$

The commutative and associative laws combined say that it doesn't matter in what order or in what groupings we add the given numbers. We could even choose to add the 1 and the 5 first.

$$1 + 3 + 5 = (1 + 5) + 3$$
$$= 6 + 3$$
$$= 9$$

We get the same result no matter which way we order and group the numbers.

Subtracting Whole Numbers

The keystrokes for subtraction are the same as for addition, except that the $\boxed{-}$ key is used instead of $\boxed{+}$.

Example 23

Evaluate $8361 - 6582$.

Solution: The keystrokes are

📱	8361	$\boxed{-}$	6582	$\boxed{\text{ENTER}}$
				1779

> **COMMON ERROR** Many calculators have a $\boxed{(-)}$ or $\boxed{(+/-)}$ key, which looks almost like the $\boxed{-}$ key. The $\boxed{(-)}$ or $\boxed{(+/-)}$ key is used to enter negative numbers. Do not use it for subtraction.

The commutative law *does not apply* to subtraction. In other words, the order in which we subtract terms *is* important.

Example 24

In addition, $3 + 2$ is equal to $2 + 3$
In subtraction, $3 - 2$ is *different* from $2 - 3$

The associative law also does not apply to subtraction. In other words, the way in which terms are grouped for subtraction *is* important.

Example 25
Show that $7 - (5 - 2)$ is *different* from $(7 - 5) - 2$.

Solution: As we said earlier, we evaluate the quantities inside the parentheses first. So

$$7 - (5 - 2) = 7 - 3 = 4$$

but

$$(7 - 5) - 2 = 2 - 2 = 0$$

Example 26
Evaluate the expression $16 - [(7 - 2) - (6 - 3)]$.

Solution: We start by removing the inner parentheses.

$$16 - [(7 - 2) - (6 - 3)] = 16 - [5 - 3]$$
$$= 16 - 2$$
$$= 14$$

Parentheses on the Calculator

To evaluate an expression containing parentheses on the calculator, we use the $[(]$ key and the $[)]$ key.

Example 27
Evaluate $82 - (45 - 27)$ by calculator.

Solution: The keystrokes are

Additive Identity

The sum of a number and the addition or subtraction of zero is that number.

Example 28
(a) $7 + 0 = 7$
(b) $7 - 0 = 7$

Applications

Problems in mathematics are often stated in verbal form. We discuss word problems in detail in Chapter 11, but we can start now to sharpen our skills by doing simple verbal problems in arithmetic. When doing the exercise, look for words that indicate addition (*sum, increase, gain, accumulation,* etc.), for words that indicate subtraction (*difference, loss, decrease, deduction,* etc.), and for words that stand for equality (*equals, is, are, amounts to,* etc.).

Example 29

A bank account having a balance of $5832 received two deposits of $359 and $38, followed by a withdrawal of $836. Find the new balance.

Solution: Do not hesitate to use a dictionary if you are unfamiliar with the banking terms *balance, deposit,* and *withdrawal.* It should be clear that we must add the deposits to the initial balance, and subtract the withdrawal.

$$\text{New balance} = \$5832 + \$359 + \$38 - \$836$$
$$= \$5393$$

SECTION 1-2 Exercises Adding and Subtracting Whole Numbers

Adding Whole Numbers

Add by calculator.

1. 874
    647

2. 364
    542

3. 837
    894

4. 855
    627
    642

5. 6773
    9354
    659

6. 945
    399
    231

7. 7764
    5056
    693
    943
    701
    109
    791

8. 85 326
    8 654
    754
    193
    490
    2 100
    923

9. 63 048
    84 336
    5 487
    902
    1 084
    74
    9 372

10. 76 109
    8 533
    863
    12 959
    10 263
    1 472
    62 923

11. 74 352
    6 475
    496
    379
    5 258
    974

12. 7325
    539
    734
    94
    3749
    741

13. 537 + 829 + 274 + 826

14. 923 + 263 + 724

15. 382 + 730 + 263 + 274

16. 482 + 428 + 284 + 836

Subtracting Whole Numbers

Subtract by calculator.

17. 724
    −384

18. 8263
    −7253

19. 294
    −163

20. 724
    −267

21. 737 − 274

22. 823 − 243

23. 982 − 230 − 263

24. 782 − 328 − 184

Symbols of Grouping

Combine as indicated.

25. 3957 − (284 − 163)

26. 737 − (406 − 264) + 517

27. (634 − 372) − (364 − 164)

28. (847 − 274) − 472

29. (36 − 27) + 83 − (83 − 53)

30. 92 − (34 − 27) + (27 − 18)

1-2 Adding and Subtracting Whole Numbers

Applications

31. A house and lot sold together for $50 523, which was $8394 more than its original cost. What was its original cost?

32. Find the perimeter (the distance around) of a four-sided figure (a quadrilateral) if the lengths of the four sides are 17 m, 16 m, 23 m, and 8 m.

33. The sum of the sides of a triangle is 734, and two of the sides are 372 and 263. Find the length of the third side.

34. Four cartons, having heights of 34 in., 51 in., 29 in., and 38 in., are placed on top of each other on a skid having a height of 4 in. What is the total height of cartons and skid?

1-3 MULTIPLYING WHOLE NUMBERS

Factors and Product

The numbers we multiply to get a product are called factors.

$$3 \times 5 = 15$$

factors ⎯⎯⎯↑ ↑   ↑⎯⎯⎯ product

Symbols for Multiplication

The symbols used for multiplication are the times symbol ($\times$), the multiplication dot ($\cdot$), and parentheses.

Example 30

The product of 3 and 5 can be written

$$3 \times 5, \quad 3 \cdot 5, \quad 3(5), \quad (3)5, \quad \text{or} \quad (3)(5)$$

Notice that the multiplication dot looks almost like a decimal point, and that the times symbol can be mistaken for an x. We avoid these whenever there is a chance of confusion. It is recommended under SI and current engineering practices that the multiplication dot be used only between units of measure such as kW·h or N·m, or in the British Engineering System ft·lb.

To show the multiplication of two letter quantities (numbers represented by letters) like a and b, or a number and a letter quantity, we simply write the two quantities with no symbol between them.

Example 31

The expression ab means a times b.
The expression $5a$ means 5 times a.

Multiplying by Calculator

Many calculators use an $\boxed{\times}$ key for multiplication, while some use a $\boxed{*}$ key. The keystrokes used to multiply the factors 13 and 27 are

12 CHAPTER 1 / Arithmetic with Whole Numbers

Commutative Law

If a and b are two numbers, the commutative law for multiplication states that the order of the factors in multiplication is not important.

| COMMUTATIVE LAW FOR MULTIPLICATION | $ab = ba$ | 2 |

Example 32

It is no surprise that
$$2 \times 3 = 3 \times 2$$

Associative Law

The associative law for multiplication allows us to group the numbers to be multiplied in different ways.

| ASSOCIATIVE LAW FOR MULTIPLICATION | $a(bc) = (ab)c$ | 4 |

Example 33

Show that $2(3 \times 4)$ equals $(2 \times 3)4$ and also equals $(2 \times 4)3$.

Solution:
$$2(3 \times 4) = 2(12) = 24$$
$$(2 \times 3)4 = (6)4 = 24$$
$$(2 \times 4)3 = (8)3 = 24$$

This last calculation makes use of the commutative law as well as the associative law. We see that the result is the same regardless of which factors we multiply first.

Distributive Law

The distributive law tells us that

| DISTRIBUTIVE LAW | $a(b + c) = ab + ac$ | 5 |

Example 34

Evaluate $5(3 + 6)$.

Solution: We will do the calculation two different ways.

(a) Let us combine the numbers inside the parentheses first.
$$5(3 + 6) = 5(9) = 45$$

(b) The distributive law enables us to do the same computation in a different way. We can multiply both the 3 and the 6 by 5 and then add.
$$5(3 + 6) = 5(3) + 5(6) = 15 + 30$$
$$= 45 \quad \text{as before}$$

The Multiplicative Identity

Multiplying any whole number by one gives that number as a product.

Example 35

(a) $5 \times 1 = 5$
(b) $1 \times 9 = 9$

Multiplying by Zero

The product of zero and any number is zero.

Example 36

(a) $56(0) = 0$
(b) $0(385) = 0$
(c) $274(0)(837) = 0$

Applications

Example 37

The power in watts (W) delivered to a device is equal to the product of the current in the device in amperes (A), and the voltage across the device in volts (V). Find the power consumed by a motor that draws 3 A at a voltage of 115 V.

Solution: Multiplying,

$$\text{Power} = \text{current} \times \text{voltage}$$
$$= (3 \text{ A})(115 \text{ V}) = 345 \text{ W}$$

SECTION 1-3 Exercises Multiplying Whole Numbers

Multiply by calculator.

1. 4853×429
2. 780×347
3. 752×32
4. $863 \times 796 \times 534$
5. 96×4278
6. 322×64
7. 42×862
8. 392×83
9. $49 \times (6 \times 3) \times 8$
10. $(9 \times 89) \times (48 \times 8)$
11. $(836 + 726) \times 14$
12. $74 \times (47 + 53) \times 7$

Applications

13. If 26 tonnes of gravel are needed for each kilometre of roadway, how many tonnes will be required for 200 kilometres?
14. Two trucks carry 3 tonnes of asphalt each and a third carries 4 tonnes. What is the value of the whole, at $33 per tonne?
15. What will the total mass of 455 brackets be, if each has a mass of 13 grams?
16. If there are 360 degrees per revolution, how many degrees are there in 9 revolutions?

17. What is the cost of 23 gallons of paint at $13 per gallon?

1-4 DIVIDING WHOLE NUMBERS

Definitions

The *dividend*, when divided by the *divisor*, gives us the *quotient*:

$$\text{dividend} \div \text{divisor} = \text{quotient}$$

or

$$\frac{\text{dividend}}{\text{divisor}} = \text{quotient}$$

A quantity *a/b* is also a *fraction*. The fraction line indicates division. The quantity *a/b* can also be referred to as the *ratio* of *a* to *b*. Fractions and ratios are treated in Chapter 2.

Dividing by Calculator

To divide by calculator, enter the dividend, then press $\boxed{\div}$, then enter the divisor and press $\boxed{\text{ENTER}}$, $\boxed{=}$, or $\boxed{\text{EXE}}$, depending on your particular calculator.

Example 38

To divide 1305 by 145 on a calculator, the keystrokes are

When we added, subtracted, or multiplied two integers, we always got an integer for an answer. This is not always the case when dividing.

Example 39

When we divide 1 by 3, we get 0.33333333333. . . . This is a repeating decimal which can also be written $0.\overline{3}$. Here we must choose how many digits we wish to retain. We must also *round* our answer to the chosen number of digits. We cover rounding in Chapter 3.

Division Involving 1

(a) Any whole number except zero divided by itself is 1.
(b) Any whole number divided by 1 is just that number.

Example 40

(a) $7 \div 7 = 1$
(b) $9 \div 1 = 9$

Zero

Zero divided by any quantity (except zero) is zero.

Example 41
(a) $0 \div 48 = 0$
(b) $0 \div (824 + 293) = 0$

However, division by zero is not defined. It is an illegal operation in mathematics.

Example 42
(a) $48 \div 0$ is undefined
(b) $(284 + 194) \div 0$ is undefined
(c) $0 \div 0$ is undefined

Example 43
Using your calculator, divide 5 by zero.

Solution: The keystrokes are

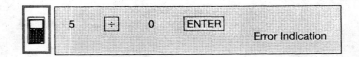

SECTION 1-4 Exercises Dividing Whole Numbers

Divide by calculator.

1. $2496 \div 832$
2. $2002 \div 91$
3. $456 \div 38$
4. $13\,893 \div 421$
5. $2715 \div 543$
6. $5550 \div 75$
7. $0 \div 7$
8. $0 \div 2734$
9. $72 \div 0$
10. $734 \div 0$

Applications

11. How many cubic metres of concrete are needed to make a counterweight with a mass of 6921 kg if the density of concrete is 2307 kg/m³ mass? (Hint: *Volume = mass ÷ density*)

12. A 2247 hectare parcel of land is to be divided into seven equal lots. Find the area of each lot.

13. At the rate of 540 kilometres in 6 hours, how many kilometres would a person drive in 9 hours? (Hint: *Rate = distance ÷ time*)

14. If a factory can assemble 1704 computers in 8 hours, how many are assembled in 3 hours? (Hint: *Work rate = amount done ÷ time worked*)

15. A voltage of 124 V is measured across a 31 Ω resistor. Find the current through the resistor. (Hint: *Current = voltage ÷ resistance*)

16. A steel flywheel (density = 450 lb/ft³) weighs 225 lb. Find the volume of the flywheel.

1-5 POWERS AND ROOTS OF WHOLE NUMBERS

Definitions

In the expression

$$2^4$$

The number 2 is called the *base* and the number 4 is called the *exponent*. The expression is read "two to the fourth power." Its value is the product of four 2's.

$$2^4 = (2)(2)(2)(2) = 16$$

Example 44

$$4^3 = (4)(4)(4) = 64$$

Powers by Calculator

To square a number on the calculator, simply enter the number and press the $\boxed{x^2}$ key. Some calculators will give the answer immediately, while on others you will need to press $\boxed{\text{ENTER}}$, $\boxed{=}$, or $\boxed{\text{EXE}}$, as in the following example.

Example 45

To square 5, the keystrokes are

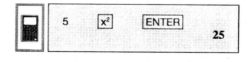

Many calculators have an $\boxed{x^3}$ key for cubing a number. If you do not have one, or to raise a number to other powers, use the *power key,* marked $\boxed{\wedge}$, $\boxed{y^x}$, or $\boxed{x^y}$. We'll use $\boxed{\wedge}$ in the examples. Enter the base, press the power key, then enter the exponent. Finally press the $\boxed{\text{ENTER}}$ or $\boxed{\text{EXE}}$ key to get the result.

Example 46

The keystrokes for finding 5^4 are

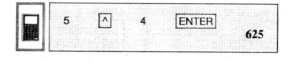

The exponents we have shown are all positive integers. A number may also be raised to a negative exponent like -3, a fractional exponent like $\frac{2}{3}$, or a decimal exponent like 1.53. We explain the meaning of such exponents and show how to calculate with them in later chapters.

Applications

Powers must often be calculated in technical work. Here we do an example from electrical technology.

Example 47

The power (in watts) developed in a resistor is equal to the product of the resistance R (in ohms) and the square of the current I (in amperes). Find the power when a current of 17 amperes flows through a resistance of 3 ohms.

Solution: The power P is equal to

$$P = RI^2$$
$$= (3)(17)^2 = 867 \text{ watts}$$

Root of a Whole Number

If $a^n = b$, then

$$\sqrt[n]{b} = a$$

which is read "the nth root of b equals a." The symbol $\sqrt{}$ is a *radical sign*, b is the *radicand,* and n is the *index* of the radical.

If no index is written, it is assumed to be 2, indicating a *square root.* Thus

$$\sqrt{b} \text{ is the same as } \sqrt[2]{b}$$

Example 48

(a) $\sqrt{9} = 3$ because $3^2 = 9$
(b) $\sqrt[3]{27} = 3$ because $3^3 = 27$
(c) $\sqrt[4]{16} = 2$ because $2^4 = 16$

Roots by Calculator

To find square roots on some calculators, we press the $\boxed{\sqrt{x}}$ (or $\boxed{\sqrt{}}$) key first, enter the number, then press $\boxed{\text{ENTER}}$. On other calculators we enter the number and then press the $\boxed{\sqrt{x}}$ or $\boxed{\sqrt{}}$ key.

Example 49

Find $\sqrt{625}$ by calculator.

Solution: The keystrokes are

or

Many calculators have a $\boxed{\sqrt[3]{}}$ key for finding cube roots. If your calculator does not have that key, or to find other roots, we use a *root key*, marked $\boxed{\sqrt[x]{y}}$, $\boxed{\sqrt[y]{x}}$, or something similar. We will use $\boxed{\sqrt[y]{}}$ in the examples.

Example 50
Find $\sqrt[4]{81}$, the fourth root of 81.

Solution: The keystrokes are

Example 51
A cubical block of concrete has a volume of 6859 cubic centimetres (cm^3). Find the length of one side of the block by taking the cube root of the volume.

Solution: By calculator,

$$\text{Length of one side} = \sqrt[3]{6859 \text{ cm}^3} = 19 \text{ cm}.$$

SECTION 1-5 Exercises Powers and Roots of Whole Numbers

Powers by Calculator

Square by calculator.

1. 3^2
2. 4^2
3. 2^2
4. 8^2
5. 11^2
6. 14^2
7. 23^2
8. 21^2

Cube by calculator.

9. 4^3
10. 6^3
11. 5^3
12. 7^3

Evaluate each power by calculator.

13. 5^4
14. 3^5
15. 2^7
16. 4^5
17. 16^2
18. 13^3
19. 8^3
20. 4^7

Roots by Calculator

Evaluate each square root by calculator.

21. $\sqrt{16}$
22. $\sqrt{36}$
23. $\sqrt{2025}$
24. $\sqrt{3721}$

Evaluate each root by calculator.

25. $\sqrt[3]{27}$
26. $\sqrt[3]{8}$
27. $\sqrt[4]{16}$
28. $\sqrt[4]{81}$
29. $\sqrt[5]{32}$
30. $\sqrt[5]{1024}$

Applications

31. Find the power developed by a current of 23 amperes in a resistance of 5 ohms (see Example 47).

32. The number of feet traveled by a falling object in t seconds is approximately equal to $16t^2$. Find the distance traveled in 11 seconds.

33. Find the length of the side of a cube of steel whose volume is 2744 cubic centimetres (cm^3). See Example 51.

34. The period (time in seconds for one swing) of a certain pendulum is equal to $\sqrt{289}$. Find this period.

1-6 SCIENTIFIC NOTATION

Scientific notation provides a convenient way to represent and to calculate with very large or very small numbers. In this section we give a brief introduction only. We will learn how to convert numbers to and from scientific notation. Our main treatment will come in Chapter 3, where we will do computations in scientific notation.

Powers of Ten

The number 10 raised to a power is called a *power of ten*. Powers of ten are found the same way as powers of any other numbers.

Example 52

(a) $10^3 = 10 \times 10 \times 10 = 1000$
(b) $10^4 = 10 \times 10 \times 10 \times 10 = 10\,000$

Summarizing some of the powers of ten in a table:

Powers of Ten	
$10^6 =$	1 000 000
$10^5 =$	100 000
$10^4 =$	10 000
$10^3 =$	1 000
$10^2 =$	100
$10^1 =$	10
$10^0 =$	1

Scientific Notation

The numbers needed for technical work are often very large or very small. For example, a certain computer might have a memory of 200 000 000 bits, and might do a certain operation in 0.0000000004 seconds. Numbers such as these, clumsy to express in decimal notation, are written more easily in *scientific notation*. Calculators and computers switch into scientific notation as needed, or when made to do so.

In this section we learn how to write whole numbers in scientific notation. We'll see that the first number above can be written 2×10^8 bits, and in Chapter 6 we show how to write the second number as 4×10^{-10} seconds.

A number is said to be in *scientific notation* when it is written as a number between 1 and 10, multiplied by a power of ten.

Example 53

Some numbers written in scientific notation are

(a) 7×10^6 (b) 5×10^{15} (c) 6×10^3 (d) 3×10^5

Converting Numbers to Scientific Notation

To convert a number to scientific notation:

1. Write the given number as the product of a number between 1 and 10, times another number that is a power of ten.
2. Rewrite the power of ten as 10^n.

Example 54
Convert 30 000 to scientific notation.

Solution:

1. We write 30 000 as the product of 3 and 10 000.
$$30\ 000 = 3 \times 10\ 000$$
2. Then we rewrite 10 000 as a power of ten.
$$10\ 000 = 10^4$$
so
$$30\ 000 = 3 \times 10^4$$

Example 55
Convert 1 840 000 to scientific notation.

Solution:

1. We write 1 840 000 as the product of 1.84 and 1 000 000.
$$1\ 840\ 000 = 1.84 \times 1\ 000\ 000$$
2. Then we rewrite 1 000 000 as a power of ten.
$$1\ 000\ 000 = 10^6$$
so
$$1\ 840\ 000 = 1.84 \times 10^6$$

Converting from Scientific Notation

To convert from scientific notation, simply reverse the process.

Example 56
Write 9×10^6 as a whole number.

Solution: Using the fact that
$$10^6 = 1\ 000\ 000$$
we get
$$9 \times 10^6 = 9 \times 1\ 000\ 000$$
$$= 9\ 000\ 000$$

SECTION 1-6 Exercises Scientific Notation

Powers of Ten

Evaluate each power of ten.

1. 10^5
2. 10^2
3. 10^1
4. 10^0
5. 10^4
6. 10^6
7. 10^7
8. 10^9

Write each number as a power of ten.

9. 1000
10. 100 000
11. 10 000
12. 100

Converting Numbers to Scientific Notation

Convert each number to scientific notation.

13. 30 000   14. 900 000   15. 2 000 000
16. 800   17. 7000   18. 30 000 000

Converting from Scientific Notation

Convert each number from scientific notation.

19. 3×10^3 20. 8×10^5 21. 2×10^2
22. 9×10^4 23. 7×10^6 24. 5×10^7

1-7   COMBINED OPERATIONS

Order of Operations

If the expression to be evaluated does not contain parentheses, perform the operations in the following order:

     1st   powers and roots, in any order
     2nd   multiplications and divisions, from left to right
     3rd   additions and subtractions, from left to right

We first show a problem containing both addition and multiplication.

Example 57

Evaluate $7 + 3 \times 4$ first without a calculator, and then by calculator.

Solution: The multiplication is done before the addition

$$7 + 3 \times 4 = 7 + 12 = 19$$

A calculator will automatically perform the multiplication before the addition, so we may key the numbers in as written.

Next we do a calculation having both a power and multiplication.

Example 58

Evaluate 5×3^2.

Solution: We raise to the power before multiplying:

$$5 \times 3^2 = 5 \times 9 = 45$$

A calculator will do the squaring operation on the 3 only, so there is no need to enter the 3^2 operation first. The keystrokes are

Parentheses

When an expression contains parentheses, first evaluate the expression within the parentheses and then the entire expression.

Example 59

Evaluate $(7 + 3) \times 4$.

Solution:
$$(7 + 3) \times 4 = 10 \times 4 = 40$$

The keystrokes are

Note that the use of parentheses was essential here.

If the sum or difference of more than one number is to be raised to a power, those numbers must be enclosed in parentheses.

Example 60

Evaluate $(5 + 2)^2$.

Solution: We combine the numbers inside the parentheses before squaring.
$$(5 + 2)^2 = 7^2 = 49$$

There are two ways to do this on a calculator. The first way is to use the $\boxed{\text{ENTER}}$, $\boxed{=}$, or $\boxed{\text{EXE}}$ key to combine the 5 and the 2, before squaring. The keystrokes are

A second way is to use parentheses. The keystrokes are

On some calculators the result will be displayed immediately after pressing the $\boxed{x^2}$ key.

Example 61

Evaluate $(2 + 6)(7 + 9)$.

Solution: The two quantities in parentheses are evaluated before multiplying:
$$(2 + 6)(7 + 9) = 8 \times 16 = 128$$

The keystrokes are

On some calculators you do not need to use the ⊠ key between the two expressions.

Example 62

Evaluate $\dfrac{8+4}{9-3}$.

Solution: *Here the fraction line acts like parentheses,* grouping the 8 and the 4, as well as the 9 and the 3. Written on a single line, this problem would be

$$(8 + 4) \div (9 - 3)$$

or $12 \div 6 = 2$.
 The calculator keystrokes are almost the same as for the preceding example. Enter the numerator first and use the ÷ key instead of the ⊠ key.

SECTION 1-7 Exercises Combined Operations

Perform each computation by calculator.

1. $(37)(28) + (36)(64)$
2. $(22)(53) - (586)(4) + (47)(59)$
3. $(63 + 36)(97 - 37)$
4. $(89 - 74 + 95)(87 - 49)$
5. $\dfrac{219}{73} + \dfrac{194}{97}$
6. $\dfrac{228}{38} - \dfrac{78}{26} + \dfrac{364}{91}$
7. $\dfrac{648 + 699}{337 + 112}$
8. $\dfrac{809 - 463 + 1858}{958 - 364 + 508}$
9. $(5 + 6)^2$
10. $(422 + 113 - 533)^4$
11. $(423 - 420)^3$
12. $\left(\dfrac{853 - 229}{874 - 562}\right)^2$
13. $\left(\dfrac{141}{47}\right)^3$
14. $\sqrt{434 + 466}$
15. $\sqrt{(8)(72)}$
16. $\sqrt[3]{657 + 553 - 1085}$
17. $\sqrt[4]{(27)(768)}$
18. $\sqrt{\dfrac{2404}{601}}$
19. $\sqrt[4]{\dfrac{1136}{71}}$
20. $\sqrt{961} + \sqrt{121}$
21. $\sqrt[4]{625} + \sqrt{961} - \sqrt[3]{216}$
22. $\sqrt[4]{256} \times \sqrt{49}$

Chapter 1 REVIEW EXERCISES

Perform each computation by calculator.

1. 21^2
2. $137 + 29 + 474 + 226$
3. $(947 - 324) - 172$
4. 52×12
5. $237 - 174$
6. 5^3
7. $147 \div 21$
8. $(13 + 96)(57 - 23)$
9. $0 \div 4$
10. 3^4
11. $537 - (206 - 164) + 317$
12. $(934 - 242) - (564 - 364)$
13. 2^5
14. 4^6
15. 25^2
16. $(81)^2$
17. $\sqrt{49}$
18. $19 \times (5 \times 6) \times 3$
19. $\sqrt[3]{125}$
20. $\sqrt[3]{8}$
21. $822 - 130 - 133$
22. $5757 - (584 - 233)$
23. $36 \times 11 \times 53$
24. $(4 \times 29) \times (22 \times 5)$
25. $2552 \div 232$
26. $1245 \div 83$
27. $44 \div 0$
28. $(7 + 2)^2$
29. $(553 - 546)^3$
30. $(22)(74) + (24)(74)$
31. $\dfrac{176}{22} + \dfrac{64}{16}$
32. $\sqrt{2209}$
33. $\sqrt{(9)(49)}$
34. $\sqrt[4]{625} \times \sqrt{81}$
35. $\dfrac{947 + 513}{137 + 228}$

Evaluate each power of ten.

36. 10^3
37. 10^4
38. 10^5
39. State the place value of the 5 and the 6 in the number 62 571.

Convert each number from scientific notation.

40. 73×10^3
41. 4×10^5
42. 2×10^4

Place each of the following numbers on a number line.

43. $\dfrac{5}{8}$
44. -9
45. $\sqrt{7}$

Convert each number to scientific notation.

46. 50 000
47. 300 000
48. 7 000 000

Insert the proper sign of equality or inequality (=, <, >) between each pair of numbers.

49. 2 and 3
50. -7 and -8
51. $\dfrac{6}{3}$ and 2

Write each number as a power of ten.

52. 1000
53. 100 000
54. 1 000 000

55. How many cubic metres of steel are needed to make a casting with a mass of 21 624 kg if the density of steel is 7208 kg/m³? (Hint: *Volume = mass ÷ density*)
56. A 417 hectare parcel of land is to be divided into three equal lots. Find the area of each lot.
57. At the rate of 525 kilometres in 7 hours, how many kilometres would a person drive in 4 hours? (Hint: *Rate = distance ÷ time*)
58. If a person can make 21 lamps in 7 days, how many can be made in 5 days? (Hint: *Work rate = amount done ÷ time worked*)
59. A voltage of 219 V is measured across a 73 Ω resistor. Find the current through the resistor. (Hint: *Current = voltage ÷ resistance*)

60. Find the perimeter of a four-sided figure (a quadrilateral) if the lengths of the four sides are 22 cm, 15 cm, 33 cm, and 18 cm.
61. The sum of the sides of a triangle is 224, and two of the sides are 77 and 103. Find the length of the third side.
62. Four boards having lengths of 134 cm, 31 cm, 229 cm, and 308 cm are placed end to end. What is the total length?
63. What is the cost of 47 tons of gravel at $16 per ton?
64. How much will 37 fans weigh if each weighs 4 lb?
65. If there are 360 degrees per revolution, how many degrees are there in 11 revolutions?
66. The number of feet traveled by a falling object in t seconds is approximately equal to $16t^2$. Find the distance traveled in 21 seconds.
67. Find the length of the side of a cubical shipping container whose volume is 216 cubic feet.

WRITING*

68. Suppose someone tells you, "Of all the kinds of numbers, it's most important to understand the difference between exact and approximate numbers." Do you agree? If so, why is this distinction so important? And just what is the difference between exact and approximate numbers? Write a paragraph or so in which you discuss these questions, and file it with your notes on this chapter.

*Many of the writing questions in this book are from P. Calter, *Technical Mathematics, Third Edition* (Upper Saddle River, NJ: Prentice Hall, 1995).

Common Fractions

Outcomes

When you have completed this chapter, you will be able to

1. Use the terminology of common fractions.
2. Convert a common fraction into equivalent fractions.
3. Convert between mixed numbers and improper fractions.
4. Multiply and divide common fractions.
5. Add and subtract common fractions.
6. Simplify complex fractions.
7. Combine operations involving common fractions.

While the use of calculators and computers, which both work with decimal numbers, has reduced the use of common fractions, it is still important to study them. Fractions are used in many places, such as measurements to fractions of an inch, ratio, and proportion. But more importantly, it will help prepare us for a study of algebraic fractions in Chapter 17, which cannot be handled by most calculators or computers.

2-1 TERMINOLOGY

Numerator and Denominator

Every fraction has a *numerator*, a *denominator*, and a *fraction line*. The numerator is written above the fraction line and the denominator is written below the fraction line.

---- **Example 1** ----

In the fraction $\frac{2}{3}$,

the numerator is 2 and the denominator is 3. You will often see fractions in printed text written with a slanted fraction line, to save space. Thus the fraction in this example may be printed as 2/3. However, we suggest that your handwritten fractions should always be of the *built-up* type, with a horizontal fraction line, for clarity.

We may think of a fraction as a way to specify a portion of a whole quantity. Thus the fraction $\frac{2}{3}$ says, "*Divide the whole into 3 equal parts, and take 2 of them.*"

Here the denominator 3 *denominates* or *names* the size of each part, here thirds.

The numerator 2 *enumerates* or *counts* the number of thirds to be taken.

FIGURE 2-1
Three slices of pizza.

Example 2
We may think of $\frac{3}{8}$ of a pizza, Fig. 2-1, as three pieces of the pizza after it has been divided into 8 equal pieces.

A Fraction as Indicating Division

Another way of thinking of a fraction is as the indicated division of two quantities.

Example 3
The fraction $\frac{5}{9}$ may be thought of as 5 divided by 9. It is also called the *quotient* of 5 divided by 9. Thought of in this way, 5 is the *dividend*, 9 is the *divisor*, and the fraction line is a symbol for division.

Fraction Line as a Symbol of Grouping

We have mentioned symbols of grouping such as parentheses ( ), brackets [ ], and braces { }. A *fraction line* is also a symbol of grouping.

Example 4
In the fraction

$$\frac{9}{2+5}$$

the 9 in the numerator cannot be divided by the 2 or by the 5 separately, because they are *grouped* by the fraction line. The 2 and the 5 must be added together before dividing.

Zero Denominator or Numerator

Recall from Chapter 1 that division by zero is not defined. That means that the denominator of a fraction may not be zero.

Example 5
The fraction

$$\frac{5}{0}$$

is not defined.

A zero *numerator,* of course, gives a fraction that is equal to zero.

Example 6
The value of the fraction

$$\frac{0}{5}$$

is zero.

Common Fractions

A *common fraction* is one in which numerator and denominator are both *integers*.

Example 7

The fractions

$$\frac{3}{5}, \quad -\frac{9}{2}, \quad \text{and} \quad \frac{121}{125}$$

are common fractions. The fraction $\frac{2.97}{5}$ is *not* a common fraction because numerator and denominator are not both integers. The fractions

$$\frac{x}{2}, \quad \frac{7}{y}, \quad \text{and} \quad \frac{a}{b}$$

are *not* common fractions. They are called *algebraic* fractions. We study them in Chapter 17.

Proper and Improper Fractions

To define proper and improper fractions, we must first say what we mean by the absolute value. The *absolute value* of a number is, geometrically, the distance of the number from zero on the number line, regardless of direction. Thus, +4 and −4 are both 4 units from zero in Fig. 2-2. Thus the absolute value of +4 is 4, and the absolute value of −4 is 4.

A *proper fraction* is one whose numerator is smaller in absolute value than its denominator, while an *improper fraction* is one whose numerator is larger in absolute value than or equal to its denominator.

Example 8

Some proper fractions are

$$\frac{2}{3}, \quad \frac{71}{-92}, \quad \text{and} \quad \frac{14}{15}$$

and some improper fractions are

$$\frac{3}{2}, \quad \frac{-32}{21}, \quad \frac{5}{4}, \quad \text{and} \quad \frac{15}{14}$$

Mixed Numbers

A *mixed number* is the sum of a whole number and a proper fraction. Such a sum is written without the + sign. Thus,

$$2 + \frac{3}{4} \quad \text{is written as} \quad 2\frac{3}{4}.$$

■ **FIGURE 2-2**
Both +4 and −4 have an absolute value of 4, because both are 4 units from zero.

Example 9
Some mixed numbers are

$$3\frac{1}{2}, \quad 72\frac{22}{35}, \quad \text{and} \quad 14\frac{3}{4}$$

Similar Fractions
Similar or *like* fractions are fractions having the same denominator.

Example 10
$$\frac{8}{9} \quad \text{and} \quad \frac{2}{9} \quad \text{are similar fractions}$$

while

$$\frac{3}{5} \quad \text{and} \quad \frac{3}{7} \quad \text{are } not \text{ similar fractions}$$

Equivalent (Equal) Fractions
Fractions are called *equivalent* (or equal) if they represent the same value.

Example 11
The fractions

$$\frac{2}{3} \quad \text{and} \quad \frac{6}{9}$$

are equivalent fractions. If you ate $\frac{2}{3}$ of a pie or $\frac{6}{9}$ of that pie (Fig. 2-3), it would be the same amount of pie.

$\frac{2}{3}$ of a pie

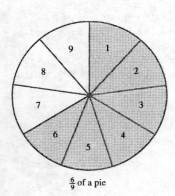
$\frac{6}{9}$ of a pie

■ FIGURE 2-3
Two-thirds of a pie.

Complex Fractions
A *complex fraction* is one with more than one fraction line.

Example 12
Some complex fractions are

$$\frac{\frac{3}{5}}{12}, \quad \frac{5}{1+\frac{4}{7}}, \quad \text{and} \quad \frac{\frac{5}{6}-\frac{7}{8}}{\frac{2}{3}+\frac{4}{5}}$$

SECTION 2-1 Exercises Terminology

Label each as a common fraction, algebraic fraction, proper fraction, improper fraction, mixed number, or complex fraction. More than one name may apply.

1. $\dfrac{4}{5}$ 2. $\dfrac{2}{7}$ 3. $\dfrac{6}{5}$ 4. $1\dfrac{3}{4}$

5. $\dfrac{x}{y}$ 6. $\dfrac{\frac{5}{7}}{3}$ 7. $\dfrac{12}{7}$ 8. $\dfrac{z}{2}$

9. $\dfrac{\frac{w}{3}+5}{\frac{x}{4}}$ 10. $\dfrac{5}{2}$ 11. $7\dfrac{5}{6}$ 12. $\dfrac{9}{4}$

State whether the following pairs of fractions are similar or equivalent.

13. $\dfrac{3}{7}$ and $\dfrac{5}{7}$ 14. $\dfrac{3}{7}$ and $\dfrac{6}{14}$ 15. $\dfrac{5}{3}$ and $\dfrac{15}{12}$ 16. $\dfrac{21}{22}$ and $\dfrac{13}{22}$

2-2   CHANGING THE FORM OF A FRACTION

Fundamental Principle of Fractions

In Sec. 2-1 we said that equivalent fractions $\left(\text{such as } \dfrac{2}{3} \text{ and } \dfrac{6}{9}\right)$ are those that represent the same value. We will now practise writing equivalent fractions. Our tool for doing this is called *the fundamental principle of fractions.*

> If the numerator and the denominator of a fraction are multiplied or divided by the same (nonzero) number, the value of the fraction remains unchanged.

Example 13

If we multiply numerator and denominator of the fraction $\dfrac{5}{8}$, say, by 3,

$$\dfrac{5}{8} = \dfrac{5}{8} \cdot \dfrac{3}{3} = \dfrac{15}{24}$$

we get the fraction $\dfrac{15}{24}$, which is equal to $\dfrac{5}{8}$. You can verify this by getting a decimal value for each on your calculator.

Reducing to Lowest Terms

One common use for the fundamental principle of fractions is to "simplify" a fraction, or *reduce a fraction to lowest terms*. By this we mean to find an equivalent fraction with the smallest possible denominator.

Example 14

Reduce the fraction $\frac{26}{65}$ to lowest terms.

Solution: By trial and error, we find that both numerator and denominator are divisible by 13. Dividing,

$$\frac{26}{65} = \frac{26 \div 13}{65 \div 13} = \frac{2}{5}$$

Canceling

As a shortcut, we can reduce a fraction by striking out any factor contained in both numerator and denominator. Recall from Sec. 1-3 that the factors of some number n are those numbers that we multiply to get n. Thus the factors of 12 are 1, 2, 2, and 3.

Example 15

To reduce the fraction $\frac{26}{65}$, from Example 14, let us first factor numerator and denominator.

$$\frac{26}{65} = \frac{(2)(13)}{(5)(13)}$$

We then strike out the 13 from numerator and denominator, because $\frac{13}{13}$ equals 1.

$$\frac{26}{65} = \frac{(2)\cancel{(13)}}{(5)\cancel{(13)}} = \frac{2}{5}$$

This is called *canceling*. But be careful when canceling. Any factor to be canceled must be a factor of the *entire* numerator and the *entire* denominator.

Reducing by Prime Factoring

Any fraction can be reduced to lowest terms by the use of prime numbers. A prime number is any natural number that can only be divided evenly by itself and one (unity). The prime numbers between zero and twenty are emphasized in the following list.

0, 1, **2**, **3**, 4, **5**, 6, **7**, 8, 9, 10, **11**, 12, **13**, 14, 15, 16, **17**, 18, **19**, 20

Example 16

Reduce the fraction $\frac{980}{2058}$ to lowest terms.

Solution: We convert the numerator and the denominator to *prime factors*

$$\frac{980}{2058} = \frac{(2)(2)(5)(7)(7)}{(2)(3)(7)(7)(7)}$$

Then simplifying or "canceling," we end up with the following

$$\frac{980}{2058} = \frac{\cancel{(2)}(2)(5)\cancel{(7)}\cancel{(7)}}{\cancel{(2)}(3)\cancel{(7)}\cancel{(7)}(7)} = \frac{(2)(5)}{(3)(7)} = \frac{10}{21}$$

This is always a reduction to the lowest term.

Example 17

In the fraction

$$\frac{5+7}{(7)(8)}$$

there is a 7 in both numerator and denominator. However, 7 is *not a factor* of the numerator so 7 *cannot be canceled*.

Writing a Fraction with a Different Denominator

Another use for the fundamental principle of fractions is to change a fraction to an equivalent fraction with a different denominator. The purpose of this is, of course, to be able to combine (add or subtract) that fraction with another one having the same denominator, as we'll show in Sec. 2-5.

Example 18

Rewrite the fraction $\frac{7}{8}$ with a denominator of 40.

Solution: We must multiply the denominator (8) by 5 to make it 40. But the fundamental principle of fractions says that we must multiply the numerator by the same number to avoid changing the value of the fraction. Multiplying gives

$$\frac{7}{8} = \frac{(7)(5)}{(8)(5)} = \frac{35}{40}$$

We sometimes want to write a whole number as a fraction with a given denominator, so that it may be combined with a given fraction.

Example 19

Rewrite the number 3 as a fraction with a denominator of 9.

Solution: We first write 3 as a fraction, and then multiply by $\frac{9}{9}$.

$$3 = \frac{3}{1} = \frac{(3)(9)}{(1)(9)} = \frac{27}{9}$$

Converting an Improper Fraction to a Mixed Number

Recall from Sec. 2-1 that an *improper fraction* $\left(\text{like } \frac{9}{8}\right)$ is one whose numerator is larger than or equal to its denominator, and that a *mixed number* $\left(\text{such as } 4\frac{1}{3}\right)$ is the sum of a whole number and a proper fraction.

To convert an improper fraction to a mixed number, simply divide the numerator by the denominator. Express any remainder as the numerator of a fraction whose denominator is the same as in the original fraction.

Example 20

Write the improper fraction $\frac{13}{5}$ as a mixed number.

Solution: We divide 13 by 5 and get

$$\frac{13}{5} = 13 \div 5 = 2\frac{3}{5}$$

We show how to reverse this process and convert a mixed number to an improper fraction in Sec. 2-5.

SECTION 2-2 Exercises Changing the Form of a Fraction

Reduce to lowest terms.

1. $\frac{4}{12}$ 2. $\frac{5}{25}$ 3. $\frac{42}{21}$ 4. $\frac{9}{3}$

5. $\frac{44}{99}$ 6. $\frac{30}{42}$ 7. $\frac{28}{21}$ 8. $\frac{40}{72}$

Rewrite each fraction with the denominator indicated.

9. $\frac{3}{4}$ with denominator 12   10. $\frac{5}{7}$ with denominator 42

11. $\frac{8}{3}$ with denominator 21   12. $\frac{4}{5}$ with denominator 25

Change each improper fraction to a mixed number.

13. $\frac{12}{5}$ 14. $\frac{22}{7}$ 15. $\frac{15}{6}$

16. $\frac{103}{25}$ 17. $\frac{82}{7}$ 18. $\frac{19}{3}$

2-3 MULTIPLYING FRACTIONS

To multiply a quantity by a fraction means to take that fractional part *of* the quantity.

Example 21

(a) To multiply $\frac{1}{2}$ and 4,

$$\frac{1}{2} \times 4$$

is the same as *taking one half of four*. We know that half of four is two, so

$$\frac{1}{2} \times 4 = \frac{1}{2} \text{ of } 4 = 2$$

as shown in Fig. 2-4.

■ FIGURE 2-4

$\frac{1}{2}$ of 4 = 2

FIGURE 2-5

FIGURE 2-6

(b) $\frac{1}{2} \times \frac{1}{2}$ is the same as half of one half, or $\frac{1}{4}$ as in Fig. 2-5.

(c) $\frac{1}{3} \times \frac{3}{4}$ means one third of three quarters, or $\frac{1}{4}$ as in Fig. 2-6.

In the examples given, we knew the answers from experience. We could also have gotten the same result by using the following rule.

> **RULE** The product of two fractions is a fraction whose numerator is the product of the original numerators and whose denominator is the product of the original denominators.

Example 22

We now repeat Example 21, using the rule. We also reduce our answers to lowest terms.

(a) $\frac{1}{2} \times 4 = \frac{1}{2} \times \frac{4}{1} = \frac{(1)(4)}{(2)(1)} = \frac{4}{2} = 2$

(b) $\frac{1}{2} \times \frac{1}{2} = \frac{(1)(1)}{(2)(2)} = \frac{1}{4}$

(c) $\frac{1}{3} \times \frac{3}{4} = \frac{(1)(3)}{(3)(4)} = \frac{3}{12} = \frac{1}{4}$

It is usually easier to do the reducing before actually multiplying out.

Example 23

Multiply

$$\frac{5}{7} \times \frac{7}{8}$$

Solution:

$$\frac{5}{7} \times \frac{7}{8} = \frac{(5)(7)}{(7)(8)}$$

Now *before* actually multiplying, let us cancel any factor contained in both numerator and denominator.

$$\frac{5}{7} \times \frac{7}{8} = \frac{(5)(\overset{1}{\cancel{7}})}{(\underset{1}{\cancel{7}})(8)} = \frac{5}{8}$$

The method is the same if there are *more than two* fractions to be multiplied.

Example 24

$$\frac{2}{9} \times \frac{1}{2} \times \frac{9}{11} = \frac{(\overset{1}{\cancel{2}})(1)(\overset{1}{\cancel{9}})}{(\underset{1}{\cancel{9}})(\underset{1}{\cancel{2}})(11)} = \frac{1}{11}$$

Our next example shows canceling when one of the factors to be canceled is a multiple of the other factor to be canceled.

Example 25

Multiply $\frac{5}{6}$ and $\frac{8}{9}$.

Solution: We see that both the 8 in the numerator and the 6 in the denominator are divisible by 2.

$$\frac{5}{6} \times \frac{8}{9} = \frac{5}{\underset{3}{\cancel{6}}} \times \frac{\overset{4}{\cancel{8}}}{9} = \frac{(5)(4)}{(3)(9)} = \frac{20}{27}$$

Multiplying Mixed Numbers

We may multiply any combination of fractions, mixed numbers, and whole numbers with the same rule for multiplying fractions. However, before multiplying, convert each mixed number to an improper fraction, and write each whole number as a fraction with 1 in the denominator. If you cancel before multiplying, it simplifies the work and reduces the chance for error.

Example 26

$$\frac{1}{5} \times 6 \times \frac{3}{8} \times 5\frac{1}{3} = \frac{1}{5} \times \frac{6}{1} \times \frac{3}{8} \times \frac{16}{3}$$

$$= \frac{(1)(6)(\overset{1}{\cancel{3}})(\overset{2}{\cancel{16}})}{(5)(1)(\underset{1}{\cancel{8}})(\underset{1}{\cancel{3}})} = \frac{12}{5} = 2\frac{2}{5}$$

Example 27

A current of $2\frac{4}{5}$ amperes (A) flows in a resistor having a resistance of $21\frac{3}{4}$ ohms (Ω). Find the voltage across the resistor by multiplying the current by the resistance.

Solution: Multiplying,

$$Voltage = current \times resistance = 2\frac{4}{5} \times 21\frac{3}{4}$$

$$= \frac{\overset{7}{\cancel{14}}}{5} \times \frac{87}{\underset{2}{\cancel{4}}} = \frac{7}{5} \times \frac{87}{2}$$

$$= \frac{609}{10} = 60\frac{9}{10} \text{ volts}$$

CHAPTER 2 / Common Fractions

SECTION 2-3 Exercises Multiplying Fractions

Multiply. Reduce your result to lowest terms.

1. $\dfrac{2}{3} \times \dfrac{3}{7}$
2. $\dfrac{1}{7} \times \dfrac{3}{5}$
3. $\dfrac{5}{8} \times \dfrac{2}{3}$
4. $\dfrac{5}{12} \times \dfrac{4}{5}$
5. $\dfrac{2}{5} \times \dfrac{3}{4}$
6. $\dfrac{5}{8} \times \dfrac{16}{17}$
7. $\dfrac{7}{15} \times \dfrac{5}{9}$
8. $\dfrac{7}{12} \times \dfrac{8}{11}$
9. $\dfrac{1}{2} \times \dfrac{3}{8} \times \dfrac{2}{3}$
10. $\dfrac{7}{12} \times \dfrac{2}{3} \times \dfrac{4}{5}$
11. $\dfrac{2}{5} \times \dfrac{3}{4} \times \dfrac{1}{3}$
12. $\dfrac{1}{2} \times \dfrac{3}{8} \times \dfrac{4}{5}$
13. $4\dfrac{2}{3} \times \dfrac{3}{4}$
14. $\dfrac{5}{8} \times 3\dfrac{1}{2}$
15. $2\dfrac{4}{15} \times \dfrac{5}{9}$
16. $\dfrac{5}{12} \times 3\dfrac{8}{11}$
17. $3\dfrac{1}{3} \times 2\dfrac{1}{4}$
18. $2\dfrac{3}{8} \times 2\dfrac{1}{4}$
19. $3\dfrac{4}{15} \times 4\dfrac{1}{9}$
20. $1\dfrac{7}{12} \times 4\dfrac{4}{11}$
21. $\dfrac{1}{2} \times 5\dfrac{2}{3} \times \dfrac{3}{4}$
22. $\dfrac{3}{8} \times \dfrac{2}{3} \times 2\dfrac{1}{2}$
23. $1\dfrac{7}{15} \times \dfrac{5}{8} \times \dfrac{1}{2}$
24. $\dfrac{7}{12} \times \dfrac{3}{5} \times 2\dfrac{6}{11}$
25. $3\dfrac{1}{2} \times 3\dfrac{2}{3} \times \dfrac{1}{4}$
26. $\dfrac{7}{8} \times 5\dfrac{2}{3} \times 3\dfrac{1}{3}$
27. $2\dfrac{1}{12} \times \dfrac{7}{8} \times 3\dfrac{1}{2}$
28. $\dfrac{5}{12} \times 1\dfrac{2}{5} \times 3\dfrac{2}{11}$

Applications

29. A gear in a certain mechanism turns at a rate of $3\dfrac{5}{8}$ rev/s. How many revolutions will it make in $2\dfrac{3}{5}$ seconds?

30. If a certain stock costs $\$55\dfrac{7}{8}$ per stock lot, how much will $6\dfrac{2}{3}$ stock lots cost?

31. A municipal landfill is subdivided into cells. It takes an average of $33\dfrac{1}{6}$ truck loads to fill one cell. How many loads are required to fill $4\dfrac{3}{5}$ cells?

♛ 32. A certain car can go $33\dfrac{1}{3}$ miles on a gallon of gasoline. How far will it go on $11\dfrac{7}{8}$ gallons?

2-4 DIVIDING FRACTIONS

Reciprocals

To find the *reciprocal* of a fraction, simply *invert* the fraction, so that the numerator and denominator change places.

Example 28

$\frac{2}{3}$ is the reciprocal of $\frac{3}{2}$

$\frac{5}{3}$ is the reciprocal of $\frac{3}{5}$

Dividing Fractions

Let us divide a number, say 3, by 2. We get

$$3 \div 2 = \frac{3}{2}$$

But by our rule for multiplying fractions, we know that

$$\frac{3}{2} = 3 \times \frac{1}{2}$$

In other words,

3 *divided by* 2 is the same as 3 *multiplied by* $\frac{1}{2}$

But 2 and $\frac{1}{2}$ are reciprocals. So we may say that

3 *divided by* 2 is the same as 3 *multiplied by the reciprocal* of 2.

Of course, this does not work only for the numbers 3 and 2. In more general language, we have,

> To divide by a number is the same as to multiply by the reciprocal of that number.

Thus, *to divide by a fraction, invert the fraction and multiply.*

Example 29

Divide $\frac{1}{5} \div \frac{4}{7}$.

Solution:

$$\frac{1}{5} \div \frac{4}{7} = \frac{1}{5} \times \frac{7}{4} = \frac{7}{20}$$

As with multiplication, try to cancel before multiplying. Don't forget to reduce your answer to lowest terms.

Example 30

$$\frac{3}{28} \div \frac{6}{7} = \frac{\overset{1}{\cancel{3}}}{\underset{4}{\cancel{28}}} \times \frac{\overset{1}{\cancel{7}}}{\underset{2}{\cancel{6}}} = \frac{1}{8}$$

Dividing Whole and Mixed Numbers

To find the reciprocal of a mixed number, first rewrite it as an improper fraction, then invert. To find the reciprocal of a whole number, simply write it as a fraction with 1 in the denominator, and invert as before.

Example 31

Find the reciprocals of (a) 5 and (b) $3\frac{3}{4}$.

Solution:

(a) The reciprocal of 5 is $\frac{1}{5}$.

(b) The mixed number $3\frac{3}{4}$ is equal to the improper fraction $\frac{15}{4}$. The reciprocal of $\frac{15}{4}$ is $\frac{4}{15}$.

If the divisor or the dividend is a mixed number, change it to an improper fraction, and multiply as before. Cancel before multiplying out, and reduce your answer to lowest terms.

Example 32

Divide and simplify.

(a) $6 \div \frac{2}{7} = \frac{6}{1} \div \frac{2}{7} = \frac{\overset{3}{\cancel{6}}}{1} \times \frac{7}{\underset{1}{\cancel{2}}} = \frac{21}{1} = 21$

(b) $\frac{5}{9} \div 3 = \frac{5}{9} \div \frac{3}{1} = \frac{5}{9} \times \frac{1}{3} = \frac{5}{27}$

(c) $2\frac{1}{3} \div \frac{1}{5} = \frac{7}{3} \div \frac{1}{5} = \frac{7}{3} \times \frac{5}{1} = \frac{35}{3} = 11\frac{2}{3}$

(d) $\frac{3}{8} \div 4\frac{1}{2} = \frac{3}{8} \div \frac{9}{2} = \frac{\overset{1}{\cancel{3}}}{\underset{4}{\cancel{8}}} \times \frac{\overset{1}{\cancel{2}}}{\underset{3}{\cancel{9}}} = \frac{1}{12}$

(e) $9 \div 3\frac{1}{2} = \frac{9}{1} \div \frac{7}{2} = \frac{9}{1} \times \frac{2}{7} = \frac{18}{7} = 2\frac{4}{7}$

(f) $1\frac{9}{10} \div 2\frac{4}{5} = \frac{19}{10} \div \frac{14}{5} = \frac{19}{\underset{2}{\cancel{10}}} \times \frac{\overset{1}{\cancel{5}}}{14} = \frac{19}{28}$

Example 33

A kennel owner has $28\frac{3}{8}$ cans of dog food left in stock. If each dog is fed an average of $\frac{4}{5}$ of a can of dog food mixed in with dry food, how many servings does he have left? Find the number of servings by dividing the number of cans by the serving size.

Solution: Dividing

$$\text{Number of servings} = \frac{\text{number of cans}}{\text{serving size}} = 28\frac{3}{8} \div \frac{4}{5}$$

$$= \frac{227}{8} \times \frac{5}{4} = \frac{1135}{32}$$

$$= 35\frac{15}{32} \text{ servings}$$

2-4 Dividing Fractions

SECTION 2-4 Exercises Dividing Fractions

Reciprocals Find the reciprocal of each number or fraction.

1. 2
2. $\frac{1}{7}$
3. $\frac{2}{3}$
4. $\frac{4}{5}$
5. $\frac{12}{13}$
6. 16

Dividing Fractions Divide. Reduce your result to lowest terms.

7. $\frac{2}{3} \div \frac{5}{7}$
8. $\frac{2}{7} \div \frac{3}{5}$
9. $\frac{3}{8} \div \frac{2}{3}$
10. $\frac{7}{12} \div \frac{4}{5}$
11. $\frac{1}{5} \div \frac{3}{4}$
12. $\frac{5}{8} \div \frac{13}{17}$
13. $\frac{1}{15} \div \frac{5}{9}$
14. $\frac{7}{22} \div \frac{8}{11}$
15. $14 \div \frac{2}{7}$
16. $25 \div \frac{2}{5}$
17. $9 \div \frac{2}{3}$
18. $15 \div \frac{3}{5}$
19. $\frac{1}{5} \div 12$
20. $\frac{7}{8} \div 24$
21. $\frac{2}{15} \div 21$
22. $\frac{5}{12} \div 3$
23. $1\frac{2}{3} \div \frac{3}{4}$
24. $\frac{3}{8} \div 3\frac{1}{2}$
25. $3\frac{4}{15} \div \frac{7}{9}$
26. $\frac{7}{12} \div 3\frac{8}{11}$
27. $2\frac{1}{3} \div 5\frac{1}{4}$
28. $5\frac{1}{8} \div 2\frac{1}{4}$
29. $5\frac{4}{15} \div 3\frac{1}{9}$
30. $2\frac{5}{12} \div 4\frac{4}{11}$
31. $28 \div 3\frac{3}{7}$
32. $22 \div 2\frac{3}{5}$
33. $4 \div 2\frac{2}{3}$
34. $18 \div 4\frac{4}{5}$
35. $3\frac{2}{5} \div 16$
36. $2\frac{5}{8} \div 24$
37. $3\frac{7}{15} \div 24$
38. $3\frac{7}{12} \div 3$
39. $34\frac{2}{3} \div 6\frac{3}{4}$
40. $2\frac{5}{8} \div 8\frac{1}{2}$
41. $3\frac{4}{15} \div 6\frac{5}{9}$
42. $2\frac{5}{12} \div 6\frac{8}{11}$

Applications

43. The annual interest on a loan is $6\frac{3}{8}$%. For monthly interest, the rate is divided equally over 12 months. What is the monthly interest rate?

44. A developer has built 82 houses and has completed $\frac{1}{3}$ of the next house. The total time to date is $21\frac{2}{3}$ months. Find the average number of months it takes to build a house by dividing the number of houses by the number of months.

45. The power to a certain resistor is $82\frac{1}{3}$ watts when a current of $3\frac{4}{5}$ amperes flows through the resistor. Find the voltage across the resistor by dividing the power by the current.

 46. A force of $28\frac{3}{4}$ lb will stretch a certain spring a distance of $1\frac{5}{16}$ in. from its unstretched or "free" length, as in Fig. 2-7. Find the spring constant by dividing the force by the distance.

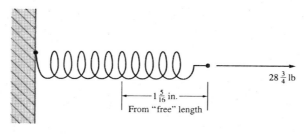

FIGURE 2-7
A spring stretched by a force.

2-5 ADDING AND SUBTRACTING FRACTIONS

In this section we show how to add and subtract fractions with the same denominators or with different denominators. We also show how to add and subtract fractions with whole or with mixed numbers. When we add or subtract fractions (or other quantities as well), we say that we *combine* them.

Adding and Subtracting Similar Fractions

Recall that *similar fractions,* or *like* fractions, are those having the same denominator. To add or subtract similar fractions, simply add or subtract the numerators over the common denominator.

Example 34

Combine the fractions

$$\frac{5}{8} + \frac{1}{8} - \frac{3}{8}$$

Solution: The denominators are the same for the three fractions, so we combine the numerators over that denominator.

$$\frac{5}{8} + \frac{1}{8} - \frac{3}{8} = \frac{5 + 1 - 3}{8} = \frac{3}{8}$$

Least Common Denominator

A *common denominator* for two or more fractions is a common multiple of the individual denominators.

Example 35

A common denominator for the fractions

$$\frac{3}{5} \quad \text{and} \quad \frac{1}{10}$$

is 50, because 50 is a multiple of both 5 and 10. Other common denominators of these fractions are 10, 20, 30, 40, 60, and so on.

The *least common denominator,* or LCD, of two or more fractions is the smallest of the common denominators of those fractions.

Example 36
The LCD in Example 35 is 10, the smallest of the common denominators.

Adding and Subtracting Fractions with Different Denominators

The first step is to rewrite each given fraction as an equivalent fraction so that both have a common denominator. Any common denominator will work. However, if we use the least common denominator, our result will not have to be reduced later.

Example 37
Add $\frac{3}{5}$ and $\frac{1}{10}$.

Solution: Here the denominators are different, 5 and 10. We saw in Example 36 that the LCD for these fractions is 10. So we rewrite the first fraction as an equivalent fraction having a denominator of 10, and combine as above.

$$\frac{3}{5} + \frac{1}{10} = \frac{(3)(2)}{(5)(2)} + \frac{1}{10} = \frac{6}{10} + \frac{1}{10} = \frac{6+1}{10} = \frac{7}{10}$$

It is often necessary to rewrite *both* given fractions to obtain equivalent fractions with a common denominator.

Example 38
Add $\frac{3}{4}$ and $\frac{1}{3}$.

Solution: The denominators are different here also. This time, though, we cannot change one to equal the other; we must change *both*. The smallest multiple of both 4 and 3 is 12, so 12 is the least common denominator. We thus rewrite each fraction as an equivalent fraction having a denominator of 12.

$$\frac{3}{4} + \frac{1}{3} = \frac{(3)(3)}{(4)(3)} + \frac{(1)(4)}{(3)(4)} = \frac{9}{12} + \frac{4}{12} = \frac{9+4}{12} = \frac{13}{12}$$

or $1\frac{1}{12}$, as a mixed number.

When combining fractions with larger denominators, the LCD can more easily be found by separating each denominator into its prime factors, as shown in the following example.

Example 39
Combine

$$\frac{5}{12} + \frac{1}{18} - \frac{7}{30}$$

Solution: Let us separate each denominator into its prime factors.

$$\frac{5}{12} + \frac{1}{18} - \frac{7}{30} = \frac{5}{(2)(2)(3)} + \frac{1}{(2)(3)(3)} - \frac{7}{(2)(3)(5)}$$

A common denominator must contain each of the prime factors from each of the three denominators, or

$$\text{LCD} = (2)(2)(3)(3)(5) = 180$$

Notice that the LCD has *two* 2's. That is because the first denominator, 12, has two 2's as factors. Similarly, our LCD has two 3's, because the denominator 18 has two 3's in its factors. Now we multiply each fraction by the factors needed to make the denominator equal to 180.

$$\frac{5}{(2)(2)(3)} + \frac{1}{(2)(3)(3)} - \frac{7}{(2)(3)(5)}$$

$$= \frac{5}{(2)(2)(3)} \times \frac{(3)(5)}{(3)(5)} + \frac{1}{(2)(3)(3)} \times \frac{(2)(5)}{(2)(5)} - \frac{7}{(2)(3)(5)} \times \frac{(2)(3)}{(2)(3)}$$

$$= \frac{5}{12} \times \frac{15}{15} + \frac{1}{18} \times \frac{10}{10} - \frac{7}{30} \times \frac{6}{6}$$

$$= \frac{75}{180} + \frac{10}{180} - \frac{42}{180} = \frac{43}{180}$$

Combining Fractions and Whole Numbers

To add a fraction and a whole number, simply express them as the parts of a mixed number.

---- **Example 40** ----

$$5 + \frac{1}{8} = 5\frac{1}{8}$$

One way to *subtract* a fraction from a whole number is to *borrow* from the whole number, as shown here.

---- **Example 41** ----

Subtract $\frac{29}{32}$ from 5.

Solution: We borrow 1 from the 5, and call it $\frac{32}{32}$.

$$5 - \frac{29}{32} = 4 + \frac{32}{32} - \frac{29}{32} = 4 + \frac{32-29}{32} = 4 + \frac{3}{32} = 4\frac{3}{32}$$

Converting a Mixed Number to an Improper Fraction

To convert a mixed number to an improper fraction, write the integer part as a fraction with the same denominator as the fraction part, and add.

Example 42

Write the mixed number $4\frac{2}{3}$ as an improper fraction.

Solution: We rewrite the integer part (4) as a fraction $\left(\frac{12}{3}\right)$.

$$4\frac{2}{3} = \frac{4}{1} + \frac{2}{3} = \frac{4(3)}{3} + \frac{2}{3}$$

$$= \frac{12}{3} + \frac{2}{3} = \frac{12 + 2}{3} = \frac{14}{3}$$

Adding and Subtracting Fractions and Mixed Numbers

To add or subtract mixed numbers, or mixed numbers and fractions, we combine the whole numbers and fractions *separately*.

Example 43

Combine $6\frac{3}{4} + 2\frac{1}{2} - 3\frac{1}{8}$.

Solution: We group the whole numbers and fractions separately.

$$6\frac{3}{4} + 2\frac{1}{2} - 3\frac{1}{8} = (6 + 2 - 3) + \left(\frac{3}{4} + \frac{1}{2} - \frac{1}{8}\right) = 5 + \left(\frac{3}{4} + \frac{1}{2} - \frac{1}{8}\right)$$

The LCD for the three fractions is 8, so

$$5 + \left(\frac{3}{4} + \frac{1}{2} - \frac{1}{8}\right) = 5 + \left(\frac{6}{8} + \frac{4}{8} - \frac{1}{8}\right) = 5 + \frac{9}{8} = 6\frac{1}{8}$$

When *subtracting* mixed numbers, we must sometimes *borrow* from the whole number part, as we did when subtracting a fraction from a whole number.

Example 44

Subtract $7\frac{1}{5} - 2\frac{4}{5}$.

Solution: We see that we cannot subtract $\frac{4}{5}$ from $\frac{1}{5}$, so we borrow 1 from the 7 and call it $\frac{5}{5}$.

$$7\frac{1}{5} - 2\frac{4}{5} = 6 + \frac{5}{5} + \frac{1}{5} - 2\frac{4}{5}$$

$$= (6 - 2) + \left(\frac{5}{5} + \frac{1}{5} - \frac{4}{5}\right) = 4 + \frac{2}{5} = 4\frac{2}{5}$$

We must sometimes add and subtract fractions, whole numbers, and mixed numbers, as shown here.

---- **Example 45** ----

Combine $9 + 3\frac{5}{6} - 5\frac{2}{5} + \frac{2}{3}$.

Solution: We combine the whole numbers and fractions separately.

$$9 + 3\frac{5}{6} - 5\frac{2}{5} + \frac{2}{3} = (9 + 3 - 5) + \left(\frac{5}{6} - \frac{2}{5} + \frac{2}{3}\right)$$

We combine, using an LCD of 30 for the fractions.

$$(9 + 3 - 5) + \left(\frac{5}{6} - \frac{2}{5} + \frac{2}{3}\right) = 7 + \frac{25}{30} - \frac{12}{30} + \frac{20}{30}$$

$$= 7 + \frac{33}{30} = 7\frac{11}{10} = 8\frac{1}{10}$$

---- **Example 46** ----

A doorway (Fig. 2-8) measures $43\frac{3}{4}$ in. to the outside of the door frame. Each frame has a width of $5\frac{7}{32}$ in. Find the distance between the frames.

Solution: The two frames have a combined width of

$$5\frac{7}{32} + 5\frac{7}{32} = 10\frac{14}{32} = 10\frac{7}{16}$$

■ **FIGURE 2-8**
A doorway.

2-5 Adding and Subtracting Fractions

We now subtract this amount from $43\frac{3}{4}$.

$$43\frac{3}{4} - 10\frac{7}{16} = 43\frac{12}{16} - 10\frac{7}{16}$$
$$= 33\frac{5}{16} \text{ in.}$$

SECTION 2-5　Exercises　Adding and Subtracting Fractions

Similar Fractions　　Combine and reduce to lowest terms.

1. $\frac{2}{7} + \frac{3}{7}$
2. $\frac{5}{11} + \frac{3}{11}$
3. $\frac{5}{8} + \frac{1}{8}$
4. $\frac{3}{16} + \frac{5}{16}$
5. $\frac{9}{13} - \frac{6}{13}$
6. $\frac{7}{11} - \frac{5}{11}$
7. $\frac{8}{21} - \frac{7}{21}$
8. $\frac{7}{15} - \frac{4}{15}$
9. $\frac{2}{5} + \frac{8}{5} - \frac{6}{5}$
10. $\frac{11}{15} - \frac{2}{15} - \frac{4}{15}$
11. $\frac{4}{13} - \frac{1}{13} + \frac{5}{13} - \frac{3}{13}$
12. $\frac{21}{22} - \frac{9}{22} - \frac{3}{22} + \frac{5}{22}$

Finding the LCD　　Find the least common denominator for the given fractions.

13. $\frac{5}{8}$ and $\frac{3}{16}$
14. $\frac{2}{5}$ and $\frac{4}{15}$
15. $\frac{2}{3}$ and $\frac{3}{4}$
16. $\frac{5}{11}$ and $\frac{4}{33}$
17. $\frac{3}{5}$ and $\frac{4}{7}$
18. $\frac{4}{9}$ and $\frac{3}{5}$

Fractions with Different Denominators　　Combine and simplify.

19. $\frac{2}{3} + \frac{1}{6}$
20. $\frac{5}{11} + \frac{3}{22}$
21. $\frac{3}{8} + \frac{1}{16}$
22. $\frac{3}{16} + \frac{5}{8}$
23. $\frac{9}{11} - \frac{5}{22}$
24. $\frac{5}{6} - \frac{7}{12}$
25. $\frac{7}{24} - \frac{1}{12}$
26. $\frac{7}{15} - \frac{11}{30}$
27. $\frac{2}{5} + \frac{8}{15} - \frac{11}{30}$
28. $\frac{11}{24} - \frac{1}{12} - \frac{5}{48}$
29. $\frac{4}{13} - \frac{1}{26} + \frac{5}{39} - \frac{3}{52}$
30. $\frac{3}{22} + \frac{9}{11} - \frac{3}{11} + \frac{5}{33}$

Fractions and Whole Numbers

Combine the fractions and whole numbers.

31. $\dfrac{2}{3} + 6$

32. $7 + \dfrac{3}{22}$

33. $6 + \dfrac{3}{16}$

34. $\dfrac{5}{11} + 9$

Fractions and Mixed Numbers

Convert each mixed number to an improper fraction.

35. $3\dfrac{4}{5}$

36. $5\dfrac{2}{3}$

37. $1\dfrac{3}{4}$

38. $7\dfrac{2}{7}$

Combine the fractions and mixed numbers.

39. $\dfrac{1}{3} + 5\dfrac{1}{6}$

40. $3\dfrac{5}{11} + \dfrac{5}{22}$

41. $7\dfrac{3}{8} + \dfrac{3}{16}$

42. $\dfrac{5}{16} + 3\dfrac{3}{8}$

43. $5\dfrac{7}{11} - \dfrac{4}{22}$

44. $3\dfrac{5}{12} - \dfrac{1}{6}$

45. $2\dfrac{1}{24} - \dfrac{7}{12}$

46. $3\dfrac{1}{15} - \dfrac{13}{30}$

47. $5\dfrac{3}{5} + \dfrac{7}{15} - 1\dfrac{13}{30}$

48. $2\dfrac{1}{24} - \dfrac{11}{12} - 1\dfrac{1}{48}$

49. $6\dfrac{2}{13} - \dfrac{3}{26} + \dfrac{7}{39} - 1\dfrac{5}{52}$

50. $8\dfrac{5}{22} - 2\dfrac{7}{11} - 1\dfrac{3}{11} + 2\dfrac{7}{33}$

Applications

51. The total markup on shoes at a retail store is $71\dfrac{3}{4}$%. The markup is a combination of fixed overhead which is $35\dfrac{7}{16}$%, unexplained losses that amount to $3\dfrac{3}{32}$%, and finally profit. What percent of the markup is profit?

52. A stock whose price was $\$55\dfrac{1}{2}$ gained $\$\dfrac{7}{8}$ and then lost $\$1\dfrac{3}{4}$. What was the final price?

53. A production time inventory system is used to estimate the amount of time it takes to safely produce a item. Assembly takes $3\dfrac{4}{5}$ minutes, packaging takes $1\dfrac{1}{4}$ minutes and shipping preparation takes $\dfrac{2}{3}$ minute. What is the total estimated time for the production of this item?

54. A car trip took $5\dfrac{3}{4}$ hours, including a rest stop of $\dfrac{37}{60}$ of an hour. What was the total driving time?

55. A rectangular garden has a length of $32\dfrac{7}{8}$ ft and a width of $9\dfrac{3}{5}$ ft. Find the perimeter (the distance around) the garden.

56. A cast part having a length of $4\dfrac{35}{64}$ in. has $\dfrac{7}{32}$ in. ground off one end, Fig. 2-9. That end is then plated with a thickness of $\dfrac{1}{64}$ in. of nickel. Find the final length of the part (Fig. 2-9).

FIGURE 2-9
A cast part.

$4\dfrac{35}{64}$ in.

Grind off $\dfrac{7}{32}$ in.

Then plate $\dfrac{1}{64}$ in. thick

2-5 Adding and Subtracting Fractions

2-6 COMPLEX FRACTIONS

Recall from Sec. 2-1 that a complex fraction is one that has more than one fraction line.

Example 47

Some complex fractions are

$$\frac{\frac{2}{3}}{17}, \quad \frac{2}{3+\frac{5}{7}}, \quad \text{and} \quad \frac{\frac{7}{2}-\frac{3}{4}}{\frac{5}{3}+\frac{1}{6}}$$

To *simplify* a complex fraction, we perform the indicated operations to obtain a common fraction or a mixed number.

Example 48

Simplify the complex fraction

$$\frac{\frac{2}{3}}{\frac{5}{7}}$$

Solution: Here we have one fraction $\left(\frac{2}{3}\right)$ divided by another $\left(\frac{5}{7}\right)$. So we perform the indicated division.

$$\frac{\frac{2}{3}}{\frac{5}{7}} = \frac{2}{3} \div \frac{5}{7} = \frac{2}{3} \times \frac{7}{5} = \frac{14}{15}$$

Example 49

Simplify the complex fraction

$$\frac{\frac{3}{5}+\frac{2}{3}}{\frac{1}{3}}$$

Solution: We add the fractions in the numerator, and then divide by the fraction in the denominator.

$$\frac{\frac{3}{5}+\frac{2}{3}}{\frac{1}{3}} = \frac{\frac{9}{15}+\frac{10}{15}}{\frac{1}{3}}$$

$$= \frac{19}{15} \div \frac{1}{3} = \frac{19}{\cancel{15}_{5}} \times \frac{\cancel{3}^{1}}{1} = \frac{19}{5} = 3\frac{4}{5}$$

Remember that a fraction line is also a symbol of grouping, so all the fractions grouped by a fraction line must be combined first.

Example 50

Simplify the complex fraction

$$\frac{\frac{3}{8} - \frac{1}{4}}{\frac{1}{2} + \frac{2}{3}}$$

Solution: We start by combining the fractions in the numerator, and those in the denominator.

$$\frac{\frac{3}{8} - \frac{1}{4}}{\frac{1}{2} + \frac{2}{3}} = \frac{\frac{3}{8} - \frac{2}{8}}{\frac{3}{6} + \frac{4}{6}} = \frac{\frac{1}{8}}{\frac{7}{6}} = \frac{1}{8} \div \frac{7}{6} = \frac{1}{\underset{4}{8}} \times \frac{\overset{3}{6}}{7} = \frac{3}{28}$$

Alternate Solution: Here, we start by multiplying numerator and denominator of the original fraction by the LCD of all individual fractions, 24.

$$\frac{\frac{3}{8} - \frac{1}{4}}{\frac{1}{2} + \frac{2}{3}} \cdot \frac{24}{24} = \frac{9 - 6}{12 + 16} = \frac{3}{28}$$

Example 51

A steel plate, Fig. 2-10, is $10\frac{7}{8}$ in. long. We want to shorten it by $2\frac{11}{16}$ in. and then drill a hole at one quarter of the new length from the end of the plate. The distance d to the hole can be expressed by the complex fraction

$$\frac{10\frac{7}{8} - 2\frac{11}{16}}{4}$$

Simplify this complex fraction.

■ **FIGURE 2-10**
A steel plate.

Solution:

$$\frac{10\frac{7}{8} - 2\frac{11}{16}}{4} = \frac{\frac{87}{8} - \frac{43}{16}}{4} = \frac{\frac{174}{16} - \frac{43}{16}}{4}$$

$$= \frac{131}{16} \times \frac{1}{4} = \frac{131}{64} = 2\frac{3}{64} \text{ in.}$$

SECTION 2-6 Exercises Complex Fractions

Simplify.

1. $\dfrac{\frac{1}{4}}{\frac{4}{5}}$ 2. $\dfrac{\frac{2}{5}}{\frac{2}{15}}$ 3. $\dfrac{\frac{4}{5}}{\frac{5}{6}}$ 4. $\dfrac{\frac{2}{3}}{\frac{3}{4}}$

5. $\dfrac{\frac{5}{7}}{\frac{2}{3}}$ 6. $\dfrac{\frac{2}{5}}{\frac{1}{7}}$ 7. $\dfrac{\frac{1}{3}+\frac{1}{2}}{\frac{1}{4}}$ 8. $\dfrac{\frac{2}{3}+\frac{5}{6}}{\frac{1}{3}}$

9. $\dfrac{\frac{2}{5}}{\frac{1}{5}+\frac{3}{5}}$ 10. $\dfrac{\frac{3}{4}}{\frac{1}{5}+\frac{3}{10}}$ 11. $\dfrac{2}{3+\frac{5}{7}}$ 12. $\dfrac{3}{5+\frac{2}{5}}$

13. $\dfrac{\frac{1}{2}-\frac{1}{4}}{\frac{1}{3}+\frac{1}{6}}$ 14. $\dfrac{\frac{7}{2}-\frac{3}{4}}{\frac{5}{3}+\frac{1}{6}}$

Chapter 2 REVIEW EXERCISES

Perform the indicated operations and reduce to lowest terms.

1. $\dfrac{1}{3} \div \dfrac{5}{8}$ 2. $\dfrac{3}{7} \div \dfrac{3}{8}$ 3. $\dfrac{5}{8} + \dfrac{7}{16}$

4. $\dfrac{5}{16} + \dfrac{5}{9}$ 5. $\dfrac{1}{2} \times \dfrac{5}{8} \times \dfrac{1}{3}$ 6. $\dfrac{5}{12} \times \dfrac{2}{5} \times \dfrac{4}{5}$

7. $3\dfrac{7}{15} \times \dfrac{2}{9}$ 8. $\dfrac{7}{12} \times 3\dfrac{7}{11}$ 9. $\dfrac{5}{12} - \dfrac{1}{24} + \dfrac{3}{24} - \dfrac{3}{48}$

10. $\dfrac{5}{22} + \dfrac{7}{11} - \dfrac{3}{11} + \dfrac{7}{33}$ 11. $4 + \dfrac{7}{16}$ 12. $\dfrac{7}{11} + 8$

13. $\dfrac{1}{4} \times 3\dfrac{2}{3} \times \dfrac{1}{4}$ 14. $\dfrac{5}{8} \times \dfrac{2}{3} \times 5\dfrac{1}{2}$ 15. $\dfrac{2}{3} + 7\dfrac{1}{6}$

16. $2\dfrac{5}{11} + \dfrac{7}{22}$ 17. $\dfrac{3}{15} \div \dfrac{7}{9}$ 18. $\dfrac{5}{22} \div \dfrac{7}{11}$

19. $\dfrac{3}{5} \div 22$ 20. $\dfrac{7}{8} \div 44$ 21. $1\dfrac{1}{44} - \dfrac{9}{22}$

22. $5\dfrac{2}{15} - \dfrac{11}{30}$ 23. $3\dfrac{2}{5} + \dfrac{2}{15} - 3\dfrac{11}{30}$ 24. $5\dfrac{1}{8} - \dfrac{7}{12} - 2\dfrac{1}{48}$

25. $7\dfrac{2}{5} \div \dfrac{8}{5}$ 26. $\dfrac{5}{22} \div 4\dfrac{8}{11}$ 27. $30 \div 5\dfrac{2}{7}$

28. $14 \div 6\dfrac{3}{5}$ 29. $8\dfrac{2}{15} \div 14$ 30. $5\dfrac{5}{12} \div 4$

31. $\dfrac{7}{8} \times \dfrac{1}{3}$ 32. $\dfrac{7}{12} \times \dfrac{3}{5}$ 33. $\dfrac{3}{8} + \dfrac{5}{8}$

34. $\dfrac{7}{16} + \dfrac{3}{16}$ 35. $\dfrac{\frac{1}{3}}{9}$ 36. $\dfrac{\frac{3}{5}}{7}$

50 CHAPTER 2 / Common Fractions

37. $\dfrac{\frac{2}{3}+\frac{3}{5}}{\frac{1}{6}}$
38. $\dfrac{\frac{2}{7}}{\frac{3}{7}+\frac{5}{7}}$
39. $\dfrac{6}{3+\frac{1}{5}}$
40. $\dfrac{\frac{2}{3}-\frac{1}{5}}{\frac{1}{3}+\frac{4}{5}}$

Find the reciprocal.

41. 7
42. $\dfrac{1}{3}$

Convert each mixed number to an improper fraction.

43. $5\dfrac{1}{3}$
44. $2\dfrac{4}{7}$

Convert each improper fraction to a mixed number.

45. $\dfrac{22}{7}$
46. $\dfrac{17}{3}$

47. It takes $124\dfrac{3}{5}$ hours to plow a farm. This farm is subdivided into 7 equal fields. How much time should it take to plow each field?

48. It was found that the aerial distribution of $14\dfrac{2}{5}$ airplane-loads of bait, containing antirabies vaccine for foxes, covers $\dfrac{3}{8}$ of a township designated as the test area to be treated. Find the density of coverage for the whole township. (Hint: To find the density of coverage for the total test area, divide the number of loads by the size of the test area already covered.)

49. A conveyor belt makes $43\dfrac{1}{5}$ rotations in one minute. How many rotations will it make in 5 minutes and 20 seconds? Hint: The time in minutes and seconds can be simplified as follows.

$$5 \text{ min } 20 \text{ s} = 5\dfrac{20}{60} \text{ min} = 5\dfrac{1}{3} \text{ min}$$

50. A cylinder has volume of 546 cm³. A machinist turns it to a cone. The volume of the cone is 182 cm³. What fraction of the cylinder is the cone?

51. A variable loan had three interest rate changes during the year. For the first 4 months, the rate was $6\dfrac{1}{4}$%, during the second 4 month period the rate was $7\dfrac{1}{2}$%, and in the final 4 months it dropped to $5\dfrac{7}{8}$%. What was the average interest rate for the full year?

52. A forest fire crew is paid on the basis of $104 per day. In one week, Fred worked $6\dfrac{3}{4}$ days, Joanne worked $5\dfrac{7}{8}$ days, and George worked $3\dfrac{1}{2}$ days. What was the total number of days that had to be paid and the total cost for the week?

53. A machine with a weight of $73\dfrac{3}{5}$ lb is placed in a box with a weight of $11\dfrac{3}{4}$ lb. The machine is surrounded by $4\dfrac{1}{3}$ lb of packing material. What is the combined weight?

54. A rectangular courtyard has a length of $7\frac{3}{8}$ yd and a width of $4\frac{4}{5}$ yd. Find the perimeter (the distance around) of the courtyard.

55. A force of $8\frac{3}{5}$ lb will stretch a certain spring a distance of $\frac{3}{16}$ in. Find the spring constant by dividing the force by the distance.

56. A pressure hose coupling cross section has an area of $12\frac{1}{3}$ square inches. A splitter coupling is attached with 3 new smaller same size outlets. In order to maintain pressure the total cross sectional area of the new outlets must be the same. What is the area of one outlet?

WRITING

57. Suppose that a vocal member of your local school board says that the study of fractions is no longer important now that we have computers and insists that it be cut from the curriculum to save money.

    Write a short letter to the editor of your local paper in which you agree or disagree. Give your reasons for eliminating or retaining the study of fractions.

TEAM PROJECT*

You will find a team project suggested at the end of most chapters. These projects are generally longer and more complicated than the usual exercise, and are meant to be done by several people working as a team over a few days.

If your class decides to do one or more of these, you should divide into teams of about three to six students. Then each team should:

    Study the problem.
    Estimate the answer(s).
    List possible tools and methods for solution. Choose one.
    Solve the problem.
    Check the answer.
    Write a brief report.

Students now have more computational tools at their disposal than anyone has had before. Depending on the problem, you might choose from the following:

    Manual computation
    Trial and error
    Scientific calculator
    Graphing calculator
    Computer, using BASIC, a spreadsheet like Lotus, mathematical computation software like *MathCad*, or a computer algebra program like *Derive, Maple,* or *Mathematica*.

*Many of the team projects in this book are from P. Calter, *Technical Mathematics, Third Edition* (Upper Saddle River, NJ: Prentice Hall, 1995).

You might choose *more than one* tool, perhaps doing the computation by calculator and checking the results by computer. For some projects you may want to borrow measuring equipment like a tape measure or a transit, or consult reference books. Sometimes it may be useful to make a model. Be sure to consider *any* aids that may help you.

Your first team project is about the ancient Egyptians' method of decomposing fractions.

58. The ancient Egyptians wrote each fraction $\left(\text{except for } \frac{2}{3}\right)$ as the sum of unit fractions (a fraction with a numerator of 1). Thus,

$$\frac{3}{5} \text{ would be written as } \frac{1}{2} + \frac{1}{10}$$

and

$$\frac{5}{7} \text{ would be written as } \frac{1}{2} + \frac{1}{7} + \frac{1}{14}$$

This is called the *decomposition* of a fraction. Repetitions were not allowed. Thus $\frac{3}{5}$ would not be written as

$$\frac{1}{5} + \frac{1}{5} + \frac{1}{5}$$

Decompose each of the following proper fractions into unit fractions.

$$\frac{3}{4} \quad \frac{2}{5} \quad \frac{4}{5} \quad \frac{5}{6} \quad \frac{2}{7} \quad \frac{3}{7} \quad \frac{4}{7} \quad \frac{6}{7}$$

$$\frac{3}{8} \quad \frac{5}{8} \quad \frac{7}{8} \quad \frac{2}{9} \quad \frac{4}{9} \quad \frac{5}{9} \quad \frac{7}{9} \quad \frac{8}{9}$$

Review Exercises

Arithmetic with Decimal Numbers

Outcomes

When you have completed this chapter, you will be able to

1. Define decimal numbers and the nature of significant digits in rounding.
2. Add and subtract decimal numbers.
3. Multiply and divide decimal numbers.
4. Calculate values of powers and roots of decimal numbers.
5. Write decimal numbers in scientific notation.
6. Enter decimal numbers in scientific notation form into the scientific calculator.
7. Perform calculations with numbers in scientific notation.

We did arithmetic with whole numbers in Chapter 1, and arithmetic with fractions in Chapter 2. Here we do arithmetic with decimal numbers. In this chapter we will do arithmetic with *approximate numbers,* the kind of numbers most common in technical work, as well as with exact numbers.

We will learn to distinguish between exact and approximate numbers, to calculate with them, and to round our answers properly. We expand our work with reciprocals, powers, roots, and scientific notation to approximate numbers. Finally, we cover combined operations on the calculator.

We do not yet cover arithmetic with negative numbers; we will do those in Chapter 6.

3-1 DECIMAL NUMBERS

Reading Decimal Numbers

In technical work, numbers are usually written in the familiar *decimal system.*

Example 1

Some decimal numbers are

$$3.50, \quad 137.3, \quad 1.0000, \quad \text{and} \quad 2746$$

We say that the decimal system uses a *base of 10* because it takes 10 units in any place to equal 1 unit in the next higher place.

The names of the places relative to the decimal point are shown in Fig. 3-1. We use those names when reading a decimal number.

FIGURE 3-1
Values of the positions in a decimal number.

> **Example 2**
> Read the numbers
> **(a)** 16 374.2 **(b)** 0.347
>
> **Solution:** We read
> **(a)** *sixteen thousand, three hundred seventy-four, and two tenths*
> **(b)** *three hundred forty-seven thousandths*

Significant Digits

When we do calculations with approximate numbers, we must be careful about how many digits we keep in our answer. If we keep too few, we may be losing accuracy. If we keep too many, we may be falsely stating the accuracy of our result, and we would certainly be doing extra work by carrying along all those digits.

We will soon see that the number of digits we keep is, of course, dependent on the number of digits in our original numbers. In this section we show how to determine how many digits in our original numbers will have an effect on any calculations we do with those numbers. Such digits are called *significant digits*.

All nonzero digits in a number are, as you might expect, significant digits.

> **Example 3**
> The numbers 831.1, 32.62, 8321, and 2.441 each have four significant digits.

Zeros are sometimes significant and sometimes not. In general, zeros not used as placeholders, just to locate the decimal point, *are* significant. This includes zeros located between two nonzero digits. They are *always significant.*

> **Example 4**
> The numbers 330.1, 60.02, 8021, and 2.001 each have four significant digits.

Final zeros to the right of the decimal point *are significant.*

> **Example 5**
> The numbers 73.30, 2.620, and 3.000 each have four significant digits. The zeros here are not needed to locate the decimal point. Ask yourself why someone bothered to write such a zero in. If it is not needed to locate the decimal point, its only purpose must be to show that that digit is significant and has the value of zero.

3-1 Decimal Numbers

Zeros used as placeholders, to locate the decimal point, are *not significant*. These include any zeros to the left of the first nonzero digit.

Example 6

The numbers 0.0274 and 0.000 923 each have three significant digits. The zeros in these numbers serve only to locate the decimal point.

Final zeros to the left of the decimal point, such as those in the number 5000, are *not* significant as written. If you want to indicate that some such zeros are significant, place an overscore over the last trailing zero that is significant.

Example 7

The numbers 825 $\overline{0}$00, 1 372 000, and 5 00$\overline{0}$ each have four significant digits.

Rounding

We will see, in the next few sections, that the numbers we get from a computation often contain worthless digits that must be thrown away. Whenever we do this, we must *round* our answer.

We'll see later that we sometimes round to a certain number of *decimal places*, and sometimes round to a certain number of *significant digits*. The reasons for doing one or the other will come later; for now we'll just get some practice in rounding.

Round down (do not change the last retained digit) when the first discarded digit is 4 or less.

Example 8

Round to one decimal place (to the nearest tenth): 372.8499.

Solution: The first discarded digit is 4, so we round down and get 372.8.

Example 9

Round to three significant digits: 45.736.

Solution: The first discarded digit is 3, so we round down and get 45.7.

Round up (increase the last retained digit by one) when the first discarded digit is 6 or more, or a 5 followed by a nonzero digit in any of the decimal places to the right.

Example 10

Round to two decimal places (to the nearest hundredth): 372.825 001.

Solution: The first discarded digit is a 5, followed eventually by a nonzero digit (1), so we round up and get 372.83.

Example 11

Round to four significant digits: 4.039 62.

Solution: The first discarded digit is 6, so we round up and get 4.040.

As we said, we sometimes round to a certain decimal place, and sometimes to a certain number of significant digits. The following example shows some numbers rounded to a given number of decimal places, and to a given number of significant digits.

Example 12

Number	Rounded to three decimal places (nearest thousandth)	Rounded to three significant digits
3.3857	3.386	3.39
0.3656	0.366	0.366
12.2846	12.285	12.3
735.7294	735.729	736

When the first discarded digit is 5 followed by all zeros, it does not usually matter whether you round up or down. However, it is important when adding or subtracting a long column of figures. If, when discarding a 5, you always rounded up, you could bias the result in that direction. To avoid that, you want to round up about as many times as you round down, and a simple way to do that is always to *round to the nearest even digit*. This is just a convention. We could just as easily round to the nearest odd digit. To be consistent in this book, we will always round to the nearest even digit if the first discarded digit is 5 followed by all zeros. This is shown in the following example.

Example 13

Number	Rounded to three decimal places	Rounded to three significant digits
3.3855	3.386	3.39
3.3845	3.384	3.38
12.2855	12.286	12.3
12.2845	12.284	12.3

SECTION 3-1 Exercises Decimal Numbers

State the number of decimal places and of significant digits in each approximate number.

1. 62.5
2. 5847
3. 9.007
4. 6800
5. 700 000
6. 9000.0
7. 0.854 97
8. 8.000 00

Round each number to two decimal places.

9. 58.309
10. 2.997
11. 86.03954
12. 39.958
13. 689.0495
14. 8.678
15. 9876.760
16. 730.875

Round each number to one decimal place.

17. 45.97
18. 364.69
19. 9.7503
20. 0.845
21. 3984.68
22. 84.209
23. 4358.38
24. 0.385

Round each number to the nearest hundred.

25. 58 029 **26.** 9440 **27.** 8 475 309 **28.** 394 203

Round each number to three significant digits.

29. 7346 **30.** 3745 **31.** 0.047 58 **32.** 475 203
33. 230 578 **34.** 0.076 543 **35.** 43.555 **36.** 79.650 001

3-2 ADDING AND SUBTRACTING DECIMAL NUMBERS

Adding and Subtracting Approximate Numbers

We showed how to add and subtract whole numbers in Chapter 1. We assumed them to be exact. But now let us tackle the problem mentioned earlier. How many digits do we keep in our answer when adding or subtracting approximate numbers? We use the following rule:

> **RULE** For each number to be added or subtracted, note the place value of the last significant digit. Identify which of these has the greatest value. Round your answer to that place value.

Example 14

Add 16.8 + 21.4125 + 6.667.

Solution: We use the calculator keystrokes for addition as shown in Chapter 1, and get an unrounded sum of 44.8795. We then note the place value of the last significant digit in each given number; tenths in 16.8, ten-thousandths in 21.4125, and thousandths in 6.667. The greatest of these place values is tenths, so we round our answer to the nearest tenth.

```
  16.8 |
  21.4 | 125
   6.6 | 67
  44.8 | 795
       └──→ discard
```

So the sum, rounded up to the nearest tenth, is 44.9.

Example 15

Add 354 100 + 2745.

Solution: Our unrounded sum is 356 845. Lacking any other information, we can only assume that the zeros in 354 100 are not significant, so that this number is "good" only to the nearest hundred. Thus we round our answer to the nearest hundred, getting 356 800.

Adding and Subtracting Exact and Approximate Numbers

Sometimes we must add an exact number and an approximate number. How do we round the answer? It's simple. Treat the exact number as if it had more decimal places than the approximate number.

Example 16

Find the sum, in minutes, of 3 hours and 15.7 minutes.

Solution: Here, the number 3 is exact, and the number 15.7 is approximate. First we convert 3 hours to minutes and get 180 minutes. This is an exact number. To it we must add the approximate number 15.7.

$$\begin{array}{r} 180 \text{ min} \\ + 15.7 \text{ min} \\ \hline 195.7 \text{ min} \end{array}$$

Since 180 is *exact*, we *do not* round our answer to the nearest 10 minutes, but retain as many decimal places as in the approximate number. Our answer is thus 195.7 min.

Applications

Example 17

The sides of a triangular window have lengths of 3.075 m, 273 m, and 2.5125 m. Find the perimeter (the distance around) of the window.

Solution: The perimeter is obtained by adding the lengths of the sides.

$$3.075 \text{ m} + 2.73 \text{ m} + 2.5125 \text{ m} = 8.3175 \text{ m}$$

The answer 8.3175 m must be rounded to two decimal places (because 2.73 m has only two decimal places). The perimeter's total length has a rounded value of 8.31 m.

SECTION 3-2 Exercises Adding and Subtracting Decimal Numbers

Combine as indicated. Round your answers properly.

1. 638.2
    +723.4

2. 723.6
    +284

3. 11.283
    − 6.26

4. 155.6
    +273.91
    − 18.3

5. 17.3
    +21.93
    − 1.9

6. 9.5
    −9.82
    +2.1

7. 7327 + 7.5
8. 19.69 + 5.8
9. 75.33 − 53.189
10. 936.23 − 362.6
11. 0.005 385 + 0.000 974 − 0.005 27
12. 6322.7 − 13.70 + 39.43 − 7.269
13. 9.273 − 6.3853 − 2.83 + 39.282 − 3.42
14. 54.63 − 23.066 − 23.97 + 133.885 − 7.221

Applications

15. A tractor sold for $30 183, which was $9925 more than its original cost. What was its original cost?

16. Find the perimeter of a four-sided figure (a quadrilateral) if the lengths of the four sides are 22.7 m, 28.16 m, 25.27 m, and 9.556 m.

17. A company distributed $255 000 to its four partners. One partner got $83 274, a second got $49 222, and a third got $52 000. How much, to the nearest dollar, did the fourth partner get?

18. Three resistors, having values of 836 ohms (Ω), 264.75 Ω, and 2844 Ω, are wired in series. Find the total resistance by adding the three individual resistances.

19. A circular oil drum has an inside radius of 5.500 ft and a wall thickness of 0.0625 ft. What is the outside diameter of the drum?

20. A batch of concrete is made by mixing 625 lb of aggregate, 264 lb of sand, 153 lb of cement, and 59.25 lb of water. Find the total weight of the mixture.

3-3 MULTIPLYING DECIMAL NUMBERS

Multiplying Approximate Numbers

The calculator keystrokes for multiplying approximate numbers are the same as when we multiplied whole numbers in Chapter 1. The difference is that we must now round our product to the proper number of digits. We use the following rule:

> **RULE** When multiplying two or more approximate numbers, round the product to as many significant digits as in the factor having the fewest significant digits.

If both factors have the same number of significant digits, we simply keep that same number of digits in our product.

Example 18

$$23.1 \times 18.8 = 434$$

three digits   three digits   three digits

When the factors have *different* numbers of significant digits, keep the same number of significant digits in your answer as is contained in the factor that has the fewest significant digits.

Example 19

Multiply 23.1 by 18.89 and round the product to the proper number of significant digits.

Solution: Our calculator gives an unrounded product of 436.359. Since the number 23.1 has only three significant digits, we round our answer to three significant digits.

$$23.1 \times 18.89 = 436$$

three digits   four digits   keep three digits

---

60   CHAPTER 3 / Arithmetic with Decimal Numbers

> **COMMON ERROR** Do not confuse significant digits with decimal places. The number 274.56 has five significant digits and two decimal places.
> Decimal places determine how we round after adding or subtracting. Significant digits determine how we round after multiplying, and, as we will soon see, after dividing, raising to a power, or taking roots.

Multiplying Exact and Approximate Numbers

When using *exact* numbers in a computation, treat them as if they had *more significant digits* than any of the approximate numbers in that computation.

Example 20

If a complete desktop computer has a mass of 27.7 kg, what will be the mass of 4 such identical computers?

Solution: Multiplying, the total mass is obtained as follows,

$$4 \times 27.7 \text{ kg} = 110.8 \text{ kg}$$

Since the 4 is an exact number, we retain as many significant digits as contained in 27.7 and round our answer to 111 kg.

SECTION 3-3 Exercises Multiplying Decimal Numbers

Multiplying Approximate Numbers

Multiply. Retain the proper number of digits in your answer.

1. 1.84×2.35
2. 8.67×1.26
3. 22.35×84.16
4. 1.035×2.943
5. 77.283×12.652
6. 8.2873×11.847
7. 8.74×7.183
8. 30.5×7.265
9. 1.4853×14.29
10. 3.800×24.7
11. $35.2 \times 0.006\ 238\ 8$
12. $2.63 \times 8.96 \times 3.34$
13. $6.40 \times 6.78 \times 1.882$
14. $1.7322 \times 16.4 \times 0.003\ 35$
15. 14.300×186.92
16. 1395×3.958
17. $4.39 \times (6.38 \times 3.58) \times 0.883$
18. $(9.28 \times 8.39) \times (4.38 \times 0.853)$
19. $(8.36 + 9.26) \times 14.3$
20. $7.43 \times (4.67 + 5.03) \times 7.59$

Multiplying Exact and Approximate Numbers

Assume the integers in these problems to be exact.

21. 34×1.93
22. 4.003×6
23. 364×2.38
24. $23 \times 0.473 \times 3$
25. $2.836 \times 4 \times 3.72$
26. $2 \times 8 \times 1.004$

Applications

Work the financial calculations to the nearest penny, regardless of the significant digits in the original numbers.

27. If 30.57 m of telephone cable are required to hook up one house, how many metres of cable will be required to hook up 145 houses? What will the total cost of the cable be if the price is $0.0489 per metre?
28. If there are 360 degrees per revolution, how many degrees are there in 0.175 revolutions?
29. The current to a motor is 2.552 A when the line voltage is 115 V. Find the power consumed in the motor by multiplying the current by the voltage.
30. A flywheel rotates 73.5 times per minute. How many revolutions will it make in 2.28 minutes?

31. One inch equals exactly 2.54 cm. Convert 15.285 in. to cm.

32. What will be the cost of building a wall 10.4 m long, at $584 per m?

33. What is the cost of 25 cans of paint if each costs $17.23?

 34. Two joggers start from the same place on a straight road, and run in opposite directions. One runs at the rate of 8.30 mi/h and the other at 11.4 mi/h. How far apart will they be at the end of 0.750 hour?

3-4 DIVIDING DECIMAL NUMBERS

Dividing Approximate Numbers

The rule for rounding in division is almost the same as with multiplication:

> **RULE** After dividing approximate numbers, round the quotient to as many significant digits as there are in the original number having the fewest significant digits.

Example 21
Divide 18.63 by 2.15 by calculator.

Solution: The keystrokes are the same as those we used for dividing whole numbers in Chapter 1, and we get a quotient of 8.665116279. Since 2.15 has three significant digits, we round our quotient up to 8.67.

Example 22
A storm travels a distance of 288 kilometres in 9.0 hours. Find the speed of the storm.

Solution: The speed is equal to the distance divided by the time. Dividing,

$$Speed = \frac{distance}{time} = \frac{288 \text{ km}}{9.0 \text{ h}} = 32 \text{ km/h}$$

Reciprocal of a Decimal Number

In Chapter 2 we defined the reciprocal of any number as *1 divided by the number.*

Example 23
The reciprocal of 2 is $\frac{1}{2}$.

To find reciprocals by calculator, simply enter the number and press the $\boxed{x^{-1}}$ key or the $\boxed{1/x}$ key. Keep as many significant digits in your answer as there are significant digits in the original number.

CHAPTER 3 / Arithmetic with Decimal Numbers

Example 24

Find the reciprocal of 3.74 by calculator.

Solution: The keystrokes are

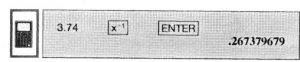

Since 3.74 has three significant digits, we round our answer to 0.267.

Example 25

When solving a certain circuit for the equivalent resistance R, we find that R in ohms (Ω) is equal to the reciprocal of 0.285. Find R.

Solution: Taking the reciprocal,

$$R = \frac{1}{0.285} = 3.51 \ \Omega$$

SECTION 3-4 Exercises Dividing Decimal Numbers

Divide. Round your answer to the proper number of digits. For this set of problems only, assume that the whole numbers are exact.

1. $6.34 \div 1.25$
2. $7.47 \div 2.46$
3. 2.645×1.463
4. $41.35 \div 4.837$
5. $183 \div 2.352$
6. $21.53 \div 7.47$
7. $1.574 \div 5.23$
8. $2.705 \div 1.21$
9. $397.3 \div 1832.2$
10. $9.782 \div 0.078\ 19$
11. $162.6 \div 89.3$
12. $37.4 \div 6.214$
13. $86.74 \div 5$
14. $8 \div 4.495$
15. $723.1 \div 9$

Reciprocals of a Decimal Number

Find the reciprocal of each number, retaining the proper number of digits in your answer.

16. 843
17. 0.000 458 0
18. 893 926
19. 93.78
20. 0.00375
21. 937.8
22. 6.338
23. 9.62
24. 93.2
25. 562
26. 7.001
27. 9.52

Applications

28. If 735 bonds are worth $25 836, what is the value of each bond?
29. At the rate of 861.9 km in 1.335 hours, how many kilometres would an aircraft fly in 10.25 hours? (*Distance = rate × time*)
30. A voltage of 163.4 V is measured across a 1745 Ω resistor. Find the current through the resistor. (*Current = voltage ÷ resistance*)
31. How long will it take 9 people to do the same work that 17 people can do in 8.50 days?
32. If two straight lines are perpendicular, the slope of one line is the negative reciprocal of the slope of the other. If the slope of a line is −1.73, find the slope of a perpendicular to that line.

33. When an object is placed 284 cm in front of a certain thin lens having a focal length f, the image will be formed 173 cm from the lens. The distances are related by

$$\frac{1}{f} = \frac{1}{284} + \frac{1}{173}$$

Find f.

34. The sine of an angle θ (written sin θ) is equal to the reciprocal of the cosecant of θ (written csc θ). Find sin θ if csc $\theta = 5.26$.

35. Find the equivalent resistance of a 451 Ω resistor and a 879 Ω resistor, connected in parallel. (Find the sum of the reciprocals of the given resistances. Then take the reciprocal of that sum.)

 36. A steel casting (density = 455 lb/ft^3) has a weight of 629.3 lb. Find the volume of the casting. (*Volume = weight ÷ density*)

37. How many cubic feet of water will a weight of 8632 lb occupy, if the density of water is 62.4 lb/ft^3?

38. A steel girder 25.82 ft long is divided into three equal sections. Find the length of each section.

3-5 POWERS AND ROOTS OF DECIMAL NUMBERS

Powers

As we showed in Chapter 1, to square a number on the calculator, simply enter the number and press the $\boxed{x^2}$ key, and to raise a number to other powers, use the power key, marked $\boxed{\wedge}$, $\boxed{y^x}$, or $\boxed{x^y}$. In our examples we'll use $\boxed{\wedge}$. When finding the power of an approximate number, round your result to the number of significant digits contained in the original number (the base).

Example 26

Find $(2.93)^3$ by calculator.

Solution: Here, the base (2.93) is approximate, and has three significant digits. The keystrokes are

We round the result to 25.2.

Fractional Exponents

We will soon see that a fractional exponent is another way of indicating a radical. For now, we just evaluate fractional exponents on the calculator. We enter the base, press the power key, then enter *the exponent in parentheses*, as shown in the following example.

Example 27

Find $8^{2/3}$ by calculator.

Solution: The keystrokes are

To raise an approximate number to a fractional exponent we use the same keys as in the preceding example. However, now we round our answer to as many significant digits as contained in the base.

Example 28

Evaluate $(62.4)^{2/3}$.

Solution: The keystrokes are

which we round to 15.7.

Roots

In Sec. 1-5 we took roots of whole numbers. Now we will take roots of approximate numbers. We keep as many significant digits in our answer as contained in the number of which we are taking the root.

To find *square* roots or *cube* roots, enter the number and then press the $\boxed{\sqrt{x}}$ key or the $\boxed{\sqrt[3]{}}$ key. To find other roots, use the root key, marked $\boxed{\sqrt[y]{}}$, $\boxed{\sqrt[x]{y}}$, or $\boxed{\sqrt[y]{x}}$. We'll indicate the root key by $\boxed{\sqrt[y]{}}$ in the examples. (See the following section if you have no root key.) Retain as many significant digits in your answer as there are significant digits in the original number.

Example 29

Find $\sqrt[4]{97.6}$ (the fourth root of 97.6).

Solution: The keystrokes are

which we round to 3.14.

Roots and Powers Related

If your calculator does not have a root key, you can find roots by using the power key. For this, you need to know the relationship between roots and exponents.

To take the nth root of a number x is the same as raising x to the $1/n^{th}$ power.

In Chapters 7 and 20 we will explain this idea in much more detail. For now, we will use it only for computing roots.

Example 30

Find $\sqrt[5]{94.7}$ using the power key.

Solution: Taking the 5th root of a number is equal to raising that number to the 1/5th power.

$$\sqrt[5]{94.7} = (94.7)^{1/5}$$

On the calculator:

which we round to 2.48.

Example 31

The volume of a certain cube is 73.5 cubic centimetres. Find the side of the cube by taking the cube root of the volume.

Solution: Taking the cube root,

$$\sqrt[3]{73.4 \text{ cm}^3} = 4.188\ 859 \text{ cm}$$

which we round to three significant digits, getting 4.19 cm.

SECTION 3-5 Exercises Powers and Roots of Decimal Numbers

Powers

Evaluate each expression. Retain the correct number of digits in your answer.

1. $(2.13)^2$
2. $(1.34)^3$
3. $(3.22)^5$
4. $(1.08)^2$
5. $(0.511)^5$
6. $(0.291)^4$
7. $(1.002)^9$
8. $(21.5)^3$
9. $(6.24)^5$
10. $(7.003)^3$
11. $(4.89)^4$
12. $(3.33)^3$
13. $(8.53)^{4/5}$
14. $(5.83)^{2/3}$
15. $(6.94)^{3/4}$
16. $(5.397)^{1/2}$
17. $(76.9)^{5/7}$
18. $(7.534)^{3/5}$
19. $(6.53)^{3/4}$
20. $(2.094)^{2/5}$
21. $(8.365)^{5/7}$
22. $(8.65)^{3/2}$
23. $(9265)^{1/3}$
24. $(7.52)^{2/3}$

Roots

Evaluate each radical by calculator, retaining the proper number of digits in your answer.

25. $\sqrt{84.2}$
26. $\sqrt{7.527}$
27. $\sqrt[3]{53.0}$
28. $\sqrt{492}$
29. $\sqrt{8426}$
30. $\sqrt[3]{4973}$
31. $\sqrt[3]{843}$
32. $\sqrt[3]{53.8}$
33. $\sqrt[3]{87.36}$

Applications

34. Evaluate the geometric mean B between 14.3 and 25.6 if
$$B = \sqrt{(14.3)(25.6)}$$

35. Find the volume of a cube with a side length of 8.24 cm by cubing this dimension.

36. The volume of a sphere with a radius of 2.85 m is $(4/3)\pi(2.85)^3$ m³. Find this volume.

37. The distance traveled by a falling body, starting from rest, is equal to $4.9t^2$, where t is the elapsed time. The distance fallen in 5.25 seconds is thus $4.9(5.25)^2$ m. Evaluate this quantity.

38. An investment of $8263 at a compound interest rate of 7.5%, left for 5.25 years, will be worth $8263(1.075)^{5.25}$ dollars. Find this amount to the nearest dollar.

39. Find the magnitude of the impedance Z in a circuit having a resistance of 375 Ω and a reactance of 624 Ω if
$$Z = \sqrt{(375)^2 + (624)^2} \quad \Omega$$

40. The power dissipated by a current I flowing through a resistance R is equal to I^2R. Therefore, the power in a 1829 Ω resistor carrying a current of 1.263 A is $(1.263)^2(1829)$ W. Evaluate this expression.

 41. Find the period T (time for one swing) of a simple pendulum 1.88 ft long, if
$$T = 2\pi \sqrt{\frac{1.88}{32.2}} \quad \text{seconds}$$

3-6 SCIENTIFIC NOTATION

In Sec. 1-6 we introduced scientific notation with whole numbers and showed, for example, that the number 30 000 can be written as 3×10^4. We apply scientific notation to decimal numbers here and do some calculations in scientific notation. We will not calculate powers of ten with negative exponents yet. We will save those for the chapter on signed numbers.

Converting from Decimal Form to Scientific Notation

Rewrite the given number with just one digit to the left of the decimal point, and discard any nonsignificant zeros. Then multiply by that multiple of ten that makes it equal to the given number. Finally write the multiple of ten as a power of ten.

---- **Example 32** ----

Convert 53 100 to scientific notation.

Solution: We place the decimal point between the 5 and the 3. The two zeros are not significant (see Sec. 3-1) so we discard them, getting 5.31. Finally, the 5.31 must be multiplied by 10 000 so that our new expression is equal to the given number.
$$53\ 100 = 5.31 \times 10\ 000$$
We now express 10 000 as a power of ten (10^4).
$$53\ 100 = 5.31 \times 10\ 000 = 5.31 \times 10^4$$

To convert from scientific notation to decimal form, simply reverse the process.

Example 33

Convert 3.84×10^5 to decimal form.

Solution:
$$3.84 \times 10^5 = 3.84 \times 100\ 000$$
$$= 384\ 000$$

Entering Numbers in Scientific Notation in a Calculator

A calculator display will, as we have seen, switch to scientific notation by itself if the display gets too long for it to show. We can also force a calculator to display in scientific notation by pressing the key that switches the display between fixed notation and scientific notation. Markings for this key vary, but some are $\boxed{\text{FE}}$, $\boxed{\text{FSE}}$, and $\boxed{\text{SCI}}$. Most newer calculators have a third type of notation. This is evident with the $\boxed{\text{FSE}}$ key, where "F" is for *F*ixed decimal, "S" is for *S*cientific notation and "E" is for *E*ngineering notation. Engineering notation is similar to scientific notation and is used extensively with the metric system. A full description will be included in Chapter 6. Most calculators allow you to select the kind of display you want.

We can enter numbers directly in scientific notation by means of the ENTER EXPONENT key, usually marked $\boxed{\text{EE}}$, $\boxed{\text{EXP}}$, or $\boxed{\text{EEX}}$. We will use $\boxed{\text{EE}}$ in our examples.

Example 34

The keystrokes to enter the number 8.25×10^8 are

With the calculator in "scientific" mode, the display will show something like 8.25E8 or 8.25 08, depending on the calculator.

We now show how to add, subtract, multiply, and divide numbers in scientific notation.

Adding and Subtracting Numbers in Scientific Notation

If two or more numbers to be added have the *same power of ten,* simply add the numbers and keep the same power of ten, as shown in the following example.

Example 35

Add (3.14×10^4) and (5.27×10^4).

Solution:
$$(3.14 \times 10^4) + (5.27 \times 10^4) = (3.14 + 5.27) \times 10^4$$

When adding 3.14 and 5.27, we follow the rule for adding approximate numbers and retain two decimal places in our answer.

$$(3.14 + 5.27) \times 10^4 = 8.41 \times 10^4$$

Example 36

$$(4.86 \times 10^5) - (2.63 \times 10^5) + (3.63 \times 10^5) = (4.86 - 2.63 + 3.63) \times 10^5$$
$$= 5.86 \times 10^5$$

Example 37

Add (5.83×10^5) and (1.85×10^5) by calculator.

Solution: The keystrokes are

with the calculator in Scientific Notation mode, or 7.68×10^5.

If the powers of ten are different, they must be made equal before the numbers can be combined. A shift of the decimal point of one place to the left will increase the exponent by 1. Conversely, a shift of the decimal point one place to the right will decrease the exponent by 1.

Example 38

$$(3.41 \times 10^5) + (6.5 \times 10^4) = (3.41 \times 10^5) + (0.65 \times 10^5)$$
$$= 4.06 \times 10^5$$

Example 39

$$(2.80 \times 10^4) - (1.6 \times 10^3) + (3 \times 10^2) = (2.80 \times 10^4) - (0.16 \times 10^4) + (0.03 \times 10^4)$$
$$= 2.67 \times 10^4$$

By calculator, the keystrokes are

Multiplying Numbers in Scientific Notation

We multiply powers of ten by *adding their exponents*. We are really using one of the laws of exponents that we will study in detail in Chapter 7.

Example 40

$$10^2 \times 10^4 = 10^{2+4} = 10^6$$

To multiply two numbers in scientific notation, multiply the decimal parts and the powers of ten separately.

Example 41

Multiply (4.15×10^4) by (2.63×10^3).

Solution: We multiply the decimal parts by calculator, and round the product to three significant digits $(4.15 \times 2.63 = 10.9)$, and multiply the powers of ten by adding their exponents.

$$(4.15 \times 10^4)(2.63 \times 10^3) = (4.15 \times 2.63)(10^4 \times 10^3)$$
$$= 10.9 \times 10^{4+3} = 10.9 \times 10^7$$
$$= 1.09 \times 10^8$$

Dividing Numbers in Scientific Notation

We divide powers of ten by *subtracting their exponents:*

Example 42:

$$\frac{10^7}{10^2} = 10^{7-2} = 10^5$$

As with multiplication, we divide the decimal parts and the powers of ten *separately*. When dividing the decimal parts, we follow the rules for rounding for division.

Example 43

$$\frac{9.62 \times 10^7}{3.83 \times 10^2} = \frac{9.62}{3.83} \times \frac{10^7}{10^2} = 2.51 \times 10^{7-2} = 2.51 \times 10^5$$

Computing in Scientific Notation by Calculator

Simply follow the same calculator procedures as you would for decimal numbers and let the calculator worry about the powers of ten.

Example 44

Multiply (7.42×10^6) by (5.639×10^3).

Solution: The keystrokes are

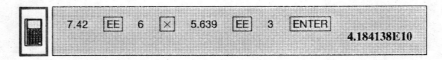

We round to three significant digits, getting 4.18×10^{10}.

FIGURE 3-2
A steel wire in tension.

Example 45

The elongation ΔL of a steel wire having a length of $L_0 = 2.90$ m and a cross-sectional area of $A = 8.19 \times 10^{-6}$ m², with a tensile load of $F = 6.99 \times 10^3$ N and an elastic modulus of $E = 2.00 \times 10^{11}$ N/m² is given by

$$\Delta L = \frac{F L_0}{E A} = \frac{(6.99 \times 10^3)(2.90)}{(2.00 \times 10^{11})(8.19 \times 10^{-6})} \quad \text{m}$$

Find ΔL. Fig. 3-2

Solution: The keystrokes are

Which is rounded to 0.0124 m or 1.24 cm.

SECTION 3-6 Exercises Scientific Notation

Converting to and from Scientific Notation

Write each number in scientific notation. Be sure to retain the proper number of significant digits.

1. 2373
2. 56 632
3. 7260
4. 123.9
5. 1 920 000
6. 673 900
7. 5228
8. 64 991
9. 7010
10. 325 800 000

Convert each number from scientific notation to decimal notation.

11. 9.023×10^2
12. 8.9902×10^4
13. 4.01×10^3
14. 7.100×10^4
15. 9.520×10^5
16. 2.94×10^6

Perform the following computations in scientific notation. Combine the powers of ten by hand or with your calculator.

Adding and Subtracting Numbers in Scientific Notation

17. $(8.40 \times 10^3) + (7.90 \times 10^4)$
18. $(42.0 \times 10^3) + 4900$
19. $(4.8862 \times 10^3) + (6.0 \times 10^2)$
20. $995 - (7.00 \times 10^2)$
21. $(8.30 \times 10^2) + (5.30 \times 10^2)$
22. $(9.50 \times 10^3) - (7.45 \times 10^3)$

Multiplying Numbers in Scientific Notation

23. $10^3 \times 10^4$
24. $10^5 \times 10^3$
25. $10^3 \times 10^6$
26. $(7.39 \times 10^6)(4.62 \times 10^5)$
27. $(5.03 \times 10^3)(6.17 \times 10^8)$
28. $(5.49 \times 10^5)(4.91 \times 10^3)$
29. $(3.86 \times 10^2)(1.28 \times 10^7)$
30. $(3.92 \times 10^3)(4.15 \times 10^5)$
31. $(4.23 \times 10^6)(5.94 \times 10^2)$

Dividing Numbers in Scientific Notation

32. $10^{10} \div 10^6$
33. $10^8 \div 10^5$
34. $10^9 \div 10^7$
35. $(3.25 \times 10^2) \div 0.00416$
36. $45\,400 \div (5.25 \times 10^3)$
37. $(8 \times 10^7) \div (4 \times 10^3)$
38. $(2 \times 10^6) \div (4 \times 10^4)$
39. $(6 \times 10^9) \div 200\,000\,000$

FIGURE 3-3
Resistors in series.

40. $(8.17 \times 10^8) \div 9\,020\,000$

41. Four resistors, having resistances of 3.52×10^5 Ω, 2.89×10^3 Ω, 5.60×10^7 Ω, and 9.52×10^4 Ω, are wired in series, Fig. 3-3. Find the total resistance by adding the four resistances.

42. Find the equivalent resistance if the four resistors in Exercise 41 are wired in parallel, Fig. 3-4. (Add the reciprocals of the four resistors. Then take the reciprocal of that sum.)

43. Find the current through a resistor if it dissipates 7.82×10^3 watts when the voltage drop across it is 3.74×10^3 V. (Divide the power by the voltage drop.)

44. The voltage across a 3.92×10^4 Ω resistor is 7.39×10^2 V. Find the power dissipated in the resistor. (Divide the square of the voltage by the resistance.)

45. A tensile load of 3.96×10^4 N is supported by a member having a cross-sectional area of 2.34×10^{-3} m². Find the stress in the member. (Divide the load by the area.)

46. A bar having a cross-sectional area of 0.97 in.² and a length of 3.62×10^2 in. is subjected to a tensile load of 5.27×10^3 lb. The elongation is 0.0446 in. Find the modulus of elasticity E of the material. (Multiply the load by the length, and divide by the area and by the elongation.)

3-7 COMBINED OPERATIONS

In Sec. 1-7 we showed how to do combined operations with whole numbers on the calculator. Combined operations with approximate numbers are done the same way. However, we must round our answer properly using the rules given earlier in this chapter.

Example 46

Evaluate the expression

$$\left(\frac{118.8 + 4.23}{\sqrt{136}}\right)^3$$

Solution: The keystrokes are

How many digits should we keep? In the numerator, we added a number with one decimal place to another with two decimal places, so we are allowed to keep just one. Thus the numerator, after addition, is good to one decimal place, or, in this case, four significant digits. The denominator, however, has just three significant digits, so we round our answer to three significant digits, getting 1170.

■ FIGURE 3-5
A rectangular courtyard.

Example 47

A rectangular courtyard (Fig. 3-5) having sides of 6.64 m and 10.28 m has a diagonal measurement x given by the expression

$$x = \sqrt{(6.64)^2 + (10.28)^2}$$

Evaluate the expression to find x.

Solution: The keystrokes are

or

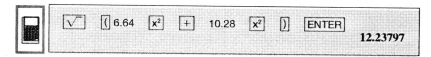

In either case, our answer is rounded to three significant digits, because 6.64 has only three, and therefore $x = 12.2$ m.

SECTION 3-7 Exercises Combined Operations

Perform the computations, keeping the proper number of digits in your answer.

1. $(7.37)(3.28) + (8.36)(9.64)$
2. $(522)(9.53) - (586)(4.70) + (847)(7.59)$
3. $(63.5 + 83.6)(8.37 \times 10^2 - 1.72 \times 10^2)$
4. $(8.93 - 3.74 + 9.05)(68.7 - 64.9)$
5. $\dfrac{583}{473} + \dfrac{946}{907}$
6. $\dfrac{6.73}{8.38} - \dfrac{5.97}{8.06} + \dfrac{8.63}{1.91}$
7. $\dfrac{6.47 + 8.604}{3.37 + 90.8}$
8. $\dfrac{809 - 463 + 744}{758 - 964 + 508}$
9. $(5.37 + 8.36)^2$
10. $(4.25 + 4.36 - 5.24)^4$
11. $(6.423 + 9.05)^3$
12. $\left(\dfrac{45.3 - 8.34}{8.74 - 5.62}\right)^{2.5}$
13. $\left(\dfrac{8.90}{4.70}\right)^2$
14. $\sqrt{4.34 \times 10^3 + 5.66 \times 10^2}$
15. $\sqrt[3]{657 + 553 - 842}$
16. $\sqrt{(28.1)(5.94 \times 10^2)}$
17. $\sqrt[5]{(9.06)(4.86)(7.93)}$
18. $\sqrt{\dfrac{653}{601}}$
19. $4\sqrt{\dfrac{4.50}{7.81}}$
20. $\sqrt{9.74} + \sqrt{12.5}$
21. $\sqrt[4]{528} + \sqrt{94.2} - \sqrt[3]{284}$
22. $\sqrt[4]{653} \times \sqrt{55.3}$

Chapter 3 REVIEW EXERCISES

Find the reciprocal of each number, retaining the proper number of digits in your answer.

1. 328
2. 21.78
3. 2.358
4. 3.82
5. 53.2
6. 8.62

Write each number in scientific notation.

7. 2300
8. 3000
9. 83.9×10^4
10. 24 491
11. 3020
12. 34 000

Convert each number from scientific notation to decimal notation.

13. 3.023×10^2
14. 1.9902×10^8
15. 8.01×10^3

Evaluate each expression, retaining the correct number of digits in your answer.

16. $(7.33)^3$
17. $(2.53)^4$
18. $(7.83)^2$
19. $(2.94)^3$
20. $(6.327)^2$
21. $(4.69)^5$
22. $\sqrt{34.2}$
23. $\sqrt{3.527}$
24. $\sqrt[3]{23.0}$
25. $\sqrt{892}$
26. $(3.2534)^3$
27. $(2.53)^4$
28. $(0.0440)^{0.552}$
29. $(3.365)^{3.7}$
30. $(4.65)^{3/2}$
31. $(3261)^{1/3}$
32. $\sqrt{3626}$
33. $\sqrt[3]{8373}$
34. $\sqrt[5]{837}$
35. $3\sqrt{92.8}$
36. $(322)(7.53) - (186)(7.70) + (347)(259)$
37. $(7.93 - 8.74 + 3.05)(18.7 - 4.9)$
38. $\dfrac{383}{173} + \dfrac{746}{307}$
39. $\dfrac{3.47 + 4.604}{7.37 + 30.8}$
40. $\dfrac{309 - 263 + 444}{458 - 264 + 808}$
41. $(8.25 + 2.36 - 9.24)^4$
42. $\left(\dfrac{75.3 - 2.34}{5.74 - 4.62}\right)^{2.5}$
43. $\sqrt{7.34 + 3.66}$
44. $\sqrt{(68.1)(3.94)}$
45. $\sqrt{\dfrac{553}{301}}$
46. $\sqrt{6.74} + \sqrt{22.5}$
47. $\sqrt[4]{253} \times \sqrt{75.3}$
48. $(4 \times 10^4)(2 \times 10^2)$
49. $(3 \times 10^3)(5 \times 10^2)$
50. $(2.39 \times 10^6)(8.60 \times 10^5)$
51. $(1.03 \times 10^3)(3.17 \times 10^8)$
52. $(2.95 \times 10^2)^3$
53. $\dfrac{4.39 \times 10^8}{3.22 \times 10^7} - 8.56$
54. $(2.52 \times 10^3)^3$
55. $\dfrac{(3.24 \times 10^6) - (1.35 \times 10^3)}{(2.12 \times 10^7) + (1.58 \times 10^2)}$
56. $(2.52 \times 10^3)^2$
57. $(8.20 \times 10^3)^4$
58. $(2.40 \times 10^3) + (4.90 \times 10^4)$
59. $(62.5 \times 10^3) + 1920$
60. $(3.25 \times 10^5) - (2.30 \times 10^4) + (4.00 \times 10^3)$
61. $(1.35 \times 10^2) + (2.34 \times 10^3) - (3.16 \times 10^4)$
62. $(2.26 \times 10^6) + (9.17 \times 10^4)$
63. A new memory unit containing 5.34×10^5 bits of memory is added to a computer that already has a memory of 8.84×10^7 bits. Find the total number of bits after the new unit is installed.

64. The wind pressure against the side of a certain building is 6.90 kN/m². If the area of the side of the building is 4.26×10^2 m², find the total force of the wind.

65. How long will it take a spacecraft traveling at a rate of 7.8×10^7 m/h to travel the 3.8×10^8 m from the earth to the moon?

WRITING

66. Suppose you have submitted a report that contains calculations in which you have rounded the answers according to the rules given in this chapter. Jones, your company rival, has sharply attacked your work, calling it "inaccurate" because you did not keep enough digits, and your boss seems to agree.

    Write a memo to your boss defending your rounding practices. Point out why it is misleading to retain too many digits. Do not write more than one page. You may use numerical examples to prove your point.

67. Study your calculator and its manual specifically on the different display formats (normal, scientific notation, and so forth). List the different formats available to you and explain the differences between them. Write a few lines explaining how to switch from one format to another.

TEAM PROJECT

68. Starting with the template in Fig. 3-6, a new template is to be made by increasing dimension A by 25.0%, and decreasing dimension B by 30.0%. Find the dimension x (in millimetres) for the new template.

■ **FIGURE 3-6**
A template, not drawn to scale.

Units of Measurement

Outcomes

When you have completed this chapter, you will be able to

1. Use the International System of Units (SI) (metric system of units).
2. Evaluate and use the SI prefixes.
3. Convert within the SI (metric) system of units.
4. Convert within the British/American customary system.
5. Convert between the British and metric units.
6. Perform calculations with denominate numbers.

When we measure any physical quantity, such as the length of a board, we make use of *units of measure,* such as the metre or in some cases the foot. In general use today is the SI or *metric* system of units. SI stands for Le Système International d'Unités or the *International System of Units.* In lesser use today is the *British/ American Customary System of Weights and Measure.*

In this chapter we learn about different units of measure and how to convert between them. We also learn how to add, subtract, multiply, and divide *denominate numbers,* those that have units (2.73 cm is a denominate number).

4-1 THE SI (METRIC) SYSTEM OF UNITS

The SI metric system evolved from a decimal system first proposed in 1670 by Gabriel Mouton (1618–1694) and adopted in France by decree in 1795. The modern *International System of Units* was adopted in 1960 by the International Committee of Weights and Measures.

The SI is a system using a coherent set of units. There are two classes of SI units, the *base units* and the *derived units.* The decimal multiples and sub-multiples are formed by the use of *prefixes.*

Base units are defined as those seven units that have well-defined physical standards. They include the metre (m), the kilogram (kg), the second (s), the kelvin temperature scale (K), the ampere (A), the candela (cd), and the mole (mol).

Derived units are formed by combining base units, for example the metre per second (m/s). Some derived units are given their own name such as the newton, which is equal to the kilogram-metre per second squared or $N = \frac{kg \cdot m}{s^2}$.

SI (Metric) Prefixes

Converting between metric units of different size is made easy because larger and smaller metric units are related to the nominal units by *factors of ten.* These larger and smaller units are created by placing a *prefix* before the basic unit. A prefix is

TABLE 4-1 Metric Prefixes

Prefix	Symbol	Factor	Base 10 exponential form	Pronunciations
yotta	Y	1 000 000 000 000 000 000 000 000	10^{24}	yot'ta
zetta	Z	1 000 000 000 000 000 000 000	10^{21}	zet'ta
exa	E	1 000 000 000 000 000 000	10^{18}	ex'a
peta	P	1 000 000 000 000 000	10^{15}	pe'ta
tera	T	1 000 000 000 000	10^{12}	ter'a
giga	G	1 000 000 000	10^{9}	ji'ga
mega	M	1 000 000	10^{6}	meg'a
kilo	k	1 000	10^{3}	kil'o
hecto	h	100	10^{2}	hek'to
deca (deka, US)	da	10	10^{1}	dek'a
		1	10^{0}	
deci	d	0.1	10^{-1}	des'i
centi	c	0.01	10^{-2}	sen'ti
milli	m	0.001	10^{-3}	mil'i
micro	μ	0.000 001	10^{-6}	mi'kro
nano	n	0.000 000 001	10^{-9}	nan'o
pico	p	0.000 000 000 001	10^{-12}	pē'co
femto	f	0.000 000 000 000 001	10^{-15}	fem'to
atto	a	0.000 000 000 000 000 001	10^{-18}	at'to
zepto	z	0.000 000 000 000 000 000 001	10^{-21}	zep'to
yocto	y	0.000 000 000 000 000 000 000 001	10^{-24}	yok'to

a group of letters placed at the beginning of a word to modify the meaning of that word. For example, the prefix *kilo* means 1000 or 10^3. Thus a kilogram is 1000 grams.

Example 1

(a) A kilometre (km) is a thousand metres, because kilo means one thousand.

$$1 \text{ km} = 1000 \text{ m}$$

(b) A centimetre (cm) is one hundredth of a metre, because centi means one hundredth.

$$1 \text{ cm} = 0.01 \text{ m}$$

(c) A millimetre (mm) is one thousandth of a metre, because milli means one thousandth.

$$1 \text{ mm} = 0.001 \text{ m}$$

The metric system has some units of measure that are used universally along with SI units. (See Table 4-2.)

Converting SI Units

In this section, we will convert from one SI unit to another.

In the SI system, Table 4-1 has the conversion factors required.

TABLE 4-2 List of Common Units Outside the SI That Are Accepted for Use with the SI

Name	Symbol	Value in SI units
minute (time)	min	1 min = 60 s
hour	h	1 h = 60 min = 3600 s
day	d	1 d = 24 h = 86 400 s
degree (angle)	°	$1° = (\pi/180)$ rad
minute (angle)	′	$1′ = (1/60)° = (\pi/10\ 800)$ rad
second (angle)	″	$1″ = (1/60)′ = (\pi/648\ 000)$ rad
litre	L	$1\ L = 1\ dm^3 = 10^{-3}\ m^3$
tonne	t	$1\ t = 10^3\ kg = 1\ Mg$
are	a	$1\ a = 1\ dam^2 = 10^2\ m^2$
hectare	ha	$1\ ha = 1\ hm^2 = 10^4\ m^2$
degree Celsius	°C	$t°C = tK - 273.15$

Converting between units is accomplished by multiplying by a conversion factor. Conversion usually involves one or two conversion factors. The procedure outlined here will also work between measurement systems and within the British system.

Example 2

Convert 2.56 kilometres to metres.

Solution: From Table 4-1, we find the following factor for kilo and apply it to metre

$$1\ km = 1000\ m$$

> **COMMON ERROR** When the letter "m" is used alone as in 3.5 m, it not the symbol for the prefix milli, it is the symbol for a metre, used without a prefix.

Dividing both sides by 1 km, (the unit we are converting *from*) we get the conversion factor:

$$1 = \frac{1000\ m}{1\ km}$$

Multiplying yields

$$2.56\ km = 2.56\ km \times \frac{1000\ m}{1\ km} = 2560\ m$$

> **TIP:** We can multiply an expression by a conversion factor because the conversion factor is always equal to 1.

Using More Than One Conversion Factor

Sometimes you may not be able to use a single multiplying factor to complete the conversion.

4-1 The SI (Metric) System of Units

Example 3

Convert 3.785 kilolitres to centilitres.

Solution: The named unit in question is the litre with two different prefixes. Therefore two conversion factors are necessary. From Table 4-1,

(a) 1 kL = 1000 L
(b) 1 cL = 0.01 L

To obtain the factor in the first conversion, dividing both sides by 1 kL (the unit we are converting *from*) we get the conversion factor to obtain litres:

$$1 = \frac{1000 \text{ L}}{1 \text{ kL}}$$

The second conversion factor from litres to centilitres is obtained by dividing both sides by 0.01 L (the unit we are converting *from* in the second step). This division gives us the conversion factor to obtain centilitres:

$$\frac{1 \text{ cL}}{0.01 \text{ L}} = 1$$

Therefore

$$3.785 \text{ kL} = 3.785 \text{ kL} \times \frac{1000 \text{ L}}{1 \text{ kL}} \times \frac{1 \text{ cL}}{0.01 \text{ L}} = 378\,500 \text{ cL}$$

TIP: A simpler procedure for converting within the metric system will be outlined in the section involving signed numbers.

Example 4

Convert 2.75×10^5 dynes to newtons (N).

Solution: These units of force do not have any base units in their names, nor any prefixes. Thus we cannot tell just from their names how they are related to each other. However, from the Table of Conversion Factors (Appendix B) we find that

$$1 \text{ dyne} = 10^{-5} \text{ newtons.}$$

Converting in the usual way, *dividing the equation by the unit we are converting from,* produces the conversion factor

$$1 = \frac{10^{-5} \text{ newtons}}{1 \text{ dyne}}$$

Then:

$$2.75 \times 10^5 \text{ dynes} \times \frac{10^{-5} \text{ newtons}}{\text{dynes}} = 2.75 \text{ newtons} = 2.75 \text{ N}$$

SECTION 4-1 Exercises Converting Between Metric Units

Convert the following

1. 22.3 km to m
2. 34.5 m to Mm
3. 47 835 mm to m
4. 6.75 L to cL
5. 45.2 km to Mm
6. 67.8 MW to GW
7. 3.476 kJ to J
8. 2 393 267 μm to km
9. 3.45 μmol to nmol
10. 42.5 TW to MW

Applications

11. The four sides of a quadrilateral measure 52 cm, 93 cm, 67 cm, and 89 cm. What is the perimeter of the quadrilateral in metres?
12. Three resistors connected in series have the following rated resistance, 320 cΩ, 4.8 kΩ, and 0.045 MΩ. What is the total resistance in kΩ?
13. A room has a length of 609 cm. Rugs are sold by the metre. How long is the room in metres?
14. Sonja ran three mornings a week to keep in shape. Monday she ran 4562 m, Wednesday she ran 6363 m, and Friday 5726 m. What was the average number of kilometres run per day?
15. A hiker wants to average 16 km per day for five days. In the first four days he travels 17 250 m, 14 655 m, 18 550 m, and 16 505 m. How many metres must he walk on the last day to maintain his average?

4-2 BRITISH/AMERICAN CUSTOMARY SYSTEM OF WEIGHTS AND MEASURE

This system of measures developed through the ages. Many of the units originally came from the Roman empire and subsequently from customary use and various trade practices. Standardization over time came through laws and the enforcement of even stricter laws. Currently, the base units are defined in terms of SI units.

The relationship between various units within the British system is based on a set of conversion factors, for example some of those for linear measurement.

Linear Conversion factors for a mile.

1 mile = 8 furlongs = 80 chains = 1760 yards = 5280 feet = 63 360 inches

The type of units a person chooses to use in any application depends on the nature of the work being performed.

Converting Within the British/American System

In this section we will convert from one British/American unit to another. Conversion within the British/American system requires the use of conversion factors.

Example 5

Convert 1.532 miles to feet

Solution: From the Table of Conversion Factors, Appendix B, we find the relationship between miles and feet

$$5280 \text{ feet} = 1 \text{ mile}$$

Dividing both sides by 1 mile (*the unit we are converting from*) we get the conversion factor

$$\frac{5280 \text{ feet}}{1 \text{ mile}} = 1$$

Multiplying yields

$$1.532 \text{ miles} = 1.532 \text{ miles} \times \frac{5280 \text{ feet}}{1 \text{ mile}} = 8088.96 \text{ ft} = 8089 \text{ ft}$$

Note that we have rounded our answer to four significant digits, because the number used in the calculation (1.532) is an approximate number with four significant digits. The number 5280 is exact.

Significant Digits

You should try to use a conversion factor that is exact, or one that contains at least one more significant digit than your original number. Once you have completed the conversion, you should round your answer to as many significant digits as in the original number. The Table of Conversion Factors shows whether a conversion factor is exact or approximate.

Example 6

Convert 934 acres to square miles (mi^2).

Solution: From the Table of Conversion Factors we find the equation

$$1 \text{ mi}^2 = 640 \text{ acres}$$

where 640 is an exact number. Dividing both sides by 640 acres, we obtain the proper conversion factor, which is

$$\frac{1 \text{ mi}^2}{640 \text{ acres}} = 1$$

Multiplying, we obtain

$$934 \text{ acres} = 934 \text{ acres} \times \frac{1 \text{ mi}^2}{640 \text{ acres}} = 1.459375 \text{ mi}^2$$

The conversion factor used here is exact, so we round our answer to three significant digits, that is, 1.46 mi^2.

Example 7

Convert 27 375 yards to nautical miles.

Solution: In the Table of Conversion Factors we do not find any conversion factor between nautical miles and yards, but we see that

$$1 \text{ nautical mile} = 6076 \text{ ft} \quad \text{(approximately)}$$
$$3 \text{ ft} = 1 \text{ yd} \quad \text{(exactly)}$$

Therefore

$$27\,375 \text{ yd} = 27\,375 \text{ yd} \times \frac{3 \text{ ft}}{1 \text{ yd}} \times \frac{1 \text{ nautical mile}}{6076 \text{ ft}} = 13.51629361 \text{ nautical miles}$$

Rounding to four significant digits, because the conversion factor of nautical miles to feet is approximate to four significant digits, we get

$$13.52 \text{ nautical miles.}$$

SECTION 4-2 Exercises British/American Customary System of Weights and Measure

Convert.

1. 152 inches to feet
2. 6.53 miles to yards
3. 762.0 feet to inches
4. 627 feet to yards
5. 29 tons to pounds
6. 88.90 pounds to ounces
7. 89 600 pounds to short tons
8. 8552 ounces to pounds
9. 1.25 miles to nautical miles
10. 2.63 bushels to cubic feet
11. 375 cubic inches to cubic feet
12. 34.5 board feet to cubic inches
13. 3.75 cubic feet to board feet
14. 88.2 cubic yards to cubic feet
15. 2.88 US gallons to cubic inches
16. 5.82 cubic feet to US gallons
17. 87.2 cubic inches to imperial gallons
18. 392 acres to square miles
19. 8834 square feet to acres
20. 994 square yards to acres

Applications

21. A certain farm has an area of 93 850 square feet. How many acres is that?
22. Convert all the dimensions for the part in Fig. 4-1 to inches.
23. The jet fuel tank, Fig. 4-2, has a volume of 15.7 ft³. How many US gallons of jet fuel will it hold?
24. A certain circuit board has a weight of 0.176 lb. Find its weight in oz.

FIGURE 4-1

FIGURE 4-2
A jet fuel tank.

4-3 CONVERTING BETWEEN BRITISH/AMERICAN AND METRIC UNITS

We convert between British/American and metric units in the same way we converted within each system.

Example 7
Convert 2.84 gallons (imperial) to litres.

Solution: From the Table of conversion Factors we find that

$$1 \text{ gallon (imperial)} = 4.546 \text{ litres}$$

Converting,

$$2.84 \text{ gal} = 2.84 \text{ gal} \times \frac{4.546 \text{ L}}{1 \text{ gal}} = 12.9 \text{ L}$$

rounded to three significant digits.

SECTION 4-3 Exercises Converting Between Systems

1. 364 kilometres to miles
2. 4.325 nautical miles to metres
3. 10.5 gallons (US) to litres
4. 10.5 gallons (imperial) to litres
5. 180 knots to miles per hour
6. 95 km/h to mi/h
7. 4 cords to cubic metres
8. 98 ft·lbf to N·m
9. 60 mi/h to ft/s
10. 32 psi (lbf/in.^{2}) to kPa
11. 364 000 m to ft
12. 6.83 in. to mm
13. 735 900 lb to kg
14. 1.27 hp to W

FIGURE 4-3

15. 4.66 gallons to litres
16. 825×10^4 dynes to newtons
17. 3.94 yards to metres
18. 83.4 cm³ to gal
19. 0.275 pounds to grams
20. 2.840 km to yd

Applications

21. An air conditioning unit has an output of 20 $\overline{0}$00 BTU. What is its output in kilojoules?
22. A certain laptop computer has a mass of 6.35 kg. What is its mass in pounds?
23. A stream has a flow rate of 5.3×10^3 ft³/s. What is the flow rate in kilolitres per second.
24. Convert all the dimensions in Fig. 4-3 to centimetres.

4-4 ARITHMETIC WITH NUMBERS HAVING UNITS OF MEASURE

In this section we show how to add, subtract, multiply, and divide *denominate* numbers (numbers having units of measure). We will also find powers and roots of denominate numbers.

Adding and Subtracting Denominate Numbers

If the numbers to be added or subtracted have the same units, simply combine them as usual. The answer will have the same units as the original numbers. Be sure to keep the proper number of decimal places.

If the numbers have *different* units, they must be converted such that all have the same units, before combining.

Example 8

Add 3.823 in. and 9.48 cm, and express the result in centimetres.

Solution: We convert 3.823 in. to cm, knowing that 1 in. = 2.54 cm.

$$3.823 \text{ in.} = 3.823 \text{ in.} \times \frac{2.54 \text{ cm}}{1 \text{ in.}} = 9.710 \text{ cm}$$

Adding,

$$9.710 \text{ cm} + 9.48 \text{ cm} = \underline{19.19 \text{ cm}} \quad \text{rounded}$$

Multiplying Denominate Numbers

Simply multiply the two numbers, keeping the proper number of significant digits. The units in the product will be *the product of the units in the original numbers*.

Example 9
(a) 7.29 N $\times$ 0.500 m = 3.65 N·m
(b) 10.2 cm $\times$ 2.05 cm = 20.9 cm²

Dividing Denominate Numbers

Divide the given numbers as usual. The units in the quotient will be the quotient of the units in the original numbers.

Example 10
Divide 394 kilometres by 7.25 hours.

Solution: Dividing,

$$\frac{394 \text{ kilometres}}{7.25 \text{ hours}} = 54.3 \text{ km/h}$$

If the units of dividend and divisor *are the same,* we get a number with *no units.* Such *dimensionless quantities* are of great importance in technical work.

Example 11
An arc 6.23 cm in length is cut off by two radii in a circle with a radius of 5.86 cm, Fig. 4-4. Divide the arc length by the radius.

Solution: Dividing,

$$\frac{6.23 \text{ cm}}{5.86 \text{ cm}} = 1.06$$

The centimetres cancel, leaving the quotient 1.06 without units. (We'll see later that this dimensionless quotient is the *radian measure* of the angle A shown in the figure.)

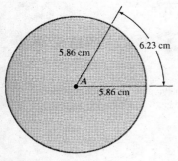

FIGURE 4-4
An arc of a circle.

Powers of Denominate Numbers

When raising a denominate number to a power, *be sure to raise the units* to the same power.

Example 12
A certain square board is 5.34 cm on a side. Square that number to get the area of the board.

Solution: Squaring both the number and the units gives

$$(5.34 \text{ cm})^2 = (5.34)^2 (\text{cm})^2$$
$$= 28.5 \text{ cm}^2$$

Roots of Denominate Numbers

When taking the root of a denominate number, we must also take the same root of the units of measure.

Example 13
A certain cube has a volume of 837 cubic centimetres (cm^3). Find the side of that cube by taking the cube root of the given number.

Solution: Taking the cube root of both the number and the units,

$$\sqrt[3]{837 \text{ cm}^3} = \sqrt[3]{837} \sqrt[3]{\text{cm}^3} = 9.42 \text{ cm}$$

SECTION 4-4 Exercises Arithmetic with Numbers Having Units of Measure

Be sure to include the proper units with your numerical results.

Adding and Subtracting Denominate Numbers
1. Add 64.8 cm and 34.1 in. Give the result in centimetres.
2. Add 43.1 yd and 32.9 m. Give the result in metres.
3. Add 33.6 acres and 41.5 hectares. Give the result in hectares.
4. Add 66.3 ft^3 and 21.7 m^3. Give the result in cubic metres.

Multiplying Denominate Numbers
5. Multiply 53.4 N and 5.93 m.
6. Multiply 34.8 in. and 3.72 cm. Give the result in square centimetres.
7. Multiply 3.41 cm, 4.93 cm, and 1.82 cm.
8. Multiply 2.73 m, 1.83 yd, and 5.22 ft. Give the result in cubic metres.

Dividing Denominate Numbers
9. Divide 83.2 kilometres by 2.84 hours.
10. Divide 823 m by 34.7 seconds.
11. Divide 56.2 metres by 34.1 seconds.
12. Divide 88.8 kilometres by 2.44 seconds.
13. Divide 567 cm by 22.5 cm.
14. Divide 83.2 cm by 30.2 in.
15. Divide 294 yd by 162 m.

Powers of Denominate Numbers
16. Find the square of 2.83 cm.
17. Find the square of 5.23 cm.
18. Find the square of 3.75 m.
19. Find the cube of 44.3 mm.
20. Find the cube of 1.66 m.
21. Find the cube of 4.72 m.

Roots of Denominate Numbers
22. Find the square root of 622 m^2.
23. Find the square root of 25.8 mm^2.
24. Find the cube root of 73.6 cubic metres.
25. Find the cube root of 255 cubic centimetres.

Applications
26. A landing strip, Fig. 4-5, is 355 m long and 48.5 ft wide. Find the area of the strip in square metres by multiplying the length by the width and doing the proper conversion of units.
27. A cylindrical storage tank, Fig. 4-6, has a volume of 3740 ft^3. The area of one end is 185 m^2. Find the height h of the tank, in metres, by dividing the volume by the area of the end and making the proper unit conversions.

■ FIGURE 4-5
A landing strip.

■ FIGURE 4-6
Cylindrical storage tank.

■ FIGURE 4-7
Computer chip.

■ FIGURE 4-8
Two building lots.

■ FIGURE 4-9
Ceramic cube.

28. A square computer chip, Fig. 4-7, has a face area of 1.74 mm². Find the length L of a side by taking the square root of the area.

29. A building lot, Fig. 4-8, has an area of 2.74 acres. An adjoining lot containing 15 400 ft² is obtained. How many acres are in the combined lots?

30. A ceramic cube, Fig. 4-9, measures 2.45 in. on a side. Find its volume by cubing the length of the side.

4-5 CONVERTING AREAS AND VOLUMES

If the conversion factor needed to convert an area or a volume can be found in a table of conversion factors, do the conversion as was shown for other units. If the conversion factor needed is not in the table, it can often be obtained by squaring or cubing a conversion factor for length.

Example 14
Convert 52 450 cubic millimetres (mm³) to cubic decimetres (dm³).

Solution: From the table of Metric Prefixes we find that

$$1 \text{ mm} = 0.001 \text{ m}$$

and

$$1 \text{ dm} = 0.1 \text{ m}$$

First we convert these equations to the cubic version by cubing both sides of each equation, resulting in

$$(1 \text{ mm})^3 = (0.001 \text{ m})^3$$

and

$$(1 \text{ dm})^3 = (0.1 \text{ m})^3$$

or

$$1 \text{ mm}^3 = 0.000\,000\,001 \text{ m}^3$$

and

$$1 \text{ dm}^3 = 0.001 \text{ m}^3$$

The next step is dividing the first cubed equation 1 mm³ to get the conversion factor from cubic millimetres to cubic metres

$$1 = \frac{0.000\,000\,001 \text{ m}^3}{1 \text{ mm}^3}$$

and then dividing the second cubed equation by 0.001 m³ to get the second conversion from cubic metres to cubic decimetres.

$$\frac{1 \text{ dm}^3}{0.001 \text{ m}^3} = 1$$

Then

$$52\,450 \text{ mm}^3 = 52\,450 \text{ mm}^3 \times \frac{0.000\,000\,001 \text{ m}^3}{1 \text{ mm}^3} \times \frac{1 \text{ dm}^3}{0.001 \text{ m}^3} = 0.052\,450 \text{ dm}^3$$

Finally, since 1 dm³ = 1 L, we could write the last answer as 0.052 45 L rounded.

SECTION 4-5 Exercises Converting Areas and Volumes

Convert the following areas.

1. 0.4327 ha to m²
2. 1636 m² to ha
3. 24.8 ft² to m²
4. 59.2 in.² to cm²
5. 557 mm² to in.²
6. 372.0 m² to ft²
7. 982 km² to acres
8. 5.95 acres to ha
9. 8845 ft² to yd²
10. 7.66 in.² to mm²

Convert the following volumes.

11. 7360 ft³ to in.³
12. 4.83 m³ to yd³
13. 73.8 yd³ to m³
14. 19.2 in.³ to cm³
15. 267 mm³ to in.³
16. 4523 L to m³
17. 112 litres to gallons (US)
18. 1.43 cm³ to m³
19. 2640 ft³ to yd³
20. 1.36 in.³ to mm³

Applications

21. The surface area of a certain lake, Fig. 4-10, is 7360 yd². Convert this to square metres.
22. A solar collector, Fig. 4-11, has an area of 8834 in.² Convert this to square metres.

■ FIGURE 4-10
A lake.

■ FIGURE 4-11
Solar collector.

FIGURE 4-12
Hot air balloon.

FIGURE 4-13
Gasoline tank.

23. The volume of a balloon, Fig. 4-12, is 8360 ft³. Convert this to cubic inches.
24. The volume of a certain gasoline tank, Fig. 4-13, is 9274 cm³. Convert this to litres.

4-6 CONVERTING RATES

A *rate* is the amount of one quantity expressed per unit of another quantity.

---- **Example 15** ----
Some typical rates are

(a) Flow rate (L/min)
(b) Rate of travel (km/h)
(c) Application rate (kg/h)
(d) Unit price (dollars/kg)

The units given here are, of course, not the only ones that could be used. Flow rate, for example, could be in litres per hour, or any unit of volume per unit of time.

A good clue that you are dealing with a rate is the word *per,* as in dollars *per* kilogram. Note that each rate contains *two* units of measure.

---- **Example 16** ----
Kilometres per hour has kilometres in the numerator and hours in the denominator.

It may be necessary to convert either or both of those units to other units. Sometimes a single conversion factor can be found (such as 1 km/h = 0.2778 m/s), but more often you will have to convert each unit with a separate conversion factor.

Example 17

Chromium is to be plated onto steel automobile parts at the rate of 1.75 pounds of chromium per square yard of steel. Convert this rate to kilograms of chromium per square metre of steel.

Solution: We write the original quantity as a fraction, and multiply by the appropriate factors, themselves written as fractions.

$$1.75 \text{ lb/yd}^2 = \frac{1.75 \text{ lb}}{\text{yd}^2} \times \frac{1 \text{ yd}^2}{0.8361 \text{ m}^2} \times \frac{0.4536 \text{ kg}}{1 \text{ lb}} = 0.949 \text{ kg/m}^2$$

When rates involve unit pricing, English speaking North America usually writes a price in this format, $3.99/lb, with the monetary symbol leading the denominate number. Most of the world writes the unit of currency in the conventional position for units of measure, that is, after the denominate number. North American custom is inconsistent, in that amounts quoted in cents, follow standard unit style rules, typically, "The price of an apple is 50¢." The usual form in most of the world would be 3.99 $/lb. This form should be used for rate conversions involving currency.

Example 18

Gasoline costs $1.24 per gallon in parts of the United States. What is the cost in Canadian dollars per litre? The exchange rate at the time of the purchase is 1 $(Can.) = 0.6774 $(US).

Solution: We write the original price in proper unit format, as a fraction, $\frac{1.24 \ \$(US)}{\text{gal}}$, and multiply the appropriate conversion factors, themselves written as fractions.

$$\$1.24/\text{gal (U.S.)} = \frac{1.24 \ \$(US)}{\text{gal}} \times \frac{1 \ \$(Can.)}{0.6774 \ \$(US)} \times \frac{1 \text{ gal.}}{3.785 \text{ L}} = 0.4836 \ \$(Can.)/L$$

The rounded answer would be $0.48 per litre in Canadian currency.

SECTION 4-5 Exercises Converting Rates

Convert units on the following time rates.

1. 5.86 ft/s to miles per hour
2. 777 gal/min to cubic metres per hour
3. 66.2 mi/h to kilometres per hour
4. 52.0 knots to km/min
5. 953 births/year to births per week
6. 8336 Btu/h to kilojoules per minute
7. 88.9 m²/min to ha per year

Convert units on the following unit prices.

8. $1.25/g to dollars per kilogram
9. $800/acre to cents per square metre
10. $3.54/lb to cents per ounce
11. $4720/ton to cents per pound

Applications

12. An airplane is cruising at a speed of 785 mi/h. Convert this speed to km/h.
13. Insecticide is applied to a crop at the rate of 5.00 gal/acre. How many litres per hectare is this?
14. A certain ore costs $25.35 per short ton. Find the cost in lire (Italy) per metric ton, if the exchange rate is 1800 lire to the dollar.

Chapter 4 REVIEW EXERCISES

Convert.

1. 3646 Btu/h to kJ/s
2. 22.3 miles to kilometres
3. 2.28×10^9 ohms to megohms
4. $7360/ton to cents per pound
5. 7250 picofarads to microfarads
6. 274 acres to square metres
7. 6253 cm² to square inches
8. 2.74 yd³ to cubic metres
9. 9.26 m²/min to hectares per year
10. 385×10^4 N to kilonewtons
11. 228 kg to slugs
12. 72.2 short tons to kilograms
13. 83.2 m to kilometres
14. 49.2 mi/h to km/h
15. 283 200 g to kilograms
16. 34.9×10^{-5} kW to watts
17. $995/acre to cents per square metre
18. 28 300 nanoseconds to milliseconds
19. 635 lb to newtons
20. 88.2 hp to kilowatts
21. 28.2 ft/s to miles per hour
22. $9.85/g to dollars per kilogram
23. A car travels at a speed of 55.5 mi/h. Convert this speed to km/h.
24. A farm has an area of 474 acres. Another parcel containing 0.225 square miles is obtained. How many acres are in the combined parcels?
25. A driveway is 22.5 ft long and 3.10 m wide. Find the area of the driveway in square metres by multiplying the length by the width and doing the proper conversion of units.
26. A cylindrical bar has a volume of 0.372 ft³. The length of the bar is 1.75 m. Find the area of one end of the bar, in square centimetres, by dividing the volume by the length and making the proper units conversions.
27. A cubical crate measures 6.45 m on a side. Find its volume by cubing the length of the side.
28. A square tile has an area of 94.6 cm². Find the length of a side by taking the square root of the area.

WRITING

29. Your company is agonizing over whether to convert to metric or stay with the English system of measurements. Write a letter to your company president giving specific reasons why you think you should "go metric."
30. Suppose that you work in the same company described in Exercise 29. Write a letter to your company president giving specific reasons why you think you should *not* convert to metric.

TEAM PROJECT

31. The Chinese have a different system of weights than the English or the metric system. These are the *tael*, the *catty* (16 taels), and the *picul* (100 catties). One catty is equal to 0.605 kg. Determine
    (a) how many ounces are equal to one tael.
    (b) how many piculs are in one ton.
    (c) how many catties are needed to make 7.00 kg.

5 Percentage

Outcomes

When you have completed this chapter, you will be able to

1. Define with respect to percent problems: base, amount, and rate.
2. Convert fractions or mixed numbers to a percent.
3. Convert percent to decimals.
4. Convert a percent to a common fraction.
5. Solve percentage problems, for various missing elements such as amount, base, or rate.
6. Evaluate a measurement when percent error, percent concentration, or percent efficiency are given.

Of all the mathematical topics we cover in this text, probably the one most used in everyday life is percentage. The daily news usually has several references to percentage; the inflation rate is now such-and-such percent, the unemployment rate is such-and-such percent, and so on. In this chapter we define *percent,* show how to convert numbers and fractions to and from percents, and work numerous types of percentage problems.

5-1 CONVERTING TO AND FROM PERCENT

Definition

The word *percent* means by the hundred, or per hundred. A percent thus gives the number of parts *in every hundred*.

Example 1

If we say that a certain alloy is 25% zinc by weight, we mean that 25 out of every 100 lb of alloy will be zinc.

Percent as a Rate

The word *rate* is often used to indicate a percent, or percent rate, as in "rejection rate," "rate of inflation," or "growth rate."

> **Example 2**
> A germination rate for seeds of 94% means that, on the average, 94 seeds out of every 100 would be expected to germinate.

Percent as a Fraction

Percent is another way of expressing a fraction having 100 as a denominator.

> **Example 3**
> If we say that a mason has finished 60% of a chimney, we mean that $\frac{60}{100}$ (or $\frac{3}{5}$) of the chimney is finished.

Converting Decimals to Percent

We will now get some practice in converting decimals and fractions to percent, and vice versa. This will give us the skills needed to do the percentage problems coming later.

To convert decimals to percent, simply move the decimal point two places to the right and write the percent symbol (%) following the number.

> **Example 4**
> (a) $0.50 = 50\%$
> (b) $2.70 = 270\%$
> (c) $0.006 = 0.6\%$

Converting Fractions or Mixed Numbers to Percent

First write the fraction or mixed number as a decimal, and then proceed as above.

> **Example 5**
> (a) $\frac{1}{5} = 0.20 = 20\%$
> (b) $\frac{7}{4} = 1.75 = 175\%$
> (c) $2\frac{1}{2} = 2.5 = 250\%$

With a fraction that yields a repeating decimal, we must either be told how many significant digits to retain or decide how many from the other numbers in the problem.

> **Example 6**
> Change $\frac{1}{3}$ to a percent, keeping four significant digits.
>
> **Solution:**
> $$\frac{1}{3} = 0.33333\ldots = 33.33\%$$

Converting Percent to Decimals

Move the decimal point two places to the left and remove the percent sign.

Example 7
(a) 22% = 0.22
(b) 2.3% = 0.023
(c) 235% = 2.35

Converting Percent to a Fraction

Write a fraction with 100 in the denominator and the percent in the numerator. Remove the percent sign and reduce the fraction to lowest terms.

Example 8
(a) $25\% = \dfrac{25}{100} = \dfrac{1}{4}$

(b) $62.5\% = \dfrac{62.5}{100} = \dfrac{625}{1000} = \dfrac{5}{8}$

(c) $120\% = \dfrac{120}{100} = \dfrac{6}{5} = 1\dfrac{1}{5}$

(d) $12\dfrac{1}{3}\% = \dfrac{12\dfrac{1}{3}}{100} = 12\dfrac{1}{3} \div 100 = \dfrac{37}{3} \div \dfrac{100}{1} = \dfrac{37}{3} \times \dfrac{1}{100} = \dfrac{37}{300}$

SECTION 5-1 Exercises Converting to and from Percent

Convert each decimal to a percent.

1. 6.29
2. 0.377
3. 0.0046
4. 0.748
5. 0.653
6. 0.0827
7. 3.959
8. 0.594

Convert each fraction to a percent. Round to three significant digits.

9. $\dfrac{3}{5}$
10. $\dfrac{1}{4}$
11. $\dfrac{9}{10}$
12. $\dfrac{7}{9}$
13. $\dfrac{2}{7}$
14. $\dfrac{7}{11}$
15. $\dfrac{1}{25}$
16. $\dfrac{4}{9}$
17. $\dfrac{1}{3}$
18. $\dfrac{3}{19}$
19. $\dfrac{7}{15}$
20. $\dfrac{11}{60}$

Convert each percent to a decimal.

21. 45%
22. 3.64%
23. $365\dfrac{1}{2}\%$
24. $18\dfrac{1}{3}\%$
25. 7.4%
26. 13%
27. 52%
28. 305%

Convert each percent to a fraction.

29. 12.5%
30. $37\dfrac{1}{2}\%$
31. 225%
32. 7%
33. 62.5%
34. 120%
35. 45%

5-2 SOLVING PERCENTAGE PROBLEMS

Amount, Base, and Rate

Percentage problems always involve three quantities:

1. The *percent rate*
2. The *base,* the quantity we are taking the percent of
3. The *amount* we get when we take the percent of the base, also called the *percentage*

The rate, base, and amount are related by the following equation:

PERCENTAGE	Amount = base × rate where the rate is expressed as a decimal.	13

In the percentage problems that follow, you will be given *two* of these three quantities (amount, base, or rate) and be required to find the third.

Finding the Amount When the Base and Rate Are Known

We simply substitute the given base and rate into the percentage equation and solve for the amount.

Example 9

What is 45 percent of 60?

Solution: In this problem the rate is 45%, so

$$\text{Rate} = 0.45$$

But is 60 the amount or the base?

> **TIP** If you have trouble telling which number is the base and which number is the amount, look for the key phrase "percent of." The quantity following this phrase is always the base.

We look for the key phrase "percent of."

What is 45 *percent of* 60?

The quantity following *percent of* is the base, so the base here is 60.

$$\text{Base} = 60$$

Then,

$$\text{Amount} = \text{base} \times \text{rate} = 60(0.45) = 27$$

> **COMMON ERROR** Don't forget to convert the percent rate to a decimal when using the percentage equation.

Example 10

Find 0.5% of 7000.

Solution: We substitute with

$$\text{Base} = 7000 \quad \text{and} \quad \text{Rate} = 0.5\% = 0.005 \text{ (not 0.5)}$$

So

$$\text{Amount} = \text{base} \times \text{rate} = 7000(0.005) = 35$$

Finding the Base When a Percent of It Is Known

We see from the percentage equation that the base equals the amount divided by the rate (expressed as a decimal), or

$$\text{Base} = \frac{\text{amount}}{\text{rate}}$$

Example 11

20% of what number is 55?

Solution: Finding the key phrase,

$$20 \text{ percent of [what number] is } 55?$$

The quantity following *percent of* is "what number," so it is clear that we are looking for the base. So

$$\text{Amount} = 55 \quad \text{and} \quad \text{Rate} = 0.20$$

Then,

$$\text{Base} = \frac{\text{amount}}{\text{rate}} = \frac{55}{0.20} = 275$$

Example 12

525 is 75% of what number?

Solution: From the percentage equation,

$$\text{Base} = \frac{\text{amount}}{\text{rate}} = \frac{525}{0.75} = 700$$

Finding the Percent One Number Is of Another Number

From the percentage equation, the rate equals the amount divided by the base, or

$$\text{Rate} = \frac{\text{amount}}{\text{base}}$$

Example 13

65 is what percent of 125?

Solution: From the percentage equation, with amount = 65 and base = 125,

$$\text{Rate} = \frac{\text{amount}}{\text{base}} = \frac{65}{125} = 0.52 = 52\%$$

Example 14
What percent of 1.8 is 1.35?

Solution: From the percentage equation,

$$\text{Rate} = \frac{\text{amount}}{\text{base}} = \frac{1.35}{1.8} = 0.75$$
$$= 75\%$$

SECTION 5-2 Exercises Solving Percentage Problems

Finding the Amount

Find:
1. 15% of 500 tonnes
2. 45% of 300 m
3. 70% of 1400 kg
4. $66\frac{1}{3}$% of 915 litres
5. 6% of 25 m³
6. $16\frac{2}{3}$% of $42
7. 25% of 505 km
8. 20% of 625 litres
9. 55% of 962 ha

Finding the Base

Find the number of which:
10. 58 is $12\frac{1}{2}$%
11. 75 is 1.5%
12. $\frac{2}{5}$ is $5\frac{1}{2}$%
13. 100 is $33\frac{1}{3}$%
14. 54 is 54%
15. 54 is 75%
16. $\frac{1}{4}$ is 25%
17. $\frac{3}{7}$ is $16\frac{2}{3}$%
18. $\frac{6}{10}$ is $66\frac{2}{3}$%

Finding the Rate

What percent of:
19. 32 is 16?
20. 64 is 8?
21. 35 is 7?
22. 960 men are 480 men?
23. 650 h are 195 h?
24. 600 t are 250 t?
25. 125 L are 12.5 L?
26. 75 kg are 15 kg?
27. 60 cm are 4 cm?

Profit and Loss

28. Find the profit or loss:
    (a) on land that cost $15 600 and was sold at a loss of 30%.
    (b) on goods that cost $7830 and were sold at a gain of 23%.

Applications

29. A resistance, now 6050 Ω, is to be increased by 35%. How much resistance should be added?

30. It is estimated that $\frac{1}{2}$% of the earth's surface receives more energy than the total projected needs for the year 2000. Assuming the earth's surface area to be 5.10×10^8 km², find the required area in ha.

31. As an incentive to install solar equipment, a tax credit of 35% of the first $1000 and 20% of the next $6400 spent on solar equipment is proposed. How much credit would a homeowner get when installing $4500 worth of solar equipment?

32. A woman receives a salary of $32 000 a year. She pays $20\frac{1}{2}$% of it for board, 9% of it for clothing, and 14% for incidentals. What are her yearly expenses?

33. A girder with a length of 15.4 m expands 0.15% when subjected to summer temperatures. Find the expansion in centimetres.
34. It is estimated that the United States has tar sands containing 3.3×10^{10} barrels of oil and that 90% of the tar sands are in Utah. If it is possible to extract 9% of the oil from the sands, how much oil could be extracted from the Utah tar sands?
35. How much metal will be obtained from 590 tonnes of ore if the metal is $20\frac{1}{4}$% of the ore?
36. If a person rents a house for $4300 a year, which is 16% of the value of the property, what is its value?
37. A Department of Energy, Mines, and Resources report on an experimental electric car gives the range of the car as 160 km and states that this is "50% better than on earlier electric vehicles." What was the range of earlier electric vehicles?
38. A man withdrew 40% of his bank deposits and spent 20% of the money drawn in the purchase of a radio worth $325. How much money did he have in the bank?
39. A merchant sold 75% of his goods for $2751. How much were the goods worth?
40. Solar panels provide 65% of the heat for a certain building. If $700/year is now being spent for heating oil, how much would be spent if the solar panels were not used?
41. A fire destroyed $16\frac{2}{3}$% of a stock of goods, and the value of the goods destroyed in the fire was $500. What was the value of the entire stock?
42. The distance between two stations on a certain railroad is 243.6 km, which is $83\frac{1}{3}$% of the total length of the road. What is the total length of the road?
43. If the United States imports 8.62 million barrels of oil per day and if this is 47.3% of its needs, how much oil is needed per day?
44. A tank with a capacity of 45 050 litres contains 7320 litres of water. Express the amount of water in the tank as a percent of the total capacity.
45. In a journey of 4590 km, a person traveled 1020 km by car and the rest by rail. What percent of the distance was traveled by rail?
46. A power supply has a dc output 200.0 V with a ripple of 6.0 V peak to peak. Express the ripple as a percent of the dc output voltage.
47. The construction of a factory cost $380 000 for materials and $490 000 for labour. What percent of the total was the labour cost?
48. If the United States needs 18.9 million barrels of oil per day and imports 9.14 million barrels per day, what percent of its oil needs are imported?
49. By insulating, a homeowner's yearly fuel oil consumption dropped from 3575 litres to 2305 litres. His present oil consumption is what percent of the former?
50. A city lot was sold for $159 000 at a profit of 12%. What did it cost?
51. A merchant gained $45 500 in one year on goods sold at a profit of 25%. What was the cost of the goods?
52. A machine was sold for $187, which was a loss of 15%. What was the cost?
53. An architect charged $\frac{1}{4}$% for plans and specifications and $1\frac{3}{8}$% for superintending a building that cost $32 000. What was the amount of her fee?
54. Find the selling price of an item having a list price of $348.99, discounted 14%.
55. A person paid $5789.70 for a car, including a 4% sales tax. Find the price of the car without the tax.

5-3 PERCENT CHANGE

Percents are often used to *compare* two quantities.

Example 15

If you wanted to compare the price of oil from last year to this year, you might say

The price of oil this year is 8% greater than last year.

Finding the Percent Change

To get an equation for percent change, we start with the percentage equation

$$\text{Amount} = \text{base} \times \text{rate}$$

When the two numbers being compared involve a change from one to the other, the *original value* is usually taken as the base. We then replace "Amount" with "new value − original value," since this is the amount by which the quantity changes.

$$(\text{New value} - \text{original value}) = (\text{original value}) \times \text{rate}$$

Dividing by "original value" gives the rate, or percent change. Then multiplying by 100 and affixing the percent sign, we get the percent change expressed as a percent.

$$\text{Percent change} = \frac{\text{new value} - \text{original value}}{\text{original value}} \times 100 \qquad 14$$

Example 16

A certain price rose from \$1.35 to \$1.50. Find the percent change in price.

Solution: We use the original value, \$1.35, as the base. From the percent change equation,

$$\text{Percent change} = \frac{1.50 - 1.35}{1.35} \times 100 = 11.1\% \text{ increase}$$

Be sure to show the *direction* of change with a positive or negative sign, or with a word like *increase* or *decrease*.

Finding the New Value

We often want to find the new value when the original value is changed by a given percent. We see from the equation for percent change that

$$\text{New value} = \text{original value} + (\text{original value} \times \text{percent change})$$

Example 17

Find the cost of a \$250 radio after the price increases by 5%.

Solution: The original value is \$250 and the percent change, expressed as a decimal, is 0.05. So

$$\text{New value} = 250 + 250(0.05) = \$262.50$$

Thus the new value in Example 17 is equal to 100% of the old value, plus 5% of the old value, or 105% of the old value. In general, to increase a quantity by P percent, simply multiply that quantity by $(1 + P)$, where P is in decimal form.

Example 18

The price of the $250 radio of Example 17 after increasing by 5%, is

$$\text{New value} = 250 \times 1.05 = \$262.50$$

as before.

Similarly, to find the new value after a decrease of P percent, multiply the original value by $(1 - P)$, where again, P is in decimal form.

Example 19

What will be the value of a $4750 machine after a decrease of 12% in value?

Solution: The new value will be 12% less than the original, or 88% of the original. Thus,

$$\text{New value} = (\text{original value}) \times 0.88$$
$$= (4750)0.88 = \$4180$$

Finding the Original Value

Finally, we do a problem where we have the new value and the percent change, and we want to find the original value.

Example 20

What number, increased by 15% of itself, is 364?

Solution: The percent change is 15%, so the new value is 1.15 times as great as the original value. If we call the original value x, then

$$364 = 1.15x$$

We get x simply by dividing both sides by 1.15. Thus

$$x = \frac{364}{1.15} = 317 \quad (\text{rounded})$$

SECTION 5-3   Exercises   Percent Change

Finding Percent Change Find the percent change when a quantity changes:

1. from 43.4 to 85.8
2. from 96 to 34
3. from 556 to 763
4. from 0.00 904 to 0.00 793

Finding the New Value Find the new value when

5. 273 is increased by 2.45%
6. 4.83 is increased by 34%
7. 8475 is decreased by 2%
8. 8.36 is decreased by 52%

Finding the Original Amount

What number increased by:
9. 15% of itself = 414?
10. 19% of itself = 595?
11. $5\frac{1}{2}$% of itself = 422?
12. $12\frac{1}{2}$% of itself = 900?

What number decreased by:
13. 10% of itself = 18?
14. 40% of itself = 12?
15. 20% of itself = 620?
16. 90% of itself = 1.84?

Applications

17. The temperature in a building rose from 294.15 K (21°C) to 296.15 K (23°C) during the day. Find the percent change in temperature.
18. A casting with an initial mass of 205 kg has 13% of its material shaved off. Find its final mass.
19. The profits of a mill in two years were $66 990, and the profits for the second year were 3% greater than the profits the first year. What were the profits each year?
20. An automobile gas turbine is expected to provide a 30% improvement in fuel economy over a conventional gas piston engine. If a car gets 14.1 km/L, what fuel efficiency would you expect if its conventional engine is replaced by a gas turbine of the same power rating?

> **TIP:** To obtain the metric efficiency rating that is currently used (litres per 100 kilometres), the following procedure is used
> Find the reciprocal of the km/L rating and multiply top and bottom of the fraction by 100.
> $$14.1 \text{ km/L converts to } \frac{1 \text{ L}}{14.1 \text{ km}} = \frac{0.0709 \text{ L}}{1 \text{ km}} \times \frac{100}{100} = \frac{7.09 \text{ L}}{100 \text{ km}}$$

21. A certain common stock rose from a value of 23\frac{1}{2}$ per share to 24\frac{3}{8}$ per share. What was the percent change in value?
22. A house that costs $598 per year to heat has insulation installed in the attic, causing the fuel bill to drop to $501 per year. Find the percent change in the cost.
23. A mechanic who received $84 per day had her salary raised by 7%. What were her daily wages then?
24. A crew paved 1350 m of roadway on one day and 1520 m the following day. What is the percent change in the amount they paved?
25. A certain country's energy consumption of 37 million barrels of oil equivalent per day is expected to climb to 48 million in six years. Find the percent increase in consumption.
26. The population of a certain town is 6024, which is $16\frac{2}{3}$% more than it was 4 years ago. What was the population then?

27. The temperature of a room rose from 524.67°R (65°F) to 531.67°R (72°F). Find the percent increase.
28. A train running at 24 mi/h increases its speed by $37\frac{1}{2}$%. How fast does it then go?
29. If the energy consumption next year will be 3.6% higher than the 37 million barrels per day consumed now, how many barrels per day will be needed next year?

5-4 PERCENT ERROR, PERCENT CONCENTRATION, AND PERCENT EFFICIENCY

Percent Error

Percent error is a common way of expressing the accuracy of a measurement. The percent error is the difference between the measured value and the known or "true" value, expressed as a percent of the known value.

$$\text{Percent error} = \frac{\text{measured value} - \text{known value}}{\text{known value}} \times 100 \quad \boxed{15}$$

Example 21

A gauge block that is certified to be 2.50 cm long is measured with a caliper. The reading is 2.53 cm. What is the percent error in the reading?

Solution: From the percent error equation,

$$\text{Percent error} = \frac{2.53 - 2.50}{2.50} \times 100 = +1.2\%$$

As with percent change, be sure to specify the *direction* of the error, either with words or with a plus or minus sign.

Percent Concentration

In a mixture of two or more ingredients, the *percent concentration* of one ingredient is defined as

$$\text{Percent concentration of ingredient } A = \frac{\text{amount of } A}{\text{amount of mixture}} \times 100 \quad \boxed{16}$$

Example 22

A certain antifreeze mixture contains 22 litres of antifreeze and 112 litres of water. Find the percent of antifreeze in the mixture.

Solution: The total amount of mixture is

$$22 + 112 = 134 \text{ litres}$$

So, by the equation for percent concentration,

$$\text{Percent antifreeze} = \frac{22}{134} \times 100 = 16.4\%$$

COMMON ERROR The denominator in the equation for percent concentration must be the *total amount* of mixture, or the sum of all the ingredients. Do not use just one of the ingredients.

Percent Efficiency

With any device, such as a motor or an amplifier, power is put into the device and power is taken out of the device, often in a different form.

---- **Example 23** ----

A motor takes power in electrical form and delivers power in mechanical form.

It is inevitable that the power output of any device is *always less* than the power input, because of power losses within the device. The *efficiency* of the device is a measure of those losses. The percent efficiency is defined as the output divided by the input, expressed as a percent.

$$\text{Percent efficiency} = \frac{\text{output}}{\text{input}} \times 100 \quad \boxed{17}$$

---- **Example 24** ----

A certain electric motor consumes 981 W and has an output of 1.21 hp. Find the efficiency of the motor (1 hp = 746 W).

Solution: Since output and input must be in the same units, we must convert either to horsepower or to watts. Converting the output to watts, we obtain

$$\text{Output} = 1.21 \text{ hp} \left(\frac{746 \text{ W}}{\text{hp}}\right) = 903 \text{ W}$$

Then

$$\text{Percent efficiency} = \frac{903}{981} \times 100 = 92.0\%$$

SECTION 5-4   Exercises   Percent Error, Percent Concentration, and Percent Efficiency

Percent Error

1. A certain quantity is measured at 237 units but is known actually to be 241 units. What is the percent error in the measurement?
2. A shaft is known to have a diameter of 43.000 cm. It is measured at 42.872 cm. Find the percent error in the measurement.
3. A certain capacitor has a working voltage of 150 V dc with an error factor of (−20%, +50%). Within what two voltages would the actual working voltage lie?
4. A resistor is labeled as 42500 Ω with a tolerance of ±6%. Between what two values is the actual resistance expected to lie?
5. A voltmeter is checked by measuring the EMF of a standard cell, known to be 1.5831 V, and the reading obtained is 1.5893 V. What is the percent error in the reading?

Percent Concentration

6. A solution is made by mixing 125 litres of alcohol and 275 litres of water. What is the percent concentration of alcohol?
7. How many litres of alcohol are contained in 400 litres of a gasohol mixture that is 4% alcohol by volume?

8. How many litres of gasoline are there in 345 gal (US) of a methanol-gasoline blend that is 15% methanol by volume?

9. A scheme for transporting powdered coal from a mine to a power station calls for mixing the powdered coal with oil so that the resulting slurry is 20% oil by weight. While traveling through the pipeline, 70% of the powdered coal becomes liquefied. What percent of the slurry reaching the power plant will be liquid?

10. Twelve cubic feet of cement is contained in a concrete mixture that is 16% cement by volume. What is the volume of the total mixture?

Percent Efficiency

11. A certain device consumes 11.9 kW and delivers 7.46 kW. What is its efficiency?

12. An electric motor consumes 2350 W. Find the horsepower it can deliver if it is 70% efficient. (1 hp = 746 W)

13. A certain speed reducer delivers 2.3 kW with a power input of 3.1 kW. Find the percent efficiency of the speed reducer.

14. A certain motor consumes 2200 W and has an efficiency of 65%. What is the output in horsepower?

15. A water pump requires an input of $1\frac{1}{2}$ hp and delivers 20 000 lb of water per hour to a house 36 ft above the pump. Find the percent efficiency. Hint: Multiply the flow rate times the height to get foot-pounds per hour, and convert that to horsepower.

Chapter 5 REVIEW EXERCISES

1. Find the percent change when a quantity changes:
   (a) from 23.4 to 37.8
   (b) from 66 to 54

2. A certain quantity is measured at 127 units but is known actually to be 131 units. What is the percent error in the measurement?

3. Find the new value when:
   (a) 773 is increased by 8.75%
   (b) 24.5 is decreased by 7%

4. A solution is made by mixing 775 litres of alcohol and 365 litres of water. What is the percent concentration of alcohol?

5. What number decreased by:
   (a) 15% of itself is 263
   (b) 30% of itself is 452

6. A concrete mixture that is 18% cement by volume contains 23.6 m³ of cement. What is the volume of the total mixture?

7. Convert each percent to a fraction.
   (a) 52.5%
   (b) $47\frac{1}{2}$%
   (c) 325%
   (d) 9%

8. How many litres of alcohol are contained in 522 litres of a gasohol mixture that is 5% alcohol by volume?

9. Convert each fraction to a percent. Round to three digits.
   (a) $\frac{1}{7}$
   (b) $\frac{7}{15}$
   (c) $\frac{3}{25}$
   (d) $\frac{5}{7}$

10. How many litres of gasoline are there in 145 gallons (US) of a methanol–gasoline blend that is 14% methanol by volume?

11. Convert each decimal to a percent.
    (a) 8.24   (b) 0.627   (c) 0.0027   (d) 0.358
12. The temperature in a building rose from 298.15 K (25°C) to 301.15 K (28°C) during the day. Find the percent change in temperature.
13. What number increased by:
    (a) 25% of itself is 214?   (b) 29% of itself is 95?
14. A tank with a capacity of 23 200 litres contains 17 600 litres of water. Express the amount of water in the tank as a percent of total capacity.
15. Find:
    (a) 48% of 2830 kg   (b) $26\frac{1}{3}$% of 362 kJ
    (c) 45% of 736 km   (d) 16% of 837 litres
16. Find the dollar loss on land that cost $65 200 and was sold at a loss of 3%.
17. Find the number of which:
    (a) 937 is $13\frac{1}{3}$%   (b) 93 is 24%   (c) 92 is 25%
18. A merchant gained $82 300 in one year on goods sold at a profit of 12%. What was the cost of the goods?
19. Convert each percent to a decimal.
    (a) 15%   (b) 7.64%   (c) $65\frac{1}{2}$%   (d) $28\frac{1}{3}$%
20. A certain speed reducer delivers 5.3 hp with a power input of 6.1 hp. Find its percent efficiency.
21. What percent of:
    (a) 736 men is 263 men?   (b) 826 hours is 315 hours?
    (c) 25 gal is 13 gal?   (d) 37 in. is 9 in.?

WRITING

22. This chapter starts by saying, "Of all the mathematical topics we cover in this text, probably the one most used in everyday life is percentage." Do you agree? Write a few paragraphs saying if you agree or not, and back your reasons up with specific examples from personal experience.

TEAM PROJECT

23. Write an hour exam that will test for the most important skills covered in this chapter on percentage. Include at least two word problems. Be sure that it is not too long and hard, and not too short and easy. Then solve the exam and make revisions if necessary. Produce an answer key, to be handed in with the final version of the exam.

6 Arithmetic with Signed Numbers

Outcomes

When you have completed this chapter, you will be able to

1. Evaluate signed numbers.
2. Add and subtract signed numbers.
3. Multiply and divide signed numbers.
4. Simplify roots and powers of signed numbers.
5. Perform calculations with combined operations involving signed numbers.

Up to this point we have done arithmetic with positive numbers and zero only. But positive numbers alone are not enough to enable us to do the calculations we may need to do. Just the simple notion of indicating a loss on a balance sheet requires a *negative* number. Negative temperatures are familiar to everyone. In technical work we encounter negative forces, negative voltages, negative pressures, and so on.

In this chapter we extend our arithmetic to handle negative numbers by learning a few rules of signs. We will follow the same order we used with whole numbers, fractions, decimal numbers, and denominate numbers. After some definitions, we will do addition and subtraction, then multiplication, division, and finally powers and roots.

We learn here how to calculate using positive and negative numbers, but that is not our only goal. We are also laying the groundwork for studying *algebraic* quantities in the following chapters.

6-1 SIGNED NUMBERS

Negative Numbers

■ **FIGURE 6-1** The number line.

Recall that a positive number is a number that is greater than zero. Now we define a *negative number* as a number that is *less* than zero. On the number line (Fig. 6-1) they are all the numbers lying to the left of zero. Negative numbers may be integers or fractions, rational or irrational. To distinguish negative numbers from positive numbers we always place a negative sign $(-)$ in front of a negative number.

Example 1

Some negative numbers are

$$-5, \quad -6.293, \quad -\frac{2}{3}, \quad -2\frac{7}{8}, \quad \text{and} \quad -\sqrt{5}$$

We usually omit writing the positive sign (+) in front of a positive number. Thus a number without a sign is always assumed to be positive. In this chapter we will often write in a (+) sign for emphasis.

Example 2

Some positive numbers are

$$5, \ +5, \ \frac{2}{3}, \ +\frac{2}{3}, \ \text{and} \ \sqrt{5}$$

The Opposite of a Number

The *opposite* of a number n is that number which, when added to n, gives a sum of zero.

Example 3

The opposite of 2 is -2.
The opposite of -6 is $+6$.

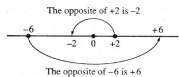

FIGURE 6-2
The opposite of a number.

Geometrically, the opposite of a number n lies on the opposite side of the zero point of the number line from n, and at an equal distance from the origin (Fig. 6-2).

Signs of Equality and Inequality

In Chapter 1 we introduced the signs of equality and inequality:

$$a = b \quad a \neq b \quad a > b \quad a < b \quad a \leq b \quad a \geq b \quad a \approx b$$

We used these signs to show the relative position of positive numbers on the number line. Here we use them for negative numbers as well. Notice that the arrow (>) always points to the number which is more to the left on the number line.

Example 4

Place the proper equality or inequality sign between -2.15 and -2.83.

Solution: Since -2.15 lies to the right of -2.83 on the number line (Fig. 6-3), then -2.15 is *greater than* -2.83. So we write

$$-2.15 > -2.83$$

FIGURE 6-3

Absolute Value

The *absolute value* of a number is, geometrically, the distance of that number from the zero point of the number line, regardless of direction. Thus the absolute value of a number is never negative.

Example 5

The numbers 4 and -4 both have an absolute value of 4, because both are 4 units from the zero point of the number line (Fig. 6-4).

FIGURE 6-4

Absolute value is indicated by enclosing the number between *vertical bars*.

Example 6
(a) $|-4| = 4$ (b) $|+4| = 4$ (c) $|5 - 8| = |-3| = 3$

Negative Numbers on the Calculator

Note that the $(-)$ sign is used for *two different things:*

1. To indicate a negative quantity
2. For the operation of subtraction

This difference is clear on most calculators, which usually have separate keys for these two functions. The $\boxed{-}$ key is used for subtraction but is not used to enter a negative number.

Some calculators have a $\boxed{(-)}$ key (sometimes called a *negation key*) for entering a negative number. Notice that the negative sign is in parentheses on this key, while it is not in parentheses on the subtraction key. Some calculators use a $\boxed{-x}$ key for the same purpose. To enter a negative number on such a calculator, press the $\boxed{(-)}$ key *first,* and then enter the number.

On other calculators, the change-sign key is used to enter a negative number. The change-sign key is marked differently on different calculators, with $\boxed{+/-}$ or $\boxed{CHS}$ being common. Check your manual to see how this key is marked on your calculator. To enter a negative number on a calculator having a change-sign key, simply enter the number first, without sign, and then press the change-sign key.

SECTION 6-1 Exercises Signed Numbers

Negative Numbers

Trace or photocopy the number line of Fig. 6-1, or make your own. On it locate the following numbers.

1. -3 2. 6 3. $-\dfrac{3}{2}$ 4. -2.74

Signs of Equality and Inequality

Place the proper equality or inequality sign between each pair of numbers.

5. 5 and 1   6. 5 and -1 7. -5 and 1
8. -5 and -1 9. -12.2 and -12.3 10. -15.4 and -16.8

Absolute Value

Evaluate each expression.

11. $|-6|$ 12. $|-24|$ 13. $|3 - 6|$
14. $|6 - 3|$ 15. $|15 - 16|$ 16. $|1 - 7|$

Negative Numbers on the Calculator

Enter each negative number on your calculator.

17. -23 18. -82 19. -3.72 20. -4.836

6-2 ADDING AND SUBTRACTING SIGNED NUMBERS

Adding Signed Numbers

Let's say that we have a shoe box (Fig. 6-5) into which we toss all our uncashed cheques and unpaid bills until we have time to deal with them. Let's further assume that the total cheques minus the total bills in the shoebox is $500.

We can think of the amount of a cheque as a positive number, because it increases our wealth, and the amount of a bill as a negative number because

■ **FIGURE 6-5**
A shoe box with bills and cheques.

it decreases our wealth. We thus represent a cheque for $100 as (+100) and a bill for $200 as (−200).

Now, let's *add a cheque* for $100 to the box. If we had $500 at first, we must now have $600.

$$500 + (+100) = 600$$

or

$$500 + 100 = 600$$

Here we have added a positive number, and our total has increased by the amount. That is easy to understand. But what does it mean to *add a negative number*?

To find out, let us now *add a bill* for $100 to the box. If we had $500 at first, we must now have $100 less, or $400. Representing the bill by (−100) we have

$$500 + (-100) = 400$$

This means that to *add a negative number* is no different than to *subtract the absolute value* of that number.

$$500 + (-100) = 500 - 100$$

This gives us our rule of signs for addition. If a and b are two positive quantities, then

| RULE OF SIGNS FOR ADDITION | $a + (-b) = a - b$ | 6 |

The operation of adding a negative quantity (−b) to the quantity a is equivalent to the operation of subtracting the positive quantity b from a.

---- **Example 7** ----
(a) $7 + 2 = 9$
(b) $7 + (-2) = 7 - 2 = 5$
(c) $-7 + 2 = -5$
(d) $-7 + (-2) = -7 - 2 = -9$

Subtracting Signed Numbers

Let us return to our shoebox problem. But now instead of adding cheques or bills to the box, we will *subtract* (remove) cheques or bills from the box.

First we remove (subtract) a cheque for $100 from the box. If we had $500 at first, we must now have $400.

$$500 - (+100) = 400$$

or

$$500 - 100 = 400$$

Here we have *subtracted a positive number*, and *our total has decreased* by the amount subtracted, as expected.

Now let us see what it means to *subtract a negative number*. We will remove (subtract) a bill for $100 from the box. If we had $500 at first, we must now have $100 more, or $600, since we have removed a bill. Representing the bill by (-100) we have

$$500 - (-100) = 600$$

Thus *to subtract a negative number* is the same as to *add the absolute value* of that number.

$$500 - (-100) = 500 + 100$$

Thus, if a and b are positive quantities, our rule of signs for subtraction is

| RULE OF SIGNS FOR SUBTRACTION | $a - (-b) = a + b$ | 7 |

The operation of subtracting a negative quantity $(-b)$ from the quantity a is equivalent to the operation of adding the positive quantity b to a.

Example 8

(a) $7 - 2 = 5$
(b) $7 - (-2) = 7 + 2 = 9$
(c) $-7 - 2 = -9$
(d) $-7 - (-2) = -7 + 2 = -5$

Example 9

A surveyor places a stake at A, Fig. 6-6, and calls its elevation 0 m. She then goes downhill and places stake B, which she labels -5.85 m. She then places stake C 3.77 m below B. What is the elevation of C?

■ **FIGURE 6-6**
Placement of surveyor's stakes.

Solution: We must subtract 3.77 m from -5.85 m.

$$\text{Elevation of } C = -5.85 - 3.77 = -9.62 \text{ m}$$

SECTION 6-2 Exercises Adding and Subtracting Signed Numbers

Adding Signed Numbers

Add.
1. $(+5) + (-3)$
2. $(+4) + (-11)$
3. $(+77) + (-24)$
4. $(-7) + (+2)$
5. $(-9) + (+11)$
6. $(-23) + (+82)$
7. $(-9) + (-3)$
8. $(-4) + (-41)$
9. $(-35) + (-54)$
10. $(+5.52) + (-3.14)$
11. $(+4.26) + (-11.82)$
12. $(+72.5) + (-27.2)$
13. $(-7.82) + (+2.26)$
14. $(-9.77) + (+11.26)$
15. $(-23.2) + (+82.7)$
16. $(-9.33) + (-3.25)$
17. $(-4.94) + (-41.22)$
18. $(-35.5) + (-54.9)$

Subtracting Signed Numbers

Subtract.
19. $(+7) - (-3)$
20. $(+9) - (-61)$
21. $(+77) - (-49)$
22. $(-7) - (+4)$
23. $(-5) - (+66)$
24. $(-43) - (+84)$
25. $(-9) - (-3)$
26. $(-2) - (-96)$
27. $(-37) - (-79)$
28. $(+7.74) - (-3.69)$
29. $(+9.46) - (-16.84)$
30. $(+74.7) - (-47.2)$
31. $(-7.84) - (+4.26)$
32. $(-9.77) - (+61.46)$
33. $(-43.2) - (+84.7)$
34. $(-9.33) - (-3.47)$
35. $(-9.94) - (-96.42)$
36. $(-37.7) - (-54.9)$

Applications

37. A balance sheet showed a balance of −$3482.77, and another bill came in for $1735.83. Find the new balance.
38. During a certain month the temperature dropped from +4.8°C to −11.4°C. Find the difference in the temperatures.
39. A submarine, Fig. 6-7, was at an elevation of −238 m, and then descended another 177 m. Find the new elevation.

■ FIGURE 6-7
A diving submarine.

40. The voltage in a certain circuit was -33.8 V. It was then reduced by 11.3 V. Find the new voltage.

6-3 MULTIPLYING SIGNED NUMBERS

Rules of Signs

To get our rules of signs for multiplication, we use the idea of *multiplication as repeated addition*. For example, to multiply 3 by 4 means to add four 3's (or three 4's).

$$3 \times 4 = 3 + 3 + 3 + 3$$

or

$$3 \times 4 = 4 + 4 + 4$$

Let us return to our shoebox example. Recall that it contains uncashed cheques and unpaid bills. Let's first add 5 cheques ($+5$), each worth \$100 ($+100$), to the box. The value of the contents of the box then increases by \$500 ($+500$). Multiplying,

(Number of cheques) $\times$ (value of one cheque) = change in value of contents

$$(+5)(+100) = +500$$

Thus a positive number times a positive number gives a positive product. This is nothing new.

Now let's add 5 *bills* to the box, thus decreasing its value by \$500. To show this multiplication, we use ($+5$) for the number of bills, (-100) for the value of each bill, and (-500) for the change in value of the box contents.

$$(+5)(-100) = -500$$

Here, a positive number times a negative number gives a negative product.

Next we *remove* 5 cheques from the box, thus decreasing its value by \$500.

$$(-5)(+100) = -500$$

Here again, the product of a positive number and a negative number is negative. Thus it doesn't matter whether the negative number is the first or the second. This, of course, is what you would expect from the commutative property, which applies to negative numbers as well as to positive numbers.

Finally, we remove 5 bills from the box, causing its value to increase by \$500.

$$(-5)(-100) = +500$$

Here, the product of two negative numbers is positive.

We summarize these findings to get our *rules of signs for multiplication*. If a and b are two positive numbers, then

| RULE OF SIGNS FOR MULTIPLICATION | $(+a)(+b) =$
 $(-a)(-b) = +ab$ | 9 |

The product of two numbers of like sign is positive.

and

| RULE OF SIGNS FOR MULTIPLICATION | $(+a)(-b) =$
 $(-a)(+b) = -ab$ | 10 |

The product of two numbers of opposite sign is negative.

Example 10

(a) $(3)(-4) = -12$
(b) $(-3)(4) = -12$
(c) $(-3)(-4) = 12$
(d) $(-9.641)(-8.31) = 80.1$ (rounded)

Multiplying a String of Numbers

When we multiply two negative numbers we get a positive product. So when multiplying a string of numbers, if an even number of factors are negative, the answer will be positive.

Example 11

(a) $(4)(-3)(-2)(-1) = -24$
(b) $(4)(-3)(-2)(1) = 24$
(c) $(-1.38)(-2.73)(-1.84) = -6.93$ (rounded)

Example 12

The velocity v of an object thrown upward or downward is found by adding the initial velocity v_0 expressed in metres per second, to the product of the acceleration a due to gravity of 9.81 m/s² and the time t expressed in seconds, or

$$v = v_0 + at$$

Find the velocity after 1.90 seconds of a ball thrown upward with an initial velocity of 15.6 m/s.

Solution: If we take the upward direction as positive, Fig. 6-8, then the initial velocity is positive because the ball is thrown upwards, and the acceleration due to gravity is negative because it acts downwards. Then

$$v_0 = 15.6 \text{ m/s} \quad \text{and} \quad a = -9.81 \text{ m/s}^2$$

Substituting gives

$$v = \frac{+15.6 \text{ m}}{\text{s}} + \left(\frac{-9.81 \text{ m}}{\text{s}^2}\right)(1.90 \text{ s}) = -3.04 \text{ m/s}$$

Here the negative sign indicates that the velocity is in the negative direction, that is, the ball is now moving downwards. In Fig. 6-8, when the ball hits the ground it will have a velocity of -15.6 m/s. It has the same speed as it had initially, but because it is moving in the opposite direction the final velocity is the negative of the initial velocity.

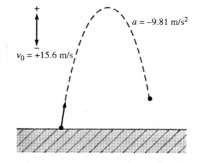

■ **FIGURE 6-8**
A ball thrown upward.

SECTION 6-3 Exercises   Multiplying Signed Numbers

Multiply. Treat all integers as exact. Treat all decimal fractions as approximate, and round your answer to the proper number of significant digits.

1. $(+8)(-4)$
2. $(+3)(-2)$
3. $(+7)(-11)$
4. $(-2)(+3)$
5. $(-9)(+3)$
6. $(-5)(-8)$

7. $(-3)(-5)$
8. $(-6)(-8)$
9. $(-2)(-23)$
10. $(+2.18)(-3.24)$
11. $(+82.3)(-15.2)$
12. $(+1.57)(-4.11)$
13. $(-412)(+2.43)$
14. $(-1.29)(+723)$
15. $(-4.85)(-4.28)$
16. $(-173)(-3.25)$
17. $(-5.26)(-618)$
18. $(-722)(-1.23)$
19. $(-2)(+3)(-7)$
20. $(-3)(-9)(+3)$
21. $(+6)(-5)(-8)$
22. $(-6)(-3)(-5)$
23. $(+3)(-6)(-8)$
24. $(-4)(-2)(-23)$
25. $(+1.52)(+3.48)(-1.74)$
26. $(-2.31)(+32.4)(-25.5)$
27. $(-1.72)(+4.67)(-2.81)$
28. $(+2.61)(-602)(+1.63)$
29. $(-1.84)(-2.22)(+523)$
30. $(+2.81)(-7.35)(-1.25)$

Applications

31. Find the velocity after 1.25 seconds of a stone thrown upward with a velocity of 10.9 m/s. See Example 12.

32. A broker bought 500 shares of stock for $35.25 per share, whose price then rose $1.25 per share and later dropped $0.42 per share. Find the final value of the stock.

33. A beam, Fig. 6-9, has three forces acting on it. If we consider the upward forces to be positive, the three forces are +15 500 N, +7260 N, and −12 200 N. The *moment* or torque τ of each force about p is equal to the product of the force and the distance that this force is applied from p. Thus the sum of the individual moments is

$$\tau_{total} = (+15\ 500\ \text{N})(1.39\ \text{m}) + (+7260\ \text{N})(2.97\ \text{m}) + (-12\ 200\ \text{N})(2.39\ \text{m})$$

Evaluate τ_{total}.

34. The total voltage drop across a resistor is equal to the resistance times the current in a resistor. A 285 Ω resistor, Fig. 6-10, has two currents flowing through it. If we take the flow direction from a to b as positive, the currents are +1.02A and −2.11A. The sum of the voltage drops across it is then

$$V = (285)(1.02) + (285)(-2.11) \text{ volts}$$

Find V.

FIGURE 6-9
A beam.

FIGURE 6-10
Voltage across a resistor.

6-4 DIVIDING SIGNED NUMBERS

Rules of Signs

We will use the rules of signs for multiplication to get the rules of signs for division. We know that the product of a negative number and a positive number is negative. Thus,

$$(-2)(+3) = -6$$

If we divide both sides of this equation by $(+3)$ we get

$$-2 = \frac{-6}{+3}$$

From this we see that *a negative number divided by a positive number gives a negative quotient.*

Again starting with

$$(-2)(+3) = -6$$

we divide both sides by (-2) and get

$$+3 = \frac{-6}{-2}$$

Here we see that *a negative number divided by a negative number gives a positive product.*

We also know that the product of two negative numbers is positive. Thus

$$(-2)(-3) = +6$$

Dividing both sides by (-3) we get

$$-2 = \frac{+6}{-3}$$

Thus *a positive number divided by a negative number gives a negative quotient.*

We combine these findings with the fact that the quotient of two positive numbers is positive and get our *rules of signs for division.*

| RULE OF SIGNS FOR DIVISION | $\dfrac{+a}{+b} = \dfrac{-a}{-b} = +\dfrac{a}{b}$ | 11 |

The quotient is positive when dividend and divisor have the same sign.

and

| RULE OF SIGNS FOR DIVISION | $\dfrac{+a}{-b} = \dfrac{-a}{+b} = -\dfrac{a}{b}$ | 12 |

The quotient is negative when dividend and divisor have opposite signs.

Example 13

(a) $8 \div (-4) = -2$

(b) $(-8) \div 4 = -2$

(c) $(-8) \div (-4) = 2$

(d) $\dfrac{125}{-25} = -5$

Example 14

Divide 85.4 by -2.5386 and round to the proper number of significant digits.

Solution: By calculator,

$$85.4 \div 2.5386 = 33.6405893$$

which we round to three significant digits and determine the sign by inspection, getting

$$85.4 \div (-2.5386) = -33.6.$$

Example 15

An airplane descending for a landing has a change in altitude of -250 m every 5.0 minutes. Find the rate of change of altitude in metres per minute.

Solution: Dividing,

$$\text{Rate of change} = \dfrac{-250 \text{ m}}{5.0 \text{ min}} = -50 \text{ m/min}$$

SECTION 6-4 Exercises Dividing Signed Numbers

Divide. Treat all integers as exact and round the quotient to four decimal places. Treat all decimal fractions as approximate, and round your answer to the proper number of significant digits.

1. $(+16) \div (-4)$
2. $(+6) \div (-2)$
3. $(+4) \div (-12)$
4. $(-9) \div (+3)$
5. $(-6) \div (+3)$
6. $(-7) \div (-8)$
7. $(-8) \div (-5)$
8. $(-3) \div (-8)$
9. $(-8) \div (-23)$
10. $(+3.16) \div (-7.14)$
11. $(+22.3) \div (-45.2)$
12. $(+7.57) \div (-3.11)$
13. $(-515) \div (+8.43)$
14. $(-135) \div (+323)$
15. $(-1.86) \div (-2.28)$
16. $(-153) \div (-2.25)$
17. $(-6.26) \div (-118)$
18. $(-612) \div (-6.23)$

Applications

19. A current in a certain circuit is changing at a rate of -173 volts every 15.0 minutes. What is the rate of change in volts per minute?

20. Descending a certain flight of 25 steps, we change in elevation by -4.75 m. How much do we change in elevation in one step?

6-5   CONVERTING WITHIN SI

Previously, in Sec. 4-1, we used conventional methods when converting between multiples and submultiples of metric units. In this section we are going to perform the same conversion using the properties of positive and negative numbers, as they appear in the base ten exponent equivalent of metric prefixes.

In Example 3 of Chapter 4, we converted 3.785 kL to 378 500 cL. If you observe any conversion of one size of metric unit to another, only the decimal place changes while the significant digits remain the same.

We will use the value of the prefixes to calculate the change in the position of the decimal marker.

Example 16

Convert 3.785 kilolitres to centilitres.

Solution: In Table 4-1, we locate the prefix kilo which gives a multiplier 1000 or 10^3, and the prefix centi which gives us a multiplier of 0.01 or 10^{-2}. Using only the exponents of these two multipliers, we subtract the second or target (centi) exponent from the initial (kilo) exponent.

$$\text{Decimal change} = (\text{initial exponent} - \text{target exponent})$$
$$= 3 - (-2) = +5$$

The $+5$ indicates that the decimal moves 5 places to the right and

$$3.785 \text{ kL} = 378\,500 \text{ cL}$$

Example 17

Convert 234 000 micrometres (μm) to metres (m)

Solution: From Table 4-1 we find that micro (μ) has multiplier 10^{-6} and metre does not have a prefix, and therefore has a multiplier value of one (no prefix) or 10^0.

Subtracting the initial exponent from the target exponent, we get

$$\text{Decimal change} = (\text{initial exponent} - \text{target exponent})$$
$$= (-6) - (0) = -6$$

The -6 indicates that the decimal moves 6 places to the left and

$$234\,000 \text{ }\mu\text{m} = 0.234 \text{ m}$$

Areas and Volumes

The same basic procedure with a variation can be applied to areas and volumes. Area is measured using a square unit, or in other words, a unit with an exponent of two on the unit. An example of this is square centimetres (cm^2). Similarly, volume is a cubed unit such as cubic decimetres (dm^3). The procedure is simply to apply the same procedure as for linear measurements and double or triple the decimal place change accordingly.

Example 18

Convert 34 000 000 cm² to square hectametres (hm²)

Solution: From Table 4-1 we find that centi (c) has multiplier 10^{-2} and hecto has a multiplier value of 10^2. Subtracting the initial exponent from the target exponent, we get

$$\text{Decimal change} = (\text{initial exponent} - \text{target exponent})$$
$$= (-2) - (2) = -4$$

Or 4 places left. But, because this is an area, we multiply the decimal place change by 2 therefore:

$$(-4)(2) = -8$$

The total number of places that the decimal marker will move is 8 places left and

$$34\ 000\ 000\ \text{cm}^2 = 0.34\ \text{hm}^2$$

Further, if you recall, we know that 1 hm² = 1 ha, therefore we could also write 0.34 ha.

SECTION 6-5   Exercises   Converting Within SI

Convert

1. 23.67 TW to MW
2. 467.8 kHz to MHz
3. 0.0845 µF to pF
4. 0.00342 ms to ps
5. 55 230 g to Mg
6. 4.5 Tg to kg
7. 44 598 m² to hm²
8. 5634 cm² to m²
9. 4.8 m³ to dm³
10. 34 000 000 m³ to km³

11. A large corporation is subdividing an undeveloped commercial property that is for sale at a rate of $23 600 per hectare. The rectangular property you want to buy measures 2345 m long by 256 m wide. What is the total number of hectares, and what will be the cost? (Hint: *Area = length × width*, and 1 hm² = 1 ha)

12. A barrel has an end radius of 45 cm and the height of the barrel is 100 cm. What is the volume of this barrel measured in litres? (Hint: *Volume = $\pi r^2 h$*, and 1 L = 1 dm³)

13. A lake has a surface area of 36 250 000 m² and a maximum depth of 35 m. A reliable estimate of the volume of water can be achieved by the equation

$$\text{Volume} = \frac{1}{3} \times \text{surface area} \times \text{depth}.$$

Calculate the water volume in cubic kilometres (km³).

14. A microchip has an area of 588 mm². During production, these are etched on a large sheet measuring 0.5 m². How many microchips are produced per sheet?

15. A ship primer paint has a rate of coverage of 12.5 m²/L. If a barrel contains 0.5 m³ of this paint, how many square metres will this barrel of paint cover?

6-6 POWERS AND ROOTS OF SIGNED NUMBERS

We did powers and roots of positive whole numbers in Sec. 1-5 and powers and roots of positive approximate numbers in Sec. 3-5. Here we extend those operations to *negative* numbers, both exact and approximate.

Powers of a Negative Number

Let us raise a negative number, say (-5), to an even power, say 2.

$$(-5)^2$$

Recalling that $(-5)^2$ is a shorthand way of writing (-5) times (-5), we get

$$(-5)^2 = (-5)(-5) = +25$$

Since the product of an even number of negative numbers is always positive, we see that *a negative base raised to an even power gives a positive number.*

Similarly, if we raise (-5) to an odd power, say 3, we get

$$(-5)^3 = (-5)(-5)(-5) = -125$$

Since the product of an odd number of negative numbers is negative, we find that *a negative base raised to an odd power gives a negative number.*

Example 19

(a) $(-2)^2 = (-2)(-2) = 4$
(b) $(-2)^3 = (-2)(-2)(-2) = -8$
(c) $(-1)^{24} = 1$
(d) $(-1)^{25} = -1$

> **COMMON ERROR** Be careful how you read and write powers of negative numbers. The exponent does not apply to the negative sign unless that sign is included inside parentheses. Thus,
>
> $$(-3)^2 = (-3)(-3) = +9$$
>
> but
>
> $$-3^2 = -(3)(3) = -9$$

Powers of Negative Numbers by Calculator

Recall that to raise a number to any power, we use a *power key* marked something like $\boxed{\wedge}$, $\boxed{x^y}$, or $\boxed{y^x}$, depending on your calculator. To raise a *negative* number to a power, enclose the number, including the negative sign, in parentheses.

Example 20

To find $(-2.95)^3$ by calculator, the keystrokes are

which we round to -25.7.

If you try to find the power of a negative number by your calculator, you might get an error indication. Some calculators will not accept a negative base, even though this is a valid operation.

Then how do you do it? Simply enter the base as positive, find the power, and determine the sign by inspection.

Example 21
Find $(-1.45)^5$.

Solution: From the calculator,

$$(+1.45)^5 = 6.41$$

Since we know that a negative number raised to an odd power is negative, we write

$$(-1.45)^5 = -6.41$$

Negative Exponent

A number can be raised to a negative exponent on the calculator using the same key we used for positive exponents.

Example 22
Evaluate $(3.85)^{-3}$.

Solution: On a calculator having a negation key $\boxed{+/-}$, the keystrokes are

which we round to 0.0175.

Principal Root

The principal root of a positive number is defined as the *positive root*. Thus, $\sqrt{4} = +2$, not ± 2.

The principal root is negative only when we take an odd root of a negative number.

Example 23

$$\sqrt[3]{-27} = -3$$

Because $(-3)(-3)(-3) = -27$.

Odd Roots of Negative Numbers by Calculator

An even root of a negative number (such as $\sqrt{-4}$) is not a real number. It is what we call an *imaginary number*. We will study these in Chapter 29. But an *odd* root of a negative number is *not* imaginary. It is a real, negative number.

Example 24

$$\sqrt[3]{-8} = -2$$

because

$$(-2)^3 = -8$$

Most calculators have a key for taking the nth root of a number. If your calculator does not have such a key, you can take the nth root of a number by raising the number to the $1/n$th power.

Example 25

Find the fifth root of -357.

Solution: The keystrokes are

giving -3.24, rounded.

As with powers, some calculators will not accept a negative radicand. Fortunately, we can outsmart these calculators and take odd roots of negative numbers anyway.

Example 26

Find $\sqrt[7]{-1054}$.

Solution: We know that an odd root of a negative number is real and negative. So we take the seventh root of $+1054$, by calculator,

$$\sqrt[7]{+1054} = 2.703 \text{ (rounded)}$$

and we have only to append the minus sign.

$$\sqrt[7]{-1054} = -2.703$$

SECTION 6-6 Exercises Powers and Roots of Signed Numbers

Treat all integers as exact. Treat all decimal fractions as approximate, and round your answer properly.

Powers of a Negative Number

Raise each number to the power indicated.

1. $(-2)^5$ 2. $(-3)^3$ 3. $(-5)^2$ 4. $(-4)^3$
5. $(-6)^2$ 6. $(-4)^4$ 7. $(-2)^6$ 8. $(-5)^3$
9. $(-1.24)^5$ 10. $(-2.13)^3$ 11. $(-2.35)^2$ 12. $(-3.14)^3$
13. $(-2.36)^2$ 14. $(-2.24)^4$ 15. $(-2.11)^6$ 16. $(-1.25)^3$

Negative Exponent

Raise each number to the power indicated. Round to three significant digits.

17. $(3)^{-4}$ 18. $(4)^{-3}$ 19. $(2)^{-3}$ 20. $(5)^{-3}$
21. $(7)^{-2}$ 22. $(2)^{-4}$ 23. $(3)^{-6}$ 24. $(7)^{-3}$
25. $(3.14)^{-5}$ 26. $(8.23)^{-3}$ 27. $(5.55)^{-2}$ 28. $(4.14)^{-3}$
29. $(7.16)^{-2}$ 30. $(5.27)^{-4}$ 31. $(6.21)^{-6}$ 32. $(8.25)^{-3}$

Odd Roots of Negative Numbers

Evaluate each root.

33. $\sqrt[3]{-8}$ 34. $\sqrt[3]{-64}$ 35. $\sqrt[5]{-32}$ 36. $\sqrt[3]{-125}$
37. $\sqrt[5]{-243}$ 38. $\sqrt[3]{-27}$ 39. $\sqrt[3]{-8.4}$ 40. $\sqrt[3]{-27.6}$
41. $\sqrt[3]{-64.5}$ 42. $\sqrt[3]{-125.8}$ 43. $\sqrt[3]{-23.1}$ 44. $\sqrt[3]{-7.42}$
45. $\sqrt[3]{-9.52}$ 46. $\sqrt[3]{-16.3}$ 47. $\sqrt[3]{-53.1}$ 48. $\sqrt[3]{-9.12}$
49. $\sqrt[5]{-5.12}$ 50. $\sqrt[5]{-96.3}$

6-7 SCIENTIFIC AND ENGINEERING NOTATION

In Sec. 1-6 we introduced scientific notation with whole numbers, and in Sec. 3-6 we applied scientific notation to decimal numbers. Now we will do scientific notation with negative exponents.

Negative Powers of Ten

For the sake of clarity, let us call a number such as 1000 a *multiple* of ten, and a number such as 10^3 a *power* of ten. In Sec. 1-6 we gave a table of multiples and powers of ten. We repeat that table here, extended to include negative exponents.

Powers of Ten

10^6 =	1 000 000
10^5 =	100 000
10^4 =	10 000
10^3 =	1 000
10^2 =	100
10^1 =	10
10^0 =	1
10^{-1} =	0.1
10^{-2} =	0.01
10^{-3} =	0.001
10^{-4} =	0.000 1
10^{-5} =	0.000 01
10^{-6} =	0.000 001

This table extends indefinitely, for both positive and negative exponents.

Converting from Decimal Form to Scientific Notation

To convert a number from decimal form to scientific notation, rewrite the given number with just one digit to the left of the decimal point, and discard any nonsignificant zeros. Then multiply by that multiple of ten that makes it equal to the given number. Finally write the multiple of ten as a power of ten.

Example 27

Convert 0.00531 to scientific notation.

Solution: We place the decimal point between the 5 and the 3. The three zeros are not significant (see Sec. 3-1) so we discard them, getting 5.31. Finally, the

5.31 must be multiplied by 0.001 so that our new expression is equal to the given number.
$$0.005\,31 = 5.31 \times 0.001$$
We now express 0.001 as a power of ten (10^{-3}).
$$0.005\,31 = 5.31 \times 0.001 = 5.31 \times 10^{-3}$$

To convert from scientific notation to decimal form, simply reverse the process.

Example 28
Convert 3.84×10^{-5} to decimal form.

Solution:
$$3.84 \times 10^{-5} = 3.84 \times 0.000\,01$$
$$= 0.000\,038\,4$$

Entering Numbers in Scientific Notation in a Calculator

We can enter numbers directly in scientific notation by means of the ENTER EXPONENT key, usually marked $\boxed{\text{EE}}$, $\boxed{\text{EXP}}$, or $\boxed{\text{EEX}}$. We will use $\boxed{\text{EE}}$ in our examples.

Example 29
The keystrokes to enter the number 7.25×10^5 are

or 725 000.

The way we enter a *negative* exponent depends on whether our calculator uses a $\boxed{+/-}$ negation key or a change-sign key.

Example 30
The keystrokes to enter the number 8.25×10^{-2} on a calculator that has a $\boxed{+/-}$ negation key are

or 0.0825. On a calculator that has a change-sign key, the keystrokes are

Note that we enter the exponent and then change its sign.

6-7 Scientific and Engineering Notation

We add, subtract, multiply, and divide numbers in scientific notation that have negative exponents exactly the way we did with those having positive exponents. See Sec. 3-6.

Engineering Notation: A Variation on Scientific Notation

On some calculators, a variation of scientific notation occurs called Engineering Notation. This notation is used with the SI in engineering applications. In engineering applications, only prefixes that are multiples of 1000 are used. Therefore, prefixes such as centi, deci, deca, and hecto do not appear. On an SI engineering drawing, we would not see 28 cm, but would see 280 mm. Engineering notation is designed such that the power of ten part of scientific notation is always a multiple of 1000 and can be replaced by an SI prefix.

Engineering notation can have *one* to *three digits* before the decimal point. On your calculator, when it is set in engineering notation, values are entered using the same methods as for scientific notation.

Example 31

Convert 0.000 034 2 Ω to engineering notation.

Solution: As in scientific notation, we move the decimal right, but *in groups of three places* resulting in a decimal place between the 4 and the 2. In order to maintain the value, we now write

$$34.2 \times 10^{-6} \ \Omega$$

From the SI prefix table, we find that 10^{-6} can be replaced by the prefix micro, and can be written

$$34.2 \ \mu\Omega$$

SECTION 6-7 Exercises Scientific And Engineering Notation

Write each number in scientific and engineering notation. Be sure to retain the proper number of digits.

1. 0.373
2. 0.006 32
3. 0.0260
4. 0.000 239

5. 0.002 000 0
6. 0.000 395 0
7. 0.0228
8. 0.000 049 1

Convert each number from scientific notation to decimal notation.

9. 2.323×10^{-2}
10. 3.2902×10^{-4}
11. 7.01×10^{-3}

12. 2.950×10^{-4}
13. 3.370×10^{-5}
14. 5.14×10^{-6}

Perform the following computations in scientific notation. Combine the powers of ten by hand or with your calculator.

15. $(18.4 \times 10^{-3}) + (17.9 \times 10^{-4})$
16. $(13.2 \times 10^{-3}) + 0.0234$

17. $(28.3 \times 10^{-2}) + (15.3 \times 10^{-2})$
18. $(26.5 \times 10^{-3}) - (17.4 \times 10^{-3})$

19. $(3.59 \times 10^{-6})(6.23 \times 10^{-5})$
20. $(1.03 \times 10^{-3})(8.27 \times 10^{-3})$

21. $(6.19 \times 10^5)(9.11 \times 10^{-3})$
22. $(7.16 \times 10^{-2})(9.28 \times 10^7)$
23. $(23.2 \times 10^2) \div 0.00824$
24. $72\,400 \div (2.82 \times 10^{-3})$
25. $(3.18 \times 10^{-7}) \div (9.34 \times 10^{-3})$
26. $(11.2 \times 10^3) \div (7.34 \times 10^{-4})$

Find answers in scientific notation.

Applications

27. Find the current through a resistor if it dissipates 2.32×10^{-2} watts when the voltage drop across it is 6.72×10^{-3} V. (Divide the power by the voltage drop.)

28. The voltage across a 6.14×10^4 Ω resistor is 2.27×10^{-2} V. Find the power dissipated in the resistor. (Divide the square of the voltage by the resistance.)

29. A tensile load of 6.19×10^3 N is supported by a wire having a cross-sectional area of 3.95×10^{-5} m². Find the stress in the member. (Divide the load by the area.) Convert to engineering notation.

30. A wire 9.19×10^2 cm long when loaded is seen to stretch 9.66×10^{-2} cm. Find the strain in the wire. (Divide the amount stretched by the length.)

31. A rod having a cross-sectional area of 8.72×10^{-2} in.² and a length of 5.82×10^2 in. is subjected to a tensile load of 7.14×10^3 lb. The elongation is 9.26×10^{-2} in. Find the modulus of elasticity E of the material. (Multiply the load by the length, and divide by the area and by the elongation.) (Answer only in scientific notation.)

6-8 COMBINED OPERATIONS

In Sec. 1-7 we showed how to do combined operations with whole numbers on the calculator, and in Sec. 3-7 we showed the same for decimal numbers. Now we include negative numbers.

Example 32

Evaluate the expression

$$\left(\frac{-41.8 + 4.23}{-21.3 \sqrt[3]{-136}} \right)^2$$

Solution: On a calculator having parentheses and a $\boxed{+/-}$ negation key, we can do the computation in a single line. The keystrokes are

which we round to three significant digits, getting 0.118.

However, many students would prefer to do the computation in stages, using the $\boxed{\text{ENTER}}$ key to obtain the intermediate values. Then pressing the $\boxed{\text{ANS}}$ key will make that intermediate result available for the next step.

Example 33
Repeating Example 32,

as before.

SECTION 6-8 Exercises Combined Operations

Perform the computations, keeping the proper number of digits in your answer.

1. $(-2.67)(7.18) + (-3.34)(-2.64)$
2. $(312)(5.53) - (186)(-7.70) + (-247)(-3.39)$
3. $(-13.7 + 33.6)(2.37 \times 10^{-2} - 7.72 \times 10^{-2})$
4. $\dfrac{381}{-623} - \dfrac{-446}{317}$
5. $\dfrac{-2.13}{5.32} - \dfrac{-7.97}{3.06} + \dfrac{3.23}{-1.51}$
6. $\dfrac{3.27 + (-6.204)}{(-7.27) + 68.8}$
7. $\dfrac{463 - (-284) + 274}{274 - 750 + (-284)}$
8. $(3.57 - 7.36)^{-2}$
9. $(7.45 + 9.36 - 2.74)^{-4}$
10. $(2.723 - 3.55)^{-3}$
11. $\sqrt{7.24 \times 10^{-3} + 3.86 \times 10^{-2}}$
12. $\sqrt[3]{-284 - 926 - 264}$
13. $\sqrt{(-88.1)(-3.54 \times 10^{-2})}$
14. $\sqrt[5]{(3.06)(-8.36)(4.93)}$
15. $\sqrt{\dfrac{-385}{-973}}$
16. $\sqrt[4]{\dfrac{-7.50}{-3.61}}$
17. $\sqrt[4]{278} - \sqrt{25.2} - \sqrt[3]{-836}$

Chapter 6 REVIEW EXERCISES

Place the proper equality or inequality sign between each pair of numbers.
1. -3 and -1
2. -14.4 and -14.5
3. -13.2 and -18.6

Write each number in scientific and engineering notation. Be sure to retain the proper number of digits.
4. 0.595
5. 0.008 54
6. 0.000 573 0

Evaluate each expression.
7. $|8 - 5|$
8. $|13 - 18|$
9. $|1 - 9|$

Convert each number from scientific notation to decimal notation.
10. 4.545×10^{-4}
11. 5.4704×10^{-2}
12. 4.730×10^{-2}

Perform each computation, rounding your answer properly.
13. $(16.2 \times 10^{-5}) + (19.7 \times 10^{-2})$
14. $(15.4 \times 10^{-5}) + 0.0452$
15. $(48.3 \times 10^{-5}) - (19.2 \times 10^{-5})$
16. $(1.05 \times 10^{-5})(6.49 \times 10^{-5})$
17. $(+4.16)(-5.42)$
18. $(+1.39)(-2.11)$
19. $(-214)(+4.25)$
20. $(-1.42)^3$
21. $(-4.15)^5$
22. $(-4.53)^4$
23. $(-5.12)^5$
24. $(5.12)^{-3}$
25. $(6.45)^{-5}$
26. $(3.33)^{-4}$
27. $(2.12)^{-5}$
28. $\sqrt[5]{-6}$
29. $\sqrt[5]{-82}$
30. $\sqrt[3]{-54}$
31. $\sqrt[5]{-143}$
32. $\sqrt[3]{-425}$
33. $\sqrt[5]{-55}$
34. $\sqrt[3]{-6.8}$
35. $\sqrt[5]{-49}$
36. $(-195)(-5.43)$
37. $(-3.48)(-816)$
38. $(-944)(-1.45)$
39. $(+1.34)(+5.26)(-1.92)$
40. $(+5.18) \div (-9.12)$
41. $(+9.39) \div (-5.11)$
42. $(-313) \div (+6.25)$
43. $(-153) \div (+545)$
44. $(-45.4) + (+64.9)$
45. $(-7.55) + (-5.43)$
46. $(-2.72) + (-21.44)$
47. $(-4.89)(9.16) + (-5.52)(-4.82)$
48. $(514)(3.35) - (168)(-9.90) + (-429)(-5.57)$
49. $(-15.9 + 55.8)(4.59 \times 10^{-4} - 9.94 \times 10^{-4})$
50. $(5.39 - 9.58)^{-4}$
51. $(9.23 + 7.58 - 4.92)^{-2}$
52. $(4.945 - 5.33)^{-5}$
53. $(+9.92) - (-5.87)$
54. $(+7.28) - (-18.62)$
55. $(-9.62) - (+2.48)$
56. $(-7.99) - (+81.28)$
57. $(-7.55) - (-5.29)$
58. $(-7.72) - (-78.24)$
59. $(-135) \div (-4.43)$
60. $(-8.48) \div (-116)$
61. $(-1.94)(+2.89)(-4.61)$
62. $(+4.81)(-804)(+1.85)$
63. $(-1.62)(-4.44)(+345)$
64. $(+4.61)(-9.53)(-1.43)$

65. A wire 7.17×10^4 cm long when loaded is seen to stretch 7.88×10^{-4} cm. Find the strain in the wire. (Divide the amount stretched by the length.)

66. A current in a certain circuit is changing at a rate of -195 volts every 13 minutes. What is the rate of change in volts per minute?

67. A broker bought 300 shares of stock for $84.26 per share, which then rose $1.43 per share and later dropped $0.24 per share. Find the final value of the stock.

68. The total voltage drop across a resistor is equal to the sum of the product of the currents and the resistance in a given resistor. A resistor with a rating of 463 Ω has two currents flowing through it. The currents are +1.04 A and −4.11 A. The sum of the voltage drops across it is then,

$$V = (463 \text{ } \Omega)(1.04 \text{ A}) + (463 \text{ } \Omega)(-4.11 \text{ A}) = \qquad \text{volts}$$

Find V.

69. A balance sheet showed a balance of −$5264.99, and another bill came in for $1953.65. Find the new balance.

70. A tensile load of 6.99×10^5 N is supported by a wire having a cross-sectional area of 5.25×10^{-7} m². Find the stress in the member. (Divide the load by the area.)

71. During a certain month the temperature dropped from −13.1°C to −34.6°C. Find the difference in the temperatures.

72. The voltage in a certain circuit was −55.6 V. It was then reduced by 11.5 V. Find the new voltage.

73. Find the current through a resistor if it dissipates 4.54×10^{-4} watts when the voltage drop across it is 8.94×10^{-5} V. (Divide the power by the voltage drop.)

74. The voltage across a 8.12×10^2 Ω resistor is 4.49×10^{-4} V. Find the power dissipated in the resistor. (Divide the square of the voltage by the resistance.)

75. A submarine was at an elevation of −456 ft, and then descended another 199 ft. Find the new elevation.

76. A rod having a cross-sectional area of 6.94×10^{-4} in.⁴ and a length of 3.64×10^4 in. is subjected to a tensile load of 9.12×10^5 lb. The elongation is 7.48×10^{-4} in. Find the modulus of elasticity E of the material. (Multiply the load by the length, and divide by the area and by the elongation.)

WRITING

77. Does your calculator have two separate keys marked with a minus sign? Why? Is there any difference between them, and if so, what? When would you use each? Write a paragraph or two explaining these keys and answering these questions.

Introduction to Algebra

Outcomes

When you have completed this chapter, you will be able to

1. Define the following words: variable, expression, and term.
2. Identify and define an equation.
3. Separate a term into variables and constants.
4. Identify the symbols of grouping.
5. Add and subtract polynomials.
6. Use the law of exponents in multiplication and division.

In Chapter 1 we showed how to raise a number to a power. For example, 3^2, or 3 raised to the power 2, means

$$3^2 = (3)(3)$$

In a similar way, x^2, or x raised to the power 2, means

$$x^2 = (x)(x)$$

where x can stand for any number, not just 3. Going further, we can represent the exponent by a symbol, say n. Thus

$$x^n = (x)(x)(x)(x) \ldots (x)$$

n factors

Thus, 3^2 is an arithmetic expression, while x^n is an *algebraic* expression. In this chapter we will give definitions which apply to algebraic expressions and learn how to add and subtract algebraic expressions, how to remove symbols of grouping, and how to handle expressions with exponents.

7-1 ALGEBRAIC EXPRESSIONS

Mathematical Expressions

A grouping of mathematical symbols, such as signs of operation, numbers, and letters is called a *mathematical expression*.

Example 1
Some mathematical expressions are
(a) $x^2 + 3x - 4$ (b) $2 \tan 5x$ (c) $3 + 8 \ln x$

Algebraic Expressions

A mathematical expression that contains only numbers, algebraic symbols (such as x and y), and operations (addition, subtraction, multiplication, division, roots, and powers) is called an *algebraic expression*. Example 1(a) above shows an algebraic expression. The others are called transcendental expressions. We will study those in later chapters.

Equations

When one expression is set equal to another expression, we get an *equation*.

──── **Example 2** ────

Some equations are
(a) $x + 8 = 3x - 1$
(b) $4x^2 - 2x + x = 0$
(c) $y = x^2 + 9$

Constants

A quantity that does not change value in a particular problem is called a *constant*. It can be a number, such as 3, 9.36, or π, or a letter. Letters that represent constants are usually chosen from the beginning of the alphabet (a, b, c, etc.).

──── **Example 3** ────

The constants in the expression

$$2x^2 + 3x + 4$$

are 2, 3, and 4.

An expression in which the constants are represented by letters is called a *literal expression*.

──── **Example 4** ────

The constants in the literal expression

$$ax^2 + bx + c$$

are a, b, and c.

Variables

A quantity that may change in a particular problem is called a *variable*. Letters that represent variables are usually chosen from the end of the alphabet (x, y, z, etc.).

──── **Example 5** ────

In the expression

$$ax + by + cz$$

the letters a, b, and c would usually represent constants, and x, y, and z would usually represent variables.

Symbols of Grouping

Parentheses ( ), brackets [ ], and braces { }, are called *symbols of grouping*. As their name implies, these are used to group parts of the expression together, and affect the meaning of the expression.

Example 6

The expression
$$\{[(2 - x^2) + 4] + (5 - 3x)\}$$
is an example of the use of symbols of grouping.

We will show how to remove symbols of grouping in Sec. 7-2. We will also see in Sec. 8-6 that symbols of grouping may be used to indicate multiplication.

Terms

The plus and minus signs separate an expression into *terms*.

Example 7

The expression $2x^2 + 5x + 3$ has three terms.

$$\underset{\text{first term}}{2x^2} \quad \underset{\text{second term}}{5x} \quad \underset{\text{third term}}{3}$$

An exception is when the plus or minus sign is within a symbol of grouping.

Example 8

The expression $[(2x + 4x^2) - 5] + (3x + 4)$ has two terms.

$$\underset{\text{first term}}{[(2x + 4x^2) - 5]} \quad \underset{\text{second term}}{(3x + 4)}$$

Factors

A divisor of a term is called a *factor* of that term.

Example 9

Some factors of ax^2 are a, x, and x^2, because a, x, and x^2 divide evenly into ax^2.

We do not usually state the separate factors of a number as factors of a term.

Example 10

The factors of $6ax$ are usually given as 6, a, and x, *not* as 2, 3, a, and x.

7-1 Algebraic Expressions

Coefficient

In elementary algebra, the *coefficient* of a term is the numerical part of the term. It is usually written before the variables.

Example 11

The coefficient of xy in the term $4xy$ is 4. It is also correct to say that $4x$ is the coefficient of y, but we will not use the word *coefficient* in that way.

When we say "coefficient," it will always refer to the constant part of the term, also called the *numerical coefficient*. It will include *any constants in literal form*.

Example 12

If b is understood to be a constant in the term $2bxy^2$, the coefficient is then $2b$.

If there is no numerical coefficient written before a term, it is understood to be 1, since

$$x \equiv 1x$$

Example 13

In the expression $-x^3$, the coefficient is -1.

COMMON ERROR Don't forget to include the negative sign with the coefficient.

Degree

The power to which the variable in a term is raised is called the *degree* of that term.

Example 14

(a) $4y$ is a term of degree 1, also called a first-degree term.
(b) $2x^2$ is a term of degree 2, or a second-degree term.
(c) $3z^8$ is a term of degree 8, or an eighth-degree term.

The term *degree* is used *only* when the exponents are positive integers.

Example 15

You would not say that $y^{1/2}$ is a "half-degree" term.
You would not say that x^{-2} is a "negative-two-degree" term.

If a term has more than one variable, *add their powers* to obtain the degree of the term.

Example 16

(a) $5x^3y^3$ is a term of degree 6, since the sum of the exponents 3 and 3 is 6.
(b) $4x^2yz$ is a term of degree 4, since $4x^2yz = 4x^2y^1z^1$, and the sum of the exponents is $2 + 1 + 1 = 4$.

The *degree of an expression* is the same as that of the highest degree term in that expression.

Example 17

$4x^2 - 5x + 1$ is an expression of degree 2.

SECTION 7-1 Exercises Algebraic Expressions

Mathematical Expressions

Which of the following mathematical expressions are also algebraic expressions?
1. $x + 2y$
2. $y - \log x$
3. $3 \sin x$
4. $x^2 - z^3$

Algebraic Expressions

Which of the following algebraic expressions are literal expressions?
5. $5xy - 2x$
6. $ax + by$
7. $2az - 3bx$
8. $4x^2 + 4y^2$

Terms

How many terms are there in each expression?
9. $x^3 - 2x$
10. $(5y + y^2) - 5$
11. $(z - 9)(z + 4)$
12. $5(1 + x) + 3(4 - 3x)$

Factors

Write the factors of each expression.
13. $3ax$
14. $9xyz$
15. $7x^2y^3$
16. $6a^2bx$

Coefficient

Write the coefficient of each term. Assume that letters from the beginning of the alphabet $(a, b, c, \ldots)$ are constants.
17. $6x^2$
18. x
19. $-x$
20. $3cx^3$
21. $\dfrac{ay^2}{2}$
22. $(1/3)(x + 1)(b)$
23. $(c - 2)3x^4$
24. $\dfrac{a}{y + 1}$

Degree

State the degree of each term.
25. $3x$
26. $4y^2$
27. $3xy$
28. $5x^2y^3$

State the degree of each expression.
29. $3x + 4$
30. $5 - xy$
31. $3x^2 - 2x + 5$
32. $2xy^2 + xy - 4$

7-2 ADDING AND SUBTRACTING POLYNOMIALS

Polynomials

An algebraic expression in which the power of every variable is a positive integer is called a *polynomial*.

Example 18

Some polynomials are

$$3x^4, \quad 4x^2 - 5x - 4, \quad \text{and} \quad 2xy + y^3$$

Some expressions that are *not* polynomials are

$$1/x, \quad 3x^{-4}, \quad 4x^2 - 5\sqrt{x} - 4, \quad \text{and} \quad 2xy + y^{1/3}$$

Combining Like Terms

Terms that differ only in their coefficients are called *like* terms. Their variable parts are the same.

Example 19

$4xz$ and $-5xz$ are like terms.

We add and subtract algebraic expressions by *combining like terms*. Like terms are added by adding their coefficients. We also call this *collecting terms*.

Example 20

(a) $3y + 4y = 7y$
(b) $18z - 9z = 9z$

Don't forget that any term with no numerical coefficient has an unwritten coefficient of 1.

Example 21

(a) $4x + x = 5x$
(b) $-3y + y = -2y$

Commutative Law of Addition

We used the commutative law (Eq. 1) when adding numbers in Chapter 1. We saw that it means that the order in which we add the terms does not affect the sum. Thus $3 + 4 = 4 + 3$. The commutative law applies, of course, when the expression contains letters. Thus $b + c = c + b$. This law enables us to rearrange the terms of an expression to make it easier to simplify.

Example 22

Simplify the expression

$$8y + 5x - 5z + 2z - x - 4y$$

Solution: We use the commutative law to rearrange the expression to get like terms together. We can then easily collect terms.

$$8y + 5x - 5z + 2z - x - 4y = 8y - 4y + 5x - x - 5z + 2z$$
$$= 4x + 4y - 3z$$

With practice, you will soon be able to omit the first step and collect terms by inspecting the original expression.

The procedure is no different when the terms have *decimal coefficients*. Just be sure to round properly.

Example 23

$$3.92x - 4.02y - 2.24x + 1.85y = 3.92x - 2.24x - 4.02y + 1.85y$$
$$= 1.68x - 2.17y$$

Removing Symbols of Grouping

To add or subtract expressions we must first remove any parentheses or other symbols of grouping, such as brackets and braces. If the parentheses are preceded only by a (+) sign (or by no sign), they may simply be removed.

Example 24

(a) $(x - y + z) = x - y + z$
(b) $(a - b) + (c + d) = a - b + c + d$

When the parentheses are preceded only by a (−) sign, change the sign of every term within the parentheses before removing them.

Example 25

$$-(x - y + z) = -x + y - z$$

An expression may contain braces or brackets as well as parentheses. If a group of terms is inside another group, simplify the innermost group first.

Example 26

Simplify the expression

$$3[x - 2(x + y)]$$

Solution: We start removing the parentheses by multiplying $(x + y)$ by (-2),

$$3[x - 2(x + y)] = 3[x - 2x - 2y]$$
$$= 3[-x - 2y]$$
$$= -3x - 6y$$

Example 27

Simplify the expression

$$3a + \{d - [2e + [-3b - (7c + 4f)]]\}$$

Solution: The braces and one set of brackets are preceded by a (+) sign so we may remove them immediately, getting

$$3a + d - [2e - 3b - (7c + 4f)]$$

Next we remove parentheses within the remaining square brackets. Since these are preceded by a (−) sign, we change the signs of the terms.

$$3a + d - [2e - 3b - 7c - 4f]$$

We remove the brackets, changing the signs of the terms as we do.

$$3a + d - 2e + 3b + 7c + 4f$$

Finally, it is common practice to write the terms in alphabetical order. Our simplified expression is then

$$3a + 3b + 7c + d - 2e + 4f$$

Once the parentheses are gone, like terms may be combined.

Example 28

$$-(a - b) - (a + b) = -a + b - a - b = -2a$$

Combining More Than Two Expressions

The procedure is no different when more than two expressions are combined.

Example 29

$$(3a + 5b - 2c) - (2a - b - 4c) - (a + 3b + c)$$
$$= 3a + 5b - 2c - 2a + b + 4c - a - 3b - c$$
$$= 3a - 2a - a + 5b + b - 3b - 2c + 4c - c$$
$$= 3b + c$$

Vertical Addition and Subtraction

It is often easier to arrange the expressions *vertically,* with like terms in the same column.

Example 30

Combine

$$(3x - 2y + 4z) - (7y + 2w - 5x) - (3w + 2z + 4y - x)$$

Solution: We remove parentheses and write the expressions one above the other, with like terms in the same vertical column, and collect terms.

$$\begin{array}{r} 3x - 2y + 4z \\ 5x - 7y - 2w \\ \underline{x - 4y - 2z - 3w} \\ 9x - 13y + 2z - 5w \end{array}$$

Instructions Given Verbally

In preparation for verbal problems to come later, we give some problems in verbal form.

Example 31

Find the sum of $x - 2y + z$ and $3x + y + z$.

Solution: We have

$$(x - 2y + z) + (3x + y + z)$$
$$= x - 2y + z + 3x + y + z$$
$$= x + 3x - 2y + y + z + z$$
$$= 4x - y + 2z$$

Example 32

Subtract $5a - 2b + 6c$ from the sum of $8a - 4b - 3c$ and $a - b + 6c$.

Solution: We add the last two expressions,

$$(8a - 4b - 3c) + (a - b + 6c) = 9a - 5b + 3c$$

Then we subtract the first expression.

$$9a - 5b + 3c - (5a - 2b + 6c)$$
$$= 9a - 5b + 3c - 5a + 2b - 6c$$
$$= 4a - 3b - 3c$$

SECTION 7-2 Exercises Adding and Subtracting Polynomials

Combining Like Terms

Combine as indicated, and simplify.

1. $8y + 2y$
2. $6x - 8x$
3. $38.2a - 17.2a$
4. $2.94z + 5.37z$
5. $5ac + 9ac - 20ac$
6. $9xyz - 2xyz + 7xyz$
7. $7.39y - 6.62y + 1.94y$
8. $23.9ab + 54.9ab - 65.1ab$
9. $5x + 2x - 8x - x$
10. $9a - 2a + 7a - 3a$
11. $88x + 23y - 17z + 68y - 36x + 39z$
12. $a - 6b + 2c + a + 6b + n - 2c$
13. $1.95x - 4.38z + 2.83a - 5.21z - 9.27x$
14. $33.9ab - 82.4ac + 29.3ad - 84.2ac + 73.2ab$

Removing Symbols of Grouping

Remove symbols of grouping and collect terms.

15. $(4z + 2) + (z - 5)$
16. $(2x + 5) + (x - 2)$

17. $(c - 5) - (6 + 3c)$
18. $(2x + 6a) - (4x - a)$
19. $(1.24z + 2.22) + (9.24z - 1.25)$
20. $(24.2x + 8.25) + (83.6x - 2.92)$
21. $(0.826c - 5.37) - (2.76 + 0.273c)$
22. $(7262x + 1.26a) - (2844x - 8.23a)$
23. $(5 - 2a^3 + 3a^4) + (4a^4 - 6a^3 - 7)$
24. $(y - z - b) - (b + y + z)$
25. $(5bc + 6c - a) - (8c - 2a + 3bc)$
26. $(2z + 5c - 3a) - (6a + 2c - 4z)$
27. $(2.25c - 9.28b + 3.82a) - (-3.92a - 1.72b - 8.33c)$
28. $(23y^2 + 4y^3 - 12) - (11y^3 - 8y^2 + y)$
29. $24ab - (16ab - 3x^2 + 7z - 2y^2)$
30. $(18y^2 - 12xy) - (6y^2 + xy - a)$
31. $(4y + 2y^2) - (3y - 6b + 4y^2 + 5)$
32. $(-6x - z) - \{3y + [7x - (3z + 8y + x)]\}$
33. $-2a + \{d - [6c + [-5b - (4a + 9b)] - (2d - a + 3c)]\}$
34. $-6x - \{9y + [5x - (-2y - 8b)] - [2b + a]\}$

Combining More Than Two Expressions

Combine and simplify. Use either horizontal or vertical addition and subtraction.

35. $(4ab + 6bc + 8cd) + (-6ab - 3ab) + (4cd - 6bc)$
36. $(2a + 3b - 1) + (2c + d - b) - (3a - 4c + 5 - 6d)$
37. $(3b - 7p + 4r) + (3s - 11p - 19r) - (-3p + r - 10s) - (-5b + 2p - 2r + 4s)$
38. $(2bx + 9x + 4a) - (4bx - a) + (3b + a)$
39. $(-a + 5x) + (a^3 + 3a - 11x^2 + 2a^2 - 3) - (7a^2 + 3a^3 - 6x^3 + 12)$
40. $(2y - x + 3z - 13) - (4x + 2 - 5y) + (6z + 8)$

Instructions Given Verbally

41. Add $a - c + b$ and $b + c - a$.
42. Find the sum of $6bc + n^2 + 3p$ and $-5x + 3n^2$.
43. Subtract $5a + 7d - 4b + 6c$ from $8b - 10c + 3a - d$.
44. Subtract $2xy + 4y - 3x$ from $5x - 2xy + 8y$.
45. What is the sum of $24by^5 - 14bx^4$, $-72bx^5 + 2by^5 - 3bx^4$, and $9bx^4 + 23by^4 - 21by^5$?
46. Add $3b(y^2 - z) - 6nx^2$ to the sum of $2b(z + y^2) + 3nx^2$ and $4b(-z - y^2) - 2nx^2$.
47. What is the sum of $5a^2 - 8c + 5bd$, $4a^2 - 3bd + 3c$, $3bd - 6c + 12a^2$, and $5a^2 - 3c$?
48. Add $3(x + z) - 5(y + z)$, $2(x + y) + 12(z + y)$, and $4(x + y) - 6(x + z)$.
49. Subtract $4(b + c)$ from the sum of $6(b + c)$ and $3(b + c)$.
50. From the sum of $4a - 6b + c$ and $7d - 2a + 8b$, subtract the sum of $3d + 4c - 2a$ and $-5c + 6b - 2a$.

7-3 LAWS OF EXPONENTS

Definitions

We have done some work with powers of numbers in Chapter 1. Now we will expand those ideas to include powers of algebraic expressions. Here we will deal only with expressions that have integers (positive or negative, and zero) as exponents.

Example 33

We study such expressions as

$$x^4, \quad b^{-3}, \quad a^0, \quad a^3b^2d^{-1}, \quad \text{and} \quad (x+y)^3$$

Recall from Chapter 1 that a positive exponent shows how many times the base is to be multiplied by itself.

Example 34

In the expression 2^5, the base is 2 and the exponent is 5.

Its meaning is

$$2^5 = (2)(2)(2)(2)(2) = 32$$

In general,

POSITIVE INTEGRAL EXPONENT	x^n means $\underbrace{(x)(x)(x)(x)\ldots(x)}_{n \text{ factors}}$	18

COMMON ERROR An exponent applies only to the symbol directly in front of it. Thus,

$$5y^3 = 5(y^3)$$

but

$$5y^3 \neq 5^3 y^3$$

Multiplying Powers

Let us multiply x^3 by x^4, two quantities that have the same base x. From the definition of an exponent we have

$$x^3 = x \cdot x \cdot x$$

and

$$x^4 = x \cdot x \cdot x \cdot x$$

Multiplying gives

$$x^3 \cdot x^4 = (x \cdot x \cdot x)(x \cdot x \cdot x \cdot x)$$
$$= x \cdot x \cdot x \cdot x \cdot x \cdot x \cdot x$$
$$= x^7$$

We see that the exponent in the result is the sum of the two original exponents.

$$x^3 \cdot x^4 = x^{3+4} = x^7$$

This gives our first law of exponents.

| PRODUCT | $x^m x^n = x^{m+n}$ | 19 |

When multiplying powers having the same base, add the exponents.

Example 35

$$x^2(x^5) = x^{2+5} = x^7$$

Every quantity has an exponent of 1, even though it is not usually written. Thus

$$x = x^1$$

Example 36

(a) $(x)(x^2) = (x^1)(x^2) = x^{1+2} = x^3$
(b) $(b)(b^3)(b^2) = (b^1)(b^3)(b^2) = b^{1+3+2} = b^6$

Quotients

Let us divide x^4 by x^2. By Eq. 18,

$$\frac{x^4}{x^2} = \frac{(x)(x)(x)(x)}{(x)(x)} = \frac{(\cancel{x})(\cancel{x})}{(\cancel{x})(\cancel{x})}(x)(x)$$
$$= (x)(x) = x^2$$

The same result could have been obtained by subtracting exponents,

$$\frac{x^4}{x^2} = x^{4-2} = x^2$$

We state it as another law of exponents:

| QUOTIENT | $\dfrac{x^m}{x^n} = x^{m-n}$
$(x \neq 0)$ | 20 |

When dividing powers having the same base, subtract the exponents.

Example 37

(a) $\dfrac{x^2 y^5}{x y^3} = x^{2-1} y^{5-3} = x y^2$

(b) $\dfrac{x^{a+b}}{x^{a-b}} = x^{a+b-(a-b)} = x^{2b}$

Power Raised to a Power

Now let us take a quantity which itself is raised to a power and raise that expression to another power. Let us raise y^3 to the power 2.

$$(y^3)^2$$

By Eq. 18,

$$(y^3)^2 = (y^3)(y^3)$$
$$= (y)(y)(y)(y)(y)(y)$$
$$= y^6$$

a result that could have been obtained by multiplying the exponents.

$$(y^3)^2 = y^{3(2)} = y^6$$

In general,

| POWERS | $(x^m)^n = x^{mn}$ | 21 |

When a power is raised to a power, multiply the exponents.

Example 38
(a) $(n^4)^3 = n^{4(3)} = n^{12}$
(b) $(10^2)^4 = 10^{2(4)} = 10^8$

Product Raised to a Power

Let us now find a rule for raising a product, such as ab, to a power, say 2.

$$(ab)^2$$

By Eq. 18

$$(ab)^2 = (ab)(ab)$$
$$= (a)(b)(a)(b)$$
$$= (a)(a)(b)(b)$$
$$= a^2b^2$$

In general,

| PRODUCT RAISED TO A POWER | $(xy)^n = x^n y^n$ | 22 |

When a product is raised to a power, each factor may be separately raised to the power.

Example 39
(a) $(abc)^4 = a^4 b^4 c^4$
(b) $(3y)^3 = 3^3 y^3 = 27y^3$
(c) $(4.23 \times 10^2)^3 = (4.23)^3 \times (10^2)^3 = 75.7 \times 10^6$
(d) $(2y^4 z^n)^3 = 2^3 (y^4)^3 (z^n)^3 = 8y^{12} z^{3n}$
(e) $(-x^2 z)^5 = (-1)^5 (x^2)^5 (z)^5 = -x^{10} z^5$

7-3 Laws of Exponents

> **COMMON ERROR** There is no similar rule for the sum of two quantities raised to a power.
>
> $$(x + y)^n \neq x^n + y^n$$

Quotient Raised to a Power

We can show, in a similar way to that used for products, that

$$\left(\frac{x}{y}\right)^2 = \left(\frac{x}{y}\right)\left(\frac{x}{y}\right) = \frac{x^2}{y^2}$$

Or, in general,

| QUOTIENT RAISED TO A POWER | $\left(\frac{x}{y}\right)^n = \frac{x^n}{y^n}$ $(y \neq 0)$ | 23 |

When a quotient is raised to a power, the numerator and denominator may be separately raised to the power.

Example 40

(a) $\left(\dfrac{2}{a}\right)^3 = \dfrac{2^3}{a^3} = \dfrac{8}{a^3}$

(b) $\left(\dfrac{2x}{4y}\right)^2 = \dfrac{2^2 x^2}{4^2 y^2} = \dfrac{4x^2}{16y^2}$

(c) $\left(\dfrac{3c^2}{4d^4}\right)^3 = \dfrac{3^3(c^2)^3}{4^3(d^4)^3} = \dfrac{27c^6}{64d^{12}}$

(d) $\left(-\dfrac{x^2}{y}\right)^4 = \dfrac{(-1)^4(x^2)^4}{y^4} = \dfrac{x^8}{y^4}$

Zero Exponent

If we divide x^n by itself we get, by Eq. 20

$$\frac{x^n}{x^n} = x^{n-n} = x^0$$

But any expression divided by itself equals 1, so

| ZERO EXPONENT | $x^0 = 1$ $(x \neq 0)$ | 24 |

Any expression (except 0) raised to the zero power equals 1.

--- **Example 41** ---

(a) $(9626)^0 = 1$
(b) $(abc)^0 = 1$
(c) $(4x^2 + 9x - 35)^0 = 1$
(d) $7a^0 = 7(1) = 7$

Negative Exponent

We now divide x^0 by x^a. By Eq. 20

$$\frac{x^0}{x^a} = x^{0-a} = x^{-a}$$

Since $x^0 = 1$, we get

| NEGATIVE EXPONENT | $x^{-n} = \dfrac{1}{x^n}$ $x \neq 0$ | 25 |

--- **Example 42** ---

(a) $6^{-1} = \dfrac{1}{6^1} = \dfrac{1}{6}$

(b) $x^{-2} = \dfrac{1}{x^2}$

(c) $2x^{-3} = 2\left(\dfrac{1}{x^3}\right) = \dfrac{2}{x^3}$

(d) $\dfrac{3}{x^{-2}} = 3x^2$

(e) $4x^{-2} + 2y^{-3} = \dfrac{4}{x^2} + \dfrac{2}{y^3}$

SECTION 7-3　Exercises　Laws of Exponents

Definitions

Evaluate each expression.

1. 3^3
2. $(3)^5$
3. $(-2)^4$
4. $(-2)^5$
5. $(0.001)^3$
6. $(-5)^3$

Multiplying Powers

Multiply.

7. $(x^4)(x^2)$
8. $(y^b)(y^3)$
9. $(a^3)(a^6)$
10. $(10^5)(10^9)$
11. $(10^2)(10^6)$
12. $(z^{11})(z^2)$

Quotients Divide. Write your answers without negative exponents.

13. $\dfrac{a^6}{a^4}$
14. $\dfrac{2^4}{2^3}$
15. $\dfrac{y^{a+1}}{y^{a-2}}$
16. $\dfrac{10^6}{10^2}$
17. $\dfrac{10^{b-1}}{10^{a-3}}$
18. $\dfrac{10^4}{10^3}$
19. $\dfrac{b^{-4}}{b^{-5}}$
20. $\dfrac{x^{-6}}{y}$

Power Raised to a Power Simplify.

21. $(a^2)^4$
22. $(2^5)^2$
23. $(z^c)^a$
24. $(y^{-1})^{-3}$
25. $(a^{x-1})^3$

Product or Quotient Raised to a Power Raise to the power indicated.

26. $(ac)^3$
27. $(3a)^2$
28. $(4a^3c^2)^4$
29. $(2xyz)^5$
30. $\left(\dfrac{2}{3}\right)^3$
31. $\left(-\dfrac{2}{5}\right)^3$
32. $\left(\dfrac{a}{b}\right)^3$
33. $\left(\dfrac{4x^2}{3y^2}\right)^2$
34. $\left(\dfrac{2ab^3}{3c^2d}\right)^3$

Zero Exponent Evaluate.

35. $(2x^2 - 8x + 32)^0$
36. $108a^3c^0$
37. $\dfrac{82}{y^0}$
38. $\dfrac{c}{y^0}$
39. $\dfrac{(z^{-n})(z^2)}{z^{2-n}}$
40. $4\left(\dfrac{a^7}{y^4}\right)^0$

Negative Exponent Write each expression with positive exponents only.

41. x^{-1}
42. $(-b)^{-3}$
43. $\left(\dfrac{2}{x}\right)^{-4}$
44. $ab^{-5}c^{-2}$
45. $\left(\dfrac{4x^3}{3y^2}\right)^{-3}$
46. $a^{-5}bc^{-2}$
47. $4w^{-4} - 3z^{-2}$
48. $\left(\dfrac{a}{b}\right)^{-4}$

Express without fractions, using negative exponents where needed.

49. $\dfrac{1}{a}$
50. $\dfrac{5}{x^3}$
51. $\dfrac{c^2}{d^3}$
52. $\dfrac{w^4z^{-2}}{x^{-3}}$
53. $\dfrac{y^2}{x^{-4}}$
54. $\dfrac{c^{-1}d^{-4}}{b^{-2}c^{-3}}$

Chapter 7 REVIEW EXERCISES

Evaluate each expression.

1. $(-3)^4$
2. $(0.002)^3$
3. $(-4)^3$
4. Subtract $2w + 3z - 4x + 6y$ from $4x - 11y + 3w - z$.
5. What is the sum of $7w^2 - 2y + 3xz$, $3w^2 - 5xz + 3y$, $8xz - 3y + 15w^2$, $3y + 2w^2 - 2xz$, and $8w^2 - 3y$?

Divide. Write your answers without negative exponents.

6. $\dfrac{w^6}{w^4}$
7. $\dfrac{b^{w+1}}{b^{w+2}}$
8. $\dfrac{10^{a+1}}{10^{w-3}}$
9. $\dfrac{x^{-4}}{x^{-5}}$

Simplify.

10. $(w^2)^4$
11. $(1^5)^2$
12. $(x^y)^w$
13. $(b^{-1})^{-3}$

Multiply.
14. $(a^4)(a^2)$ 15. $(b^x)(b^3)$ 16. $(z^{w-2})(z^{1-w})$ 17. $(10^5)(10^9)$

Raise to the given power.
18. $(ay)^3$ 19. $(4x^3y^2)^4$ 20. $(2abz)^5$
21. $\left(\dfrac{1}{3}\right)^3$ 22. $\left(-\dfrac{2}{7}\right)^2$ 23. $\left(\dfrac{3ax^3}{4y^2z}\right)^3$

Combine as indicated and simplify.
24. $8b + 2b$ 25. $5a + 2a - 8a - a$ 26. $5wy + 9wy - 20wy$
27. Add $w - 6x + 2y$ and $w + 6x + y - 2z$.
28. What is the sum of $4ax + 6xy - 8yz$, $-6wx - 3ax$, $4yz - 6xy$, and $5ax - 3xy$?

Write without fractions, using negative exponents where needed.
29. $\dfrac{1}{w}$ 30. $\dfrac{w^4 z^{-2}}{a^{-3}}$ 31. $\dfrac{y^{-1}z^{-4}}{x^{-2}y^{-3}}$

Evaluate.
32. $(2a^2 - 8a + 32)^0$ 33. $108w^3y^0$
34. $\dfrac{32}{a^0}$ 35. $7\left(\dfrac{w^3}{z^2}\right)^0$

How many terms are there in each expression?
36. $a^3 - 2a$ 37. $(5b + b^2) - 5$ 38. $(z - 9)(z + 4)$

Write each expression without negative exponents.
39. a^{-1} 40. $(-x)^{-3}$ 41. $wx^{-5}y^{-2}$
42. $\left(\dfrac{2}{a}\right)^{-4}$ 43. $\left(\dfrac{4a^3}{3b^2}\right)^{-3}$ 44. $\left(\dfrac{w}{x}\right)^{-4}$

Remove symbols of grouping and collect terms.
45. $(4z + 2) + (z - 5)$ 46. $2a^3 - (4a - w + b)$
47. $13b^2 - 14ab - (2b^2 + ab - w)$
48. $(5xy + 2y - w) - (6w + 3xy) + (-4y - 7w)$
49. $(z + 6w - 5y) - (3x + 4z) - (w - 3y) + (-x - 2z)$
50. $-5a - \{3b + [2a - (-6b - 3x)] - [8x + w]\}$
51. Add $-w + 5a$, $w^3 + 3w - 11a^2 + 2w^2 - 3$, $7w^2 + 3w^3 - 6a^3 + 12$, and $9w^2 - 5w^3 + 2a^2 - 5 + 7a$.
52. Subtract $23b^2 + 4b^3 - 12$ from $11b^3 - 8b^2 + b$.

WRITING

53. Suppose your friend is sick and has missed the introduction to algebra, and is still out of class. Write a note to your friend explaining in your own words what you think algebra is, and how it is related to the arithmetic that you both have just finished studying.

8 Multiplication of Algebraic Expressions

Outcomes

When you have completed this chapter, you will be able to

1. Multiply two or more monomials.
2. Multiply a monomial by a multinomial.
3. Multiply two binomials.
4. Multiply two or more multinomials.
5. Raise a multinomial to a power.
6. Simplify an expression by removing grouping symbols.

In Chapter 7 we learned how to add and subtract algebraic expressions. Here we learn how to multiply them, and in Chapter 9 we divide algebraic expressions.

We start by multiplying an expression having just one term (monomial) by another monomial. We then go on to multiply a monomial by an expression having more than one term (called a multinomial). Finally, we multiply a multinomial by another multinomial. In this chapter we will multiply only polynomials, those multinomials whose exponents are all positive integers, but the methods we will learn apply equally well to any multinomials.

8-1 PRODUCT OF TWO OR MORE MONOMIALS

Symbols and Definitions

Multiplication is indicated in several ways: by the usual × symbol, by a dot, or by parentheses, brackets, or braces. Thus, the product of b and d could be written

$$b \times d \quad b \cdot d \quad b(d) \quad (b)d \quad (b)(d)$$

Most common of all is to use no symbol at all. The product of b and d would usually be written bd. Avoid using the × symbol because it could get confused with the letter x.

We get a product when we multiply two or more factors.

$$(\text{factor})(\text{factor})(\text{factor}) = \text{product}$$

Rules of Signs

When we multiply two factors that have the *same* sign, we get a product that is *positive*. When we multiply two factors that have *opposite* signs, we get a product that is *negative*. If a and b are positive quantities, we have

RULES OF SIGNS	$(+a)(+b) = (-a)(-b) = +ab$	9
	$(+a)(-b) = (-a)(+b) = -(+a)(+b) = -ab$	10

The product of two factors of like signs is positive, of unlike signs is negative.

Example 1
(a) $(+x)(+z) = xz$
(b) $(+x)(-z) = -xz$
(c) $(-x)(+z) = -xz$

When we multiply *more* than two factors, every pair of negative factors will give a positive product. Thus, if there is an even number of negative factors, the final result will be positive; if there is an odd number of negative factors, the result will be negative.

Example 2
(a) $(a)(-b)(-c) = abc$
(b) $(-p)(-q)(-r) = -pqr$
(c) $(-w)(-x)(-y)(-z) = wxyz$

Commutative Law

When we multiplied *numbers*, we saw that the order of multiplication did not make any difference, since 2×3 gives the same product as 3×2.
  In symbols,

COMMUTATIVE LAW FOR MULTIPLICATION	$ab = ba$	2

Example 3
Multiply $-2y$ by $4x$.

Solution:

$$(-2y)(4x) = -2(y)(4)(x)$$
By Eq. 2,
$$= -2(4)(y)(x)$$
$$= -8yx$$
or
$$= -8xy$$

We usually write the letters in alphabetical order.

Multiplying Monomials

Recall that a monomial is an algebraic expression having one term, such as the expressions $3y^2$ and $(8x)^3$. To multiply monomials, we use the laws of exponents and the rules of signs.

Example 4

$$(-5xy^2)(2x^3y^2) = (-5)(2)(x)(x^3)(y^2)(y^2)$$
$$= -10x^{1+3}y^{2+2}$$
$$= -10x^4y^4$$

It is no harder to multiply three or more monomials than to multiply two monomials.

Example 5

Multiply $-3a$, $-2a^2b$, and ab^2c^3.

Solution: By Eq. 9,

$$-3a(-2a^2b)(ab^2c^3) = -3(-2)(1)(a)(a^2)(a)(b)(b^2)(c^3)$$
$$= 6a^{1+2+1}b^{1+2}c^3$$
$$= 6a^4b^3c^3$$

If there are *literals* in the exponents, simply combine them using the laws of exponents we have already studied.

Example 6

Multiply $5a^nx^2$ by $-3a^3$.

Solution: By Eq. 10,

$$5a^nx^2(-3a^3) = 5(-3)(a^n)(a^3)(x^2)$$
$$= -15a^{n+3}x^2$$

The procedure is no different when the quantities to be multiplied are approximate numbers. We must, however, retain the proper number of digits in our answer. Recall from Sec. 3-3 that *when multiplying approximate numbers we retain as many significant digits in our product as contained in the factor having the fewest significant digits.*

Example 7

$$(3.848x^2)(5.24xy^2) = 20.2x^3y^2$$

Since one of our numerical factors has four significant digits and the other has only three, we round our product to three significant digits.

SECTION 8-1 Exercises Product of Two or More Monomials

Multiply the following monomials, and simplify.

1. $(x)(-y)$
2. $(-w)(-z)$
3. $(a)(-b)(c)$
4. $(-x)(-y)(z)$
5. $(2.82x)(3.26y)$
6. $(33.5a)(3.72ab)$
7. $(-8xy)(-2x)$
8. $(-3ab)(-5ac)$

9. $a^2(a^5)$
10. $xy^2(5x^2)$
11. $(3.72x^2a)(-4.26xa^3)$
12. $(-6.72ab^4)(5.27a^2bc)$
13. $(5x^2y)(-3xyz^2)(2x^2y^3z)$
14. $(-3a^2bc)(4abc^3)(-6ab^2)$
15. $8x(xy)(5yz)(-x^2y^3z)$
16. $(-5c^2)(-2bc)(-4a^2c)(-6ab^3)(-3b^2c)$
17. $(3.11a^mx^2)(2.94a^3)$
18. $(6.82b^3x^3)(5.16b^2x)$
19. $(x^my^n)(3x^3y^2)$
20. $(2w^3y^3)(7w^ay^b)$

8-2 PRODUCT OF A MULTINOMIAL AND A MONOMIAL

A *multinomial* is an algebraic expression having more than one term.

Example 8

Some multinomials are

(a) $6x - 8$
(b) $7x^2 + 2x - 1$
(c) $x^{-2} + 5$

Recall that a *polynomial* is a monomial or a multinomial in which the powers to which the variable is raised are all positive integers. The first two expressions in the preceding example are polynomials, but the third is not. The examples in this chapter will show multiplication of polynomials only, but the methods we show are valid for any multinomials. We have not yet covered the rules needed for multiplying other multinomials, such as those containing radicals, negative exponents, logarithms, and so forth. In later chapters we will show multiplication of such expressions.

A *binomial* is a polynomial with two terms, and a *trinomial* is a polynomial having three terms. In Example 8, the first expression is a binomial and the second is a trinomial.

To multiply a multinomial and a monomial, we use the distributive law (Eq. 5):

| DISTRIBUTIVE LAW | $a(b + c) = ab + ac$ | 5 |

Example 9

$$2x(x + 1) = 2x(x) + 2x(1) = 2x^2 + 2x$$

Be especially careful when multiplying negative quantities.

Example 10

$$-y(y^2 - 3) = -y(y^2) - y(-3) = -y^3 + 3y$$

Example 11

$$5x(-2x^2 - 1) = 5x(-2x^2) + 5x(-1)$$
$$= -10x^3 - 5x$$

We can extend the distributive law, which was written for a monomial times a binomial, to a multinomial having any number of terms. We simply multiply every term in the multinomial by the monomial.

Example 12

Multiply $2y^2 + 5y - 8$ by $-4y^2$.

Solution:
$$-4y^2(2y^2 + 5y - 8) = -4y^2(2y^2) + (-4y^2)(5y) - (-4y^2)(8)$$
$$= -8y^4 - 20y^3 + 32y^2$$

SECTION 8-2 Exercises Product of a Multinomial and a Monomial

Multiply the multinomial by the monomial, and simplify.

1. $3(-2 - x)$
2. $x(b + 2)$
3. $2(a + 3b)$
4. $x(x - 5)$
5. $3.83b(b^2 + 1.27)$
6. $2.03x(1.27x - 2.36)$
7. $3x(-7 - 10x)$
8. $b^4(b^2 + 8)$
9. $a^2b(2a + b - ab)$
10. $-6x^3y^3(3xy + 5x^2y^3 - 2xy^2)$
11. $2ab(9a^2 + 6ab - 3b^2)$
12. $-4.27m^2(2.83m^4 + 6.82m^2 - 3.25m^3 + 2.47)$
13. $-5.16xy(1.23x^2y - 5.83xy^2 + 4.27x^2y^2 - 2.94xy)$
14. $6mn^2(5m^3n + 4mn^2 - 2m^2n + mn)$

8-3 PRODUCT OF TWO BINOMIALS

To multiply any multinomials, we multiply every term in one multinomial by each term in the other multinomial, and combine like terms. Here we apply this rule to binomials and later use it for multinomials having any number of terms.

Example 13
$$(x - 2)(x + 3) = (x)(x) + (x)(3) + (-2)(x) + (-2)(3)$$
$$= x^2 + 3x - 2x - 6$$
$$= x^2 + x - 6$$

We can also multiply two multinomials by using the distributive rule.

Example 14

Repeating Example 13 using the distributive rule gives,
$$(x - 2)(x + 3) = x(x + 3) - 2(x + 3)$$
$$= x^2 + 3x - 2x - 6$$
$$= x^2 + x - 6$$

as before.

FOIL Rule

One way to keep track of the terms when multiplying binomials is by the FOIL rule. Multiply the **F**irst terms, then **O**uter, **I**nner, and **L**ast terms. These products can, of course, be done in any order, but we can avoid getting mixed up if we always follow the same FOIL order.

Example 15

Repeating Example 13,

$$\begin{array}{c} \quad\;\; \mathbf{F} \quad\; \mathbf{O} \quad\;\; \mathbf{I} \quad\;\, \mathbf{L} \\ \quad\;\; \downarrow \quad\;\; \downarrow \quad\;\; \downarrow \quad\;\; \downarrow \\ (x-2)(x+3) = x^2 + 3x - 2x - 6 \\ = x^2 + x - 6 \end{array}$$

COMMON ERROR Be careful, the FOIL Rule applies only to the multiplication of two binomials. Do not try to apply it to any other combination.

You might prefer to arrange the expressions *vertically*.

Example 16

Repeating Example 13, we get,

$$\begin{array}{r} x - 2 \\ x + 3 \\ \hline 3x - 6 \\ x^2 - 2x \quad\;\; \\ \hline x^2 + x - 6 \end{array}$$

Product of the Sum and Difference of Two Terms

The product of two binomials gives us expressions that will occur often enough later for us to take special note of them here. These are called *special products*. We will need them especially when we study factoring.

Let us multiply the binomials $(x - y)$ and $(x + y)$. We get

$$(x - y)(x + y) = x^2 + xy - xy - y^2$$

The sum of the two middle terms is zero, so we get

DIFFERENCE OF TWO SQUARES	$(x-y)(x+y) = x^2 - y^2$	31

Thus, the product is the *difference of two squares:* the squares of the original numbers.

Example 17

$$(x + 4)(x - 4) = x^2 - 4^2 = x^2 - 16$$

Example 18
$$(a + b)(a - b) = a^2 - b^2$$

Example 19
$$(2x + 3y)(2x - 3y) = (2x)^2 - (3y)^2 = 4x^2 - 9y^2$$

Example 20
$$(7a^2x + 3by^2)(7a^2x - 3by^2) = (7a^2x)^2 - (3by^2)^2$$
$$= 49a^4x^2 - 9b^2y^4$$

Product of Two Binomials Having the Same First Term

When we multiply $(x + a)$ and $(x + b)$, we get

$$(x + a)(x + b) = x^2 + ax + bx + ab$$

Combining like terms, we obtain a trinomial in which the coefficient of the first term (called the *leading coefficient*) is equal to 1.

TRINOMIAL WITH LEADING COEFFICIENT = 1	$(x + a)(x + b) = x^2 + (a + b)x + ab$	35

Example 21
$$(x - 3)(x + 5) = x^2 + (-3 + 5)x + (-3)5$$
$$= x^2 + 2x - 15$$

General Quadratic Trinomial

When we multiply two binomials of the form $ax + b$ and $cx + d$, we get

$$(ax + b)(cx + d) = acx^2 + adx + bcx + bd$$

Combining like terms, we obtain

GENERAL QUADRATIC TRINOMIAL	$(ax + b)(cx + d) = acx^2 + (ad + bc)x + bd$	36

Example 22
$$(4x - 2)(5x - 3) = 4(5)x^2 + [(4)(-3) + (-2)(5)]x + (-2)(-3)$$
$$= 20x^2 - 22x + 6$$

SECTION 8-3 Exercises Product of Two Binomials

Multiply the following binomials, and simplify.
1. $(x + y)(x + z)$
2. $(4a - 3)(a + 2)$
3. $(4m + n)(2m^2 - n)$
4. $(y + 2)(y - 2)$
5. $(2x - y)(x + y)$
6. $(a^2 - 3b)(a^2 + 5b)$
7. $(4xy^2 - 3a^3b)(3xy^2 + 4a^3b)$
8. $(2m^2 - 2n^2)(2m^2 + 2n^2)$
9. $(a - 7x)(2a + 3x)$
10. $(3x - z^2)(4x - 3z^2)$
11. $(ax - 5b)(ax + 5b)$
12. $(5y^2 + 3z)(5y^2 - 3z)$
13. $(2.93x - 1.11y)(x + y)$
14. $(2.84a^2 - 3.82b)(a^2 + 5.11b)$
15. $(4.03y^2 - 3.92a^3b)(3.26y^2 + 4.73a^3b)$
16. $(2.83m^2 - 2.12n^2)(2.83m^2 + 2.12n^2)$

Applications

17. A rectangle has its length L increased by 2 units and its width W decreased by 3 units. Write an expression for the area of the rectangle, and multiply out.
18. A car traveling at a speed v for a time t will go a distance equal to vt. If the speed is decreased by 8.5 km/h and the time is increased by 2.4 h, write an expression for the new distance traveled, and multiply it out.

8-4 PRODUCT OF TWO MULTINOMIALS

We multiply multinomials in the same way that we multiplied other expressions; *multiply every term in one multinomial by every term in the other.* Then combine like terms.

Example 23

$$(a + 1)(a^2 + a + 1) = a(a^2) + a(a) + a(1) + 1(a^2) + 1(a) + 1(1)$$
$$= a^3 + a^2 + a + a^2 + a + 1$$
$$= a^3 + 2a^2 + 2a + 1$$

Example 24

$$(x + 2)(x^2 + 4x - 3) = x(x^2) + x(4x) + x(-3) + 2(x^2) + 2(4x) + 2(-3)$$
$$= x^3 + 4x^2 - 3x + 2x^2 + 8x - 6$$
$$= x^3 + 6x^2 + 5x - 6$$

Example 25

$$(w - x^2)(w^3 + aw^2 + bx^2 - x^3)$$
$$= w^4 + aw^3 + bwx^2 - wx^3 - w^3x^2 - aw^2x^2 - bx^4 + x^5$$

Example 26

Multiply $(3 + x)(4 - x)(5 - x)(6 + x)$.

Solution: Let us first multiply one pair of binomials, say, $(5 - x)(6 + x)$.

$$(5 - x)(6 + x) = 30 + 5x - 6x - x^2 = 30 - x - x^2$$

Then let us multiply that product by $(4 - x)$.

$$(4 - x)[(5 - x)(6 + x)] = (4 - x)(30 - x - x^2)$$
$$= 120 - 4x - 4x^2 - 30x + x^2 + x^3$$
$$= 120 - 34x - 3x^2 + x^3$$

Finally we multiply that product by $(3 + x)$.

$$(3 + x)[(4 - x)(5 - x)(6 + x)] = (3 + x)(120 - 34x - 3x^2 + x^3)$$
$$= 360 - 102x - 9x^2 + 3x^3 + 120x - 34x^2 - 3x^3 + x^4$$
$$= 360 + 18x - 43x^2 + x^4$$

Example 27

$$(x^2 + 3x - 1)(x^2 - 3x + 2) = x^4 - 3x^3 + 2x^2 + 3x^3 - 9x^2 + 6x - x^2 + 3x - 2$$
$$= x^4 - 8x^2 + 9x - 2$$

SECTION 8-4 Exercises Product of Two Multinomials

Multiply the following binomials and trinomials, and simplify.
1. $(x - 3)(x + 4 - y)$
2. $(a - d)(a - 2d + 5)$
3. $(w^2 + w - 5)(4w - 2)$
4. $(a^2 - 5)(3a^2 - 7a - 4)$
5. $(b^7 - 2.82b^5 + 4.27b^3)(b + 2.93)$
6. $(x + 3.88)(x^3 - 2.15x - 6.03)$
7. $(1 + c^2)(4c^2 + 7c - 3)$
8. $(2x^2 - 6xy + 3y^2)(3x + 3y)$
9. $(b^2 - bx + x^2)(b + x)$
10. $(a^2 + 2a - 2)(a + 1)$
11. $(b^2 + bx + x^2)(b - x)$
12. $(b^2 + bx - x^2)(b - x)$

Multiply the following binomials and polynomials, and simplify.
13. $(a^2 + 5a - xy)(a + z)$
14. $(c^2 - cm + cn + mn)(c - m)$
15. $(y^2 - x^2)(y^3 + ay^2 - abxy + bx^2 - x^3)$

Multiply the following monomials and binomials, and simplify.
16. $4(5b + 3c)(5b + 3c)$
17. $(4w - 5z)(4w - 5z)d$
18. $z(22.1a - 3.03b)(2.26a + 38.2b)$
19. $(d - 4.11)(d - 4.93)(d + 2.26)$
20. $(m + 3)(m + 3)(m - 5)$
21. $(c + m)(c + n)(c - p)$
22. $(z + 5)(z - 2)(z - 5)(z + 2)$
23. $(2 + y)(2 - y)(2 - y)(4 + y^2)$

Multiply the following trinomials, and simplify.
24. $(x - y - z)(x + y + z)$
25. $(5x^3 + 2xy^2 - 2x)(5x^2 - 2x)$
26. $(5x - y + 2x)(4x - y + 6)$
27. $(x + y - z)(x - y - z)$
28. $(a^2 - 5.93a + 31.4)(a^2 - 5.37a + 4.03)$
29. $(m^3 - 4.83 + 32.4m)(m^2 - 3.37m + 2.26)$
30. $(am - ym + yx)(am + ym - yx)$

8-5 POWERS OF MULTINOMIALS

We see from Eq. 18 that raising an expression to a power is the same as multiplying the expression by itself the proper number of times, provided that the power is a positive integer.

Example 28

$$(x + 2)^2 = (x + 2)(x + 2)$$
$$= x^2 + 2x + 2x + 4$$
$$= x^2 + 4x + 4$$

Perfect Square Trinomial

Let us square a binomial, such as $(x + y)$. We get

$$(x + y)^2 = (x + y)(x + y) = x^2 + xy + xy + y^2$$

Collecting terms, we get an expression that is called a *perfect square trinomial*.

| PERFECT SQUARE TRINOMIAL | $(x + y)^2 = x^2 + 2xy + y^2$ | 37 |

The first and last terms of the perfect square trinomial are the squares of the two terms of the binomial; its middle term is twice the product of the terms of the binomial. Similarly, when we square $(x - y)$, we get

| PERFECT SQUARE TRINOMIAL | $(x - y)^2 = x^2 - 2xy + y^2$ | 38 |

Example 29

$$(2x - 3)^2 = (2x)^2 - 2(2x)(3) + 3^2$$
$$= 4x^2 - 12x + 9$$

Cube of a Binomial

The following example shows how to cube a binomial. The same method can, of course, be extended to cube any multinomial, or to raise a multinomial to any positive integer power.

Example 30

Cube $(x + 1)$.

Solution: By the definition of an exponent, we get

$$(x + 1)^3 = (x + 1)(x + 1)(x + 1)$$

Let us first multiply out $(x + 1)(x + 1)$.

$$(x + 1)^3 = (x + 1)(x^2 + x + x + 1)$$
$$= (x + 1)(x^2 + 2x + 1)$$

Then by the distributive law,

$$(x + 1)^3 = x(x^2 + 2x + 1) + 1(x^2 + 2x + 1)$$
$$= x^3 + 2x^2 + x + x^2 + 2x + 1$$
$$= x^3 + 3x^2 + 3x + 1$$

Example 31

$$(x + 2)^3 = (x + 2)(x + 2)(x + 2)$$
$$= (x + 2)(x^2 + 4x + 4) \text{ (from Example 28)}$$
$$= x^3 + 4x^2 + 4x + 2x^2 + 8x + 8$$

Combining like terms,

$$(x + 2)^3 = x^3 + 6x^2 + 12x + 8$$

SECTION 8-5 Exercises Powers of Multinomials

Square each binomial.

1. $(x + y)$
2. $(m + n)$
3. $(a - d)$
4. $(z - w)$
5. $(B + D)$
6. $(C - D)$
7. $(4.92y + 3.12z)$
8. $(2.45d - 1.93x)$
9. $(5n + 6x)$
10. $(d^3 + d^2)$
11. $(1 - w)$
12. $(cy^2 - c^3y)$
13. $(b^3 - 13)$
14. $(a^n + b^3)$

Square each trinomial.

15. $(x + y + z)$
16. $(x - y - z)$
17. $(a + b - 1)$
18. $(5a^3 - 3a + 16)$
19. $(c^2 - cd + d^2)$
20. $(w^2 - 5w + 2)$

Cube each binomial.

21. $(x - y)$
22. $(1.93 - 3.24a)$
23. $(3.02m + 2.16n)$
24. $(a^2 + 1)$
25. $(c + d)$
26. $(4p - q)$

Applications

27. A square of side x has each side increased by 2 units. Write an expression for the area of the new square, and multiply out.
28. When a current I flows through a resistance R, the power in the resistance is I^2R. If the current is increased by 2.5 amperes, write an expression for the new power, and multiply out.
29. If the radius r of a sphere is decreased by 2 units, write an expression for the new volume of the sphere, and multiply out. (Volume = $\frac{4}{3}\pi r^3$)

8-6 REMOVING SYMBOLS OF GROUPING

We learned earlier that symbols of grouping indicate that the enclosed expression is to be taken as a whole. Here we see that they also *indicate multiplication*.

Example 32

(a) $x(y)$ is the product of x and y.
(b) $x(y + 4)$ is the product of x and $(y + 4)$.
(c) $(x - 3)(y + 1)$ is the product of $(x - 3)$ and $(y + 1)$.
(d) $-(y + 1)$ is the product of -1 and $(y + 1)$.

After the multiplication has been performed, we simply remove the parentheses.

Example 33

(a) $x(y - 3) = xy - 3x$

(b) $8(2y + 1) - 3 = 16y + 8 - 3 = 16y + 5$

COMMON ERROR Don't forget to multiply every term within the grouping by the preceding factor.

$$-(3x + 4) \neq -3x + 4$$

COMMON ERROR Multiply the terms within the parentheses only by the factor directly preceding it.

$$a - 3(x + y) \neq (a - 3)(x + y)$$

If there are groupings within groupings, start simplifying with the innermost groupings.

Example 34

(a) $4[2(x + 1) + 3] = 4[2x + 2 + 3]$
$= 4[2x + 5]$
$= 8x + 20$

(b) $a + 4[3 + 2(b - 3)] = a + 4[3 + 2b - 6]$
$= a + 4[2b - 3]$
$= a + 8b - 12$

(c) $2\{[(1 - m) + (n + 3)] - 5\} - 4 = 2\{[1 - m + n + 3] - 5\} - 4$
$= 2\{[4 - m + n] - 5\} - 4$
$= 2\{-m + n - 1\} - 4$
$= -2m + 2n - 2 - 4$
$= -2m + 2n - 6$

SECTION 8-6 Exercises Removing Symbols of Grouping

Remove symbols of grouping, and simplify.

1. $a + (b + a)$
2. $x + (2 - x)$
3. $x + (x - y)$
4. $-(a + 3.92) - (a - 4.14)$
5. $x - 2.66[y - 3.02(x + 6.22y) + 4.98y]$
6. $[(5x - 4y) - (2x - 5y)][(-3x - 6y) - (2x + 4y)]$
7. $\{[3 - (x + 7)] - 3x\} - (x + 7)$
8. $[a(a + b) - a^2](a^2 + b^2)(a^3 - 4ab + b^3)$
9. $6p - \{3p + [2q - ((5p + 4q) + p) - (3p + 2)] - 2p\}$
10. $[(5a - 2)(2a + 3) - 6a^2][(3a - 4)(2a - 4) + 5a^2]$

Chapter 8 REVIEW EXERCISES

Multiply and simplify.

1. $(3ay^2)(-4a^3y)$
2. $(3x^2y)(-5xyz^2)(4x^2y^3z)$
3. $(a - 3)(a + 4 - b)$
4. $(2 + x)(2 - x)(2 - x)(4 + x^2)$
5. $(x - y)(x - 2.45y + 5.83)$
6. $[z^2 + (x - y)z - xy](x + z)$
7. $(x^2 - 3y)(x^2 + 5y)$
8. $4(5p + 3q)(5p + 3q)$
9. $(4a + 2b)(2a^2 - b)$
10. $(4.45x - 5.92y)(4.23x - 5.86y)z$
11. $(c + 3)(c + 3)(c - 5)$
12. $(a - b - c)(a + b + c)$
13. $(p^2 - px + x^2)(p + x)$
14. $(x^2 + 2x - 2)(x + 1)$
15. $(5a^3 + 2ab^2 - 2a)(5a^2 - 2a)$
16. $(a + b)(a + c)$
17. $(p + q - r)(p - q - r)(p + q + r)$
18. $x^2y(2x + y - xy)$
19. $(3m^3 + m^2 - 5m + 3)(m^2 - 5m + 4)$
20. $(a + 2)(a - 2)$
21. $(a^2 - b^2)(a^2 + b^2 - ab)$
22. $x^4(x^2 + 8)$
23. $(bm - by)(bm + by + nz)$
24. $3a(-2 - 10a)$
25. $(-5z^2)(-2yz)(-4x^2z)(-6xy^3)(-3y^2z)$
26. $ab^2(5a^2)$

Square each binomial.

27. $(2p + 3q)$
28. $(x - 5.03y)$
29. $(x^3 + x^2)$

Square each trinomial.

30. $(x^3 - 3.33x + 1.84)$
31. $(x^2 - xy + y^2)$
32. $(a^2 - 5a + 2)$

Remove symbols of grouping, and simplify.

33. $x - 2[y - 3(x + 2y) + 3y]$
34. $\{[4 - (x + 4)] - 2x\} - (x + 6)$
35. $2x - \{3y + [2x - (4y + 2x) + y - (4x + 2)] - 2y\}$

Cube each binomial.

36. $(a - 2b)$
37. $(4x + 3y)$
38. $(x^2 + 2)$

WRITING

39. Your friend refuses to learn the FOIL rule. "I want to learn math, not memorize a bunch of tricks!" he declares. What do you think? Write a paragraph or so giving your opinion on the value or harm in learning devices like the FOIL rule.

TEAM PROJECT

40. Figure 8-1 shows a geometric representation of $(a + b)^2$, the square of the binomial $(a + b)$. Use the figure to evaluate $(a + b)^2$. Then make a similar sketch to evaluate each of the following:

(a) $(2a + b)^2$
(b) $(a + 2)(a + 3)$
(c) $(a + b)(c + d)$
(d) $(a + b + 1)^2$

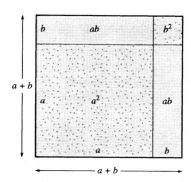

■ **FIGURE 8-1**
A geometric representation of $(a + b)^2$.

9 Division of Algebraic Expressions

Outcomes

When you have completed this chapter, you will be able to

1. Divide a monomial by a monomial.
2. Divide a polynomial by a monomial.
3. Divide a polynomial by a polynomial.

This chapter completes our study of the four basic operations with algebraic expressions; we have already covered addition, subtraction, and multiplication of algebraic expressions, and now we do division. As with multiplication, we start with monomials and gradually work up to longer expressions.

9-1 QUOTIENT OF TWO MONOMIALS

Symbols for Division

The following symbols are commonly used to indicate division.

$$x \div y \qquad \frac{x}{y} \qquad x/y$$

We get a *quotient* when we divide the *dividend* by the *divisor*.

$$\text{quotient} = \frac{\text{dividend}}{\text{divisor}}$$

The quantity x/y $\left(\text{or } \dfrac{x}{y}\right)$ is also called a *fraction*, where x is the *numerator* and y is the *denominator*. It is also spoken of as the *ratio* of x to y.

Division by Zero

Division by zero is not permitted.

Example 1

In the fraction

$$\frac{x + 8}{x - 3}$$

x cannot equal three, or the illegal operation of division by zero will result. We say $x \neq 3$.

Rules of Signs

If the dividend and the divisor have the same sign, the quotient is positive.

$$\frac{+a}{+b} = \frac{-a}{-b} = \frac{a}{b}$$

If the dividend and divisor have the opposite signs, the quotient is negative.

$$\frac{+a}{-b} = \frac{-a}{+b} = -\frac{a}{b}$$

The fraction itself has an algebraic sign which, when negative, reverses the sign of the quotient. These three ideas are summarized in the following rules:

RULES OF SIGNS FOR DIVISION		
$\frac{+a}{+b} = \frac{-a}{-b} = -\frac{-a}{+b} = -\frac{+a}{-b} = \frac{a}{b}$	11
$\frac{+a}{-b} = \frac{-a}{+b} = -\frac{-a}{-b} = -\frac{a}{b}$	12

These rules show that any pair of negative signs may be removed without changing the value of the fraction.

Example 2

Simplify $-\frac{y}{-x}$.

Solution: Removing the pair of negative signs, we obtain

$$-\frac{y}{-x} = \frac{y}{x}$$

Dividing a Monomial by a Monomial

The quotient of any quantity (except zero) and itself equals one.

Example 3

$$\frac{a}{a} = 1 \quad (a \neq 0)$$

Thus if the same factor appears both in the divisor and the dividend, they may be eliminated because their quotient is 1. This is called *canceling*.

Example 4

Divide $8xy$ by $2x$.

Solution:

$$\frac{8xy}{2x} = \frac{8}{2} \cdot \frac{\cancel{x}}{\cancel{x}} \cdot y = 4y$$

9-1 Quotient of Two Monomials

When the quantities to be divided have *exponents,* we make use of the law of exponents for division, Eq. 20.

| QUOTIENT | $\dfrac{x^m}{x^n} = x^{m-n}$ | 20 |

Example 5

Divide y^7 by y^5.

Solution: By Eq. 20,

$$\frac{y^7}{y^5} = y^{7-5} = y^2$$

Any numerical coefficients are divided separately.

Example 6

Divide $10y^5$ by $2y^2$.

Solution:

$$\frac{10y^5}{2y^2} = \frac{10}{2} \cdot \frac{y^5}{y^2}$$
$$= 5y^{5-2} = 5y^3$$

When there is more than one unknown in the expressions to be divided, treat each unknown separately.

Example 7

Divide $21x^2y^4z^3$ by $7xy^2z^2$.

Solution:

$$\frac{21x^2y^4z^3}{7xy^2z^2} = \frac{21}{7} \cdot \frac{x^2}{x} \cdot \frac{y^4}{y^2} \cdot \frac{z^3}{z^2}$$
$$= 3xy^2z$$

When dividing expressions it is common to get a quotient containing negative exponents.

Example 8

$$\frac{5y^4}{y^6} = 5y^{4-6} = 5y^{-2}$$

You may leave the answer in this form, or eliminate the negative exponent by using Eq. 25. Thus,

$$5y^{-2} = \frac{5}{y^2}$$

The process of dividing a monomial by a monomial is also referred to as *simplifying* a fraction, or *reducing a fraction to lowest terms*.

Example 9

Simplify the fraction

$$\frac{4xy^4z}{12x^3yz^2}$$

Solution: The procedure is no different than if we had been asked to divide $4xy^4z$ by $12x^3yz^2$.

$$\frac{4xy^4z}{12x^3yz^2} = \frac{4}{12}x^{1-3}y^{4-1}z^{1-2}$$

$$= \frac{1}{3}x^{-2}y^3z^{-1}$$

or,

$$= \frac{y^3}{3x^2z}$$

When the expressions to be divided have negative exponents, we simply apply Eq. 25, as before.

Example 10

Divide $16x^{-3}y^2z^{-3}$ by $4x^2y^4z^{-5}$.

Solution: Proceeding as before, we obtain

$$\frac{16x^{-3}y^2z^{-3}}{4x^2y^4z^{-5}} = \frac{16}{4}x^{-3-2}y^{2-4}z^{-3-(-5)}$$

$$= 4x^{-5}y^{-2}z^2$$

$$= \frac{4z^2}{x^5y^2}$$

Alternate Solution: We may choose to eliminate the negative exponents first.

$$\frac{16x^{-3}y^2z^{-3}}{4x^2y^4z^{-5}} = \frac{16y^2z^5}{4x^2(x^3)(y^4)(z^3)} = \frac{4z^2}{x^5y^2}$$

SECTION 9-1 Exercises Quotient of Two Monomials

Divide the following monomials.

1. x^7 by x^4
2. $21a^4$ by $3a^2$
3. $5xyz$ by xy
4. m^3n by mn
5. $4a^2d$ by $-2ad$
6. $-360x^4y^2$ by $-30x^2y$
7. $31ab^9$ by ab^8
8. $54x^2z$ by $-9x$
9. $-99ad$ by $3a$
10. $49xyz$ by $-7y$
11. $42p^5q^4r^2 \div 7p^3qr$
12. $50x^3y^5z^3 \div -10xy^3z^2$
13. $-32m^2nx \div 4mx$
14. $48cd^2z^3 \div -24cd$
15. $-36a^4b^2c \div 9ab$
16. $45m^2q$ by $-5mq$

17. $-24m^3n^3z$ by $4m^3z$
18. $-27a^5p^2$ by $-9a^3q^2$
19. $25a^4bcxyz$ by $5a^2bcxz$
20. $18d^3f^2$ by $3d^2f$
21. $32a^2bc \div -8ab$
22. $-12m^2n^3 \div 4mn^2$
23. $-36a^2by^2 \div 12a^2y$
24. $44a^2b^3c^4 \div 11a^2bc$
25. $64x^2y^2$ by $8xy$
26. $24pq^2r^3s$ by $8r$
27. $x^2y^6z^2$ by x^2z^2
28. $(a-x)^5$ by $(a-x)^2$
29. a^x by a^y
30. $66c^2dy^3 \div (-22cy)$
31. $-35a^3b^2z \div 7ab^2$
32. $19e^2m^2n^2 \div (-em^2n)$
33. $95abc \div 5a^2b^3c$
34. $45a^2b^2d^2 \div 15abd$

9-2 DIVIDING A POLYNOMIAL BY A MONOMIAL

To divide a polynomial by a monomial, we simply divide each term of the polynomial by the monomial.

Example 11

Divide $6x^3 - 3x^2$ by $3x$.

Solution:

$$\frac{6x^3 - 3x^2}{3x} = \frac{6x^3}{3x} - \frac{3x^2}{3x} = 2x^2 - x$$

Example 12

Divide $8x^2 - 4x + 4$ by $2x$.

Solution:

$$\frac{8x^2 - 4x + 4}{2x} = \frac{8x^2}{2x} - \frac{4x}{2x} + \frac{4}{2x}$$

This is really a consequence of the distributive law, Eq. 5.

$$\frac{8x^2 - 4x + 4}{2x} = \frac{1}{2x}(8x^2 - 4x + 4) = \left(\frac{1}{2x}\right)(8x^2) - \left(\frac{1}{2x}\right)(4x) + \left(\frac{1}{2x}\right)(4)$$

Each of these terms is now simplified as in the preceding section.

$$\frac{8x^2}{2x} - \frac{4x}{2x} + \frac{4}{2x} = 4x - 2 + \frac{2}{x}$$

COMMON ERROR Do not to forget to divide *every* term of the polynomial by the monomial.

$$\frac{4x^2 - 2x + 5}{2x} \neq 2x - 1 + 5$$

COMMON ERROR There is no similar rule for dividing a monomial by a polynomial.

$$\frac{a}{b+c} \neq \frac{a}{b} + \frac{a}{c}$$

Example 13

Divide $4xy - 8x^2y + 3xy^2 - 2x^2y^2$ by $16xy$.

Solution:

$$\frac{4xy - 8x^2y + 3xy^2 - 2x^2y^2}{16xy} = \frac{4xy}{16xy} - \frac{8x^2y}{16xy} + \frac{3xy^2}{16xy} - \frac{2x^2y^2}{16xy}$$

$$= \frac{1}{4} - \frac{x}{2} + \frac{3y}{16} - \frac{xy}{8}$$

We may also combine these terms over the common denominator 16, getting

$$\frac{4 - 8x + 3y - 2xy}{16}$$

Alternate Solution: We can get the same result by canceling, noting that every term in both numerator and denominator contains an x and a y.

$$\frac{4\cancel{x}\cancel{y} - 8x^2\cancel{y} + 3\cancel{x}y^2 - 2x^2y^2}{16\cancel{x}\cancel{y}} = \frac{4 - 8x + 3y - 2xy}{16}$$

SECTION 9-2 Exercises Dividing a Polynomial by a Monomial

Divide the polynomial by the monomial.

1. $15x^3 + 3x^2$ by x
2. $42m^6 - 2m^3$ by $2m$
3. $36d^5 - 6d^2$ by $3d$
4. $22x^2 + 11y^5$ by 11
5. $48c^4 + 36c^5$ by $12c^2$
6. $27n^3 - 9n^2$ by $-3n$
7. $39p^2 + 52p^3$ by $-13p^2$
8. $40a^5 - 20a^2$ by $10a$
9. $-25x^3 - 15x^2$ by $5x$
10. $-55a^3d + 22a^2$ by $-a^2$
11. $-bm^3n^2 - 3bm^2n$ by $-m^2n$
12. $-5a^5b + 5ab^2$ by $-ab$
13. $8a^3b^2 + 4a^2b^3$ by $4ab^2$
14. $10x^3y - 5xy^4$ by $-5xy$
15. $x^2y^3z - xy^4z^2$ by $-xy^2z$
16. $16a^3bc^2 + 12a^2b^5c$ by $-4abc$
17. $m^2n^2 + m^3n^2 - m^2n^3$ by $-mn$
18. $p^5q^2 - p^2q^5 - p^3q^3$ by p^2q^2
19. $x^3y^3 - x^4y + xy^4$ by $-xy$
20. $-a^3b^3 - a^2b^2 - ab$ by $-ab$
21. $c^3 - 4c^2d^2 + d^3$ by cd^2
22. $r^4s^3 - r^2s^2 + r^4s^2$ by $-r^2s$
23. $a^4 + 2a^2b^2 - b^4$ by a^2b^2
24. $m^5n^2 + m^2n^2 - m^2n^4$ by $-m^2n$
25. $4x^3z + 2xz^2 - 3z^4$ by $-xz$
26. $ab^3 + a^3c^2 - b^2c^4$ by $-abc$
27. $p^3q^3 + pq^2r^3 - p^2r^4$ by $-p^2r$
28. $7m^4n^2 + 7m^3n^3 - 7m^2n^4$ by $-7m^2n^3$
29. $4c^4d + 3c^2d^3 - cd^5$ by $-cd^2$
30. $3a^2x + 5a^3x^3 - 2ax^2$ by $-a^2x^2$
31. $8b^2c^4 + 4b^2c - 12b^3c^3$ by $-b^3c^2$

9-3 QUOTIENT OF TWO POLYNOMIALS

To divide one polynomial by another polynomial, follow these steps.

1. Write the divisor and the dividend in the order of descending powers of the variable.
2. Supply any missing terms, using coefficients of zero.
3. Set up the division in long-division form, as in the following example.

Note that this method is used only for polynomials, expressions in which the exponents are all positive integers.

Example 14

Divide $(4x + x^2 + 3)$ by $(x + 1)$.

Solution:

1. Write the dividend in descending order of the powers.

$$x^2 + 4x + 3$$

2. There are no missing terms, so we go on to the next step.
3. Set up in long-division format.

$$(x + 1)\overline{\smash{)}x^2 + 4x + 3}$$

4. Divide the first term in the dividend (x^2) by the first term in the divisor (x). The result (x) is written above the dividend, in line with the term having the same power. It is the first term of the quotient.

$$(x + 1)\overline{\smash{)}x^2 + 4x + 3}^{\,x}$$

5. Multiply the divisor by the first term of the quotient. Write the result below the dividend. Subtract it from the dividend.

$$\begin{array}{r} x \\ (x+1)\overline{\smash{)}x^2 + 4x + 3} \\ -\underline{(x^2 + x)} \\ 3x + 3 \end{array}$$

6. Repeat Steps 4 and 5 until the degree of the remainder is *less* than the degree of the divisor.

$$\begin{array}{r} x + 3 \\ (x+1)\overline{\smash{)}x^2 + 4x + 3} \\ -\underline{(x^2 + x)} \\ 3x + 3 \\ -\underline{(3x + 3)} \\ 0 \end{array}$$

The result is written

$$\frac{x^2 + 4x + 3}{x + 1} = x + 3$$

We now try a harder example.

Example 15

Divide $(2w^2 + 6w^4 - 2)$ by $(w + 1)$.

Solution:

1. Write the dividend in descending order of the powers.

$$6w^4 + 2w^2 - 2$$

2. Supply the missing terms with coefficients of zero.

$$6w^4 + 0w^3 + 2w^2 + 0w - 2$$

3. Set up in long-division format.

$$(w + 1)\overline{\smash{)}6w^4 + 0w^3 + 2w^2 + 0w - 2}$$

4. Divide the first term in the dividend ($6w^4$) by the first term in the divisor (w). The result ($6w^3$) is written above the dividend, in line with the term having the same power. It is the first term of the quotient.

$$\begin{array}{r} 6w^3 \\ (w+1)\overline{\smash{)}6w^4 + 0w^3 + 2w^2 + 0w - 2} \end{array}$$

5. Multiply the divisor by the first term of the quotient. Write the result below the dividend. Subtract it from the dividend.

$$\begin{array}{r} 6w^3 \\ (w+1)\overline{\smash{)}6w^4 + 0w^3 + 2w^2 + 0w - 2} \\ -\,(6w^4 + 6w^3) \\ \hline -6w^3 + 2w^2 + 0w - 2 \end{array}$$

6. Repeat Steps 4 and 5, each time using the new dividend obtained, until the degree of the remainder is less than the degree of the divisor.

$$\begin{array}{r} 6w^3 - 6w^2 + 8w - 8 \\ (w+1)\overline{\smash{)}6w^4 + 0w^3 + 2w^2 + 0w - 2} \\ -\,(6w^4 + 6w^3) \\ \hline -6w^3 + 2w^2 + 0w - 2 \\ -\,(-6w^3 - 6w^2) \\ \hline 8w^2 + 0w - 2 \\ -\,(8w^2 + 8w) \\ \hline -8w - 2 \\ -\,(-8w - 8) \\ \hline 6 \end{array}$$

The result is written

$$\frac{6w^4 + 2w^2 - 2}{w + 1} = 6w^3 - 6w^2 + 8w - 8 + \frac{6}{w + 1}$$

COMMON ERROR Errors are often made during the subtraction step (Step 5 in the preceding example).

$$\begin{array}{r} 6w^3 \\ (w+1)\overline{\smash{)}6w^4 + 0w^3 + 2w^2 + 0w - 2} \\ -\,(6w^4 + 6w^3) \\ \hline 6w^3 + 2w^2 + 0w - 2 \end{array}$$

No. Should be $-6w^3$

9-3 Quotient of Two Polynomials

Example 16

Divide $(2x^2y - y^3 + x^3 - 2xy^2)$ by $(x - y)$.

Solution: Let us write each expression in the order of descending powers of x, that is, $x^3, x^2, \ldots$, and proceed as before.

$$
\begin{array}{r}
x^2 + 3xy + y^2 \\
x - y \overline{\smash{\big)}\, x^3 + 2x^2y - 2xy^2 - y^3} \\
\underline{-(x^3 - x^2y)} \\
3x^2y - 2xy^2 \\
\underline{-(3x^2y - 3xy^2)} \\
xy^2 - y^3 \\
\underline{-(xy^2 - y^3)} \\
0
\end{array}
$$

SECTION 9-3 Exercises Quotient of Two Polynomials

Divide each polynomial by the given binomial.

1. $a^2 + 15a + 56$ by $a + 7$
2. $27x^3 - 8y^3$ by $3x - 2y$
3. $a^2 + a - 56$ by $a - 7$
4. $4x^2 + 23x + 15$ by $4x + 3$
5. $2x^2 + 11x + 5$ by $2x + 1$
6. $6x^2 - 7x - 3$ by $2x - 3$
7. $a^2 - 15a + 56$ by $a - 7$
8. $a^2 - a - 56$ by $a + 7$
9. $3x^2 - 4x - 4$ by $2 - x$
10. $a^4 + 3a^2 + 2$ by $a^2 + 1$
11. $a^8 - 3a^4 + 2$ by $a^4 - 1$
12. $1 - x^3y^3$ by $1 - xy$
13. $a^3 - 8a - 3$ by $a - 3$

■ Chapter 9 REVIEW EXERCISES

Divide.

1. $p^2q^3r - pq^4r^2$ by $-pq^2r$
2. $16x^3yz^2 + 12x^2y^5z$ by $-4xyz$
3. $m^2 - m - 56$ by $m + 7$
4. $3n^2 - 4n - 4$ by $2 - n$
5. $p^2 - 2pq + q^2 - r^2$ by $p - q - r$
6. $a^2b^6c^2$ by a^2c^3
7. $(u - v)^6$ by $(u - v)^3$
8. $a^2b^2 + a^3b^2 - a^2b^3$ by $-ab$
9. $x^5y^2 - x^2y^5 - x^3y^3$ by x^2y^2
10. $75abc \div 5a^3b^4c$
11. $-44ab^2 \div 4a^2b$
12. $1 - m^3n^3$ by $1 - mn$
13. $p^2 + 3pq + 2q^2$ by $p + q$
14. $m^3 - 2m - 3$ by $m - 3$
15. $x^2 - xy - 2y^2$ by $x + 2y$
16. $p^2 - q^2 + 2q - p$ by $p - q$
17. $a^4 + a^3 - 3a^2 + a$ by $a^2 - a$
18. $x^4 + 3x^2 + 2$ by $x^2 + 1$
19. $z^8 + z^4 + 1$ by z^4
20. $p^3 - 4p^2q^2 + p$ by pq^2
21. $x^4y^3 - x^2y^2 + x^4y^2$ by $-x^2y$

WRITING

22. In simplifying a fraction we are careful not to change its value. If that's the case, and the new fraction is "the same" as the old, then why bother? Write a few paragraphs explaining why you think it's valuable to simplify fractions (and other expressions as well) or if there is some advantage to just leaving them alone.

23. In the following chapter we will solve simple equations, something you must have done before. Without peeking ahead, write down in your own words whatever you remember about solving equations. You may give a list of steps, or give a description in paragraph form.

TEAM PROJECT

24. Write an hour exam that will test for the most important skills covered in this chapter. Be sure that it is not too long and hard, and not too short and easy. Then work the test and produce an answer key, to be handed in with the exam.

10 Simple Equations

Outcomes

When you have completed this chapter, you will be able to

1. Define an equation.
2. Substitute numerical values into equations.
3. Substitute denominate numbers into a formula.
4. Identify the degree of an equation.
5. Solve a simple equation.
6. Solve simple literal equations.

An *equation* is obtained when two mathematical expressions are set equal to each other. For example, when the mathematical expression $2(x - 3)$ is set equal to the expression $5x + 3$, we get the equation

$$2(x - 3) = 5x + 3$$

In this chapter we will first learn how to substitute numerical values into an equation, and then go on to solve simple equations.

It is important to be able to solve equations because they are used to describe so many things in the physical world. In the following chapter we'll learn how to write an equation to describe a simple situation, but for now we'll just learn how to solve a given equation. In later chapters we will solve more complex equations.

10-1 SUBSTITUTING INTO EQUATIONS AND FORMULAS

Before learning how to substitute into and to solve equations, we will first learn some terms that we will use in talking about equations.

Equations

An equation has two sides and an equal sign.

$$2x^2 - 5 = 3x + 2$$

left side $\quad\uparrow\quad$ right side
equal sign

Identities

An equation that is true for *any value* of the variable is called an *identity*.

Example 1

$$x(x - 3) = x^2 - 3x$$

is an identity, because it is true for any value of x.

Conditional Equations

A *conditional equation* is one whose sides are equal only for certain values of the variable.

Example 2

The equation

$$x + 1 = 4$$

is a conditional equation because the sides are equal only when $x = 3$.

When we say *equation* in this textbook, we will mean a *conditional equation*.

Substituting into Equations

To *substitute into an equation* means to replace the letter quantities in an equation with their given numerical values, and to perform the indicated computation. This is different from, and usually simpler than, solving an equation.

Example 3

Substitute the values

$$a = 4, \quad b = 2, \quad \text{and} \quad c = 6$$

into the equation

$$x = \frac{2a + b}{c}$$

Solution: Substituting, we obtain

$$x = \frac{2(4) + 2}{6} = \frac{10}{6} = 1\frac{2}{3}$$

When substituting integers, as in the preceding example, treat them as exact numbers and give your answer in exact form. When substituting approximate numbers, be sure to round your answer to the proper number of digits. Recall the rules for rounding from Chapter 3:

> When we *add* or *subtract* approximate numbers, we round our answer to the same number of *decimal places* as in the original number having the fewest decimal places.
>
> When we *multiply, divide, raise to powers,* or *take roots,* we round our answer to the same number of *significant digits* as in the original number having the fewest significant digits.

10-1 Substituting into Equations and Formulas

Example 4

Substitute the values

$$r = 12.8, \quad m = 1.04, \quad \text{and} \quad p = 33.7$$

into the equation

$$x = \frac{p+r}{m} + \frac{p-r}{7}$$

Solution: We substitute, treating the 7 as an exact number,

$$x = \frac{33.7 + 12.8}{1.04} + \frac{33.7 - 12.8}{7}$$

$$= \frac{46.5}{1.04} + \frac{20.9}{7}$$

$$= 44.7 + 2.99 = 47.7$$

Note that in the last step of the addition we were justified in keeping only one decimal place.

Substituting into Formulas

The equation for the area of a circle of radius r:

$$A = \pi r^2$$

is also called a *formula* for the area of a circle. A formula is an equation which expresses some general mathematical or physical fact. To substitute into a formula we use the same procedure as for substituting into equations. But now we carry units along with the numerical values. Sometimes we will need a conversion factor to make the units cancel properly, so that the answer will be in the desired units.

Example 5

Find the area of a circle of radius 3.75 cm.

Solution: We substitute 3.75 cm into the formula for the area of a circle given above, and get

$$A = \pi r^2 = \pi (3.75 \text{ cm})^2$$

$$= 44.2 \text{ cm}^2$$

Here we have used an accurate value of π from our calculator, so we round our answer to the number of digits in the given radius.

Often when substituting into a formula, some conversion of units must be done to make those units cancel properly. Such conversions may be done after substituting into the formula or before, as shown in the next example.

Example 6

The velocity v of a moving object starting from rest is equal to the acceleration a times the elapsed time t.

$$v = at$$

Find the velocity, in metres per second, for an object that starts from rest and accelerates at 18.8 ft/s² for 0.138 minutes.

Solution: In this example we will convert units *before* substituting. Since 1 ft = 0.3048 m we get,

$$\frac{18.8 \text{ ft}}{\text{s}^2} \cdot \frac{0.3048 \text{ m}}{\text{ft}} = 5.73 \text{ m/s}^2$$

and since 1 minute = 60 seconds, we get,

$$0.138 \text{ minute} \cdot \frac{60 \text{ seconds}}{\text{minute}} = 8.28 \text{ seconds}$$

Substituting into the given formula,

$$v = at = \frac{5.73 \text{ m}}{\text{s}^2} \cdot 8.28 \text{ s} = 47.4 \text{ m/s}$$

rounded to three significant digits.

COMMON ERROR Students often neglect to include units when substituting into a formula, with the result that the units often do not cancel properly.

In our next example, the units to be used in a certain formula are specified. Here it is necessary to convert all quantities to those specified units before substituting into the formula.

Example 7

An engineering handbook gives the following formula for the flow of water over a dam (Fig. 10-1) as

$$Q = 0.727 \, Lh \sqrt{2gh}$$

where

Q = volumetric flow rate,
L = width of the dam,
h = height of water over the dam,
g = acceleration due to gravity. ($g = 9.81$ m/s^2 = 32.2 ft/s^2).

Find the rate of flow, in metres per second squared, over a dam with a width of 4.72 metres, if the height of water is 8.75 inches.

FIGURE 10-1
Water over a dam.

Solution: To use the formula we must substitute appropriate measurements. (All lengths are in metres for the SI, and in inches for the British system. Time is always measured in seconds.)

$$L = 4.72 \text{ m}$$

$$h = 8.75 \text{ in.} \left(\frac{0.0254 \text{ m}}{\text{in.}}\right) = 0.222\ 25 \text{ m} \qquad \text{Rounding will be performed in the final answer.}$$

Substituting,

$$Q = 0.727\ Lh\sqrt{2gh}$$

$$= 0.727(4.72 \text{ m})(0.222\ 25 \text{ m})\sqrt{2\left(9.81\ \frac{\text{m}}{\text{s}^2}\right)(0.222\ 25 \text{ m})}$$

$$= 1.59\ \frac{\text{m}^3}{\text{s}}$$

SECTION 10-1 Exercises Substituting into Equations and Formulas

Substituting

Substitute the given integers into each equation and evaluate y. Do not round your answer.

1. $y = 3x + 3 \quad (x = 4)$
2. $y = 3m^2 + 2m - 6 \quad (m = -3)$
3. $y = 4a^3 - 3x \quad (a = -2, x = -3)$
4. $y = 4x^3 + 5x^2 - x - 12 \quad (x = -2)$
5. $y = 4c + 3d^2 - 5e^3 \quad (c = 5, d = -4, e = 2)$
6. $y = \frac{x^2}{z} - \frac{w^2}{x} + \frac{x}{z^2} \quad (x = -3, w = 2, z = 5)$

Substitute the given approximate numbers into each equation and evaluate y. Assume that all the given integers are exact numbers. Round your answer to the proper number of digits.

7. $y = 6 - 4x \quad (x = -3.445)$
8. $y = -4d^3 - 2e^2 \quad (d = -6.375, e = 7.995)$
9. $y = a^3 - a + 9 \quad (a = 8.01)$
10. $x = \sqrt[3]{7y - 5z} \quad (y = 5.39, z = -3.27)$
11. $y = \sqrt[3]{x^3 + 5x} \quad (x = 4.52)$
12. $y = (2z + 3x)^{2.3} \quad (z = 3.7, x = 4.8)$

Applications

13. An amount of money P deposited for t years at a simple interest rate r (expressed as a decimal) will grow to a new amount A, where $A = P(1 + rt)$. Find the amount to which $2500 will accumulate in 4 years at a simple interest rate of 5.5%.

14. In time t, an object having initial velocity v_0 and acceleration a will travel a distance s, where $s = v_0 t + \frac{1}{2}at^2$. Find the displacement after 2.2 seconds of a body thrown downward with a speed of 3 m/s. Use $a = 9.81$ m/s².

15. To convert a temperature F given in degrees Fahrenheit to a temperature C given in degrees Celsius, we use the formula $C = \frac{5}{9}(F - 32)$. Convert 105.0°F to degrees Celsius.

16. The torque T, in newton-metres delivered by a motor of power rating P, in watts, rotating at N revolutions per minute is given by the working formula

$$T = \frac{60 P}{2\pi N} \text{ N·m}$$

Find the torque delivered by a motor rated at 497 W rotating at 1500 rev/min.

17. Find the torque delivered by a motor rated at 2.05 kW rotating at 120 rad/s. (Hint: the working formula requires rotations per *minute*).

18. A bar of original length L equal to 30.4 m and with a cross-sectional area A of 32.8 cm² is subject to a tensile load F of 19 000 N. The elongation ΔL is 2.45 mm. Use the equation $E = \dfrac{FL}{A \, \Delta L}$ to find the modulus of elasticity E in newtons per square metre or pascals. Note: Pa = N/m²

19. Use the equation, $A = P(1 + r)^t$, to find the amount A obtained when an amount $P = \$7950$ is allowed to accumulate for $t = 4$ years at a compound interest rate of $r = 5\frac{1}{4}\%$.

20. A metal having a resistance R_1 at temperature t_1 will have a new resistance R at temperature t, where $R = R_1[1 + \alpha(t - t_1)]$, and α is the temperature coefficient of resistance. If the resistance of a copper coil is 775 Ω at 20°C, and the temperature coefficient of resistance is 0.003 93 at 20°C, find the resistance at 80°C.

 21. The formula for the pressure loss h in a pipe is

$$h = \frac{6270 \, fLQ^2}{D^5} \text{ ft}$$

Where f is the friction factor, L the length of the pipe in feet, Q the flow rate in cubic feet per second, and D the pipe diameter in inches. Calculate the pressure drop in a pipe with a length of 250 ft and a diameter of 3.25 in. In this pipe, $f = 0.019$ and the flow rate is 122 gal/min.

10-2 SOLVING SIMPLE EQUATIONS

The Solution of an Equation

A value of the variable that makes the sides of an equation equal is called a *solution* of that equation.

Example 8

The solution of the equation

$$x + 1 = 4$$

is $x = 3$. The value 3 is also called the *root* of the equation. We also say that it *satisfies* the equation, since it makes the equation true.

Checking

We can check a solution by substituting it back into the original equation.

Example 9

Is 12 a solution of the equation $3(x - 4) = 2x$?

Solution: Substituting 12 for x in the equation, we get

$$3(12 - 4) \stackrel{?}{=} 2(12)$$
$$3(8) \stackrel{?}{=} 24$$
$$24 = 24 \quad \text{checks}$$

So the value $x = 12$ checks, and 12 is a solution of the given equation.

COMMON ERROR Check your solution only in the original equation. Later versions may already contain errors.

First-Degree Equations

Recall that a first-degree term is one in which the variable is raised to the power 1. An equation in which all terms containing the variable are of first degree is called a *first-degree equation*. A first-degree equation is also called a *linear* equation.

Example 10

The equations

$$5 + 9x = 4 - 2x, \quad 5x - 5z = 7z, \quad \text{and} \quad 4\frac{x}{3} - 8 = 12 - 3x$$

are all of the first degree.

The equation $x^2 - 3x + 2 = 0$ is not of the first degree since it contains a second-degree term (x^2).

Solving an Equation

When we *solve* an equation, our object is to get the variable standing alone on one side of the equation. To accomplish this, *we perform the same mathematical operation on both sides of the equation.* That is, we may add the same quantity to both sides, subtract the same quantity from both sides, multiply both sides by the same quantity, and so forth. This will be made clear by examples.

Example 11

Solve the equation $5x = 4x + 7$.

Solution: Subtracting $4x$ from both sides, we obtain

$$5x - 4x = 7 + 4x - 4x$$

Combining like terms yields

$$x = 7$$

Check:
Substituting 7 for x in the original equation gives

$$5(7) \stackrel{?}{=} 4(7) + 7$$
$$35 \stackrel{?}{=} 28 + 7$$
$$35 = 35 \quad \text{checks}$$

Example 12

Solve the equation $2x - 3 = 7 - 3x$.

Solution: Adding $3x + 3$ to both sides,

$$\begin{array}{rcl} 2x - 3 &=& 7 - 3x \\ \underline{3x + 3} & & \underline{3 + 3x} \\ 5x &=& 10 \end{array}$$

Dividing both sides by 5, we obtain

$$x = 2$$

Check:
Substituting 2 for x in the original equation yields

$$2(2) - 3 \stackrel{?}{=} 7 - 3(2)$$
$$4 - 3 \stackrel{?}{=} 7 - 6$$
$$1 = 1 \quad \text{checks}$$

Symbols of Grouping

We showed earlier that parentheses, brackets, and braces are often used to group quantities within an expression. When the equation contains such symbols of grouping, remove them early in the solution.

Example 13

Solve the equation $2 - 4(x + 2) = 5 - 3(2x + 1)$.

Solution: Removing the parentheses, we obtain

$$2 - 4x - 8 = 5 - 6x - 3$$

Combining like terms,

$$-4x - 6 = -6x + 2$$

Adding $6x + 6$ to both sides,

$$\begin{array}{rcl} \underline{6x + 6} &=& \underline{6x + 6} \\ 2x &=& 8 \end{array}$$

Dividing by 2,

$$x = 4$$

Check:

$$2 - 4(4 + 2) \stackrel{?}{=} 5 - 3(8 + 1)$$
$$2 - 24 \stackrel{?}{=} 5 - 27$$
$$-22 = -22 \quad \text{checks}$$

Simple Fractional Equations

An equation that contains one or more fractions is called a *fractional equation*.

Example 14

Some fractional equations are

$$\frac{x}{5} = 3x \quad \text{and} \quad \frac{2x}{5} = \frac{2}{7}$$

10-2 Solving Simple Equations

Here we will solve some very simple fractional equations, and in Chapter 18 we will learn how to solve more complex types.

If an equation contains a single fraction, that fraction can be eliminated by multiplying both sides by the denominator of the fraction.

Example 15

Solve
$$\frac{x}{2} - 4 = 7$$

Solution: Multiplying both sides by 2,
$$2\left(\frac{x}{2} - 4\right) = 2(7)$$
$$x - 8 = 14$$

Adding 8 to both sides,
$$x = 8 + 14 = 22$$

Check:
$$\frac{22}{2} - 4 \stackrel{?}{=} 7$$
$$11 - 4 = 7 \quad \text{checks}$$

When an equation has two or more fractions, we can clear the fractions by multiplying both sides by the least common denominator.

Example 16

Solve
$$\frac{x}{2} - 3 = \frac{x}{5}$$

Solution: Multiplying by 10, the product of the denominators 2 and 5,
$$10\left(\frac{x}{2} - 3\right) = 10\left(\frac{x}{5}\right)$$
$$10\left(\frac{x}{2}\right) - 10(3) = 10\left(\frac{x}{5}\right)$$
$$5x - 30 = 2x$$

Rearranging,
$$5x - 2x = 30$$
$$3x = 30$$

Dividing both sides by 3
$$x = 10$$

Check:
$$\frac{10}{2} - 3 \stackrel{?}{=} \frac{10}{5}$$
$$5 - 3 \stackrel{?}{=} 2$$
$$2 = 2 \quad \text{checks}$$

Strategy

Remember that our object when solving an equation is to get the variable by itself on one side of the equation. To do this we use any valid mathematical operation, applied to both sides of the equation. It is not possible to give a procedure that will work for every equation, but the following tips should help. You may have to do these operations in a different order than that shown.

> **TIPS**
> 1. Eliminate any fractions by multiplying both sides by the least common denominator.
> 2. Remove any parentheses by performing the indicated multiplication.
> 3. Move all terms containing x to one side of the equation, and move all other terms to the other side.
> 4. Combine like terms on the same side of the equation at any stage of the solution.
> 5. Remove any coefficient of x by dividing both sides by that coefficient.
> 6. Check the answer by substituting it back into the original equation.

> **COMMON ERROR** The mathematical operations you perform must be done to both sides of the equation in order to preserve the equality.

Example 17

Solve

$$\frac{x-2}{3} = \frac{2x-4}{2}$$

Solution: We eliminate the fractions by multiplying by 6,

$$6\left(\frac{x-2}{3}\right) = 6\left(\frac{2x-4}{2}\right)$$

$$2(x-2) = 3(2x-4)$$

Removing parentheses,

$$2x - 4 = 6x - 12$$

We get all x terms on one side by adding 12 and subtracting $2x$ from both sides,

$$-4 + 12 = 6x - 2x$$

Combining like terms and switching sides,

$$4x = 8$$

Dividing by the coefficient of x,

$$x = 2$$

Check:

$$\frac{2-2}{3} \stackrel{?}{=} \frac{2(2)-4}{2}$$

$$0 = 0 \quad \text{checks}$$

SECTION 10-2 Exercises Solving Simple Equations

Solve and check each equation. Treat the constants in these equations as exact numbers. Leave your answers in fractional, rather than decimal, form.

1. $x + 9 = 16$
2. $3x - 2 = 10$
3. $30 + 5x = 20x$
4. $7x - 29 = 6$
5. $4x + 9 = 11x - 3x$
6. $20 - y = 13$
7. $x + 9 = 5$
8. $-6y - 4 = 2y$
9. $5x + 8 = 9x$
10. $4x = 3x - 6$
11. $10 - 5y = 1 - y$
12. $x + 3 = 10$
13. $x - 5 = 6$
14. $8y - 6 = 5y$
15. $7x - 3 = 5x + 1$
16. $25x - 5 = 3x + 6$
17. $y - 4 = 0$
18. $21x = -8 + 5x$
19. $4x - 5 = 10x - 2$
20. $7x + 15 = 8$
21. $6x + 4 = 3x + 19$
22. $16y - 10 = 10y + 14$
23. $\dfrac{1}{3a} = 7$
24. $\dfrac{y}{5} = 4$
25. $17 - 14x = 8 - 11x$
26. $3y - 15 + 4y = 6$
27. $5x - 10 + 13 = 18$
28. $44 - 11x + 4x = -5$
29. $49 - 5y = 3y - 7$
30. $5 + y = 1 + 15y$
31. $4x - 111 = 21 + 14x$
32. $47x - 84 = 2x + 6$
33. $16 = x + 3$
34. $15x - 9 = 7x - 5$
35. $3 - (2b + 4) = 5b - 2$
36. $x + 16 = 1 - 4(3x + 1)$
37. $6x - 25 = 4x + 7$
38. $80 - 7x = 5x - 16$
39. $3(y - 5) + 2(3 + y) = 21$
40. $5(3x - 13) = 10$
41. $4(y - 5) = 2(y - 10)$
42. $2x + 1 = -2(2x + 5)$
43. $3(15 - x) = 3(5 + x)$
44. $4(3x + 20) = 2(x - 5)$
45. $3(2x + 1) = 7$
46. $3(x - 5) = 4(x - 1)$
47. $5x + 6 = 6(x - 3) + 2$
48. $6(2x - 3) + 2(6x + 2) = 5(x + 1)$
49. $19y - 60 = 4y + 15$
50. $5(2x - 8) = 2(10 - 5x)$
51. $5 + 3(x - 2) = 58$
52. $3(3x + 13) = 5(x - 1)$
53. $4(x - 5) - 2(6x + 3) = 22$
54. $5(y - 1) + 4(3y + 3) = 3(4y - 6)$
55. $2(4x - 3) = 3(5x + 2)$
56. $3(4x - 2) = 9x$
57. $5a = -2(13 - 9a)$
58. $5 - 3(x - 2) = 4(3x - 1)$
59. $6x + 5 = 7(x - 3)$
60. $3(7 - 2x) + 2 = 3(4x - 1)$
61. $\dfrac{x}{4} - 3 = 11$
62. $5 + \dfrac{7x}{3} = \dfrac{x}{2}$
63. $\dfrac{5x}{4} = 2x - 3$
64. $\dfrac{y}{5} - \dfrac{y}{2} + \dfrac{y}{4} = 5$
65. $\dfrac{2x - 4}{7} = \dfrac{2 - 2x}{4}$

10-3 SIMPLE EQUATIONS WITH APPROXIMATE COEFFICIENTS AND LITERALS

Equations with Approximate Numbers

In technical work we usually must solve equations that have approximate numbers, rather than coefficients and constants that are integers. The method is no different than before. Just remember to round your answer to the proper number of significant digits.

Example 18

Solve for x: $4.23x + 1.33 = 2.16x + 5.82$.

Solution: Collecting terms,

$$4.23x - 2.16x = 5.82 - 1.33$$
$$2.07x = 4.49$$

Dividing,

$$x = \frac{4.49}{2.07} = 2.17$$

rounded to three significant digits.

Example 19

Solve for x: $2.93(x + 4.28) = 24.2 - 5.82x$.

Solution: Removing parentheses gives

$$2.93x + 12.5 = 24.2 - 5.82x$$

Collecting terms,

$$2.93x + 5.82x = 24.2 - 12.5$$
$$8.75x = 11.7$$

Dividing,

$$x = \frac{11.7}{8.75} = 1.34$$

rounded to three significant digits.

Simple Literal Equations

An equation in which some or all of the coefficients are represented by letters is called a *literal* equation.

Example 20

The equation

$$5x + 3c = d$$

is a literal equation.

We often want to isolate a particular literal quantity on one side of the equal sign. When we do this we say we are *solving* for that quantity. To solve for a literal quantity, we use the same methods as before.

Example 21

Solve the equation of Example 20 for x.

Solution: Subtracting $3c$ from both sides,

$$5x = d - 3c$$

Dividing by 5 gives

$$x = \frac{d - 3c}{5}$$

We will solve more complex literal equations in Chapter 18.

SECTION 10-3 Exercises Simple Equations with Approximate Coefficients and Literals

Approximate Coefficients

Solve for x. Round your answer to the proper number of significant digits.

1. $8.27x = 4.82$
2. $24.8x - 28.4 = 0$
3. $2.84x + 2.83 = 83.7$
4. $3.82 = 29.3 + 3.28x$
5. $3.82x - 3.28 = 5.29x + 5.82$
6. $382x + 827 = 625 - 846x$
7. $2.94(x + 8.27) = 3.27$
8. $9.38(5.82 + x) = 23.8$
9. $2.34(x - 4.27) = 5.27(x + 3.82)$
10. $92.1(x - 2.34) = 82.7(x - 2.83)$
11. $2.83x + 12.1(x - 6.54) = 2.47(x - 12.43)$
12. $1.58(8.22 + x) + 42.8 = 6.44(x - 6.17) - 1.87(x + 9.22)$

Literal Equations

Solve for x.

13. $dx + 5 = 8$
14. $8 + bx = d$
15. $ax + b = c + d$
16. $dx - 5 = cx - 2$
17. $a(x + 1) = b(x - 2)$
18. $3(bx + c) = 5$

Chapter 10 REVIEW EXERCISES

Solve and check each equation. Treat the constants in these equations as exact numbers. Leave your answers in fractional, rather than decimal, form.

1. $x + 3 = 15$
2. $2x - 6 = 12$
3. $10 + 2x = 30x$
4. $5x - 19 = 7$
5. $3x + 9 = 12x$
6. $25 - y = 17$
7. $x + 2 = 51$
8. $-8y - 2 = 2y$
9. $3x + 8 = 5x$

10. $2x = 3x - 7$
11. $12 - 5y = 6$
12. $x + 8 = 16$
13. $x - 2 = 61$
14. $3y - 5 = 5y$
15. $2x - 6 = 3x + 2$
16. $15x - 3 = 5x$
17. $y - 3 = 8$
18. $11x = -8 + 2x$
19. $5x - 5 = 16x$
20. $4x + 18 = 83$
21. $3x + 8 = 4x + 12$
22. $2x + 2 = 8(x - 4)$
23. $6(2 - 2x) + 6 = 7(3x - 1)$
24. $\dfrac{x}{2} - 6 = 21$
25. $2 + \dfrac{3x}{4} = \dfrac{8x}{3}$
26. $\dfrac{2x}{3} = 5x - 3$
27. $\dfrac{y}{3} - \dfrac{y}{2} + \dfrac{y}{4} = 7$

Substitute the given integers into each equation. Do not round your answer.

28. $y = 4x + 7$ ($x = 8$)
29. $y = 2m^2 + 6m - 2$ ($m = -7$)
30. $y = 2a^3 - 7x$ ($a = -3, x = -7$)
31. $y = 6x^3 + 2x^2 - x - 1$ ($x = -3$)
32. $y = 7c + 2d^2 - 8e$ ($c = 2, d = -1, e = 6$)

Substitute the given approximate numbers into each equation. Round your answer to the proper number of digits.

33. $y = 4 - 8x$ ($x = -1.34$)
34. $y = -2d^3 - 5e^2$ ($d = -2.335, e = 3.195$)
35. $y = a^3 + a + 2$ ($a = 4.61$)

Use the formulas given in Section 10-1 Exercises for problems 36–44.

36. Find the amount to which $7420 will accumulate in 6 years at a simple interest rate of 3.8%.
37. A ball is thrown downward with a speed of 6.7 m/s. Find the distance the ball has traveled after 5.6 seconds (i.e., find the displacement of the ball at 5.6 s).
38. Convert 26°F to degrees Celsius.
39. Find the torque delivered by a motor rated at 1.12 kW rotating at 2740 rev/min.
40. Find the torque delivered by a motor rated at 1.16 kW rotating at 225 rad/s.
41. A bar measuring 51.4 m long having a cross-sectional area of 24.5 cm² is subject to a tensile load of 21 500 N. The elongation is 1.55 mm. Find the modulus of elasticity in newtons per square metre.
42. Find the amount S obtained when an amount $P = \$3354$ is allowed to accumulate for $t = 7$ years at a compound interest rate of $n = 6.75\%$.
43. The resistance of a copper coil is 827 Ω at 18°C and the temperature coefficient of resistance is 0.004 11 at 18°C. Find the resistance at 75°C.

44. Calculate the pressure drop in a pipe with a diameter of 2.84 in., and a length of 124 ft. In this pipe, $f = 0.022$ and the flow rate is 184 gal/min.

Solve for x.

45. $8.24(x + 1.17) = 6.67$
46. $2.48(3.12 + x) = 53.2$
47. $4.24(x - 6.77) = 2.67(x + 7.72)$
48. $12.1(x - 5.34) = 22.7(x - 5.83)$
49. $5.23x + 62.1(x - 2.14) = 7.37(x - 22.13)$
50. $4.18(3.22 + x) + 21.8 = 3.84(x - 2.47) - 4.17(x + 3.62)$
51. $cx + 5 = d$
52. $d + ax = c$
53. $mx + b = a + b$
54. $cx - d = 2x - a$

WRITING

55. You probably have some idea what field you later plan to enter: business, engineering, and so forth. Look at a book from that field and find at least one formula commonly used. Write it out, describe what it is used for, and explain all the quantities it contains, and their units.

Solving Word Problems

Outcomes

When you have completed this chapter, you will be able to

1. Solve word problems by translating them into mathematical expressions.
2. Estimate results to a word problem.
3. Solve uniform motion problems.
4. Solve basic financial problems.
5. Solve mixture problems.

In this chapter we are asked to read a short problem statement (the "word problem") and to decide what is being asked for. We then write an equation which enables us, by solving that equation, to find the quantity we seek.

Keep in mind that we are not solving these problems for their own sake. We don't really care how long it takes runner A to reach city Q. What we are trying to do is to help you to read a technical statement, to extract the important information, and to use simple mathematics to find a missing quantity. These skills will also help you to deal with the mountains of technical material in written form with which we are all faced: instruction manuals, textbooks, specifications, contracts, insurance policies, building codes, handbooks, and so forth.

11-1   SOLVING WORD PROBLEMS

It is not possible to give a step-by-step procedure that will enable someone to solve *any* word problem, but we can give some good tips that are almost always helpful. We first give some general tips on how to approach any word problem and later show how to set up specific kinds of problems.

Study the Problem

Read the problem carefully, more than once if necessary. Be sure you understand the words used. A technical problem will often have unfamiliar terms. Look up their meanings in a dictionary, handbook, or textbook.

Example 1

Word problems may contain words or phrases like

      simply supported beam   angular velocity   groundspeed
      concentrated load   centre of gravity   kinetic energy

It goes without saying that you must know the meanings of these words before you can solve a problem in which they appear.

Picture the Problem

Try to visualize the situation described in the problem. Form a picture in your mind. Draw a diagram showing as much of the given information as possible.

Example 2

Suppose that part of a word problem states that "a car goes from city P to city Q, which is 378 kilometres away, at an average rate of 90 kilometres per hour."

At this point you should see not just dead words on paper, but a car moving along a road (perhaps a red sports car cruising down the highway, stereo playing, you at the wheel). Make a sketch something like Fig. 11-1, showing the road as a line with cities P and Q at either end, 378 kilometres apart. Show the car with an arrow giving the direction of travel and the speed.

■ FIGURE 11-1

Of course, we have not solved the problem yet (it isn't even fully stated), but at least we have a good grip on the information given so far.

Identify the Unknown(s)

Look for a sentence that asks a question (it may end in a question mark). Find a sentence that starts with *Find* . . . or *Calculate* . . . or *What is* . . . or *How much* . . . or similar phrases. Such a sentence will usually contain the unknown quantity, the thing we want to solve for. Then label this unknown with a statement such as:

Let x = cost of each part, dollars

Be sure to include units of measure when defining the unknown.

Example 3

Suppose one sentence in a word problem says, "Find the speed of train B." You may be tempted to label the unknown in one of the following ways:

Let x = train B	This is too vague. Are we talking about the speed of train B or the time it takes for train B or the distance traveled by train B or something else?
Let x = speed of the train	Which train?
Let x = speed of train B	Units are missing.
Let x = speed of train B, km/h	Good

Estimate the Answer

A valuable skill to develop is the ability to *estimate* the answer before starting the problem. Then you will be able to see if the answer you get is reasonable or not.

One way to estimate an answer is to *make simplifying assumptions*.

Example 4

Suppose we want to find the volume of the spindle in Fig. 11-2. To get an estimate, we can find the volume of a cylinder of the same length as the spindle, but with a diameter somewhere between the largest and smallest diameters of the spindle, say, 17 mm. This will not give us the exact answer, of course, but will put us in the ballpark. If our final answer is very different than our estimate, we would check our work.

FIGURE 11-2
A spindle.

Another technique is to find two numbers that *bracket* the answer.

Example 5

We can *bracket* the volume of the spindle by noting that it must be less than that of a cylinder 19.00 mm in diameter and 35.00 mm in length, but greater than that of a cylinder 15.00 mm in diameter and 35.00 mm in length.

Estimation is not always easy, and it takes some time to develop this skill, but it is well worth it. We will show estimation in most of the word problems in this chapter.

Write and Solve an Equation

We must now write an equation to relate the given quantities to the unknown quantity. Sometimes the equation will be a formula from mathematics, such as the relationship between the volume of a sphere and its radius. At other times you will need a formula from technology, such as the one relating the strength of a steel rod to its diameter. The relationships you will need for the problems in this book can be found in Appendix A, "Summary of Facts and Formulas." Problems that use some of these formulas are treated later in this chapter. For now, we will solve simple number problems, in which the formula is given verbally right in the problem statement.

When you have an answer, label it with the same words you used in defining the unknown.

Don't Give Up

Be persistent. If you don't "get it" at first, keep trying. Take a break and return to the problem later. Read the problem again. Perhaps try a different approach. You know that these "school problems" you encounter in a mathematics class all have solutions, so keep at it.

Number Problems

These problems may not have much practical value, but they have great value in giving us practice in setting up and solving word problems. In such problems, mathematical operations are often indicated with words such as

> *. . . the difference of . . .*
> *. . . is equal to . . .*
> *. . . the quotient of . . .*

Locate any such words and replace them with mathematical symbols.

Example 6

If two times a number is increased by seven, the result is equal to five less than four times the number. Find the number.

Solution: Let x = the number. Then,

$$2x = \text{two times the number}$$

From the problem statement,

$$2x + 7 = 4x - 5$$

Solving for x,

$$7 + 5 = 4x - 2x$$
$$2x = 12$$
$$x = 6$$

Check Your Answer

First see if your answer is reasonable. Did you come up with a bridge cable 1 mm in diameter or a person walking at a rate of 89 km/h? Does your answer agree, within reason, with your estimate? Then you should check your answer in the original problem *statement*. Do *not* check the answer in your equation, which may already contain an error.

Example 7

Let us check the answer from Example 6. Using the words of the problem statement, we get,

Twice six increased by seven is: $2(6) + 7 = 12 + 7 = 19$

Five less than four times six is: $4(6) - 5 = 24 - 5 = 19$ checks

> **COMMON ERROR** Checking an answer by substituting into the equation is *not good enough*. The equation may already contain an error. Check your answer in the problem statement.

Define Other Unknowns

If there is a second unknown, you will often be able to define it in terms of the original unknown, rather than to introduce a new symbol.

Example 8

Find two numbers whose sum is 50 and whose difference is 30.

Solution: This problem has two unknowns. We could choose to label each with a separate symbol,

$$\text{Let } x = \text{larger number}$$

and

$$y = \text{smaller number}$$

but a better way would be to label the second unknown in terms of the first:

$$\text{Let } x = \text{larger number}$$

then,

$$50 - x = \text{smaller number}$$

Solving the problem, we get the difference by subtracting the smaller number from the larger.

$$x - (50 - x)$$

We are told that this equals 30, so our equation is,

$$x - (50 - x) = 30$$

Removing parentheses gives

$$x - 50 + x = 30$$

Collecting like terms we get,

$$2x - 50 = 30$$

Adding 50 to both sides gives

$$2x = 80$$

$$x = 40 = \text{the larger number}$$

$$50 - x = 10 = \text{the smaller number}$$

The usual steps followed in solving a word problem are summarized here.

1. **Study the Problem.** Look up unfamiliar words. Make a sketch. Try to visualize the situation in your mind.
2. **Identify the Unknown(s).** Give it a symbol, such as *let x = ...* If there is more than one unknown, try to label the others in terms of the first. Include units.
3. **Estimate the Answer.** Make simplifying assumptions or try to bracket the answer.
4. **Write and Solve an Equation.** Look for a relationship between the unknown and the known quantities that will lead to an equation. Write and solve that equation for the unknown. Include units in your answer. If there is a second unknown be sure to find it also.
5. **Check Your Answer.** See if the answer looks reasonable. See if it agrees with your estimate. Be sure to do a numerical check in the problem statement itself.

SECTION 11-1 Exercises Solving Word Problems

Identify an unknown and rewrite each expression as an algebraic expression.
1. Ten more than three times a number.
2. Two numbers whose sum is 17.
3. Two numbers whose difference is 42.
4. The amounts of antifreeze and of water in 4 gallons of an antifreeze-water solution.
5. A fraction whose denominator is 4 more than 6 times its numerator.
6. The angles in a triangle, if one angle is three times the other.
7. The number of litres of antifreeze in a radiator containing x litres of a mixture that is 11% antifreeze.
8. The distance traveled in x hours by a car going 78 km/h.
9. If x equals the length of a rectangle, write an expression for the width of the rectangle if (a) the perimeter is 64; (b) the area is 320.

Solve each problem for the required quantity.
10. Four less than 6 times a number is 33. Find the number.
11. Find a number such that the sum of 16 and that number is 3 times that number.
12. Six less than 5 times a number is 19. Find the number.
13. The sum of 45 and some number equals 6 times that number. Find it.
14. Ten less than three times a certain number is 29. Find the number.
15. When three is added to seven times a number, the result is equal to the same as when we subtract seven from nine times that number. Find that number.
16. The denominator of a fraction is 2 less than 4 times its numerator, and the reduced value of the fraction is $\frac{4}{15}$. Find the numerator.

11-2 UNIFORM MOTION PROBLEMS

Number problems were good for a warm-up, but now we will do some types of problems that have more application to technology. We will start with uniform motion problems. The ideas here should be familiar; you already know that if you walk at a rate of 5 kilometres per hour for 2 hours you will travel 10 kilometres. The problems in this section are based on that one simple idea.

Uniform Motion

Motion is called *uniform* when the speed does not change. The distance traveled at constant speed is related to the speed (the *rate* of travel) and the elapsed time by

$$\text{rate} \times \text{time} = \text{Distance}$$

or

$$RT = D$$

Be careful not to use this formula for anything but uniform motion.

Example 9

A plane flies for 2.45 h at 325 km/h. How far does it travel?

Solution: From the equation above,

$$D = 325\,(2.45) = 796 \text{ km}$$

We now do a typical motion problem, in which we will organize our information in table form, and use the equation (rate × time = distance) to help in the solution.

Example 10

A truck leaves a city traveling at a speed of 72.5 km/h, and a car leaves the same city 1.25 hour later to overtake the truck. If the car's speed is 108.0 km/h, how long will it take for the car to overtake the truck?

Estimate
The truck has a head start of 1.25 hour and goes at 72.5 km/h, so it is about 91 km ahead, when the car starts. But the car goes about 36 km/h faster than the truck, so each hour the truck's lead is cut by 36 km. Thus it would take between 2 and 3 hours to reduce the 91 km lead to zero.

Solution: Let us follow the list of steps suggested for solving any word problem.

Make a Sketch
A simple sketch, Fig. 11-3, shows the city, the road, the meeting point, and the rates of the truck and the car.

Identify the Given Information
Let us write down the formula for uniform motion,

$$\text{rate} \times \text{time} = \text{Distance}$$

■ FIGURE 11-3

Truck, 72.5 km/h

City ———————————————— Meeting point

Car, 108 km/h
(1.25 h later)

Under each of the three quantities we write the given value of that quantity, both for the truck and for the car. Reading the problem statement, we see that the speed of the truck and of the car are given.

	Rate (km/h)	×	Time (h)	=	Distance (km)
Truck	72.5				
Car	108.0				

Define the Unknown
What are we looking for? We seek the time for the car to overtake the truck, so we write

Let t = time traveled by car (hours)

and enter this information into our table

	Rate (km/h)	×	Time (h)	=	Distance (km)
Truck	72.5				
Car	108.0		t		

Define the Other Unknowns
We have three empty boxes in our table. First, the time for the truck can be written in terms of t by noting that the truck travels for 1.25 hour longer than the car.

	Rate (km/h)	×	Time (h)	=	Distance (km)
Truck	72.5		$t + 1.25$		
Car	108.0		t		

We complete the table by noting that distance = rate × time.

	Rate (km/h)	×	Time (h)	=	Distance (km)
Truck	72.5		$t + 1.25$		$72.5(t + 1.25)$
Car	108.0		t		$108.0t$

Write and Solve an Equation
An equation says that something is equal to something else. What quantities are equal in this problem? Since, at the instant that the car overtakes the truck, they have both gone the same distance, we set the distance gone by the truck equal to the distance gone by the car.

$$108.0t = 72.5(t + 1.25)$$

Solve the Equation
Removing the parentheses,

$$108.0t = 72.5t + 72.5(1.25)$$
$$108.0t - 72.5t = 90.63$$
$$35.5t = 90.63$$
$$t = 2.55 \text{ hours} = \text{time for car}$$

Check Your Answer

We first note that the answer (2.55 h for the car) agrees with our estimate, which was between 2 and 3 hours. Next, for a more accurate check, let's find the distance traveled by the truck and by the car.

$$\text{Time for truck} = t + 1.25 = 3.80 \text{ h}$$

$$\text{Truck distance} = (72.5 \text{ km/h})(3.80 \text{ h}) = 275.5 \text{ km}$$

$$\text{Car distance} = (108.0 \text{ km/h})(2.55 \text{ h}) = 275.4 \text{ km}$$

These distances agree to three significant digits, the precision to which we are working in this problem. Since most of our original numbers are known only to three significant digits, we cannot expect our answers to check to more than three significant digits.

In the preceding example we got our equation from the fact that both vehicles traveled the same distance. This, of course, will not be true for *every* motion problem, and you may have to look for other relationships in order to write an equation.

SECTION 11-2 Exercises Uniform Motion Problems

1. Two planes start from the same city at the same time. One travels at 403 km/h and the other at 426 km/h, in the opposite direction. How long will it take for them to be 2800 km apart?

2. The pointer of a certain meter can travel to the right at the rate of 10.0 cm/s. What must be the return rate if the total time for the pointer to traverse the full 12.0 cm scale and return to zero must not exceed 2.0 seconds?

3. A train travels from P to Q at a rate of 22.5 km/h. After it has been gone 2.75 hours, an express train leaves P for Q traveling at 85.5 km/h, and reaches Q 1.50 hours ahead of the first train. Find the distance from P to Q, and the time taken by the express train.

4. Two submarines start from the same spot and travel in opposite directions, one at 115 km/day and the other at 182 km/day. How long will it take for the submarines to be 1470 km apart?

5. A bus travels 87.5 km to another town at a speed of 72.0 km/h. What must be its return rate if the total time for the round trip is to be 2.50 hours?

6. Spacecraft A is over Houston at noon on a certain day and traveling at a rate of 275 km/h. Spacecraft B, attempting to overtake and dock with A, is over Houston at 1:15 P.M. and is traveling in the same direction as A, at 444 km/h. At what time will B overtake A? At what distance from Houston?

7. A freight train leaves A for B, 175 miles distance and travels at the rate of 31.5 mi/h. After 1.50 hours, a train leaves B for A, traveling at 21.5 mi/h. How many miles from B will they meet?

8. An oil slick from a runaway offshore oil well is advancing toward a beach 354 miles away at the rate of 10.5 mi/day. Two days after the spill, cleanup ships leave the beach and steam toward the slick at a rate of 44.6 mi/day. At what distance from the beach will they reach the slick?

9. A certain shaper has a forward cutting speed of 115 ft/min and a stroke of 10.5 in. It is observed to make 429 cuts (and returns) in 4.0 minutes. What is the return speed?

11-3   FINANCIAL PROBLEMS

We all deal with money, so word problems involving money are probably the easiest for us to understand. We will need to use percentage often here, so you may want to review Chapter 5 quickly before starting these problems.

Example 11

A roofer needs $1700 worth of materials, excluding waste, to roof a certain house. Of his total purchase, 8% will end up as waste. Find the total cost of materials rounded up to the nearest ten dollars, if he buys enough to allow for waste. (Normal business practice is to round up cost estimates).

Estimate
We know that he must buy more than $1700 worth of materials, but not too much more. Let's guess at $2000 worth. The waste would then be 8% of $2000 or $160 worth, leaving $1840 actually used. This is *more* than is needed. So our answer must be between $1700 and $2000.

Solution: We will let x stand for the quantity we seek, so

$$\text{Let } x = \text{cost of roofing material, dollars}$$

From that amount we subtract the 8.0% of that amount for waste, and the result is the $1700 needed. That gives us our equation

$$x - 0.080x = 1700$$

Solving for x,

$$0.920x = 1700$$

$$x = \frac{1700}{0.092} = \$1847.83 = \$1850$$

This number lies within the values we bracketed in our estimate.

Example 12

A federal income tax table states that if your taxable income is over $55 100 but not over $115 000, your tax is $12 470 plus 31% of the amount over $55 100. What was your taxable income if your tax bill was $26 834? Work to the nearest dollar.

Estimate
For an income of $55 100, the tax is $12 470. For an income of $115 000, the tax is $12 470 + 0.31(115 000 − 55 100) = $31 039. Our tax ($26 834) lies between these values, but closer to the higher value than the lower. To get closer, let's guess at an income of $100 000, with a tax of $12 470 + 0.31(100 000 − 55 100), or $26 389, a bit lower than our actual tax. Thus, our income must be a little over $100 000 but not more than $115 000.

Solution:

$$\text{Let } x = \text{taxable income}$$

Your income then exceeds $55 100 by $(x - 55\,100)$, so your tax is

$$\text{Tax} = 12\,470 + 31\% \text{ of } (x - 55\,100)$$

But your tax bill was $26 834 so our equation is

$$12\,470 + 0.31(x - 55\,100) = 26\,834$$

Solving for x,

$$0.31(x - 55\,100) = 26\,834 - 12\,470$$
$$0.31x - 0.31(55\,100) = 14\,364$$
$$x - 55\,100 = 46\,335$$
$$x = 46\,335 + 55\,100 = \$101\,435$$

This agrees well with our estimate.

Example 13

A person had $24 924.85 in a bank, part in a savings account that earned an interest of 3.55% per year, and the remainder in a money market account earning 5.75% per year. (Assume that both are *simple interest;* that is, the interest is computed only once per year.) The interest earned from both accounts combined was $1125.84. How much was in each account?

Estimate

Let's assume that *half the money* (about $12 500) was put into each account. The interest at 3.55% would be $444, and at 5.75% would be $719, for a total of $1163. This is more than was actually earned. We conclude that *more than half* was invested at the lower rate.

Solution: Here we are looking for *two* things, the amount in each of two accounts. Let us set x equal to one of them. So,

Let x = amount deposited in the savings account

Then

$24\,924.85 - x$ = amount deposited in the money market account

The interest earned on each account is then

3.55% of x = interest earned by savings account

and

5.75% of $(24\,924.85 - x)$ = interest earned by money market account

Setting the sum of the two earnings equal to $1125.84 gives us our equation.

$$0.0355x + 0.0575(24\,924.85 - x) = 1125.84$$

Solving for x,

$$0.0355x + 1433.18 - 0.0575x = 1125.84$$
$$1433.18 - 1125.84 = 0.0575x - 0.0355x$$
$$0.0220x = 307.34$$
$$x = \$13\,970.00$$

That is the amount in the savings account. In the money market account we have

$$\$24\,924.85 - x = \$10\,954.85$$

We see that more than half the money was put into the lower-interest savings account, as we predicted in our estimate.

SECTION 11-3 Exercises Financial Problems

1. How much must a person earn to have $45 824 after paying 28% in taxes?
2. A carpenter estimates that a certain deck needs $4285 worth of lumber, if there were no waste. How much should she buy, if she estimates the waste at 7%?
3. The labour costs for a certain brick wall were $1118 per day for 10 masons and helpers. If a mason earns $125/day and a helper $92/day, how many masons were on the job?
4. According to a tax table, for your filing status, if your taxable income is over $38 000 but not over $91 850, your tax is $5700 plus 28% of the amount over $38 100. If your tax bill was $8126, what was your taxable income? Work to the nearest dollar.
5. A company has $173 924 in bonds and from them earns $13 824 in simple interest annually. Part of the money is invested at 6.75% and the remainder at 8.24%. How much is invested at each rate?
6. A company had $528 374 invested, part in stocks that earned 9.45% per year, and the remainder in bonds that earned 6.12% per year, both simple interest. The amount earned from both investments combined was $42 852. How much was in each investment?
7. A student sold a computer and printer for a total of $995, getting $1\frac{1}{2}$ times as much for the printer as for the computer. What was the price of each?

11-4 MIXTURE PROBLEMS

Before starting a typical mixture problem, let's look at some basic ideas about mixtures.

Basic Relationships

If a mixture contains two ingredients, A and B, the total amount of the mixture is equal to the sum of the amounts of the ingredients:

Total amount of mixture = amount of A + amount of B

If there are more than two ingredients, the total amount, of course, equals the sum of the individual amounts. If more of a particular ingredient, say A, is added to the mixture and/or some is removed, then,

Final amount of A = initial amount of A + amount of A added − amount of A removed

--- Example 14 ---

A soldering machine for circuit boards contains 124 kg of solder, half lead and half zinc. From this, 22 kg of solder is removed, then 35 kg of lead is added. How much lead is contained in the final mixture?

Solution:

Initial mass of lead = 0.5(124) = 62 kg

Amount of lead removed = 0.5(22) = 11 kg

Amount of lead added = 35 kg

By the equation above,

Final amount of lead = 62 − 11 + 35 = 86 kg

Percent Concentration

Recall from Sec. 5-4 that the percent concentration of each ingredient is

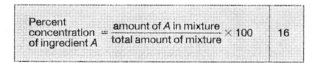

Example 15
The final percent concentration of lead in the preceding example is, by Eq. 16

$$\text{Percent lead} = \frac{86}{137} \times 100 = 63\%$$

Example 16
A tank contains 88.6 litres of gasohol that is 12.2% alcohol. How many litres of alcohol are in the tank?

Solution: Again using Eq. 16, but in the form,

Amount of A in mixture = percent concentration of A × total amount of mixture

$$\text{Litres of alcohol in tank} = 0.122 \text{ of } 88.6 \text{ litres}$$
$$= 10.8 \text{ litres}$$

Example 17
Find the percent concentration of alcohol in the tank of Example 16 if 2.6 litres of pure alcohol are added.

Solution: The amount of alcohol in the tank increases by 2.6 litres, so

$$\text{Total amount of alcohol} = 10.8 + 2.6 = 13.4 \text{ litres}$$

The total amount of mixture also increases by 2.6 litres, so

$$\text{Total amount of mixture} = 88.6 + 2.6 = 91.2 \text{ litres}$$

Then by Eq. 16,

$$\text{Percent concentration of alcohol} = \frac{13.4}{91.2} \times 100 = 14.7\%$$

Two Mixtures

When two mixtures are combined into a third mixture, the amount of any ingredient A in the final mixture is

Final amount of A = amount of A in first mixture + amount of A in second mixture

Example 18
A vat of 125 litres of gasohol containing 11.2% alcohol is mixed with 225 litres of gasohol containing 7.75% alcohol. How many litres of alcohol are in the final mixture?

Solution: By the equation above,

$$\text{Final amount of alcohol} = 11.2\% \text{ of } 125 + 7.75\% \text{ of } 225$$
$$= 0.112(125) + 0.0775(225)$$
$$= 31.4 \text{ litres}$$

Example 19

How much steel containing 5.25% nickel must be combined with another steel containing 2.84% nickel to make 3.25 tonnes of steel containing 4.15% nickel?

Estimate

The final steel needs 4.15% of 3.25 tonnes of nickel, about 0.135 tonnes. If we assume that *equal amounts* of each steel were used, the amount of nickel would be 5.25% of 1.625 tonnes, or 0.085 tonnes, from the first alloy, and 2.84% of 1.625 tonnes, or 0.046 tonnes, from the second alloy. This gives a total of 0.131 tonnes of nickel in the final steel. This is not enough (we need 0.135 tonnes). Thus, *more than half* of the final alloy must come from the higher-nickel alloy, or between 1.625 tonnes and 3.25 tonnes.

Solution: Let x = tonnes of 5.25% steel needed. Fig. 11-4 shows the three alloys, and the amount of nickel in each, with the nickel drawn as if it were separated from the rest of the steel. By the equation above, the mass of the 2.84% steel is $3.25 - x$. The mass of nickel it contains is, by Eq. 16,

$$0.0284(3.25 - x)$$

The mass of nickel in x tonnes of 5.25% steel is

$$0.0525x$$

The sum of these must give the mass of nickel in the final mixture:

FIGURE 11-4

Final amount of A =
$$\text{amount of } A \text{ in first mixture and amount of } A \text{ in second mixture}$$
$$0.0525x + 0.0284(3.25 - x) = 0.0415(3.25)$$

Clearing parentheses,
$$0.0525x + 0.0923 - 0.0284x = 0.1349$$
$$0.0241x = 0.0426$$
$$x = 1.77 \text{ tonnes of } 5.25\% \text{ steel}$$
$$3.25 - x = 1.48 \text{ tonnes of } 2.84\% \text{ steel}$$

Check

First we see that more than half the final alloy comes from the higher-nickel steel, as predicted in our estimate. Now let's see if the final mixture has the proper percentage of nickel.

$$\text{Final tonnes of nickel} = 0.0525(1.77) + 0.0284(1.48)$$
$$= 0.135 \text{ tonnes}$$
$$\text{Percent nickel} = \frac{0.135}{3.25} \times 100 = 0.0415$$

or 4.15%, as required.

Alternate Solution: This problem can also be set up in table form, as follows.

	Percent Nickel	×	Amount (tons)	=	Amount of Nickel (tons)
Steel with 5.25% Ni	5.25%		x		$0.0525x$
Steel with 2.84% Ni	2.84%		$3.25 - x$		$0.0284(3.25 - x)$
Final Steel	4.15%		3.25		$0.0415(3.25)$

We then equate the sum of the amounts of nickel in the original steels with the amount in the final steel, getting

$$0.0525x + 0.0284(3.25 - x) = 0.0415(3.25)$$

as before.

COMMON ERROR If you wind up with an equation that looks like this,

$$(\ )t \text{ nickel} + (\ )t \text{ iron} = (\ )t \text{ iron}$$

you know something must be wrong. When using the equation

Final amount of A = amount of A in first mixture + amount of A in second mixture

all the terms must be for the same ingredient.

SECTION 11-4 Exercises Mixture Problems

1. From 55.5 litres of wine containing 12.0% alcohol, 10.0 litres is removed. Then 5.50 litres of alcohol is added. How many litres of alcohol are contained in the final mixture?

2. A casting that has a mass of 875 kg is made of a brass alloy that contains 87.5% copper. How many kilograms of copper are in the casting?

3. Find the percent concentration of copper in the casting of the preceding problem, if it is melted with 125 kg of pure copper.

4. There are 235 litres of fuel for a two-cycle engine that contains 3.15% oil. To this is added 186 litres of fuel containing 5.05% oil. How many litres of oil are in the final mixture?

5. How much steel containing 1.15% chromium must be combined with another steel containing 1.50% chromium to make 8.00 tonnes of steel containing 1.25% chromium?

6. Two gasohol mixtures are available: one with 6.35% alcohol and the other with 11.28% alcohol. How many litres of the 6.35% mixture must be added to 25.0 litres of the other mixture to make a final mixture containing 8.00% alcohol?

7. How many tonnes of tin must be added to 2.75 tonnes of bronze to raise the percentage of tin from 10.5% to 16.0%?

8. How many pounds of nickel silver containing 12.5% zinc must be melted with 248 lb of nickel silver containing 16.4% zinc to make a new alloy containing 15.0% zinc?

9. How many pounds of candy costing $3.55 per pound must be mixed with 10 pounds of candy costing $5.25 per pound to make a mixture that costs $4.00 per pound?

10. How many quarts of olive oil costing $4.86 per quart must be mixed with 136 quarts of corn oil costing $2.75 per quart to make a blend costing $3.00 per quart?

■ Chapter 11 REVIEW EXERCISES

1. A company has $223 812 invested in two separate accounts and earns $14 824 in simple interest annually from these investments. Part of the money is invested at 5.94% and the remainder at 8.56%. How much is invested at each rate?

2. An airplane left an airport at 1:00 P.M. traveling at a rate of 387 km/h. After traveling for some distance, one engine failed, and the airplane returned to the field at a reduced rate of 184 km/h, arriving at 3:15 P.M. At what distance from the field did it turn around?

3. From 12.5 litres of coolant containing 13.4% antifreeze, 5.50 litres is removed. If 5.50 litres of antifreeze is then added, how much antifreeze will be contained in the final mixture?

4. A person had $125 815 invested, part in a mutual fund that earned 10.25% per year and the remainder in a bank account that earned 4.25% per year, both simple interest. The amount earned from both investments combined was $11 782. How much was in each investment?

5. How much must a person earn to have $50 000 after paying 27% in taxes?

6. In a group of 105 persons, composed of men, women, and children, there are four times as many children as men, and twice as many women as men. How many are there of each?

7. The labour costs for a certain project were $3971 per day for 15 technicians and helpers. If a technician earns $325/day and a helper $212/day, how many technicians were on the job?

8. A person walks a certain distance at a rate of 6.1 km/h. Another person takes 2.1 hours longer to walk the same distance, at a rate of 5.2 km/h. What distance did they walk?
9. If $5x + 3$ stands for 38, for what number does $2x - 4$ stand?
10. A certain submarine can travel at a rate of 31.5 km/h submerged and 42.3 km/h on the surface. How far can it travel submerged, and return to the starting point on the surface, if the total round trip is to take 8.25 hours?
11. A mixture is made where 15.6 gallons of cleaner containing 17.8% acid is added to 25.5 gallons of cleaner containing 10.3% acid. How many gallons of acid are in the final mixture?
12. A certain space probe traveled at a speed of 800 mi/h, and then, by firing retro-rockets, slowed to 595 mi/h. The total distance traveled was 47 100 miles in a time of 63.5 hours. Find the distance from the starting point at which the retro-rockets were fired.
13. How many pounds of copper are in an ingot that weighs 1550 lb and is made of a brass that contains 73.2% copper?
14. A ship leaves port at a rate of 15.3 mi/h. After 8.25 hours a launch leaves the same port to overtake the ship and travels at a rate of 22.0 mi/h. Find the time it will take for the launch to overtake the ship and the distance from port at which they will meet.
15. How much brass containing 75.5% copper must be combined with another brass containing 86.3% copper to make 5250 lb of brass containing 80.0% copper?

WRITING

16. Make up a word problem. It can be similar to one given in this chapter or, better, very different. Try to solve it yourself. Then swap with a classmate and solve each other's problem. Note down where the problem may be unclear, unrealistic, or ambiguous. Finally, each of you should rewrite your problem if needed.

12 Functions and Graphs

Outcomes

When you have completed this chapter, you will be able to

1. Define a function.
2. Define dependent and independent variables.
3. Identify a function as an equation, a formula, or as an ordered pair.
4. Use functional notation.
5. Introduce graphing and rectangular coordinates.
6. Graph ordered pairs.
7. Graph linear functions.
8. Graph quadratic and other functions.

In this chapter, we introduce functions and functional notation. We use functions to solve applications including the volume of a cylinder and the velocity of a uniformly accelerating object. Then we introduce the rectangular coordinate system and study the graph of a function. We pay particular attention to linear and quadratic functions. We explore the use of a graphing calculator, with its ability to trace and zoom in on the graph of a function.

12-1 FUNCTIONS

The equations we solved in the last two chapters were equations in one variable, such as

$$3(x - 2) = 5(x + 1) - 4$$

■ FIGURE 12-1
Area and radius of a circle.

But many equations and formulas involve two or more variables. For example, the formula for the area of a circle, $A = \pi r^2$, involves two variables, area A and radius r. As the radius r of the circle changes, the area A changes also (Fig. 12-1). The variables r and A are related. In this section we study equations that express relationships between variables.

The idea of a *function* provides us with a different way of speaking about mathematical relationships. It may become apparent as you study this chapter that these same problems could be solved without ever introducing the idea of a function. Does the function concept, then, merely give us new jargon for the same old ideas? Not really. A new way of *speaking* about something can lead to a new way of *thinking* about it, and so it is with functions. It also will lead to powerful and convenient functional notation, which will be especially useful as you continue your study of mathematics.

In this section, we define a function and study the different forms of a function. Many formulas are functions and express functional relationships between variables.

Example 1

The equation

$$A = s^2$$

is the formula for the area A of a square with side s (Fig. 12-2). This equation expresses the relationship between the two variables, s and A. It expresses one quantity, A, in terms of another quantity, s. In fact, if $s = 3$ cm,

$$\text{then } A = 9 \text{ cm}^2$$

We say that *the area A is a function of the side s*. Given any side s of a square, we can determine the area A of that square, using the equation above.

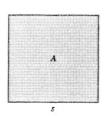

■ FIGURE 12-2
Area and side of a square.

Definition of a Function

If we have an equation that enables us to find exactly one y for any given x, then we say *y is a function of x*. In Example 1, A is a function of s, since for each value of s, there is exactly one value of A.

Example 2

In the equation $y = 2x - 3$,

$$\text{if } x = 4$$

then, by substitution,

$$y = 2(4) - 3$$
$$= 8 - 3$$
$$= 5$$

Similarly,

$$\text{if } x = 5$$

then

$$y = 2(5) - 3 = 7$$

In fact, for each value of x, there is exactly one value of y, and y is a function of x.

Example 3

In the equation $x^2 + y^2 = 25$,

$$\text{if } x = 3$$

then

$$3^2 + y^2 = 25$$
$$y^2 = 25 - 9$$
$$y^2 = 16$$

or

$$y = \pm 4$$

Since y has *two* values for $x = 3$, the equation $x^2 + y^2 = 25$ is *not* a function.

Explicit Form

An equation is said to be in *explicit form* when one variable is isolated on one side of the equal sign. In the equation $y = 2x - 3$, the variable y is expressed *explicitly* in terms of x. That is, y has been isolated on one side of the equal sign.

---- **Example 4** ----

The equations

$$y = x^2 - 1, \quad A = s^2, \quad \text{and} \quad x = 3y + z$$

are all functions in explicit form.

Implicit Form

When a variable is *not* isolated on one side of the equal sign, the equation is said to be in *implicit form*.

---- **Example 5** ----

The equations

$$x + y = 1, \quad x = y^2 + 3x, \quad \text{and} \quad x^2 + y^2 = 25$$

are all in implicit form.

Dependent and Independent Variables

In the explicit equation $y = 2x - 3$, the variable y has been isolated on the left. If we know any value of x, then we can easily determine the value of y by substitution. In this equation, the variable y is called the *dependent* variable, because its value *depends* on the value of the other variable x. The variable x is called the *independent* variable. Only equations in explicit form have dependent and independent variables.

---- **Example 6** ----

In the equation

$$y = x - 3$$

y is the dependent variable and x is the independent variable. If we solve that same equation for x, we get

$$x = y + 3$$

Now x is the dependent variable and y is the independent variable. We could also write that same equation as

$$x - y = 3$$

This is an implicit form of the equation and neither x nor y is dependent or independent.

Types of Functions

There are many different types of functions and many have special names. For example,

$y = 2x - 3$	is called a *linear* function.
$y = 4x^2 - 3x + 5$	is called a *quadratic* function.
$y = x^3$	is called a *power* function.
$y = \cos 3x$	is called a *trigonometric* function.
$y = 2^x$	is called an *exponential* function.
$y = \log x$	is called a *logarithmic* function.

We will study all these types of functions in this book.

Functions in Verbal Form

A function may be expressed in verbal form by a sentence. We may then express that sentence in equation form.

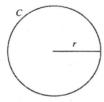

■ **FIGURE 12-3**
Circumference and radius of a circle.

─── Example 7 ───
The circumference of a circle is twice the product of π times the radius. Express the circumference as a function in equation form.

Solution: Letting C represent the circumference of the circle and r represent the radius (Fig. 12-3), the equation form is

$$C = 2\pi r$$

A Function in the Form of a Formula

Many formulas are functions. The most common formulas are listed in Appendix A, "Summary of Facts and Formulas," beginning on page 693.

─── Example 8 ───
Express the volume V of a sphere as a function of its radius r.

Solution: This is another way of saying: "Write an equation for the volume of a sphere in terms of its radius." The formula for the volume of a sphere is found in Appendix A, "Summary of Facts and Formulas," Eq. 95.

$$V = \frac{4}{3}\pi r^3$$

A Function as a Set of Ordered Pairs

We must often deal with pairs of numbers rather than with single values. An *ordered pair* of numbers is a pair of numbers in which the order of the numbers is important. An ordered pair of numbers is written with the numbers separated by a comma and enclosed within parentheses. Since order counts, the ordered pair (2, 3) is different from the ordered pair (3, 2). A set of ordered pairs is a function if for each first number there is only *one* second number.

Example 9

The set of ordered pairs

$$(-2, 4), \quad (-1, 1), \quad (0, -2), \quad (1, -5), \quad (2, -8)$$

is a function, because each first number has exactly one second number associated with it.

This function could also be described by a table of values:

First	-2	-1	0	1	2
Second	4	1	-2	-5	-8

Example 10

The set of ordered pairs

$$(2, -4), \quad (0, 0), \quad (1, 1), \quad (2, 4)$$

is not a function, because there are two second numbers, -4 and 4, for the first number 2.

Example 11

In this table, the quantity r represents the radius (in cm) of a circle and C represents the circumference (in cm) of the circle.

r	2.3	3.5	4.0	5.6	8.1	9.9
C	14.5	22.0	25.1	35.2	50.9	62.2

This table of values defines a function, since for each value of the radius, there is only one value of the circumference. We say that the circumference C is a function of the radius r.

This function may also be expressed as the set of ordered pairs (r, C): (2.3, 14.5), (3.5, 22.0), (4.0, 25.1), (5.6, 35.2), (8.1, 50.9), and (9.9, 62.2).

SECTION 12-1   Exercises   Functions

Definition of a Function

Find the value y of the given function for the specified value of x.

1. $y = 3x - 2, x = 1$
2. $y = 2x^2 + x - 1, x = 3$
3. $y = x^3, x = -4$
4. $y = 2^x, x = 3$

State whether the equation is in implicit or explicit form. If explicit, name the dependent and the independent variables.

5. $y = 4x - 1$
6. $x = xy + 2$
7. $x^2 + y^2 = 25$
8. $s = 16.1 t^2$

Functions in Verbal Form

Express in equation form.

9. y is equal to 5 times the square of x.
10. y is the amount by which 3 exceeds x.
11. y is the sum of x and its reciprocal.
12. y is twice the square root of x.

13. The surface area A of a sphere is 4π times the square of the radius r.
14. The area A of a circle is π times the square of the radius r.
15. The velocity v is the product of the acceleration a and time t.
16. The power P dissipated in a resistor is the quotient of the square of the voltage V divided by the resistance R.

A Function in the Form of a Formula

Refer to Appendix A to find the formula, if necessary.

17. A car is traveling on the highway at 90 km/h. Express the distance d traveled as a function of the time t traveled.
18. If work is the product of force and distance, express the amount of work W done in lifting a block with a mass of 50 kg as a function of the distance lifted. A force of

$$F = mass \times acceleration = 50 \text{ kg} \times 9.81 \frac{\text{m}}{\text{s}^2} = 491 \text{ N}$$

is required to lift a 50 kg block.

19. The first minute of a telephone call costs $0.75 and each additional minute is $0.53. Express the cost c as a function of time t.
20. An empty spool for storing wire cable has a mass of 66.4 kg. Each metre of cable has a mass of 13 kg. Find the mass m of the filled spool as a function of the length x of cable on it.
21. Express the area A of a triangle with a base of 10 cm as a function of its height h.
22. Express the base b of a rectangle as a function of the height h, if the area A is 4.7 cm².
23. Express the area A of a square as a function of its perimeter p.
24. Express the circumference C of a circle as a function of its diameter D.
25. Express the area A of a circle as a function of its circumference C.
26. Express the volume V of a cube as a function of its side s.

A Function as a Set of Ordered Pairs

27. Express this function as a set of ordered pairs:

x	-2	-1	0	1	2
y	-5	-2	1	4	7

28. Make a table of values for this function: $(-1, 1)$, $(0, 0)$, $(1, 1)$, $(2, 4)$, $(3, 9)$.

12-2 FUNCTIONAL NOTATION

In this section we learn to use functional notation, determine the domain and range of a function, and find the value of a function by substitution.

Functional Notation

Sometimes we use a special notation with functions, a notation which specifies the rule, or operations on a variable. If y is a function of x, we write

$$y = f(x)$$

and we read this as *y equals f of x*.

Just as we use the symbol x to represent a number without saying which number we are specifying, we use the notation $y = f(x)$ to represent a function f without having to specify which particular function we are referring to. The letter

f is used as the *name* of the function, to identify the rule or operations to be performed on the independent variable x.

> **COMMON ERROR** The equation $y = f(x)$ does *not* mean "y equals f times x."
> $$y \neq f \times x$$

Example 12
If $y = 2x - 3$, then y is a function of x. We write $y = f(x)$, where $f(x) = 2x - 3$. In other words,

$$y = 2x - 3 \quad \text{and} \quad f(x) = 2x - 3$$

are two *different* ways of describing the *same* function.

Example 13
The volume V of a cube is a function of the length s of a side (Fig. 12-4), such that

$$V = s^3$$

In functional notation, we write

$$V = f(s) \quad \text{or} \quad f(s) = s^3$$

Note that the independent variable is s and the dependent variable is V.

FIGURE 12-4
Volume and side of a cube.

Example 14
The distance traveled by a freely falling body is a function of the elapsed time. We can express this relationship in functional notation by the equation

$$s = f(t)$$

where s represents the distance fallen, and t represents the elapsed time. From physics we learn that

$$s = v_0 t + \frac{1}{2} a t^2$$

where v_0 is the initial velocity and a is the acceleration due to gravity. Note that the independent variable is t, v_0 and a are constants, and s is the dependent variable.

The letter f is usually used to represent a function, but other letters such as g and h may be used. Subscripts, as in f_1 and f_2, are also used to distinguish one function from another. For example, we could say $y = g(x)$, $y = h(x)$, $y = f_1(x)$, or $y = f_2(x)$ to describe different functions.

Domain and Range

The *domain* of the function $y = f(x)$ is the set of all x values, and the *range* of the function is the set of all y values of the function. If the domain of a function is not given, then it is assumed to be the set of *all possible real values* of x that yield *real values* of y.

Example 15

Find the domain and range of the function

$$y = x^2$$

Solution: We can square any real number so x can be any real number. Thus the domain is all real numbers. Any real number squared will be nonnegative, so the range is all nonnegative real numbers.

>   Domain: all real numbers x
>   Range: $y \geq 0$

Example 16

Find the domain and range of the function

$$y = \sqrt{x - 2}$$

Solution: Since we cannot take the square root of a negative number, then x must be greater than or equal to 2:

$$x \geq 2$$

Since the square root of any number is nonnegative, then

$$y \geq 0$$

So, we get

>   Domain: $x \geq 2$
>   Range: $y \geq 0$

In terms of dependent and independent variables, the domain is the set of all possible values of the independent variable, and the range is the set of all possible values of the dependent variable.

Example 17

A body, falling freely from rest, falls a distance s feet in t seconds according to the formula

$$s = 16t^2$$

Find the domain and range of this function.

Solution: Since $s = 16t^2$, t is the independent variable and s is the dependent variable. Since t represents the elapsed time, t cannot be negative, so

$$t \geq 0$$

The distance s is the product of 16 times a nonnegative number, t^2, so s is also nonnegative,

$$s \geq 0$$

So, we have

>   Domain: $t \geq 0$
>   Range: $s \geq 0$

Substituting into Functions

Functional notation is often used to designate the value of the function for a specific value of the independent variable. The value of the function $f(x)$ when $x = a$ is denoted as $f(a)$. Thus, $f(3)$ is the value of the function $f(x)$ when $x = 3$.

Example 18

Given $f(x) = x^2 - 5$, find $f(3)$ and $f(-2)$.

Solution: The notation $f(3)$ represents the value of the function $f(x)$ when $x = 3$. Therefore, we substitute 3 for x in the equation of the function.

$$f(x) = x^2 - 5$$
$$f(3) = 3^2 - 5$$
$$= 9 - 5$$
$$= 4$$

Similarly,

$$f(-2) = (-2)^2 - 5$$
$$= 4 - 5$$
$$= -1$$

We may substitute a literal value, such as a, into the equation of the function in the same way.

Example 19

Given $g(x) = 2x - 3$, find $g(a)$ and $g(-5b)$.

Solution: Substituting a for x in the equation of the function, we get

$$g(a) = 2(a) - 3$$
$$= 2a - 3$$

Substituting $-5b$ for x, we get

$$g(-5b) = 2(-5b) - 3$$
$$= -10b - 3$$

Example 20

If $g(x) = 2x - 3$, find $3g(-4)$.

Solution: The notation $3g(-4)$ means 3 times $g(-4)$. First we find $g(-4)$.

$$g(-4) = 2(-4) - 3$$
$$= -8 - 3$$
$$= -11$$

Now we multiply by 3:

$$3g(-4) = 3(-11)$$
$$= -33$$

SECTION 12-2 Exercises Functional Notation

1. Rewrite "y is a function of x such that $y = 2x - 1$" using functional notation.
2. Rewrite "v is a function of t, such that $v = 24 + 32t$" using functional notation.
3. The Fahrenheit temperature F is 32 more than $\frac{9}{5}$ of the Celsius temperature C. Express this in functional notation.
4. Work W is defined as force F times distance d. Write work as a function of distance.

Domain and Range

Find the domain and range of each function in questions 5–10.

5. $y = 3x - 2$ 6. $f(x) = x^2 + 1$ 7. $g(x) = \frac{1}{x}$ 8. $y = \sqrt{x - 1}$

9. The work W done by a constant force F acting on an object and moving it through a distance d is given by

$$W = Fd$$

10. If a body has a constant acceleration of 9.8 m/s² and initial velocity of 1.4 m/s then its velocity v after time t is given by

$$v = 1.4 + 9.8t$$

Substituting into Functions

11. Given $f(x) = 3x - 2$, find $f(1)$, $f(-2)$, $f(a)$, and $2f(3)$.
12. Given $g(x) = 5x + 2$, find $g(2)$, $g(-3)$, $g(b)$, and $3g(-1)$.
13. Given $h(x) = 3x^2 + 1$, find $h\left(\frac{1}{2}\right)$, $h(0)$, $h(c)$, and $h(1) - h(2)$.
14. Given $f_1(x) = 4 - x^2$, find $f_1(0.2)$, $f_1(0)$, $f_1(p)$, and $\frac{f_1(1)}{f_1(3)}$.

Round to three significant digits.

Applications

15. The power (in watts) dissipated in a resistor is given by

$$g(R) = \frac{V^2}{R}$$

where V is the voltage in volts and R is the resistance in ohms. If the voltage is 115 volts, find $g(7.00)$, $g(72.0)$, and $g(4750)$.

16. The velocity of a uniformly accelerating body at time t is given by

$$f(t) = v_0 + at,$$

where a is the acceleration and v_0 is the initial velocity. If v_0 is 8.00 ft/s and $a = 32.2$ ft/s², find $f(5.00)$, $f(10.0)$, and $f(20.0)$.

12-3 RECTANGULAR COORDINATES

In this section we define the rectangular coordinate system, and learn to graph points in rectangular coordinates. Graphing is a skill that is indispensable in mathematics. We often collect data, and it is essential for us to be able to graph that data and interpret the graph. Most physical phenomena can be represented algebraically as functions (equations), and the graph is a "picture" of that function. Graphs can

FIGURE 12-5
Real number line.

FIGURE 12-6
Rectangular coordinate system.

be used to show the relationship between two quantities, such as time and velocity, diameter and area, and voltage and current. Lastly, we must be able to read information from a graph.

The Rectangular Coordinate System

In Chapter 1 we saw that we can graphically represent every real number as a point on a line. We called that line the *real number line* (Fig. 12-5).

But how can we graph an *ordered pair* of numbers? To graph an ordered pair, we need two number lines, placed at right angles to each other, intersecting each other at the zero mark. We call this a *rectangular coordinate system*.

The horizontal number line is called the *abscissa* and in this case is assigned as the x axis and the vertical number line is called the *ordinate* and in this case is assigned to be the y axis. They intersect at a point called the *origin*. These two axes divide the plane into four regions called *quadrants*. The quadrants are numbered counterclockwise, using Roman numerals, as shown in Fig. 12-6.

Example 21
In Fig. 12-7, name the quadrant in which each point lies.

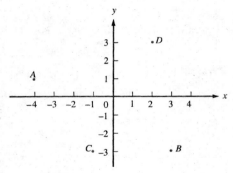

FIGURE 12-7

Solution: A: quadrant II; B: quadrant IV; C: quadrant III; D: quadrant I.

Graph of a Point

Each ordered pair of numbers can be graphed as a point. Fig. 12-8 shows a point P in quadrant I. Its horizontal distance from the origin, called the x *coordinate* of the point, is 2 units; thus, $x = 2$. Its vertical distance from the origin, called the y *coordinate* of the point, is 3 units; thus, $y = 3$. The numbers in the ordered pair (2, 3) are called the *rectangular coordinates,* or simply *coordinates,* of the point P. They are always written in the same order, with the x coordinate first, and the y coordinate second. A letter identifying the point is sometimes written before the coordinates, as in $P(2, 3)$.

To plot any ordered pair (h, k), simply place a point at a distance h units from the y axis and k units from the x axis. Points with positive x coordinates are located to the *right* of the y axis and points with negative x coordinates are located to the *left* of the y axis. Points with positive y coordinates are located *above* the x axis and points with negative y coordinates are located *below* the x axis.

FIGURE 12-8
Graph of a point.

FIGURE 12-9

Example 22

Graph the points

(a) $P(3, 4)$
(b) $Q(-3, 2)$
(c) $R(4, -3)$
(d) $S(-2, -4)$

Solution: We draw the coordinate axes, label each axis, and label the scale on each axis.

(a) The point $P(3, 4)$ is located 3 units to the right of the y axis and 4 units above the x axis. So, we start at the origin, move 3 units to the right (on the x axis), and then move 4 units straight up. We label the point $P(3, 4)$, as in Fig. 12-9.

(b) The point $Q(-3, 2)$ is located 3 units to the left of the y axis and 2 units above the x axis.

(c) The point $R(4, -3)$ is located 4 units to the right of the y axis and 3 units below the x axis.

(d) The point $S(-2, -4)$ is located 2 units to the left of the y axis and 4 units below the x axis.

Graphing a Set of Ordered Pairs

To graph a *set* of ordered pairs, we simply plot each ordered pair. If the ordered pairs represent a function, we usually connect the points with a smooth curve, unless we have reason to believe that there are sharp corners, breaks, or gaps in the graph.

Example 23

Graph this set of ordered pairs and connect them in order with a smooth curve that seems to fit the curve best: $P_1(-4, 4)$, $P_2(-3, 0)$, $P_3(-2, -3)$, $P_4(-1, -4)$, $P_5(0, -3)$, $P_6(1, 0)$, $P_7(2, 4)$.

12-3 Rectangular Coordinates

Solution: The points are graphed in Fig. 12-10 and then connected with a smooth curve.

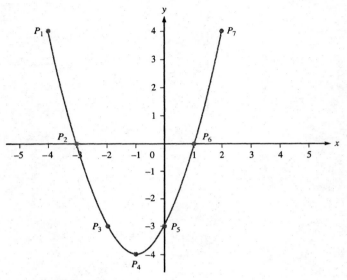

■ **FIGURE 12-10**
Graph of a set of ordered pairs.

Graphing Data

Ordered pairs of numbers may be given in a chart or a table. This information, called *data,* may be the result of an experiment or calculations. The ordered pairs are graphed as points, and the points are connected by a smooth curve.

Example 24

Graph the following data, and connect the points with a smooth curve. The data shows the relationship between the diameter D and the circumference C of several circles.

D (in cm)	4.6	7.0	8.0	11.2	16.2	19.8
C (in cm)	14.5	22.0	25.1	35.2	50.9	62.2

Solution: We choose to graph the diameter D on the horizontal axis and the circumference C on the vertical axis. Thus, C is the dependent variable and D is the independent variable; C is being graphed as a function of D. We draw and label the axes as the D (horizontal) and C (vertical) axes. Since D ranges from 0 to 20, a convenient scale on the D axis is 2. Since C ranges from 0 to 70, a convenient scale on the C axis is 10. Each pair of numbers represents a point (D, C). The points (4.6, 14.5), (7.0, 22.0), (8.0, 25.1), (11.2, 35.2), (16.2, 50.9), and (19.8, 62.2) are plotted and connected, as shown in Fig. 12-11. In this case the graph is a straight line.

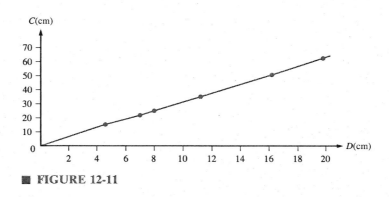

FIGURE 12-11

SECTION 12-3 Exercises Rectangular Coordinates

Graph of a Point

Write the coordinates of each point in Fig. 12-12.
1. A
2. B
3. C
4. D
5. E
6. F
7. G
8. H

Graph the following points.
9. (3, 4)
10. (5, 3)
11. (−1, 2)
12. (−2, 4)
13. (4, −3)
14. (5, −2)
15. (−2, −3)
16. (−3, −4)
17. (−3, 0.5)
18. (1.6, −3)
19. (0, −2)
20. (−3, 0)
21. (−4, 0)
22. (0, −4)
23. $\left(-\frac{1}{2}, 3\frac{3}{4}\right)$
24. $\left(-2\frac{1}{2}, -\frac{3}{4}\right)$
25. (−3.7, −2.3)
26. (1.4, −3.7)
27. Which quadrant contains points having a positive x coordinate and a negative y coordinate?
28. In which quadrant does a point (h, k) lie if $h < 0$ and $k < 0$?
29. In which quadrants is the x coordinate negative?
30. In which quadrants is the y coordinate positive?

FIGURE 12-12

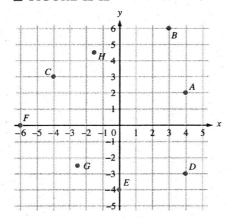

31. What is the x coordinate of any point on the y axis?

32. What is the y coordinate of any point on the x axis?

33. In which quadrants is the y coordinate divided by the x coordinate negative?

34. In which quadrants is the x coordinate divided by the y coordinate positive?

Graphing a Set of Ordered Pairs

Graph each set of points. Connect them in order with line segments and identify the geometric figure formed.

35. $P_1(-3, 1)$, $P_2(2, 2)$, $P_3(5, -2)$, $P_4(0, -3)$

36. $P_1(-3.6, 4.4)$, $P_2(0, 6.2)$, $P_3(3.1, 0)$, $P_4(-0.5, -1.8)$

Graph each set of points. Connect them in order with a smooth curve that seems to fit the data best.

37. $P_1(-4, -9)$, $P_2(-3, -2)$, $P_3(-2, -1)$, $P_4(-1, 0)$, $P_5(0, 7)$

38. $P_1(-1, -7)$, $P_2(0, -2)$, $P_3(1, 1)$, $P_4(2, 2)$, $P_5(3, 1)$, $P_6(4, -2)$, $P_7(5, -7)$

Graphing Data

Graph the following data. Label the graph and connect the points with a smooth curve.

39. C represents the temperature in degrees Celsius, and F represents the temperature in degrees Fahrenheit.

C	−20.5	0	20.5	49.6	82.7	100
F	−4.9	32	68.9	121.3	180.9	212

40. Let t represent the time (in years) a sum of money is left in a savings account, and let I represent the interest (in dollars) earned during that time.

t	0.5	1	2.5	3.5	5	7.5	10
I	27.50	55.00	137.50	192.50	275.00	412.50	550.00

12-4 LINEAR FUNCTIONS

In this section we study linear functions. We learn to graph a linear function, find its intercepts and slope, and solve some applications involving linear functions.

Linear Function

A *linear function* is a function defined by an equation having the form

$$y = mx + b$$

where m and b are constants.

─── **Example 25** ───
The equation $y = 3x - 2$ defines a linear function in which $m = 3$ and $b = -2$.

Graph of a Function

The graph of a function defined by an equation consists of all the points whose coordinates *satisfy* the equation—that is, the equation is true when the coordinate values of the point are substituted in.

Example 26

Show that the point (2, 4) satisfies the equation

$$y = 3x - 2$$

Solution: If $x = 2$ and $y = 4$, the equation becomes

$$4 = 3(2) - 2$$

or

$$4 = 4$$

Since this is a true equation, the coordinates of the point (2, 4) satisfy the equation. Therefore the point (2, 4) is on the graph of the equation $y = 3x - 2$, as we will see in Example 27.

Table of Values

The graph of the function

$$y = f(x)$$

consists of *all* points whose coordinates (x, y) satisfy the equation. Since the value of y depends on the value of x, then y is the dependent variable and x is the independent variable. Some points can be determined by choosing specific values for x and calculating y for each value of x.

To organize our work we may make a table of values that consists of the coordinates of several points on the graph of the function. To graph the function, we plot those points and connect them with a curve.

Example 27

Make a table of values for the linear function

$$y = 3x - 2$$

for x from -2 to 2 and graph the function.

Solution: We choose integral values for x and calculate y by substituting these x values into the equation. The table of values is shown on the right.

■ FIGURE 12-13

$y = 3x - 2$

$y = 3x - 2$
if $x = -2$ then $y = 3(-2) - 2 = -8$
if $x = -1$ then $y = 3(-1) - 2 = -5$
if $x = 0$ then $y = 3(0) - 2 = -2$
if $x = 1$ then $y = 3(1) - 2 = 1$
if $x = 2$ then $y = 3(2) - 2 = 4$

x	y
-2	-8
-1	-5
0	-2
1	1
2	4

Note: A horizontal arrangement of this table looks like this:

x	-2	-1	0	1	2
y	-8	-5	-2	1	4

Each ordered pair represents one point of the graph. The horizontal axis is the axis of the independent variable x, and the vertical axis is the axis of the dependent variable y. We plot these points and connect them with a smooth curve (Fig. 12-13). As we see, the graph of the equation $y = 3x - 2$ is a straight line.

In fact, *the graph of every linear function is a straight line*. If we know any *two* points on a line, then we can draw that line. So our table of values can be *very small* for a linear function. We can "get by" with plotting only two points, but it's wise to plot an extra point as a check.

Intercepts

The point where the graph crosses the x axis is called the *x intercept*. To find the x intercept, we note that *the y coordinate of any point on the x axis is 0*. So, to find the x intercept, we substitute 0 for y in the equation and solve for x.

The point where the graph crosses the y axis is called the *y intercept*. To find the y intercept, we note that *the x coordinate of any point on the y axis is 0*. So, to find the y intercept, we substitute 0 for x in the equation and solve for y.

— **Example 28** —

Find the intercepts of the function

$$y = 3x - 2$$

Solution: To find the x intercept, we substitute 0 for y in the equation to get

$$0 = 3x - 2$$

FIGURE 12-14
Intercepts.

$y = 3x - 2$

and solve for x:

$$x = \frac{2}{3}$$

So the x intercept is $\left(\frac{2}{3}, 0\right)$, as shown in Fig. 12-14. Sometimes we refer to the x coordinate as the x intercept; so we may say *the x intercept is* $\frac{2}{3}$, instead of saying *the x intercept is* $\left(\frac{2}{3}, 0\right)$.

To find the y intercept, we substitute 0 for x in the equation to get

$$y = 3(0) - 2$$

and solve for y:

$$y = -2$$

So the y intercept is $(0, -2)$, as shown in Fig. 12-14. Sometimes we refer to the y coordinate as the y intercept; so we may say *the y intercept is* -2, instead of saying *the y intercept is* $(0, -2)$.

— **Example 29** —

Find the y intercept of the general linear function

$$y = mx + b$$

Solution: To find the y intercept, we substitute 0 for x in the equation:

$$y = m(0) + b$$

or

$$y = b$$

So, *the y intercept of the linear function $y = mx + b$ is the constant b*.

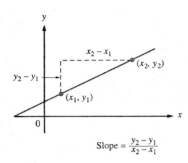

FIGURE 12-15
Slope of a line.

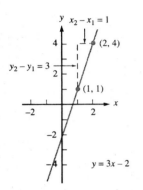

FIGURE 12-16

Slope

A very important concept in the study of linear functions is *slope*. Slope is a measure of the direction and steepness of a line. We can calculate the slope of a line if we know the coordinates of two points. If (x_1, y_1) and (x_2, y_2) are two points on a line (Fig. 12-15), the slope is defined as

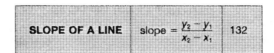

SLOPE OF A LINE   slope $= \dfrac{y_2 - y_1}{x_2 - x_1}$ 132

Example 30

In Example 27, we found that the points (1, 1) and (2, 4) are on the graph of

$$y = 3x - 2$$

Use these points to calculate the slope of the line.

Solution: We use Eq. 132 to find the slope of the line, with $(x_1, y_1) = (1, 1)$ and $(x_2, y_2) = (2, 4)$ (Fig. 12-16). So, we let $x_1 = 1$, $y_1 = 1$, $x_2 = 2$, and $y_2 = 4$,

$$\text{Slope} = \frac{4 - 1}{2 - 1}$$
$$= \frac{3}{1}$$
$$= 3$$

So the slope of the line $y = 3x - 2$ is 3, which is also the coefficient of x.

In fact, *the slope of the line $y = mx + b$ is always the coefficient m.*

Slope-Intercept Form

The form of a linear function

$$y = mx + b$$
     ↕   ↕
   slope y intercept

is called the *slope-intercept form,* because m is the slope of the line and b is the y intercept.

SLOPE-INTERCEPT FORM $y = mx + b$ 137

Example 31

Find the slope and the y intercept of the linear function

$$y = 2x + 3$$

Solution: The equation $y = 2x + 3$ is in slope-intercept form.

Since the slope m is the coefficient of the x term, then $m = 2$ and the slope is 2. The y intercept b is the constant term 3, so $b = 3$ and the y intercept is 3.

12-4 Linear Functions

■ FIGURE 12-17

Example 32

Graph and find the equation of the line passing through the points $(-1, 3)$ and $(2, -3)$.

Solution: To graph the line, we plot the two points and draw the line through them (Fig. 12-17). To find the equation of the line, we need to know the slope and the y intercept of the line. We use Eq. 132 to find the slope of the line, with $(x_1, y_1) = (-1, 3)$ and $(x_2, y_2) = (2, -3)$,

$$m = \text{slope} = \frac{-3 - 3}{2 - (-1)}$$
$$= \frac{-6}{3}$$
$$= -2$$

So, substituting $m = -2$ into Eq. 137,

$$y = -2x + b$$

Since the point $(-1, 3)$ is on the line, its coordinates satisfy the equation. We let $x = -1$ and $y = 3$,

$$3 = -2(-1) + b$$

Solving for b,

$$b = 1$$

So

$$y = -2x + 1$$

Horizontal and Vertical Lines

Now we consider some properties of horizontal and vertical lines.

Example 33

Find the slope and the y intercept of the linear function

$$y = 2$$

Graph the function.

Solution: We write the equation in slope-intercept form:

$$y = 0x + 2$$

So the slope m is 0 and the y intercept is 2. We make a table of values to include at least three points, in which y is always 2:

x	0	1	2
y	2	2	2

We graph these points in Fig. 12-18 and get a horizontal line.

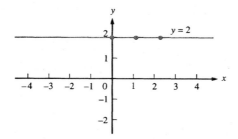

■ FIGURE 12-18
Graph of a horizontal line.

The slope of every horizontal line is 0. The equation of the horizontal line with y intercept b is

| HORIZONTAL LINE | $y = b$ | 135 |

Example 34

Find the equation of the vertical line that passes through $(4, 3)$ and $(4, -1)$.

Solution: We plot these two points and draw the line through them (Fig. 12-19).

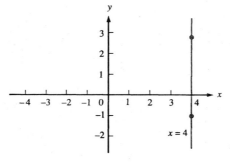

■ FIGURE 12-19
Graph of a vertical line.

Using Eq. 132 to find the slope,

$$m = \frac{-1 - 3}{4 - 4}$$
$$= \frac{-4}{0}$$

Now $\frac{-4}{0}$ is undefined, so *the slope is undefined.* From the graph, we observe that there is no y intercept. This equation cannot be expressed in slope-intercept form. We note that the x coordinate of every point on the graph is 4, so the equation of this line is

$$x = 4$$

This equation, however, does *not* describe a *function*, since all points have the same x coordinate.

The slope of a vertical line is undefined. The equation of the vertical line with x intercept a is

| VERTICAL LINE | $x = a$ | 136 |

Applications

Many applications involve formulas that are linear functions. A *formula* is an equation in which letters represent physical or geometric quantities. These letters may be either variables or constants. We graph a formula the same way we graph other functions.

Example 35

The velocity v of an object t seconds after being thrown down from the top of a building with an initial velocity v_0, is given by the formula

$$v = v_0 + gt$$

where g is the acceleration due to gravity, 9.8 m/s². If a ball is thrown with an initial downward velocity of 5.0 m/s, the equation becomes

$$v = 5.0 + 9.8t$$

Graph this equation for t from 0.0 to 10.0 seconds.

Solution: In this equation, t is the independent variable and v is the dependent variable. To make the table of values, we calculate the values for v by substituting the t values in the given equation and make a table of values

t	0.0	2.0	4.0	6.0	8.0	10.0
v	5.0	24.6	44.2	63.8	83.4	103.0

Since this is a linear function, we know the graph is a line. Since t goes from 0.0 to 10.0, we choose a scale of 2 for the t axis. Since v goes from 5.0 to 103.0, we choose a scale of 25 for the v axis. We use different scales for the two axes, since the ranges are so different. We plot the points and connect them with a straight line (Fig. 12-20).

■ **FIGURE 12-20**

$v = 5.0 + 9.8t$

SECTION 12-4 Exercises Linear Functions

For each equation,

(a) complete this table of values;

x	-2	-1	0	1	2
y					

(b) graph the equation;
(c) find the x and y intercepts; and
(d) find the slope.

1. $y = x$
2. $y = x + 2$
3. $y = 2x + 3$
4. $y = 3x + 1$
5. $y = 1 - 3x$
6. $y = 3 - 2x$
7. $y = \dfrac{x}{2} - 1$
8. $y = 2 - \dfrac{x}{3}$
9. $y = 3$
10. $y = -1$

For each equation,

(a) graph the equation;
(b) find the x and y intercepts; and
(c) find the slope.

11. $x = -1$
12. $x = 2$

Graph and find the equation of the linear function passing through the points.

13. $(1, 2)$ and $(3, 6)$
14. $(-1, 4)$ and $(3, -4)$
15. $(-1, -1)$ and $(4, -2)$
16. $(-2, -3)$ and $(2, -1)$

Applications

17. The circumference C of a circle is given by

$$C = 2\pi r$$

where r is the radius of the circle. Graph this function for values of r from 0 to 10 cm.

18. If an amount of money p (the principal) is invested at an annual interest rate r for t years, the simple interest earned is given by the formula

$$I = prt$$

Write the equation for the interest if $5000 is invested at the rate of 6.5%. Graph the function for t from 0 to 5 years.

19. The voltage V in a circuit with constant resistance R is a linear function of the current I, given by

$$V = IR$$

In a certain circuit the resistance is 4.5 ohms. Write the equation for the voltage as a function of the current. Graph the function for I from -2.5 to 2.5 amps.

20. The work W done by a constant force F acting on an object is a linear function of the distance d moved by the object, given by

$$W = Fd$$

A force of 100 N is applied to move a block. Write the equation for the work W done on the block as a function of the distance d. Graph the function for d from 0 to 12 m.

21. The mass m of an object of density D is a linear function of its volume V, given by

$$m = DV$$

The density of iron is 7.87 g/cm³. Write the equation for the mass m of an iron object as a function of its volume V. Graph the function for V from 0 to 400 cm³.

22. The velocity v of sound (in m/s) is a linear function of the temperature T (in °C), given by

$$v = 0.607T + 332$$

Graph the velocity v of sound as a function of the temperature T for T from −30°C to 30°C.

23. The total cost C of producing x units is a linear function:

$$C = mx + C_0$$

where m is the cost per unit and C_0 is the fixed (or initial) cost. A computer company determines that it costs $100 to produce one keyboard and that the fixed costs were $250 000. Express the cost C as a function of x and graph the function for x from 0 to 1000 keyboards.

24. The total cost C of producing x units is a linear function:

$$C = mx + C_0$$

where m is the cost per unit and C_0 is the fixed (or initial) cost. A new company invests $750 000 for equipment to produce laser disks. The unit cost per laser disk is $35. Express the cost C as a function of x and graph the function for x from 0 to 5000 disks.

 25. A moving company buys a 35 ft van, which has a useful lifetime of 6 years. After x months of use, the value y of the van is estimated by the equation

$$y = 37\,000 - 500x$$

Graph y as a function of x for x from 0 to 6 years.

12-5 QUADRATIC AND OTHER FUNCTIONS

In this section, we study quadratic functions. We learn to graph a quadratic function, find its vertex, and solve some applications of quadratic functions. We then graph some other functions that are neither linear nor quadratic.

Quadratic Function

A quadratic function has the form

$$y = ax^2 + bx + c$$

where a, b, and c are constants, and $a \neq 0$.

---- **Example 36** ----

The equation

$$y = x^2 - 2x - 3$$

defines a quadratic function in which $a = 1$, $b = -2$, and $c = -3$.

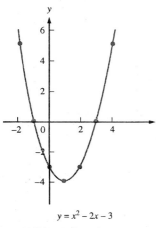

FIGURE 12-21

Graph of a Quadratic Function

The graph of a quadratic function has some very special properties which we now consider.

―― **Example 37** ――
Graph the quadratic function
$$y = x^2 - 2x - 3$$
for integral values of x from -2 to 4.

Solution: By substitution, we obtain the table of values,

x	-2	-1	0	1	2	3	4
y	5	0	-3	-4	-3	0	5

These ordered pairs are plotted and connected with a smooth curve in Fig. 12-21. Note that there are *two* x intercepts, $(-1, 0)$ and $(3, 0)$. The y intercept is $(0, -3)$.

> **COMMON ERROR** Be especially careful when substituting negative values into an equation. It is easy to make an error.

Parabola

The graph of a quadratic function is a curve called a *parabola*. Every parabola resulting from a quadratic function of the form $y = ax^2 + bx + c$ has a point that is either a high point on the curve or a low point. This point is called the *vertex*. Low points on a curve are called *minimum points*. High points on a curve are called *maximum points*. The vertex is either a minimum or a maximum point, depending on whether the curve opens up or down.

The graph of a quadratic function generally includes the vertex. Without proof we give this equation for the x coordinate of the vertex:

x COORDINATE OF THE VERTEX	$x = \dfrac{-b}{2a}$	**139**

Once we know the x coordinate of the vertex, we can calculate the y coordinate from the equation.

―― **Example 38** ――
Find the coordinates of the vertex of
$$y = x^2 - 2x - 3$$

Solution: Using Eq. 139 for the x coordinate, with $a = 1$ and $b = -2$,
$$x = \frac{-(-2)}{2(1)}$$
$$= 1$$

Substituting 1 for x in the equation,
$$y = 1^2 - 2(1) - 3 = -4$$

So the vertex of the parabola is $(1, -4)$. This is verified in Fig. 12-21, in which $(1, -4)$ is the lowest point of the graph.

12-5 Quadratic and Other Functions

> **TIP** When graphing a quadratic function, be sure to plot the vertex and points to the left and right of the vertex.

Example 39

Find the vertex and graph the function

$$y = -x^2 - 4x - 2$$

Solution: Using Eq. 139 for the x coordinate of the vertex, with $a = -1$ and $b = -4$,

$$x = \frac{-(-4)}{2(-1)} = -2$$

We make a table of values, choosing several x values to the left of -2 and several to the right of -2.

				vertex ↓			
x	-5	-4	-3	-2	-1	0	1
y	-7	-2	1	2	1	-2	-7

These points are plotted and connected with a smooth curve in Fig. 12-22.

FIGURE 12-22

Graphs of Other Functions

So far, we have graphed only functions that are linear or quadratic. There are also many other important types of functions. Some of these, such as trigonometric and exponential functions, will be covered in later sections. Here we consider a few other examples of functions.

FIGURE 12-23

Example 40

Graph the function

$$y = x^3 - 3x$$

for x from -3 to 3.

Solution: By substitution, we obtain the table.

x	-3	-2	-1	0	1	2	3
y	-18	-2	2	0	-2	2	18

We plot these points, and connect them with a smooth curve to get the graph (Fig. 12-23).

Example 41

Graph the function

$$y = \frac{1}{x}$$

for x from -3 to 3.

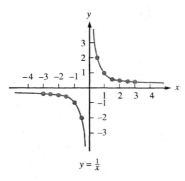

FIGURE 12-24

Solution: By substitution, we obtain the table, taking x increments of $\frac{1}{2}$.

x	-3	$\frac{-5}{2}$	-2	$\frac{-3}{2}$	-1	$\frac{-1}{2}$	0	$\frac{1}{2}$	1	$\frac{3}{2}$	2	$\frac{5}{2}$	3
y	$\frac{-1}{3}$	$\frac{-2}{5}$	$\frac{-1}{2}$	$\frac{-2}{3}$	-1	-2	undef.	2	1	$\frac{2}{3}$	$\frac{1}{2}$	$\frac{2}{5}$	$\frac{1}{3}$

We graph this function in Fig. 12-24. Note that something very special is happening near $x = 0$. First, x cannot equal zero, since $\frac{1}{0}$ is not defined; thus 0 is not in the domain of the function. But, for positive values of x, as x gets smaller (closer to 0), y gets larger. The graph approaches the y axis but never touches it! And, as negative values of x approach 0, y gets smaller (more negative). Again, the graph approaches the y axis but never touches it! A line which the graph approaches but never touches is called an *asymptote* of the graph. Here, the y axis is an asymptote. Similarly, we observe that the x axis is another asymptote of this graph—on the left and the right, the graph approaches the x axis, but never touches it.

─── **Example 42** ───

Graph the function

$$y = \sqrt{x}$$

for x from 0 to 9.

Solution: By substitution, we make the table of values, choosing perfect square values of x.

x	0	1	4	9
y	0	1	2	3

FIGURE 12-25

We plot these points and connect them in Fig. 12-25. For the function $y = \sqrt{x}$, the domain and range are

Domain: $x \geq 0$
Range: $y \geq 0$

So the graph of the function lies entirely in the first quadrant.

SECTION 12-5 Exercises Quadratic and Other Functions

The Graph of a Quadratic Function

Fill in the missing values and graph the function.

1. $y = x^2 - 3$

x	-3	-2	-1	0	1	2	3
y							

2. $y = 6x - x^2 - 5$

x	0	1	2	3	4	5	6
y							

12-5 Quadratic and Other Functions

Find the vertex and graph the function.

3. $y = -x^2$
4. $y = 1 - x^2$
5. $y = x^2 + 2x + 3$
6. $y = -x^2 + 4x - 1$
7. $y = \dfrac{x^2}{4}$
8. $y = 5x - x^2$

Graphs of Other Functions

Fill in the missing values and graph the function.

9. $y = x^3$

x	-2	-1	$-\frac{1}{2}$	0	$\frac{1}{2}$	1	2
y							

10. $y = \dfrac{x^3}{3}$

x	-2	-1	$-\frac{1}{2}$	0	$\frac{1}{2}$	1	2
y							

For each function, make a table of ordered pairs, using integral values of x from -3 to 3, rounding y values to tenths. Graph the function.

11. $y = \dfrac{-x^3}{3}$
12. $y = 2x^3$
13. $y = x^3 + 3x^2$
14. $y = 3x - x^3$
15. $y = -\dfrac{1}{x}$
16. $y = \dfrac{1}{x - 1}$
17. $y = \dfrac{1}{x + 1}$
18. $y = 1 + \dfrac{1}{x}$

For each function, make a table of ordered pairs, and graph the function.

19. $y = -\sqrt{x}$
20. $y = \sqrt{x + 1}$
21. $y = \sqrt{1 - x}$
22. $y = 1 + \sqrt{x}$

Applications

23. It has been shown that blood near the edge of an artery flows more slowly than blood near the centre of the artery. The velocity v (in cm/s) of the blood x cm from the centre of an artery is given by

$$v = 1000(0.04 - x^2)$$

Graph v as a function of x, for x from 0.0 to 0.2 cm.

24. The revenue R received by a company for selling a CD player at a price of p dollars is given by the revenue equation:

$$R = 8000p - 40p^2$$

Graph the revenue R as a function of the price p, for p from 0 to 200 dollars.

25. The farm population in the United States as a percentage of the total population can be expressed as a hyperbolic function:

$$y = \dfrac{200}{2t - 3895}$$

where y represents the percentage and t represents the year. Graph y as a function of t from the year 1950 to 2000.

12-6 READING GRAPHS

In this section we learn how to interpret a graph so that we can read information from it.

Example 43

An object is thrown down from the top of a cliff with an initial velocity of 6.5 m/s. Its velocity v, after t seconds, is given by the equation

$$v = 6.5 + 9.8t$$

(a) Graph the function for t from 0.0 to 10.0 seconds.
(b) From the graph, estimate the velocity of the object after $t = 7.0$ seconds.

Solution: First we make a table of values for the function

$$v = 6.5 + 9.8t$$

Using t values from 0.0 to 10.0, we calculate the v values from the equation. Since this is a linear function, we need to plot only three points.

t (seconds)	0.0	5.0	10.0
v (m/s)	6.5	55.5	104.5

We plot the three points and draw the line (Fig. 12-26). To estimate the velocity from the graph we need to draw some extra lines. First, locate the point on the horizontal (time) axis where $t = 7.0$. Draw a vertical line up to where it meets the graph. Next, from the point where the vertical line intersects the graph, draw a horizontal line to the vertical v axis and read the value of the velocity. We estimate the velocity is about 75 m/s.

■ **FIGURE 12-26**

Example 44

If an amount of money a is invested at an annual interest rate r for t years, and the interest is compounded annually, the accumulated amount y is given by the formula

$$y = a(1 + r)^t$$

Suppose you have $100 which you invest at an annual interest rate of 5.5%, compounded annually.

(a) Write the formula for the accumulated amount.
(b) Graph the formula for $t = 0$ to 10 years.
(c) Estimate how many years it would take to accumulate to $150.

Solution:

(a) The interest rate r, expressed as a decimal, is

$$r = 5.5\% = 0.055$$

Substituting $a = 100$ and $r = 0.055$ in the given equation,

$$y = 100(1 + .055)^t$$

or

$$y = 100(1.055)^t$$

FIGURE 12-27

(b) We make the table of values, and substitute into the equation to find y, rounding to the nearest dollar.

t (years)	0	2	4	6	8	10
y (dollars)	100	111	124	138	153	171

We plot these points and connect them with a smooth curve (Fig. 12-27).

(c) To estimate how many years it would take to accumulate to $150, we draw a horizontal line at $y = 150$. Then we draw the vertical line from the point of intersection to the t axis and estimate that the time when the accumulated amount will be $150 is about 7.6 years.

Example 45

The height s (in m) of an object t seconds after it is thrown from the top of a building is given by

$$s = -4.9t^2 + 9.8t + 39.2$$

(a) Graph s as a function of t, for t from 0.0 to 4.0 seconds.
(b) Find the time at which the object hits the ground.
(c) From the graph, estimate the time at which it reaches its greatest height.
(d) Estimate the greatest height the object reaches.

Solution:

(a) Since s is a quadratic function of t, the graph is a parabola. We make the table of values, for t from 0.0 to 4.0 seconds.

t (seconds)	0.0	1.0	2.0	3.0	4.0
s (metres)	39.2	44.1	39.2	24.5	0.0

FIGURE 12-28

We plot these points (Fig. 12-28) and draw the parabola.

(b) The object hits the ground when $s = 0$. This happens when the value for $t = 4.0$ seconds. In this specific case, the object hits the ground after 4.0 seconds as calculated for the table.

(c) The object reaches its greatest height at the vertex. From the graph, we can estimate that the vertex is (1, 44). We check it analytically, using Eq. 136 for the x coordinate of the vertex, with $a = -4.9$ and $b = 9.8$,

$$x = \frac{-9.8}{2(-4.9)} = 1$$

The object reaches its greatest height at $t = 1.0$ seconds.

(d) Its greatest height is the s coordinate of the vertex, 44.1 m.

Example 46

The force F acting on a spring is a function of the distance x the spring is stretched. This data was taken in an experiment.

x (cm)	0.0	1.0	2.0	3.0	4.0
F (N)	0.0	23.8	47.6	71.4	95.2

FIGURE 12-29

(a) Graph the force as a function of the displacement.
(b) From the graph, estimate the force required to stretch the spring 5.0 cm.
(c) Estimate the amount of stretch when the spring is acted upon by a force of 80 N.

Solution:

(a) We plot the points in the data table. They are linear, so we draw the line through the points (Fig. 12-29).
(b) Although we don't have an equation for this function, we can estimate values from the graph. To estimate the force required to stretch the spring 5.0 cm, we must extend the graph to the right to $x = 5.0$. We draw the vertical line at $x = 5.0$ up to the function line, draw the horizontal line over to the F axis and read the F value. We estimate it's about at $F = 120$. So the necessary force is about 120 N.
(c) Similarly, to estimate the distance the spring is stretched when a force of 80 N is applied, we find $F = 80$ on the vertical F axis, draw the horizontal line across to the function line, draw the vertical line down to the horizontal x axis, and read the x value. We estimate it's about $x = 3.4$. So a force of 80 N would stretch this spring about 3.4 cm.

Graphical Solution of Equations

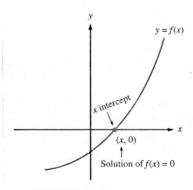

FIGURE 12-30
Graphical solution of the equation $f(x) = 0$.

We can use our knowledge of graphing functions to *solve equations* of the form $f(x) = 0$. The graph of the function $y = f(x)$ touches the x axis at an x intercept (Fig. 12-30), where the y coordinate is zero. Substituting $y = 0$ into the equation $y = f(x)$, we get $f(x) = 0$. So the x value of an x intercept is also a *solution* (or *root*) of the equation $f(x) = 0$.

Thus, to solve an equation graphically, we put the equation into the form

$$f(x) = 0$$

and graph the function

$$y = f(x)$$

The x intercepts are the solutions of the equation $f(x) = 0$. These solutions can be *approximated* from the graph.

The graphical solution of equations is a very powerful method of solving equations. It can be used to solve *any* equation, as long as the associated function can be graphed.

Example 47

Solve the equation graphically. Approximate the solution(s) to the nearest tenth.

$$6x^2 - 4 = 5x$$

Solution: We put the equation into the form

$$f(x) = 0$$

by bringing all terms to one side of the equal sign:

$$6x^2 - 5x - 4 = 0$$

So,

$$f(x) = 6x^2 - 5x - 4$$

We make a table of values,

x	-2	-1	0	1	2
y	30	7	-4	-3	10

graph the function (Fig. 12-31), and read the x intercepts from the graph. We estimate (to the nearest tenth) that the intercepts are at $x = -0.5$ and $x = 1.3$. So the approximate solutions of the equation
$$6x^2 - 4 = 5x$$
are -0.5 and 1.3.

Check: We substitute into the original equation,

$6(-0.5)^2 - 4 \stackrel{?}{=} 5(-0.5)$ $6(1.3)^2 - 4 \stackrel{?}{=} 5(1.3)$

$-2.5 = -2.5$ check $6.14 = 6.5$ check

We see from the check that the solution -0.5 is exact, and that the solution 1.3 is approximately correct (the check is accurate to 1 significant digit, if we round 6.5 back to 6).

FIGURE 12-31

$y = 6x^2 - 5x - 4$

SECTION 12-6   Exercises   Reading Graphs

Estimate to two significant digits.

1. The displacement s for an object thrown upward with an initial velocity v_0 after t seconds is given by
$$s = v_0 t - \frac{1}{2}gt^2$$
where g is 9.8 m/s². An object is thrown upward with an initial velocity of 50 m/s.
   (a) Write the equation for displacement as a function of time.
   (b) Graph the function for t from 0 to 10 seconds.
   (c) From the graph, estimate the displacement after 7.5 seconds.

2. According to Ohm's law, the current I (in amps) is equal to the voltage V (in volts) divided by the resistance R (in ohms),
$$I = \frac{V}{R}$$
The voltage in a certain circuit is 100 volts.
   (a) Write the equation for the current in this circuit as a function of resistance.
   (b) Graph the function for R from 10 to 300 ohms.
   (c) From the graph, estimate the resistance when the current is 5 amps.

3. The surface area A of a cylinder with radius r and height h is given by
$$A = 2\pi rh + 2\pi r^2$$
A cylinder has a radius of 6.0 cm.
   (a) Write an equation for the surface area as a function of the height.
   (b) Graph the function for h from 0.0 to 8.0 cm.
   (c) From the graph, estimate the height if the surface area is 500 cm².
   (d) Estimate the surface area if the height is 4.0 cm.

4. The surface area A of a cylinder with radius r and height h is given by
$$A = 2\pi rh + 2\pi r^2$$
The height of the cylinder is 3.0 m.
  (a) Write the equation for the surface area as a function of the radius.
  (b) Graph the function for r from 0.0 to 5.0 m.
  (c) From the graph, estimate the radius when the surface area is 75 m^2.

5. The velocity v of a rocket after t seconds is given by
$$v = v_0 + at$$
where v_0 is the initial velocity and a is the acceleration. A rocket has an initial velocity of 5.0 m/s and an acceleration of 2.0 m/s^2.
  (a) Write an equation for its velocity v as a function of time t.
  (b) Graph the function for t from 0.0 to 10.0 seconds.
  (c) From the graph, estimate the velocity after 5.5 seconds.
  (d) Estimate the time at which the velocity is 20.0 m/s.

6. The volume of a cone with radius r and height h is given by
$$V = \frac{1}{3}\pi r^2 h$$
A certain conical sand pile is 2.7 m high.
  (a) Write the equation for the volume of this sand pile as a function of the radius.
  (b) Graph the function for r from 0.0 to 3.0 m.
  (c) From the graph, estimate the radius of the sand pile if the volume is 8.0 m^3.

7. The height s (in m) of an object t seconds after it is thrown from the top of a building is given by
$$s = -9.8t^2 - 24.5t + 122.5$$
  (a) Graph s as a function of t, for t from 0.0 to 3.0 seconds.
  (b) Estimate the time at which the object hits the ground.

8. The height s (in metres) of an object t seconds after it is thrown upward from the ground is given by
$$s = 33.6t - 4.8t^2$$
  (a) Graph the function for t from 0.0 to 8.0 seconds.

From the graph, estimate
  (b) the time when the object hits the ground;
  (c) the time when the object reaches its maximum height;
  (d) the maximum height of the object; and
  (e) the time(s) when the object is 30 metres above the ground.

9. The length l of a spring is a linear function of the force F applied to it. This data was taken in an experiment.

Force (N)	0	2	4	6	8	10
Length (cm)	5.00	5.14	5.28	5.42	5.56	5.70

  (a) Graph the length l as a function of the force F.
  (b) Estimate the force necessary to stretch the spring to 6 cm.

12-6 Reading Graphs

10. The following is a table of data which relates the force F required to pull a block along a rough surface and the normal force N.

N (lb)	500	600	700	800	900
F (lb)	225	270	315	360	405

   (a) Graph the force F as a function of the normal force N.
   (b) Estimate the force required to pull a block with normal force of 500 lb.

11. The velocity v of a projectile fired upward is a linear function of time t. This data was taken in an experiment.

t (seconds)	2	4	6
v (ft/s)	32	-32	-96

   (a) Graph v as a function of t.
   (b) The velocity of a projectile when $t = 0$ is called the initial velocity. Find the initial velocity.
   (c) Estimate the time at which the velocity is 0.

12. The linear velocity v of a point on a rotating circle of radius r is a function of its angular velocity ω. The following data were taken in an experiment.

ω (rev/s)	1600	1800	2000	2200	2400
v (ft/s)	64 000	72 000	80 000	88 000	96 000

   (a) Graph the linear velocity v as a function of the angular velocity ω.
   (b) Estimate the angular velocity of a point moving with a linear velocity of 70 000 ft/s.
   (c) Estimate the linear velocity of a point moving with an angular velocity of 1000 rev/s.

13. The simple interest I earned on a savings account is a linear function of time t. After 1 year, the account earned \$35 in interest, and after 5 years, it earned \$175 in interest.
   (a) Graph I as a function of t, for t from 0 to 5 years.
   (b) From the graph estimate the interest after 4 years.

14. The volume V of a cone is a linear function of the height h if the radius is constant. For this function, $V = 1.57$ cm^3 when $h = 1.50$ cm and $V = 2.36$ cm^3 when $h = 2.25$ cm.
   (a) Graph V as a function of h, for h from 0.00 to 3.00 cm.
   (b) From the graph estimate the volume when $h = 2.00$ cm.

Graphical Solution of Equations

Solve the equation graphically. Approximate the solution(s) to the nearest tenth.

15. $x^2 - 3 = 4x$ 16. $x^2 + x = 3$ 17. $-x^2 = 4 + 6x$
18. $1 - x^2 = 3x$ 19. $x^3 - 4x = 4$ 20. $x^3 + x = 1$

12-7 GRAPHING CALCULATORS

A graphing calculator helps us to graph functions quickly. Since there are many graphing calculators on the market, our discussion is general, and not calculator-specific. In this section we graph functions using a graphing calculator. We use the TRACE and ZOOM features of the graphing calculator to determine the x and y intercepts and the maximum and minimum points of a graph.

Graph of a Function

We set the calculator *graph mode* for graphing functions in the rectangular coordinate system. We set the *window* (the displayed portion of the coordinate plane) by defining the minimum and maximum values of x and y that will appear in the window. We enter the function and graph. The steps are simple!

Graphing a Function on a Graphing Calculator

1. Set the window.
2. Enter the equation.
3. Graph.

Example 48

Use a graphing calculator to graph the function

$$y = x^2$$

for $-10 \leq x \leq 10$ and $0 \leq y \leq 100$.

Solution: If we look at the screen on the calculator where the window variables are defined, we see a screen similar to this:

```
Xmin=
Xmax=
Xscl=
Ymin=
Ymax=
Yscl=
```

These six quantities are called the *window variables*. In this case, we're given that $-10 \leq x \leq 10$ and $0 \leq y \leq 100$, so we let Xmin = -10, Xmax = 10, Ymin = 0, and Ymax = 100. The values of Xscl and Yscl determine the spacing between the tick marks on the axes. We choose the scales so that there are neither too many nor too few tick marks on an axis. A scale of 5 on the x axis and a scale of 10 on the y axis are appropriate choices for this window. So we set Xscl = 5 and Yscl = 10. Our window variables look like this:

```
Xmin= -10
Xmax= 10
Xscl= 5
Ymin= 0
Ymax= 100
Yscl= 10
```

In the screen where the function is defined, we enter the function

$$y = x^2$$

Now we can display the graph (Fig. 12-32). We see that the graph is a parabola with the vertex at $(0, 0)$.

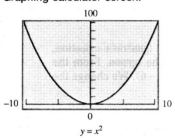

■ **FIGURE 12-32**
Graphing calculator screen.

Note: When a figure includes a graphing calculator screen, *we will show the values of Xmin, Xmax, Ymin, and Ymax, although they do not appear on the screen.*

Complete Graph

We are often *given* the domain of a function we wish to graph. But if we're not told which x values to use, we must determine an appropriate domain. We want to include those x values that show the important features of the graph, such as asymptotes, intercepts, and maximum and minimum points. A graph that shows all the important features is called a *complete* graph.

If we don't know an appropriate domain and range which will display a complete graph of the function, we may need to *explore* with different windows. On the graphing calculator, a good starting point may be a window in which $-5 \leq x \leq 5$ and $-3 \leq y \leq 3$. In this window, the origin is in the centre of the screen and a scale of 1 is appropriate on both the x and y axes. Many graphing calculators have built-in window settings (standard, initial, decimal, trig, etc.), which use predefined domains and ranges. These settings are very handy and convenient to use. Check your calculator manual for specifics.

FIGURE 12-33

Example 49

Graph the function and estimate the important features of the graph:

$$y = x^3 - 3x^2 + 2$$

Solution: We set the window so that

$$-5 \leq x \leq 5 \quad \text{and} \quad -3 \leq y \leq 3$$

with x and y scales of 1. We enter and graph the function (Fig. 12-33). The complete graph of this function is displayed in the window. From the graph we estimate that the y intercept is 2, the x intercepts are about -0.7, 1, and 2.7, and maximum and minimum points are at about $(0, 2)$ and $(2, -2)$.

Changing the Window on the Graphing Calculator

Sometimes the screen does not show a complete graph of the function, or it may not show the portion of the graph that we're interested in. After graphing with one window setting, if the important features of the graph are not shown on the screen, the window may be changed to a more appropriate one.

FIGURE 12-34

Example 50

Make a complete graph of the function

$$y = 2x^2 + 3x - 5$$

Solution: We really don't know what an appropriate window for this function is, so we try

$$-5 \leq x \leq 5 \quad \text{and} \quad -3 \leq y \leq 3$$

We enter and graph the function (Fig. 12-34). Since this is a quadratic equation, the graph should be a parabola. But the vertex is *not* on the screen. From the portion shown, we estimate that the vertex is at about $(-1, -6)$. We change the window so that

$$-4 \leq x \leq 2 \quad \text{and} \quad -8 \leq y \leq 2$$

and graph the function again (Fig. 12-35). Since this graph shows the vertex of the parabola, and the x and y intercepts, it is now a complete graph. From the graph, we estimate that the x intercepts are -2.5 and 1, the y intercept is -5, and the vertex is at *about* $(-1, -6)$.

FIGURE 12-35

Using the TRACE Feature of the Graphing Calculator

The TRACE feature on a graphing calculator is used to move the cursor from one point to the next, while displaying the coordinates of the point on the screen.

Example 51

Find the approximate coordinates of the vertex of the parabola

$$y = 3x^2 - 7x + 5$$

Solution: Using a window so that $-5 \leq x \leq 5$ and $-3 \leq y \leq 3$, we graph the function

$$y = 3x^2 - 7x + 5$$

FIGURE 12-36

(Fig. 12-36). With the graph on the screen, we turn on the TRACE feature. A blinking point, called the *cursor*, appears on the graph. The coordinates of that point are shown at the bottom of the screen. We use the arrow keys to move the cursor along the curve until it is as close to the vertex as we can get. Since this parabola opens up, the vertex is the point which has the smallest y value. We read the coordinates of the point (Fig. 12-37). The coordinates of the vertex are about $x = 1.1702128$ and $y = 0.91670439$. *Not all of these digits are significant.* In fact, we can conclude only that the vertex is *approximately* at (1.2, 0.9).

FIGURE 12-37
Tracing on a graph.

COMMON ERROR When using the TRACE feature on a graphing calculator, the coordinates of the points are *approximations*, and many digits may *not* be significant.

FIGURE 12-38

Note: Many calculators have a feature to find the minimum value of a function. Using this feature, we find that the vertex (a minimum point) is at (1.167, 0.917) rounded to thousandths place.

Using the ZOOM Feature of the Graphing Calculator

The ZOOM feature on a graphing calculator allows us to enlarge or reduce a portion of a graph. To enlarge a small portion of a graph, we can "zoom in" on a region around a point or we can "box" in a portion of the window. To display a larger portion of the graph on the screen, we can "zoom out"—this is useful if only part of the complete graph is shown. The extent of the enlargement or reduction depends on the "zoom factors."

FIGURE 12-39
Zooming out.

Example 52

Graph the function

$$y = 5x - x^3$$

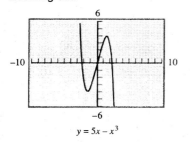

Solution: We clear the graph screen and set the window so that $-5 \leq x \leq 5$ and $-3 \leq y \leq 3$. We graph the function (Fig. 12-38). The maximum and minimum points are not shown on this graph, so we would like to see a larger portion of the graph. We set the zoom factors to 2 for both the x and the y axes—this *doubles* the length of both the x and y axes when we zoom out. We "zoom out" to see a larger portion of the graph (Fig. 12-39). We use the TRACE feature to find that the x intercepts are at about $(-2.2, 0)$ and $(2.2, 0)$, and the minimum and maximum points are at about $(-1.3, -4.3)$ and $(1.3, 4.3)$.

12-7 Graphing Calculators

Sketching a Graphing Calculator Graph on Graph Paper

Often we need to sketch a graph on graph paper after we've graphed it on the calculator. To sketch the graph, *we must still plot points and connect them.* We use the graph on the calculator to determine the coordinates of important points, such as the x and y intercepts and maximum or minimum points. We plot these points and connect them with a smooth curve. If asymptotes exist, we show them. Finally we label the x and y axes and label the scales on each axis.

Graphical Solution of Equations by Graphing Calculator

In Sec. 12-6 we showed how to solve an equation graphically. Any equation can be solved graphically if we can graph the associated function, and the graphing calculator certainly simplifies this process. After we graph the function, we find the x intercepts using the root feature of the graphing calculator. The "roots" are the solutions of the original equation.

FIGURE 12-40
Finding a root.

Example 53

Use a graphing calculator to solve the equation graphically. Approximate the solutions to three significant digits.

$$1.07x^2 + 2.93x = 6.75$$

Solution: We bring all terms to one side of the equal sign,

$$1.07x^2 + 2.93x - 6.75 = 0$$

set the left side equal to y,

$$y = 1.07x^2 + 2.93x - 6.75$$

and graph the function. Using the root feature (Fig. 12-40), we find that one solution is

$$x = -4.23$$

Again, using the root feature (Fig. 12-41), we find that the other solution is

$$x = 1.49$$

Note that the y value at this x intercept is not *exactly* 0,

$$y = -2 \cdot 10^{-13}$$

but it is very close to 0! Again, we're only finding *approximations*, and this is close enough.

FIGURE 12-41
Another root.

SECTION 12-7 Exercises Graphing Calculators

Graph of a Function

Set the window so that $-5 \leq x \leq 5$ and $-3 \leq y \leq 3$ with a scale of 1 for both x and y. Clear the graph screen and graph each function on the calculator. Sketch the graph on graph paper.

1. $y = 3x - 2$
2. $y = 1 - 2x$
3. $y = x^2 - 2x - 1$
4. $y = x^2 + 3x + 1$
5. $y = x^3 - x$
6. $y = 2x - x^3$

Changing the Window on the Graphing Calculator

Graph these functions for $-5 \leq x \leq 5$ and $-3 \leq y \leq 3$, or use a similar window on your calculator. The complete graphs of these functions may not appear on the screen. Change the window as needed, so that you display the complete graph. Sketch on graph paper, using the same x and y scales.

7. $y = 2x^2 + 3x - 6$
8. $y = 1 - 5x - 2x^2$
9. $y = 2 + 3x - x^3$
10. $y = x^3 - 3x^2 - 2$

Using the TRACE Feature of the Graphing Calculator

Graph each function and use the TRACE feature to approximate important features of the graph, such as the x and y intercepts and maximum and minimum points. Round to the nearest tenth. Sketch.

11. $y = 1 - 2x - x^2$
12. $y = x^2 - 7x + 10$
13. $y = 3x^2 - 4 - x^3$
14. $y = 4 - 3x^2 - x^3$

Using the ZOOM Feature

Graph each function and use the TRACE, ZOOM, and other calculator features to approximate important features of the graph. Round to the nearest hundredth. Sketch.

15. $y = -x^2 + 2x + 5$
16. $y = 5 - 4x - 3x^2$
17. $y = -x^3 + 2x^2 - 1$
18. $y = x^3 - 4x^2 - 2$

Graphical Solution of Equations

Use a graphing calculator to solve the equation graphically. Approximate the solutions to three significant digits.

19. $1.32x^2 + 1.95x = 4.02$
20. $2.56x^2 - 6.75 = 1.17x$
21. $2.56x^3 - 6.75 = 7.25x^2$
22. $1.36x + 3.75x^3 = 2.89$

Chapter 12 REVIEW EXERCISES

Express in equation form:
1. y is 3 times the square of x.
2. y is the amount by which 4 exceeds x.
3. Find the domain and range of the function
$$y = \sqrt{x - 3}$$
4. Given $f(x) = 2x^2 - 1$, find $f(-1)$, $f(2)$, $f(a)$, and $3f(0)$.
5. Given $g(x) = x - x^3$, find $g(1)$, $g(-2)$, $g(b)$, and $2g\left(\dfrac{1}{2}\right)$.
6. In which quadrant does a point (h, k) lie if $h > 0$ and $k < 0$?
7. Fill in the missing values and graph the equation.

$y = 3x - 2$

x	-2	-1	0	1	2
y					

State whether the equation is in implicit form or explicit form. If explicit, name the dependent and the independent variables.

8. $y = x^2 - 1$
9. $2x + 3y = 5$

10. Graphically solve the equation
$$x^3 - x^2 = 2x - 4$$
on a graphing calculator, rounding to the nearest tenth. Sketch on graph paper.

11. Make a table of ordered pairs, taking values of x from -3 to 3, and graph the function
$$y = x^2 + x - 6$$

12. Graph the linear function
$$y = 4 - 2x$$
Find the x intercept, the y intercept, and the slope of the line.

13. Graph the function
$$y = 3 - 5x - 2x^2$$
on a graphing calculator. Use the TRACE and ZOOM features to approximate the x and y intercepts and maximum and minimum points of the graph, rounding to the nearest hundredth. Sketch on graph paper.

14. The force F_r required to pull a block at a constant speed along a rough surface with a coefficient of friction μ_k is a linear function of the normal force F_N, expressed as follows
$$F_r = \mu_k K_N$$
A block of wood is being pulled across a wood surface with a coefficient of friction of 0.20. Graph the force F_r as a function of the normal force F_N, for F_N from 0 N to 35 N.

15. The power company determines your monthly charge as follows: The customer is charged \$6.65 for the first 200 kW·h of electricity used and an additional \$0.16 per kW·h for electricity used over 200 kW·h. Express the monthly charge C as a function of the electricity used x.

16. The centripetal force F of a body of mass m revolving in a circle of radius r with velocity v is given by
$$F = \frac{mv^2}{r}$$
A mass of 0.30 kg is revolving around a circle of radius 0.20 m. Graph F as a function of v, for v with a domain of 0 m/s to 3.0 m/s. From the graph, estimate the velocity of the mass when the centripetal force is $6.0\frac{\text{kg·m}}{\text{s}^2} = 6.0$ N.

17. When an amount of money a is invested at an annual interest rate r for t years, the accumulated amount is given by
$$y = a(1 + r)^t$$
if the interest is compounded annually. Write a formula for the amount accumulated when \$300 is invested, if the interest is compounded annually for two years. Find the accumulated amounts for interest rates of 6.5%, 7.5%, and 10.2%, rounding to the nearest dollar.

18. If an object is dropped from a height h (in m), its velocity v (in m/s) when it hits the ground is given by
$$v = \sqrt{2ah}$$
where a is 9.81 m/s². Write the equation for velocity and graph v as a function of h, for h from 0.0 to 20.0 m.

19. The quantity r represents the radius of the base of a cone and V represents the volume of the cone with radius r. Data were collected in an experiment:

r (cm)	1.5	2.6	3.5	4.4
V (cm³)	10.5	31.4	57.0	90.0

Graph the volume V as a function of the radius r.

20. The resistance R of a wire having resistivity ρ and cross-sectional area A is a linear function of the length L of the wire, given by

$$R = \rho \frac{L}{A}$$

An aluminum wire has a diameter of 10.0 mm. The resistivity of aluminum is 2.63×10^{-8} ohm-metres. Graph the resistance R in the wire as a function of its length L, for L from 0.0 to 8.0 m.

 21. The amount of work required to lift a weight is defined as the product of the weight and the distance lifted. Graph the work required to lift a 75 lb weight from 0.0 to 4.00 ft in increments of 0.5 feet. From the graph, estimate the work required to lift the weight 3.8 feet.

WRITING

22. State, in your own words, what is meant by a "function." What are the domain and range of a function and how are they found? Describe the graph of a function, list the important features, and how you find them. You may use examples to help describe these terms.

TEAM PROJECT

23. Use a graphing calculator to graph the function

$$y = x^3 - 2x^2 + x - 30$$

Find, to the nearest hundredth, the x and y intercepts and maximum and minimum points.

13 Plane Geometry

Outcomes

When you have completed this chapter, you will be able to

1. Introduce the basic concepts of plane geometry.
2. Define properties of lines and angles.
3. Define the measurement of angles.
4. Work with the properties of triangles.
5. Work with the properties of quadrilaterals.
6. Work with the properties of circles.

In this chapter we present the basics of plane geometry, and applications of these concepts. They are presented informally, without proof, and the emphasis is on their use in applications. We examine lines, angles, polygons, triangles, quadrilaterals, and circles.

13-1 LINES AND ANGLES

In this section we study lines and angles. We look at two ways of defining an angle and three ways of measuring angles. We examine different types of angles and different relationships between angles.

Lines, Segments, and Rays

■ FIGURE 13-1
Line AB.

■ FIGURE 13-2
Line segment AB.

■ FIGURE 13-3

■ FIGURE 13-4
Ray AB.

A *line* (Fig. 13-1) is often denoted by a single letter, such as L. If A and B are two points on a line, then the line is denoted by the symbol $\overleftrightarrow{AB}$, or simply line AB.

A *line segment* (Fig. 13-2) is part of a line that lies between two points on the line. The two points are called the *endpoints* of the line segment. If A and B are the endpoints of the line segment, then the segment is sometimes denoted by the symbol $\overline{AB}$, or just segment AB. The *length* of the line segment is denoted by the symbol AB, where AB is a distance and represents a positive number.

— **Example 1** —
In Fig. 13-3, $\overline{PQ}$ is a line segment whose endpoints are P and Q. The length of the segment is PQ.

A *ray* (Fig. 13-4) is part of a line that lies entirely on one side of a point. The point is called the *endpoint* of the ray. If A is the endpoint of the ray and B is any

■ FIGURE 13-5

other point of the ray, then the ray is denoted by the symbol $\vec{AB}$, or simply ray AB, with the endpoint always given first.

---- **Example 2** ----
In Fig. 13-5, $\vec{PQ}$ is a ray; its endpoint is P.

Angles

■ FIGURE 13-6
Angle.

An *angle* (Fig. 13-6) is formed by two rays with a common endpoint. The common endpoint (the point of intersection of the two rays) is called the *vertex,* and the rays are called the *sides* of the angle. Often we give the angle a special name, such as θ or α (these are Greek letters). An angle can also be named using the angle symbol and the vertex, as in $\angle B$. If a point on each side of the angle is given, you may denote the angle by $\angle ABC$ or $\angle CBA$, taking special care to write the vertex as the middle point. So the angle in Fig. 13-6 can be written in any of the following ways:

$$\theta \quad \angle B \quad \angle ABC \quad \angle CBA$$

---- **Example 3** ----
Name the angle in Fig. 13-7 four different ways.

Solution: The angle can be written as α, $\angle Y$, $\angle XYZ$, or $\angle ZYX$.

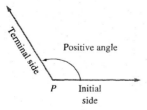

■ FIGURE 13-7

An Angle Formed by a Rotation

Sometimes it is useful to think of an angle as being formed by the *rotation* of a ray. If we begin with a ray (Fig. 13-8) in some *initial position* (the *initial side*) and rotate the ray about its endpoint to a new position (the *terminal side*), then we have formed an angle. The angle is defined as the amount of rotation of the ray. We use a curved arrow to indicate the direction of the rotation. We usually consider angles formed by rotation in the *counterclockwise* direction as *positive,* and angles formed by *clockwise* rotation as *negative.*

■ FIGURE 13-8
Angle formed by a rotation.

■ FIGURE 13-9
Positive and negative angles.

---- **Example 4** ----
In Fig. 13-9, $\angle P$ is a positive angle, since it is formed by a counterclockwise rotation, and $\angle Q$ is a negative angle, since it is formed by a clockwise rotation.

Measure of an Angle

Angles are generally measured using one of three units: revolutions, degrees, or radians.

Angles are measured in SI using the radian, but the degree (a non-SI unit acceptable for use with the International System) is the most commonly used angular unit of measure. A natural unit of angular measure, the revolution is used extensively in engineering applications.

If we rotate a ray counterclockwise so that it returns to its initial position, we call the rotation one *revolution.*

Angles are also measured in degrees. We define one *degree* (1°) to be equal to $\frac{1}{360}$ of a revolution. Thus *there are 360 degrees in one revolution.*

FIGURE 13-10
Protractor.

We often use a *protractor* (Fig. 13-10) to measure angles in degrees. If the direction of rotation is not indicated, the angle is assumed to be positive.

> **TIP** In geographic mapping, the initial position of a ray is at the North position and rotates clockwise with east being at 90 degrees. One full rotation is also equal to 360 degrees.

Example 5

Estimate the measure of the angles in Fig. 13-11 and then use a protractor to measure them to the nearest degree.

Solution: Since the direction of the rotation is not indicated, the angles are assumed to be positive. The measure of angle A is 30°. The measure of angle B is 90°. The measure of angle C is 150°.

> **TIP:** We will frequently omit the words "measure of angle" when discussing angle measurements. For example, we may say $A = 30°$ instead of the more correct, but wordy, *the measure of angle A is 30°*.

FIGURE 13-11

1 rad ≈ 57.3°

FIGURE 13-12
Angle of 1 radian.

Fractional parts of a degree may be expressed as decimal degrees (tenths, hundredths, thousandths, etc.) or as minutes and seconds. A *minute* (′) is equal to $\frac{1}{60}$ of a degree and a *second* (″) is equal to $\frac{1}{60}$ of a minute, or $\frac{1}{3600}$ of a degree. Surveyors must frequently be accurate to the nearest second.

In some fields, angles are measured in *radians*. The angle in Fig. 13-12 has a measure of 1 radian; its vertex is the centre of a circle, and it "cuts off" a part of the circle equal in length to the radius. There are 2π radians in one revolution, where $\pi \approx 3.1416$. If the unit of an angle measurement is not given, the angle is assumed to be in radians. One radian is about 57.3°. (This topic will be covered in more detail in Sec. 24-1.)

You may find on your calculator a unit of angle measurement called a *grad*. There are 400 grads in one revolution. The grad is popular in Europe. We will not use grads in this book.

Conversions

We convert angles using the same method we used to convert other units in Chapter 4. When converting between units of measurement, *retain as many significant digits in the conversion as are given in the original measurement*. A table of conversions is given in Appendix B, starting on page 702. A summary of angle conversions follows:

ANGLE CONVERSIONS	1 rev = 360° 1° = 60′ 1′ = 60″ 1 rev = 2π rad ≈ 6.28 rad 360° = 2π rad ≈ 6.28 rad	79

246   CHAPTER 13 / Plane Geometry

Example 6

Convert 34°23'56" to decimal degrees.

Solution:

$$34°23'56" = 34° + 23' + 56"$$

$$= 34° + 23'\left(\frac{1°}{60'}\right) + 56"\left(\frac{1'}{60"}\right)\left(\frac{1°}{60'}\right)$$

$$= \left(34 + \frac{23}{60} + \frac{56}{3600}\right)° = 34.3989°$$

We round to six significant digits, since there were six digits in the original measurement.

Many scientific calculators have special keys for converting degrees, minutes, and seconds into decimal degrees, and vice versa. Consult your calculator manual for specific instructions.

Example 7

Convert 67.2765° to degrees, minutes, and seconds.

Solution:

$$67.2765° = 67° + 0.2765°$$

Converting to degrees and minutes:

$$= 67° + 0.2765°\left(\frac{60'}{1°}\right)$$

$$= 67° + 16.59'$$

Converting to degrees, minutes, seconds:

$$= 67° + 16' + 0.59'$$

$$= 67° + 16' + 0.59'\left(\frac{60"}{1'}\right)$$

$$= 67° + 16' + 35" \text{ (rounded)}$$

$$= 67°16'35"$$

Example 8

Convert to radians.
(a) 1.50 rev   **(b)** 245°

Solution:

(a) $1.50 \text{ rev} = (1.50 \text{ rev})\left(\frac{2\pi \text{ rad}}{1 \text{ rev}}\right)$

$= 9.42 \text{ rad}$

(b) $245° = 245°\left(\frac{2\pi \text{ rad}}{360°}\right) = 4.28 \text{ rad}$

Alternate Solution:

Often conversions from degrees to radians are reported as a product of a fraction and π. This form is considered to be exact.

$$245° = 245°\left(\frac{2\pi \text{ rad}}{360°}\right) = \frac{49\pi}{36} \text{ rad}$$

■ FIGURE 13-13

■ FIGURE 13-14
Types of angles.

─── **Example 9** ───
Convert to revolutions.
(a) 7.25 rad (b) 52°12′15″

Solution:

(a) $7.25 \text{ rad} = (7.25 \text{ rad})\left(\dfrac{1 \text{ rev}}{2\pi \text{ rad}}\right) = 1.15 \text{ rev}$

(b) First we change to decimal degrees.

$$52°12′15″ = 52° + \dfrac{12°}{60} + \dfrac{15°}{3600} = 52.20417°$$

Converting to revolutions:

$$52°12′15″ = 52.20417° \left(\dfrac{1 \text{ rev}}{360°}\right)$$
$$= 0.145012 \text{ rev}$$

─── **Example 10** ───
Convert to degrees.
(a) 1.72 rev (b) 3.90 rad (c) $\dfrac{4}{3}\pi$ rad.

Solution:

(a) $1.72 \text{ rev} = 1.72 \text{ rev}\left(\dfrac{360°}{1 \text{ rev}}\right) = 619°$

(b) $3.90 \text{ rad} = (3.90 \text{ rad})\left(\dfrac{360°}{2\pi \text{ rad}}\right) = 223°$

(c) $\left(\dfrac{4}{3}\pi \text{ rad}\right)\left(\dfrac{360°}{2\pi \text{ rad}}\right) = 240°$

Note that when an angle is considered a rotation, there is no limit to the size of an angle (Fig. 13-13). The measure of the angle is signed—it may be negative, zero, or positive, or even very large, like 900°.

Types of Angles

A *right* angle is defined as a 90° angle, or $\dfrac{1}{4}$ revolution. It is usually denoted by a small square at the vertex, as in Fig. 13-14. An *acute* angle is an angle between 0° and 90°. An *obtuse* angle is an angle between 90° and 180°.

─── **Example 11** ───
In Fig. 13-11, angle A is acute, since $A = 30°$; angle B is a right angle, since $B = 90°$; angle C is an obtuse angle, since $C = 150°$.

Special Relationships Between Angles

Two angles are called *adjacent* if they have a common side and a common vertex and they lie on opposite sides of their common side (Fig. 13-15). Two angles are

(a) Adjacent angles

(b) Complementary angles

(c) Supplementary angles

FIGURE 13-15
Special relationships between angles.

called *complementary* if their sum is 90° (Fig. 13-15). Two angles are *supplementary* if their sum is 180° (Fig. 13-15). Note that complementary and supplementary angles are not always adjacent.

---- **Example 12** ----

In Fig. 13-16, find θ.

FIGURE 13-16

Solution: The two angles form a right angle and thus are complementary, so

$$\theta + 23.1° = 90°$$
$$\theta = 90° - 23.1° = 66.9°$$

---- **Example 13** ----

In Fig. 13-17, find θ.

FIGURE 13-17

Solution: Since θ is supplementary to the 44.2° angle,

$$\theta = 180° - 44.2° = 135.8°$$

When two lines intersect they form two pairs of opposite angles. The angles within each pair are called *vertical* angles (Fig. 13-18). A very important theorem in geometry is

FIGURE 13-18
Vertical angles.

$A = B$

FIGURE 13-19

| Vertical angles are equal. | 60 |

---- **Example 14** ----

In Fig. 13-19, $\theta = 41.6°$, since vertical angles are equal.

13-1 Lines and Angles

Another very important theorem is

| When two lines intersect, the adjacent angles are supplementary. | 61 |

FIGURE 13-20
Supplementary adjacent angles.

Thus, when two lines intersect, the sum of adjacent angles is 180° (Fig. 13-20). So, $A + B = 180°$.

Example 15

In Fig. 13-21, find θ.

FIGURE 13-21

Solution: When two lines intersect, the adjacent angles are supplementary and their sum is 180°, so

$$\theta + 65.2° = 180°$$
$$\theta = 180° - 65.2° = 114.8°$$

Two lines that intersect at a right angle are called *perpendicular* (Fig. 13-22). Two lines in a plane that never intersect are called *parallel* (Fig. 13-23).

A *transversal* is a line that intersects two or more lines. In Fig. 13-24, two parallel lines, L_1 and L_2, are intersected by a transversal, T. Angles A, B, G, and H are called *exterior* angles, and angles C, D, E, and F are called *interior* angles. Angles A and E are called *corresponding* angles. Other corresponding pairs of angles are C and G, B and F, and D and H. Angles C and F are called *alternate*

FIGURE 13-22
Perpendicular lines.

FIGURE 13-23
Parallel lines.

FIGURE 13-24
Transversal T.

interior angles, as are angles *D* and *E*. Refer to Fig. 13-24 to verify another important theorem in geometry:

> If two parallel lines are intersected by a transversal, corresponding angles are equal, and alternate interior angles are equal. | 62

Example 16
Find θ in Fig. 13-25.

■ FIGURE 13-25   ■ FIGURE 13-26

Solution: We need to introduce another angle, α (Fig. 13-26), which is a corresponding angle to 53.7°. Thus, $\alpha = 53.7°$. Now, θ is supplementary to α, so

$$\theta = 180° - \alpha = 180° - 53.7° = 126.3°$$

SECTION 13-1 Exercises Lines and Angles

Lines, Segments, and Rays
1. Draw a segment *RS*, so that *RS* = 3 cm.
2. Draw a ray *QP*. What is the endpoint of ray *QP*?

Angles
3. Name the angle in Fig. 13-27 four different ways.
4. Name the angle in Fig. 13-28 four different ways.

■ FIGURE 13-27   ■ FIGURE 13-28

Measure of an Angle
5. Estimate the measure of each angle in Fig. 13-29 and then measure with a protractor to the nearest degree.

■ FIGURE 13-29

(a)   (b)   (c)

■ FIGURE 13-30

6. Estimate the measure of each angle in Fig. 13-30 and then measure with a protractor to the nearest degree.
7. What is a radian?
8. Draw an angle whose measure is 2 rads.

Conversions

Round all conversions appropriately.

Convert to decimal degrees.
9. 32°18′   10. 140°37′30″   11. 93°47′29″
12. 67°42′   13. 235°22′30″   14. 69°36′27″

Convert to degrees, minutes, and seconds.
15. 56.25°   16. 172.125°   17. 12.345°
18. 74.7°   19. 263.875°   20. 164.985°

Convert to radians.
21. 2.50 rev   22. 15.0 rev   23. 60.0°
24. 75.0°   25. 125°50′   26. 215°32′15″

Convert to revolutions.
27. 3.50 rad   28. 1.25 rad   29. 301°
30. 335°   31. 75°24′55″   32. 146°35′51″

Convert to decimal degrees.
33. 4.50 rev   34. 2.1345 rad

Convert to degrees, minutes, and seconds.
35. 5.123556 rad   36. 3.61000 rev

Sketch the following angles and show the rotation:
37. 135°   38. −60°   39. 300°   40. −540°
41. 225°   42. −45°   43. 150°   44. −840°

Special Relationships Between Angles

45. Draw two complementary angles that are not adjacent.
46. Draw two supplementary angles that are adjacent.
47. Find θ in each part of Fig. 13-31.
48. Find θ in each part of Fig. 13-32.
49. Find angles A, B, C, D, E, F, and G in Fig. 13-33, given that L_1 and L_2 are parallel.
50. Find angles A, B, C, D, E, F, and G in Fig. 13-34, given that L_1 and L_2 are parallel.

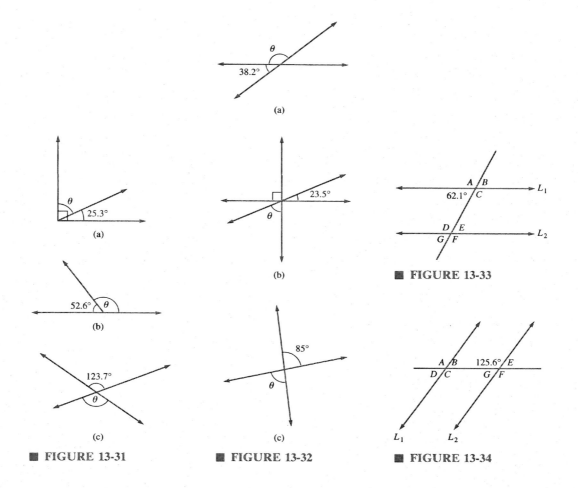

FIGURE 13-31 **FIGURE 13-32** **FIGURE 13-33** **FIGURE 13-34**

Applications

51. An escalator in a large store (Fig. 13-35) makes an angle of 34.8° with the first floor. Find the other angles A, B, and C.

52. A ray of light makes a 35.3° angle with the plane of a mirror. What angle does the ray of light make with a line perpendicular to the mirror?

53. A ship is traveling due west. If the captain turns 90° to the right, what is its new direction?

54. A support for a wall (Fig. 13-36) makes an angle of 32.5° with the floor at a corner. What angle does it make with the wall?

FIGURE 13-35

FIGURE 13-36

13-1 Lines and Angles   253

13-2 TRIANGLES

In this section we define the polygon and the triangle and look at different types of triangles. We discuss the concepts of perimeter and area and relate those concepts to the triangle. Lastly, we look at some properties of special triangles.

Polygons

A *polygon* (Fig. 13-37) is a plane figure formed by three or more line segments, called the *sides* of the polygon. The sides are joined at their endpoints to form a closed figure. A point where the two sides meet is called a *vertex* (plural, *vertices*). If all the sides and angles of a polygon are equal, the polygon is called a *regular* polygon.

■ FIGURE 13-37
Polygon.

─── **Example 17** ───

The polygons in Fig. 13-38 are all regular. The square and the equilateral triangle are the most common regular polygons. The polygons in Fig. 13-39 are *not* regular polygons, since their sides are not all equal.

■ FIGURE 13-39
Polygons (not regular).

■ FIGURE 13-38
Regular polygons.

Perimeter of a Polygon

The *perimeter* of a polygon is simply the sum of its sides. It is the distance around the polygon.

■ FIGURE 13-40

■ FIGURE 13-41

■ FIGURE 13-42
Types of triangles.

Scalene

Equilateral

Isosceles

Example 18

The perimeter p of the polygon in Fig. 13-40 is the sum of its sides. We start at the lower left corner.

$$p = 1\text{ cm} + 2\text{ cm} + 2\text{ cm} + 1\text{ cm} + 3\text{ cm} = 9\text{ cm}$$

Area of a Polygon

The *area* of a polygon is a measure of the interior of the polygon. It is usually expressed as a number of square units.

Example 19

The area A of the polygon in Fig. 13-41 can be found by counting the square units. Thus

$$A = 6 \text{ square units}$$

Types of Triangles

A *triangle* is a polygon with three sides. There are many different types of triangles (Fig. 13-42). A *scalene* triangle has no equal sides and an *equilateral* triangle has three equal sides. An *isosceles* triangle has two equal sides. In an isosceles triangle, the angle between the two equal sides is called the *vertex angle* and the side opposite the vertex angle is called the *base*.

An *acute* triangle (Fig. 13-43) has three acute angles, and an *obtuse* triangle has one obtuse angle. A *right* triangle (Fig. 13-43) has one right angle, and an *oblique* triangle (Fig. 13-43) has no right angle.

Sum of the Angles

A very important theorem in geometry concerns the angles of a triangle: *The sum of the angles of any triangle is* 180° (Fig. 13-44).

| SUM OF THE ANGLES OF A TRIANGLE | $A + B + C = 180°$ | 103 |

This enables us to find the third angle of a triangle if the other two angles are known.

■ FIGURE 13-43
Types of triangles.

■ FIGURE 13-44
Sum of the angles of a triangle.

$A + B + C = 180°$

13-2 Triangles

Example 20

Find angle *A* in a triangle if the other two angles are 53.2° and 46.5°.

Solution: By Eq. 103, the sum of the angles must be 180°. Therefore,

$$A = 180° - (53.2° + 46.5°)$$
$$A = 80.3°$$

Altitude and Base of a Triangle

An *altitude* of a triangle (Fig. 13-45) is a line segment which extends from a vertex perpendicular to the opposite side (called the *base*), or to an extension of that side. Every triangle has three altitudes, one from each vertex. *The three altitudes of a triangle intersect in a single point.*

■ **FIGURE 13-45**
Altitude and base of a triangle.

Example 21

Draw the three altitudes in the given triangle and show that they intersect in a single point.

Solution: An altitude is drawn from each vertex (Fig. 13-46) so that it is perpendicular to the opposite side. We see that all three altitudes intersect in a single point.

■ **FIGURE 13-46**
Altitudes.

Median

A *median* of a triangle (Fig. 13-47) is a line segment joining a vertex to the midpoint of the opposite side. In Fig. 13-47, the markings on the base indicate equal line segments.

Every triangle has *three* medians, one from each vertex. *The three medians intersect in a single point* (but not usually the same point at which the altitudes intersect).

■ **FIGURE 13-47**
Median of a triangle.

■ **FIGURE 13-48**
Medians.

Example 22

Draw the three medians in an acute triangle and show that they intersect in a single point.

Solution: A median is drawn from each vertex to the midpoint of the opposite side (Fig. 13-48). We see that the three medians intersect in a single point.

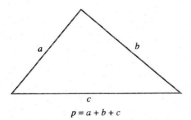

FIGURE 13-49
Perimeter of a triangle.

Perimeter of a Triangle

The perimeter of a triangle having sides a, b, and c is the sum of a, b, and c (Fig. 13-49).

| PERIMETER OF A TRIANGLE | $p = a + b + c$ | 100 |

Example 23

Find the perimeter of the triangle in Fig. 13-50.

Solution: To find the perimeter, use Eq. 100.

$$p = a + b + c$$
$$= 3.21 \text{ cm} + 6.62 \text{ cm} + 5.83 \text{ cm}$$
$$= 15.66 \text{ cm}$$

FIGURE 13-50

Area of a Triangle

One way to find the area of a triangle is to use the familiar formula: *The area of a triangle equals one-half the product of the base b and the altitude h to that base* (Fig. 13-51).

| AREA OF A TRIANGLE WITH BASE b AND HEIGHT h | $A = \frac{1}{2}bh$ | 101 |

FIGURE 13-51
Area of a triangle.

Example 24

Use Eq. 101 to find the area of the triangle in Fig. 13-50.

Solution: To find the area, use Eq. 101

$$A = \frac{1}{2}bh$$
$$= \frac{1}{2}(6.62 \text{ cm})(2.83 \text{ cm})$$
$$= 9.37 \text{ cm}^2$$

Hero's Formula

There is another method of finding the area of a triangle. If a, b, and c are the lengths of the sides of a triangle, then the area is given by *Hero's* (or *Heron's*) formula.

| HERO'S FORMULA FOR THE AREA OF A TRIANGLE | $A = \sqrt{s(s-a)(s-b)(s-c)}$ where $s = \frac{a+b+c}{2}$ | 102 |

Example 25

Use Hero's formula to find the area of the triangle in Fig. 13-50.

Solution: Using Eq. 102,

$$s = \frac{3.21 + 6.62 + 5.83}{2} = 7.83$$

$$A = \sqrt{s(s-a)(s-b)(s-c)}$$
$$= \sqrt{7.83(7.83 - 3.21)(7.83 - 6.62)(7.83 - 5.83)} = 9.36 \text{ cm}^2$$

Note that this result agrees with the answer in Example 24, since there is always some uncertainty in the last significant digit.

We use Hero's formula to find the area of a triangle when we know all three sides of the triangle. If we know a base and a height, then we would use the familiar formula, $A = \frac{1}{2}bh$.

Example 26

A triangular parking lot is bounded by three streets (Fig. 13-52). Find the cost of paving the area at a rate of $15.50/m², if the sides of the triangle are 1371 cm, 1067 cm and 762 cm.

■ **FIGURE 13-52**

Solution: We calculate the area using Hero's formula.

$$s = \frac{1371 + 1067 + 762}{2} = 1600$$

$$A = \sqrt{1600(229)(533)(838)}$$
$$= 404\,541.7477$$
$$= 405\,000 \text{ cm}^2$$

Converting cm² to m²,

$$A = 405\,000 \text{ cm}^2 \left(\frac{1 \text{ m}^2}{10\,000 \text{ cm}^2}\right) = 40.5 \text{ m}^2$$

Calculating the cost,

$$\text{Cost} = \left(\frac{\$15.50}{\text{m}^2}\right)(40.5 \text{ m}^2) = \$628$$

SECTION 13-2 Exercises Triangles

Types of Triangles

In Exercises 1 through 6 (Fig. 13-53), (a) find the measure of angle A, and (b) name all the terms from this list that apply to the triangle: scalene, isosceles, equilateral, right, acute, obtuse, oblique.

■ FIGURE 13-53

1.
2.
3.
4.
5.
6.

7. Sketch a right triangle, an equilateral triangle, and a scalene triangle. Measure the three angles of each triangle with a protractor. What is the sum of the angles of each triangle?

8. Sketch an isosceles triangle, an acute triangle, and an obtuse triangle. Measure the three angles of each triangle with a protractor. What is the sum of the angles of each triangle?

Sum of the Angles

Find the third angle in the triangle when the other two angles are given.

9. 62.7°, 75.9°
10. 32.6°, 87.4°
11. 47°20′, 67°35′
12. 95°45′, 38°30′

Altitude

13. Draw an obtuse triangle and its three altitudes. Do they intersect in a single point?

14. Draw an acute triangle and its three altitudes. Do they intersect in a single point?

Median

15. Draw an acute triangle and its three medians. Do they intersect in a single point?

16. Draw an obtuse triangle and its three medians. Do they intersect in a single point?

17. Draw an isosceles triangle and the median from the vertex angle. Is the vertex angle bisected? Is the base bisected? Is the median perpendicular to the base?

18. Draw an equilateral triangle and the medians. What can you say about the angles? What can you say about the medians and sides?

■ FIGURE 13-54

■ FIGURE 13-55

■ FIGURE 13-56

■ FIGURE 13-57

Perimeter and Area

19. Find the perimeter and area of the triangle in Fig. 13-54.
20. Find the perimeter and area of the triangle in Fig. 13-55.
21. Find the perimeter and area of the triangle in Fig. 13-56.
22. Find the perimeter and area of the triangle in Fig. 13-57.

Applications

23. A wedge is in the shape of an isosceles triangle. If the vertex angle is 22.8°, what are the two base angles?
24. A vertical flagpole is supported by a wire that forms an angle of 42.5° with the pole. Find the angle that the wire makes with the ground.
25. Three stakes, A, B, and C, are placed in the ground so that ∠ABC is 43.8°, and ∠BCA is 62.3°. Make a sketch and find ∠CAB.
26. Two angles of a triangular lot measure 48.5° and 74.6°. Find the third angle.

Round to the nearest dollar.

27. Find the cost of fencing in a triangular pasture, whose sides are 191 m, 139 m, and 115 m, if fence costs $1.65 per metre.
28. A garden is made in the shape of a regular pentagon, 12.0 m on a side. Find the cost of putting a brick border around the garden, if the brickwork costs $18.75 per metre.
29. What is the cost of a triangular piece of land with a base of 139 m and with an altitude of 68.6 m, if the price is at a rate of $2410 per hectare? (1 ha = 10 000 m²)
30. A soil analysis indicates that a farmer's field should be fertilized at a rate of 1120 kg/ha. A 20 kg bag of fertilizer costs $7.50. Find the cost of fertilizing a triangular field whose base is 175 metres and altitude is 149 metres.
31. At $4.50 a square yard, find the cost of paving a triangular court whose sides are 35.0 ft, 30.0 ft, and 25.0 ft.
32. If indoor-outdoor carpeting costs $24.00 a square yard, what is the cost of carpeting a triangular deck whose sides are 24.0 ft, 15.0 ft, and 15.0 ft?

13-3 RIGHT TRIANGLES

In this section, we study the right triangle, which you recall is a triangle with one right angle. We define the hypotenuse and legs and learn how to find them by using the Pythagorean theorem. Finally, we consider properties of two special triangles: the 30-60-90 right triangle and the 45-45-90 right triangle.

■ FIGURE 13-58
Right triangle.

Definitions

In a right triangle (Fig. 13-58), the side that is opposite the right angle is called the *hypotenuse*. The two other sides are called *legs*. Note that the hypotenuse is always the longest side of a right triangle.

Sum of the Angles

The right angle of a right triangle is, of course, 90°. Since the sum of the angles of any triangle is 180°, then two other angles must total 90°. These two angles (A and B in Fig. 13-59) are therefore complementary. Each one must be less than 90°, and therefore both are acute.

■ FIGURE 13-59
Acute angles of a right triangle.

| The acute angles of a right triangle are complementary. | $A + B = 90°$ | 106 |

Example 27
Find angle A in Fig. 13-60.

■ FIGURE 13-60

Solution: Since angle A and angle B are the acute angles of a right triangle, they are complementary.

$$A + B = 90°$$

So,

$$A = 90° - B = 90° - 27.3° = 62.7°$$

Pythagorean Theorem

The sides of a right triangle are related by the Pythagorean theorem. This theorem was named for the Greek mathematician and philosopher, Pythagoras, who lived

FIGURE 13-61
Pythagorean Theorem.

about 2500 years ago. Pythagoras proved that *the square of the hypotenuse of a right triangle is equal to the sum of the squares of the two legs* (Fig. 13-61).

| PYTHAGOREAN THEOREM | $c^2 = a^2 + b^2$ | 107 |

We use the Pythagorean theorem to find a missing side of a right triangle if we know the other two sides. If the missing side is the hypotenuse c, then we take the square root of both sides of Eq. 107 to get

| HYPOTENUSE OF A RIGHT TRIANGLE | $c = \sqrt{a^2 + b^2}$ | 108 |

Example 28
A right triangle has legs of length 4.37 cm and 6.59 cm. Find the length of the hypotenuse.

Solution: We sketch the triangle in Fig. 13-62 and let $a = 4.37$ and $b = 6.59$ in Eq. 108. Then c is the length of the hypotenuse.

$$c = \sqrt{4.37^2 + 6.59^2}$$
$$c = \sqrt{62.525} = 7.91 \text{ cm}$$

FIGURE 13-62

> **COMMON ERROR** Remember that the Pythagorean theorem applies only to *right triangles*. We need trigonometry (Chapter 22) to find the sides and angles of other (oblique) triangles.

Finding a Leg of a Right Triangle

FIGURE 13-63
Leg of a right triangle.

Suppose you need to find a leg, say a, of a right triangle (Fig. 13-63). We will solve Eq. 107 for the leg a.

$$c^2 = a^2 + b^2$$

Interchanging sides,

$$a^2 + b^2 = c^2$$

Subtracting b^2 from both sides,

$$a^2 = c^2 - b^2$$

Taking the square root of both sides,

$$a = \sqrt{c^2 - b^2}$$

| LEG OF A RIGHT TRIANGLE | $a = \sqrt{c^2 - b^2}$ | 109 |

CHAPTER 13 / Plane Geometry

■ FIGURE 13-64

Example 29

A ladder 9.00 m long reaches to the top of a building when its foot stands 4.00 m from the building (Fig. 13-64). How high is the building?

Solution: We see that the unknown side of the right triangle is a leg. We use Eq. 109, letting a = the length of the missing leg.

$$a = \sqrt{9.00^2 - 4.00^2} = 8.06 \text{ m}$$

COMMON ERROR You *add* the squares of the two legs to get the square of the hypotenuse. If the missing side is a leg, then you must *subtract* the square of the given leg from the square of the hypotenuse.

The 30-60-90 Right Triangle

When we draw a median in an equilateral triangle, the median bisects the angle and is perpendicular to the opposite side (Fig. 13-65). We consider just one of these

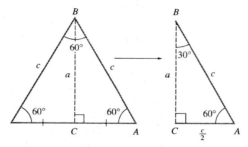

■ FIGURE 13-65

right triangles. Since $\angle B$ was 60° and this angle was bisected, then $\angle ABC$ in the right triangle is 30°. Angle C is a right angle, and $\angle A$ is still 60°. The three angles of the triangle are 30°, 60°, and 90°. Therefore, we call this triangle a *30-60-90 right triangle*. Since the base c of the equilateral triangle is bisected, the base of the right triangle is $\frac{c}{2}$. We use Eq. 109 to find the missing leg a.

$$a = \sqrt{c^2 - \left(\frac{c}{2}\right)^2} = \sqrt{\frac{4c^2}{4} - \frac{c^2}{4}} = \sqrt{\frac{3c^2}{4}} = \frac{c}{2}\sqrt{3}$$

So, in a 30-60-90 right triangle (Fig. 13-66), *the shorter leg is half the length of the hypotenuse and the longer leg is $\sqrt{3}$ times the shorter leg.*

30-60-90 RIGHT TRIANGLE RELATIONSHIPS

110

■ FIGURE 13-66

When we solve problems involving 30-60-90 right triangles, we assume the angle measurements are exact, and so we round the results to the number of significant digits in the given side.

Example 30

The hypotenuse of a 30-60-90 right triangle is 3.74 cm long. Find the two legs.

Solution: We sketch the triangle in Fig. 13-67.

$$b = \frac{3.74}{2} = 1.87 \text{ cm}$$

$$a = (1.87)(\sqrt{3}) = 3.24 \text{ cm}$$

FIGURE 13-67

Example 31

The altitude of an equilateral triangle is 7.46 cm. Find the side c.

Solution: We sketch the equilateral triangle in Fig. 13-68, and draw an altitude to the base. Angle B is 60°, since all angles of an equilateral triangle are 60°. Thus, the altitude is the longer leg of a 30-60-90 right triangle. The side c of the equilateral triangle is the hypotenuse of the 30-60-90 right triangle. We are given the longer leg and need to find the hypotenuse c.

$$\frac{c}{2}\sqrt{3} = 7.46$$

Solving for c,

$$c = \frac{(2)(7.46)}{\sqrt{3}} = 8.61 \text{ cm}$$

FIGURE 13-68

The 45-45-90 Right Triangle

A 45-45-90 right triangle (Fig. 13-69) has two 45° angles and one right angle. It is also called an *isosceles right triangle*. In an isosceles right triangle, the two legs are equal, and the third side is the hypotenuse. One angle is a right angle and the other two angles are equal, since the base angles of an isosceles triangle are equal. Therefore each base angle must be 45°.

If we are given the leg a of an isosceles right triangle (Fig. 13-70), we can find the hypotenuse c using Eq. 108

$$c = \sqrt{a^2 + a^2} = \sqrt{2a^2} = a\sqrt{2}$$

FIGURE 13-69
Isosceles right triangles.

Thus, in an isosceles right triangle, the hypotenuse c is the length of the leg a times $\sqrt{2}$.

ISOSCELES RIGHT TRIANGLE RELATIONSHIPS

■ FIGURE 13-70

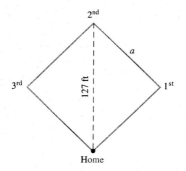

■ FIGURE 13-71

Example 32

Find the hypotenuse of an isosceles right triangle whose legs are 5.00 cm each.

Solution:
$$c = a\sqrt{2} = (5.00)(\sqrt{2}) = 7.07 \text{ cm}$$

Example 33

A baseball diamond is a square. If it is 127 ft from home plate to second base (Fig. 13-71), how far is it between the bases? (Round to the nearest foot.)

Solution: The diagonal of a square separates the square into two isosceles right triangles. The diagonal is the hypotenuse of each triangle. So,
$$a\sqrt{2} = 127$$
$$a = \frac{127}{\sqrt{2}} = 90 \text{ ft}$$

SECTION 13-3 Exercises Right Triangles

Finding the Sides of a Right Triangle

1. What is wrong with this definition: "The hypotenuse is the longest side of a triangle"?
2. Name the hypotenuse and the legs of the triangles in Fig. 13-72.

■ FIGURE 13-72

(a)

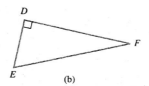

(b)

13-3 Right Triangles   265

3. Find the missing sides in Fig. 13-73.
4. Find the missing sides in Fig. 13-74.

■ FIGURE 13-73

■ FIGURE 13-74

30-60-90 and 45-45-90 Right Triangles

5. Find the missing sides in Fig. 13-75.
6. Find the missing sides in Fig. 13-76.

■ FIGURE 13-75

■ FIGURE 13-76

7. Find the altitude and the area of an equilateral triangle 4.75 metres on a side.
8. Find the side of an equilateral triangle whose altitude is 8.50 cm.
9. Find the diagonal of a square whose side is 56.0 cm.
10. Find the diagonal of a square whose perimeter is 13.0 cm.
11. Find the perimeter of a square whose diagonal is 18.25 cm.
12. Find the side of a square whose diagonal is 3.56 cm.

■ FIGURE 13-77

■ FIGURE 13-78

13. The length of a rectangle is 54.2 mm. The diagonal is 67.3 mm long. Find the perimeter and area of the rectangle.

14. Find the perimeter and area of a 30-60-90 right triangle whose longer leg is 11.0 mm.

15. A vertical pole 35.0 m high is supported by a guy wire attached to the top. The wire reaches the ground 45.0 m from base of the pole. How long is the wire?

16. How high is a loading platform if the bottom of a 4.57 m ramp is 3.96 m from the base of the platform?

17. A ladder 14.0 m long reaches to the top of a building when its foot stands 5.00 m from the building. How high is the building?

18. Two streets, one 15.6 m wide and the other 29.8 m wide, cross at right angles. What is the distance between opposite corners?

19. A rectangular room is 7.13 m by 7.32 m. What is the distance from one corner to its opposite corner?

 20. A ladder 28.0 ft long stands flat against the side of a building. How many feet must it be pulled out at the bottom so that the top may be lowered 5.0 ft?

21. A house is 36.0 ft wide, the ridge is 10.0 ft higher than the side walls, and the rafters project 2.00 ft beyond the sides of the house (Fig. 13-77). How long are the rafters?

22. A carpenter must build a ramp for disabled persons to enter a door 3.50 ft above the ground level. How long will the ramp be if it must begin to rise 25.0 ft from the base of the building?

23. A hex head bolt measures 0.750 in. across the flats (Fig. 13-78). Find the distance between opposite corners. (Hint: Draw 6 equilateral triangles using diagonals, and then separate those into 12 equal right triangles.)

24. A room is 18.0 ft long, 15.0 ft wide, and 8.00 ft high. What is the distance from one of the lower corners to the opposite upper corner?

13-4   QUADRILATERALS

■ FIGURE 13-79
Quadrilateral.

A *quadrilateral* (Fig. 13-79) is a polygon that has 4 sides. The trapezoid, parallelogram, rectangle, rhombus, and square are quadrilaterals.

Sum of the Angles

The sum of the angles of a quadrilateral is 360°.

SUM OF THE ANGLES OF A QUADRILATERAL	$A + B + C + D = 360°$	63

FIGURE 13-80
Trapezoid.

Trapezoid

A *trapezoid* (Fig. 13-80) is a quadrilateral in which only one pair of opposite sides are parallel. The two parallel sides are called the *bases, a* and *b*. The perpendicular distance between the bases is called the *altitude, h*. (The altitude is sometimes called the *height*.)

The perimeter *p* of a trapezoid is the sum of its sides, just as it is for any polygon.

PERIMETER OF A TRAPEZOID	$p = a + b + c + d$	64

The area *A* of a trapezoid is the product of the average of the bases, $\frac{a+b}{2}$, times the altitude *h*.

$$A = \left(\frac{a+b}{2}\right)h$$

AREA OF A TRAPEZOID	$A = \frac{(a+b)h}{2}$	65

If the two nonparallel sides are equal, the trapezoid is called an *isosceles* trapezoid (Fig. 13-81). In the figure, the marks on the two nonparallel sides indicate that the sides are equal.

FIGURE 13-81
Isosceles trapezoid.

FIGURE 13-82

─── **Example 34** ───

Find the cost of flooring the trapezoidal deck in Fig. 13-82, if the flooring costs $23.99 per square metre. Round up to the nearest dollar.

Solution: Using Eq. 65 to find the area of the deck,

$$A = \frac{(a+b)h}{2} = \frac{(658 + 396)323}{2}$$

$$A = 170\,221$$

or $\qquad A = 17\overline{0}\,000 \text{ cm}^2$

Converting to m²,

$$17\overline{0}\,000 \text{ cm}^2 \left(\frac{1 \text{ m}^2}{10\,000 \text{ cm}^2}\right) = 17.0 \text{ m}^2$$

$$\text{Cost} = \left(\frac{\$23.99}{\text{m}^2}\right)(17.0 \text{ m}^2) = \$407.83 = \$408$$

FIGURE 13-83
Parallelogram.

Parallelogram

A *parallelogram* (Fig. 13-83) is a quadrilateral in which both pairs of opposite sides are parallel. In a parallelogram, both pairs of opposite sides are equal, and opposite angles are equal. In the figure, equal angles are marked with the same number of arcs.

FIGURE 13-84
Diagonals of a parallelogram.

A *diagonal* of a parallelogram is a line segment joining two opposite vertices of the parallelogram. Every parallelogram has two diagonals and the diagonals bisect each other (Fig. 13-84).

The perimeter of a parallelogram is the sum of its four sides (Fig. 13-83).

$$p = a + b + a + b$$

Combining terms,

$$p = 2a + 2b$$

Factoring out a 2 on the right, we get the formula,

PERIMETER OF A PARALLELOGRAM	$p = 2(a + b)$	66

FIGURE 13-85
Area of a parallelogram.

The area A of a parallelogram is equal to the product of its base b and its altitude h (Fig. 13-85).

AREA OF A PARALLELOGRAM	$A = bh$	67

--- **Example 35** ---

Find (a) the perimeter and (b) the area of the parallelogram in Fig. 13-86.

Solution:

(a) Using Eq. 66 for the perimeter of a parallelogram,

$$p = 2(a + b) = 2(36.2 + 45.6)$$
$$p = 163.6 \text{ cm}$$

FIGURE 13-86

(b) To find the area, first we draw an altitude to complete a 30-60-90 triangle. We see that the altitude is the short leg of the 30-60-90 right triangle and that the 36.2 cm side is the hypotenuse. The short leg is half the hypotenuse, so the altitude is 18.1 cm. Using Eq. 67 to find the area,

$$A = bh = (45.6)(18.1)$$
$$A = 825 \text{ cm}^2$$

FIGURE 13-87
Rectangle.

Rectangle

A *rectangle* (Fig. 13-87) is a parallelogram with four right angles. The diagonals of a rectangle bisect each other and are equal (Fig. 13-88).

In a rectangle, we call the base b and the height a. To find the perimeter, we use Eq. 66 to get the familiar formula,

PERIMETER OF A RECTANGLE	$p = 2(a + b)$	68

FIGURE 13-88
Diagonals of a rectangle.

To find the area of a rectangle, we use Eq. 67 to get the familiar formula,

AREA OF A RECTANGLE	$A = ab$	69

Example 36

How much can a farmer expect to earn from a rectangular corn field measuring 191 m by 267 m, if a hectare yields $556 on the average.

Solution: First we find the area of the field. Using Eq. 69,

$$A = ab = (191 \text{ m})(267 \text{ m})$$
$$A = 50\,997 \text{ m}^2 = 51\,\overline{0}00 \text{ m}^2$$

Now we convert the square metres into hectares.

$$A = 51\,\overline{0}00 \text{ m}^2 = (51\,\overline{0}00 \text{ m}^2)\left(\frac{1 \text{ ha}}{10\,000 \text{ m}^2}\right) = 5.10 \text{ ha}$$

Multiplying the number of hectares by $556 per hectare we have

$$\text{Earnings} = (5.10 \text{ ha})(\$556/\text{ha}) = \$2835.60 = \$2840$$

So the farmer could expect to earn approximately $2840 from the corn in the field.

Rhombus

FIGURE 13-89 Rhombus.

A *rhombus* (Fig. 13-89) is a parallelogram in which all four sides are equal. The diagonals of a rhombus bisect each other, are perpendicular to each other, and bisect the angles of the rhombus (Fig. 13-90).

Since all four sides in a rhombus are equal, the perimeter is just four times the length of a side a.

| PERIMETER OF A RHOMBUS | $p = 4a$ | 70 |

FIGURE 13-90 Diagonals of a rhombus.

The area of a rhombus is the product of the base a and the altitude h (Fig. 13-89).

| AREA OF A RHOMBUS | $A = ah$ | 71 |

FIGURE 13-91

Example 37

Find (a) the perimeter and (b) the area of the rhombus in Fig. 13-91.

Solution:
(a) To find the perimeter of the rhombus, we use Eq. 70 with $a = 4.37$,

$$p = 4a = 4(4.37)$$
$$p = 17.48 \text{ cm}$$

(b) To find the area, we must first find the altitude h. Using Eq. 110 for the long leg of a 30-60-90 triangle, we have

$$h = \left(\frac{4.37}{2}\right)\sqrt{3} = 3.785 \text{ cm}$$

Thus, the area of the rhombus (Eq. 71) is

$$A = ah = (4.37)(3.785)$$
$$A = 16.5 \text{ cm}^2$$

Square

A *square* (Fig. 13-92) is a rectangle with four equal sides. The diagonals of a square bisect each other, are equal, are perpendicular to each other, and bisect the angles of the square (Fig. 13-93).

■ **FIGURE 13-92**
Square.

■ **FIGURE 13-93**
Diagonals of a square.

Since all four sides in a square are equal, the perimeter is just four times the length of a side a.

| PERIMETER OF A SQUARE | $p = 4a$ | 72 |

The area A of a square is equal to the side a squared (Fig. 13-92).

| AREA OF A SQUARE | $A = a^2$ | 73 |

Example 38

Find the area of a square plot of land 352 m on a side, and express it in hectares (ha).

Solution: Using Eq. 73 for the area of a square,

$$A = a^2 = (352)^2$$
$$A = 123\,900 \text{ m}^2$$

Converting to hectares,

$$A = 123\,900 \text{ m}^2 = 123\,900 \text{ m}^2 \left(\frac{1 \text{ ha}}{10\,000 \text{ m}^2}\right)$$

$$A = 12.4 \text{ ha}$$

Name	Figure	Perimeter	Area
Trapezoid		$p = a + b + c + d$	$A = \dfrac{(a+b)h}{2}$
Parallelogram		$p = 2(a + b)$	$A = bh$
Rectangle		$p = 2(a + b)$	$A = ab$
Rhombus		$p = 4a$	$A = ah$
Square		$p = 4a$	$A = a^2$

■ **FIGURE 13-94**
Quadrilaterals.

Since rectangles, rhombi, and squares are all parallelograms, the perimeter and area formulas for the parallelogram may be used for the rectangle, rhombus, and square.

Information on the quadrilaterals is summarized in Fig. 13-94.

SECTION 13-4 Exercises Quadrilaterals

Quadrilaterals

1. Draw a parallelogram. Measure the angles with a protractor and add them. Is the sum close to 360°?

2. Draw a trapezoid. Measure the angles with a protractor and add them. Is the sum close to 360°? Do the same for a rhombus.

3. Draw a parallelogram and its two diagonals. State four properties concerning parallelograms and/or their diagonals.

4. Draw a rectangle and its two diagonals. State six properties concerning rectangles and/or their diagonals.

5. Draw a rhombus and its two diagonals. State six properties concerning rhombi and/or their diagonals.

6. Draw a square and its two diagonals. State seven properties concerning squares and/or their diagonals.

FIGURE 13-95

FIGURE 13-96

FIGURE 13-97

FIGURE 13-98

7. Find the perimeter and area of the quadrilaterals in Fig. 13-95.
8. Find the perimeter and area of the parallelogram in Fig. 13-96.
9. Find the perimeter and area of the rhombus in Fig. 13-97.
10. Find the perimeter and area of the trapezoid in Fig. 13-98.

Round to the nearest dollar.

Applications

11. Find the cost of shingling both sides of a two sided roof if each side measures 12.04 m by 5.95 m. The cost of shingling is $13.44 per square metre.
12. What is the cost of paving a sidewalk 86.87 m long and 1.68 m wide if the cost is $17.34 per square metre?
13. How many square tiles with an edge length of 20 cm will it take to cover a rectangular floor 7.20 m by 10.80 m?
14. What will it cost to concrete a cellar floor 10.67 m long and 9.60 m wide at a cost of $4.84 per square metre?
15. Find the cost of paving with asphalt a road 20.1 km long and 11.4 m wide if the cost is $1.375 per square metre.
16. A baseball diamond is square, 27.4 m (90 ft) on a side. What part of a hectare does a baseball diamond cover? (1 ha = 10 000 m²)
17. Find the area of a garden, in acres, that is 150.0 ft by 75.0 ft.
18. Find the cost of painting a vertical wall of a house 32.0 ft by 36.0 ft, if a gallon of paint covers 375 ft² and costs $12.95.
19. What is the cost of plastering the walls and ceiling of a room 15.2 ft long, 11.8 ft wide, and 8.75 ft high, at $9.50/yd², deducting 124 ft² for doors and windows?
20. Find the cost of lining a topless rectangular tank 54.3 in. long, 48.9 in. wide, and 36.5 in. deep, with zinc, weighing 5.21 lb per sq. ft, at $1.65 per lb installed.

13-5 CIRCLES

In this section we discuss the circle and the parts of a circle. We show how to calculate the diameter, circumference, and area of a circle.

Definitions

A *circle* (Fig. 13-99) is the set of all points in a plane that are equidistant from a given point C, called the *centre* of the circle. The distance from the centre to any

FIGURE 13-99
Circle.

FIGURE 13-100
Compass.

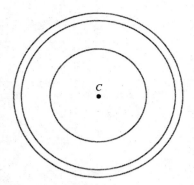

■ **FIGURE 13-101**
Concentric circles.

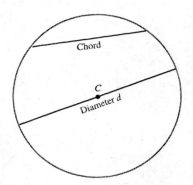

■ **FIGURE 13-102**
Chord and diameter of a circle.

point on the circle is called the *radius*. The term *radius* may also refer to a line segment whose endpoints are the centre and a point on the circle. All the *radii* (plural of *radius*) of a circle, of course, are equal.

A circle may be drawn using an instrument called a *compass* (Fig. 13-100). Circles that have the same centre (Fig. 13-101) are called *concentric*.

A *chord* (Fig. 13-102) is a line segment whose endpoints are on the circle. If the chord passes through the centre of the circle, then the chord is called a *diameter*. The diameter d of a circle is equal to twice its radius r.

DIAMETER AND RADIUS OF A CIRCLE	$d = 2r$ or $r = \dfrac{d}{2}$	74

Example 39

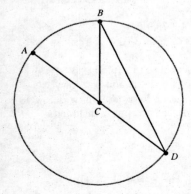

■ **FIGURE 13-103**

If C is the centre of the circle in Fig. 13-103, name the diameter, three radii, and two chords. If the diameter is 3.72 cm, find the radius.

Solution: The diameter is $\overline{AD}$; three radii are $\overline{AC}$, $\overline{CD}$, and $\overline{BC}$; two chords are $\overline{AD}$ and $\overline{BD}$. Using Eq. 74, with $d = 3.72$,

$$r = \frac{d}{2} = \frac{3.72}{2} = 1.86 \text{ cm}$$

Circumference

The perimeter of a circle is the distance around a circle and is called its *circumference*. The ratio (quotient) of the circumference C of a circle to (divided by) its diameter d is always a constant. This constant is called *pi* (a Greek letter pronounced *pie*), and is written π.

$$\frac{\text{Circumference}}{\text{Diameter}} = \frac{C}{d} = \pi$$

The constant π is an irrational number and has no exact fractional or decimal representation. It is approximately equal to 3.14, rounded to 2 decimal places. The value of π, rounded to 25 decimal places, is

$$\pi \approx 3.141\ 592\ 653\ 589\ 793\ 238\ 462\ 643\ 4 \text{ (rounded)}$$

All scientific calculators have a special key for π. You use the π key on your calculator when π is in a calculation, because the value the calculator gives for π has *many* significant digits and you won't be introducing rounding errors into the calculations.

We see that the circumference C of a circle is equal to π times the diameter d. We can also express the circumference in terms of the radius:

$$C = \pi d$$

Since $d = 2r$

$$C = \pi(2r) = 2\pi r$$

CIRCUMFERENCE OF A CIRCLE	$C = \pi d$ or $C = 2\pi r$	75

Example 40

Find the circumference of a circular water main 94.6 cm in diameter.

Solution: By Eq. 75,

$$C = \pi d = \pi(94.6 \text{ cm}) = 297 \text{ cm}$$

Area of a Circle

The area A of a circle is equal to π times the square of the radius r.

AREA OF A CIRCLE	$A = \pi r^2$	76

COMMON ERROR Some students have trouble distinguishing the formulas for the circumference and area of a circle. Remember that units of area are square units and the formula for the area of a circle has a square (πr^2) in it.

Example 41

Find the cross-sectional area of a pipe 3.26 cm in diameter.

Solution: By Eq. 74,

$$r = \frac{d}{2} = \frac{3.26 \text{ cm}}{2} = 1.63 \text{ cm}$$

By Eq. 76,

$$A = \pi r^2 = \pi(1.63 \text{ cm})^2 = 8.35 \text{ cm}^2$$

Arcs, Sectors, Segments, and Tangents

Any two points on a circle divide the circle into two parts, called *arcs* (Fig. 13-104). The shorter of the two arcs is named by its two endpoints A and B, and is denoted by the symbol $\widehat{AB}$, or simply arc AB. A *semicircle* is an arc which is "cut off" by a diameter of the circle. The endpoints of the semicircle are the endpoints of the diameter. In Fig. 13-104, $\widehat{DE}$ is a semicircle. A semicircle is half a circle.

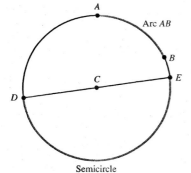

FIGURE 13-104
Arcs of a circle.

13-5 Circles

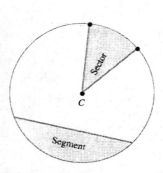

FIGURE 13-105
Sector and segment of a circle.

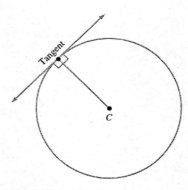

FIGURE 13-106
Tangent of a circle.

A *sector* of a circle (Fig. 13-105) is a region bounded by two radii and an arc. It is shaped like a piece of pie.

A *segment* of a circle (Fig. 13-105) is a region bounded by a chord and an arc.

A *tangent* (Fig. 13-106) is a line that intersects a circle in exactly one point. A tangent is always perpendicular to the radius drawn to the point of contact. We can often use that fact to solve problems involving tangents of circles.

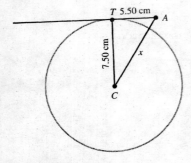

FIGURE 13-107

Example 42

From a point A outside a circle of radius 7.50 cm, a tangent is drawn to the circle (Fig. 13-107). The line segment $\overline{TA}$ is 5.50 cm long. Find the distance from A to the centre of the circle.

Solution: Since $\overline{AT}$ is a tangent, it is perpendicular to the chord $\overline{CT}$. Since triangle ATC is a right triangle, we can use the Pythagorean theorem to find the hypotenuse.

$$x = \sqrt{7.50^2 + 5.50^2} = 9.30 \text{ cm}$$

So, the point A is 9.30 cm from the centre of the circle.

SECTION 13-5 Exercises Circles

Definitions

1. Identify each of the following in Fig. 13-108: centre, radius, diameter, chord, arc, and semicircle.
2. Use a compass to draw a circle with all of the following and label each item: centre, radius, diameter, chord, arc, semicircle, and sector.

Circumference and Area

3. Find the circumference and area of a circle whose radius is 3.60 cm.
4. Find the circumference and area of a circle whose diameter is 7.56 metres.
5. Find the area of a circle whose circumference is 12.75 cm.
6. Find the circumference of a circle whose area is 15.4 m².
7. Draw a circle with a tangent. Draw the radius to the point of contact. What can you conclude?

FIGURE 13-109

FIGURE 13-108

FIGURE 13-110

Applications

8. Find the distance *TS* in Fig. 13-109, if the diameter is 12.2 metres, *SC* is 18.6 metres, $\overline{CT}$ is a radius, and $\overline{ST}$ is a tangent.
9. What is the circumference of a circular pond 25.0 m in diameter?
10. What length of fence is needed to enclose a circular garden 15.8 m in radius?
11. The orbit of the earth around the sun is nearly circular. Find the length of that orbit, assuming the earth is $15\overline{0}\,000\,000$ km away from the sun.
12. A certain car tire is 76.2 cm in diameter. How far will the car move forward with one revolution of the wheel?
13. Find the area of the face of a cylindrical piston 9.53 cm in diameter.
14. A circular fountain in a park has a basin that is 17.5 m in circumference. What is the diameter of the basin? What is the area of the basin?
15. The area of the bottom of a circular plate is 325 cm². Find its diameter.
16. Find the cost of cleaning up a circular oil spill 1.75 km in diameter, if it costs $6050/km² for cleanup.
17. A running track (Fig. 13-110) consists of a square with a semicircle at each end. Find the length of the track if the radius of the semicircles is 42.8 metres.
18. A bicycle tire has a radius of 30.0 cm. How many revolutions will it make in going one kilometre?

19. An automobile tire is 26.4 inches in diameter and the wheel makes 8.00 revolutions per second. How far (in miles) will the car move in 1.00 hour?
20. Two ventilating pipes, each 20.0 inches in diameter, are joined to form one single large pipe having the same capacity (cross-sectional area) as the combined capacity of the two single pipes. What is the diameter of the single large pipe?
21. A circle 16.0 inches in diameter has a chord 9.00 inches long. Find the perpendicular distance from the chord to the center of the circle.
22. A circular path is 1.20 yd wide and surrounds a circular garden 214.0 yd in diameter. Find the area of the path.
23. A round washer has an outside diameter of 1.25 inches. It has a round hole in the center 0.50 inches in diameter. Find the area of the face of the washer.

13-5 Circles   **277**

Chapter 13 REVIEW EXERCISES

(a)

(b)

FIGURE 13-111

FIGURE 13-112

(a)

(b)

FIGURE 13-113

(a)

(b)

FIGURE 13-114

FIGURE 13-115

FIGURE 13-116

1. Find θ in each part of Fig. 13-111.
2. Find angles A, B, C, D, E, F, and G in Fig. 13-112, given that L_1 and L_2 are parallel.
3. Three stakes, A, B, and C, are placed in the ground so that $\angle ABC$ is 55.3° and $\angle ACB$ is 43.9°. Make a sketch and find $\angle BAC$.
4. Find the perimeter and area of the triangles in Fig. 13-113.
5. A vertical pole is supported by a cable that forms an angle of 38.7° with the pole. Find the angle that the cable makes with the ground.
6. Find the cost of a triangular plot of land whose sides measure 160 m, 109 m and 99 m. The cost per hectare is $2595.
7. The rafters of a certain roof form an isosceles triangle with the horizontal beam. What are the two base angles, if the vertex angle is 124°?
8. Find the missing sides of the triangles in Fig. 13-114.
9. Find the altitude and area of an equilateral triangle 6.54 cm on a side.
10. The perimeter of a square dance floor is 53.5 m. Find the distance from one corner to the opposite corner.
11. Find the perimeter and area of a 30-60-90 right triangle whose shortest leg is 7.67 cm.
12. A ladder 10 m long stands flat against the side of a building. How many feet must it be pulled out at the bottom so that the top may be lowered 2 m?
13. Draw a parallelogram, a rectangle, a rhombus, and a square. Draw their diagonals. For each figure, state several properties that are true.
14. Find the perimeter and area of the shape in Fig. 13-115.
15. Find the perimeter and area of the parallelogram in Fig. 13-116.

16. What will it cost to carpet a rectangular floor, 6.50 m by 5.40 m, at $12.50 per square metre?
17. A rectangular field is 65.8 rods long and 52.9 rods wide. How many acres does it contain? (1 acre = 160 rod²)
18. Find the circumference and area of a circle whose diameter is 12.4 in.
19. A ranger measured the distance around a circular lake to be 2.15 miles. Find its diameter in ft and its area in acres. (1 acre = 43 560 ft²)
20. Find the diameter of a circular garden whose area is 1.00 acre. (1 acre = 43 560 ft²)
21. A circular water main 28.0 inches in diameter branches off into four smaller equal circular mains with the same total combined capacity as the larger main. Find the diameter of the smaller mains.

WRITING

22. Plane geometry is everywhere in the real world! Write an essay on plane geometry in either art, nature, or architecture.

TEAM PROJECT

23. In "The Musgrave Ritual" Sherlock Holmes calculates the length of the shadow of an elm tree that is no longer standing. He does know that the elm was 64 ft high and that the shadow was cast at the instant that the sun was grazing the top of a certain oak tree. Holmes held a 6.0 ft fishing rod vertical and measured the length of its shadow at the proper instant. It was 9.0 feet long. He then said, "Of course the calculation now is a simple one. If a rod of six feet threw a shadow of nine, a tree of sixty-four feet would throw one of _____." How long was the shadow of the tree?

   Now use Sherlock Holmes' method to find the height of a flagpole, building, or tree on your campus. Find a way to check your result.

14 Solid Geometry

Outcomes

When you have completed this chapter, you will be able to

1. Identify the physical properties of solid objects, such as volume and area as described in solid geometry.
2. Calculate the area and volume of various shaped prisms.
3. Calculate the area and volume of cylinders.
4. Calculate the area and volume of pyramids and cones.
5. Calculate the area and volume of the frustum of a pyramid or cone.
6. Calculate the area and volume of a sphere.

In this chapter we introduce you to the study of solid geometric figures. We define the prism, cylinder, pyramid, cone, and sphere and find the volumes and surface areas of these solids.

    Students need to have a working knowledge of solid geometry. Auto manufacturers and consumers use the volume of cylinders to indicate the power of an engine; technicians calculate volumes of tanks and pipes, weights of steel balls, surface areas of domes, and so forth. All these involve concepts of solid geometry.

14-1

PRISMS

In this section we consider the prism. We define various terms, find the volume and surface area of prisms, and solve applications.

■ FIGURE 14-1
Right prism.

Definitions

A *right prism* (Fig. 14-1) is a solid that is bounded by two equal parallel polygons (called the *bases*) and a number of rectangles (called the *lateral faces*). The lateral faces are perpendicular to the bases. In Fig. 14-1, the two triangles are the bases; they are equal and parallel. The three rectangles are the lateral faces. In this book, the term *prism* refers to a right prism.

    Prisms occur in many forms, depending on the shape of the base. Figure 14-2 shows several examples of prisms: a brick with a rectangular base, an optical prism with a triangular base, and an I beam with an I-shaped base.

(a)   (b)   (c)

■ **FIGURE 14-2**
Examples of right prisms.

Volume of a Prism

The volume of a solid is a measure of the space it occupies or encloses. It is measured in cubic units (cm^3, ft^3, m^3, etc.). The volume of a prism is equal to the product of the area B of the base and the altitude h (Fig. 14-3).

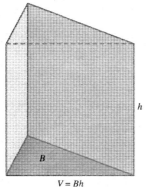

$V = Bh$

■ **FIGURE 14-3**
Volume of a prism.

| VOLUME OF A PRISM | $V = Bh$ | 80 |

Example 1

Find the volume of the triangular prism in Fig. 14-4.

■ **FIGURE 14-4**

Solution: In Fig. 14-4, the bases of the prism are the two triangles on the left and right ends of the prism. We find the area of the triangular base B,

$$B = \frac{1}{2}(2.45)(3.13)$$

$$B = 3.834 \text{ cm}^2$$

Using Eq. 80 to find the volume of the prism, with $h = 5.38$ cm,

$$V = Bh = (3.834)(5.38)$$

Rounding to three significant digits,

$$V = 20.6 \text{ cm}^3$$

■ **FIGURE 14-5**
Lateral area of a prism.

Lateral Area of a Prism

The area of all the lateral faces is called the *lateral area* (Fig. 14-5). To find the lateral area, we just add the areas of the rectangular lateral faces.

Example 2

Find the lateral area of the prism in Fig. 14-4.

Solution: The prism in Fig. 14-4 has three rectangular lateral faces. To calculate these areas, we first need to find the hypotenuse c of the right triangle:

$$c = \sqrt{3.13^2 + 2.45^2} = 3.975$$

Now we add the areas of the three rectangular faces:

$$L = (3.13)(5.38) + (2.45)(5.38) + (3.975)(5.38)$$

Rounding to three significant digits,

$$L = 51.4 \text{ cm}^2$$

Now let's take a look at an alternative way to calculate the lateral area L. To visualize this, we cut along the dotted edges of the prism (Fig. 14-6) and "unfold" the prism. The "unfolded" lateral area is a rectangle whose length is the perimeter

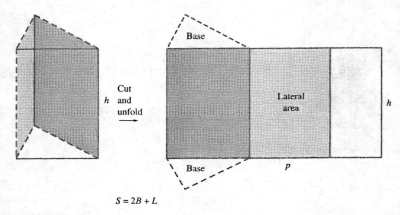

■ **FIGURE 14-6**
Surface area of a prism.

p of the base and whose width is the height h of the rectangular solid. Thus, the lateral area L of a prism is the product of the perimeter p of the base and the height h.

| LATERAL AREA OF A PRISM | $L = ph$ | 81 |

Example 3

Use Eq. 81 to find the lateral area of the prism in Fig. 14-4.

Solution: In Example 2, we found that the hypotenuse of the triangular base was 3.975 cm. Now we can find the perimeter p of the triangular base:

$$p = 3.13 + 2.45 + 3.975$$
$$p = 9.555 \text{ cm}$$

Substituting into Eq. 81 with $h = 5.38$ cm,

$$L = ph = (9.555)(5.38)$$
$$L = 51.4 \text{ cm}^2$$

This agrees with our answer in Example 2—we just used a different method to find it.

Surface Area of a Prism

The surface area of a solid is the total area of the entire surface of the solid. The surface area S of a prism (Fig. 14-6) is the sum of the area of the two bases, $2B$, and the lateral area, L.

| SURFACE AREA OF A PRISM | $S = 2B + L$ | 82 |

Example 4

Find the surface area of the prism in Fig. 14-4.

Solution: In Example 1, we found the area B of the base,

$$B = 3.834 \text{ cm}^2$$

In Example 3, we found the lateral area L,

$$L = 51.4 \text{ cm}^2$$

Substituting into Eq. 82,

$$S = 2B + L = 2(3.834) + 51.4$$
$$S = 59.1 \text{ cm}^2$$

FIGURE 14-7 Rectangular solid.

Rectangular Solid

A *rectangular solid,* or box (Fig. 14-7), is a prism with a rectangular base. There are simple formulas for the volume and surface area of a rectangular solid.

To find the volume, we use the area B of the rectangular base:

$$B = lw$$

Substituting into Eq. 80,

$$V = Bh = (lw)h = lwh$$

| VOLUME OF A RECTANGULAR SOLID | $V = lwh$ | 83 |

The surface of a rectangular solid consists of six rectangles, two of each size: l by w, h by w, and l by h. Thus,

| SURFACE AREA OF A RECTANGULAR SOLID | $S = 2(lw + hw + lh)$ | 84 |

FIGURE 14-8

Example 5

Find the volume and surface area of the rectangular solid in Fig. 14-8.

Solution: We use Eq. 83 for the volume of a rectangular solid, letting $l = 4.89$, $w = 3.26$, and $h = 2.10$,

$$V = lwh = (4.89)(3.26)(2.10)$$
$$V = 33.5 \text{ cm}^3$$

We use Eq. 84 for the surface area of a rectangular solid,

$$S = 2(lw + hw + lh) = 2[(4.89)(3.26) + (2.10)(3.26) + (4.89)(2.10)]$$

Rounding to three significant digits,

$$S = 66.1 \text{ cm}^2$$

FIGURE 14-9
Cube.

Cube

A *cube* (Fig. 14-9) is a prism, in which all six faces are equal squares. There are simple formulas for its volume and surface area.

To find the volume of a cube with side a, we use the area B of the square base

$$B = a^2$$

Substituting into Eq. 80 with $h = a$,

$$V = Bh = (a^2)a$$

| VOLUME OF A CUBE | $V = a^3$ | 85 |

The surface of a cube with side a consists of six squares, each having area a^2. Thus

| SURFACE AREA OF A CUBE | $S = 6a^2$ | 86 |

Example 6

Find the volume and surface area of a cube with side 1.75 cm.

Solution: To find the volume, we use Eq. 85 with $a = 1.75$ cm,

$$V = a^3 = (1.75 \text{ cm})^3$$
$$V = 5.36 \text{ cm}^3$$

To find the surface area, we use Eq. 86 with $a = 1.75$ cm,

$$S = 6a^2 = 6(1.75 \text{ cm})^2$$
$$S = 18.4 \text{ cm}^2$$

Applications

Applications involve finding areas and volumes of tanks, beams, boxes, rooms, wedges, and so forth.

■ FIGURE 14-10

■ FIGURE 14-11

Example 7

Find the number of cubic metres of steel necessary to make the I beam in Fig. 14-10.

Solution: The base of the I beam is shaped like an I. We sketch the I, and separate the regions into rectangles so we can calculate its area (Fig. 14-11). The area of the base B is

$$B = (15.0 \text{ cm})(8.0 \text{ cm}) + (8.0 \text{ cm})(10.0 \text{ cm}) + (15.0 \text{ cm})(8.0 \text{ cm}) = 320 \text{ cm}^2$$

The beam is 4.25 m long so $h = 4.25$ m which we convert to cm.

$$h = 4.25 \text{ m} = (4.25 \text{ m})\left(\frac{100 \text{ cm}}{1 \text{ m}}\right) = 425 \text{ cm}$$

Using Eq. 80 for the volume,

$$V = Bh = (320 \text{ cm}^2)(425 \text{ cm})$$
$$V = 130\,000 \text{ cm}^3$$

Converting to cubic metres,

$$V = 136\,000 \text{ cm}^3 \left(\frac{1 \text{ m}^3}{100^3 \text{ cm}^3}\right) = 0.136 \text{ m}^3$$

$$V = 0.14 \text{ m}^3 \quad \text{(rounding to two significant digits)}$$

This table summarizes the formulas found in this section.

	Lateral Area	Surface Area	Volume
Prism	$L = ph$	$S = 2B + L$	$V = Bh$
Rectangular solid		$S = 2(lw + hw + lh)$	$V = lwh$
Cube		$S = 6a^2$	$V = a^3$

SECTION 14-1 Exercises Prisms

Prisms

Find the volume and surface area of the following prisms.

1.

■ FIGURE 14-12

2.

■ FIGURE 14-13

3.

4.

■ FIGURE 14-14

■ FIGURE 14-15

Rectangular Solids

Find the volume and surface area of the rectangular solids with these dimensions.

5. $l = 4.60$ cm, $w = 5.80$ cm, $h = 2.50$ cm
6. $l = 1.55$ m, $w = 2.75$ m, $h = 4.25$ m
7. $l = 5.25$ cm, $w = 2.56$ cm, $h = 4.58$ cm
8. $l = 3.24$ m, $w = 5.38$ m, $h = 1.47$ m

Cubes

Find the volume and surface area of the cubes with the given side.

9. $a = 3.52$ m
10. $a = 4.58$ cm
11. $a = 0.549$ m
12. $a = 0.375$ cm

Applications

13. A brick measures 8.85 cm by 5.75 cm by 21.1 cm. Find the volume of the brick.
14. A classroom is 15.0 m long, 12.0 m wide, and 4.0 m high. If there are 36 people in the room, find the number of cubic metres of air for each person in the room.
15. White light is separated into a spectrum by passing through a prism which is 4.25 cm in height and whose base is a right isosceles triangle whose equal sides are 2.68 cm long. Find its volume.
16. Find the volume of a gas tank that is a rectangular prism 33.0 cm by 39.4 cm by 91.4 cm.
17. A rectangular tank is to be lined with stainless steel sheets. Find the area to be lined (bottom and sides) if the tank is 12.5 m long, 8.75 m wide, and 4.25 m high.
18. A rectangular storage shed is to be painted. Find the area to be painted (top and sides) if the shed is 5.20 m long, 4.50 m wide, and 6.50 m high.
19. Find the number of cubic metres of concrete required to make the concrete retaining wall in Fig. 14-16. Find the total cost of the concrete at $87.50 per cubic metre.
20. Find the volume of the solid steel wedge in Fig. 14-17. Find its mass (in kg) if the steel has a density of 7.82 grams per cubic centimetre.
21. Find the number of ft³ of steel necessary to make the I beam in Fig. 14-18.

■ FIGURE 14-16

■ FIGURE 14-17

■ FIGURE 14-18

22. How many cubic yards of gravel can a truck carry if the trailer is 6.35 yd long, 2.75 yd wide, and 1.45 yd high?

23. How many loads of gravel will be needed to cover a driveway 75.2 yd long and 12.5 ft wide, to a depth of 15.0 in., if one truckload contains 8.00 yd³ of gravel?

24. A cube has a surface area of 57.8 in.². Find the length of its side and its volume.

25. A rectangular prism with a square base has a volume of 892 cubic inches. The height is 8.00 inches. Find the total surface area.

14-2 CYLINDERS

In this section we define the cylinder, and find its volume and surface area. We study liquid capacity and conversion of units, and solve applications involving automobile cylinders, pipes, and cylindrical tanks.

A *circular cylinder* (Fig. 14-19) is a geometric solid which has two equal parallel *circular* bases. It is similar to a prism and can be thought of as a "circular prism." Although there are non-circular cylinders, all the cylinders that we deal with in this book are circular cylinders. Whenever we say "cylinder" we are referring to a circular cylinder.

■ FIGURE 14-19
Cylinder.

Volume of a Cylinder

Since a cylinder can be thought of as a circular prism, we use Eq. 80 for the volume of a prism to determine the equation of the volume of a cylinder having height h and base radius r (Fig. 14-20):

$$V = Bh$$

Since the base is a circle, its area B is

$$B = \pi r^2$$

Substituting, we get

$$V = (\pi r^2)h$$

■ FIGURE 14-20
Volume of a cylinder.

| VOLUME OF A CYLINDER | $V = \pi r^2 h$ | 87 |

■ FIGURE 14-21

Example 8
Find the volume of a cylindrical soup can (Fig. 14-21) whose diameter is 6.60 cm and height is 9.75 cm.

Solution: We use Eq. 87 for the volume,

$$V = \pi r^2 h$$

First we need the radius,

$$r = \frac{d}{2} = \frac{6.60}{2} = 3.30 \text{ cm}$$

So,

$$V = \pi(3.30)^2(9.75) = 334 \text{ cm}^3$$

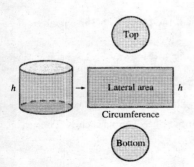

FIGURE 14-22
Surface area of a cylinder.

Surface Area of a Cylinder

The surface area of a cylinder can be visualized by "unfolding" the surface of the cylinder. Think of a soup can; if we remove the top and bottom "lids" and then slit the can up one side and spread out the lateral area (Fig. 14-22), we get two circles and a rectangle. By Eq. 76, the area B of each circle is

$$B = \pi r^2$$

The lateral area is a rectangle, whose length is the circumference C of the base and whose width is the height h of the can. (Note that this also follows from Eq. 81 that the lateral area of a prism is the perimeter of the base times the height.)

So,

$$L = Ch$$

Since

$$C = 2\pi r$$

We substitute and get

$$L = (2\pi r)h = 2\pi rh$$

Substituting in Eq. 82, we get

$$S = 2B + L = 2(\pi r^2) + 2\pi rh$$

SURFACE AREA OF A CYLINDER	$S = 2\pi r^2 + 2\pi rh$	88

Example 9

Find the surface area of the can in Example 8 (Fig. 14-21).

Solution: We use Eq. 88

$$S = 2\pi r^2 + 2\pi rh$$

Since $r = 3.30$ cm and $h = 9.75$ cm, we substitute and get

$$S = 2\pi(3.30)^2 + 2\pi(3.30)(9.75)$$
$$= 271 \text{ cm}^2$$

Liquid Capacity

In solving volume problems, we are often asked to find the volume of liquids in gallons, litres, quarts, and so forth. This type of volume is called *liquid capacity*. These problems are solved by finding the volume in cubic units and then converting to liquid capacity units.

Example 10

A hot water tank (Fig. 14-23) has a diameter of 38.0 cm and a height of 1.69 m. How many litres does it hold?

Solution: We find the radius from the diameter

$$r = \frac{d}{2} = \frac{38.0}{2} = 19.0 \text{ cm}$$

FIGURE 14-23

The units of r and h must agree, and so we convert $h = 1.69$ m to centimetres.

$$1.69 \text{ m} = (1.69 \text{ m})\left(\frac{100 \text{ cm}}{1 \text{ m}}\right) = 169 \text{ cm}$$

Using Eq. 87 to calculate V

$$V = \pi r^2 h = \pi (19.0)^2(169)$$
$$V = 191\ 700 \text{ cm}^3$$

We must now convert cm^3 to litres. One litre (L) is 1000 cm^3.
So,

$$V = (191\ 700 \text{ cm}^3)\left(\frac{1 \text{ L}}{1000 \text{ cm}^3}\right)$$
$$V = 192 \text{ L}$$

Hollow Cylinders

A *hollow cylinder* is the shell between two concentric cylinders, like a pipe or tube. To find the volume of a hollow cylinder, we subtract the volume of the small cylinder from the volume of the large cylinder. If the radius of the small cylinder is r and the radius of the large cylinder is R, then

$$V = \pi R^2 h - \pi r^2 h$$

Example 11

Find the volume of metal in a 4.75 m length of the hollow iron pipe shown in Fig. 14-24.

Solution: The large radius R is

$$R = \frac{4.86}{2} = 2.43 \text{ cm}$$

The small radius r is

$$r = \frac{3.64}{2} = 1.82 \text{ cm}$$

The length of the cylinder h, in cm, is

$$h = 4.75 \text{ m} = 475 \text{ cm}$$

So,

$$V = \pi R^2 h - \pi r^2 h$$
$$= \pi(2.43)^2(475) - \pi(1.82)^2(475)$$
$$= 3870 \text{ cm}^3$$

FIGURE 14-24

SECTION 14-2 Exercises Cylinders

Volume and Surface Area

1. Find the volume of a cylinder whose radius is 3.46 m and whose height is 9.52 m.
2. Find the surface area of the cylinder in Exercise 1.
3. Find the surface area of a cylinder whose diameter is 5.65 cm and whose height is 8.75 cm.
4. Find the volume of the cylinder in Exercise 3.

Liquid Capacity

5. Find the liquid capacity, in litres, of the tank in Fig. 14-25.
6. Find the liquid capacity, in mL, of a can 6.75 cm in diameter and 10.3 cm high. (1 mL = 1 cm³)

Hollow Cylinders

7. Find the volume of the hollow cylinder in Fig. 14-26.
8. Find the volume of a hollow cylinder with an inside radius of 5.25 cm, an outside radius of 7.75 cm, and a height of 6.50 cm.
9. Find the amount of steel in a steel shaft 35.0 cm long with a diameter of 2.35 cm.

Applications

10. Find the volume of silage that can be stored in a cylindrical silo 8.0 m in diameter and 37.8 m high.
11. Find the volume of copper necessary to make one kilometre of copper wire 0.7938 cm in diameter.
12. Find the amount of wood (in m³) in a cylindrical cedar post 12.5 cm in diameter and 2.50 m long.
13. A stainless steel cylindrical milk tank 2.51 m in diameter and 38.9 m long is being constructed. Find the amount, in m², of stainless steel needed, allowing 10.0% for seams and waste.
14. The cylindrical tank of an electric hot water heater is 54.6 cm in diameter and 148.6 cm in height. How many square metres of insulation are needed to wrap the lateral area and top?
15. In automotive engineering, the volume (or displacement) of a cylinder in a piston engine is measured in cubic centimetres. The diameter of the cylinder is called the bore and the height is called the stroke. What is the displacement of a cylinder with a bore of 72.0 mm and a stroke of 85.0 mm?

■ FIGURE 14-25

16. What is the engine displacement in cubic centimetres of an 8-cylinder engine if each cylinder has a bore of 65.0 mm and a stroke of 88.0 mm? (Refer to Exercise 15.)
17. An oil tank is 24.0 m long and has a diameter of 10.5 m. How many litres will it hold?
18. An artesian well driller had to drill down 287 m before striking water. If the pipe is 8.00 cm in diameter, how much (in litres) is the "head" (water stored in the pipe)?
19. A certain pipe sleeve is a hollow cylinder with an outer diameter of 16.0 in., an inner diameter of 10.0 in., and a length of 24.0 in. Compute the volume of the sleeve in cubic feet. (1 ft³ = 1728 in.³)

■ FIGURE 14-26

20. Find the amount of iron in a 10.0 ft length of the hollow pipe in Fig. 14-27.
21. A gasoline storage tank has the shape of a cylinder with a diameter of 60.0 ft and a height of 25.5 ft. How many imperial gallons does it hold? (1 ft³ = 6.226 gal (Imp.))
22. A water main is 16.0 inches in diameter. If the water flows at 1.56 ft per second, how much water (in U.S. gallons) flows in one hour? (1 gal (U.S.) = 231 in.³)

■ FIGURE 14-27

23. An engine piston has a height of 3.90 in. and a volume of 21.5 cubic inches. Find the diameter of the piston.
24. A cylindrical quart can is 4.50 in. high. What is its diameter? (1 gal (U.S.) ≈ 231 in.³)

14-3 PYRAMIDS AND CONES

In this section we define pyramids and cones and find their volumes and surface areas. We also discuss weight, density, and other applications.

■ FIGURE 14-28
Pyramid.

■ FIGURE 14-29
Pyramids.

Pyramid

A *pyramid* (Fig. 14-28) is a solid whose base is a polygon and whose lateral faces are triangles with a common vertex. Figure 14-28 shows a pyramid whose base is a square. The *altitude h* (or *height*) is the perpendicular distance from the vertex to the base.

A *regular* pyramid is a pyramid whose base is a regular polygon and whose vertex is centred over the base. All pyramids we consider in this chapter will be regular pyramids. The faces of a regular pyramid are equal isosceles triangles.

Pyramids are named by the shape of their base. Figure 14-28 shows a square pyramid, and Fig. 14-29 shows a triangular pyramid and a hexagonal pyramid.

Volume of a Pyramid

The volume of a pyramid is *one-third* the volume of a prism of the same base and height. Recall from Section 14-1 that the volume of a prism is the product of the area of the base B and the height h. So, the volume V of a pyramid (Fig. 14-30) is one-third the area of the base B times the height h.

| VOLUME OF A PYRAMID | $V = \frac{1}{3}Bh$ | 89 |

Example 12

Find the volume of a square pyramid (Fig. 14-30) if the base is 23.8 cm on a side and the height is 32.5 cm.

Solution: For this pyramid,

$$h = 32.5 \text{ cm}$$

Since the base is a square, its area B is

$$B = (23.8)^2 = 566.4 \text{ cm}^2$$

Substituting into Eq. 89, we get

$$V = \frac{1}{3}Bh = \frac{1}{3}(566.4)(32.5)$$

$$V = 6140 \text{ cm}^3$$

■ FIGURE 14-30
Volume of a pyramid.

$V = \frac{1}{3}Bh$

14-3 Pyramids and Cones

FIGURE 14-31
Regular pyramid.

Lateral and Surface Area of a Regular Pyramid

In a regular pyramid (Fig. 14-31), each lateral face is an isosceles triangle; the altitude s of the isosceles triangle is called the *slant height* of the pyramid. The area A of one lateral face (one of the isosceles triangles) is one-half the product of the base b of the triangle times the slant height s.

$$A = \frac{1}{2}bs$$

If we multiply this area A by the number n of lateral faces, then we have the lateral area L of the pyramid.

$$L = nA$$

Substituting,

$$L = n\left(\frac{1}{2}bs\right)$$

Rearranging the factors,

$$L = \frac{1}{2}(nb)s$$

We note that nb is the perimeter p of the base of the pyramid, so,

$$nb = p$$

Substituting,

$$L = \frac{1}{2}ps$$

So, the lateral area of a pyramid is one-half the perimeter p of the base times the slant height s of the pyramid.

LATERAL AREA OF A PYRAMID	$L = \frac{1}{2}ps$	90

The surface area of a pyramid includes both the area B of the base and the lateral area L.

SURFACE AREA OF A PYRAMID	$S = B + L$	91

FIGURE 14-32

Example 13

Find the lateral area and the surface area of a square pyramid with a base 4.76 cm on a side and a slant height of 6.23 cm.

Solution: We sketch the pyramid in Fig. 14-32. For this pyramid, the base is a square of side 4.76 cm, so

$$p = 4(4.76 \text{ cm}) = 19.04 \text{ cm}$$

and

$$s = 6.23 \text{ cm}$$

Substituting into Eq. 90 to find the lateral area of a pyramid,

$$L = \frac{1}{2}ps = \frac{1}{2}(19.04 \text{ cm})(6.23 \text{ cm})$$

$$L = 59.31 \text{ cm}^2$$

To find the surface area, we need to calculate the area B of the square base,

$$B = s^2 = (4.76 \text{ cm})^2$$

$$B = 22.66 \text{ cm}^2$$

Substituting into Eq. 91 for the surface area of the pyramid,

$$S = B + L = 22.66 \text{ cm}^2 + 59.31 \text{ cm}^2$$

$$S = 82.0 \text{ cm}^2$$

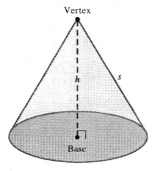

■ FIGURE 14-33
Right circular cone.

Cone

A *circular cone* (Fig. 14-33) is a geometric solid that has a circular base and a lateral surface which tapers from the base to a point called the *vertex*. The *altitude h* (or *height*) of a cone is the perpendicular distance from the vertex to the base. The *slant height* of a cone is the distance from the vertex to a point on the circular base. In a *right* circular cone, the line from the vertex to the centre of the base is perpendicular to the base.

Volume of a Cone

We can think of a cone as a "circular pyramid." Since the volume of a pyramid is one-third the volume of the prism with the same dimensions, then the volume of a cone is one-third the volume of the cylinder which has the same radius and height (Fig. 14-34), or

$$V = \frac{1}{3}(\pi r^2 h)$$

■ FIGURE 14-34
Volume of a cone.

where r is the radius of the base and h is the height of the cone.

VOLUME OF A CONE	$V = \frac{1}{3}\pi r^2 h$	92

■ FIGURE 14-35

—— **Example 14** ——

Find the volume of a cone whose diameter is 8.50 cm and height is 14.0 cm.

Solution: We sketch the cone in Fig. 14-35. For this cone, $d = 8.50$ cm and $h = 14.0$ cm. We find the radius,

$$r = \frac{d}{2} = \frac{8.50}{2} = 4.25 \text{ cm}$$

Substituting into Eq. 92 for the volume of the cone,

$$V = \frac{1}{3}\pi r^2 h = \frac{1}{3}\pi(4.25)^2(14.0)$$

$$V = 265 \text{ cm}^3$$

Lateral and Surface Area of a Cone

The lateral area of a cone is determined using Eq. 90

$$L = \frac{1}{2}ps$$

where s is the slant height of the cone and the perimeter p is the circumference of the circular base. Thus,

$$p = 2\pi r$$

Substituting into Eq. 90, we have

$$L = \frac{1}{2}ps = \frac{1}{2}(2\pi r)s$$

or

$$L = \pi rs$$

| LATERAL AREA OF A CONE | $L = \pi rs$ | 93 |

The total surface area is

$$S = B + L$$

Substituting $B = \pi r^2$ and $L = \pi rs$, we have

$$S = \pi r^2 + \pi rs$$

| SURFACE AREA OF A CONE | $S = \pi r^2 + \pi rs$ | 94 |

Example 15

Find the lateral and surface areas of a cone (Fig. 14-36) with diameter 5.50 cm and slant height 8.75 cm.

Solution: For this cone, $d = 5.50$ cm and $s = 8.75$ cm. First we find r,

$$r = \frac{5.50}{2} = 2.75 \text{ cm}$$

Substituting into Eq. 93 to find the lateral area L,

$$L = \pi rs = \pi(2.75)(8.75)$$

$$L = 75.6 \text{ cm}^2$$

Substituting into Eq. 94 to find the surface area S,

$$S = \pi r^2 + \pi rs = \pi(2.75)^2 + 75.6$$

$$S = 99.4 \text{ cm}^2$$

FIGURE 14-36

Mass

Often we need to calculate the mass of an object. Mass is a function of density and volume. To calculate the mass of an object, we must know the mass per unit volume of the material. The mass per unit volume is called the *density*. Typical units of density are kg/m³ or lb(mass)/ft³. The mass m of an object is the product of the density ρ of the material and the volume V of the object.

$$\text{Mass} = (\text{density})(\text{Volume})$$

or

$$m = \rho V$$

The densities of common materials can be found in a density table of most engineering handbooks. In this book, we provide the densities as needed.

Example 16

Find the mass of the cone in Example 14 (Fig. 14-35), if it is made of bronze with a density of $9.20 \times 10^3 \, \frac{\text{kg}}{\text{m}^3}$.

Solution: The density is the mass per unit volume, so it is given that the density is

$$\rho = 9.20 \times 10^3 \, \frac{\text{kg}}{\text{m}^3}$$

Using the volume from Example 14, and converting it to cubic metres,

$$V = 265 \text{ cm}^3 = (265 \text{ cm}^3)\left(\frac{1 \text{ m}^3}{100^3 \text{ cm}^3}\right) = 265 \times 10^{-6} \text{ m}^3$$

Substituting into the mass equation,

$$m = \rho V = \left(9.20 \times 10^3 \, \frac{\text{kg}}{\text{m}^3}\right)(265 \times 10^{-6} \text{ m}^3)$$

$$m = 2.438 \text{ kg} = 2.44 \text{ kg} \quad \text{(Rounded)}$$

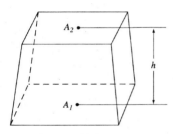

■ FIGURE 14-37

Frustum of a Cone or Pyramid

Many common objects are truncated cones or pyramids. In everyday encounters, we see coffee cups, tree trunks, waste baskets, and many other items.

The Volume of a Frustum

The volume of these objects is found using the following common formula.

$$V = \frac{1}{3}h(A_1 + A_2 + \sqrt{A_1 A_2})$$

where A_1 and A_2 are the areas of the two bases and h is the altitude.

| VOLUME OF A FRUSTUM | $V = \frac{1}{3}h(A_1 + A_2 + \sqrt{A_1 A_2})$ | 95 |

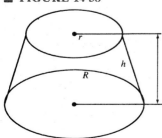

■ FIGURE 14-38

When dealing with the frustum of a cone, Eq. 95 may be used, or a simplified version Eq. 96 can also be used.

$$V = \frac{1}{3}\pi h(R^2 + r^2 + Rr)$$

| VOLUME OF THE FRUSTUM OF A CONE | $V = \frac{1}{3}\pi h(R^2 + r^2 + Rr)$ | 96 |

14-3 Pyramids and Cones

Example 17

A log cut from a maple tree has the shape of a frustum of a cone. The diameter at the bottom cut is 76 cm, the top cut is 43 cm and the length of the log is 914 cm. How many cubic metres of maple lumber are in this log prior to sawing into planks?

Solution: First we change all the measurements into metres by the simple procedure of moving the decimal place left by two places

$$\text{Bottom diameter} = 76 \text{ cm} = 0.76 \text{ m}$$
$$\text{Top diameter} = 43 \text{ cm} = 0.43 \text{ m}$$
$$\text{Length} = 914 \text{ cm} = 9.14 \text{ m} = h$$

Therefore

$$R = \frac{0.76 \text{ m}}{2} = 0.38 \text{ m}$$

and

$$r = \frac{0.43 \text{ m}}{2} = 0.215 \text{ m}$$

Substituting in Eq. 96

$$V = \frac{1}{3}\pi(9.14 \text{ m})((0.38 \text{ m})^2 + (0.215 \text{ m})^2 + (0.38 \text{ m})(0.215 \text{ m}))$$

$$V = 2.6 \text{ m}^3 \quad \text{Rounded}$$

Example 18

A map of Fish Lake shows a contour at a depth of 10 metres. The surface area of the lake has been measured at 10.6 km² and the 10 metre contour area has been measured at 2.7 km². If the lake were to shrink during a hot spell to the 10 m contour, how many cubic metres of water would be gone? (Fig. 14-39)

Solution: We first convert the areas into square metres by simply moving the decimal right by six places.

$$\text{Surface Area} = 10.6 \text{ km}^2 = 10\,600\,000 \text{ m}^2$$
$$\text{Area of 10 metre contour} = 2.7 \text{ km}^2 = 2\,700\,000 \text{ m}^2$$
$$h = 10 \text{ m, the elevation difference between the two surfaces.}$$

Substituting into Eq. 95, the general frustum formula, we get

$$V = \frac{1}{3}h(A_1 + A_2 + \sqrt{A_1 A_2})$$

$$V = \frac{1}{3}(10)(10\,600\,000 + 2\,700\,000 + \sqrt{(10\,600\,000)(2\,700\,000)})$$

$$V = 62\,000\,000 \text{ m}^3 \quad \text{(Rounded)}$$

So during a dry spell, 62 000 000 m³ of water will be lost due to various causes. Note: This calculation assumes that the lake bottom slopes uniformly down to the 10 m depth contour.

Fish Lake

FIGURE 14-39

The Lateral Area of the Frustum of a Cone or a Regular Pyramid

To calculate the lateral area LA of the frustum of a cone or a regular pyramid we work with the average of the perimeters of the upper and lower surfaces P_1 and P_2 multiplied by the slant height s.

$$LA = s\left(\frac{P_1 + P_2}{2}\right) \quad \text{Or} \quad LA = \frac{s}{2}(P_1 + P_2)$$

LATERAL AREA OF A FRUSTUM OF A CONE OR PYRAMID	$LA = \frac{s}{2}(P_1 + P_2)$	97

Example 19

Find the lateral area of a regular square frustum in Fig. 14-40 which has a top edge of 2.3 cm and a bottom edge of 3.7 cm. The slant height is 5.3 cm.

Solution: First we find the perimeter of the top and bottom squares

$$P_1 = 4 \times 2.3 \text{ cm} = 9.2 \text{ cm}$$
$$P_2 = 4 \times 3.7 \text{ cm} = 14.8 \text{ cm}$$

Using Eq. 97

$$A = \frac{5.3}{2}(9.7 + 14.8) = 64.93 \text{ cm}^2$$

or

$$A = 65 \text{ cm}^2 \quad \text{Rounded}$$

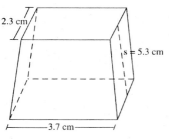

FIGURE 14-40

SECTION 14-3 Exercises Pyramids and Cones

Volume

1. Find the volume of a square pyramid 6.21 cm on a side and 7.36 cm in height.
2. Find the volume of a regular triangular pyramid 8.94 cm on edge and with an altitude of 15.4 cm.
3. Find the volume of a cone whose radius is 35.2 mm and height is 87.6 mm.
4. Find the volume of a cone whose diameter is 0.0781 m and height is 0.0459 m.

Surface Area

5. Find the surface area of a regular triangular pyramid whose base is 6.56 m on a side and slant height is 9.25 m.
6. Find the surface area of a square pyramid whose base is 14.25 m on a side and slant height is 8.75 m.
7. Find the surface area of a cone whose diameter is 15.0 cm and slant height is 8.50 cm.
8. Find the surface area of a cone whose radius is 2.50 m and slant height is 4.75 m.
9. A cone has a diameter of 12.0 cm, a height of 18.0 cm, and a slant height of 19.0 cm. Find the surface area of the cone.
10. A square pyramid has a base 24.0 cm on a side, a height of 35.0 cm, and a slant height of 37.0 cm. Find the surface area of the pyramid.

Mass

11. Find the mass of the pyramid in Exercise 1 if it is made of gold with a density of $\rho \text{ (gold)} = 19.3 \times 10^3 \frac{\text{kg}}{\text{m}^3}$.

12. Find the mass of the pyramid in Exercise 2 if it is made of bronze with a density of $\rho \text{ (bronze)} = 8.78 \times 10^3 \frac{\text{kg}}{\text{m}^3}$.

13. Find the mass of the cone in Exercise 3 if it is made of silver with a density of $\rho \text{ (silver)} = 10.5 \times 10^3 \frac{\text{kg}}{\text{m}^3}$.

14. Find the mass of the cone in Exercise 4 if it is made of steel with a density of $\rho \text{ (steel)} = 7.84 \times 10^3 \frac{\text{kg}}{\text{m}^3}$.

15. A sand pile is in the shape of a cone, 8.54 m in diameter and 5.56 m in height. Find the mass of the sand pile if the density of sand is $\rho \text{ (sand)} = 1.63 \times 10^3 \frac{\text{kg}}{\text{m}^3}$.

16. A granite monument is in the shape of a tall square pyramid. It is 27.0 m tall and the base is 2.25 m on a side. Find the mass in tonnes of the monument if the density of granite is $\rho \text{ (granite)} = 2.61 \times 10^3 \frac{\text{kg}}{\text{m}^3}$.

17. Find the mass of a conical brass casting which is 46.7 cm high and 12.9 cm in diameter. The density of brass is $\rho \text{ (brass)} = 8.65 \times 10^3 \frac{\text{kg}}{\text{m}^3}$.

Frustum

18. A large pot for planting has the shape of a frustum of a cone. The top has a diameter of 60 cm and the bottom has a diameter of 45 cm. The pot has a total height of 30 cm. How many litres of soil will be required to fill it?

19. A swimming pool at a rural home is kidney shaped. The water filling it is considered for fire protection in insurance fee evaluation. The top has an area of 49 m² while the deepest portion has an area of 17 m². The depth of the pool at the deepest portion is 3.2 m. Using the frustum formula, calculate the number of cubic metres of water you have in reserve for fire protection.

20. A disposable flat bottom round drinking cup is made of coated paper. How many square centimetres of coated paper are used in making this cup assuming no waste? The top diameter is 5.7 cm, the bottom diameter is 3.8 cm and the slant height is 5.9 cm.

21. A circular hole needs to be dug 3.5 m wide at the top, 2.0 m wide at the bottom, and 1.8 m deep. If the density of the soil to be removed is $\rho \text{ (soil)} = 1.29 \times 10^3 \frac{\text{kg}}{\text{m}^3}$, how many tonnes will be removed?

Applications

22. A drinking cup is in the shape of a cone, 7.65 cm in diameter and 8.75 cm high. Find its volume. Will the drinking cup hold *more* than or *less* than $\frac{1}{2}$ cup of water? (1 cup = 250 cm³).

23. A canvas tent is in the shape of a square pyramid. It is 1.40 m tall with a slant height of 1.69 m, and the base is 1.98 m on a side. How much canvas was needed to make the tent (floor and sides)?

24. A steeple, in the shape of a hexagonal pyramid, is to be painted. It is 30.0 m high with a slant height of 30.2 m, and its base is 4.00 m on a side. A litre of paint will cover 11.8 m². How much will the paint cost if we use two coats? The cost of paint is $24.99 per five litre can.

25. The tin cover of a maple syrup sap bucket is conical. It is 15.0 inches in diameter and 2.50 inches in height with a slant height of 7.91 inches. Find the lateral area of the cover. Then find the mass of the cover, allowing 5% for overlap, if tin sheeting has a mass of 2.55 pounds (mass) per square foot.

26. A grain bin on a combine is conical. It is 58.0 inches high and 54.0 inches in diameter. Find the amount of metal (in ft^2) necessary to make the open bin and the volume (in ft^3) of grain it will hold. (Hint: Use the Pythagorean theorem to find the slant height.)

27. A concrete monument in the shape of a square pyramid has a mass of 15 950 lb. The concrete has a density of 138 lb/ft^3. If the base of the monument is 4.00 ft on the side, how tall is the monument?

28. A cone with slant height of 12.0 in. has a base with a diameter of 8 in. What is the lateral area of this cone?

29. Find the capacity in imperial gallons of a pail which has the shape of a frustum of a cone, if the pail has a height of 1.25 inches, a bottom diameter of 9.00 inches and a top diameter of 12.0 inches. (1 gal (Imp.) = 277.4 in.^{3})

30. Find the approximate volume, in acre-feet, of a lake which has a surface area of 4.67 acres and a depth of 32.5 feet (hint: use the cone formula). (1 acre-foot = 43 560 ft^3)

14-4 SPHERES

In this section we define a sphere and find its volume and surface area. These are essential skills in dealing with applications such as ball bearings, geodesic domes, planets, and satellites.

Definitions

A *sphere* (Fig. 14-41) is a geometric solid in which every point of the sphere is equidistant from a given point, called the *centre* of the sphere. The *radius* of a sphere is the distance from the centre to a point of the sphere. The *diameter* of a sphere is the length of a line segment through the centre with its endpoints on the sphere. Thus, the diameter is twice its radius, or the radius is half the diameter,

$$r = \frac{d}{2}$$

A *great circle* of a sphere (Fig. 14-41) is a circle cut by a plane which passes through the centre of the sphere. The radius of a great circle is the radius r of the sphere. A great circle separates the sphere into two equal half-spheres, called *hemispheres*.

FIGURE 14-41
Sphere.

Volume of a Sphere

The volume V of a sphere is the product of $\frac{4}{3}\pi$ times the cube of the radius r.

| VOLUME OF A SPHERE | $V = \frac{4}{3}\pi r^3$ | 98 |

Surface Area of a Sphere

The surface area S of a sphere is 4 times the area of the great circle of the sphere. Since the radius of a great circle is the radius r of the sphere, then the area A of a great circle is

$$A = \pi r^2$$

and the surface area S of a sphere is

$$S = 4A$$

Substituting,

$$S = 4(\pi r^2)$$

| SURFACE AREA OF A SPHERE | $S = 4\pi r^2$ | 97 |

Example 20

A baseball has a diameter of 7.52 cm. Find its volume and surface area.

Solution: To find the volume, we first find the radius, using $d = 7.52$ cm,

$$r = \frac{d}{2} = \frac{7.52}{2} = 3.76 \text{ cm}$$

Substituting into Eq. 98 for the volume,

$$V = \frac{4}{3}\pi r^3 = \frac{4}{3}\pi(3.76)^3$$

$$V = 223 \text{ cm}^3$$

To find the surface area, we use Eq. 99,

$$S = 4\pi r^2 = 4\pi(3.76)^2$$

$$S = 178 \text{ cm}^2$$

SECTION 14-4 Exercises Spheres

Volume and Surface Area of a Sphere

Find the volume and surface area of each of the following spheres.

1. Tennis ball, radius = 3.32 cm
2. Beach ball, radius = 28.4 cm
3. Earth, diameter = 12 700 km
4. Ball bearing, diameter = 0.925 mm

Applications

5. A weather balloon is filled with helium, to a diameter of 2.50 m. Find its volume.
6. Oil comes out of a sprayer in spherical droplets 1.005 mm in diameter. Find the volume (in cm³) of 1 million droplets.
7. A glass globe is filled with water. Find its capacity in litres if the diameter of the globe is 19.3 cm.
8. A hemispherical vat has a diameter of 54.1 cm. Find its capacity in litres.
9. A spherical gas storage tank, 35.4 m in diameter, is to be painted. How much paint is needed for two coats, if a 4.00 L can of paint covers 50.0 m²?
10. A company manufactures plastic hemispherical covers for plants. The covers are 42.5 cm in diameter. Find the cost of making 125 000 covers, if the plastic costs $4.75 per square metre.

11. A hemispherical dome on a concert wall has a radius of 22.1 m. Find the cost of applying a thin layer of gold to the dome at $20.77 per square metre.
12. Two fuel storage tanks each have the same capacity. One is spherical with a diameter of 3.00 m. The other is cylindrical with the same diameter and a height of 2.00 m. Which of the two tanks required more surface area material? How much more?
13. Find the mass of a golden sphere 17.8 cm in diameter. (The density of gold is 19.3×10^3 kg/m^3.)

14. Find the weight of a brass sphere 3.25 ft in diameter. The density of brass is 511 lb/ft^3.
15. For the shot put, a contestant throws (puts) an iron sphere (a shot). Find the weight of a shot 5.00 inches in diameter, if it is made of iron whose density is 491 lb/ft^3.
16. Find the weight of 1000 steel ball bearings 0.10 inch in diameter, if the density of the steel is 486 lb/ft^3.
17. A certain sphere holds 1.00 cubic foot of water. What is its diameter? What is its surface area?
18. The diameter of a softball is 3.82 inches. What is its circumference and its volume? If the weight is 0.406 lb, what is its density in pounds per cubic foot?
19. Find the volume of a sphere whose surface area is 1.00 ft^2.

Chapter 14 REVIEW EXERCISES

1. Find the volume and surface area of the rectangular prism whose length is 6.75 cm, width is 4.52 cm, and height is 5.88 cm.
2. Find the volume and surface area of the cube whose side is 3.75 cm.
3. Find the volume and surface area of the prism in Fig. 14-42.
4. An excavation for a building foundation is 15.2 m long, 9.90 m wide, and 3.65 m deep. How many cubic metres of soil are removed? How many truck-loads will be necessary if the truck has a capacity of 5.0 cubic metres?
5. An octagonal shaped silo has a base where each wall is 2.61 m wide. The height of the silo is 9.75 m. In order to paint it, how may litres of paint will be necessary if coverage is 6.31 m^2/L?
6. Find the volume and surface area of the cylinder in Fig. 14-43.
7. A steam boiler has 18 cylindrical flues, each 7.62 cm in diameter and 2.44 m long. Find the total area of heating surface (the lateral area) of the 18 flues.
8. A brass rod is 0.914 cm in diameter and 1.98 m long. Find its volume in cubic centimetres.
9. The inside diameter of a pipeline is 122 cm. How many kilolitres of crude oil will a kilometre hold?
10. A pipe made of concrete has an outside diameter of 93.2 cm and an inside diameter of 72.6 cm. The length of the pipe is 7.47 m. What is the total volume of the concrete making up the pipe?
11. Find the volume of a triangular pyramid 34.5 mm high and whose base is 27.5 mm on each side.
12. Find the total surface area of a cone whose height is 3.75 m, slant height is 4.37 m, and radius is 2.25 m.

■ FIGURE 14-42

■ FIGURE 14-43

13. One of the colossal Egyptian pyramids, the great pyramid of Cheops in Gizeh, originally rose to a height of 146.6 m and had a slant height of 186.2 m. Its square base was 228.6 m on each side. Find the volume and the total surface area of the great pyramid of Cheops.
14. A large bucket that is a frustum of a cone has a height of 35.0 cm. The bottom diameter is 25.0 cm and the top diameter is 55.0 cm. How many litres will this bucket hold?
15. Find the volume of the sawdust collector in Fig. 14-44.
16. The roof of a building (Fig. 14-45) is in the shape of a square pyramid and must be shingled. Find the lateral area of the roof. How many boxes of shingles are needed, if each box covers 50.0 ft^2?
17. A pendulum weight is a solid gold cone, 3.00 inches in diameter and 5.00 inches in height. Find its value if gold sells for $390.00 per troy ounce. (The density of gold is 1204 lb/ft^3.)
18. Find the surface area, volume, and weight of an iron ball, 3.50 inches in diameter. (The density of iron is 491 lb/ft^3.)
19. A hemispherical radome protects a radar antenna on an aircraft. If the radome is 19.2 ft in diameter, what volume is enclosed? If the radome is constructed of a sheet material with a mass of 0.371 lb/ft^2, find the mass of the radome.
20. Find the weight of a round head rivet made of steel, as illustrated in Fig. 14-46. The shaft has a diameter of 0.50 inch and a length of 1.52 inches. The head is a hemisphere with a diameter of 0.925 inch. The density of steel is 4.5 ounces per cubic inch.

■ FIGURE 14-44

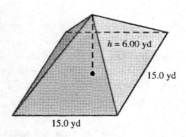

■ FIGURE 14-45

■ FIGURE 14-46

WRITING

21. In this chapter, we studied geometric solids, including the prism, rectangular solid, cube, cylinder, pyramid, cone, frustum, and sphere. In your own words, describe each of these geometric solids. Which solid is your favourite? Give an example of your favourite solid from real life.

TEAM PROJECT

22. A brick of platinum and a brick of gold both have the following dimensions, 5.0 cm thick, 10.0 cm wide and 20.0 cm long. Using a specific day, because precious metal prices vary daily, determine which brick is the more valuable. Determine the market value of each of the bricks, for that day, keeping in mind that precious metal prices are quoted in troy ounces.

15 Right Triangle Trigonometry

Outcomes

When you have completed this chapter, you will be able to

1. Define an angle in standard position.
2. Use the Pythagorean theorem using notation for rectangular coordinates.
3. Identify the trigonometric functions using notation for rectangular coordinates.
4. Find an angle using inverse trigonometric functions.
5. Solve for missing components of a right angle triangle.

One of the most important topics in mathematics is trigonometry. In this chapter, we lay the foundation for solving applications involving right triangles using trigonometry.

Trigonometry is the branch of mathematics that deals mostly with the solution of triangles and the functions derived from right triangles. The six trigonometric functions are introduced here and are used to solve right triangles. We consider the inverse trigonometric functions and some of the many applications involving right triangle trigonometry, such as problems in surveying, navigation, angles of depression and elevation, and indirect measurement.

15-1 SINE, COSINE, AND TANGENT FUNCTIONS

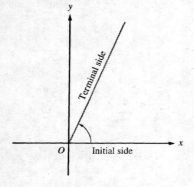

FIGURE 15-1 Standard position of an angle.

In this section we define an angle in standard position and define the three most useful trigonometric functions. We calculate the trigonometric functions, for a given angle or for a given point on the terminal side of the angle. We will use this in Sec. 15-4 to solve applications involving right triangles.

Standard Position of an Angle

In Sec. 13-1 we discussed the definition of an angle formed by a rotating ray. In this section, we extend that concept with angles drawn on the coordinate axes. If we draw an angle on the coordinate axes, with its vertex at the origin and one side along the positive x axis (initial side), then the angle is said to be in *standard position* (Fig. 15-1). All the angles referred to in this chapter will be positive acute angles and their terminal sides will be in quadrant I.

■ FIGURE 15-2

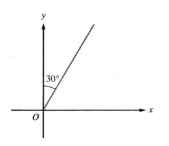

■ FIGURE 15-3
Angle *not* in standard position.

Example 1

Sketch a 30° angle in standard position.

Solution: The vertex of the angle is at the origin and the one side is along the positive x axis (Fig. 15-2). The terminal side is drawn in quadrant I, so that a 30° angle is formed. The position of the terminal side may be found using a protractor if accuracy is important.

COMMON ERROR Be sure to draw the standard position angle so that it makes the angle with the positive x axis, *not* with the y axis (Fig. 15-3).

Example 2

Sketch the positive angle θ in standard position for which the point (3, 4) is on the terminal side. Measure the angle.

Solution: We draw the vertex of the angle at the origin and draw one side along the positive x axis. We plot the point (3, 4) and draw a line from the vertex through the point (Fig. 15-4). Using a protractor, we find that the angle θ is approximately 53°.

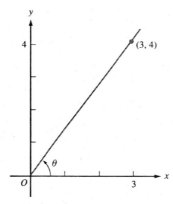

■ FIGURE 15-4

■ FIGURE 15-5

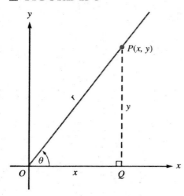

Trigonometric Functions

Now let's consider an angle θ in standard position (Fig. 15-5). We select a point P on the terminal side of θ, and let the coordinates of P be (x, y). If we drop a perpendicular from P to a point Q on the x axis, we have a right triangle. The hypotenuse of the right triangle is $\overline{OP}$, and we call its length r (for *radius*). The leg opposite θ is $\overline{PQ}$ and has length y. The leg adjacent to θ is $\overline{OQ}$ and has length x.

The hypotenuse r is found by the Pythagorean theorem.

| HYPOTENUSE | $r = \sqrt{x^2 + y^2}$ | 112 |

The right triangle has three sides (x, y, and r) which are related to each other by Eq. 112 and are functions of the angle θ. The quotients (ratios) of any two sides

15-1 Sine, Cosine, and Tangent Functions

$\left(\dfrac{y}{r}, \dfrac{x}{r}, \text{ and } \dfrac{y}{x}\right)$ are very important numbers. If we had chosen to place point P farther out on the terminal side of angle θ, then the lengths x, y, and r would all be greater than they are now. But the quotient of any two sides would be the same, regardless of where we placed P. The quotients change only if we change the angle θ. That is, the quotients of two sides are functions of the angle θ. They are called the *trigonometric functions*.

The most useful trigonometric functions are the sine, cosine, and tangent functions. The symbols for these functions are given below with their definitions.

TRIGONOMETRIC FUNCTIONS	sine $\theta = \sin \theta = \dfrac{y}{r}$	113
	cosine $\theta = \cos \theta = \dfrac{x}{r}$	114
	tangent $\theta = \tan \theta = \dfrac{y}{x}$	115

Example 3

Find the sine, cosine, and tangent of the angle in Fig. 15-4, rounded to three significant digits.

Solution: For this angle θ, $x = 3$ and $y = 4$
Using Eq. 112 to calculate r,

$$r = \sqrt{3^2 + 4^2} = \sqrt{25} = 5$$

Using Eqs. 113–115,

$$\sin \theta = \frac{y}{r} = \frac{4}{5} = 0.800$$

$$\cos \theta = \frac{x}{r} = \frac{3}{5} = 0.600$$

$$\tan \theta = \frac{y}{x} = \frac{4}{3} = 1.33$$

COMMON ERROR When writing the trigonometric functions, *do not omit the angle.*
If $\sin \theta = 0.800$, NEVER write $\sin = 0.800$

You will notice that the trigonometric functions are *pure* numbers; that is, they have no units. If the sides of a triangle have units, the units cancel out in the division.

Example 4

Sketch the angle in standard position for which the point $Q(6, 8)$ is on the terminal side. Find the sine, cosine, and tangent of the angle, rounded to three significant digits.

Solution: We sketch the angle (Fig. 15-6) so that the point $Q(6, 8)$ is on the terminal side. For point Q, $x = 6$ and $y = 8$.

■ FIGURE 15-6

Using Eq. 112 to calculate r,
$$r = \sqrt{6^2 + 8^2} = \sqrt{100} = 10$$

By Eqs. 113–115,
$$\sin \theta = \frac{y}{r} = \frac{8}{10} = 0.800$$
$$\cos \theta = \frac{x}{r} = \frac{6}{10} = 0.600$$
$$\tan \theta = \frac{y}{x} = \frac{8}{6} = 1.33$$

We see that the values of the trigonometric functions in Example 4 are the same as in Example 3. That is because the points (3, 4) and (6, 8) are both on the same terminal side (Fig. 15-7). The angle is the same for both examples.

So the trigonometric functions are not functions of the chosen point on the terminal ray. The specific point doesn't matter. Any point on the terminal side of the angle will do. The trigonometric functions are functions of the angle, and not of a particular point on the terminal side.

Trigonometric Functions by Calculator

We generally use a scientific calculator to find the trigonometric functions of a given angle. First, the calculator must be in the proper angle unit mode, either "degree" or "radian," depending on the units of the given angle. Refer to your calculator manual to determine how to change modes on your calculator and how to enter the trigonometric functions.

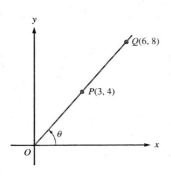

FIGURE 15-7

---- **Example 5** ----

Use your calculator to evaluate the cosine of a 30° angle, to three significant digits.

Solution: With our calculator in degree mode, we find that
$$\cos 30° = 0.8660254038$$
Rounding to three significant digits,
$$\cos 30° = 0.866$$

---- **Example 6** ----

Evaluate tan 1.25 to three significant digits.

Solution: We're to find the tangent of an angle whose measure is 1.25. Recall that an angle with no units is assumed to be in radians. So, with our calculator in radian mode, we find that
$$\tan 1.25 = 3.009569674$$
Rounding to three significant digits,
$$\tan 1.25 = 3.01$$

COMMON ERROR It is easy to forget to change to degree or radian mode as necessary. Be sure to check it each time.

SECTION 15-1 Exercises Sine, Cosine, and Tangent Functions

Standard Position of an Angle

Sketch each angle in standard position, showing the rotation. Then check your accuracy with a protractor.

1. 45°
2. 60°
3. 80°
4. 20°
5. 30°
6. 75°
7. 10°
8. 40°

Trigonometric Functions

The given point is a point on the terminal side of a positive angle θ in standard position. Sketch each angle and give the sine, cosine, and tangent for each angle rounded to three significant digits.

9. (5, 12)
10. (6, 8)
11. (15, 8)
12. (7, 24)
13. (2, 3)
14. (1, 5)
15. ($\sqrt{3}$, 1)
16. (5, $\sqrt{16}$)
17. (34.6, 79.1)
18. (502, 289)

Using a protractor, sketch the given angle in standard position on coordinate axes. Drop a perpendicular from any point on the terminal side and measure the distances, x, y, and r in cm. Use these measured distances to determine the sine, cosine, and tangent of the angle. Round to three significant digits. Check your results with the calculator. They should match approximately.

19. 60°
20. 30°
21. 45°
22. 75°

Plot the given point on coordinate axes and draw a line segment to the origin. With a protractor, measure the angle formed with the positive x axis to the nearest degree. Find the tangent of the angle two ways: (a) find the tangent of the measured angle by calculator, and (b) calculate the ratio $\frac{y}{x}$. They should be approximately equal.

23. (4, 3)
24. (2, 5)
25. (1, 6)
26. (4, 1)

27. Draw a 45° angle in standard position. Drop a perpendicular from a point on the terminal side, to form a 45-45-90° right triangle. Label the lengths of the legs 1 and 1, and the hypotenuse $\sqrt{2}$. Use these lengths to determine the sine, cosine, and tangent of 45°. Verify these functions using your calculator.

28. Draw a 60° angle in standard position. Drop a perpendicular from a point on the terminal side, to form a 30-60-90° right triangle. Label the lengths of the two legs 1 and $\frac{1}{2}$, and the hypotenuse $\frac{\sqrt{3}}{2}$. Use these lengths to determine the functions of sine, cosine, and tangent of 60°. Verify these functions using your calculator.

Evaluate, rounding to three significant digits.

29. sin 52.1°
30. cos 84.3°
31. tan 12.5°
32. sin 46.9°
33. cos 36.8°
34. tan 18.3°
35. sin 78.6°
36. cos 54.3°
37. tan 67°12′
38. sin 29°35′
39. cos 43°40′18″
40. sin 6°52′42″
41. tan (1.54 rad)
42. cos (0.367 rad)
43. cos (0.0256 rad)
44. sin (1.27 rad)

15-2 INVERSE TRIGONOMETRIC FUNCTIONS

In this section we study the inverse trigonometric functions. This involves finding the angle when a trigonometric function (sine, cosine, or tangent) is given. We also learn to use the special inverse function notation. We need this for applications involving right triangles.

Finding the Angle When the Trigonometric Function Is Given

Suppose we were given:
$$\sin \theta = 0.256$$

What is θ? To begin, θ is an angle; specifically, it is an angle whose sine is 0.256. We call the angle θ the *inverse sine* of 0.256, and we write
$$\theta = \sin^{-1} 0.256$$

The angle θ is also called the *arcsine* of 0.256, and may also be written
$$\theta = \arcsin 0.256$$

Inverse Functions by Calculator

You can use your calculator to find an angle, given one of its trigonometric functions. Consult your calculator manual for specifics on the inverse trigonometric functions.

Example 7

Use the calculator to find the angle θ such that
$$\sin \theta = 0.256$$

Solution: First we rewrite the equation
$$\sin \theta = 0.256$$
as
$$\theta = \sin^{-1} 0.256$$

With our calculator in degree mode, we find that
$$\sin^{-1} 0.256 = 14.83284792$$

Rounding to the nearest tenth, we get
$$\theta = 14.8°$$

Example 8

Show the meaning of the equation $\cos \theta = 0.345$ by completing this sentence:
_____ is the _____ whose _____ is _____.

Solution: $\underline{\theta}$ is the <u>angle</u> whose <u>cosine</u> is $\underline{0.345}$.

Example 9

Find θ to the nearest tenth of a degree if
$$\tan \theta = 3.75$$

Solution: We rewrite the equation as
$$\theta = \tan^{-1} 3.75$$

With the calculator in degree mode, we find that
$$\tan^{-1} 3.75 = 75.06858282$$

Rounding,
$$\theta = 75.1°$$

Example 10

Evaluate $\cos^{-1} 0.345$, in radians, rounded to the nearest hundredth.

Solution: With the calculator in radian mode, we find that
$$\cos^{-1} 0.345 = 1.218557542$$
Rounding,
$$\cos^{-1} 0.345 = 1.22$$

TIP If θ is an angle and a is the sine of that angle, then these equations all mean the same thing:
$$\sin \theta = a$$
$$\sin^{-1} a = \theta$$
$$\arcsin a = \theta$$
They all mean "θ is the angle whose sine is a."

There may be some confusion with the special notation used to indicate the inverse trigonometric function. In the expression
$$\sin^{-1} a$$
the raised -1 is *not* an exponent, but indicates the inverse sine (arcsin) of the number a.
$$\sin^{-1} a = \arcsin a$$
However, in the expression
$$(\sin \theta)^{-1}$$
the -1 *is* an exponent, and we saw in Chapter 7 that
$$(\sin \theta)^{-1} = \frac{1}{\sin \theta}$$

COMMON ERROR $\quad \sin^{-1} a \neq (\sin a)^{-1}$

The $\sin^{-1} a$ is *not* the reciprocal of a.

The same is true for the other trigonometric functions.

Example 11

In radian mode, show that
$$\cos^{-1} 0.123 \neq (\cos 0.123)^{-1}$$

Solution: We see that $\cos^{-1} 0.123$ is the *angle* whose cosine is 0.123. With the calculator in radian mode, we find that
$$\cos^{-1} 0.123 = 1.45$$
Since $(\cos 0.123)^{-1}$ is the *reciprocal* of $\cos 0.123$, we find that
$$(\cos 0.123)^{-1} = \frac{1}{\cos 0.123} = 1.01$$
Since $1.45 \neq 1.01$,
$$\cos^{-1} 0.123 \neq (\cos 0.123)^{-1}$$

SECTION 15-2 Exercises   Inverse Trigonometric Functions

Finding the Angle When the Trigonometric Function Is Given

Show the meaning of these equations by completing the sentence: ____ is the ____ whose ____ is ____.

1. $\sin \theta = 0.636$
2. $\cos \alpha = 0.398$
3. $\arctan 3.45 = \alpha$
4. $\arcsin 0.035 = \theta$
5. $\cos^{-1} a = 30°$
6. $\tan^{-1} b = \dfrac{\pi}{3}$
7. $\tan \dfrac{\pi}{2} = x$
8. $\sin 3 = y$
9. $\sin^{-1} a = B$
10. $\cos^{-1} x = Z$
11. $\sin 1 = 0.8415$
12. $\tan 0.5 = 0.5463$
13. $\arccos 0.399 = 66.5°$
14. $\arctan 0.222 = 12.5°$

Write the equation two other ways.

15. $\sin \alpha = 0.634$
16. $\cos \theta = 0.294$
17. $\tan \theta = 4.67$
18. $\sin^{-1} a = 45°$
19. $\cos^{-1} b = 60°$
20. $\tan^{-1} b = 30°$
21. $\arcsin 0.268 = \alpha$
22. $\arccos 0.839 = \theta$
23. $\arccos x = 0.785$
24. $\tan 27° = y$
25. $\tan 89.0° = 57.3$
26. $\arcsin 0.738 = 0.830$

Solve to the nearest tenth of a degree.

27. $\cos A = 0.707$
28. $\sin B = 0.866$
29. $\tan \theta = 4.87$
30. $\cos A = 0.197$
31. $\sin \alpha = 0.5$
32. $\tan \theta = 2$
33. $\tan \alpha = 1$
34. $\cos \alpha = 0.5$
35. $\cos B = 0.866$
36. $\sin C = 0.707$

Solve to the nearest hundredth of a radian.

37. $\sin \theta = 0.382$
38. $\tan A = 25.6$
39. $\tan B = 50.1$
40. $\cos \alpha = 0.999$

Evaluate to the nearest tenth of a degree.

41. $\arctan 6.20$
42. $\arccos 0.707$
43. $\sin^{-1} 0.707$
44. $\tan^{-1} 1$
45. $\cos^{-1} 0.5$
46. $\arcsin 0.866$

Evaluate to the nearest hundredth of a radian.

47. $\tan^{-1} 10$
48. $\arctan 35$
49. $\arccos 0.866$
50. $\sin^{-1} 0.5$

Inverse Functions by Calculator

51. Show that $\sin^{-1} 0.738 \neq (\sin 0.738)^{-1}$, in radian mode.
52. Show that $\tan^{-1} 3.00 \neq (\tan 3.00)^{-1}$, in radian mode.

15-3   SOLUTION OF RIGHT TRIANGLES

In Sec. 15-1, we defined the sine, cosine, and tangent of an angle, and saw that these functions could give us the quotient of two sides of a right triangle. But we considered only angles that were in standard position on the coordinate axes. In this section, we will see how the sine, cosine, and tangent can be used with a right triangle that is not on coordinate axes.

FIGURE 15-8
Labeling a right triangle.

Right Triangles

Figure 15-8 shows a right triangle that has been labeled in the conventional way. We will usually label a right triangle as shown in Fig. 15-8. We label the angles (vertices) with the capital letters A, B, and C, with C being the right angle. We label the sides with the lower case letters a, b, and c. Side a is opposite angle A; side b is opposite angle B; and side c (the hypotenuse) is opposite angle C (the right angle). The formulas for right triangles that we have already learned are repeated here.

PYTHAGOREAN THEOREM	$c^2 = a^2 + b^2$	107
	$c = \sqrt{a^2 + b^2}$	108
	$a = \sqrt{c^2 - b^2}$	109
SUM OF THE ACUTE ANGLES	$A + B = 90°$	106

When we solve a right triangle, it is convenient to express the trigonometric functions of the acute angles in terms of the sides. If we place triangle ABC (Fig. 15-8) on the coordinate axes, so that A is at the origin, B is in Quadrant I, and C is on the positive x axis (Fig. 15-9), then angle A is in standard position, and we can find the trigonometric functions of angle A. Using Eqs. 113–115 with $\theta = A$, $x = b$, $y = a$, and $r = c$,

FIGURE 15-9

$$\sin A = \frac{y}{r} = \frac{a}{c} = \frac{\text{leg opposite } A}{\text{hypotenuse}}$$

$$\cos A = \frac{x}{r} = \frac{b}{c} = \frac{\text{leg adjacent } A}{\text{hypotenuse}}$$

$$\tan A = \frac{y}{x} = \frac{a}{b} = \frac{\text{leg opposite } A}{\text{leg adjacent } A}$$

We can generalize for any acute angle θ in any right triangle:

RIGHT TRIANGLE TRIGONOMETRY	$\sin \theta = \dfrac{\text{opposite leg}}{\text{hypotenuse}}$	113
	$\cos \theta = \dfrac{\text{adjacent leg}}{\text{hypotenuse}}$	114
	$\tan \theta = \dfrac{\text{opposite leg}}{\text{adjacent leg}}$	115

COMMON ERROR Eqs. 113–115 above are true only in a *right* triangle.

TIP Equations 113–115 are very important since they are used so often. They should be memorized. One way to memorize these is to remember

"SOH-CAH-TOA"

which comes from

$$\underline{S}\text{in } \theta = \frac{\underline{O}\text{pposite leg}}{\underline{H}\text{ypotenuse}}$$

$$\underline{C}\text{os } \theta = \frac{\underline{A}\text{djacent leg}}{\underline{H}\text{ypotenuse}}$$

$$\underline{T}\text{an } \theta = \frac{\underline{O}\text{pposite leg}}{\underline{A}\text{djacent leg}}$$

TIP These formulas are true for a right triangle in *any* position. The angle θ need not be in standard position on the coordinate axes.

Example 12

Find the hypotenuse, the tangent of angle A, and angle A in Fig. 15-10.

Solution: We use the formula derived from the Pythagorean theorem (Eq. 108) to find the hypotenuse

$$c = \sqrt{a^2 + b^2} = \sqrt{12.0^2 + 5.00^2}$$
$$c = 13.0 \text{ cm}$$

■ FIGURE 15-10

(Note that we round to as many significant digits as are in the given numbers.)
  Hint: Always make sure your answers are reasonable as you go. Does the hypotenuse look like 13 cm (to scale)? Yes, it does.
We use Eq. 115 to find the tangent

$$\tan A = \frac{\text{opposite leg}}{\text{adjacent leg}} = \frac{12.0}{5.00}$$
$$\tan A = 2.40$$

To find angle A, we solve the equation above for A, using the inverse tangent function.

$$A = \tan^{-1} 2.40$$

With the calculator in degree mode, we find that

$$A = 67.4°$$

Once again, make sure your answers are reasonable as you go. Does angle A look like it is about 67°? Yes, it does.

Solving Right Triangles

To *solve* a triangle means to find all the missing sides and angles. In most applications we will need to find only one angle or one side. But first we will practise by finding all the missing sides and angles. We can solve any right triangle if we know at least one side and either another side or an angle. So we will know three parts (including the right angle), and we will need to find the remaining three parts.

Steps to Solve a Right Triangle

1. Make an accurate sketch (to scale) and label the known quantities.
2. Solve for the missing angles (by either subtracting a known angle from 90°, or by using an inverse trigonometric function).
3. Solve for the missing sides (by using the trigonometric ratios or the Pythagorean theorem).
4. Check.

Solving a Right Triangle When One Side and One Angle Are Known

If one side and one angle (and the right angle) are known, then we must find the other two sides and one angle.

Example 13

Solve right triangle ABC if $B = 32.5°$ and $a = 12.1$ cm.

Solution: First we make a sketch in Fig. 15-11. We make our sketch as accurate as possible (to scale), so that we can verify our results as we go. We label the known quantities, and the unknown sides and angles by letters. Next, we find the missing angle, A.

$$A = 90° - 32.5°$$
$$A = 57.5°$$

■ FIGURE 15-11

Does it look like A is about 57°? Of course.

Next, we find the missing sides, using the trigonometric ratios. We need to find b and c, and we want to use only angle B, not angle A. After all, we may have made an error in finding A, and if we used an incorrect A to find the sides, then they would be wrong also. So we want to use only the *given* quantities, B and a.

We must determine which trigonometric function to use to find b. The unknown side b is *opposite* angle B. The given side a is the side *adjacent* to angle B. The trigonometric function which relates the opposite side to the adjacent side is the tangent.

$$\tan B = \frac{\text{opposite leg}}{\text{adjacent leg}}$$

Substituting,

$$\tan 32.5° = \frac{b}{12.1}$$

Solving for b,

$$b = 12.1(\tan 32.5°)$$

With the calculator in degree mode, we find that

$$12.1(\tan 32.5°) = 7.708550156$$

So, rounding to three significant digits,

$$b = 7.71 \text{ cm}$$

(Does it look right?)

We now find side c. One way of finding c is to use the Pythagorean theorem. The problem with that is that we would need to use b, which was not given, but which we calculated. Thus, a possible error in b would be carried to c. So we choose not to use the Pythagorean theorem now. (We will use it later in the check.)

We will solve for c using a trigonometric function. We note that side c is the hypotenuse. And we will once again use the given adjacent side, a. The obvious choice for the trigonometric function is the cosine.

$$\cos B = \frac{\text{adjacent leg}}{\text{hypotenuse}}$$

Substituting,

$$\cos 32.5° = \frac{12.1}{c}$$

Solving for c, and remembering to *divide* this time,

$$c = \frac{12.1}{\cos 32.5°}$$
$$c = 14.3 \text{ cm}$$

Check:
Does it look like the hypotenuse is about 14 cm, to scale? Yes it does. We can check numerically, using the formula derived from the Pythagorean theorem:

$$\sqrt{12.1^2 + 7.71^2} \stackrel{?}{=} 14.3$$
$$14.3 = 14.3 \quad \underline{\text{checks}}$$

TIP When solving triangles using trigonometric functions, it often helps to make a table with a list of the triangle properties we already know. If you have a coloured pen or pencil, the known values are written in the box using red or green and the unknown values are filled in using regular black or blue pen or pencil. Using Example 13, we get

Angles	Sides
A = 57.5°	a = 12.1 cm
B = 32.5°	b = 7.71 cm
C = 90°	c = 14.3 cm

A quick visual check shows that the largest angle corresponds to the largest side, the intermediate angle corresponds to the intermediate side and the smallest angle corresponds to the smallest side. We will see when we study the trigonometry of oblique triangles that this relationship is still true.

COMMON ERROR Remember to use only the *given* information for each computation, rather than some quantity previously calculated. This way, any errors in the early computations will not be carried along.

Solving a Right Triangle When Two Sides Are Known

If two sides are known, then we must find one side and two angles.

Example 14

Solve right triangle ABC if $a = 42.0$ cm and $b = 48.0$ cm.

Solution: We sketch the triangle in Fig. 15-12, labeling all parts. We find the angles, A and B. To find angle A, we note that side a is opposite angle A, and side b is adjacent to angle A. We must use the tangent function.

$$\tan A = \frac{\text{opposite leg}}{\text{adjacent leg}}$$

Substituting,

$$\tan A = \frac{42.0}{48.0}$$

Solving for A,

$$A = \tan^{-1} \frac{42.0}{48.0} = 41.2°$$

Similarly,

$$\tan B = \frac{48.0}{42.0}$$

$$B = \tan^{-1} \frac{48.0}{42.0} = 48.8°$$

Do the angles look like they are about 41° and 49°? An intermediate check is that

$$41.2° + 48.8° = 90.0°$$

as it should.

Next, we find side c. We use Eq. 108:

$$c = \sqrt{42.0^2 + 48.0^2} = 63.8 \text{ cm}$$

We round to three significant digits, since the given numbers had three significant digits.

Check:

As a quick check, does it look like the hypotenuse is about 64 cm long, to scale? Yes, it does. For a numerical check, we can use one of the other trigonometric functions, say the sine, of one of the angles.

$$\sin 41.2° \stackrel{?}{=} \frac{42.0}{63.8}$$

$$0.659 = 0.658 \quad \text{checks}$$

Important: This *apparent* discrepancy in the check is due to the rounding, where there is always a little uncertainty in the last digit.

FIGURE 15-12

SECTION 15-3 Exercises Solution of Right Triangles

Right Triangles Find the third side, the tangent of angle A, and angle A.

1.

■ FIGURE 15-13

2.

■ FIGURE 15-14

3.

■ FIGURE 15-15

4.

■ FIGURE 15-16

5.

■ FIGURE 15-17

6.

■ FIGURE 15-18

7.

■ FIGURE 15-19

8.

■ FIGURE 15-20

9.

■ FIGURE 15-21

10.

■ FIGURE 15-22

Solving a Right Triangle When One Side and One Angle Are Known

Sketch each right triangle using the conventional labeling, and find all the missing parts.
11. $B = 48.0°$, $c = 35.0$ cm
12. $A = 35.0°$, $c = 42.6$ cm
13. $A = 69.0°$, $a = 4.32$ m
14. $B = 54.0°$, $b = 8.37$ cm
15. $B = 25.2°$, $a = 56.8$ cm
16. $A = 75.4°$, $b = 28.5$ cm
17. $A = 42°15'$, $a = 1.65$ m
18. $B = 64°24'$, $b = 3.75$ m

Solving a Right Triangle When Two Sides Are Known

Sketch each right triangle using the conventional labeling, and find all missing parts. Express the angles in decimal degrees.
19. $a = 4.00$ cm, $b = 7.00$ cm
20. $b = 5.00$ cm, $c = 8.00$ cm
21. $a = 4.60$ cm, $c = 6.90$ cm
22. $a = 1.50$ m, $b = 7.50$ m
23. $b = 35.70$ m, $c = 41.30$ m
24. $a = 85.6$ m, $c = 101$ m

15-4 APPLICATIONS

There are many applications involving right triangle trigonometry. In this section, we solve problems in geometry, angles of elevation and angles of depression, indirect measurements, navigation, and shop applications. In all of these applications, the key to the solution is in determining the appropriate trigonometric function to use. Each trigonometric function involves three quantities: one angle and two sides (opposite, adjacent, or hypotenuse). First we make a sketch, and determine which of the sides is given and which is needed. If they are opposite and adjacent, we use the tangent; if they are opposite and hypotenuse, we use the sine; if they are adjacent and hypotenuse, we use the cosine. Remember that each result is rounded according to the number of significant digits in the given quantities.

Geometry

Geometric problems involving polygons other than right triangles are often solved using trigonometry. In the following examples we use right triangle trigonometry to solve geometric problems.

FIGURE 15-23

Example 15

Two of the sides of an isosceles triangle have a length of 35.0 mm and the base angles are each 63.0°. Find the altitude and the base of the triangle.

Solution: We sketch the triangle in Fig. 15-23, and note that this is not a right triangle, so we cannot use the trigonometric functions directly. We draw the altitude a to the base and create two right triangles. In right triangle ABC, a is opposite the 63.0° angle and 35.0 mm is the hypotenuse, so we use the sine.

$$\sin 63.0° = \frac{a}{35.0}$$

Solving for a,

$$a = 35.0(\sin 63.0°)$$
$$a = 31.2 \text{ mm}$$

We let x be the side adjacent to the 63.0° angle. The hypotenuse is 35.0, so we use the cosine.

$$\cos 63.0° = \frac{x}{35.0}$$
$$x = 35.0(\cos 63.0°) = 15.9$$

So, the base is

$$\text{base} = 2x = 31.8 \text{ mm}$$

So, the altitude is 31.2 mm and the base is 31.8 mm.

(a) Angle of elevation

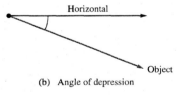

(b) Angle of depression

FIGURE 15-24

Angles of Elevation and Depression

Many right triangle trigonometry applications involve an angle made with a horizontal line (Fig. 15-24). For an object *above* the horizontal, the angle formed between the horizontal line and the line of sight from an observer to the object is called the *angle of elevation*. If the object is *below* the horizontal line, then the angle formed between the horizontal line and the line of sight from an observer to the object is called the *angle of depression*.

FIGURE 15-25

Example 16

A 38.1 m tower casts a shadow 52.4 m long. Find the angle of elevation of the sun.

Solution: In Fig. 15-25, 38.1 is opposite θ and 52.4 is adjacent to θ. Therefore we use the tangent function:

$$\tan \theta = \frac{38.1}{52.4}$$
$$\theta = \tan^{-1} \frac{38.1}{52.4} = 36.0°$$

15-4 Applications

Indirect Measurements

Distances that are difficult or impossible to measure directly are often found indirectly using trigonometry. A surveyor, for example, uses a transit to measure angles between locations, and then applies trigonometry to calculate the distances.

Example 17

A surveyor needs to measure the distance across a pond (Fig. 15-26). First she places a stake C on one shore. With a transit directly above C, she sights a stake B on the opposite shore. She then turns the transit 90° and places a third stake along that line of sight at A, 20.0 m from C. At A she measures the angle between the two stakes B and C to be 83.7°. What is the distance d across the pond?

Solution: Since d is the side opposite the 83.7° angle and 20.0 m is the adjacent side, we use the tangent.

$$\tan 83.7° = \frac{d}{20.0}$$
$$d = 20.0(\tan 83.7°)$$
$$d = 181 \text{ m}$$

So, the distance across the pond is 181 m.

■ FIGURE 15-26

Navigation

Problems in navigation often require the use of trigonometry. Before doing a navigation problem we show how compass directions are commonly designated. For example, a ship that is traveling in a compass direction of *35° east of south* is illustrated in Fig. 15-27. The direction 35° east of south means that the angle of travel is 35° in an easterly direction from due south.

■ FIGURE 15-27

Example 18

An observer at a point P (Fig. 15-28) on a coast sights a ship S in a direction of 42.0° west of north. At the same time, the ship is directly west of a point Q, which is 12.0 km due north of P. Find the distance x of the ship from point P.

■ FIGURE 15-28

Solution: First we note how the angle is drawn. From point P, a line is drawn due north. We draw the 42.0° angle on the west side of the north line. We note that x is the hypotenuse of the right triangle SPQ and 12.0 is the side adjacent to the 42.0° angle. We use the cosine:

$$\cos 42.0° = \frac{12.0}{x}$$

$$x = \frac{12.0}{\cos 42.0°}$$

$$x = 16.1 \text{ km}$$

So the distance from the ship to P is 16.1 km.

Shop Trigonometry

There are many uses of trigonometry in machine and tool shops.

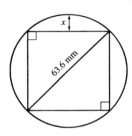

──── **Example 19** ────

Determine the depth of cut x (to the nearest tenth of a millimetre) required to mill a square on the end of a 63.6 mm circular shaft (Fig. 15-29).

Solution: The radius of the circle is 31.8 mm. We draw another radius, which is the perpendicular bisector of the top of the square. Recall from Sec. 13-4 that the diagonals of a square bisect the 90° angles of the square. So the angle θ is 45°. (We treat 45° as an exact number.) We first solve for y, and then use y to find x.

$$\sin 45° = \frac{y}{31.8}$$

$$y = 31.8(\sin 45°) = 22.486$$

$$x = 31.8 - 22.486 = 9.314$$

$$x = 9.3 \text{ mm}$$

■ **FIGURE 15-29**

The result is rounded to the tenth of a millimetre since the initial measurement was accurate only to the tenth of a millimetre. So the cut must be 9.3 mm deep on each side of the shaft.

SECTION 15-4 Exercises Applications

Geometry

Round answers to three significant digits.

1. Find the length of the altitude of an isosceles triangle with a 65° base angle and a base 145 cm long (Fig. 15-30).

2. Find the base of an isosceles triangle with a 54° vertex angle and equal sides of 43.8 mm each.

3. Find the measure of the base angles of an isosceles triangle with equal legs of 7.00 cm each and a base of 5.00 cm.

4. Find the base of an isosceles triangle with 72° base angles and an altitude 58.4 cm long.

5. The sides of a rectangle are 467 mm and 638 mm. Find the angle θ between a diagonal and the shorter side (Fig. 15-31).

FIGURE 15-30

FIGURE 15-31

FIGURE 15-33

FIGURE 15-32

FIGURE 15-34

6. The short side of a certain rectangle is 32.0 cm and the diagonal is 54.0 cm long. Find the length of the long side and the angle between the diagonal and the short side.

Angles of Elevation and Depression

7. The shadow of a tree is 15.0 m long when the angle of elevation of the sun is 68.7° (Fig. 15-32). Find the height of the tree.

8. How long a shadow will be cast by a monument that is 131 m tall, when the angle of elevation of the sun is 42.6°?

9. An airplane, at an altitude of 3810 m, is approaching an airport. If the angle of depression from the airplane to the airport is 20.3°, find the horizontal distance from the airplane to the airport.

10. The CN Tower in Toronto is 553 metres high. If you stand $100\overline{0}$ metres away, what would the angle of elevation be? (Fig. 15-33)

11. A woman at a window 18.0 m above a street is looking at a street light 13.0 m above the street. The angle of depression is 15.8°. How far is she from the street light?

12. From a boat at sea, the angle of elevation to the top of a 114 m cliff is 28.3°. How far is the boat from the base of the cliff?

13. From the top of a building, the angle of depression to a person on the street, 4.57 m in front of the building, is 65.2° (Fig. 15-34). How tall is the building?

■ FIGURE 15-35

■ FIGURE 15-36

14. From the top of a 38.1 m sea cliff, the angle of depression to a small boat sailing toward the cliff was 26.3°. Two hours later the angle of depression to the boat was 35.8°. How far had the boat traveled?

Indirect Measurements

15. A surveyor needs to measure the distance between two points B and C on opposite sides of a river (Fig. 15-35). She places a transit at a point C and sights point B across the river. She then turns the transit 90° and places a stake along that line of sight at point A 30.0 m from C. Angle CAB is measured with the transit to be 64.7°. What is the distance between points B and C?

16. To measure the height of a cliff, a surveyor measures a point 125 m from the base of the cliff and measures the angle of elevation to the top of the cliff to be 28.5°. How high is the cliff?

17. The angle of depression of a ship from the bridge of an aircraft carrier 72.2 m above water level is 10.5°. How far is the ship from the waterline of the carrier?

18. The angle of elevation from one mountain peak to the adjacent higher peak is 12°30′ (Fig. 15-36). What is the air distance between the peaks if their heights are 2576 m, and 3146 m?

19. From a point 38.1 m from the base of a building, a surveyor uses a transit to sight an angle of 36.2° to the top of a building. What is the height of the building if the sight of the transit is 1.52 m from the ground?

20. A woman wants to measure the height of an antenna on the top of a 125 m building. From a point in front of the building, she measures the angle of elevation to the top of the building to be 68.2° and the angle of elevation to the top of the antenna to be 71.0°. How tall is the antenna?

Navigation

21. An observer at a point A on a coast sights a ship in a direction of 37.0° east of south. At the same time, the ship is directly east of a point B, 19.0 km due south of A (Fig. 15-37). Find the distance of the ship from point A.

22. A ship at sea was 3.50 km due west of a lighthouse at noon one day. The ship was traveling due north, and later that day the captain sighted the lighthouse in a direction of 27° east of south. How far had the ship traveled since noon?

23. A ship traveled from a port in the direction of 57° west of north for 348 km. How far north and how far west of the port is the ship located?

24. A ship traveled from a port in the direction of 43° west of south at 26.0 knots for 3.00 hours. How far south and how far west of the port was the ship located? (A knot is 1.852 km/h.)

Structures

25. A guy wire 22.9 m long is stretched from the ground to the top of a telephone pole 17.7 m high (Fig. 15-38). Find the angle between the wire and the pole.

15-4 Applications

■ FIGURE 15-37

■ FIGURE 15-38

■ FIGURE 15-39

■ FIGURE 15-40

■ FIGURE 15-41

■ FIGURE 15-42

■ FIGURE 15-43

Shop Trigonometry

26. The roof of a building (Fig. 15-39) slopes up at an angle of 29.0° with the horizontal. The span (horizontal distance) of the roof is 9.75 m. Find the rise and the length of a roof rafter.

27. Find the angle a rafter makes with the horizontal if the rise is 3.0 m and the span is 15.0 m (Fig. 15-39). Then find the length of the rafter.

28. A stair has a rise of 12.0 cm for every 15.0 cm of run (Fig. 15-40). What is the angle of inclination θ of the stair?

29. A shelf (Fig. 15-41) is supported by a 27.0 cm brace attached to the wall 14.0 cm below the shelf. Find the angle between the brace and the wall.

30. An entrance ramp for the handicapped must be built to a door 1.45 m above ground level. If the maximum angle the ramp can make with the ground is 8°, how far from the base of the building must they begin the ramp?

31. The corner of the rectangular metal plate in Fig. 15-42 is to be cut off. Find x and θ.

32. Determine the missing dimension in the head of the screw in Fig. 15-43.

33. A machinist must make a bolt circle of radius 13.65 mm with 12 equally spaced bolt holes (Fig. 15-44). What is the straight line distance between the holes?

34. Sixteen rivets are equally spaced on the circumference of a circular plate. If the centre-to-centre distance between two adjacent rivets is 5.75 cm, what is the radius of the circle?

■ FIGURE 15-44

■ FIGURE 15-45

■ FIGURE 15-46

35. Figure 15-45 is a sketch of a dovetail wedge. Find the upper width, w, of the wedge.

36. The distance across the corners of a regular hexagonal bolt is 1.50 cm (Fig. 15-46). Find the distance across the flats.

15-5 COSECANT, SECANT, AND COTANGENT FUNCTIONS

In this section we define the cosecant, secant, and cotangent of an angle and then use the calculator to calculate them. We use these functions to express some right triangle relationships in a simpler form and also to simplify some trigonometric formulas. We didn't cover these functions earlier because we didn't need them for solving right triangles; all the calculations could be done using only the sine, cosine, and tangent functions.

Cosecants, Secants, and Cotangents as Functions

In Sec. 15-1 we defined the sine, cosine, and tangent of an angle as a ratio of coordinates or as a quotient of sides of a right triangle (for acute angles). There are three additional trigonometric functions that are similarly defined, using the triangle in Fig. 15-47.

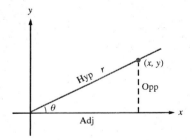
■ FIGURE 15-47

■ FIGURE 15-48
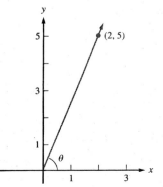

THREE MORE TRIGONOMETRIC FUNCTIONS	cosecant θ = csc θ = $\dfrac{r}{y}$ = $\dfrac{\text{hypotenuse}}{\text{opposite leg}}$	116
	secant θ = sec θ = $\dfrac{r}{x}$ = $\dfrac{\text{hypotenuse}}{\text{adjacent leg}}$	117
	cotangent θ = cot θ = $\dfrac{x}{y}$ = $\dfrac{\text{adjacent leg}}{\text{opposite leg}}$	118

Example 20

The point (2, 5) is on the terminal side of a positive angle θ in standard position (Fig. 15-48). Find the cosecant, secant, and cotangent of θ, rounded to three significant digits.

Solution: We calculate r, using Eq. 112, with $x = 2$ and $y = 5$,

$$r = \sqrt{x^2 + y^2} = \sqrt{2^2 + 5^2}$$
$$r = \sqrt{29}$$

We leave r as $\sqrt{29}$, without approximating it as a decimal, so that we don't introduce error at this point! Now we use Eqs. 116–118 to calculate the trigonometric functions.

$$\csc \theta = \frac{r}{y} = \frac{\sqrt{29}}{5} = 1.08$$

$$\sec \theta = \frac{r}{x} = \frac{\sqrt{29}}{2} = 2.69$$

$$\cot \theta = \frac{x}{y} = \frac{2}{5} = 0.400$$

── **Example 21** ──

Find the cosecant, secant, and tangent of angle A in Fig. 15-49. Round to three significant digits.

Solution: Using the formula derived from the Pythagorean theorem (Eq. 109),

$$a = \sqrt{8.92^2 - 5.27^2} = \sqrt{51.79} = 7.197$$

Here we approximate $\sqrt{51.79}$ but carry one extra digit for substituting into Eqs. 116–118,

$$\csc A = \frac{\text{hypotenuse}}{\text{opposite leg}} = \frac{8.92}{7.197} = 1.24$$

$$\sec A = \frac{\text{hypotenuse}}{\text{adjacent leg}} = \frac{8.92}{5.27} = 1.69$$

$$\cot A = \frac{\text{adjacent leg}}{\text{opposite leg}} = \frac{5.27}{7.197} = 0.732$$

■ **FIGURE 15-49**

Reciprocal Relationships

How do you determine the cosecant of a specific angle, say csc 52.6°, by calculator? You may have noticed that there is no "csc" key on your calculator.

Since

$$\csc \theta = \frac{\text{hypotenuse}}{\text{opposite leg}} \quad \text{and} \quad \sin \theta = \frac{\text{opposite leg}}{\text{hypotenuse}}$$

we can see that the cosecant and the sine are reciprocals of each other.

Therefore,

$$\csc \theta = \frac{1}{\sin \theta}$$

Similarly, the secant is the reciprocal of the cosine, and the cotangent is the reciprocal of the tangent.

RECIPROCAL RELATIONSHIPS	$\csc \theta = \dfrac{1}{\sin \theta}$	119(a)
	$\sec \theta = \dfrac{1}{\cos \theta}$	119(b)
	$\cot \theta = \dfrac{1}{\tan \theta}$	119(c)

So, to find the cosecant, secant, or cotangent of an angle, first find the sine, cosine, or tangent of the angle and then take the reciprocal.

Example 22

Find $\csc 52.6°$, rounded to three significant digits.

Solution: Since $\csc 52.6° = \dfrac{1}{\sin 52.6°}$, we evaluate with the calculator.

With the calculator in degree mode, we find that

$$\frac{1}{\sin 52.6°} = 1.258788514$$

So $\csc 52.6° = 1.26$.

Finding the Angle When the Trigonometric Function Is Given

Here we reverse the previous process. We must now use both the reciprocal and the inverse keys on the calculator.

Example 23

Given $\sec \theta = 2.25$, find θ, and round to the nearest tenth of a degree.

Solution: The secant is the reciprocal of the cosine, so

$$\sec \theta = \frac{1}{\cos \theta} \quad \text{so} \quad \frac{1}{\cos \theta} = 2.25$$

Taking reciprocals,

$$\cos \theta = \frac{1}{2.25}$$

Taking the inverse,

$$\theta = \cos^{-1}\left(\frac{1}{2.25}\right)$$

The final equation shows θ expressed in terms of an inverse trigonometric function $(\cos^{-1})$ and a reciprocal $\left(\dfrac{1}{2.25}\right)$. With the calculator in degree mode, we find that

$$\cos^{-1}\left(\frac{1}{2.25}\right) = 63.61220004°$$

So, $\theta = 63.6°$.

Example 24

Evaluate $\cot^{-1} 2.75$ in radians, and round to three significant digits.

Solution: Recall that $\cot^{-1} 2.75$ is the angle whose cotangent is 2.75. That is,

$$\cot^{-1} 2.75 = \theta$$
$$\cot \theta = 2.75$$
$$\frac{1}{\tan \theta} = 2.75$$
$$\tan \theta = \frac{1}{2.75}$$
$$\theta = \tan^{-1}\left(\frac{1}{2.75}\right)$$

With the calculator in radian mode, we find that

$$\tan^{-1}\left(\frac{1}{2.75}\right) = 0.3487710036$$

So, $\cot^{-1} 2.75 = 0.349$ rad.

SECTION 15-5 Exercises Cosecant, Secant, and Cotangent Functions

Cosecants, Secants, and Cotangents as Functions

The given point is a point on the terminal side of a positive angle θ in standard position. Sketch each angle and give the cosecant, secant, and cotangent for each angle, rounded to three significant digits.

1. (3, 4) 　2. (12, 5) 　3. (24, 7) 　4. (6, 8)
5. (1, 3) 　6. (3, 6) 　7. (3, $\sqrt{2}$) 　8. ($\sqrt{6}$, 4)
9. (27.3, 83.7) 　10. (12.5, 93.3)

Find the cosecant, secant, and cotangent of angle A (rounded to three significant digits), and the degree measure of A (rounded to the nearest tenth).

11.

■ FIGURE 15-50

12.

■ FIGURE 15-51

13.

■ FIGURE 15-52

14.

■ FIGURE 15-53

15.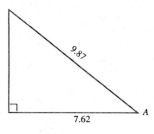

■ FIGURE 15-54

16.

■ FIGURE 15-55

Reciprocal Relationships

Evaluate, and round to three significant digits.

17. cot 63.3°   **18.** csc 74.5°   **19.** sec 28.1°
20. cot 89.6°   **21.** csc 2.39°   **22.** sec 3.58°
23. cot 52°45′   **24.** csc 73°54′   **25.** sec (1.25 rad)
26. csc (0.75 rad)

Finding the Angle When the Trigonometric Function Is Given

Find θ, and round to the nearest tenth of a degree.

27. sec θ = 1.15   **28.** csc θ = 2.25
29. cot θ = 12.6   **30.** sec θ = 1.05

Solve to the nearest thousandth of a radian.

31. csc θ = 4.50   **32.** cot θ = 8.60

Evaluate to the nearest tenth of a degree.

33. $\sec^{-1}$ 1.75   **34.** arccsc 3.50

Evaluate to the nearest thousandth of a radian.

35. arccsc 5.30   **36.** $\cot^{-1}$ 4.25

■ Chapter 15   REVIEW EXERCISES

Given a point on the terminal side of an angle in standard position,

  (a) Sketch the angle;

  (b) Give the six trigonometric functions for the angle, to three significant digits; and

  (c) Find the angle, to the nearest tenth of a degree.

1. (0.950, 4.70)   **2.** (32.5, 74.2)

Evaluate to three significant digits.

3. sin 76.8°
4. cos 5°18′
5. cot (1.25 rad)
6. csc (2.37 rad)
7. tan (12.6°)
8. sec 35°42′

Find the angle to the nearest tenth of a degree.

9. cos A = 0.866
10. cot α = 1.50
11. tan θ = 4.75
12. csc θ = 5.71

Find the angle to the nearest thousandth of a radian.

13. sin α = 0.567
14. sec θ = 1.15
15. Evaluate $\sin^{-1} 0.56$ to the nearest tenth of a degree.
16. Evaluate arcsec 1.41 to the nearest thousandth of a radian.

Find the six trigonometric functions of angle A, and the degree measure of A.

17.

■ FIGURE 15-56

18.

■ FIGURE 15-57

■ FIGURE 15-58

■ FIGURE 15-59

Sketch each right triangle and find all the missing parts.

19. $A = 37.6°$, $c = 65.7$ cm
20. $c = 4.625$ m, $b = 3.875$ m
21. $B = 0.825$ rad, $a = 3.25$ cm
22. $c = 63.3$ mm, $B = 67.8°$
23. Find the altitude of an isosceles triangle with a 57.3° vertex angle and a base 74.9 cm long (Fig. 15-58).
24. The sides of a rectangle are 15.0 cm and 25.0 cm. Find the angle between a diagonal and the longer side.
25. Find the area of a parallelogram whose sides are 48.0 cm and 36.0 cm, if one angle is 73.5°.
26. Find the diameter of a circle inscribed in a hexagon whose sides are 15.0 cm each.
27. Find the radius of the circle in Fig. 15-59.
28. How long is the shadow of a 56.0 ft tree, when the angle of elevation of the sun is 32.8°?

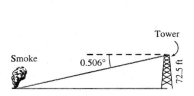

FIGURE 15-60   **FIGURE 15-61**

29. A ranger spots a fire from the top of a 72.5 ft fire tower. The angle of depression to the fire is 0.506° (Fig. 15-60). How far away from the base of the tower is the fire?
30. To measure the height of a cliff, a surveyor measures a point 902 ft from the base of the cliff and measures the angle of elevation to the top of the cliff to be 16.4°. How high is the cliff?
31. A 20.0 ft ladder leans against a vertical wall so that the base of the ladder is 4.50 ft away from the base of the wall. How high up the wall does the ladder go, and what angle does the ladder make with the ground?
32. The corner of the rectangular metal plate in Fig. 15-61 is to be cut off. Find x and θ.

WRITING

33. Write an essay on how you might use right triangle trigonometry in real-life situations on the job, in a hobby, or around the house.

TEAM PROJECT

34. Make a device for measuring angles in the horizontal plane (a "transit"), using a sighting tube of some sort and a protractor. Use your transit and a rope of known length to find some distance on your campus, such as the distance across a ball field, using the method of Example 17 in Sec. 15-4. Compare your result with the actual taped distance.

16 Factoring

Outcomes

When you have completed this chapter, you will be able to

1. Factor out a common factor.
2. Factor by grouping.
3. Factor a difference of squares.
4. Factor a simple trinomial.
5. Factor a general trinomial.
6. Factor a general trinomial by grouping.
7. Factor a perfect square trinomial.
8. Factor the sum and difference of cubes.

In Sec. 8-3 we learned how to multiply factors to get a product. In this chapter, we do the reverse: We begin with the product and find those quantities (factors) which, when multiplied together, give the original product. To *factor* expressions easily, we must be very familiar with the *special products* of algebraic multiplication. We will look at these special products first in each section.

But why must we learn to factor? We learn to factor because factoring is an essential step in simplifying expressions and solving equations, especially quadratic equations, literal equations, and formulas. Factoring is also used extensively in simplifying fractions. We will consider many factorable expressions which appear in equations that arise from applications.

16-1 COMMON FACTORS

In this section we learn to *factor* certain polynomials by factoring out a monomial which is a factor of each term of the polynomial. Recall that a monomial has only one term and a polynomial has more than one term.

Factoring an Expression

The *factors* of an expression are those quantities (numbers and variables) whose product is the original expression. In this chapter, we consider only factors with integral coefficients.

Example 1
Since
$$3a - 6b = 3(a - 2b)$$
then $\quad$ 3 and $(a - 2b)$ are factors of $3a - 6b$.

Prime Factors

If an expression has no factors, other than 1 and itself, then it is called *prime*.

Example 2
Two factors of 70 are 7 and 10. The 7 is prime because it has no factors other than 1 and 7. However, the 10 is not prime, because it can be factored into 2 times 5 (both 2 and 5 are prime). Thus the prime factors of 70 are 2, 5, and 7.

Example 3
The expressions x, $x - 2$, and $x^2 - x - 1$ are all prime, because they have no factors other than 1 and themselves.

Example 4
The expressions $x^2 - 4$ and $x^2 - 5x + 6$ are not prime, because they can be factored, as we shall see later in this chapter. We say they are *factorable*.

It's important to note that *not all* expressions are factorable. Many expressions, as in Example 3, are prime.

Common Factors

In Sec. 1-3, we discussed the distributive law

$$a(b + c) = ab + ac$$

This gives us the product of a monomial times a polynomial. If we reverse the sides of this equation, then we are *factoring* the left side.

$$ab + ac = a(b + c)$$
common factor

In this equation, the quantity a is called a *common factor,* since it is a factor *common* to both terms. It may be factored out, using the reverse of the distributive law (Eq. 5).

FACTORING OUT A COMMON FACTOR	$ab + ac = a(b + c)$	5

Example 5

Factor.

(a) $5p + 5q = 5(p + q)$
(b) $6a - 12b =$

$$6a - 6(2b) = 6(a - 2b)$$
$$\underbrace{}_{\text{common factor}}$$

The common factor 6 is *not* prime. That is, it can be factored

$$6 = 2(3)$$

However, in general, *we do not factor the numerical coefficient into primes.* So we are finished factoring the expression.

Example 6

$$6x - 2xy = \underline{2x}(3) - \underline{2x}(y) = \underline{2x}(3 - y)$$
$$\underbrace{}_{\text{common factor}}$$

Sometimes the common factor is not as obvious as it is in the examples above.

Example 7

Factor $12x^2 + 8x$.

Solution: Since 4 is a factor of both 12 and 8 and there is a common x in both terms, then $4x$ is a common factor of $12x^2 + 8x$.

$$12x^2 + 8x = \underline{4x}(3x) + \underline{4x}(2) = \underline{4x}(3x + 2)$$
$$\underbrace{}_{\text{common factor}}$$

Note that $4x$ is only *one* common factor of $12x^2 + 8x$. Other common factors are 2, 4, x, and $2x$. So the expression $12x^2 + 8x$ may be factored several different ways:

$$12x^2 + 8x = 2(6x^2 + 4x)$$

or

$$= 4(3x^2 + 2x)$$

or

$$= x(12x + 8)$$

or

$$= 2x(6x + 4)$$

But the preferred factoring is

$$12x^2 + 8x = 4x(3x + 2)$$

because $4x$ is the *greatest* common factor, as described below.

Factoring Completely

We say that an expression is factored *completely* if it is expressed as a product of its prime factors. One way to factor an expression completely is to factor out its "greatest common factor." The *greatest common factor* (also called the *gcf*) is the common factor that has the greatest numerical coefficient and the greatest degree. (Recall that the degree of a term is the power to which the variable is raised.)

But how do we find the gcf? Generally, the gcf is found by *inspection*. That is, we look at (inspect) the terms and say, "Ah ha, the gcf is $2x^2$." Specifically, the numerical coefficient of the gcf is the gcf of the coefficients of the terms, and the variable part is the highest power of each variable that is a factor of *each term*.

Example 8

Factor $10x^3 - 6x^2y$ completely.

Solution: First the gcf of the expression must be determined. The gcf of 10 and 6 is 2, the highest power of x contained in each term is x^2, and no power of y is contained in the first term. Thus, the gcf of the expression is $2x^2$.

$$10x^3 - 6x^2y = 2x^2(5x) - 2x^2(3y)$$
$$= 2x^2(5x - 3y)$$

Another method of factoring completely is to factor the expression, and keep factoring, until all the factors are prime.

Example 9

$$10x^3 - 6x^2y = 2(5x^3 - 3x^2y)$$
$$= 2x^2(5x - 3y)$$

Factors of 1

In factoring expressions, you must be especially careful with factors of 1.

Example 10

Factor $12x^2 - 6x$ completely.

Solution: By inspection we find the gcf is $6x$. Factoring $6x$ out of each term:

$$12x^2 - 6x = 6x(2x) - 6x(1)$$
$$= 6x(2x - 1)$$

Students are sometimes careless about the 1 in the factorization above. It's a common error to forget the 1, but it must be included!

> **COMMON ERROR** If the gcf of an expression is the same as one of its terms, we must include the 1 in the factored form.

Common Expressions

The common factors we have dealt with thus far have been monomials. But sometimes the common factor may be an expression, such as a binomial or a polynomial. Again, use the reverse of the distributive property:

$$ab + ac = a(b + c)$$

except that the common factor a will be an expression. Suppose the common factor was the expression $x + y$. Then the equation above would be:

$$(x + y)b + (x + y)c = (x + y)(b + c)$$

$$\underbrace{\hspace{4cm}}_{\text{common factor}}$$

Note that this is just a special case of the distributive property.

Example 11

Factor completely:

$$(x - 2)x + (x - 2)3$$

Solution: The common factor is the expression $(x - 2)$. Factoring out the expression $(x - 2)$, we get

$$(x - 2)x + (x - 2)3 = (x - 2)(x + 3)$$

Example 12

Factor completely:

$$2(p - 3) - x(p - 3)$$

Solution: The common factor is the expression $(p - 3)$. Factoring out $(p - 3)$, we get

$$2(p - 3) - x(p - 3) = (2 - x)(p - 3)$$

Factoring by Grouping

When the expression to be factored has four or more terms, these terms can sometimes be arranged in smaller groups, which can be factored separately.

Example 13

Factor $ab + ac + 2b + 2c$.

Solution: First we notice that there is no factor common to all four terms. However, the first two terms have a common factor, a, and the second two terms have a common factor, 2. We will try grouping the first two terms together and the last two terms together.

$$ab + ac + 2b + 2c = (ab + ac) + (2b + 2c)$$

Now factor out the common factor in each group:

$$= a\underbrace{(b + c)}_{} + 2\underbrace{(b + c)}_{}$$

$$\underbrace{\hspace{3cm}}_{\text{common factor}}$$

Both terms now have the common factor $(b + c)$. We now treat $(b + c)$ as a single expression and factor it out using the reverse of the distributive property. So,

$$ab + ac + 2b + 2c = (a + 2)(b + c)$$

Factoring by grouping can be quite tricky. Sometimes you must rearrange terms, factor out negative factors, and in general, use *trial and error* to see if any arrangement gives you another common factor. Sometimes no arrangement works, and the expression is not factorable.

Checking

To check if factoring has been done correctly, simply multiply the factors together. If the product is the original expression, then you were correct. This check does not, however, tell you if the expression is factored *completely*. You check that by inspecting the factors to make sure they cannot be factored further.

Example 14

Factor and check:

$$6x^2 - 3x$$

Solution: Factoring,

$$6x^2 - 3x = 3x(2x - 1)$$

To check we multiply the factors,

$$3x(2x - 1) = 3x(2x) - 3x(1)$$
$$= 6x^2 - 3x$$

So, it checks, and the factors are correct.

SECTION 16-1 Exercises   Common Factors

Factoring Completely   Factor each expression completely and check your result.

1. $xy + xz$
2. $pr - qr$
3. $3a - 3b$
4. $7x - 7y^2$
5. $x^2 - xy$
6. $ab + b^2$
7. $6x + 18$
8. $16x - 4y$
9. $24x^2 - 3$
10. $36ax + 3a$
11. $ac + a$
12. $4x - 2$
13. $4x^3 - 3x^2$
14. $5b^2 - 6b^3$
15. $24x^3y + 15x^2y^2 - 40xy$
16. $15p^3 - 10p^2 + 20p^5$
17. $4x^2y + 6x^3 - 2x$
18. $6y^3 - 12y^2 + 3y$
19. $6a^2b^3 + 4a^3b^3 - 10a^2b^4$
20. $6a^2b^3c - 9a^2bc^3 + 12a^3b^3c^2$

Common Expressions
21. $2(x + y) + x(x + y)$
22. $4(a - b) - b(a - b)$
23. $(p - 1)q - (p - 1)p$
24. $(m + n)m + (m + n)n$
25. $(c - d)d + (c - d)$
26. $(x - 3)x - (x - 3)$

27. $(x - y)^2 + (x - y)$
28. $4(a - b) - 2(a - b)^2$
29. $(3 + a)^3 - (3 + a)^2$
30. $(p - q)^3 - (p - q)^2 + 5(p - q)$

Factoring by Grouping

Group the terms in pairs and factor completely.

31. $xy + xz + 3y + 3z$
32. $pq - pr + 5q - 5r$
33. $xy - 2x + 3y - 6$
34. $ab - 3a + 4b - 12$
35. $x^2 + xy - 2x - 2y$
36. $2a - 2b - a^2 + ab$
37. $m^3 - 5m^2 + 3m - 15$
38. $1 - x + x^2 - x^3$

Applications

39. The total surface area of a cylinder is given by the formula
$$A = 2\pi r^2 + 2\pi rh$$
where r is the radius of the cylinder and h is the height (Fig. 16-1). Factor the right side of this equation.

40. If p dollars is invested for t years at an annual rate of simple interest r, the principal p would accumulate to y dollars, given by the formula:
$$y = p + prt$$
Factor the right side of this equation.

41. The total surface area of a cone is given by the formula
$$A = \pi r^2 + \pi rs$$
where r is the radius of the base and s is the slant height (Fig. 16-2). Factor the right side of the equation.

42. The total surface area A of a rectangular solid is given by the formula
$$A = 2lw + 2lh + 2wh$$
where l is the length, w is the width, and h is the height of the solid. Factor the right side of this equation.

43. In a battery, the sum of the voltage-drops across the resistors R_1, R_2, and R_3 with current i must equal the battery voltage E:
$$E = iR_1 + iR_2 + iR_3$$
Factor the right side of this equation.

44. For a simply supported beam of length L having a distributed load of w (Fig. 16-3), the bending moment M at any distance x from one end is given by
$$M = \frac{1}{2}wLx - \frac{1}{2}wx^2$$
Factor the right side of this equation.

45. An object is thrown upward with an initial velocity of 32 ft/s from a building 64 ft above the ground. The height S of the object above the ground at any time t is given by
$$S = 64 + 32t - 16t^2$$
Factor the right side of this equation.

FIGURE 16-1

FIGURE 16-2
Cone.

FIGURE 16-3
Simply supported beam.

16-2 DIFFERENCE OF TWO SQUARES

In this section we consider products that give the difference of two squares, such as

$$x^2 - y^2$$

and then we factor the difference of two squares. We learn to recognize the *pattern* for the difference of two squares, and use the pattern to factor the difference of any two even powers of expressions. We learn to group terms so that they fit into the pattern. Finally we consider applications of this type of factoring.

The Sum Times the Difference of Two Numbers

In Sec. 8-3 we saw in Eq. 31 that the sum of two numbers times the difference of the same numbers was equal to the difference of their squares.

$$\underset{\downarrow}{\text{sum}} \quad \underset{\downarrow}{\text{difference}}$$
$$(x + y)(x - y) = x^2 - y^2$$
$$\uparrow$$
$$\text{difference of two squares}$$

--- **Example 15** ---
$$(x + 3)(x - 3) = x^2 - 9$$

--- **Example 16** ---
$$(2x - 5)(2x + 5) = (2x)^2 - 5^2$$
$$= 4x^2 - 25$$

Factoring the Difference of Two Squares

If we reverse the sides of Eq. 31, we have

| FACTORING THE DIFFERENCE OF TWO SQUARES | $x^2 - y^2 = (x + y)(x - y)$ | 31 |

Once we recognize that an expression is the difference of two squares

$$x^2 - y^2$$

we can fit it into the pattern of Eq. 31.

--- **Example 17** ---
Factor $y^2 - 16$.

Solution: First we recognize that $y^2 - 16$ is a difference of two squares, since both y^2 and 16 are perfect squares. The square root of 16 is 4, so we write

$$y^2 - 16 = y^2 - 4^2$$

Thus, using Eq. 31,

$$y^2 - 16 = (y + 4)(y - 4)$$

Example 18

Factor $9x^2 - 4$.

Solution: Is $9x^2 - 4$ a difference of two squares? Does it fit into the pattern of $(x^2 - y^2)$? Yes, both $9x^2$ and 4 are perfect squares.

$$9x^2 - 4 = (3x)^2 - (2)^2$$

We fit these quantities into the pattern of the right side of Eq. 31:

$$9x^2 - 4 = (3x)^2 - (2)^2 = (3x \quad)(3x \quad)$$
$$\text{Same!}$$

$$9x^2 - 4 = (3x)^2 - (2)^2 = (3x \quad 2)(3x \quad 2)$$
$$\text{Same!}$$

$$9x^2 - 4 = (3x + 2)(3x - 2)$$
$$\quad\quad\quad\quad\quad\quad\quad \text{sum} \quad \text{difference}$$

Example 19

Factor completely:

$$2x^2 - 18$$

Solution: First we recognize that this is a difference but not of two squares. Neither $2x^2$ nor 18 are perfect squares. But there *is* a common factor, 2. We factor it out first:

$$2x^2 - 18 = 2(x^2 - 9)$$

Now we see that $(x^2 - 9)$ is the difference of two squares and can be factored further.

$$2x^2 - 18 = 2(x^2 - 9) = 2(x + 3)(x - 3)$$

TIP Always factor out the gcf *first*, before doing other types of factoring.

Difference of Two Even Powers

The difference of any two even powers can be factored as the difference of two squares, since every even power is a perfect square.

Example 20

Factor completely:

$$x^4 - y^4.$$

Solution: The expression $x^4 - y^4$ is the difference of two even powers, and therefore the difference of two squares:

$$x^4 - y^4 = (x^2)^2 - (y^2)^2$$
$$= (x^2 + y^2)(x^2 - y^2)$$

Can either of these factors be factored further? The first factor $(x^2 + y^2)$ is prime, and cannot be factored. But the second factor $(x^2 - y^2)$ is the difference of two squares, and can be factored further.

$$x^4 - y^4 = (x^2 + y^2)(x^2 - y^2)$$
$$= (x^2 + y^2)(x + y)(x - y)$$

None of these factors can be factored again, so the expression is factored completely.

In Example 20, we noted that the sum

$$x^2 + y^2$$

was prime. It *cannot* be factored, since the *sum* of two squares is prime.

Squared Expressions

Sometimes one (or both) of the perfect squares will be squares of expressions.

---- **Example 21** ----

Factor completely:

$$(a + 1)^2 - b^2$$

Solution: In this example, the expression $(a + 1)$ corresponds to the x in Eq. 31, and b corresponds to the y.

$$x^2 - y^2 = [x + y][x - y]$$

Replacing x by $(a + 1)$, $(a + 1)^2 - b^2 = [(a + 1) + b][(a + 1) - b]$

Removing the inner parentheses on the right:

$$(a + 1)^2 - b^2 = (a + 1 + b)(a + 1 - b)$$

SECTION 16-2 Exercises Difference of Two Squares

Factor completely.

1. $m^2 - n^2$
2. $c^2 - d^2$
3. $x^2 - 16$
4. $9 - x^2$
5. $25 - p^2$
6. $y^2 - 4$
7. $9y^2 - 1$
8. $36x^2 - y^2$
9. $4a^2 - 9b^2$
10. $9b^2 - 25a^2$
11. $p^4 - 4$
12. $4m^2 - n^6$
13. $3x^2 - 12y^2$
14. $18x^2 - 8y^2$
15. $4a^2b^2 - 4c^2$
16. $20x^2y^4 - 5$

Difference of Two Even Powers

17. $a^4 - b^4$
18. $x^6 - y^4$
19. $p^6 - q^8$
20. $49m^2 - 9n^4$
21. $100x^4 - 81$
22. $a^2b^4 - 4c^6$
23. $x^8 - 1$
24. $x^4 - 1$

Squared Expressions

25. $(a + b)^2 - c^2$
26. $(x - y)^2 - z^2$
27. $p^2 - (q + r)^2$
28. $m^2 - (n - p)^2$
29. $1 - (a - b)^2$
30. $(x - y)^2 - (a + b)^2$

FIGURE 16-4

Applications

31. To find the difference in elevation, h, of a slope (Fig. 16-4), a surveyor uses the equation

$$h^2 = s^2 - d^2$$

Factor the right side of this equation.

32. The area of a square picture frame (Fig. 16-5) is given by the formula

$$A = s_2^2 - s_1^2$$

Factor the right side of this equation.

33. The surface area of a washer (Fig. 16-6) is given by the formula

$$A = \pi R^2 - \pi r^2$$

Factor the right side of this equation.

34. The volume of a hollow cylinder (Fig. 16-7) is given by the formula

$$V = \pi R^2 h - \pi r^2 h$$

Factor the right side of this equation.

35. The mass of an iron pipe (Fig. 16-8) is given by the formula

$$m = \pi R^2 LD - \pi r^2 LD$$

where D is the density of iron. Factor the right side of this equation.

36. The "work-energy" principle states that in any finite motion, the work W of the resultant force on a body equals the change in its kinetic energy:

$$W = \frac{1}{2}mV^2 - \frac{1}{2}mv^2$$

Factor the right side of this equation.

FIGURE 16-5

FIGURE 16-6

FIGURE 16-7

FIGURE 16-8

16-3 SIMPLE TRINOMIALS

In this section we factor quadratic trinomials of the form $ax^2 + bx + c$, in which $a = 1$. In the next section, we factor the general quadratic trinomial which has a leading coefficient different from 1. In all cases the coefficients in this chapter will be integers. We also solve applications that require us to factor trinomials.

Recall from Sec. 8-2 that a *trinomial* is a polynomial that has three terms. A *quadratic trinomial* in x has an x^2 term, an x term, and a constant term. The coefficient of the x^2 term in a quadratic trinomial is called the *leading coefficient* and the x term is sometimes called the *middle term*.

Example 22
The expression $3x^2 - 2x + 5$ is a quadratic trinomial. The leading coefficient is 3, the coefficient of the x term is -2, and the constant term is 5.

Trinomials with a Leading Coefficient of 1

We get a quadratic trinomial with a leading coefficient of 1 when we multiply two binomials, if the coefficients of the x terms in the binomials are also 1. In Sec. 8-3, we saw that we can multiply two binomials together using the FOIL method.

Example 23
Recall how we multiply $(x + 2)(x + 3)$:

$$(x + 2)(x + 3) = x^2 + \underbrace{3x + 2x}_{} + 6$$

where the terms correspond to F + O + I + L.

So, $(x + 2)(x + 3) = x^2 + 5x + 6$

On the right, the coefficient of the x term is 5 (the sum of 2 and 3), and the constant term is 6 (the product of 2 and 3):

$$(x + 2)(x + 3) = x^2 + 5x + 6$$

(sum → 5; product → 6)

In general, by Eq. 35

$$(x + a)(x + b) = x^2 + (a + b)x + ab$$

(sum → $(a+b)$; product → ab)

If we reverse Eq. 35, then it shows us a method to factor quadratic trinomials.

FACTORING A TRINOMIAL WITH LEADING COEFFICIENT 1	$x^2 + (a + b)x + ab = (x + a)(x + b)$	35

Note that in the given trinomial, the leading coefficient is 1, the coefficient of the x term is a sum, and the constant term is a product. The key to factoring is to *find the two numbers, a and b, which have that sum and product.*

Example 24

Factor $x^2 + 3x + 2$.

Thus, $x^2 + 3x + 2 = (x + 2)(x + 1)$

It's important to note that *not all quadratic trinomials can be factored,* using only factors with integral coefficients.

Example 25

The quadratic trinomial $x^2 + x + 1$ is not factorable, since

$$x^2 + x + 1 = x^2 + 1x + 1$$

with arrows indicating sum and product,

and there are no real numbers, a and b, whose sum is 1:

$$a + b = 1$$

and whose product is 1:

$$ab = 1$$

So, the quadratic trinomial $x^2 + x + 1$ is prime.

Using the Signs to Aid Factoring

The signs of the terms of the trinomial indicate the signs of the factors.

Example 26

Factor $x^2 + 11x + 18$.

Solution: If we let $x^2 + 11x + 18 = (x + a)(x + b)$, with the $11x$ as sum and 18 as product,

then we must find two numbers, a and b, whose sum is $+11$ and whose product is $+18$:

$$a + b = +11 \quad \text{and} \quad ab = +18$$

Since both the sum $+11$ and the product $+18$ are positive, then both a and b must be positive. We consider products first because there are fewer pairs of numbers with a given product than a given sum. So, we begin with positive pairs of numbers (factors) whose products are +18. We list these pairs of factors and calculate their sums. A table is useful to organize our work:

Positive Factors of 18		Product	Sum
a	b	ab	a + b
+1	+18	+18	+19
+2	+9	+18	+11
+3	+6	+18	+9

The correct pair is +2 and +9, since the sum

$$+2 + 9 = +11$$

gives us the coefficient of the x term, +11.

Thus, $\quad\quad\quad x^2 + 11x + 18 = (x + 2)(x + 9)$

Check:

Just multiply $(x + 2)(x + 9)$ to check:

$$(x + 2)(x + 9) = x^2 + 11x + 18 \quad \text{checks}$$

Example 27

Factor $x^2 - 11x + 18$.

Solution: This trinomial is similar to the one in Example 26, except that the x term is negative. In this case we need to find a pair of numbers, a and b, whose product is +18 and whose sum is −11:

$$ab = +18 \quad \text{and} \quad a + b = -11$$

Since the product is positive, then either both numbers are positive, or both are negative. Since the sum is negative, then both must be *negative*. We make a table:

Negative Factors of +18		Product	Sum
a	b	ab	a + b
−1	−18	+18	−19
−2	−9	+18	−11
−3	−6	+18	−9

Since the sum $\quad\quad\quad -2 + -9 = -11$

and −11 is the coefficient of the x term, the result is

$$x^2 - 11x + 18 = (x - 2)(x - 9)$$

You can check by multiplying the factors, to get the original trinomial.

Example 28

Factor $t^2 - 4t - 12$.

Solution: We are looking for two numbers, a and b, whose product is −12 and whose sum is −4:

So,
$$t^2 - 4t - 12 = (t + a)(t + b)$$
where
$$ab = -12 \quad \text{and} \quad a + b = -4$$

The product is negative, so the numbers have opposite signs, one positive and one negative. There are many ways to factor -12, so we make a table to organize our work:

Factors of -12		Product	Sum
a	b	ab	$a + b$
+1	−12	−12	−11
+2	−6	−12	−4
+3	−4	−12	−1
+4	−3	−12	+1
+6	−2	−12	+4
+12	−1	−12	+11

All the pairs of factors give a product of -12, but only one pair gives the required sum of -4:
$$+2 + -6 = -4$$

Thus the numbers are $+2$ and -6.

So,
$$t^2 - 4t - 12 = (t + 2)(t - 6)$$

SECTION 16-3 Exercises Simple Trinomials

Trinomials with a Leading Coefficient of 1

Factor completely and check by multiplying.

1. $x^2 + 8x + 7$
2. $x^2 + 7x + 10$
3. $a^2 - a - 6$
4. $c^2 - 3c - 10$
5. $u^2 + 7u - 18$
6. $k^2 + k - 12$
7. $y^2 - 7y + 6$
8. $a^2 - 6a + 8$
9. $14 + 9k + k^2$
10. $21 - 10y + y^2$
11. $x^2 - 8xy + 15y^2$
12. $m^2 + 14mn + 24n^2$
13. $18 - 3x - x^2$
14. $20 - y - y^2$
15. $a^2 - 2ab - 8b^2$
16. $s^2 - 5st - 24t^2$

Applications

17. To find the width of a sheet of metal needed to make a box with a specific volume, one must solve the following equation:
$$w^2 - 9w - 60 = 0$$
Factor the left side of this equation.

18. To find the width of a brick border on a certain patio one must solve the equation
$$w^2 - 26w + 69 = 0$$
Factor the left side of this equation.

19. To find the dimensions of a rectangular field that has a perimeter of 724 m and an area of 32 400 m², one must solve the equation
$$x^2 - 362x + 32\,400 = 0$$
Factor the left side of this equation.

20. To find the dimensions of a rectangular field which is 12 m longer than it is wide and contains 448 m², one must solve the equation
$$w^2 + 12w - 448 = 0$$
Factor the left side of this equation.

21. To find the width of a certain picture frame, we must solve the equation
$$x^2 - 11x + 10 = 0$$
Factor the left side of this equation.

22. An object is thrown upward with an initial velocity of 8 m/s from a building 20 m above the ground. The height s of the object above the ground at any time t is given by
$$s = 20 + 8t - t^2$$
Factor the right side of this equation.

Factor completely and check.

23. $(x + y)^2 - 5(x + y) + 4$ 24. $(y - 3)^2 - 3(y - 3) + 2$
25. $t^4 - 2t^2 + 1$ 26. $x^4 - 10x^2 + 9$

16-4 GENERAL TRINOMIALS

In this section we learn how to factor the general quadratic trinomial where the leading coefficient is not 1. In Sec. 8-3, we saw that we get such a trinomial when we multiply two binomials, Eq. 36:

$$(ax + b)(cx + d) = acx^2 + (ad + bc)x + bd$$

In the trinomial on the right,

ac = the leading coefficient,
$ad + bc$ = the x term coefficient, and
bd = the constant term

Example 29

If we multiply the binomials below, we can see the relationship among the coefficients. We know that

$$(2x - 1)(3x + 4) = 6x^2 + 5x - 4$$

We compare this to

$$(ax + b)(cx + d) = acx^2 + (ad + bc)x + bd$$

In the trinomial,

the leading coefficient 6 is the product ac, $2(3) = 6$
the x coefficient 5 is the sum, $ad + bc$, $(2)(4) + (-1)(3) = 5$
the constant -4 is the product bd $(-1)(4) = -4$

When we factor, we reverse the process of multiplication. Reversing Eq. 36, we get

| FACTORING A GENERAL TRINOMIAL | $acx^2 + (ad + bc)x + bd = (ax + b)(cx + d)$ | 36 |

To factor a trinomial, we look for the four numbers a, b, c, and d, which satisfy the equation above.

Example 30

Factor $2x^2 + 7x + 3$.

Solution: The leading coefficient is 2 and the constant is 3. We can "set up" the problem like this:

$$2x^2 + 7x + 3 = (\_x + \_)(\_x + \_)$$

Product of coefficients must be 2.
↓
The only factors of 2 are 2 and 1.

Product of constants must be 3.
↓
The only factors of 3 are 3 and 1.

Since all the coefficients are positive, we need only consider positive factors. We "fill in the blanks," listing all the possibilities.

$$2x^2 + 7x + 3 \stackrel{?}{=} (2x + 3)(x + 1)$$

Reversing the constants, $\stackrel{?}{=} (2x + 1)(x + 3)$

Reversing the coefficients, $\stackrel{?}{=} (x + 3)(2x + 1)$

Reversing the constants, $\stackrel{?}{=} (x + 1)(2x + 3)$

Note that the third possibility is identical to the second and that the fourth is identical to the first. So we can eliminate the third and fourth possibilities. In fact, when you consider possible factorizations, *you need not reverse the factors of the leading coefficient as long as you do reverse the factors of the constant.*

Now we check the middle terms to see which, if either, give us the required $7x$.

Factors	Middle term	
$(2x + 3)(x + 1)$	$2x + 3x$	$= 5x$
$(2x + 1)(x + 3)$	$6x + x$	$= 7x$

The second possibility gives the correct middle term, so

$$2x^2 + 7x + 3 = (2x + 1)(x + 3)$$

Example 30 was simple because all terms were positive and the numbers had few factors. We will now factor some more "interesting" trinomials.

Example 31

Factor $2x^2 + 5x - 12$.

Solution: We write down the setup:

$$2x^2 + 5x - 12 = (\_x \ \_)(\_x \ \_)$$

Product must be 2   Product must be −12

In general, the leading coefficient is usually positive, and we need consider only its positive factors. In this case, the leading coefficient is 2, and the only positive factors of 2 are 2 and 1. So, we can fill in the coefficients of x:

$$2x^2 + 5x - 12 = (\underline{2}x \ \_) \ (\underline{1}x \ \_)$$

The constant term is −12. Since it is negative, the constant factors have opposite signs, one positive and one negative. The positive factors of 12 are

     1 and 12   2 and 6   3 and 4

For each pair of factors, we use opposite signs, reverse the signs, reverse the position of the factors, and reverse those signs! For the factors 1 and 12, we have 4 possible factorizations:

$$(2x + 1)(x - 12)$$

Reversing the signs:

$$(2x - 1)(x + 12)$$

Reversing the factors:

$$(2x + 12)(x - 1)$$

Reversing the signs again:

$$(2x - 12)(x + 1)$$

A similar set of possibilities is found for the factors 2 and 6, and another set of 4 possibilities is found for the factors 3 and 4. This list shows all possibilities:

$$2x^2 + 5x - 12 \stackrel{?}{=} (2x + 1)(x - 12)$$
$$(2x - 1)(x + 12)$$
$$(2x + 12)(x - 1)$$
$$(2x - 12)(x + 1)$$
$$(2x + 2)(x - 6)$$
$$(2x - 2)(x + 6)$$
$$(2x + 6)(x - 2)$$
$$(2x - 6)(x + 2)$$
$$(2x + 3)(x - 4)$$
$$(2x - 3)(x + 4)$$
$$(2x + 4)(x - 3)$$
$$(2x - 4)(x + 3)$$

By trial and error we determine which combination (if any) will produce a middle term of $+5x$. *We calculate the middle term for each pair of factors* until we find that $(2x - 3)(x + 4)$ has the required middle term of $+5x$. So, the solution is

$$2x^2 + 5x - 12 = \underline{(2x - 3)(x + 4)}$$

16-4 General Trinomials

Note: You will rarely have to test *all* the possibilities. In fact, with a little practice, you will, in most cases, quickly find the right combination by trial and error.

Test for Factorability

As we noted before, not all quadratic trinomials can be factored. We test for factorability as follows.

TEST FOR FACTORABILITY	The trinomial $ax^2 + bx + c$ (where a, b, and c are constants) is factorable if $b^2 - 4ac$ is a perfect square.	34

Example 32

Test these trinomials for factorability:

(a) $2x^2 + 5x - 12$ (b) $3x^2 - 2x + 1$

Solution:

(a) This quadratic is from Example 31, so we already know it is factorable. We verify this by calculating $b^2 - 4ac$, where $a = 2$, $b = 5$, and $c = -12$:

$$b^2 - 4ac = 5^2 - 4(2)(-12) = 121$$

The number 121 is a perfect square, so the quadratic $2x^2 + 5x - 12$ is factorable.

(b) For $3x^2 - 2x + 1$, we calculate $b^2 - 4ac$, where $a = 3$, $b = -2$, and $c = 1$:

$$b^2 - 4ac = (-2)^2 - 4(3)(1) = -8$$

Since -8 is not a perfect square, the quadratic $3x^2 - 2x + 1$ is *not* factorable.

Factoring Trinomials by Grouping

Factoring becomes more complicated when the leading coefficient can also be factored several ways. But there is an alternative method of factoring trinomials that reduces the amount of trial and error and uses the grouping method from Sec. 16-1. We will illustrate this method in the next example.

Example 33

Factor, using the grouping method:

$$6x^2 - 5x - 4$$

Solution: Letting

$$6x^2 - 5x - 4 = ax^2 + bx + c$$

then

$$a = 6, \quad b = -5, \quad \text{and} \quad c = -4$$

First testing for factorability,

$$b^2 - 4ac = (-5)^2 - 4(6)(-4) = 121 = 11^2$$

This trinomial is indeed factorable.

To use the grouping method, we must find two numbers whose product is ac and whose sum is b.

Since
$$ac = 6(-4) = -24$$
and
$$b = -5$$

then we need to find two numbers whose product is -24 and whose sum is -5. Since the product is negative, we must use opposite signs, and since the sum must be negative, the number with largest magnitude must also be negative.

Setting up a table

Factors of -24 (largest magnitude is negative)	Product -24	Sum -5
$(1)(-24)$	-24	-23
$(2)(-12)$	-24	-10
$(3)(-8)$	-24	-5

The factors 3 and -8 are the only two that fit the requirements.

We replace the middle term $-5x$ with $3x + (-8)x$ or $3x - 8x$ and substitute the original trinomial which becomes as follows

$$6x^2 - 5x - 4 = 6x^2 + 3x - 8x - 4$$

Grouping terms
$$= (6x^2 + 3x) + (-8x - 4)$$

factoring each group
$$= 3x(2x + 1) - 4(2x + 1)$$

factoring out $(2x + 1)$,
$$= (2x + 1)(3x - 4)$$

Therefore,
$$6x^2 - 5x - 4 = (2x + 1)(3x - 4)$$

This grouping method is useful because it eliminates most of the trial and error aspect of factoring trinomials.

Common Factors

If the trinomial has any common factors, it is wise to factor these common factors out first, because the coefficients will then be smaller and more manageable.

Example 34
Factor $3x^2 - 21x + 30$.

Solution: The trinomial $3x^2 - 21x + 30$ has a common factor 3. Factor out the 3 first, and then factor the new trinomial.

$$3x^2 - 21x + 30 = 3(x^2 - 7x + 10)$$
$$= 3(x - 5)(x - 2)$$

16-4 General Trinomials

Example 35

Factor $12x^2 + 14x - 6$.

Solution: The trinomial $12x^2 + 14x - 6$ has a common factor 2. Factor out the 2 and then factor the new trinomial.

$$12x^2 + 14x - 6 = 2(6x^2 + 7x - 3)$$
$$= 2(3x - 1)(2x + 3)$$

Trinomials with More Than One Variable

Two (or more) variables may appear in some factoring problems. The process of factoring is the same, but remember to include all the variables in the factors.

Example 36

Factor $9x^2 - 15xy - 14y^2$.

Solution:

$$9x^2 - 15xy - 14y^2 = (3x - 7y)(3x + 2y)$$

Remember the y's!

Example 37

Factor $4a^3 - 10a^2b + 6ab^2$.

Solution: We see that the trinomial has a common factor, $2a$, and we factor that out first:

$$4a^3 - 10a^2b + 6ab^2 = 2a(2a^2 - 5ab + 3b^2)$$

Now we factor the new trinomial by trial and error:

$$= 2a(2a - 3b)(a - b)$$

Remember the b's!

Thus,

$$4a^2 - 10a^2b + 6ab^2 = 2a(2a - 3b)(a - b)$$

SECTION 16-4   Exercises   General Trinomials

Factor completely.

1. $2x^2 + 3x + 1$
2. $2y^2 + 5y + 3$
3. $5y^2 - 7y + 2$
4. $6n^2 - 11n + 3$
5. $6x^2 - 13x - 5$
6. $4y^2 - 4y - 15$
7. $3a^2 + 7a - 6$
8. $8r^2 + 2r - 3$
9. $15 + 11x - 12x^2$
10. $12 - y - 6y^2$
11. $6t^2 + 25st + 14s^2$
12. $36u^2 - 5uv - 24v^2$
13. $3b^2 - 15b - 18$
14. $7x^2 - 14x - 21$
15. $2ab^3 - 14ab^2 + 24ab$
16. $3x^3y - 21x^2y + 30xy$
17. $15k^2 - 6k - 21$
18. $10mn^2 - 6mn - 4m$

Applications

19. An object is thrown upward with an initial velocity of 5 m/s from a building 30 m above the ground. The height s of the object above the ground at any time t is given by

$$s = 30 + 5t - 5t^2$$

Factor the right side of this equation.

20. An object is thrown into the air with an initial velocity of 15 m/s. To find the time it takes for the object to reach a height of 10 m, we must solve the quadratic equation

$$5t^2 - 15t + 10 = 0$$

Factor the left side of this equation.

21. An open box is made from a rectangular sheet of cardboard by cutting out square corners and folding up its sides (Fig. 16-9). To find the dimensions of the box, we must solve the equation

$$6x^2 - 54x - 540 = 0$$

■ FIGURE 16-9

Factor the left side of this equation.

22. A rectangular patio (Fig. 16-10) is surrounded by a brick border of width w. To find the width of the border, we must solve the equation

$$4w^2 + 80w - 276 = 0$$

■ FIGURE 16-10

Factor the left side of this equation.

23. $6(y - 2)^2 + (y - 2) - 2$ **24.** $8(x - y)^2 - 14(x - y) - 15$
25. $15 - 8(x - y) + (x - y)^2$ **26.** $6a^2 + a(b + c) - (b + c)^2$
27. $6(m + n)^2 + 7(m + n)(m - n) + (m - n)^2$

16-5 PERFECT SQUARE TRINOMIALS

In this section we define the perfect square trinomial, show which products give a perfect square trinomial, and learn how to factor a perfect square trinomial.

Perfect Square Trinomials

When a binomial is squared, we get a *perfect square trinomial*.

Example 38

$$\begin{aligned}(3x - 4)^2 &= (3x - 4)(3x - 4) \\ &= 9x^2 - 12x - 12x + 16 \\ &= 9x^2 - 24x + 16 \leftarrow \text{perfect square trinomial}\end{aligned}$$

In a perfect square trinomial, the first and last terms of the trinomial are the squares of the first and last terms of the binomial,

$$(3x - 4)^2$$
square   square
$$9x^2 \qquad 16$$

and the middle term is twice the product of the terms of the binomial.

$$(3x - 4)^2$$

twice the product $\quad 2(3x)(-4) = -24x$

Thus, $\quad (3x - 4)^2 = 9x^2 - 24x + 16$

Recall the formulas, Eqs. 37 and 38, for the perfect square trinomials from Sec. 8-5,

$$(x + y)^2 = x^2 + 2xy + y^2$$
$$(x - y)^2 = x^2 - 2xy + y^2$$

Factoring a Perfect Square Trinomial

To factor a perfect square trinomial you must first *recognize the form of the trinomial as a perfect square:* The first and last terms must both be positive perfect squares and the middle term must be twice the product of the square roots of the first and last terms. Then to factor, simply reverse the process described above. Reversing Eqs. 37 and 38, we get

FACTORING PERFECT SQUARE TRINOMIALS	$x^2 + 2xy + y^2 = (x + y)^2$	37
	$x^2 - 2xy + y^2 = (x - y)^2$	38

Example 39

Factor $x^2 + 10x + 25$.

Solution: First recognize the form.
First and last terms are positive perfect squares:

$$x^2 + 10x + 25$$
$$(x)^2 \qquad 5^2$$
$$2(x)(5) = 10x$$

The middle term is twice the product of x and 5:
So, factoring

$$x^2 \;+\; 10x \;+\; 25$$
square root — same sign — square root

We get, $\quad (x + 5)^2$

Example 40

Factor $9x^2 - 24x + 16$.

Solution: Is this a perfect square trinomial? The first and last terms are positive perfect squares and the middle term is twice the product of the square roots of the first and last terms:

$$9x^2 - 24x + 16$$
$$(3x)^2 \quad (4)^2$$
$$2(3x)(4) = 24x$$

So, yes, this is a perfect square trinomial. Since the middle term is negative, we refer to Eq. 38 and fit the trinomial into the pattern in Eq. 38:

$$9x^2 - 24x + 16 = (3x)^2 - 2(3x)(4) + (4)^2$$
$$= (3x - 4)^2$$

Of course, you would get the same result using the methods in Sec. 16-4 to factor a general trinomial. But, if you learn to recognize a perfect square trinomial, the factoring will be easier for you.

Note that *a perfect square trinomial must have a middle term*. The sum of two squares *cannot* be factored as a perfect square.

| COMMON ERROR | $x^2 + y^2 \neq (x + y)^2$ |

SECTION 16-5 Exercises Perfect Square Trinomials

Factoring a Perfect Square Trinomial

Factor completely and check.

1. $x^2 - 4x + 4$
2. $25 + 10x + x^2$
3. $y^2 + 6y + 9$
4. $b^2 - 8b + 16$
5. $36 - 12s + s^2$
6. $1 - 4x + 4x^2$
7. $9a^2 + 12ab + 4b^2$
8. $25x^2 - 60xy + 36y^2$
9. $1 - 22y + 121y^2$
10. $9a^2 - 42a + 49$
11. $100 + 60x + 9x^2$
12. $4x^2 + 20x + 25$
13. $81y^2 + 72y + 16$
14. $9x^2 - 24x + 16$
15. $16t^2 - 24tuv + 9u^2v^2$
16. $4x^2y^2 - 12xyz + 9z^2$
17. $7y^3 + 14y^2 + 7y$
18. $24x + 24x^2 + 6x^3$
19. $20xy^2 - 60xy + 45x$
20. $4x^4 - 8x^2y^2 + 4y^4$
21. $(x + y)^2 + 2(x + y) + 1$
22. $9 - 6(a - b) + (a - b)^2$
23. $4 - 4(a - b) + (a - b)^2$
24. $4(x - 2)^2 - 12(x - 2)(y - 1) + 9(y - 1)^2$
25. $16(m + n)^2 + 40(m + n)(m - n) + 25(m - n)^2$

16-6 SUM OR DIFFERENCE OF TWO CUBES

In this section we define the sum of two cubes, $x^3 + y^3$, and the difference of two cubes, $x^3 - y^3$. We find out what factors will give a product that is the sum or the difference of two cubes, and then reverse the process to find the factors of the sum or the difference of two cubes.

If we divide the sum of two cubes, $x^3 + y^3$, by $x + y$, we find that there is no remainder and therefore $x + y$ is a factor of $x^3 + y^3$:

$$\begin{array}{r}
x^2 - xy + y^2 \\
x+y \overline{\smash{)}x^3 + y^3} \\
\underline{x^3 + x^2y } \\
-x^2y \\
\underline{-x^2y - xy^2 } \\
+xy^2 + y^3 \\
\underline{+xy^2 + y^3}
\end{array}$$

Thus, the other factor of $x^3 + y^3$ is $x^2 - xy + y^2$.

FACTORING THE SUM OF TWO CUBES	$x^3 + y^3 = (x + y)(x^2 - xy + y^2)$	32

So the sum of two cubes, $x^3 + y^3$, is always divisible by the sum, $x + y$. Similarly, we can show that the *difference* of two cubes, $x^3 - y^3$, is divisible by the *difference*, $x - y$. By division, we find

FACTORING THE DIFFERENCE OF TWO CUBES	$x^3 - y^3 = (x - y)(x^2 + xy + y^2)$	33

The two equations above should be learned, if you don't want to have to divide it out each time!

Factoring the Sum or Difference of Two Cubes

Once we recognize an expression as the sum or the difference of two cubes, we apply Eq. 32 or Eq. 33.

Example 41
Factor $p^3 + 8$.

Solution: First we recognize this as the sum of two cubes:

$$p^3 + 8 = p^3 + 2^3$$

Now we apply Eq. 32:

$$p^3 + 2^3 = (p + 2)(p^2 - 2p + 2^2)$$
same sign — opposite signs — always positive

So,

$$p^3 + 8 = (p + 2)(p^2 - 2p + 4)$$

Example 42

Factor $8x^3 - 27y^3$.

Solution: First we recognize this as the difference of two cubes:

$$8x^3 - 27y^3 = (2x)^3 - (3y)^2$$

Applying Eq. 33:

$$(2x)^3 - (3y)^3 = [2x - 3y][(2x)^2 + (2x)(3y) + (3y)^2]$$
$$= [2x - 3y][4x^2 + 6xy + 9y^2]$$

So,

$$8x^3 - 27y^3 = (2x - 3y)(4x^2 + 6xy + 9y^2)$$

same sign opposite signs always positive

COMMON ERROR The middle term of the trinomials in Eq. 32 and Eq. 33 is often mistaken as 2xy.

$$x^3 + y^3 \neq (x + y)(x^2 - 2xy + y^2)$$
$$\downarrow$$
$$\text{no!}$$

SECTION 16-6 Exercises Sum or Difference of Two Cubes

Factor completely.

1. $a^3 - 1$
2. $x^3 + 1$
3. $x^3 + 27$
4. $a^3 - 8$
5. $b^3 - 64$
6. $y^3 - 27$
7. $1 - 8a^3$
8. $27p^3 + 125$
9. $1000 + 27x^3$
10. $343 - 8c^3$
11. $64a^3 - 125b^3$
12. $729c^3 + 64d^3$
13. $3y^3 - 24$
14. $4x^3 + 32$
15. $54p^4 + 2p$
16. $250q - 2q^4$
17. $(a + b)^3 - x^3$
18. $(x - y)^3 + 27z^3$
19. $8x^3 - (a - b)^3$
20. $64a^3 + (x + y)^3$
21. $(a - b)^3 + (x + y)^3$
22. $(a - 1)^3 - (b - 2)^3$
23. $x^6 + 1$
24. $1 - y^6$
25. $a^9 - 1$
26. $27x^9 + 512$

Applications

27. The volume of a hollow spherical shell having an inside radius of r and an outside radius of R is

$$V = \frac{4}{3}\pi R^3 - \frac{4}{3}\pi r^3$$

Factor the right side of this equation.

28. A concrete container for holding radioactive material is in the shape of a hollow cube whose inside dimension is s and whose outside dimension is S. If it is made of concrete of density d, its weight is

$$W = dS^3 - ds^3$$

Factor the right side of this equation.

29. Factor $a^6 - b^6$ in two different ways:

   (a) As the difference of two cubes, and
   (b) As the difference of two squares.

   Which way gives you the complete factorization more easily?

Chapter 16 REVIEW EXERCISES

Factor completely.

1. $15y^2 - 27y$
2. $25xy + 5y$
3. $15a^2 - 21ab + 3ac$
4. $(y - 2)y - (y - 2)$
5. $3x - 3y + xy - y^2$
6. An item costing P dollars is reduced in price by 12%. The resulting price C is

$$C = P - 0.12P$$

   Factor the right side of this equation.

7. $4x^2 - 25$
8. $9x^2 - 16a^2b^2$
9. $27 - 12y^2$
10. $(c - d)^2 - e^2$
11. $m^2 - (n - 1)^2$
12. $x^4 - 1$
13. Due to a drop in temperature, a spherical weather balloon shrinks from radius R to radius r. The change in surface area is given by the formula

$$S = 4\pi R^2 - 4\pi r^2$$

   Factor the right side of this equation.

14. $x^2 + 8x - 9$
15. $m^2 + m - 90$
16. $y^2 - 4y - 12$
17. $b^2 - 11b + 28$
18. $x^2 + x - 30$
19. $b^2 - 20b + 36$
20. $6a^2 + 19a + 15$
21. $24y^2 + 17y - 20$
22. $4y^2 - 25y + 6$
23. $15x^2 + x - 40$
24. $4a^2 - 12a + 9$
25. $y^2 + 16y + 64$
26. $49 - 14p + p^2$
27. $16x^2 + 72x + 81$
28. The area A of a rectangle is represented by the following equation

$$A = 2x^2 + 11x + 15$$

   If the standard formula is *Area = length × width*, which of the factors of the right hand side of the original equation would represent the length and which would represent the width. (Bonus, explain your answer.)

Factor completely.

29. $1 - y^3$
30. $a^3 - 125$
31. $27x^3 + 8y^3$
32. $2a^4 - 16ab^3$
33. $a^3 - (b - 2)^3$
34. $8 + a^6$

WRITING

35. We have studied the factoring of several different types of expressions in this chapter. List at least six types and give an example of each. State in words how to recognize each type and how to tell one from the other. List two other expressions that are not examples of the given types.

TEAM PROJECT

36. The area of a rectangle is the product of its sides, so we can think of the product
$$(x + 1)(x + 2) = x^2 + 3x + 2$$
as the area of a rectangle of sides $(x + 1)$ and $(x + 2)$ (Fig. 16-11).

Similarly, to factor an expression using areas of rectangles, we can cut rectangles of areas equal to the terms in the trinomial. You may use graph paper and choose any length for x, but avoid making it an integer value. The expression $x^2 + 3x + 2$ can be represented by rectangles of area x^2, $3x$, and 2 (Fig. 16-12). Now arrange the individual rectangles to form one large rectangle, as in Fig. 16-11. There will be only one way to do this. The sides $(x + 1)$ and $(x + 2)$ of this rectangle will be the factors of the trinomial. Use this method to factor these trinomials:

(a) $x^2 + 5x + 6$
(b) $2x^2 + 5x + 2$
(c) $x^2 + 2x - 3$

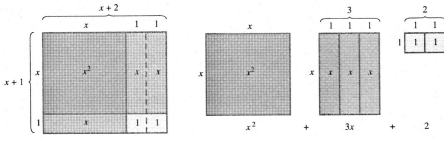

■ FIGURE 16-11   ■ FIGURE 16-12

17 Algebraic Fractions

Outcomes

When you have completed this chapter, you will be able to

1. Define an algebraic fraction.
2. Identify and manipulate equivalent algebraic fractions.
3. Reduce an algebraic fraction to lowest terms.
4. Multiply and divide algebraic fractions.
5. Add and subtract algebraic fractions.
6. Simplify complex fractions.

Because the calculator and the metric system both use decimal notation, the importance of common fractions has been somewhat reduced. However, common fractions are still widely used in everyday situations and algebraic fractions are just as important as ever. We must be able to handle algebraic fractions in order to solve fractional equations and the applications from which they arise.

In this chapter we study equivalent fractions. We learn to add, subtract, multiply, and divide fractions. We simplify complex fractions and solve fractional equations, formulas with fractions, and applications involving fractions.

17-1 EQUIVALENT FRACTIONS

In this section we define algebraic fractions, manipulate fractions with signs, and express fractions in equivalent forms by using the fundamental principle of fractions.

Algebraic Fractions

An *algebraic fraction* is a fraction whose numerator and/or denominator contain literal quantities (variables).

---- Example 1 ----
Some examples of algebraic fractions are

$$\frac{1}{x}, \quad \frac{x+y}{2}, \quad \frac{y}{x-1}, \quad \frac{5x^2 b}{7y^5}, \quad \text{and} \quad \frac{x^2 - 3x + 4}{x^2 - 9}$$

Manipulation with Signs

A fraction may have as many as three signs associated with it: the sign of the numerator, the sign of the denominator, and the sign of the fraction itself. Any two of the three signs of a fraction may be changed without changing the value of the fraction.

Example 2

The fraction $\frac{a}{b}$ may be written as $+\frac{+a}{+b}$ with three positive signs. It can also be expressed as $+\frac{-a}{-b}$ or $-\frac{+a}{-b}$ or $-\frac{-a}{+b}$ by changing two of the three positive signs to negative signs.

Fractions preceded by a negative sign are treated similarly.

Example 3

The fraction $-\frac{a}{b}$ may be written as $-\frac{+a}{+b}$ with one negative sign. It can also be expressed as

$$-\frac{-a}{-b} \text{ or } +\frac{+a}{-b} \text{ or } +\frac{-a}{+b}$$

by changing two of the three signs.

Generally, if a number is preceded by a + sign, then the + sign is omitted.

Example 4

The fraction $+\frac{-a}{+b}$ is generally written as $\frac{-a}{b}$.

Examples 2, 3, and 4 can be summarized as the rules of signs.

RULES OF SIGNS		
$\frac{a}{b} = \frac{-a}{-b} = -\frac{a}{-b} = -\frac{-a}{b}$	11
$-\frac{a}{b} = -\frac{-a}{-b} = \frac{-a}{-b} = \frac{-a}{b}$	12

It is useful to eliminate as many negative signs as possible when you simplify fractions.

17-1 Equivalent Fractions

Example 5

Use the rules of signs to simplify each of the following fractions:

(a) $\quad -\dfrac{-x}{y}$ (b) $\quad -\dfrac{-p}{-q}$

Solution:

(a) $\quad -\dfrac{-x}{y} = \dfrac{x}{y}$

(b) $\quad -\dfrac{-p}{-q} = -\dfrac{p}{q}$

A negative sign in front of the quantity can be interpreted as a factor of -1. So,

$$-(a - b) = -1(a - b) = -a + b = b - a$$

| $-(a-b) = b-a$ | 8 |

This is often used to simplify fractions or to express a fraction differently.

Example 6

Show that $\dfrac{x-3}{y-x} = \dfrac{3-x}{x-y}$.

Solution: Using the rules of signs,

$$\dfrac{x-3}{y-x} = \dfrac{-(x-3)}{-(y-x)}$$

Using Eq. 8,

$$= \dfrac{3-x}{x-y}$$

Example 7

Simplify the fraction $-\dfrac{p-1}{q-p}$.

Solution: Using the rules of signs, Eq. 12,

$$-\dfrac{p-1}{q-p} = \dfrac{-(p-1)}{q-p}$$

Using Eq. 8,

$$= \dfrac{1-p}{q-p}$$

Alternate Solution:

$$-\dfrac{p-1}{q-p} = \dfrac{p-1}{-(q-p)} = \dfrac{p-1}{p-q}$$

So, we see that the given fraction can be simplified by associating the negative sign with either the numerator or the denominator. The resulting simplified fractions look different, but both are equal to the original equation.

Equivalent Fractions

Recall that two fractions are *equivalent* if they have the same value. When we calculate with a fraction, we frequently have to use a different yet equivalent form of that fraction. We use the *fundamental principle of fractions* to find equivalent fractions.

FUNDAMENTAL PRINCIPLE OF FRACTIONS	If the numerator and the denominator of a fraction are multiplied or divided by the same (nonzero) number, then the value of the fraction remains unchanged.

In symbols, if $\frac{a}{b}$ is a fraction, and c is any nonzero number, then the fundamental principle of fractions gives us:

EQUIVALENT FRACTIONS	$\frac{a}{b} = \frac{ac}{bc}$	39
	$\frac{a}{b} = \frac{a \div c}{b \div c}$	40

Example 8

Multiply the numerator and denominator of the fraction $\frac{3a}{5x}$ by $2x$ to obtain an equivalent fraction.

Solution:
$$\frac{3a}{5x} = \frac{(3a)(2x)}{(5x)(2x)} = \frac{6ax}{10x^2}$$

Fractions Involving Polynomials

The fundamental principle of fractions also applies to fractions involving polynomials.

Example 9

Multiply the numerator and denominator of the fraction $\frac{1}{x-2}$ by $(x+2)$ to obtain an equivalent fraction.

Solution:
$$\frac{1}{x-2} = \frac{1(x+2)}{(x-2)(x+2)} = \frac{x+2}{x^2-4}$$

> **COMMON ERROR** We may *not add or subtract* the same quantity in the numerator and denominator of a fraction to obtain an equivalent fraction, because this will change the value of the fraction. For example,
> $$\frac{x}{y} \neq \frac{x+1}{y+1}$$

Similarly,

> **COMMON ERROR** We may *not take the square root* of both the numerator and denominator of a fraction to obtain an equivalent fraction, because this will change the value of the fraction. For example,
> $$\frac{9a^2}{4b^2} \neq \frac{3a}{2b}$$

Reducing Fractions to Lowest Terms

You will recall from Sec. 2-2 that a common fraction is reduced to lowest terms, or simplified, when it has the smallest possible denominator. An algebraic fraction is simplified when the numerator and the denominator have no common factor other than 1. We can cancel common factors in algebraic fractions, just like we did with common fractions.

| REDUCING FRACTIONS | $\dfrac{ac}{bc} = \dfrac{\cancel{ac}}{\cancel{bc}} = \dfrac{a \cdot 1}{b \cdot 1} = \dfrac{a}{b}$ | 39 |

Example 10

Reduce $\dfrac{9x^2y}{6x^3y}$ to lowest terms.

Solution: We cancel common factors, one by one, until there are no common factors left.

$$\frac{9x^2y}{6x^3y} = \frac{\overset{3}{\cancel{9}} \cdot \overset{1}{\cancel{x^2}} \cdot \overset{1}{\cancel{y}}}{\underset{2}{\cancel{6}} \cdot \underset{x}{\cancel{x^3}} \cdot \underset{1}{\cancel{y}}} = \frac{3}{2x}$$

Alternate Solution: Sometimes it is easier to factor the greatest common factor out of the numerator and the denominator and cancel. The gcf is $3x^2y$. We factor and cancel.

$$\frac{9x^2y}{6x^3y} = \frac{\overset{1}{\cancel{(3x^2y)}}(3)}{\underset{1}{\cancel{(3x^2y)}}(2x)} = \frac{3}{2x}$$

Example 11

Reduce the fraction $\dfrac{2(x-3)}{(x-3)(x+3)}$ to lowest terms.

Solution: A common factor is $(x-3)$. We cancel $(x-3)$ in the numerator and in the denominator:

$$\frac{2(x-3)}{(x-3)(x+3)} = \frac{2}{x+3} \cdot \frac{\cancel{x-3}}{\cancel{x-3}} = \frac{2}{x+3}$$

To reduce a fraction in which the numerator and/or denominator are not factored, we must first factor them and then cancel any factors common to both.

Example 12

Reduce to lowest terms:

$$\frac{3x-6}{4x-8}$$

Solution:

$$\frac{3x-6}{4x-8} = \frac{3\cancel{(x-2)}}{4\cancel{(x-2)}} = \frac{3}{4}$$

Reducing more complicated fractions involves more factoring.

Example 13

Reduce $\dfrac{6a^2 - a - 2}{3 + 5a - 2a^2}$ to lowest terms.

Solution:

$$\frac{6a^2 - a - 2}{3 + 5a - 2a^2} = \frac{(3a-2)\cancel{(2a+1)}}{\cancel{(1+2a)}(3-a)} = \frac{3a-2}{3-a}$$

Note that

$$2a + 1 = 1 + 2a$$

and that the terms of identical factors need not be written in the same order. But take care to cancel only common *factors*.

> **COMMON ERROR** Remember to cancel only common *factors*. You cannot cancel common *terms*. For example,
>
> $$\frac{3x-1}{6x-5} \neq \frac{x-1}{2x-5} \quad \text{and} \quad \frac{x^2-4}{x^2-9} \neq \frac{4}{9}$$

Simplifying Fractions by Changing Signs

Sometimes fractions can be reduced by manipulating the algebraic signs.

17-1 Equivalent Fractions

--- **Example 14** ---

Reduce to lowest terms: $\dfrac{y - x}{x - y}$

Solution: Since, by Eq. 8,

$$y - x = -(x - y)$$

we have

$$\frac{y - x}{x - y} = \frac{-\cancel{(x - y)}^{1}}{\cancel{(x - y)}_{1}} = \frac{-1}{1} = -1$$

--- **Example 15** ---

Reduce $\dfrac{9 - x^2}{x^2 - 5x + 6}$ to lowest terms.

Solution: First we factor,

$$\frac{9 - x^2}{x^2 - 5x + 6} = \frac{(3 + x)(3 - x)}{(x - 3)(x - 2)}$$

Replacing $(x - 3)$ with $-(3 - x)$,

$$= \frac{(3 + x)\cancel{(3 - x)}^{1}}{-\cancel{(3 - x)}_{1}(x - 2)}$$

Canceling the common factors,

$$= \frac{3 + x}{-(x - 2)}$$

Simplifying,

$$\frac{9 - x^2}{x^2 - 5x + 6} = \frac{3 + x}{2 - x}$$

SECTION 17-1 Exercises Equivalent Fractions

Algebraic Fractions and Manipulation with Signs

Use the rules of signs to simplify the fractions.

1. $\dfrac{-r}{-s}$ 2. $-\dfrac{-m}{n}$ 3. $-\dfrac{x}{-2}$ 4. $-\dfrac{-3}{-y}$

Equivalent Fractions

Write an equivalent fraction.

5. $\dfrac{a - b}{b - 1}$ 6. $\dfrac{p - q}{p - 3}$ 7. $-\dfrac{r - s}{r - 1}$ 8. $-\dfrac{x - 2}{x - y}$

Multiply the numerator and denominator of the given fractions by the given factor to obtain an equivalent fraction.

9. $\dfrac{3x}{5y}$, by $4x$ 10. $\dfrac{2}{p - 3}$, by $(p + 3)$

11. $\dfrac{a + b}{a - b}$, by $(a - b)$ 12. $\dfrac{2x - 1}{3x - 2}$, by $-3x$

CHAPTER 17 / Algebraic Fractions

Reducing Fractions to Lowest Terms

Reduce to lowest terms, without using negative exponents.

13. $\dfrac{12x}{8y}$
14. $\dfrac{6a^2}{4a^6}$
15. $\dfrac{18x^4y}{15x^2y^3}$
16. $\dfrac{15ab^4}{21a^5b}$
17. $\dfrac{81p^6qr^4}{45pq^5r^4}$
18. $\dfrac{49xy^3z^5}{56x^6y^3z}$
19. $\dfrac{4(x-2)}{(x-2)(x+2)}$
20. $\dfrac{(a+1)(a-3)}{(a+4)(a-3)}$
21. $\dfrac{8(x-1)(x-2)(x-3)}{6(x-1)(x+2)(x-3)}$
22. $\dfrac{12(p+2q)(3p+q)(2p-q)}{15(2p-q)(p+2q)(3p-q)}$
23. $\dfrac{5x-15}{3x-9}$
24. $\dfrac{4+2y}{5y+10}$
25. $\dfrac{x^2-16}{x^2-7x+12}$
26. $\dfrac{x^2-25}{x^2-10x+25}$
27. $\dfrac{x^2+14x+49}{x^2-49}$
28. $\dfrac{4x^2-9}{2x^2-9x+9}$
29. $\dfrac{x^2-5x+6}{x^2-x-6}$
30. $\dfrac{6x^2-11x+4}{9x^2-9x-4}$
31. $\dfrac{1+4x+4x^2}{6x^2-x-2}$
32. $\dfrac{9x^2-4}{9x^2-12x+4}$
33. $\dfrac{x^3+y^3}{x^2-y^2}$
34. $\dfrac{a^3-1}{a^2-1}$
35. $\dfrac{(x-y)^3}{x^3-y^3}$
36. $\dfrac{(a+b)^3}{a^3+b^3}$

Simplifying Fractions by Changing Signs

Reduce to lowest terms.

37. $\dfrac{a-b}{b-a}$
38. $\dfrac{2m-2n}{3n-3m}$
39. $-\dfrac{q-3p}{3p-q}$
40. $-\dfrac{2x-3y}{3y-2x}$
41. $\dfrac{(x-y)(a-b)}{y-x}$
42. $\dfrac{a-b}{(b-a)(b-2)}$
43. $\dfrac{(2x-1)(x-3)}{(3-x)(2+x)}$
44. $\dfrac{(1-x)(2-x)(3-x)}{(x-3)(x+2)(x-1)}$
45. $\dfrac{6-5x+x^2}{x^2-x+6}$
46. $\dfrac{4a-8a^2}{2a-1}$
47. $\dfrac{3x^2-9x}{9-x^2}$
48. $\dfrac{y^2-4y+3}{1-y^2}$
49. $\dfrac{2xy-y^2}{y^2-4x^2}$
50. $\dfrac{a^2-6a+8}{12-3a}$
51. $\dfrac{x^3-8}{4-x^2}$
52. $\dfrac{p^3-q^3}{(q-p)^3}$
53. $\dfrac{6x^2y-3x^2}{x^3-8y^3}$
54. $\dfrac{6y^2-2y}{1-27y^3}$

Applications

The following expressions are used in the given application. Simplify each expression.

55. Velocity of a bullet:

$$\dfrac{mv^2+Mv^2}{mv^2}$$

56. Electricity:

$$\dfrac{E^2R^2-E^2r^2}{(R^2+2Rr+r^2)^2}$$

17-1 Equivalent Fractions

57. Machine design:
$$\frac{R^3 - r^3}{R^2 - r^2}$$

58. Mechanical advantage of a machine:
$$\frac{4F^2 - 4f^2}{2F - 2f}$$

17-2 MULTIPLYING AND DIVIDING ALGEBRAIC FRACTIONS

In this section, we multiply and divide with algebraic fractions.

Multiplying Algebraic Fractions

Recall from Sec. 2-3 how we multiply two common fractions:

Example 16
$$\left(\frac{2}{3}\right)\left(\frac{4}{5}\right) = \frac{2 \cdot 4}{3 \cdot 5} = \frac{8}{15}$$

So, *the product of two or more fractions is a fraction whose numerator is the product of the original numerators and whose denominator is the product of the original denominators.* This rule applies to algebraic fractions as well. In symbols,

| MULTIPLICATION OF FRACTIONS | $\frac{a}{b} \cdot \frac{c}{d} = \frac{ac}{bd}$ | 41 |

Example 17
$$\left(\frac{2a}{b}\right)\left(\frac{x}{3y}\right) = \frac{(2a)(x)}{(b)(3y)} = \frac{2ax}{3by}$$

If there are common factors in the numerator and denominator, it is recommended that *we cancel the common factors before we multiply.* If this is done first, the resulting fraction will be in lowest terms.

Example 18
$$\frac{3x}{4y} \cdot \frac{6y}{7z} = \frac{3x}{\underset{2}{4y}} \cdot \frac{\overset{3}{6y}}{7z} = \frac{9x}{14z}$$

We use the same process when multiplying fractions in which the numerator or denominator is a polynomial.

Example 19

$$\frac{x+y}{x-y} \cdot \frac{x^2-y^2}{(x+y)^2} = \frac{(x+y)(x^2-y^2)}{(x-y)(x+y)^2}$$

Factoring and canceling,

$$= \frac{\cancel{(x+y)}\cancel{(x+y)}\cancel{(x-y)}}{\cancel{(x-y)}\cancel{(x+y)}\cancel{(x+y)}} = 1$$

Dividing Algebraic Fractions

We recall from Sec. 2-4 how we divided fractions:

Example 20

$$\left(\frac{2}{5}\right) \div \left(\frac{3}{4}\right) = \frac{2}{5} \cdot \frac{4}{3} = \frac{8}{15}$$

So, to divide by a fraction, we multiply by its reciprocal. The reciprocal of a fraction is found by inverting the fraction. Thus, the reciprocal of $\frac{c}{d}$ is $\frac{d}{c}$.

DIVISION OF FRACTIONS	$\dfrac{a}{b} \div \dfrac{c}{d} = \dfrac{a}{b} \cdot \dfrac{d}{c} = \dfrac{ad}{bc}$	42

Thus, *to divide by a fraction (the divisor), we invert the divisor and multiply!*

Example 21

Divide

$$\frac{3x}{4y} \div \frac{2a}{3} = \frac{3x}{4y} \cdot \frac{3}{2a} = \frac{9x}{8ay}$$

After we invert the divisor, if there are common factors, we cancel these common factors before we multiply.

Example 22

$$\frac{2a^2}{3b} \div \frac{5x}{6y^2} = \frac{2a^2}{3b} \cdot \frac{\overset{2}{\cancel{6}}y^2}{5x} = \frac{4a^2y^2}{5bx}$$

17-2 Multiplying and Dividing Algebraic Fractions

Example 23

$$\frac{x^2 - x - 6}{3x - 3} \div \frac{x^2 - 4}{x - 1} = \frac{x^2 - x - 6}{3x - 3} \cdot \frac{x - 1}{x^2 - 4}$$

Factoring and canceling,

$$= \frac{(x + 2)(x - 3)}{3(x - 1)} \cdot \frac{(x - 1)}{(x + 2)(x - 2)}$$

$$= \frac{x - 3}{3(x - 2)}$$

Generally, we leave the result in its simplified factored form.

When dividing a fraction by a polynomial, just multiply by the reciprocal of the polynomial.

Example 24

$$\frac{a^2 + 6a + 8}{a^2} \div (3a^2 + 6a) = \frac{a^2 + 6a + 8}{a^2} \cdot \frac{1}{3a^2 + 6a}$$

Factoring and canceling,

$$= \frac{(a + 2)(a + 4)}{a^2} \cdot \frac{1}{3a(a + 2)}$$

Simplifying,

$$= \frac{a + 4}{3a^3}$$

Combined Multiplication and Division

In problems involving three or more multiplications and divisions, invert only the fraction *immediately following* the division sign.

Example 25

$$\frac{6a}{7b} \div \frac{12a^2}{35c^2} \cdot \frac{5b^2}{c^4} = \frac{6a}{7b} \cdot \frac{35c^2}{12a^2} \cdot \frac{5b^2}{c^4} = \frac{25b}{2ac^2}$$

SECTION 17-2 Exercises | Multiplying and Dividing Algebraic Fractions

Multiplying Algebraic Fractions

Multiply and simplify.

1. $\dfrac{x}{y} \cdot \dfrac{2a}{3b}$
2. $\dfrac{a}{5} \cdot \dfrac{2b}{3c}$
3. $\dfrac{5x}{2y} \cdot \dfrac{8y}{x^2}$
4. $\dfrac{3p^2}{2q} \cdot \dfrac{q^5}{12p^4}$

5. $\dfrac{24a^2b^3}{5c^2} \cdot \dfrac{15c}{36ab^3}$

6. $\dfrac{5x^2}{6a^2y^3} \cdot \dfrac{27a^3y}{35x}$

7. $\dfrac{a}{a+b} \cdot \dfrac{b}{a-b}$

8. $\dfrac{p-q}{p} \cdot \dfrac{p+q}{q}$

9. $\dfrac{x-y}{a^2} \cdot \dfrac{2a}{x-y}$

10. $\dfrac{a+b}{6} \cdot \dfrac{2a}{3(a+b)}$

11. $\dfrac{3a+2b}{6a} \cdot \dfrac{3a}{2a+3b}$

12. $\dfrac{2x}{x-2y} \cdot \dfrac{2x-y}{4x}$

13. $\dfrac{x^2+x-2}{x+1} \cdot \dfrac{x+1}{x^2-3x+2}$

14. $\dfrac{x-2}{x^2+2x-3} \cdot \dfrac{x^2+4x+3}{x-2}$

Dividing Algebraic Fractions

Divide and simplify.

15. $\dfrac{5a}{6b} \div \dfrac{3x}{7y}$

16. $\dfrac{x^2}{y^3} \div \dfrac{2a}{b^2}$

17. $\dfrac{3x^3}{2y^2} \div \dfrac{x}{y}$

18. $\dfrac{a^2}{3b} \div \dfrac{a}{6b^2}$

19. $6 \div \dfrac{3}{x^2}$

20. $9a^2 \div \dfrac{a}{b^2}$

21. $\dfrac{x+2}{3} \div \dfrac{2}{x-1}$

22. $\dfrac{3}{x-3} \div \dfrac{2}{x+2}$

23. $\dfrac{a+5}{3} \div (6-a)$

24. $\dfrac{7-x}{y} \div (x+y)$

25. $\dfrac{2x-6}{x+1} \div \dfrac{3x-9}{2x+2}$

26. $\dfrac{6x-4}{3} \div \dfrac{3x-2}{2}$

27. $\dfrac{6(x+y)^2}{x-y} \div 2(x+y)$

28. $\dfrac{8(a-b)(a+b)}{a^2} \div 4(a-b)$

29. $2(a+c) \div \dfrac{ab+bc}{c}$

30. $(x-y) \div \dfrac{xz-yz}{x+y}$

31. $\dfrac{3a^2-6ab}{2ab+4b^2} \div \dfrac{9a^2-18ab}{4ab+8b^2}$

32. $\dfrac{4x^2+2xy}{6xy-3y^2} \div \dfrac{10x^2+5xy}{12xy-6y^2}$

33. $\dfrac{2x^2+3x-2}{3xy} \div \dfrac{x^3+8}{6xy}$

34. $\dfrac{3x^2-x-2}{6x^3} \div \dfrac{2x^2-5x+3}{4x^2}$

35. $\dfrac{3x^2+x-2}{2x^2-7x+6} \div \dfrac{3x^2-8x+4}{2x^2-x-3}$

36. $\dfrac{12x^2+5x-3}{6x^2+7x+2} \div \dfrac{4x^2-x-3}{2x^2-x-1}$

37. $\dfrac{15x^2-x-2}{2x-6} \div (10x-4)$

38. $\dfrac{8x^2+10x-3}{3x+3} \div (8x+12)$

Combined Multiplication and Division

Perform the indicated operations and simplify.

39. $\dfrac{5x}{6y} \div \dfrac{2a^2}{3x} \cdot \dfrac{4y^3}{a^2}$

40. $\dfrac{2a}{3b^3} \div \dfrac{5x^2}{7b} \cdot \dfrac{5a}{6x}$

41. $\dfrac{25a^2b}{8xy} \div \dfrac{10ab^2}{3cx^2y} \cdot \dfrac{4by}{9ac}$

42. $\dfrac{14x^2y}{15ab^2} \div \dfrac{9cx}{10ab} \cdot \dfrac{21ac}{8y^2}$

43. $\dfrac{2x^2+x-1}{2x^2-x-1} \cdot \dfrac{4x^2-1}{3x^2+2x-1} \div \dfrac{4x^2-4x+1}{9x^2-6x+1}$

44. $\dfrac{2a^2-2ab}{a^2+2ab+b^2} \cdot \dfrac{2a^2+2ab}{3ab-3b^2} \div \dfrac{4ab-4b^2}{3a+3b}$

45. $\dfrac{(x-1)^3}{3x+3} \div \dfrac{x^2-1}{6x-12} \cdot \dfrac{x^2+2x+1}{x^3-1}$

46. $\dfrac{x^2-4x-5}{3x^4-27x^2} \div \dfrac{3x^2+4x+1}{x^3+x^2-6x} \cdot \dfrac{9x^3-18x^2-27x}{x^2-7x+10}$

47. $\dfrac{x^2 - 2x - 3}{2x^4 - 72x^2} \div \dfrac{x^2 - 4x + 3}{x^3 - 7x^2 + 6x} \div \dfrac{2x^2 + 5x + 2}{4x^3 + 32x^2 + 48}$

48. $\dfrac{x^3 - 5x^2 - 4x}{6x^2 + 30x + 36} \div \left(\dfrac{x^4 - 16x^2}{3x^2 - 6x - 45} \cdot \dfrac{x^2 - 6x + 5}{x^3 + 8x^2 + 16x} \right)$

Applications

■ FIGURE 17-1

49. The power P dissipated in a resistor is given by

$$P = i^2 R$$

where i is the current and R is resistance. Find the formula for P in terms of the time t, if

$$i = \dfrac{3t + 1}{2t^2} \quad \text{and} \quad R = \dfrac{2t}{3t + 1}$$

50. The resistance in a simple electric circuit is the voltage divided by the current. Find the formula for the resistance R of a circuit in which the voltage V and current i are functions of time, as follows:

$$V = \dfrac{4t + 12}{2t + 3} \quad \text{and} \quad i = \dfrac{6t}{2t^2 + 9t + 9}$$

51. The density of an object is its mass divided by its volume. Write and simplify an expression for the density of a cone (Fig. 17-1) having a mass m and a volume equal to $\dfrac{\pi r^2 h}{3}$.

17-3 ADDING AND SUBTRACTING ALGEBRAIC FRACTIONS

In Sec. 2-5, we learned how to add and subtract common fractions. In this section we learn how to add and subtract algebraic fractions. We find the least common denominator, add and subtract mixed expressions, and solve applications.

Adding and Subtracting Like Fractions

Recall that like, or similar, fractions are fractions that have the same (common) denominator. *The sum of like fractions is the sum of the numerators divided by the denominator.* When we add or subtract fractions (or other quantities), we may say that we *combine* them. As usual, we simplify the result if possible.

ADDITION AND SUBTRACTION OF FRACTIONS WITH THE SAME DENOMINATOR	$\dfrac{a}{c} + \dfrac{b}{c} = \dfrac{a + b}{c}$ $\dfrac{a}{c} - \dfrac{b}{c} = \dfrac{a - b}{c}$	43

Example 26

$$\dfrac{2}{x} + \dfrac{3}{x} = \dfrac{5}{x}$$

Sometimes the numerators can be combined into a single sum or difference, as above. If the numerators cannot be combined, then the sum or difference must be indicated using the $+$ and $-$ symbols.

Example 27

$$\frac{2x}{a^2b} - \frac{3}{a^2b} = \frac{2x - 3}{a^2b}$$

Special care must be taken when subtracting fractions involving polynomials.

Example 28

Combine and simplify:

$$\frac{2y}{y-1} - \frac{2+y}{y-1}$$

Solution: We use parentheses to show that we are subtracting the entire second numerator, $2 + y$:

$$\frac{2y}{y-1} - \frac{2+y}{y-1} = \frac{2y - (2+y)}{y-1}$$

Subtracting,

$$= \frac{2y - 2 - y}{y-1}$$

Combining,

$$= \frac{y-2}{y-1}$$

COMMON ERROR When a fraction is subtracted, be sure to subtract the entire numerator.

We use Eq. 8

$$-(a - b) = b - a$$

to combine certain fractions, as shown in the following example.

Example 29

Combine and simplify:

$$\frac{1}{x-2} + \frac{x}{2-x}$$

Solution: First we rewrite the fractions with a common denominator:
Using Eq. 8,

$$\frac{1}{x-2} + \frac{x}{2-x} = \frac{1}{x-2} + \frac{x}{-(x-2)}$$

17-3 Adding and Subtracting Algebraic Fractions

Using the rules of signs,

$$= \frac{1}{x-2} - \frac{x}{x-2}$$

Combining,

$$= \frac{1-x}{x-2}$$

COMMON ERROR You may combine numerators only when adding or subtracting fractions with like denominators. NEVER add or subtract denominators:

$$\frac{a}{b} + \frac{a}{c} \neq \frac{a}{b+c} \quad \text{and} \quad \frac{a}{b} + \frac{c}{d} \neq \frac{a+c}{b+d}$$

Example 30

$$\frac{1}{x} + \frac{1}{y} \neq \frac{1}{x+y} \quad \text{and} \quad \frac{1}{x} + \frac{y}{2} \neq \frac{1+y}{x+2}$$

Least Common Denominator

To add or subtract fractions with different denominators, we must rewrite the fractions as equivalent fractions that do have the same common denominator. Any common denominator will work. However, the *least common denominator (LCD)* is the most convenient and useful. The LCD is the simplest expression that is exactly divisible by each of the denominators. The LCD is also the least common multiple of the denominators and must contain all the prime factors of each of the denominators.

Steps to Find the LCD

1. Factor each denominator into its prime factors.
2. Take the highest power of each prime factor occurring in any denominator.
3. The LCD is the product of these highest powers.

Example 31

Find the LCD of $\frac{5}{6x^2y}$ and $\frac{4}{9xy^3}$.

Solution: Factoring the denominators into primes, we have

$$6x^2y = 2 \cdot 3x^2y$$

and

$$9xy^3 = 3^2xy^3$$

Note that with monomials, only the numerical coefficients must be factored, since the variable part is already factored. The prime factors of these two denominators are 2, 3, x, and y.
The highest power of 2 that occurs is 2^1 (or 2).
The highest power of 3 that occurs is 3^2.
The highest power of x is x^2, and the highest power of y is y^3.
The LCD is the product of these powers:

$$\text{LCD} = 2 \cdot 3^2 \cdot x^2y^3 = 18x^2y^3$$

In many problems, you may be able to find the LCD by inspection. The process given above is one that we can use when the LCD is not obvious.

Example 32
Find the LCD of the fractions

$$\frac{2}{4x^2},\ \frac{3x}{2x^2-8},\ \text{and}\ \frac{5}{x^2-2x}$$

Solution: Factoring the denominators into primes, we have

$$4x^2 = 2^2 x^2$$
$$2x^2 - 8 = 2(x^2 - 4) = 2(x+2)(x-2)$$

and

$$x^2 - 2x = x(x-2)$$

The LCD is the product of the highest powers of the factors 2, x, $x+2$, and $x-2$, so

$$\text{LCD} = 2^2 x^2 (x+2)(x-2)$$
$$= 4x^2(x+2)(x-2)$$

We could multiply this to find the product, but the LCD is more useful for addition and subtraction in its factored form.

TIP It is preferable to leave the least common denominator in its factored form for ease in addition and subtraction.

Adding and Subtracting Algebraic Fractions with Different Denominators

To add or subtract algebraic fractions with different denominators, we must rewrite the fractions as equivalent fractions that do have the same common denominator. Any common denominator will work. If the denominators have no factors in common, then the common denominator is the product of the denominators.

Example 33
Combine $\dfrac{a}{b} + \dfrac{c}{d}$.

Solution: We multiply the two denominators to get the common denominator, bd. Expressing each fraction over bd:

$$\frac{a}{b} + \frac{c}{d} = \frac{a(d)}{b(d)} + \frac{c(b)}{d(b)}$$
$$= \frac{ad}{bd} + \frac{bc}{bd}$$
$$= \frac{ad + bc}{bd}$$

This brings us to the rules for adding and subtracting fractions with different denominators.

ADDITION AND SUBTRACTION OF FRACTIONS WITH DIFFERENT DENOMINATORS	$\dfrac{a}{b} + \dfrac{c}{d} = \dfrac{ad + bc}{bd}$ $\dfrac{a}{b} - \dfrac{c}{d} = \dfrac{ad - bc}{bd}$	44

Example 34

Combine and simplify:

$$\frac{x}{2y} - \frac{3}{x}$$

Solution: By inspection, the LCD is $2xy$. We use the fundamental principal of fractions to find equivalent fractions.
Expressing each fraction over $2xy$:

$$\frac{x}{2y} - \frac{3}{x} = \frac{x(x)}{2y(x)} - \frac{3(2y)}{x(2y)}$$
$$= \frac{x^2}{2xy} - \frac{6y}{2xy}$$

Combining,

$$= \frac{x^2 - 6y}{2xy}$$

Since the numerator is not factorable, this is already simplified.

Steps to Add or Subtract Fractions with Different Denominators

1. Find the LCD.
2. Change each fraction to an equivalent one whose denominator is the LCD by multiplying the numerator and denominator of the fraction by that expression which will make its denominator equal to the LCD.
3. Add the equivalent fractions by adding the numerators and writing that sum over the LCD.
4. Simplify the result if possible.

Example 35

Combine and simplify:

$$\frac{5}{6x^2y} - \frac{4}{9xy^3}$$

Solution: We found the LCD of these fractions in Example 31. It is $18x^2y^3$. We find equivalent fractions:

$$\frac{5}{6x^2y} - \frac{4}{9xy^3} = \frac{5}{6x^2y} \cdot \frac{3y^2}{3y^2} - \frac{4}{9xy^3} \cdot \frac{2x}{2x}$$
$$= \frac{15y^2 - 8x}{18x^2y^3}$$

Example 36

Combine and simplify:

$$\frac{a}{a-b} + \frac{b}{a^2 - b^2}$$

Solution: We factor the second denominator:

$$a^2 - b^2 = (a+b)(a-b)$$

Thus the LCD is $(a+b)(a-b)$.

$$\frac{a}{a-b} + \frac{b}{a^2 - b^2} = \frac{a}{(a-b)} \cdot \frac{(a+b)}{(a+b)} + \frac{b}{(a+b)(a-b)}$$

$$= \frac{a(a+b) + b}{(a+b)(a-b)}$$

$$= \frac{a^2 + ab + b}{(a+b)(a-b)}$$

Since the numerator is not factorable, this result is in simplest form.

Mixed Expressions

A mixed expression is an expression in which some, but not all, terms are fractions. To add or subtract mixed expressions, we must still find the LCD and express each term with that denominator.

Example 37

Combine and simplify:

$$x - \frac{2}{x-1}$$

Solution: First we change the non-fraction part to fractional form:

$$x = \frac{x}{1}$$

and proceed as above. The LCD is $x - 1$. (The factor of 1 need not be written.)

$$x - \frac{2}{x-1} = \frac{x}{1} - \frac{2}{x-1}$$

$$= \frac{x}{1} \cdot \frac{(x-1)}{(x-1)} - \frac{2}{x-1}$$

$$= \frac{x(x-1) - 2}{x-1}$$

$$= \frac{x^2 - x - 2}{x-1}$$

$$= \frac{(x+1)(x-2)}{x-1}$$

Since no factors cancel, the expression is in simplest form.

17-3 Adding and Subtracting Algebraic Fractions

SECTION 17-3 Exercises Adding and Subtracting Algebraic Fractions

Adding and Subtracting Like Fractions

Combine and simplify.

1. $\dfrac{5}{y} + \dfrac{2}{y}$
2. $\dfrac{7}{x} - \dfrac{3}{x}$
3. $\dfrac{6}{ab} - \dfrac{2}{ab} + \dfrac{3}{ab}$
4. $\dfrac{5}{x^2} + \dfrac{2}{x^2} - \dfrac{3}{x^2}$
5. $\dfrac{2x}{a^2b^3} - \dfrac{5y}{a^2b^3} + \dfrac{6x}{a^2b^3}$
6. $\dfrac{2a}{x^2y} + \dfrac{3}{x^2y} - \dfrac{4a}{x^2y}$
7. $\dfrac{5a}{x-y} - \dfrac{a}{x-y}$
8. $\dfrac{2x}{a+b} + \dfrac{3x}{a+b}$
9. $\dfrac{a}{a-2} - \dfrac{a-3}{a-2}$
10. $\dfrac{q}{p+1} - \dfrac{q+1}{p+1}$
11. $\dfrac{c^2}{c-1} - \dfrac{1}{c-1}$
12. $\dfrac{x^2}{x+3} + \dfrac{3x}{3+x}$
13. $\dfrac{x^2}{x+1} - \dfrac{3-2x}{1+x}$
14. $\dfrac{6}{y-2} - \dfrac{5y-y^2}{y-2}$
15. $\dfrac{1}{m-n} + \dfrac{2}{n-m}$
16. $\dfrac{3}{x-y} - \dfrac{5}{y-x}$
17. $\dfrac{a}{a-b} - \dfrac{b}{b-a} + \dfrac{a+b}{a-b}$
18. $\dfrac{3}{1-p} + \dfrac{p}{p-1} - \dfrac{1+p}{1-p}$

Adding and Subtracting Algebraic Fractions with Different Denominators

Combine and simplify.

19. $\dfrac{1}{x} + \dfrac{1}{y}$
20. $\dfrac{p}{10} - \dfrac{2p}{15}$
21. $\dfrac{7}{a} - \dfrac{5}{a^2}$
22. $\dfrac{x}{y} + \dfrac{3}{xy}$
23. $\dfrac{2a}{3x} + \dfrac{a}{2x}$
24. $\dfrac{y}{4x} - \dfrac{5y}{6x^2}$
25. $\dfrac{x}{4} - \dfrac{3y}{2} + \dfrac{2z}{3}$
26. $\dfrac{1}{r_1} + \dfrac{1}{r_2} + \dfrac{1}{r_3}$
27. $\dfrac{1}{x} - \dfrac{1}{y} + \dfrac{1}{z}$
28. $\dfrac{a-b}{4} - \dfrac{a+b}{6}$
29. $\dfrac{1}{2} - \dfrac{5}{6ab^2} + \dfrac{3}{8a^2b}$
30. $\dfrac{2}{3x^2} + \dfrac{1}{6y} - \dfrac{3}{xy^3}$
31. $\dfrac{2}{x+y} - \dfrac{3}{x-y}$
32. $\dfrac{a+1}{a-1} - \dfrac{a-1}{a+1}$
33. $\dfrac{p-1}{p+2} + \dfrac{p+2}{p-3}$
34. $\dfrac{1}{x-1} + \dfrac{2}{x-2} - \dfrac{3}{x-3}$
35. $\dfrac{2}{b+3} - \dfrac{3}{b-2} + \dfrac{1}{b+1}$
36. $\dfrac{a}{a+b} - \dfrac{2b^2}{a^2-b^2}$
37. $\dfrac{2}{x+1} - \dfrac{x-3}{x^2-1}$
38. $\dfrac{1}{2x} - \dfrac{x}{3x-6}$
39. $\dfrac{3}{x^2-x} + \dfrac{5}{2x-2}$
40. $\dfrac{4y}{4y^2-1} - \dfrac{3y+1}{1-2y} + \dfrac{2y}{1+2y}$

Mixed Expressions

Combine and simplify.

41. $a + \dfrac{1}{a}$
42. $\dfrac{3}{y} - y$
43. $c + \dfrac{3}{c-2}$
44. $\dfrac{d}{d+1} - d$
45. $\dfrac{2}{x} - 3 + \dfrac{1}{x^3}$
46. $\dfrac{1}{a+2} - a + 3$
47. $\dfrac{2x-3}{3x-2} - \dfrac{2}{x} - 1$
48. $\dfrac{2a-3}{a-2} - \dfrac{4}{a} - 3a - 4$

Applications

The following expressions are used with the given application. Combine and simplify.

49. Forces acting on a concrete slab:

$$1 - \dfrac{4c}{\pi l} + \dfrac{c^3}{3l^3}$$

50. The time t required to make a round trip by boat:
$$\frac{d}{V-v} - \frac{d}{V+v}$$

51. Resistance of a circuit (Fig. 17-2):
$$R_1 + \frac{R_2 R_3}{R_2 + R_3}$$

52. Focal length of a thin lens (Fig. 17-3):
$$\frac{1}{f} = \frac{1}{p} + \frac{1}{q} \quad \text{Combine the right side.}$$

■ **FIGURE 17-2**

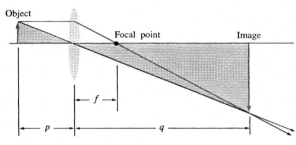

■ **FIGURE 17-3**

17-4 COMPLEX FRACTIONS

In this section we learn to simplify a complex fraction. Recall from Sec. 2-6 that a *complex fraction* is a fraction in which the numerator or denominator, or both, contains a fraction.

Example 38

Examples of complex fractions are

$$\frac{\frac{x}{y}}{z}, \quad \frac{\frac{p}{q}}{\frac{q}{r}}, \quad \frac{\frac{a}{b}}{\frac{c}{d}}, \quad \frac{1-\frac{1}{x}}{2}, \quad \frac{1-\frac{1}{x}}{1+\frac{1}{x}}, \quad \text{and} \quad \frac{x+\frac{1}{x-y}}{x-\frac{1}{x+y}}$$

The main fraction line is *significantly longer* and, if possible, in boldface type.

To simplify a complex fraction, we perform the indicated operations to obtain a simple fraction in which there is only one fraction line. In a fraction, the fraction line indicates division, so we express the fraction as a division problem, divide, and simplify the result, if possible.

---- **Example 39** ----

Simplify:

$$\frac{x}{\frac{3}{y}}$$

Solution: First we express the fraction as a division problem:

$$\frac{x}{\frac{3}{y}} = x \div \frac{3}{y}$$

Inverting and multiplying,

$$= \frac{x}{1} \cdot \frac{y}{3}$$

$$= \frac{xy}{3}$$

> **COMMON ERROR** When complex fractions are written, it is very easy to read $\frac{x}{\frac{3}{y}}$ as $\frac{\frac{x}{3}}{y}$. If there is a possibility of reading error, then a very obvious main fraction line or parentheses may be used to indicate which is the correct version of the complex fraction. The alternate forms of the complex fraction to be simplified in Example 39 should be $\frac{x}{\frac{3}{y}} = \frac{x}{\frac{3}{y}} = \frac{x}{\left(\frac{3}{y}\right)}$. If there is any possibility of confusion, the rightmost form is preferred.

If the numerator and/or denominator are more complicated, just express them as simple fractions. Then divide the fraction in the numerator by the fraction in the denominator. Recall that a fraction line is also a symbol of grouping, so all quantities grouped by a fraction line must be combined before dividing.

---- **Example 40** ----

Simplify:

$$\frac{y - \frac{1}{5}}{5 - \frac{1}{y}}$$

Solution: Express the numerator and denominator as simple fractions:

$$\frac{y - \frac{1}{5}}{5 - \frac{1}{y}} = \frac{\frac{5y - 1}{5}}{\frac{5y - 1}{y}}$$

Divide:

$$= \frac{5y - 1}{5} \div \frac{5y - 1}{y}$$

Invert and multiply:

$$= \frac{\overset{1}{\cancel{5y-1}}}{5} \cdot \frac{y}{\underset{1}{\cancel{5y-1}}}$$

$$= \frac{y}{5}$$

Alternate Solution: Multiply the numerator and denominator of the given fraction by the LCD of the fractions that appear in the numerator or denominator, and simplify. The LCD of the denominators is $5y$. We multiply the numerator and denominator by $5y$:

$$\frac{y - \frac{1}{5}}{5 - \frac{1}{y}} = \frac{\left(y - \frac{1}{5}\right)5y}{\left(5 - \frac{1}{y}\right)5y}$$

Distributing the $5y$ and multiplying,

$$= \frac{y(5y) - \frac{1}{5}(5y)}{5(5y) - \frac{1}{y}(5y)} = \frac{5y^2 - y}{25y - 5}$$

Factoring and canceling,

$$= \frac{y\overset{1}{\cancel{(5y-1)}}}{5\underset{1}{\cancel{(5y-1)}}} = \frac{y}{5}$$

We see that the two methods yield the same result, $\frac{y}{5}$.

--- **Example 41** ---

Simplify:

$$\frac{\frac{1}{9} - \frac{a}{3}}{\frac{a}{6} - \frac{1}{18}}$$

Solution: The LCD is 18. We multiply the numerator and denominator by 18:

$$\frac{\frac{1}{9} - \frac{a}{3}}{\frac{a}{6} - \frac{1}{18}} = \frac{\left(\frac{1}{9} - \frac{a}{3}\right)18}{\left(\frac{a}{6} - \frac{1}{18}\right)18}$$

Distributing,

$$= \frac{\frac{1}{9}(18) - \frac{a}{3}(18)}{\frac{a}{6}(18) - \frac{1}{18}(18)}$$

$$= \frac{2 - 6a}{3a - 1}$$

$$= \frac{2(1 - 3a)}{3a - 1}$$

$$= \frac{-2\,\overset{1}{\cancel{(3a-1)}}}{\underset{1}{\cancel{(3a-1)}}}$$

$$= -2$$

17-4 Complex Fractions

SECTION 17-4 Exercises Complex Fractions

Simplify.

1. $\dfrac{a}{\dfrac{b}{2}}$ 2. $\dfrac{\dfrac{a}{b}}{2}$

3. Show that $\dfrac{x}{\dfrac{y}{z}} \ne \dfrac{\dfrac{x}{y}}{z}$

4. Show that $\dfrac{2}{\dfrac{x}{3}} \ne \dfrac{\dfrac{2}{x}}{3}$

Simplify.

5. $\dfrac{\dfrac{a}{b}}{\dfrac{c}{d}}$ 6. $\dfrac{\dfrac{3}{x}}{\dfrac{x}{4}}$ 7. $\dfrac{x-4}{2-\dfrac{8}{x}}$ 8. $\dfrac{x-12}{1-\dfrac{12}{x}}$

9. $\dfrac{b-\dfrac{1}{4}}{4-\dfrac{1}{b}}$ 10. $\dfrac{\dfrac{1}{2}-x}{2-\dfrac{1}{x}}$ 11. $\dfrac{3+\dfrac{2}{y}}{9y-\dfrac{4}{y}}$ 12. $\dfrac{\dfrac{9}{y}-16y}{1+\dfrac{3}{4y}}$

13. $\dfrac{\dfrac{3a}{2}-1}{6a-\dfrac{8}{3a}}$ 14. $\dfrac{\dfrac{y}{3}+\dfrac{y}{4}}{\dfrac{5y}{12}+1}$ 15. $\dfrac{1+\dfrac{a}{b}}{1-\dfrac{a^2}{b^2}}$ 16. $\dfrac{\dfrac{1}{a}+\dfrac{1}{b}}{\dfrac{1}{a}-\dfrac{1}{b}}$

17. $\dfrac{1-\dfrac{7a+4}{2a^2}}{a-\dfrac{16}{a}}$ 18. $\dfrac{1-\dfrac{7b+6}{3b^2}}{b-\dfrac{9}{b}}$ 19. $\dfrac{\dfrac{x}{x-2y}-1}{1-\dfrac{x}{x+2y}}$ 20. $\dfrac{2-\dfrac{b}{2a-b}}{\dfrac{b}{2a+b}-1}$

Applications

These complex fractions are used with the given application. Simplify.

21. Resistance of two resistors in parallel (Fig. 17-4):

$$\dfrac{1}{\dfrac{1}{R_1}+\dfrac{1}{R_2}}$$

22. Focal length of a thin lens:

$$\dfrac{1}{\dfrac{p+q}{pq}}$$

■ FIGURE 17-4

FIGURE 17-5

23. Capacitance (Fig. 17-5):

$$\frac{1}{\frac{1}{C_1} + \frac{1}{C_2} + \frac{1}{C_3}}$$

24. Present value of a stock:

$$\frac{\frac{1}{1+i}}{1 - \frac{1}{1+i}}$$

Chapter 17 REVIEW EXERCISES

Reduce to lowest terms.

1. $\dfrac{x^2 - 36}{x^2 - 12x + 36}$

2. $\dfrac{x^3 + 1}{(x + 1)^3}$

3. $\dfrac{a^2 - 4b^2}{2b - a}$

4. $\dfrac{9a^2 - 24a + 16}{9a^2 - 16}$

5. $\dfrac{y^3 - x^3}{(x - y)^2}$

6. $\dfrac{a^3 + b^3}{a^2 - b^2}$

Perform the indicated operations and simplify.

7. $\dfrac{3x^2 - 75x}{2x^2 - 2x - 24} \cdot \dfrac{4x^2 - 28x + 48}{x^4 - 8x^3 + 15x^2}$

8. $\dfrac{a^3 - 2a^2b + ab^2}{4a + 4b} \div \dfrac{ab - a^2}{2a + 2b}$

9. $\dfrac{4 - n^2}{n^3 - 8} \div \dfrac{n^3 - 2n^2 - 8n}{4n^2 - 16n}$

10. $\dfrac{x^2 - 3x - 4}{3x^4 - 75x^2} \div \dfrac{2x^2 + 3x + 1}{x^3 + 3x^2 - 10x} \cdot \dfrac{6x^3 - 24x^2 - 30x}{x^2 - 6x + 8}$

Combine and simplify.

11. $\dfrac{1}{1 - a} - \dfrac{a^2}{1 - a}$

12. $\dfrac{x^2}{x + 2} - \dfrac{2 - x}{2 + x}$

13. $\dfrac{x}{x - y} + \dfrac{y}{y - x} - \dfrac{x + y}{x - y}$

14. $\dfrac{x}{12} - \dfrac{7x}{8}$

15. $\dfrac{x}{4} + \dfrac{5x}{6} - \dfrac{3x}{8}$

16. $\dfrac{x + y}{2} - \dfrac{x - y}{3}$

17. $\dfrac{x + 5}{x^2 - 16} + \dfrac{x - 3}{4 - x}$

18. $\dfrac{a - 4}{2a} + \dfrac{a + 2}{a + 1} - \dfrac{3a + 2}{a^2 + a}$

19. $\dfrac{a}{5} - 3$

20. $x - \dfrac{2}{x - 1} + 1$

21. $\dfrac{3p - 5}{p - 3} - \dfrac{5}{p} - 9$

22. $\dfrac{2x - 3}{x - 2} - \dfrac{5}{x} - 2$

Simplify.

23. $\dfrac{\dfrac{1}{3} - x}{3 - \dfrac{1}{x}}$ 24. $\dfrac{\dfrac{1}{4} - \dfrac{x}{2}}{\dfrac{x}{6} - \dfrac{1}{12}}$ 25. $\dfrac{1 - \dfrac{1}{x}}{x - \dfrac{1}{x^2}}$ 26. $\dfrac{1 - \dfrac{3x - 2}{2x^2}}{x - \dfrac{4}{x}}$

The following expressions are used with the given application. Simplify.

27. Displacement of a vibrating string:

$$\dfrac{t}{T} - \dfrac{x}{\lambda}$$

28. Focal length of a lens:

$$\dfrac{1}{\dfrac{1}{p} + \dfrac{1}{q}}$$

WRITING

29. We have seen many applications of algebraic fractions in this chapter. How are algebraic fractions used in science or in business? Give three examples, explaining what each letter represents and the units associated with each letter.

TEAM PROJECT

30. A small computer company has five employees: Ms. Adams, Ms. Black, Mr. Clark, Mr. Downing, and Mr. Ewing. They hold the positions of technician, analyst, mathematician, programmer, and manager, although not necessarily in that order. Here is some additional information about the employees:

   The analyst and the manager were same-sex roommates in college.
   The technician is a bachelor.
   Mr. Ewing and Ms. Adams have had only business contacts with each other.
   Mrs. Clark was greatly disappointed when her husband told her that the manager had refused to give him a raise.
   Mr. Downing is going to be best man when the mathematician and the analyst are married.

   What position does each person hold in the company? Explain how you came to your conclusions.

Fractional Equations 18

Outcomes

When you have completed this chapter, you will be able to

1. Solve fractional equations with numbers in the denominators.
2. Solve fractional equations with variables in the denominators.
3. Rearrange and solve literal equations and formulas for a specific variable.
4. Solve applied fractional formulas, for an unknown.

Many equations in applied mathematics involve fractions. In this chapter we solve fractional equations and formulas involving fractions. We then solve some applications with fractional equations.

18-1 SOLUTION OF FRACTIONAL EQUATIONS

In this section we solve fractional equations in which the denominators are either constants or variables.

Solving Fractional Equations

A *fractional equation* is an equation in which one or more terms is a fraction.

Example 1

Some examples of fractional equations are

$$\frac{2x}{3} = 13 - \frac{x}{5} \quad \text{and} \quad \frac{3x}{x+2} - 2 = \frac{2x-3}{2x-1}$$

A good way to begin to solve a fractional equation is to *first multiply each side of the equation by the LCD of all the fractions*. The resulting equation will contain no fractions, and can be solved using the methods described in Chapter 10. As always, it is wise to *check* solutions by substituting them into the original equation.

Example 2
Solve and check:

$$\frac{2x}{3} = 13 - \frac{x}{5}$$

Solution: The LCD of these fractions is 15. Multiplying each side of the equation by the LCD, we have

$$\left(\frac{2x}{3}\right)15 = \left(13 - \frac{x}{5}\right)15$$

$$\left(\frac{2x}{3}\right)15 = (13)15 - \left(\frac{x}{5}\right)15$$

Simplifying,

$$10x = 195 - 3x$$
$$13x = 195$$
$$x = 15$$

Check:

$$\frac{2(15)}{3} \stackrel{?}{=} 13 - \left(\frac{15}{5}\right)$$

$$10 = 10 \quad \text{checks}$$

Note: In Example 2, we could have eliminated one step by just multiplying each *term* of the equation by the LCD.

If a fraction with a polynomial in it is preceded by a minus sign, then the entire fraction must be subtracted. The wise use of parentheses is very important.

Example 3
Solve and check:

$$\frac{7y - 5}{10} - \frac{y + 1}{3} = 1$$

Solution: We multiply, term by term, by the LCD, 30,

$$\left(\frac{7y - 5}{10}\right)30 - \left(\frac{y + 1}{3}\right)30 = (1)30$$

$$(7y - 5)3 - (y + 1)10 = 30$$
$$21y - 15 - 10y - 10 = 30$$
$$11y = 55$$
$$y = 5$$

Check:

$$\frac{7(5) - 5}{10} - \frac{5 + 1}{3} \stackrel{?}{=} 1$$

$$1 = 1 \quad \text{checks}$$

Equations with the Variable in the Denominator

The procedure is the same for solving an equation in which the variable is in one or more denominators. Multiply both sides of the equation by the LCD, which in this case contains the variable. Solve the resulting equation and check.

Example 4

Solve and check:

$$\frac{1}{4x} - \frac{3}{4} = \frac{7}{x}$$

Solution: The LCD is $4x$. Multiplying term by term by $4x$, we get

$$\frac{1}{4x}(4x) - \frac{3}{4}(4x) = \frac{7}{x}(4x)$$

Simplifying

$$1 - 3x = 28$$
$$x = -9$$

Check: Substituting $x = -9$ in the original equation:

$$\frac{1}{4(-9)} - \frac{3}{4} \stackrel{?}{=} \frac{7}{-9}$$

$$-\frac{7}{9} = -\frac{7}{9} \quad \text{checks}$$

Example 5

Solve and check:

$$\frac{3x}{x+2} - 2 = \frac{2x-3}{2x-1}$$

Solution: The LCD is $(x + 2)(2x - 1)$. Multiplying term by term by the LCD,

$$\left(\frac{3x}{x+2}\right)(x+2)(2x-1) - 2(x+2)(2x-1) = \left(\frac{2x-3}{2x-1}\right)(x+2)(2x-1)$$

Simplifying,

$$3x(2x - 1) - 2(2x^2 + 3x - 2) = (2x - 3)(x + 2)$$
$$6x^2 - 3x - 4x^2 - 6x + 4 = 2x^2 + x - 6$$
$$-10x = -10$$
$$x = 1$$

Check:

$$\frac{3}{1+2} - 2 \stackrel{?}{=} \frac{2-3}{2-1}$$

$$-1 = -1 \quad \text{checks}$$

Alternate Solution: Recall that all equations can be solved graphically. To solve this equation, we rewrite it as

$$\frac{3x}{x+2} - 2 - \frac{2x-3}{2x-1} = 0$$

FIGURE 18-1

and graph the function

$$y = \frac{3x}{x+2} - 2 - \frac{2x-3}{2x-1}$$

on a graphing calculator. Using the "root" or "zero" feature of the calculator (Fig. 18-1), we find that there is an x intercept at $x = 1$, so that the solution of the given equation is

$$x = 1$$

When solving equations in which the variable is in the denominator, *the check is essential*. In the process of multiplying both sides by a variable expression (the LCD), we may introduce a false "solution," a value that does not make the original equation true. We call such a value *extraneous*. It becomes clear that a value is extraneous only in the check. If the check produces a denominator of zero in the given equation, then that value is extraneous and must be discarded. There may be *no solution*. Since we cannot predict which equations produce extraneous values, we must check all solutions.

> **COMMON ERROR** It is *essential* to check the solution of an equation in which the variable appears in a denominator, for the value may be extraneous.

─── **Example 6** ───

Solve and check:

$$\frac{2}{a-2} = \frac{a}{a-2} - 2$$

Solution: Multiplying by the LCD, $a - 2$,

$$\left(\frac{2}{a-2}\right)(a-2) = \left(\frac{a}{a-2}\right)(a-2) - 2(a-2)$$

$$2 = a - 2(a-2)$$

$$2 = a - 2a + 4$$

$$a = 2$$

Check:

$$\frac{2}{0} \stackrel{?}{=} \frac{2}{0} - 2$$

Since we have division by zero, the value 2 is not a solution; it is extraneous. Thus, there is no solution.

Fractional Solutions

If a solution is a fraction, checking by substitution may not be simple. A quick check to see if it is extraneous can be made by inspecting the denominators. A check by calculator may be easier, or you may choose to rework the problem, rather than doing a regular check.

Example 7

Check that $x = \dfrac{12}{5}$ is a solution to the equation

$$\frac{6}{x^2 - 4} - 1 = \frac{4}{x + 2} - \frac{x - 3}{x - 2}$$

Solution: A check by hand would be laborious. A quick check of the denominator does not indicate division by zero, so the value is not extraneous. We check by evaluating each side of the equation *by calculator*:

$$\frac{6}{\left(\dfrac{12}{5}\right)^2 - 4} - 1 \stackrel{?}{=} \frac{4}{\dfrac{12}{5} + 2} - \frac{\dfrac{12}{5} - 3}{\dfrac{12}{5} - 2}$$

$$2.4090909 = 2.4090909 \quad \text{checks by calculator}$$

SECTION 18-1 Exercises Solution of Fractional Equations

Solving Fractional Equations

Solve and check.

1. $\dfrac{2x}{3} = 12$
2. $\dfrac{11}{5} = \dfrac{b}{15}$
3. $a - 3 = \dfrac{2a}{5}$
4. $n + 6 = \dfrac{3n}{5}$
5. $4 - x = \dfrac{5 + x}{8}$
6. $x = \dfrac{3x}{8} - 5$
7. $\dfrac{x}{3} - \dfrac{x}{4} = 5$
8. $\dfrac{7x}{2} - \dfrac{9x}{4} = x + 6$
9. $\dfrac{x}{2} - x = \dfrac{x}{3} - 15$
10. $\dfrac{5c}{6} - \dfrac{3c}{4} = 3 - \dfrac{2c}{3}$
11. $\dfrac{4 - 5a}{3} - 2 = \dfrac{5 - 9a}{7}$
12. $\dfrac{2y - 1}{4} = \dfrac{2y + 3}{5} - 3$
13. $\dfrac{5 - a}{7} - \dfrac{a - 1}{14} = 1$
14. $\dfrac{3x - 1}{4} - \dfrac{2x + 1}{7} = 1$
15. $2 - \dfrac{x - 7}{12} = \dfrac{x + 2}{6}$
16. $\dfrac{3b - 1}{4} - \dfrac{b + 5}{3} = 1$
17. $\dfrac{5 - y}{9} - \dfrac{y + 2}{16} = \dfrac{1}{12}$
18. $\dfrac{3a - 2}{5} - \dfrac{2a - 7}{4} = 2$
19. $\dfrac{2y - 4}{3} - \dfrac{y + 1}{5} = 5$
20. $\dfrac{a - 5}{12} - \dfrac{1 - 3a}{2} = \dfrac{2}{3}$
21. $\dfrac{4y + 3}{5} - \dfrac{y + 4}{3} - \dfrac{2y - 1}{15} = 2$
22. $\dfrac{5 - 2a}{2} - \dfrac{1 - 3a}{3} = \dfrac{2a - 3}{6} + 4$

Equations with the Variable in the Denominator

Solve and check.

23. $\dfrac{1}{2x} = \dfrac{1}{6}$
24. $\dfrac{2}{3x} = \dfrac{2}{9}$
25. $\dfrac{1}{x} - \dfrac{1}{3x} = \dfrac{1}{3}$
26. $\dfrac{3x - 6}{x} + 1 = \dfrac{14}{x}$
27. $\dfrac{y - 3}{y + 3} = 7$
28. $\dfrac{x - 2}{x + 8} = 6$

29. $\dfrac{8}{x-2} = \dfrac{4x}{x-2}$

30. $\dfrac{2}{x-1} = 0$

31. $\dfrac{y}{y-5} - 2 = \dfrac{5}{y-5}$

32. $\dfrac{x+1}{2x-6} + \dfrac{3}{3-x} = \dfrac{3}{2}$

33. $\dfrac{2}{x-2} - \dfrac{9}{x} = \dfrac{1}{x}$

34. $\dfrac{4}{a-1} = \dfrac{7}{a-1} - 2$

35. $\dfrac{2}{x+1} + \dfrac{1}{x} = \dfrac{1}{x^2+x}$

36. $\dfrac{3}{a+2} = \dfrac{1}{a-2} - \dfrac{4}{a^2-4}$

37. $\dfrac{3y+4}{y+2} - \dfrac{3y-5}{y-4} = \dfrac{12}{y^2-2x-8}$

38. $\dfrac{5y}{y+3} - \dfrac{3y+2}{y-4} = \dfrac{y-3}{y^2-y-12} + 2$

39. $1.83x = \dfrac{x}{2.75} + 8.14$

40. $87.4 = \dfrac{35.6}{x} + 45.7$

41. $\dfrac{0.429}{x-0.326} - 0.567 = \dfrac{0.395}{x-0.326}$

42. $\dfrac{2.34}{x-3.89} + 4.56 = 8.46 - \dfrac{6.39}{x}$

18-2 LITERAL EQUATIONS AND FORMULAS

Recall that a literal equation is an equation in which some or all of the constants are represented by letters (literal) and that a formula is a literal equation that relates two or more mathematical or physical quantities.

Many important formulas in applications involve fractions. We look at a few of them here, and solve for different variables. The process of solving is the same as in the last section, except that now the solution will be a literal expression rather than a number.

Example 8

Solve for x:

$$\dfrac{x}{a} - b = \dfrac{x}{c}$$

Solution: We combine the two x terms on the left and bring the constant b to the right:

$$\dfrac{x}{a} - \dfrac{x}{c} = b$$

We multiply both sides of the equation by the LCD:

$$ac\left(\dfrac{x}{a} - \dfrac{x}{c}\right) = ac(b)$$

Distributing, we get

$$ac\left(\dfrac{x}{a}\right) - ac\left(\dfrac{x}{c}\right) = acb$$

Simplifying,

$$cx - ax = abc$$

Factoring,

$$(c-a)x = abc$$

Dividing,

$$x = \dfrac{abc}{c-a}$$

Example 9

The kinetic energy E of a body with mass m and velocity v is

$$E = \frac{1}{2} mv^2$$

Solve this equation for m.

Solution: Multiplying both sides by the LCD, 2, we have:

$$2E = mv^2$$

Dividing both sides by v^2, and changing sides:

$$m = \frac{2E}{v^2}$$

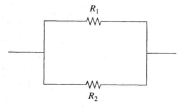

FIGURE 18-2

Example 10

If two resistors R_1 and R_2 are connected in parallel (Fig. 18-2), the total resistance R is found using the formula:

$$\frac{1}{R} = \frac{1}{R_1} + \frac{1}{R_2}$$

Solve for R_1.

Solution: The LCD is RR_1R_2. Multiplying term by term:

$$\frac{1}{R} \cdot RR_1R_2 = \frac{1}{R_1} \cdot RR_1R_2 + \frac{1}{R_2} \cdot RR_1R_2$$

Simplifying,

$$R_1R_2 = RR_2 + RR_1$$

Combining all terms involving R_1,

$$R_1R_2 - RR_1 = RR_2$$

Factoring out R_1,

$$R_1(R_2 - R) = RR_2$$

Solving for R_1,

$$R_1 = \frac{RR_2}{R_2 - R}$$

Be sure that the variable you have solved for does not also appear on the other side of the equation.

> **COMMON ERROR** When solving for a variable, be sure to combine all terms involving that variable before solving for the variable. It is *not* useful to solve for a variable *in terms of itself*.

Substituting into Formulas

In many problems we are given a formula and values for all but one variable in the formula. We substitute all known values in the formula and then solve for the unknown variable.

Example 11

The formula for the displacement s of a freely falling body having an initial velocity v and acceleration g is

$$s = vt + \frac{1}{2}gt^2$$

Find the initial velocity of an object thrust downward off the top of a 60 m building, if it hits the ground after 3.00 seconds. (Use $g = 9.81$ m/s².)

Solution: Substituting into the formula, with $s = 60$, $g = 9.81$, and $t = 3.00$, we have:

$$60 = v(3) + \frac{1}{2}(9.81)(3)^2$$

$$60 = 3v + 44.15$$

Solving,

$$v = 5.28 \text{ m/s}$$

So, the initial velocity is 5.28 m/s.

SECTION 18-2 Exercises Literal Equations and Formulas

Solve each literal equation for x.

1. $\dfrac{x}{a} + b = c$
2. $a - \dfrac{b}{x} = c$
3. $\dfrac{ax}{b} = c$
4. $a = \dfrac{bc}{x}$
5. $\dfrac{x}{a} + \dfrac{x}{b} = \dfrac{1}{c}$
6. $\dfrac{a}{x} = \dfrac{b}{x} - c$
7. $a = \dfrac{b+x}{c}$
8. $a = \dfrac{b}{x+c}$
9. $\dfrac{a}{b} = \dfrac{c}{x} + 1$
10. $a = \dfrac{bx}{b+x}$

Solve each formula for the letter in parentheses on the right.

11. $A = \dfrac{ab}{2}$ (a)
12. $V = \dfrac{1}{3}\pi r^2 h$ (h)
13. $T^2 = 4\pi^2 \dfrac{L}{g}$ (g)
14. $v^2 = \dfrac{2GM}{r}$ (r)
15. $I = \dfrac{E+e}{R}$ (e)
16. $I = \dfrac{E}{R+r}$ (r)
17. $\dfrac{1}{R} = \dfrac{1}{R_1} + \dfrac{1}{R_2} + \dfrac{1}{R_3}$ (R_1)
18. $\dfrac{1}{f} = \dfrac{1}{p} + \dfrac{1}{q}$ (p)
19. $\dfrac{E}{e} = \dfrac{R+r}{r}$ (r)
20. $\dfrac{W_1}{W_2} = \dfrac{L_2}{L_1}$ (L_1)
21. $\dfrac{P_1 V_1}{T_1} = \dfrac{P_2 V_2}{T_2}$ (T_1)
22. $C = \dfrac{C_1 C_2}{C_1 + C_2}$ (C_1)

23. $s = vt - \dfrac{1}{2}gt^2$ (g)   24. $C = \dfrac{5}{9}(F - 32)$ (F)

25. $A = \dfrac{h}{2}(a + b)$ (a)   26. $I = \dfrac{E}{r + \dfrac{R}{n}}$ (R)

Applications

27. A formula for calculating a child's dosage for medication is

$$\text{Child's dose} = \frac{\text{age of child}}{\text{age of child} + 12} \times \text{adult dose}$$

If a 10-year-old child's dose of a particular medication is 15 mg, what is the adult's dose?

28. A formula for calculating an infant's dosage for medication is

$$\text{Infant's dose} = \frac{\text{age of infant in months}}{150} \times \text{adult dose}$$

If a 12-month-old infant is to receive 8.5 mg of medication, what is the equivalent adult dose (to the nearest mg)?

29. The formula in Exercise 16 is used to find the current I (in amperes) in an electric circuit when voltage E (in volts), external resistance R (in ohms), and internal resistance r (in ohms) are known. Find R when $E = 30.5$ volts, $r = 0.756$ ohms, and $I = 1.67$ amperes.

30. The formula in Exercise 21 shows the relation among the pressure P, volume V, and temperature T (in the Kelvin temperature scale) of a gas. If the volume of oxygen is 0.0708 m^3 when the pressure is 14.8 MPa and the temperature is 27.5°C, find the volume of oxygen when the pressure is 7.94 MPa and the temperature is 34.5°C. (Hint: To convert °C to Kelvin units, add 273 to the degrees Celsius.)

31. The formula in Exercise 18 expresses the relation among the focal length f of a lens, the distance p to an object, and the distance q to the image (Fig. 18-3). Find the image distance for a lens with a focal length of 506 mm, if the object is 3250 mm from the lens.

32. What resistor must be wired in parallel with a 125 ohm resistor if the total resistance is 20.5 ohms? (Refer to Example 10).

33. The formula in Exercise 22 is used to find the capacitance C of a circuit containing two capacitances, C_1 and C_2, in series. A circuit contains two capacitances in series, one of which is 6.50 μF (Fig. 18-4). Find the second capacitance if the capacitance of the circuit is 3.50 μF.

34. The formula in Exercise 25 is used to find the area A of a trapezoid when the altitude h and the bases a and b are known. If one base is 14.4 cm, the altitude is 9.50 cm, and the area is 108 cm^2, find the second base.

■ FIGURE 18-3

■ FIGURE 18-4

18-3 APPLICATIONS

Many types of applications involve equations with fractions. These include work and uniform motion problems. The method for solving these problems is the same as that given in Chapter 11.

Work

To solve work problems, we use an equation that we know from everyday experience:

$$(\text{rate of work}) \cdot (\text{time worked}) = \text{Amount of work done}$$

or,

$$\text{rate} \cdot \text{time} = \text{Work}$$

Example 12

Suppose a person can mow a lawn in 3 hours. To find his rate of work, we solve for the rate:

$$\text{rate} \cdot \text{time} = \text{Work}$$
$$\text{rate} = \frac{\text{Work}}{\text{time}}$$

In this case, the "Work" is mowing "1 lawn" and the time worked is 3 hours. So the rate is

$$\text{rate} = \frac{1 \text{ lawn}}{3 \text{ hours}} \text{ or } \frac{1}{3} \text{ lawn/h}$$

Now we look at a problem in which two people are working together to complete a job (the work).

Example 13

One person can mow a lawn in 2.75 hours, and a second person takes 3.50 hours to mow the same lawn. How long would it take them to mow the lawn, working together?

Solution: We make a table. But first we need to find the individual rates of work for each person:

$$\text{For the 1st person: } r_1 = \frac{1 \text{ lawn}}{2.75 \text{ hours}} = \frac{1}{2.75} \text{ lawn/h}$$

$$\text{For the 2nd person: } r_2 = \frac{1 \text{ lawn}}{3.50 \text{ hours}} = \frac{1}{3.50} \text{ lawn/h}$$

We make a table:

	Rate (lawn/h)	× Time (h)	= Work (lawn)
1st person	$\frac{1}{2.75}$	t	$\frac{1}{2.75}t$
2nd person	$\frac{1}{3.50}$	t	$\frac{1}{3.50}t$
Together			1

The equation for the problem comes from the last column of the table. Their work is 1, meaning 1 lawn, because they mowed the whole (1) lawn.

1st person's work + 2nd person's work = their work together

$$\frac{1}{2.75}t + \frac{1}{3.50}t = 1$$

Multiplying through by the LCD 2.75(3.50), we get

$$2.75(3.50)\left(\frac{1}{2.75}t\right) + 2.75(3.50)\left(\frac{1}{3.50}t\right) = 2.75(3.50)(1)$$

$$3.50t + 2.75t = 9.625$$

$$6.25t = 9.625$$

$$t = 1.54 \text{ h}$$

So, it would take them 1.54 hours to mow the lawn together.

Example 14

One technician can assemble 9 instruments in 4.25 hours. A second technician can assemble 7 instruments in 3.75 hours. If they worked together, how long would it take them to assemble 25 instruments?

Solution: First we find the technicians' individual rates. Working alone, the first technician works at the rate

$$r_1 = \frac{9 \text{ instruments}}{4.25 \text{ hours}} = \frac{9}{4.25} \text{ instruments/h}$$

Working alone, the second technician works at the rate

$$r_2 = \frac{7 \text{ instruments}}{3.75 \text{ hours}} = \frac{7}{3.75} \text{ instruments/h}$$

We make a table:

	Rate (ins/h)	× Time (h)	= Work (ins)
1st technician	$\frac{9}{4.25}$	t	$\frac{9}{4.25}t$
2nd technician	$\frac{7}{3.75}$	t	$\frac{7}{3.75}t$
Together			25

18-3 Applications

Here, the technicians are assembling 25 instruments, so the total work to be completed is 25 (instruments). As in Example 13, the sum of their individual works is the total work:

$$\frac{9}{4.25}t + \frac{7}{3.75}t = 25$$

The LCD is 4.25(3.75), so multiplying by the LCD,

$$4.25(3.75)\frac{9}{4.25}t + 4.25(3.75)\frac{7}{3.75}t = 4.25(3.75)25$$

Simplifying,

$$33.75t + 29.75t = 398.4$$
$$63.5t = 398.4$$
$$t = 6.27 \text{ h}$$

So it will take them about 6.27 hours to assemble 25 instruments, working together.

Uniform Motion

Certain problems of uniform motion involve equations with fractions. In the following navigational example, we need to know that the airspeed of a plane is its speed relative to the air and the groundspeed is its speed relative to the ground. The groundspeed is equal to the sum of the airspeed and the wind velocity.

$$\text{Groundspeed} = \text{airspeed} + \text{wind velocity}$$

■ FIGURE 18-5

Example 15

A pilot flies 485 km from airport A to airport B with a tailwind of 35.2 km/h. On her return flight, she flies directly into the same wind and flies only 415 km in the same amount of time. What is the plane's airspeed?

Solution: First we make a diagram (Fig. 18-5). Let r = the plane's airspeed (in km/h). On the first flight (with the wind),

$$\text{The plane's groundspeed} = \text{airspeed} + \text{wind velocity}$$
$$= r + 35.2$$

On the return flight (against the wind),

$$\text{The plane's groundspeed} = \text{airspeed} - \text{wind velocity}$$
$$= r - 35.2$$

We make a table, filling in the time column for each flight using the rate and distance columns, and

$$\text{Time} = \frac{\text{distance}}{\text{rate}}$$

	Rate (km/h) ×	Time (h) =	Distance (km)
1st flight	$r + 35.2$	$\dfrac{485}{r + 35.2}$	485
Return flight	$r - 35.2$	$\dfrac{415}{r - 35.2}$	415

Since the times are equal, we write the equation from the time column:

$$\frac{485}{r + 35.2} = \frac{415}{r - 35.2}$$

$$485(r - 35.2) = 415(r + 35.2)$$

$$485r - 17072 = 415r + 14608$$

$$70r = 31680$$

$$r = 453 \text{ km/h}$$

So the plane is flying with an airspeed of 453 km/h.

SECTION 18-3 Exercises Applications

Solve and check. Round to three significant digits.

1. A rectangle is $\frac{3}{4}$ as wide as it is long. If the length is increased by 3.00 cm and the width decreased by 2.00 cm, then the area is 25.0 cm² more than the original area. Find the dimensions of the original rectangle.

2. The width of a rectangle is $\frac{3}{5}$ of its length and the perimeter is 48.0 cm. Find the dimensions of the rectangle.

3. One side of a triangular lot is $\frac{1}{2}$ the length of the second side. The third side is $\frac{5}{6}$ the length of the first side. Find the sides if the perimeter of the lot is 184 m.

4. After spending $\frac{3}{5}$ of her money, a woman had $24.00 left. How much did she have before?

5. A man spent $\frac{3}{4}$ of his paycheque, and then $\frac{1}{3}$ of the remainder. If he had $54.00 left, how much was his paycheque?

6. Three children were left an inheritance of which the eldest received $\frac{1}{2}$ and the second received $\frac{1}{3}$. The third child received the remainder, which was $2500. How much was the total inheritance?

Work

7. A woman can mow the lawn in 5.25 hours and her husband can mow the lawn in 4.75 hours. If each had their own mowers and mowed the lawn together, how long would it take them?

8. A holding tank can be drained through one pipe in 8.50 hours, or through another pipe in 6.75 hours. If both pipes were used, how long would it take to drain the tank?

9. One road crew can construct 525 m of road in 3.50 days. A second crew can construct 1050 m in 7.25 days. If the crews worked together, how long would it take them to construct a road one kilometre long?

10. One machine can make 25 parts in 6.75 hours. Another machine can make 18 parts in 5.25 hours. If both machines are working, how long would it take to make 100 parts?

11. A swimming pool can be filled in 15.5 hours through one opening at the top. It can be filled in 12.6 hours through a second opening. How long will it take to fill the pool if both openings are used?

12. A 200 litre tank can be filled in 8.25 minutes. It can be drained in 12.5 minutes. Five (5.00) minutes after they began filling the tank, a technician accidentally opened the drain. How much longer does it take for the tank to fill?

Uniform Motion

13. A biker rides 15.0 km down a long straight road with a wind of 8.25 km/h at his back. He turns around and finds that he can go only 7.25 km in the same time, with the wind at his front. What is his bike velocity, without wind?

14. The current in a certain stream is 4.47 km/h. A canoeist can go 7.28 km upstream in the same time it takes her to go 15.8 km downstream. What is her rate of canoeing in still water?

15. A salesman drove 298 km to a convention. Then he drove 212 km to a city at the same rate in 1.17 hours less than his first trip. How fast did the salesman drive? How long did it take him to drive to the convention?

16. Two ships leave port A at the same time and travel at the same rate. One ship sailed to port B 217 km away and the other to port C 137 km away (Fig. 18-6). The second ship arrived in port 2.25 hours before the first ship arrived at its port. How fast were the ships traveling?

■ FIGURE 18-6

17. A baseball player's batting average is found by dividing h, the number of hits, by b, the number of times at bat:

$$\frac{h}{b}$$

If the player gets a hit, then the new batting average is

$$\frac{h+1}{b+1}$$

If the player does not get a hit, then the new batting average is

$$\frac{h}{b+1}$$

Will a player's batting average always change after the player goes to bat? Under what conditions will the batting average remain the same?

■ Chapter 18 REVIEW EXERCISES

Solve and check.

1. $y - 3 = \dfrac{y+6}{7}$
2. $\dfrac{2a}{5} - \dfrac{a}{2} = a$
3. $\dfrac{2x}{5} - \dfrac{3x}{10} = 2 + \dfrac{4x}{15}$
4. $\dfrac{2b}{3} - \dfrac{3b-1}{8} = 1$
5. $\dfrac{7-y}{24} + \dfrac{3y+4}{16} = 2$
6. $\dfrac{n-5}{n-4} = \dfrac{3n}{3n+1}$
7. $2 + \dfrac{1}{b-1} = \dfrac{b}{b-1}$
8. $\dfrac{5x}{x-2} - 3 = \dfrac{4x}{2x-7}$

Solve for the variable in parentheses on the right.

9. $F = k\dfrac{M_1 M_2}{d^2}$ (M_1) **10.** $\dfrac{V_1}{T_1} = \dfrac{V_2}{T_2}$ (T_1) **11.** $F = \dfrac{9}{5}C + 32$ (C)

12. The formula,

$$I = \frac{E}{R + r}$$

is used to find the current I in an electric circuit when voltage E, external resistance R, and internal resistance r are known. Find R when $E = 33.5$ volts, $r = 0.525$ ohms, and $I = 1.25$ amperes.

13. The width of a certain rectangle is $\dfrac{2}{3}$ of the length. If the width is decreased by 3.00 cm and the length increased by 3.00 cm, the area is 36.0 cm² less than the original rectangle. Find the dimensions of the original rectangle.

14. One person can build a shed in 5.25 days and another person can build an identical shed in 4.50 days. How long should it take them to build the shed working together?

15. One mason can build 1.5 m of wall in a day, and another mason can build 1.8 m of wall in a day. How many days will it take them to build 36.5 m of wall together?

16. Two planes, flying at the same rate, are scheduled to arrive at Chicago's O'Hare Airport at the same time. One plane left its origin, 4775 km from O'Hare, 3.00 hours before the other plane left its origin, 1910 km from O'Hare. What was their groundspeed and how many hours was the longer flight?

17. Sid can type 7 pages in 2.00 hours. Shirley can type 13 pages in 3.00 hours. After Shirley has been typing a 35-page document for 2.00 hours, Sid joins her. How much longer must they work together to complete the job?

WRITING

18. When we solve a fractional equation algebraically, we sometimes get a false "solution," which is called *extraneous*. Make up a fractional equation that has an extraneous solution, solve the equation, and show that it has an extraneous solution. Where does this extraneous solution come from? How did you discover that it is extraneous?

TEAM PROJECT

19. An alphametic is an algebra puzzle in which digits are replaced by letters of the alphabet. In each puzzle each letter stands for a particular, but different, digit. For example, a familiar alphametic is

$$\begin{array}{r} \text{SEND} \\ \text{MORE} \\ \hline \text{MONEY} \end{array}$$

The solution is:

$$S = 9 \quad M = 1 \quad \text{Check:} \quad 9\,567$$
$$E = 5 \quad O = 0 \quad \phantom{\text{Check:}} \quad \underline{1\,085}$$
$$N = 6 \quad R = 8 \quad \phantom{\text{Check:}} \quad 10\,652$$
$$D = 7 \quad Y = 2$$

Solve the alphametic

$$\frac{SEVEN}{TWO} = TWO$$

in which

```
           TWO
      _____
 TWO ) SEVEN
       BOB
       ___
       JOE
       OVV
       ___
       VESN
       VESN
```

Systems of Linear Equations

Outcomes

When you have completed this chapter, you will be able to

1. Properly plot a linear equation.
2. Obtain a solution for a system of two linear equations using graphical means.
3. Obtain a solution for a system of two linear equations using algebraic addition.
4. Obtain a solution for a system of two linear equations using the substitution method.
5. Solve applications using the methods for linear equations.
6. Solve a system of three linear equations using algebraic methods.
7. Solve a system of three linear equations using determinants.

Sometimes a physical situation can be described by a single equation. For example, the motion of a car being driven down a highway at 90 km/h can be described by the equation

$$d = 90t$$

where d is the distance (in kilometres) driven during time t (in hours). However, a more complicated situation may require two or more equations for a complete description. For example, if there is a *second* car, we need *two* equations to completely describe the motion of the two cars. If the second car travels the same distance as the first car, but at 100 km/h and in two hours less time, then

$$d = 90t \qquad \text{for the first car}$$

and also

$$d = 100(t - 2) \quad \text{for the second car}$$

In this chapter we learn how to solve systems of two or more linear equations using several different methods. We show graphical and algebraic solutions. We also include applications for systems of linear equations, and solve systems involving fractional and literal equations.

19-1 GRAPHICAL METHOD

In this section we learn how to solve a system of two linear equations graphically.

Linear Equations

A *linear equation* is an equation in which the terms containing the variable(s) are all of first degree. Recall from Sec. 7-1 that the degree of a term is the power to which the variable in the term is raised. A linear equation in two variables can be written as

| LINEAR EQUATION IN TWO VARIABLES | $Ax + By = C$ | 134 |

where A, B, and C are constants, and A and B cannot both be zero. In a linear equation, no variable has an exponent other than 1.

Example 1
The equation

$$x + y = 8$$

is a linear equation in two variables, in which $A = 1$, $B = 1$, and $C = 8$ in Eq. 134.

Suppose we had to *solve* the linear equation in Example 1. We need to find all values of the variables that make the equation true. If $x = 1$, for example, then $y = 7$ would make the equation true. Denoting this as an ordered pair, we say that (1, 7) is *a solution* of the equation. Similarly, $x = 2$ and $y = 6$ would also make the equation true, so (2, 6) is a solution, and so forth. In fact, there are an *infinite* number of solutions.

Graph of a Linear Equation

We saw in Chapter 12 that these solutions can be displayed graphically, by plotting the ordered pairs as points on the coordinate plane. As you might expect, the graph of a linear equation in two variables is a *line*.

Example 2
Graph the equation:

$$x + y = 8$$

Solution: To graph the equation, we'll plot some points and connect them. To make a table of values, we solve the equation for y, so that we can easily calculate the y values from the x values:

$$y = -x + 8$$

Note that this puts the equation into the slope-intercept form of a linear function. Since the graph is a line, we will plot three points. We already have two points from the discussion of Example 1, so we make a table and calculate a third point.

x	y
1	7
2	6
3	5

We plot these points and see that the graph is a line (Fig. 19-1).

FIGURE 19-1
Graph of a linear equation.

System of Equations

If two or more equations that contain the same variables are considered together, they are called a *system of equations*. (Sometimes they are referred to as *simultaneous* equations.)

Example 3
The equations

$$x + y = 8$$
$$x - y = 4$$

are a system of two linear equations in two variables.

Solution of a System of Equations

The *solution* of a system of equations is the set of *all* values of the variables that make every equation in the system true.

Example 4
The solution of the system in Example 3 is the set of all ordered pairs (x, y) which make *both* equations true. The ordered pair (6, 2) is a solution, since the sum, $x + y$, of the numbers is 8 and the difference, $x - y$, is 4:

$$6 + 2 = 8 \quad \text{and} \quad 6 - 2 = 4$$

In fact, the ordered pair (6, 2) is the only solution, since they are the only numbers whose sum is 8 and difference is 4.

Example 5
Determine if the ordered pair $(5, -4)$ is a solution to this system:

$$2x + y = 6$$
$$3x + 4y = 4$$

Solution: We must substitute the ordered pair $(5, -4)$ into both equations. Letting $x = 5$ and $y = -4$ in the first equation, we have

$$2(5) + (-4) \stackrel{?}{=} 6$$
$$6 = 6$$

Thus the pair $(5, -4)$ satisfies the first equation. Now we substitute the values in the second equation:

$$3(5) + 4(-4) \stackrel{?}{=} 4$$
$$-1 \neq 4$$

The substitution fails in the second equation, so $(5, -4)$ is not a solution to the system.

COMMON ERROR When checking, be sure to substitute the apparent solution in *both* equations.

Graphical Solution of a System

To solve a system of two linear equations graphically, graph both equations on the same coordinate axes and find the point of intersection. The solution is given by the coordinates of the point of intersection. Any graphical solution is approximate and *must be checked in both equations*.

Example 6

Solve this system graphically and check.

$$2x - 3y = 12$$
$$x + y = 1$$

Solution: To graph the equations, we make a table of values for each equation. We solve each equation for y, and calculate at least three ordered pairs.

$$2x - 3y = 12 \qquad\qquad x + y = 1$$
$$y = \frac{2x - 12}{3} \qquad\qquad y = 1 - x$$

x	y
0	−4
3	−2
6	0

x	y
0	1
1	0
2	−1

We plot the points and draw the lines (Fig. 19-2). It looks like the lines intersect at about the point (3, −2). The apparent solution is (3, −2), but these values must be checked.

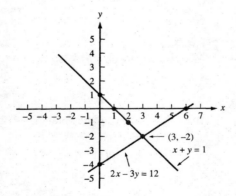

■ FIGURE 19-2

Check:
We substitute $x = 3$ and $y = -2$ into *both* of the original equations:

$$2(3) - 3(-2) \stackrel{?}{=} 12 \qquad\qquad 3 + -2 \stackrel{?}{=} 1$$
$$6 + 6 = 12 \qquad\qquad\qquad 1 = 1 \quad \text{checks}$$
$$12 = 12 \quad \text{checks}$$

The values check in both equations and therefore the ordered pair (3, −2) is the solution.

Again, a graphical solution to any problem is *approximate,* and must always be checked. If the graphical approximation happens to be exact, as in Example 6 above, the check will also be exact. Sometimes, however, the coordinates of the point of intersection are not integers, and must be approximated, say, to the nearest tenth. If the values we check are not exact, then they will not satisfy the equations exactly. The approximated values will be satisfactory, if, in the check, both sides of the equations are approximately equal.

Example 7

Solve this system graphically. Approximate the solution to the nearest tenth and check:

$$3x - y = 8$$
$$2x - 5y = 15$$

Solution: In this example, we graph the lines from the slope-intercept form of each equation. We solve each equation for y:

$$3x - y = 8 \qquad 2x - 5y = 15$$
$$y = 3x - 8 \qquad y = \frac{2}{5}x - 3$$

For the first equation, we graph a line whose slope is 3 and whose y intercept is -8. For the second, we graph a line whose slope is $\frac{2}{5}$ and whose y intercept is -3. From the graph (Fig. 19-3), we approximate the solution to be $(1.9, -2.2)$.

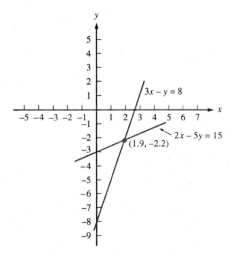

■ **FIGURE 19-3**

Check:

We check in the original equations:

$$3(1.9) - (-2.2) \stackrel{?}{=} 8 \qquad 2(1.9) - 5(-2.2) \stackrel{?}{=} 15$$
$$7.9 \approx 8 \quad \text{checks} \qquad 14.8 \approx 15 \quad \text{checks}$$

The sides of each equation are approximately equal, so the solution $(1.9, -2.2)$ is a good approximation.

■ FIGURE 19-4

Example 8

Solve this system graphically, and check:

$$x - y = 2$$
$$-3x + 3y = 9$$

Solution: We rewrite the equations in slope-intercept form:

$$y = x - 2$$
$$y = x + 3$$

and graph the equations in Fig. 19-4. We see that the lines are parallel, and there is *no solution*.

Example 9

Solve this system graphically, and check:

$$-2x + y = 5$$
$$4x - 2y = -10$$

Solution: We rewrite the equations in slope-intercept form:

$$y = 2x + 5$$
$$y = 2x + 5$$

and note that the two equations are the same. So the graphs of the equations (Fig. 19-5) are really the same line. Since all the points on the line are solutions of the system, the *solution is infinite,* and consists of

all ordered pairs (x, y) such that $-2x + y = 5$

■ FIGURE 19-5

Graphing Calculator Solution

It's very simple to solve a system of equations using a graphing calculator. Just graph the two equations on the same screen. If the point of intersection is not on the screen, zoom out or change the window until the point of intersection is on the screen. Now use the "intersect" feature to find the coordinates of the point of intersection.

Example 10

Use a graphing calculator to solve the system:

$$1.24x - 1.65y = -2.39$$
$$3.06x + 2.47y = -2.59$$

accurate to the nearest hundredth.

Solution: We solve each equation for y, so that we can enter the function in the calculator.

$$y = \frac{1.24x + 2.39}{1.65}$$

$$y = \frac{-3.06x - 2.59}{2.47}$$

We use the range $-5 \leq x \leq 5$ and $-3 \leq y \leq 3$. We graph both equations on the same screen and use the "intersect" feature to find the point of intersection (Fig. 19-6). We see that the coordinates are $(-1.25, 0.51)$, rounded to hundredths. Thus the solution is $x = -1.25$ and $y = 0.51$, rounded to hundredths.

■ FIGURE 19-6

SECTION 19-1 Exercises Graphical Method

Solution of a System of Equations

Determine if the given ordered pair is a solution to the system of equations.

1. $x + 2y = -7$
   $5x - y = 9$ $(1, -4)$
2. $2x - y = 5$
   $x - 3y = 5$ $(2, -1)$
3. $x - 2y = -3$
   $3x + y = 5$ $(-1, 1)$
4. $x + y = 3$
   $2x - 3y = 11$ $(2, 1)$
5. $2x - 3y = -5$
   $x + 6y = 2$ $(-1.6, 0.6)$
6. $2x - 5y = 8$
   $6x + y = 0$ $(0.25, -1.5)$

For each pair of lines in Fig. 19-7, name the coordinates of the point of intersection. If the coordinates are not integers, round to the nearest tenth.

7. a and b
8. a and c
9. a and d
10. b and c
11. b and d
12. c and d

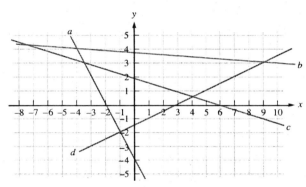

■ FIGURE 19-7

Graphical Solution of a System

Solve these systems graphically. If the solution is an ordered pair, approximate the solution to the nearest tenth and check it.

13. $x = -2$
    $y = 1$
14. $x = 3$
    $y = -2$
15. $2x - y = -5$
    $x + y = 2$
16. $3x - 2y = -4$
    $x + 2y = -4$
17. $2x + 5y = 18$
    $3x + 4y = 7$
18. $x + y = 3$
    $4x - 6y = 7$
19. $x - 2y = 6$
    $2x - 4y = 12$
20. $2x + y = -3$
    $-4x - 2y = 6$
21. $2x - 3y = 3$
    $x - 2y = -2$
22. $x - 2y = 4$
    $3x + 2y = 4$
23. $9x - 8y = 1$
    $6x + 12y = 5$
24. $4x + y = 11$
    $2x + 3y = 4$
25. $2x - 5y = 10$
    $-4x + 10y = 30$
26. $x - y = 6$
    $2x - 2y = 4$
27. $5x + 2y = 3$
    $3x - 3y = 4$
28. $5x - y = 7$
    $2x + y = 4$
29. $x + y = 2$
    $3x - y = 10$
30. $2x + y = 6$
    $3x + 4y = 4$

31. $2x - 5y = -2$
$-4x + 10y = 4$

32. $4x - 3y = -2$
$12x - 9y = 1$

33. $6.25x - 5.56y = 14.0$
$4.18x + 1.04y = 8.56$

34. $5.35x + 1.46y = 10.5$
$5.48x + 3.27y = 15.7$

35. $2x + y = -3$
$6x + 3y = 2$

36. $-2x + 3y = 4$
$4x - 6y = -8$

37. $1.67x - 0.76y - 3.18 = 0$
$2.56x + 3.82y - 11.5 = 0$

38. $2.45x + 1.73y + 1.95 = 0$
$2.38x - 1.25y - 3.45 = 0$

39. $0.2x + 0.1y - 0.1 = 0$
$-0.3x + 0.1y - 0.1 = 0$

40. $3.1x + 1.2y + 3.9 = 0$
$1.8x - 3.0y + 1.1 = 0$

Graphing Calculator Solution

Use a graphing calculator to solve the system, rounding to hundredths.

41. $x - y = 1$
$x + y = 5$

42. $x - y = 3$
$2x + 3y = 11$

43. $-2x + y = -3$
$4x - 2y = 2$

44. $-2x + 3y = 4$
$6x - 5y = 1$

45. $5.24x - 3.45y = 7.82$
$2.14x + 1.58y = 4.36$

46. $2.74x + 1.69y = -2.45$
$-6.55x - 2.33y = 4.50$

47. $2.36x + 3.47y = 4.27$
$8.46x - 12.0y = -16.7$

48. $5.36x + 1.49y = 10.4$
$9.63x + 1.03y = -20.4$

19-2   ADDITION METHOD

In the last section we learned how to approximate the solution to a system of two linear equations using a graphical method. But when the problem requires an *exact* solution, we may need to use an algebraic method to solve the system.

Algebraic Solution by the Addition Method

With any algebraic method of solving a system of equations in two variables, we try to *eliminate* one of the variables, so that we have a single equation in one variable. In the *addition method,* we eliminate one of the variables by adding the two equations together; that is, we add the left sides together, add the right sides together, and get a new equation. The new equation is in one variable and we can solve for that variable. We find the other variable by substituting back into one of the original equations.

Example 11

Solve by the addition method, and check:

$$2x - y = 3$$
$$3x + y = 2$$

Solution: If we add the two equations together, we eliminate the y terms:

$$2x - y = 3$$
$$3x + y = 2$$

Adding the two equations, we get $5x + 0 = 5$

Solving for x,

$$x = 1$$

Substituting $x = 1$ into the first of the original equations,

$$2(1) - y = 3$$

Solving for y,

$$y = -1$$

So the solution is $x = 1$ and $y = -1$. We write the solution as the ordered pair $(1, -1)$.

Check:
We substitute the values into *both* original equations.

$$2(1) - (-1) \stackrel{?}{=} 3 \quad \text{and} \quad 3(1) + (-1) \stackrel{?}{=} 2$$
$$3 = 3 \quad \text{checks} \qquad\qquad 2 = 2 \quad \text{checks}$$

In Example 11, y was eliminated when we added the original equations, because the coefficients of y in the two equations (-1 and 1) were equal in absolute value and opposite in sign. Often this is not the case. If one coefficient of a variable is a *multiple* of the other coefficient of that variable, then we may multiply each side of one equation by a factor that will make the coefficients in both equations equal in absolute value and opposite in sign.

Example 12

Solve by the addition method, and check:

$$2x + 3y = 6$$
$$x + 2y = 5$$

Solution: No pair of coefficients are equal in absolute value and opposite in sign, but one coefficient of x is a multiple of the other coefficient of x. If we multiply each side of the second equation by -2, then the x coefficients will be equal in absolute value and opposite in sign:

$$-2(x + 2y) = -2(5)$$
$$-2x - 4y = -10$$

So the new system becomes

$$2x + 3y = 6$$
$$-2x - 4y = -10$$

Adding, we get

$$-y = -4$$

Solving for y,

$$y = 4$$

Substituting $y = 4$ into the original second equation,

$$x + 2(4) = 5$$

Solving for x,

$$x = -3$$

So the solution is $(-3, 4)$.

Check:
As in Example 11, we check by substituting $x = -3$ and $y = 4$ into both equations.

$$2(-3) + 3(4) \stackrel{?}{=} 6 \qquad -3 + 2(4) \stackrel{?}{=} 5$$
$$6 = 6 \qquad\qquad\qquad 5 = 5 \quad \text{checks}$$

> **COMMON ERROR** Remember that the solution of a system of two linear equations is an ordered *pair* of numbers. After you have solved for one variable, be sure to solve for the other.

If one coefficient of a variable is not a multiple of the other coefficient of that variable, then we may have to multiply each side of *both* equations by factors that will make the coefficients of one variable opposites.

Example 13

Solve by the addition method, and check:

$$3x + 2y = 6$$
$$8x - 3y = 1$$

Solution: We choose to eliminate y, because the y coefficients are smaller than the x coefficients and they are already opposite in sign. To eliminate y, we multiply the first equation by 3 and the second by 2:

$$9x + 6y = 18$$
$$16x - 6y = 2$$

Adding, we get
$$25x = 20$$

Solving for x,

$$x = \frac{20}{25} = \frac{4}{5}$$

Substituting into the first equation,

$$3\left(\frac{4}{5}\right) + 2y = 6$$

$$2y = 6 - \frac{12}{5} = \frac{18}{5}$$

$$y = \frac{9}{5}$$

So the solution is $\left(\frac{4}{5}, \frac{9}{5}\right)$.

Check:
We substitute the solution into both equations:

$$3\left(\frac{4}{5}\right) + 2\left(\frac{9}{5}\right) \stackrel{?}{=} 6 \qquad 8\left(\frac{4}{5}\right) - 3\left(\frac{9}{5}\right) \stackrel{?}{=} 1$$

$$6 = 6 \qquad\qquad\qquad 1 = 1 \quad \text{checks}$$

Example 14

Solve, round to three significant digits, and check:

$$-3.26x + 2.04y = 15.2$$
$$5.75x + 6.42y = 3.97$$

Solution: Since it doesn't matter which variable we choose to eliminate, we choose x because its coefficients have opposite signs. We eliminate x by multiplying the first equation by 5.75 and the second equation by 3.26. We carry four (or more) significant digits and round the final answer to three significant digits at the end.

$$-18.75x + 11.73y = 87.40$$
$$18.75x + 20.93y = 12.94$$

Adding, $\quad\quad\quad\quad\quad\quad\quad 32.66y = 100.34$

$$y = 3.072$$

Substituting $y = 3.072$ into the first equation,

$$-3.26x + 2.04(3.072) = 15.2$$

Solving for x,

$$x = -2.740$$

We round to get the solution $x = -2.74$ and $y = 3.07$, and express the solution as the ordered pair $(-2.74, 3.07)$.

Check:

$$-3.26(-2.74) + 2.04(3.07) \stackrel{?}{=} 15.2 \quad\quad 5.75(-2.74) + 6.42(3.07) \stackrel{?}{=} 3.97$$
$$15.2 = 15.2 \quad\quad\quad\quad\quad\quad\quad\quad\quad 3.95 \approx 3.97 \quad \text{checks}$$

Note: Since we're dealing with rounded numbers, there's some uncertainty in the last digit.

Systems with No Solution

When solving certain systems algebraically, both variables may be eliminated. If we get an equation that is *never* true, like

$$0 = 2$$

then there is no solution. In the graph of such systems, the lines are parallel and there is no point of intersection.

Example 15

Solve by the addition method:

$$2x + y = 2$$
$$-6x - 3y = 4$$

Solution: To eliminate x, we multiply each side of the first equation by 3:

$$6x + 3y = 6$$
Adding the second equation, $\quad -6x - 3y = 4$
we get $\quad\quad\quad\quad\quad\quad\quad\quad\quad 0 = 10$

This equation is never true, and there is no solution.

Systems with Infinite Solutions

When solving certain systems algebraically, we may get an equation that is *always true*, like

$$0 = 0 \quad \text{or} \quad 2 = 2$$

In this case, there are an infinite number of solutions. In the graph of such systems, the two lines are the same, and there are an infinite number of points on the line.

Example 16

Solve by the addition method:

$$2x + 3y = 4$$
$$-6x - 9y = -12$$

Solution: To eliminate x, we multiply the first equation by 3:

$$6x + 9y = 12$$

Adding the second equation, $\quad -6x - 9y = -12$

$$0 = 0$$

This equation is always true, so the system has an infinite number of solutions. We express the solution as:

all ordered pairs (x, y) such that $\quad 2x + 3y = 4$

Solution of a System of Linear Equations by the Addition Method

This procedure may be followed to solve a system of two linear equations in two variables using the addition method:

1. If necessary, multiply each side of one or both equations by a factor, or factors, so that the coefficients of one variable are equal in absolute value and opposite in sign.
2. Add the equations, to eliminate that variable.
3. Solve the resulting equation for the remaining variable.
4. Substitute the solution for the variable found in Step 3 into an equation and solve for the remaining variable.
5. Check by substituting the solution into both original equations.

SECTION 19-2 Exercises Addition Method

Algebraic Solution by the Addition Method

Solve by the addition method, and check.

1. $x + y = 7$
   $x - y = 3$
2. $x + y = 9$
   $4x - y = 6$
3. $2x - 3y = 5$
   $4x + 3y = 7$
4. $4x + 3y = -9$
   $-4x + y = 3$
5. $-3x + 2y = 9$
   $2x + y = 1$
6. $3x - 2y = -8$
   $-x + 4y = 6$
7. $3x + y = 8$
   $x - 5y = 0$
8. $3x + 5y = 4$
   $x - y = -6$
9. $2x + 3y = 5$
   $-3x + 2y = 12$
10. $8x + 3y = 1$
    $6x + 2y = 0$
11. $5.2x + 2.3y = 3.8$
    $3.8x - 3.4y = 4.7$
12. $-5.6x + 2.8y = -4.9$
    $3.7x + 3.5y = 5.8$

13. $12x - 3y = 8$
    $4x = y + 1$
14. $2x = 3y + 6$
    $14x - 21y = -3$
15. $8x = 4y + 12$
    $2x - y = 3$
16. $x - 3y = 5$
    $2x = 6y + 10$
17. $2x + 3y = 13$
    $-3x + 2y = 0$
18. $3x - 2y = 19$
    $5x + 3y = 19$
19. $x + 5 = 3x - 2y$
    $4 - x = 2x + 4y$
20. $9 + (x - y) = 3x$
    $2(x + y) = 6x - 18 + 5y$
21. $y - 10 = -3x$
    $6x - 5 = -2x$
22. $3y = 9 + 6x$
    $2x = y - 3$
23. $1.25x - 2.24y = 1.43$
    $7.54x + 5.72y = 50.4$
24. $1.53x - 0.82y = 2.75$
    $0.97x - 1.19y = -1.42$
25. $23.6x - 37.1y = -43.7$
    $12.7x + 18.2y = 31.8$
26. $54.1x - 36.7y = 190$
    $75.8x + 41.7y = 24.1$
27. $\frac{1}{3}x - \frac{1}{2}y = -\frac{5}{6}$
    $\frac{1}{2}x + \frac{1}{3}y = 2$
28. $\frac{1}{2}x - \frac{3}{4}y = 6$
    $-\frac{2}{3}x + \frac{1}{4}y = 2$
29. $2ax + by = 3$
    $ax + 3by = 2$
30. $2x - 3y = a$
    $3x + y = b$
31. $ax + by = c$
    $dx + ey = f$

19-3 SUBSTITUTION METHOD

A second method of algebraically solving two linear equations in two variables is the *substitution method*.

Algebraic Solution by the Substitution Method

First solve either equation for one variable in terms of the other variable. Substitute this expression into the other equation to eliminate the first variable. Finish as with the addition method by solving the resulting equation for the remaining variable. Substitute that value into one of the equations and solve for the remaining variable.

Example 17

Solve by the substitution method, and check:

$$3x - 2y = 1$$
$$2x + y = 10$$

Solution: We may choose to solve either equation for either variable. Since the coefficient of y in the second equation is 1, this is the best choice:

$$y = 10 - 2x$$

We substitute the expression for y in the other (first) equation:

$$3x - 2(10 - 2x) = 1$$

Solving this equation for x:

$$3x - 20 + 4x = 1$$
$$7x = 21$$
$$x = 3$$

Now we solve for y by substituting $x = 3$ into either equation. Since y has already been solved for in the second equation, we use the equation:

$$y = 10 - 2x = 10 - 2(3)$$
$$y = 4$$

So the solution is (3, 4).

Check:

$$3(3) - 2(4) \stackrel{?}{=} 1 \qquad 2(3) + 4 \stackrel{?}{=} 10$$
$$1 = 1 \qquad\qquad 10 = 10 \quad \text{checks}$$

In general, we wish to avoid substituting a fractional expression. However, it is sometimes impossible to do so.

Example 18

Solve by substitution, and check:

$$5x + 3y = 1$$
$$3x + 4y = -6$$

Solution: In this system, no variable can be solved for without introducing a fractional expression. We choose to solve for x in the first equation:

$$x = \frac{1 - 3y}{5}$$

We substitute the right side of this equation for x into the second equation:

$$3\left(\frac{1 - 3y}{5}\right) + 4y = -6$$

To eliminate the fraction, we multiply both sides of the equation by the LCD, 5:

$$3(1 - 3y) + 20y = -30$$

Solving for y:

$$3 - 9y + 20y = -30$$
$$11y = -33$$
$$y = -3$$

Substituting $y = -3$ into the equation for x:

$$x = \frac{1 - 3y}{5} = \frac{1 - 3(-3)}{5} = \frac{10}{5}$$
$$x = 2$$

So the solution is (2, −3).

Check:

$$5(2) + 3(-3) \stackrel{?}{=} 1 \qquad 3(2) + 4(-3) \stackrel{?}{=} -6$$
$$1 = 1 \qquad\qquad\qquad -6 = -6 \quad \text{checks}$$

Solution of a System of Linear Equations by the Substitution Method

This procedure may be followed to solve a system of two linear equations in two variables using the substitution method:

1. Solve either equation for one variable in terms of the other variable.
2. Substitute this expression into the other equation.
3. Solve the resulting equation for the remaining variable.
4. Substitute the solution for the variable found in Step 3 into an equation, and solve for the remaining variable.
5. Check by substituting the solution into both original equations.

A Choice of Methods

When the method of solution for a system of linear equations is not specified, you may choose among the three methods: graphical, addition, or substitution. The addition method may be used easily with any system. The substitution method is easiest when the coefficient of one of the variables is 1 or -1. Graphical solutions are approximate and can be used when exact answers are not needed or to verify algebraic results.

SECTION 19-3 Exercises Substitution Method

Algebraic Solution by the Substitution Method

Solve by the substitution method, and check.

1. $4x - 3y = 8$
   $y = 4$
2. $x - 5 = 0$
   $3x - 2y = 1$
3. $3x - y = 10$
   $x + y = 2$
4. $-2x + 5y = 22$
   $2x + y = -10$
5. $x - y = 3$
   $2x + 3y = 11$
6. $2x + y = 2$
   $6x - y = 22$
7. $5x - 8y = 6$
   $x + 5y = -3$
8. $2x - y = 3$
   $2x + y = -1$
9. $-3x + 4y = 12$
   $3x + 5y = 15$
10. $5x - 3y = 21$
    $6x + 5y = 8$
11. $2x - 3y = 4$
    $3x - 2y = -2$
12. $2x - 3y = 3$
    $3x - 2y = -3$
13. $4x - 3y = 11$
    $5x + 4y = 6$
14. $5x - 9y = 12$
    $7x - 5y = 13$
15. $3x + 2y = -1$
    $-6x - 4y = 2$
16. $5y - 2 = 2x$
    $4x = 10y - 4$
17. $3x - 6y = 1$
    $5x - 10y = 2$
18. $3x - y = -4$
    $-3x + y = 5$
19. $y = 2x + 1$
    $y = -x + 3$
20. $y = -3x + 2$
    $y = 2x - 1$

21. $2x - y = 3a$
    $x - 2y = -2b$

22. $rx + sy = t$
    $2rx - sy = 2t$

23. $0.9x - 0.8y = 0.1$
    $0.6x + 1.2y = 0.5$

24. $1.5x - 2y = -1$
    $3x + 2.5y = 3.5$

25. $\dfrac{x}{2} + 2y = 22$
    $x - \dfrac{y}{2} = 8$

26. $\dfrac{2x}{3} + y = 18$
    $x - \dfrac{y}{4} = 6$

A Choice of Methods

Solve by any convenient method and check.

27. $y = 3x - 2$
    $y = -2x + 3$

28. $3x + 4y = 1$
    $x + 4y = -5$

29. $5x - 3y = -13$
    $x + 4y = 2$

30. $y = \dfrac{1}{3}x - 1$
    $y = -\dfrac{1}{3}x - 3$

31. $5x + 7y = 1$
    $4x - 2y = 16$

32. $3x - 2y = 7$
    $3x + 5y = 14$

Solve for x and y, by making the substitutions:

$$m = \frac{1}{x} \quad \text{and} \quad n = \frac{1}{y}$$

33. $\dfrac{12}{x} + \dfrac{6}{y} = 5$
    $\dfrac{4}{x} + \dfrac{3}{y} = 2$

34. $\dfrac{1}{x} + \dfrac{1}{y} = \dfrac{5}{6}$
    $\dfrac{1}{x} - \dfrac{1}{y} = \dfrac{13}{6}$

19-4 APPLICATIONS OF SYSTEMS OF TWO EQUATIONS

In many applications, there are two or more unknown values. To solve such problems, we must write *as many equations as there are variables.* Otherwise, it is not possible to obtain a unique numerical solution.

In general, we use the same process we used to solve applications involving one equation, except that we need two equations, since we have two variables.

Solution of Applications Involving Systems of Two Equations in Two Variables

1. Understand the written problem: Read it carefully once for a general overview and again for the details. Draw a sketch if possible.
2. Identify the unknown quantities with variables.
3. Determine any formulas you need to solve the problem. Make a table if possible, headed by the formula.
4. Write two equations relating the two variables.
5. Solve the system.
6. Answer the question(s) asked, and label your answers with the proper units.
7. Check your answer(s) in the words of the problem.

Example 19

A technician ordered 7 transistors and 4 resistors, for a total price of $3.15. The next week, she ordered 5 transistors and 6 resistors, for a total price of $3.35. Find the price of each.

Solution: We first identify and label the unknown quantities:

Let t = the price of a transistor
and r = the price of a resistor
and total prices are converted to cents.

We need to use the relationship, which we know from everyday experience, that

$$(\text{No. of items}) \times (\text{price/item}) = \text{cost}$$

We use the first sentence of the problem statement to write the first equation:

$$7t + 4r = 315$$

We use the second sentence in the problem to write the second equation:

$$5t + 6r = 335$$

We now have a system of two equations in two variables. We choose to solve this system using the addition method. To eliminate r, we multiply both sides of the first equation by 3, and we multiply both sides of the second equation by -2:

$$\begin{array}{r} 21t + 12r = 945 \\ -10t - 12r = -670 \end{array}$$

Adding, $\qquad 11t = 275$

Solving for t,

$$t = 25$$

Substituting $t = 25$ into the first equation,

$$7(25) + 4r = 315$$

Solving for r,

$$r = 35$$

So each transistor was 25¢ and each resistor was 35¢.

Check:
We check back in the words of the problem.

Seven transistors at 25¢ each comes to	$7(\$.25) = \$1.75$
and four resistors at 35¢ each comes to	$4(\$.35) = \$1.40$
for a total of	$\$3.15$ checks
Five transistors at 25¢ each comes to	$5(\$.25) = \$1.25$
and six resistors at 35¢ each comes to	$6(\$.35) = \$2.10$
for a total of	$\$3.35$ checks

You will notice that some problems in this section are similar to problems presented in Chapters 11 and 18. In these problems there are two unknown quantities, but earlier we were able to use one equation in one variable by expressing the second variable in terms of the first. It may be more natural for you to solve these problems using two equations in two variables.

19-4 Applications of Systems of Two Equations

Example 20

A solution that is 34% silver nitrate is to be mixed with a solution that is 4% silver nitrate to form 100 mL of solution that is 7% silver nitrate. How much of each is needed?

Solution:

Let x = amount of the 34% solution (in mL)
and y = amount of the 4% solution (in mL)

We use a form of Eq. 16:

% concentration of A × amount of solution = amount of A in the solution

A table is very helpful:

	% of silver nitrate (as a decimal) ×	Amount of solution (mL) =	Amount of silver nitrate (mL)
	0.34	x	$0.34x$
	0.04	y	$0.04y$
New solution	0.07	100	$0.07(100)$

We calculate the third column by multiplying the first and second columns. Then we use columns two and three to write a system of two equations:

$$x + y = 100$$
$$0.34x + 0.04y = 0.07(100)$$

We choose to solve this system using the substitution method.

Solving for y,

$$y = 100 - x$$

Substituting into the second,

$$0.34x + 0.04(100 - x) = 0.07(100)$$

To clear the equation of decimals, we multiply both sides by 100,

$$34x + 4(100 - x) = 7(100)$$

Solving for x,

$$34x + 400 - 4x = 700$$
$$30x = 300$$
$$x = 10$$

Substituting $x = 10$ into the first equation,

$$10 + y = 100$$
$$y = 90$$

So we need to mix 10 mL of the 34% solution with 90 mL of the 4% solution to make 100 mL of a 7% solution.

Check:

We check back in the words of the problem:

We have 100 mL since

$$10 + 90 = 100 \quad \text{checks}$$

We have a 7% solution since

$$(34\% \text{ of } 10) + (4\% \text{ of } 90) \stackrel{?}{=} 7\% \text{ of } 100$$

$$3.4 + 3.6 = 7 \quad \text{checks}$$

Example 21

A motorboat travels downstream for a distance of 42.6 kilometres in 2.75 hours and makes the return trip in 3.25 hours. Find the speed of the motorboat in still water and the speed of the current in the river. Round to the nearest tenth of a km/h.

Solution:

Let v = the speed of the motorboat in still water
and c = the speed of the current in the river

Then,

$$\text{Downstream speed} = v + c$$

$$\text{Upstream speed} = v - c$$

Using the equation

$$\text{speed} \times \text{time} = \text{Distance}$$

Once again we make a table:

	Speed (km/h)	×	Time (h)	=	Distance (km)
Downstream	$v + c$		2.75		$(v + c)\,2.75$
Upstream	$v - c$		3.25		$(v - c)\,3.25$

We multiply the first and second columns to find the third column. Now, using the given fact that the distance downstream is 42.6 km and the distance upstream is 42.6 km, we get a system of two equations:

$$(v + c)\,2.75 = 42.6$$
$$(v - c)\,3.25 = 42.6$$

We simplify the equations by dividing both sides of the first equation by 2.75 and both sides of the second by 3.25:

$$v + c = 15.49$$
$$v - c = 13.11$$

We can eliminate c by adding the equations,

$$2v = 28.60$$

Solving for v,

$$v = 14.30$$

Substituting to find c,

$$14.30 + c = 15.49$$
$$c = 15.49 - 14.30 = 1.19$$

Rounding to tenths, $v = 14.3$ and $c = 1.2$. So the speed of the motorboat is 14.3 km/h in still water and the speed of the current is 1.2 km/h.

Check:
Checking in the words of the problem:

Going downstream, the speed is

$$14.3 + 1.2 = 15.5 \text{ km/h}$$

After 2.75 hours, the distance traveled is

$$d = rt = (15.5)(2.75) = 42.6 \text{ km} \quad \text{checks}$$

Going upstream, the speed is

$$14.3 - 1.2 = 13.1 \text{ km/h}$$

After 3.25 hours, the distance traveled is

$$d = 13.1(3.25) = 42.6 \text{ km} \quad \text{checks}$$

SECTION 19-4 Exercises Applications of Systems of Two Equations

Write two equations in two variables, solve, and check. Round approximate answers to three significant digits.

1. The sum of the number of teeth on two gears is 73, and their difference is 17. How many teeth are on each gear?

2. A salesman sells TVs and stereos. One day he sold 3 TVs and 4 stereos for $2530. The next day he sold 4 TVs and 3 stereos for $2510. What are the prices of a TV and a stereo?

3. While wiring a new house, an electrician and her assistant worked 25.0 and 17.0 hours, respectively, and earned $671. In wiring a second house, the electrician and her assistant worked 21.0 and 13.0 hours, respectively, and earned $547. What were the hourly wages of the electrician and the assistant?

4. An airplane on a search mission flies due east from an airport, turns and flies due west back to the airport. The plane cruises at 320 km/h when flying east and 400 km/h when flying west. What is the farthest point from the airport the plane can reach if it can remain in the air for 9.00 hours?

5. A canoeist paddles downstream for a distance of 20.5 km in 4.25 hours and returns in 6.25 hours. Find the rate of the river current.

6. A fish swims 19.3 km downstream in 2.45 hours and returns in 6.75 hours. How fast does it swim in still water?

7. An airplane travels 2990 kilometres in 2.00 hours with the wind and then returns in 2.50 hours traveling against the wind. Find the speed of the airplane with respect to the air and the velocity of the wind.

8. A cyclist rode 9.25 kilometres in 15.0 minutes with the wind, and returned in 20.0 minutes against the same wind. Find her speed (in km/h) without a wind.

9. The velocity v of a uniformly accelerated particle is linearly related to the elapsed time t by the equation

$$v = v_0 + at$$

where v_0 is the initial velocity and a is the acceleration. In one trial, the velocity of a particle was 7.01 m/s after 5.00 seconds, and in another trial, the velocity was 9.25 m/s after 10.0 seconds. Find the initial velocity and the acceleration.

10. An automobile company is conducting tests on a new model of car. Starting from rest, the car accelerates at a constant rate for 11.0 s to a certain velocity. In a second test, starting at 18.0 km/h, the car accelerates at the same constant rate and reaches that certain velocity in only 8.00 s. Find that velocity and the constant acceleration (in m/s^2). (Refer to Exercise 9 if necessary.)

11. A metallurgist wishes to make 1000 kg of an alloy that is 75% copper, using a 78% copper alloy and a 63% copper alloy. How much of each does he need?

12. A creamery has a vat of milk with 2.5% butterfat, and a vat of milk with 30% butterfat. How much of each is needed to make 100 litres of milk with 3.6% butterfat?

13. Water must be added to a solution of 75% sulfuric acid in water to make 150 mL of a 20% solution of sulfuric acid in water. How much of the 75% solution should be used and how much water should be added?

14. Water must be added to a solution of 80% nitric acid to make 840 mL of a 2.0% solution. How much of the 80% solution should be used and how much water should be added?

15. A vat contains 440 litres of gasohol (gasoline and alcohol) which is 11% alcohol. How much gasohol must be drained from the vat and replaced with gasoline to produce a gasohol mixture which is 6.5% alcohol?

16. A certain brass alloy contains 60% copper, 34% zinc, and 6% lead. How many kilograms of this brass alloy and of zinc must be added to 20 kg of lead to produce a new brass alloy that is 40% zinc and 8% lead?

17. A man has $8000 invested, part at 6.0% and the remainder at 10.5%. If his total income from these two investments is $678 per year, how much has he invested at each rate?

18. A sum of $20 000 is invested, part at 8.5% and part at 7.5%. If the total interest from these two investments is $1620 per year, how much is invested at each rate?

19. A woman invested $4200, part of it in bonds bearing 9.0% interest and the remainder in stocks bearing 12.0% in dividends. If she received the same income from each investment, how much was invested in each?

20. A sum of $10 000 is invested, part at 7.0% and part at 10%. The annual income from the 7.0% investment is $65 less than from the 10% investment. How much was invested in each?

21. A carpenter and a helper can do a certain job in 12 days. If the carpenter works twice as fast as the helper, how long would it take each to do the job, working alone?

22. One person can mow a lawn twice as fast as another. Together they can mow the lawn in 6.0 hours. How long would it take the slower person to mow the lawn alone?

23. During one week two machines produce a total of 180 parts, with the faster machine working for 30 hours and the slower for 20 hours. During another week, they produce 495 parts, with the faster machine working 60 hours and the slower working for 85 hours. How many parts can each produce in an hour working alone?

24. A wind generator operates for 18 hours one day and a hydro unit operates for 20 hours; together they produce 860 kW·h. Another day the wind generator operated for 12 hours and the hydro unit operated for 15 hours; together they produced 615 kW·h that day. Find the number of kW produced by each in one hour.

 25. A piece of pipe is 21.0 ft long. It must be cut so that one piece is 3.00 ft longer than the other. What are the lengths of the two pieces of pipe?

26. A man takes a trip of 360 miles, part of the way by car at an average of 48.0 mi/h, and the rest of the way by train at an average of 64.0 mi/h. If the total travel time is 6.00 hours, how far does he travel by each mode of travel?

27. A student, traveling 655 miles on his way back to school, takes a train at an average of 70.0 mi/h part of the way, and a bus, at an average of 55.0 mi/h for the rest of the way. If the total trip was 10.0 hours, how many hours and how many miles did he travel by each method of transportation?

28. A ski lift carries a skier up a slope at the rate of 120 ft/min and she skis down from the top to the bottom on a path parallel to the lift at an average rate of 1080 ft/min. If the round trip traveling time is 20.0 minutes, how many minutes does it take her to ski down? How long is the lift?

19-5 SYSTEMS OF THREE EQUATIONS

Some applications need a system of three equations in three variables to describe them. A linear equation in three variables is an equation in three variables in which no variable is of a degree greater than 1.

Example 22

The equation

$$x + 2y + 3z = 6$$

is a linear equation in three variables, since no variable has an exponent other than 1.

Example 23

The equations

$$x - y + z = 6$$
$$x + y + 2z = 5$$
$$2x + y - z = -3$$

are a *system* of three linear equations in three variables.

Algebraic Solution

To solve a system of three equations in three variables, we first eliminate one variable and get a system of two equations in two variables. Then we eliminate a variable from *that* system to get one equation in one variable. We solve that equation for the variable, and substitute back to find the second variable. Finally we substitute both these values into one of the preceding equations and solve for the last variable. You should always check your solution in all three equations.

FIGURE 19-8

Solution of a System of Three Linear Equations in Three Variables

1. Choose a variable to be eliminated, and eliminate it from any pair of equations.
2. Eliminate the same variable from another pair of equations.
3. Solve this system of two equations in two variables.
4. Solve for the third variable by substituting these values into one of the preceding equations.
5. Check by substituting the solution into all three original equations.

A flowchart for the process is given in Fig. 19-8.

---- **Example 24** ----

Solve and check:

$$x - y + z = 6 \tag{1}$$
$$x + y + 2z = 5 \tag{2}$$
$$2x + y - z = -3 \tag{3}$$

Solution: We choose to eliminate y, since the coefficients of y in all three equations are 1 and -1. We eliminate y from equations (1) and (2) by adding them together:

$$\begin{array}{r} x - y + z = 6 \\ x + y + 2z = 5 \\ \hline 2x \quad\quad + 3z = 11 \end{array} \tag{4}$$

We eliminate y from equations (2) and (3) by multiplying both sides of the second equation by -1, and adding the equations:

$$\begin{array}{r} -x - y - 2z = -5 \\ 2x + y - z = -3 \\ \hline x \quad\quad - 3z = -8 \end{array} \tag{5}$$

Now we solve the system consisting of two new equations (4) and (5) in x and z:

$$2x + 3z = 11$$
$$x - 3z = -8$$

We solve this system by adding the equations to get:

$$3x = 3$$

Solving for x:

$$x = 1$$

We substitute $x = 1$ into equation (4):

$$2(1) + 3z = 11$$

Solving for z:

$$z = 3$$

We substitute $x = 1$ and $z = 3$ into equation (1):

$$1 - y + 3 = 6$$

Solving for y:

$$y = -2$$

So the solution is $(1, -2, 3)$.

Check:
We substitute back in all three original equations:

$$1 - (-2) + 3 \stackrel{?}{=} 6 \qquad 1 + (-2) + 2(3) \stackrel{?}{=} 5 \qquad 2(1) + (-2) - 3 \stackrel{?}{=} -3$$
$$6 = 6 \text{ checks} \qquad 5 = 5 \text{ checks} \qquad -3 = -3 \text{ checks}$$

Applications

Some applications require three equations in three variables.

Example 25

Solutions containing 25%, 50%, and 60% sulfuric acid are mixed to produce 100 mL of a 45% solution of sulfuric acid (Fig. 19-9). If four times as much of the 25% solution is used as the 50% solution, how much of each is used?

FIGURE 19-9

Solution:

Let x = number of mL of the 25% solution
y = number of mL of the 50% solution
z = number of mL of the 60% solution

The equations come from three parts of the problem statement.

1. There is a total of 100 mL:

$$x + y + z = 100$$

2. The sum of the acid in the parts is the same as the acid in the final solution:

$$0.25x + 0.50y + 0.60z = 0.45(100)$$

or simplifying,

$$25x + 50y + 60z = 4500$$

3. There is four times as much 25% solution as 50% solution:

$$x = 4y$$

So the system of three equations in three unknowns is:

$$x + y + z = 100 \qquad (1)$$
$$25x + 50y + 60z = 4500 \qquad (2)$$
$$x = 4y \qquad (3)$$

To solve, first we use (3) to substitute $4y$ for x in (1):

$$4y + y + z = 100$$

Simplifying:
$$5y + z = 100 \qquad (4)$$

Now substitute $4y$ for x in (2):
$$25(4y) + 50y + 60z = 4500$$

Simplifying:
$$150y + 60z = 4500 \qquad (5)$$

We now have a system of two equations in two unknowns, (4) and (5).

Solving for z in (4), we get
$$z = 100 - 5y \qquad (6)$$

Substituting in (5), we get
$$150y + 60(100 - 5y) = 4500$$

Solving,
$$y = 10$$

Substituting in (3),
$$x = 4(10) = 40$$

Substituting in (6),
$$z = 100 - 5(10) = 50$$

So we need to use 40 mL of the 25% solution, 10 mL of the 50% solution, and 50 mL of the 60% solution to make 100 mL of the 45% solution.

Check:
Checking back in the words of the problem:

We "produce 100 mL":
$$40 + 10 + 50 = 100$$

There is "four times as much of the 25% solution" as the 50%:
$$4(10) = 40$$

It makes a "45% solution of sulfuric acid":
$$25\%(40) + 50\%(10) + 60\%(50) \stackrel{?}{=} 45$$
$$= 45\%(100) \quad \text{checks}$$

SECTION 19-5 Exercises Systems of Three Equations

Solve and check.

1. $x + y + z = 6$
   $x - y + z = 2$
   $2x - y + 3z = 6$

2. $2x + y + z = 13$
   $x + 2y + z = 11$
   $x + 3y + 3z = 19$

3. $x + y - z = 4$
   $2x - 3y + z = 1$
   $x - 4y - 2z = -7$

4. $2x - 3y + 3z = 7$
   $3x + y - 2z = -11$
   $5x - 2y + 4z = 11$

5. $x + 2y - z = 2$
   $3x - y + 3z = -2$
   $-2x - 4y + 2z = -1$

6. $3x + y - z = -1$
   $2x - 4y + 2z = 1$
   $-x + 2y - z = -3$

7. $x + y - 3z = 7$
   $x + y + 4z = 0$
   $5x + 2z = 8$

8. $x - y = -6$
   $x + z = -9$
   $3y + z = 1$

9. $x + 3z = -2$
   $2x + y = 2$
   $7y + 4z = -4$

10. $x - y = 1$
    $y + z = 3$
    $x - z = 2$

11. $y + z = 2$
    $x - z = -1$
    $x - y = 3$

12. $x - z = 1$
    $x + y + z = 2$
    $x + y = 1$

13. $x - y + 4z = 1$
    $2x + y + 6z = 3$
    $3x + 2y - 2z = 2$

14. $x - 2y - 3z = -4$
    $2x + 3y + 4z = 8$
    $3x + 4y - 5z = 16$

Applications

■ FIGURE 19-10

Write three equations in three variables, solve, and check.

15. Three solutions contains 20%, 30%, and 40% nitric acid are mixed together to produce 100 mL of 25% solution. Twice as much of the 40% solution is used as the 30% solution. How many mL of each are used?

16. The three resistors in Fig. 19-10 have a total resistance of 2740 ohms. The second resistance is 200 more than the first resistance, and the third resistance is 100 less than twice the second resistance. Find the resistances.

17. In a certain electric circuit, the total of three voltages in series is 340 volts. The first voltage is 25 volts less than twice the second. The third is equal to the sum of the other two. Find each voltage.

18. A woman wants to invest $16 000, part in a chequing account, part in a money market account, and the rest in a certificate of deposit. She wants twice as much in the certificate of deposit as in her chequing account. She also wants $1000 more in the money market account than in the certificate of deposit. How much should be placed in each account?

19. A man has $8000 invested in three investments. On the first, he receives 8% interest, on the second, 9%, and the third, 12%. He has twice as much invested at 9% than at 8%, and receives $860 annually in interest. How much does he have in each of the three investments?

20. An assembly of a nut, a bolt, and a washer has a mass of 80 g. Two nuts and one bolt have a combined mass of 90 g. Three fastener sets each made up of a nut, a bolt, and double washers have a combined mass of 270 g. Find the mass of the individual nut, bolt, and washer.

21. A dietician is planning a meal using three foods. Each serving of food A contains 10% of the daily requirements for carbohydrates, 5% for protein, and 0% for sodium. Each serving of food B contains 10% of the daily requirements for carbohydrates, 10% for protein, and 15% for sodium. Each serving of food C contains 10% of the daily requirement for carbohydrates, 25% for protein, and 10% for sodium. How many grams of each food should be used to supply 100% of the daily requirements for each nutrient? A serving is made up of 25 g of food.

22. An athlete enters a triathlon and runs at the rate of 12 km/h, canoes at 3 km/h, and bikes at 24 km/h. She finishes the race in 3 hours and 40 minutes. Her friend who is doing the triathlon just for fun, runs at 10 km/h, canoes at 2 km/h and bikes at 16 km/h, finishing the race in 5 hours and 6 minutes. How many kilometres was each part (running, canoeing, biking), if the total course was 51 kilometres long?

19-6 SOLUTION OF SYSTEMS OF TWO LINEAR EQUATIONS BY DETERMINANTS

We have seen that systems of linear equations may be solved graphically or algebraically, using the addition method or the substitution method. But systems may also be solved using *determinants*. In this section we introduce determinants and learn how to evaluate them. Then we use determinants to solve a system of two linear equations.

To begin, our goal is to solve a general system of two linear equations:

$$a_1x + b_1y = c_1$$
$$a_2x + b_2y = c_2$$

Using the addition method, we multiply the first equation by b_2 and the second by $-b_1$:

$$a_1b_2x + b_1b_2y = b_2c_1$$
$$-a_2b_1x - b_1b_2y = -b_1c_2$$

Adding: $(a_1b_2 - a_2b_1)x = b_2c_1 - b_1c_2$

Dividing by $a_1b_2 - a_2b_1$ gives

$$x = \frac{b_2c_1 - b_1c_2}{a_1b_2 - a_2b_1}$$

Similarly, we can solve for y although we don't show the work for this. These results can be summarized as follows:

SOLUTION OF TWO LINEAR EQUATIONS		
$a_1x + b_1y = c_1$ $a_2x + b_2y = c_2$	$x = \dfrac{b_2c_1 - b_1c_2}{a_1b_2 - a_2b_1}$ $y = \dfrac{a_1c_2 - a_2c_1}{a_1b_2 - a_2b_1}$ where $a_1b_2 - a_2b_1 \neq 0$	52

These equations can be used to solve any system of two linear equations.

Definitions

The denominator, $a_1b_2 - a_2b_1$, of the fractions above can be expressed using a special arrangement of the coefficients a_1, b_1, a_2, and b_2 called a *determinant*. It is denoted by the symbol

$$\begin{vmatrix} a_1 & b_1 \\ a_2 & b_2 \end{vmatrix}$$

Each number in the determinant is called an *element*. This determinant has two *rows* and two *columns* and is called a *second-order determinant*. Every second-order determinant has two diagonals. The diagonal from the upper left to lower right is called the *principal diagonal*. The diagonal from the lower left to the upper right is called the *secondary diagonal*.

principal diagonal

secondary diagonal

Value of a Second-Order Determinant

The *value* of a second-order determinant is equal to the product of the elements on the principal diagonal minus the product of the elements on the secondary diagonal.

SECOND-ORDER DETERMINANT	$\begin{vmatrix} a_1 & b_1 \\ a_2 & b_2 \end{vmatrix} = a_1 b_2 - a_2 b_1$	54

Example 26

In the determinant $\begin{vmatrix} 2 & -3 \\ 1 & 5 \end{vmatrix}$ each of the numbers 2, 1, −3, and 5 are the elements.

The first row contains the elements 2 and −3; the second row contains the elements 1 and 5. The first column contains the elements 2 and 1; the second column contains the elements −3 and 5. The value of this determinant is equal to the product of the elements on the principal diagonal, 2(5), minus the product of the elements on the secondary diagonal, 1(−3).

$$\begin{vmatrix} 2 & -3 \\ 1 & 5 \end{vmatrix} = 2(5) - 1(-3) = 10 + 3 = 13$$

COMMON ERRORS Students sometimes *add* the numbers on a diagonal instead of *multiplying*. Don't do that.

Don't just ignore a zero element. It causes the product along its diagonal to be zero.

Example 27

$$\begin{vmatrix} 0 & 4 \\ 2 & -3 \end{vmatrix} = 0(-3) - 2(4) = -8$$

Solving a System of Two Equations by Determinants

We now return to the solution of the general system of two linear equations:

$$a_1 x + b_1 y = c_1$$
$$a_2 x + b_2 y = c_2$$

Recall that the solution for x is

$$x = \frac{b_2 c_1 - b_1 c_2}{a_1 b_2 - a_2 b_1}$$

We can replace the denominator in this expression by the determinant of the coefficients:

$$\begin{vmatrix} a_1 & b_1 \\ a_2 & b_2 \end{vmatrix} = a_1 b_2 - a_2 b_1$$

Now look at the numerator $b_2 c_1 - b_1 c_2$. This can be expressed as a determinant also:

$$\begin{vmatrix} c_1 & b_1 \\ c_2 & b_2 \end{vmatrix} = c_1 b_2 - c_2 b_1$$
$$= b_2 c_1 - b_1 c_2$$

So the numerator can also be expressed as a determinant. Note that the determinant for the numerator is the same as the determinant of the coefficients (the denominator) except that the column of x coefficients, a_1 and a_2, have been replaced by the column of constants, c_1 and c_2. So our solution for x can be expressed as

$$x = \frac{\begin{vmatrix} c_1 & b_1 \\ c_2 & b_2 \end{vmatrix}}{\begin{vmatrix} a_1 & b_1 \\ a_2 & b_2 \end{vmatrix}}$$

Similarly, we can develop an expression for y. We don't show the work for this, but we find that

$$y = \frac{\begin{vmatrix} a_1 & c_1 \\ a_2 & c_2 \end{vmatrix}}{\begin{vmatrix} a_1 & b_1 \\ a_2 & b_2 \end{vmatrix}}$$

We can now express the solution of a system of two linear equations in terms of determinants. This is called Cramer's Rule, after the Swiss mathematician Gabriel Cramer (1704–1752).

| CRAMER'S RULE | $x = \dfrac{\begin{vmatrix} c_1 & b_1 \\ c_2 & b_2 \end{vmatrix}}{\begin{vmatrix} a_1 & b_1 \\ a_2 & b_2 \end{vmatrix}}$ and $y = \dfrac{\begin{vmatrix} a_1 & c_1 \\ a_2 & c_2 \end{vmatrix}}{\begin{vmatrix} a_1 & b_1 \\ a_2 & b_2 \end{vmatrix}}$ where $\begin{vmatrix} a_1 & b_1 \\ a_2 & b_2 \end{vmatrix} \neq 0$ | 56 |

Thus, the solution for any variable is a fraction whose denominator is the determinant of the coefficients, and whose numerator is that same determinant with the column of coefficients for the variable replaced by the column of constants.

Of course the determinant of the coefficients cannot be zero or we would get division by zero. This would indicate that the system has no unique solution.

Example 28

Solve by determinants:

$$3x - 2y = 4$$
$$x + 3y = 2$$

Solution: Using Eq. 56 for x and y, we get

$$x = \frac{\begin{vmatrix} 4 & -2 \\ 2 & 3 \end{vmatrix}}{\begin{vmatrix} 3 & -2 \\ 1 & 3 \end{vmatrix}} = \frac{4(3) - 2(-2)}{3(3) - 1(-2)} = \frac{16}{11}$$

and

$$y = \frac{\begin{vmatrix} 3 & 4 \\ 1 & 2 \end{vmatrix}}{\begin{vmatrix} 3 & -2 \\ 1 & 3 \end{vmatrix}} = \frac{3(2) - 1(4)}{11} = \frac{2}{11}$$

Note that we don't need to calculate the denominator a second time—we already know it's 11.

Check:

$$3\left(\frac{16}{11}\right) - 2\left(\frac{2}{11}\right) \stackrel{?}{=} 4 \qquad \frac{16}{11} + 3\left(\frac{2}{11}\right) \stackrel{?}{=} 2$$

$$\frac{48}{11} - \frac{4}{11} \stackrel{?}{=} 4 \qquad \frac{16}{11} + \frac{6}{11} \stackrel{?}{=} 2$$

$$\frac{44}{11} \stackrel{?}{=} 4 \qquad \frac{22}{11} \stackrel{?}{=} 2$$

$$4 = 4 \quad \text{checks} \qquad 2 = 2 \quad \text{checks}$$

> **COMMON ERROR** Be sure to arrange the equations into the form
>
> $a_1x + b_1y = c_1$
>
> $a_2x + b_2y = c_2$
>
> before writing the determinants.

Calculator Solution

Many calculators can also evaluate determinants. Check your calculator manual for specific instructions. On many calculators, you first enter the *dimensions* of the determinant (the dimensions are the number of rows and columns). Then enter the elements of the determinant, and press the appropriate key to evaluate the determinant.

──── **Example 29** ────

Use a calculator to solve this system to three significant digits.

$$2.37x - 3.54y - 4.49 = 0$$
$$3.16x + 2.34 = 2.75y$$

Solution: First we rewrite the equations so they are in the correct form, with the x and y terms on the left of the equal sign and the constants on the right:

$$2.37x - 3.54y = 4.49$$
$$3.16x - 2.75y = -2.34$$

Then we enter the determinant of the coefficients in the calculator (Fig. 19-11), and evaluate the determinant (Fig. 19-12). Now we find the numerator for x by entering the constants in the first column, and evaluating that determinant by calculator.

$$\begin{vmatrix} 4.49 & -3.54 \\ -2.34 & -2.75 \end{vmatrix} = -20.6311$$

By Cramer's Rule,

$$x = \frac{-20.6311}{4.6689} = -4.42$$

Now we find the numerator for y by entering the determinant,

$$\begin{vmatrix} 2.37 & 4.49 \\ 3.16 & -2.34 \end{vmatrix} = -19.7342$$

■ **FIGURE 19-11**

```
MATRIX[A] 2 × 2
[ 2.37   -3.54      ]
[ 3.16   -2.75      ]
```

■ **FIGURE 19-12**

```
det [A]
              4.6689
```

Once again, by Cramer's Rule

$$y = \frac{-19.7342}{4.6689} = -4.23$$

rounded to three significant digits. So the solution is $(-4.42, -4.23)$.

SECTION 19-6 Exercises Solution of Systems of Two Linear Equations by Determinants

Value of a Second-Order Determinant

Evaluate each determinant.

1. $\begin{vmatrix} 2 & 1 \\ 3 & 4 \end{vmatrix}$
2. $\begin{vmatrix} 3 & -4 \\ -1 & 7 \end{vmatrix}$
3. $\begin{vmatrix} -3 & -2 \\ -5 & -3 \end{vmatrix}$
4. $\begin{vmatrix} -2 & 1 \\ -3 & 2 \end{vmatrix}$
5. $\begin{vmatrix} 2.4 & 3.7 \\ -5.3 & 1.5 \end{vmatrix}$
6. $\begin{vmatrix} 5.67 & -3.21 \\ -3.49 & -1.38 \end{vmatrix}$
7. $\begin{vmatrix} 2.43 & 0 \\ -1.25 & -3.34 \end{vmatrix}$
8. $\begin{vmatrix} 0 & 1.67 \\ 0 & 3.93 \end{vmatrix}$

Solving a System of Two Equations by Determinants

Solve each system using determinants. Check.

9. $2x - y = 9$
   $3x + y = 11$
10. $4x - 3y = 6$
    $2x - 3y = 12$
11. $5x - 3y = 7$
    $2x - 4y = 9$
12. $x - 6y = -20$
    $2x - 7y = -17$
13. $9x - y = 7$
    $5x - 3y = -11$
14. $x - 4y = -17$
    $2x - 3y = -9$
15. $5x + 2y - 20 = 0$
    $3x - 2y - 12 = 0$
16. $4 - x + 3y = 0$
    $1 + 2x + 3y = 0$
17. $y = 3 - x$
    $y = 11 - 3x$
18. $x = 2y + 3$
    $2x = 9 + 3y$

Calculator Solution

Solve each system using determinants with a calculator. Round to three significant digits.

19. $2.45x - 5.62y = 5.71$
    $3.62x - 4.49y = 12.6$
20. $5.34x = 20.5 + 4.67y$
    $2.87y = 6.37 - 3.21x$
21. $1.25x + 3.58y = 5.41$
    $6.32x - 2.90y = 1.66$
22. $4.65x - 7.75y = 9.65$
    $6.37x - 8.61y = 1.58$

Applications

Solve by determinants. Check.

23. In a certain electric circuit, two resistors, R_1 and R_2, have a total resistance of 87 ohms when connected in series. Find the resistance of the two resistors if R_1 has a resistance of 19 ohms less than R_2.

24. The sum of two capacitors is 65 microfarads and the difference between them is 29 microfarads. What is the size of each capacitor?

25. How many kilograms of candy that sells for $1.90/kg are needed to be mixed with another candy that sell for $1.50/kg in order to produce 16 kg of a mix that will sell for $1.80/kg?

26. A tea merchant prepares a blend of 12 kilograms of tea to sell for $4.30 per kilogram. For the blend, he uses two types of tea, one selling at $3.90 per kilogram, the other at $4.50 per kilogram. How much of each type should he use for the blend?

27. A dietician in a nursing home is to plan a special diet composed of two foods, A and B. Each serving of food A contains 6 units of calcium and 2 units of iron. Each serving of food B contains 5 units of calcium and 3 units of iron. How many servings of foods A and B should be used to obtain a food mix that has 52 units of calcium and 24 units of iron?

28. A certain zoo animal is kept under a strict diet. He is to receive, among other things, 40 grams of protein and 12 grams of fat. The zookeeper is able to purchase two food mixes of the following compositions: Mix M has 10% protein and 6% fat; mix N has 20% protein and 2% fat. How many grams of each mix should be used to obtain the right diet for this animal?

19-7 SOLUTION OF SYSTEMS OF THREE LINEAR EQUATIONS BY DETERMINANTS

In the last section we solved a system of two linear equations using determinants. In this section we solve a system of three linear equations using determinants.

Definitions

If we algebraically solved the system of equations:

$$a_1x + b_1y + c_1z = k_1$$
$$a_2x + b_2y + c_2z = k_2$$
$$a_3x + b_3y + c_3z = k_3$$

we would get the solution:

SOLUTION OF A SYSTEM OF THREE EQUATIONS

$$x = \frac{b_2c_3k_1 + b_1c_2k_3 + b_3c_1k_2 - b_2c_1k_3 - b_3c_2k_1 - b_1c_3k_2}{a_1b_2c_3 + a_3b_1c_2 + a_2b_3c_1 - a_3b_2c_1 - a_1b_3c_2 - a_2b_1c_3}$$

$$y = \frac{a_1c_3k_2 + a_3c_2k_1 + a_2c_1k_3 - a_3c_1k_2 - a_1c_2k_3 - a_2c_3k_1}{a_1b_2c_3 + a_3b_1c_2 + a_2b_3c_1 - a_3b_2c_1 - a_1b_3c_2 - a_2b_1c_3}$$

$$z = \frac{a_1b_2k_3 + a_3b_1k_2 + a_2b_3k_1 - a_3b_2k_1 - a_1b_3k_2 - a_2b_1k_3}{a_1b_2c_3 + a_3b_1c_2 + a_2b_3c_1 - a_3b_2c_1 - a_1b_3c_2 - a_2b_1c_3}$$

where

$$a_1b_2c_3 + a_3b_1c_2 + a_2b_3c_1 - a_3b_2c_1 - a_1b_3c_2 - a_2b_1c_3 \neq 0$$

53

Notice that the three denominators are identical.

The denominator

$$a_1b_2c_3 + a_3b_1c_2 + a_2b_3c_1 - a_3b_2c_1 - a_1b_3c_2 - a_2b_1c_3$$

can be expressed as the *determinant of the coefficients* of the three given equations, just as in Sec. 19-6. It is commonly expressed as

$$\begin{vmatrix} a_1 & b_1 & c_1 \\ a_2 & b_2 & c_2 \\ a_3 & b_3 & c_3 \end{vmatrix}$$

This determinant has three rows and three columns and is called a *third-order determinant*. The principal diagonal consists of the elements a_1, b_2, and c_3. The secondary diagonal consists of the elements a_3, b_2 and c_1.

Value of a Third-Order Determinant

An easy way to evaluate a third-order diagonal is to use its diagonals. Write the determinant and copy the first and second columns to the right of the determinant.

Now there are two more diagonals parallel to the principal diagonal to its right. There are also two more diagonals parallel to the secondary diagonal to its right.

To find the value of the determinant, first add the products of the elements of the principal diagonal and the products of the two parallel diagonals. Then from the sum, subtract the products of the elements of the secondary diagonal and the products of the two parallel diagonals.

| THIRD-ORDER DETERMINANT | $\begin{vmatrix} a_1 & b_1 & c_1 \\ a_2 & b_2 & c_2 \\ a_3 & b_3 & c_3 \end{vmatrix} = a_1b_2c_3 + a_3b_1c_2 + a_2b_3c_1 \\ - a_3b_2c_1 - a_1b_3c_2 - a_2b_1c_3$ | 55 |

Example 30

Evaluate the determinant

$$\begin{vmatrix} 1 & -2 & 3 \\ -4 & 5 & 6 \\ 7 & 8 & -9 \end{vmatrix}$$

Solution: We rewrite the determinant with the first two columns written to the right, and draw all six diagonals.

$$\begin{vmatrix} 1 & -2 & 3 \\ -4 & 5 & 6 \\ 7 & 8 & -9 \end{vmatrix} \begin{matrix} 1 & -2 \\ -4 & 5 \\ 7 & 8 \end{matrix}$$

Adding the products of the principal diagonals and subtracting the products of the secondary diagonals, we get a value of

$$(1)(5)(-9) + (-2)(6)(7) + (3)(-4)(8) - (7)(5)(3) - (8)(6)(1) - (-9)(-4)(-2)$$
$$= -45 - 84 - 96 - 105 - 48 + 72$$
$$= -306$$

COMMON ERROR Larger determinants, 4 × 4 and larger, cannot be evaluated by the method outlined in example 30. An expansion method for large determinants is included in more advanced textbooks.

Determinants by Calculator Solution

This determinant can also be evaluated by using a calculator. Check your calculator manual for specific instructions.

```
MATRIX[A] 3 × 3
[  1    -2     3    ]
[ -4     5     6    ]
[  7     8    -9    ]
```

FIGURE 19-13

```
det [A]
                -306
```

FIGURE 19-14

Example 31

Evaluate the determinant in Example 30 using a calculator.

Solution: As before, we enter the *dimensions* of the determinant (this is a 3-by-3 determinant). Then enter the elements of the determinant (Fig. 19-13), and evaluate the determinant (Fig. 19-14).

Solving a System of Three Equations by Determinants

We can solve a system of three linear equations by determinants just as we solved a system of two linear equations by determinants. We return to the equations for the solutions of the general system in Eq. 53. The numerator of each solution can be obtained from the determinant of the coefficients by replacing the coefficients of the variable in question with the constants k_1, k_2, k_3. The solution by determinants now becomes Cramer's Rule.

CRAMER'S RULE

$$x = \frac{\begin{vmatrix} k_1 & b_1 & c_1 \\ k_2 & b_2 & c_2 \\ k_3 & b_3 & c_3 \end{vmatrix}}{\begin{vmatrix} a_1 & b_1 & c_1 \\ a_2 & b_2 & c_2 \\ a_3 & b_3 & c_3 \end{vmatrix}}$$

$$y = \frac{\begin{vmatrix} a_1 & k_1 & c_1 \\ a_2 & k_2 & c_2 \\ a_3 & k_3 & c_3 \end{vmatrix}}{\begin{vmatrix} a_1 & b_1 & c_1 \\ a_2 & b_2 & c_2 \\ a_3 & b_3 & c_3 \end{vmatrix}}$$

$$z = \frac{\begin{vmatrix} a_1 & b_1 & k_1 \\ a_2 & b_2 & k_2 \\ a_3 & b_3 & k_3 \end{vmatrix}}{\begin{vmatrix} a_1 & b_1 & c_1 \\ a_2 & b_2 & c_2 \\ a_3 & b_3 & c_3 \end{vmatrix}}$$

where $\begin{vmatrix} a_1 & b_1 & c_1 \\ a_2 & b_2 & c_2 \\ a_3 & b_3 & c_3 \end{vmatrix} \neq 0$

57

Example 32

Solve by determinants, and use a calculator to evaluate the determinants.

$$3x - 7y + 3z = 6$$
$$3x + 3y + 6z = 1$$
$$5x - 5y + 2z = 5$$

Solution: In the calculator, we enter the determinant of the coefficients and evaluate it.

By Cramer's Rule, the solution is

$$x = \frac{\begin{vmatrix} 6 & -7 & 3 \\ 1 & 3 & 6 \\ 5 & -5 & 2 \end{vmatrix}}{\begin{vmatrix} 3 & -7 & 3 \\ 3 & 3 & 6 \\ 5 & -5 & 2 \end{vmatrix}} = \frac{-40}{-150} = \frac{4}{15}$$

$$y = \frac{\begin{vmatrix} 3 & 6 & 3 \\ 3 & 1 & 6 \\ 5 & 5 & 2 \end{vmatrix}}{-150} = \frac{90}{-150} = -\frac{3}{5}$$

$$z = \frac{\begin{vmatrix} 3 & -7 & 6 \\ 3 & 3 & 1 \\ 5 & -5 & 5 \end{vmatrix}}{-150} = \frac{-50}{-150} = \frac{1}{3}$$

So the solution is $x = \frac{4}{15}$, $y = -\frac{3}{5}$, and $z = \frac{1}{3}$.

Check:
We use a calculator to check in all three equations.

SECTION 19-7 Exercises Solution of Systems of Three Linear Equations by Determinants

Value of a Third-Order Determinant

Evaluate each determinant.

1. $\begin{vmatrix} 1 & 2 & 2 \\ 3 & -1 & 1 \\ -2 & 4 & -1 \end{vmatrix}$

2. $\begin{vmatrix} -2 & 1 & 3 \\ 5 & -1 & 2 \\ -3 & 2 & 4 \end{vmatrix}$

3. $\begin{vmatrix} 0 & 1 & -2 \\ 1 & 0 & -1 \\ -1 & 2 & 0 \end{vmatrix}$

4. $\begin{vmatrix} 2 & 0 & -1 \\ 0 & 1 & 3 \\ -1 & -3 & -2 \end{vmatrix}$

Determinants by Calculator Solution

Evaluate by calculator. Round to tenths.

5. $\begin{vmatrix} 1.3 & 2.3 & -3.2 \\ -3.4 & 1.7 & -2.5 \\ 5.4 & -2.4 & 0.6 \end{vmatrix}$

6. $\begin{vmatrix} 4.5 & -3.1 & 2.7 \\ 1.9 & -2.7 & -8.5 \\ -3.2 & 0.7 & 1.5 \end{vmatrix}$

Solve each system using determinants with a calculator. Check.

7. $6x - 4y - 7z = 17$
    $9x - 7y - 16z = 29$
    $10x - 5y - 3z = 23$

8. $2x + 4y - 3z = 22$
    $4x - 2y + 5z = 18$
    $6x + 7y - z = 63$

Solve by determinants. Round to tenths.

9. $2.4x - 1.7y - 3.2z = 4.5$
    $1.3x - 5.4y + 3.8z = 6.4$
    $-3.4x + 4.9y - 2.6z = 3.7$

10. $-9.8x + 7.8y + 5.7z = 9.4$
    $8.4x - 4.9y - 6.5z = 0.7$
    $5.8x - 5.8y + 3.7z = 8.5$

Solving a System of Three Equations by Determinants

Solve each system using determinants.

11. $x + y + z = 18$
    $x - y + z = 6$
    $x + y - z = 4$

12. $x + y + z = 35$
    $x - 2y + 3z = 15$
    $x - y - z = 5$

13. $x + y = 25$
    $x + z = 30$
    $y + z = 35$

14. $x - y = 6$
    $x - z = 10$
    $y + z = 14$

15. $2x + 3y + z = 2$
    $-x + 2y + 3z = -1$
    $-3x - 3y + z = 0$

16. $2x - 2y + 3z = 5$
    $2x + y - 2z = -1$
    $4x - y - 3z = 0$

Solve by determinants.

Applications

17. The three resistors in Fig. 19-15 have a total resistance of 2600 ohms when connected in series. If the resistance of R_1 is 420 ohms more than 3 times R_2, and the resistance of R_3 is 480 ohms more than R_2, find each resistance.

18. Three solutions containing 4% acid, 28% acid, and 60% acid, are to be mixed to get 1000 mL of a new solution, which is 30% acid. If the new solution is to contain twice as much of the 4% solution as of the 28% solution, how much of each should be mixed together?

FIGURE 19-15

Chapter 19 REVIEW EXERCISES

1. Determine if $(2, -1)$ is a solution of the system
$$3x - y = 7$$
$$2x + 3y = -1$$

2. Determine if $\left(\dfrac{2}{13}, -\dfrac{14}{13}\right)$ is a solution of this system:
$$5x - 3y = 4$$
$$x + 2y = -2$$

Solve these systems graphically. If the solution is an ordered pair, estimate the solution to the nearest tenth and check it.

3. $3x + y = -3$
   $x - 3y = -1$

4. $2x + 5y = 18$
   $3x + 4y = 7$

5. $2x + 3y = 5$
   $4x + 6y = 7$

6. $2x - y = 3$
   $4x - 2y = 6$

Solve by the addition method, and check.

7. $x + y = 2$
   $3x - y = 10$

8. $3x + y = -3$
   $x - 3y = -1$

9. $3x - 4y = 2$
   $5x - 7y = -5$

10. $2x + 3y = 3$
    $5x - 4y = 1$

11. $y = 5 - 2x$
    $4x = 10 - 2y$

12. $6x = 3y - 9$
    $y = 2x + 4$

Solve by the substitution method, and check.

13. $2x + y = 13$
    $3x + y = 17$

14. $2x - y = 4$
    $x + y = 5$

15. $y + 2 = 4x$
    $8x - 2y = -1$

16. $3y - 2 = 4x$
    $12x - 9y = -6$

17. $2x + 3y = 6$
    $3x + 4y = 8$

18. $3x - 5y = 7$
    $5x - 7y = 8$

Evaluate the determinant.

19. $\begin{vmatrix} -4 & 2 \\ -3 & -2 \end{vmatrix}$ **20.** $\begin{vmatrix} -2.6 & -3.4 \\ 5.3 & 4.6 \end{vmatrix}$

Solve by determinants, and check.

21. $x + 2y = 2$
$3x - 2y = -10$

22. $y = x + 1$
$y = -2x + 4$

23. $5x - 2y = 4$
$10x - 4y = 8$

24. $4.3x - 3.2y = -6.0$
$1.5x + 3.8y = -9.7$

Solve by any convenient method, and check:

25. $3x - 2y = 4$
$2y = 3x - 6$

26. $2x + 7y = 1$
$2x + 5y = 3$

27. $2x - 5y = 12$
$2x - 3y = 12$

28. $3x + 4y = 4$
$x - 2y = 0$

Write two equations in two variables, solve, and check.

29. The sum of two voltages is 86 volts and the difference is 14 volts. Find the two voltages.

30. A dietician in a hospital is to plan a special diet composed of the following: Each serving of food A contains 9 units of calcium and 4 units of iron, each serving of food B contains 5 units of calcium and 4 units of iron, and the mix of A and B must have exactly 81 units of calcium and 48 units of iron.

31. An investor bought 40 hectares of land, part at $1100 per hectare and part at $1300 per hectare. If he paid a total of $49 000, how much land was there in each part?

32. A cyclist rode 96 kilometres in 3 hours with the wind, and returned in 4 hours against the same wind. Find her speed without the wind and the speed of the wind.

33. How much solder containing 30% tin and 70% lead must be combined with another solder containing 60% tin and 40% lead to make 120 kilograms of solder containing 50% tin and 50% lead?

34. How many mL of water must be added to a solution of 85% carbolic acid to make 850 mL of a 1.5% solution?

35. How many kilograms of tea costing $1.40 per kilogram should be mixed with tea costing $3.00 per kilogram to make 16 kg of tea that sells for $2.60 per kilogram?

Solve and check.

36. $2x + y - 3z = -2$
$x + 3y - 2z = -5$
$3x + 2y - z = 7$

37. $x + y = -1$
$y + z = -2$
$2x - z = 4$

Evaluate the determinant.

38. $\begin{vmatrix} 3 & -2 & 4 \\ 0 & 5 & -1 \\ 2 & 0 & -3 \end{vmatrix}$ **39.** $\begin{vmatrix} 3.5 & -6.5 & -3.8 \\ -1.6 & 2.9 & 3.7 \\ 0 & -4.8 & -3.9 \end{vmatrix}$

Solve by determinants, and round to tenths.

40. $1.2x + 2.3y + 3.4z = 9.9$
$7.8x - 2.3y - 3.5z = 1.6$
$4.7x + 3.8y - 7.2z = 0.4$

Write three equations in three variables, solve, and check.

41. A lab technician must mix three solutions of 10%, 30%, and 60% alcohol, to make 1000 mL of a 39% solution. If the new solution must contain three times as much of the 30% solution as of the 10% solution, how much of each must she mix?

WRITING

42. We have learned four methods of solving a system of linear equations. Describe each method, including its advantages and/or disadvantages. Make up a system of two linear equations and use it to illustrate all four methods of solution. Which do you prefer, and why?

TEAM PROJECT

43. A certain nickel silver alloy contains:

  55.90% copper   0.10% lead
  31.25% zinc   12.00% nickel
  0.50% tin   0.25% manganese

How many pounds of zinc, tin, lead, nickel, and manganese must be added to 400.0 kg of this alloy to make a new "leaded nickel silver" with the following composition:

  44.50% copper   1.00% lead
  42.00% zinc   10.00% nickel
  0.50% tin   2.00% manganese

Exponents and Radicals

Outcomes

When you have completed this chapter, you will be able to

1. Change radical expressions into a fractional exponent form of expression and back.
2. Simplify a radical.
3. Rationalize a fractional expression with a radical in the denominator.
4. Add and subtract radicals.
5. Multiply radicals.
6. Divide radicals.
7. Solve radical equations.

In this chapter we study fractional exponents, rules of radicals, simplification of radical expressions, addition, subtraction, multiplication, and division of radical expressions, and radical equations. Radical expressions need to be studied because they are solutions to many equations derived from applications.

20-1 FRACTIONAL EXPONENTS

In this section we define a quantity raised to a fractional exponent and express such a quantity in both exponential and radical form. We evaluate numerical expressions with fractional exponents. Finally, we extend the laws of integral exponents from Sec. 7-3 to fractional exponents so that we can simplify algebraic expressions with fractional exponents.

To refresh your memory, we restate the laws of integral exponents:

PRODUCT	$x^m x^n = x^{m+n}$	19
QUOTIENT	$\dfrac{x^m}{x^n} = x^{m-n}$	20
POWER	$(x^m)^n = x^{mn}$	21
PRODUCT RAISED TO A POWER	$(xy)^n = x^n y^n$	22
QUOTIENT RAISED TO A POWER	$\left(\dfrac{x}{y}\right)^n = \dfrac{x^n}{y^n}$	23

ZERO EXPONENT	$x^0 = 1$	24
NEGATIVE EXPONENT	$x^{-n} = \dfrac{1}{x^n}$	25

These laws, which we developed for integral exponents, also apply to fractional exponents, providing the bases, x and y, are positive real numbers. Therefore, in this chapter, *we will assume that variable bases represent positive real numbers.*

Radical and Exponential Forms

A radical expression can be written as a quantity raised to a fractional exponent. We make the following definition:

RADICAL AND EXPONENTIAL FORMS	$\sqrt[r]{x} = x^{1/r}$	26

The expression on the left is in *radical* form. The expression on the right is in *exponential* form.

Example 1

By Eq. 26,

$$\sqrt[3]{x} = x^{1/3}$$
$$\sqrt[4]{x} = x^{1/4}, \text{ and so forth}$$

Example 2

Express $\sqrt{x}$ in exponential form.

Solution: Recall that

$$\sqrt{x} = \sqrt[2]{x}$$

Then by Eq. 26,

$$\sqrt{x} = x^{1/2}$$

Example 3

Change $y^{-1/2}$ to radical form.

Solution: By Eqs. 25 and 26,

$$y^{-1/2} = \frac{1}{y^{1/2}} = \frac{1}{\sqrt{y}}$$

The negative sign in the exponent indicates a reciprocal. Even though the exponent is negative, the expression is not negative.

COMMON ERROR If a fractional exponent is negative, it does *not* mean that the expression is negative.

Example 4
Write $x^{2/3}$ in radical form.

Solution: By Eq. 21,
$$x^{2/3} = (x^{1/3})^2 = (\sqrt[3]{x})^2$$

Or looking at it another way:
$$x^{2/3} = (x^2)^{1/3} = \sqrt[3]{x^2}$$

We see that there are two equivalent ways of writing $x^{2/3}$:
$$x^{2/3} = (\sqrt[3]{x})^2 = \sqrt[3]{x^2}$$

In general,

FRACTIONAL EXPONENTS	$x^{p/r} = (\sqrt[r]{x})^p = \sqrt[r]{x^p}$	27

Example 5
Express $x^{3/4}$ in two equivalent radical forms.

Solution:
$$x^{3/4} = (\sqrt[4]{x})^3 = \sqrt[4]{x^3}$$

It follows from Eq. 27 that

RADICAL FORMS OF x	$\sqrt[r]{x^r} = x$ and $(\sqrt[r]{x})^r = x$	28

Example 6
$$\sqrt[3]{x^3} = x \quad \text{and} \quad (\sqrt[3]{x})^3 = x$$

Evaluation of Expressions

Numerical expressions with fractional exponents can be evaluated algebraically using the laws of exponents or numerically using a calculator. We show examples of both methods of evaluation. First we show how to evaluate an expression algebraically using the laws of exponents. This yields an exact answer, with irrational answers expressed in radical form.

Example 7
Evaluate $64^{1/2}$ algebraically.

Solution:
$$64^{1/2} = \sqrt{64} = \sqrt{8^2} = 8$$

Example 8

Evaluate $(-8)^{1/3}$ algebraically.

Solution:
$$(-8)^{1/3} = \sqrt[3]{-8} = \sqrt[3]{(-2)^3} = -2$$

This illustrates that an odd root of a negative number is always negative.

Example 9

$$100000^{1/5} = \sqrt[5]{100000} = \sqrt[5]{10^5} = 10$$

Example 10

$$8^{2/3} = (\sqrt[3]{8})^2 = 2^2 = 4$$

Although an expression like this, of the form $x^{p/r}$, can be evaluated using either $(\sqrt[r]{x})^p$ or $\sqrt[r]{x^p}$, it is generally easier to *take the root first* and then raise the result to the power, using $(\sqrt[r]{x})^p$.

Example 11

$$16^{-3/4} = \frac{1}{16^{3/4}} = \frac{1}{(\sqrt[4]{16})^3} = \frac{1}{2^3} = \frac{1}{8}$$

Evaluation by Calculator

Many numerical expressions with fractional exponents represent irrational numbers and these numbers can be approximated using a calculator. Powers are generally entered using the $\boxed{\wedge}$ key. Calculators give answers in decimal form, so an irrational answer can be expressed as a decimal approximation. (Some calculators also give answers in exact radical and fraction form.) When using a calculator, the use of parentheses around the fraction is very important.

Example 12

Express in radical form and approximate to three significant digits:
$$6^{2/3}$$

Solution: Changing to radical form,
$$6^{2/3} = \sqrt[3]{6^2} = \sqrt[3]{36}$$

We use a calculator to approximate the number,
$$6^{2/3} \approx 3.30$$

rounded to three significant digits.

> **COMMON ERROR** Remember to use parentheses around a fractional exponent when entering it into the calculator.

Laws of Exponents

As stated earlier in this section, the laws of exponents also apply to fractional exponents, provided that the base is a positive real number. The following examples illustrate these laws.

Example 13

Simplify $x^{1/2}x^{1/3}$.

Solution: By Eq. 19, we add the exponents:
$$x^{1/2}x^{1/3} = x^{1/2+1/3} = x^{5/6}$$

Example 14

Simplify $\dfrac{y^{1/2}}{y^{1/3}}$.

Solution: By Eq. 20, we subtract the exponents:
$$\frac{y^{1/2}}{y^{1/3}} = y^{1/2-1/3} = y^{1/6}$$

Example 15

Simplify $(x^{1/2})^{4/5}$.

Solution: By Eq. 21, we multiply the exponents:
$$(x^{1/2})^{4/5} = x^{4/10} = x^{2/5}$$

Example 16

Expand $(3y)^{1/4}$.

Solution: By Eq. 22, each factor is raised to the $\dfrac{1}{4}$ power:
$$(3y)^{1/4} = 3^{1/4}y^{1/4}$$

Example 17

Expand $\left(\dfrac{x}{6}\right)^{1/3}$.

Solution: By Eq. 23, both the numerator and denominator are raised to the 1/3 power.
$$\left(\frac{x}{6}\right)^{1/3} = \frac{x^{1/3}}{6^{1/3}}$$

SECTION 20-1 Exercises Fractional Exponents

Radical and Exponential Forms

Change to radical form. Express in two different ways if possible.

1. $x^{1/2}$
2. $y^{3/4}$
3. $y^{1/3}$
4. $x^{2/3}$
5. $a^{-1/2}$
6. $b^{-2/3}$
7. $(x-y)^{1/2}$
8. $(a+b)^{2/3}$
9. $(-c)^{2/3}$
10. $(-d)^{-3/5}$
11. $-c^{2/3}$
12. $-d^{-3/5}$

Change to exponential form.

13. $\sqrt{x}$
14. $\sqrt[3]{a}$
15. $\sqrt[4]{y}$
16. $\sqrt{z}$
17. $\sqrt[3]{b^2}$
18. $\sqrt[5]{x^3}$
19. $\dfrac{1}{\sqrt{y}}$
20. $\dfrac{1}{\sqrt[4]{x^3}}$
21. $\sqrt{x+y}$
22. $\sqrt[3]{a-b}$

Evaluation of Expressions

Evaluate algebraically, without using a calculator.

23. $16^{1/2}$
24. $49^{1/2}$
25. $64^{1/3}$
26. $125^{1/3}$
27. $(-27)^{1/3}$
28. $(-1)^{1/3}$
29. $1^{1/5}$
30. $(10000)^{1/4}$
31. $(0.01)^{1/2}$
32. $(0.001)^{1/3}$
33. $25^{3/2}$
34. $27^{2/3}$
35. $81^{-1/2}$
36. $8^{-1/3}$
37. $64^{-2/3}$
38. $100^{-3/2}$
39. $(-32)^{4/5}$
40. $-32^{4/5}$
41. $-81^{-3/4}$
42. $-(-216)^{-2/3}$

Evaluation by Calculator

Evaluate using a calculator, and round to three significant digits.

43. $9^{3/2}$
44. $16^{3/4}$
45. $(-8)^{2/3}$
46. $(-64)^{2/3}$
47. $27^{-2/3}$
48. $125^{-2/3}$
49. $-16^{3/2}$
50. $-81^{-3/4}$
51. $5^{1/2}$
52. $(-7)^{1/3}$
53. $(-6)^{1/5}$
54. $8^{1/6}$
55. $9^{2/3}$
56. $12^{3/4}$
57. $4^{-2/3}$
58. $3^{-3/2}$
59. $(-2)^{-3/4}$
60. $-2^{-3/4}$

Laws of Exponents

Simplify using the laws of exponents. Assume all variable bases are positive. Leave your answers in exponential form, with positive exponents.

61. $a^{1/3}a^{1/4}$
62. $y^{1/2}y^{2/3}$
63. $(b^{1/2})^{6/7}$
64. $(x^{2/3})^{1/2}$
65. $(4a^2)^{1/2}$
66. $(x^3y)^{2/3}$
67. $\dfrac{x^{2/3}}{x^{1/6}}$
68. $\dfrac{b^{1/2}}{b^{1/4}}$
69. $\left(\dfrac{x}{y^2}\right)^{3/2}$
70. $\left(\dfrac{a^2}{b}\right)^{1/2}$
71. $a^{-1/3}$
72. $c^{-2/3}$
73. $x \cdot x^{1/2}$
74. $a^{2/3} \cdot a$
75. $\dfrac{a}{a^{1/3}}$
76. $\dfrac{xy}{x^{2/3}}$
77. $\dfrac{a(a^{3/2})^2}{a^{2/3}}$
78. $\dfrac{(x^{-1/2})^3 x}{x^{1/2}}$

Applications

■ FIGURE 20-1

79. A gear in the shape of Fig. 20-1 can be described by the equation

$$x^{2/3} + y^{2/3} = a^{2/3}$$

Write this equation with radicals instead of fractional exponents.

80. The surface area on the face of a cube A_f varies with the volume V as shown in the following equation

$$A_f = V^{2/3}$$

Find the surface area of a cube with a volume of 357 911 cm³.

81. If 500 bacteria double their number every hour, the equation for the number of bacteria n after t hours is

$$n = (500)(2)^t$$

Evaluate the number of bacteria after $\dfrac{2}{5}$ of an hour.

82. A yeast manufacturer finds that the yeast will triple every hour. If the manufacturer starts with 100 pounds of yeast, the number of pounds of yeast n after t hours is

$$n = (100)(3)^t$$

Evaluate the number of pounds of yeast after $\dfrac{3}{4}$ hour.

20-2 SIMPLIFICATION OF RADICALS

In this section, we use two rules of radicals, the root of a product and the root of a quotient, to simplify radical expressions.

Root of a Product

The root of a product is equal to the product of the roots:

ROOT OF A PRODUCT	$\sqrt[r]{xy} = \sqrt[r]{x}\sqrt[r]{y}$	29

Example 18

It follows from Eq. 29 that
$$\sqrt{xy} = \sqrt{x}\sqrt{y}$$
and similarly,
$$\sqrt[3]{xy} = \sqrt[3]{x}\sqrt[3]{y}, \text{ and so forth.}$$

Simplest Form of a Radical Expression

A radical expression is said to be *in simplest form* when

1. The radicand has been made as small as possible, and
2. There are no radicals in the denominator and no fractional radicands.

We now use the rule for the root of a product (Eq. 29) to simplify radical expressions. First we factor the radicand into factors that are rth powers, and then use Eq. 28 to eliminate these factors from the radicand. For example, to simplify a *square* root, we factor perfect *squares* from the radicand, apply the rule, and simplify.

Example 19

Write $\sqrt{18}$ in simplest form.

Solution: Since this is a *square* root, we factor a perfect *square* from the radicand, 18.
$$\sqrt{18} = \sqrt{9 \cdot 2}$$
By Eq. 29,
$$= \sqrt{9} \cdot \sqrt{2}$$
Simplifying,
$$= 3\sqrt{2}$$

Note that we could have factored 18 differently:
$$\sqrt{18} = \sqrt{6 \cdot 3}$$
Then, by Eq. 29,
$$\sqrt{18} = \sqrt{6} \cdot \sqrt{3}$$
But this cannot be simplified, since neither 6 nor 3 is a perfect square. In order to use Eq. 29 to simplify square roots, you must factor a *perfect square* from the radicand.

We also use the rule for the root of a product to simplify square roots of variable expressions, recalling from Eq. 28 that
$$\sqrt{x^2} = x$$
for all nonnegative x. We also use the fact that all even powers of x are perfect squares:
$$x^2 = (x)^2, \quad x^4 = (x^2)^2, \quad x^6 = (x^3)^2, \quad x^8 = (x^4)^2, \quad \text{and so forth}$$

Example 20
Simplify $\sqrt{x^8}$.

Solution:
$$\sqrt{x^8} = \sqrt{(x^4)^2} = x^4$$

Example 21
Simplify $\sqrt{8x^9}$.

Solution: Factoring out the perfect squares,
$$\sqrt{8x^9} = \sqrt{4 \cdot 2 \cdot x^8 \cdot x}$$
Rearranging, and using Eq. 29,
$$= \sqrt{4} \cdot \sqrt{x^8} \cdot \sqrt{2x}$$
Simplifying,
$$= 2 \cdot x^4 \cdot \sqrt{2x}$$
$$= 2x^4\sqrt{2x}$$

To simplify a cube root, we factor out the perfect cubes:
$$8, \quad 27, \quad 64, \quad 125, \quad 256, \ldots$$
The perfect cubes of x are
$$x^3, \quad x^6, \quad x^9, \quad x^{12}, \quad x^{15}, \ldots \quad \text{(exponents are all multiples of 3)}$$
Then we apply Eq. 28, using the fact that
$$\sqrt[3]{x^3} = x$$
for all x, and simplify.

Example 22
Simplify $\sqrt[3]{8x^6}$.

Solution:
$$\sqrt[3]{8x^6} = \sqrt[3]{8} \cdot \sqrt[3]{x^6}$$
$$= 2x^2$$

Example 23
Simplify $\sqrt[3]{16x^8}$.

Solution: Factoring out the perfect cubes, we have
$$\sqrt[3]{16x^8} = \sqrt[3]{8 \cdot 2 \cdot x^6 \cdot x^2}$$
$$= \sqrt[3]{8} \cdot \sqrt[3]{x^6} \cdot \sqrt[3]{2x^2}$$
$$= 2x^2\sqrt[3]{2x^2}$$

Example 24

Simplify $\sqrt[5]{32x^{11}}$.

Solution: Factoring out the fifth powers, we have

$$\sqrt[5]{32x^{11}} = \sqrt[5]{2^5 \cdot x^{10} \cdot x}$$
$$= \sqrt[5]{2^5} \cdot \sqrt[5]{x^{10}} \cdot \sqrt[5]{x}$$
$$= 2 \cdot \sqrt[5]{(x^2)^5} \cdot \sqrt[5]{x}$$
$$= 2 \cdot x^2 \cdot \sqrt[5]{x}$$
$$= 2x^2 \sqrt[5]{x}$$

COMMON ERROR Remember to write the root *index* for all roots other than square roots. If the index is omitted, it is assumed to be a square root.

Rationalizing the Denominator

Simplifying radical expressions also requires that we write a given expression without any radicals in the denominator. This is called *rationalizing the denominator*. To do this, we use the rule for the root of a quotient.

Root of a Quotient

The root of a quotient is equal to the quotient of the roots:

| ROOT OF A QUOTIENT | $\sqrt[r]{\dfrac{x}{y}} = \dfrac{\sqrt[r]{x}}{\sqrt[r]{y}}$ | 30 |

Note that if r is even, x must be nonnegative and y must be positive.

Example 25

$$\sqrt[3]{\frac{x}{y}} = \frac{\sqrt[3]{x}}{\sqrt[3]{y}}$$

Example 26

Simplify $\sqrt[3]{\dfrac{5}{8x^3}}$.

Solution: By Eq. 30,

$$\sqrt[3]{\frac{5}{8x^3}} = \frac{\sqrt[3]{5}}{\sqrt[3]{8x^3}} = \frac{\sqrt[3]{5}}{2x}$$

Since the denominator in the example above was a perfect cube, it was rationalized immediately after we applied Eq. 30. But frequently the radicand is *not* a perfect power, and we must make it a perfect power to rationalize it. If the denominator is a square root, multiply the numerator and denominator of the fraction by a quantity that will make the radicand in the denominator a perfect square.

Example 27

Simplify $\dfrac{3}{\sqrt{2}}$.

Solution:

$$\dfrac{3}{\sqrt{2}} = \dfrac{3}{\sqrt{2}} \cdot \dfrac{\sqrt{2}}{\sqrt{2}} = \dfrac{3\sqrt{2}}{2}$$

Here we multiply the numerator and denominator by $\sqrt{2}$ to make the denominator rational.

Example 28

Simplify $\sqrt{\dfrac{2}{3}}$.

Solution:

$$\sqrt{\dfrac{2}{3}} = \sqrt{\dfrac{2 \cdot 3}{3 \cdot 3}} = \dfrac{\sqrt{6}}{\sqrt{9}} = \dfrac{\sqrt{6}}{3}$$

Here we multiply both numerator and denominator *under* the radical sign by the smallest quantity that will make the denominator rational.

Example 29

Simplify:

$$\sqrt{\dfrac{1}{2x}} = \sqrt{\dfrac{1 \cdot 2x}{2x \cdot 2x}} = \dfrac{\sqrt{2x}}{2x}$$

To rationalize the denominator of a cube root, multiply by the *smallest* quantity that will make the denominator a perfect cube.

Example 30

$$\sqrt[3]{\dfrac{2a}{9b}} = \sqrt[3]{\dfrac{2a \cdot 3b^2}{9b \cdot 3b^2}} = \dfrac{\sqrt[3]{6ab^2}}{\sqrt[3]{27b^3}} = \dfrac{\sqrt[3]{6ab^2}}{3b}$$

In a more complicated problem we may be required to remove square factors *and* to rationalize the denominators. It is best to remove the square factors first, and then rationalize the denominator.

Example 31

$$\sqrt{\dfrac{y}{12x^3}} = \sqrt{\dfrac{y}{4 \cdot 3 \cdot x^2 \cdot x}} = \dfrac{1}{2x}\sqrt{\dfrac{y}{3x}}$$
$$= \dfrac{1}{2x}\sqrt{\dfrac{y \cdot 3x}{3x \cdot 3x}} = \dfrac{\sqrt{3xy}}{2x \cdot 3x} = \dfrac{\sqrt{3xy}}{6x^2}$$

It is important to note that while we have rules for the root of a product and of a quotient, there are *no rules* for the root of a *sum* or *difference*.

> **COMMON ERROR** The root of a sum is *not* equal to the sum of the roots.
> $$\sqrt[n]{x+y} \neq \sqrt[n]{x} + \sqrt[n]{y}$$

Example 32

(a) $\sqrt{x+y} \neq \sqrt{x} + \sqrt{y}$
(b) $\sqrt[3]{x-y} \neq \sqrt[3]{x} - \sqrt[3]{y}$
(c) $\sqrt{x^2-4} \neq x - 2$

Example 33

Show that
$$\sqrt{16+9} \neq \sqrt{16} + \sqrt{9}$$

Solution: The left side,
$$\sqrt{16+9} = \sqrt{25} = 5$$

but the right side
$$\sqrt{16} + \sqrt{9} = 4 + 3 = 7$$

Since $5 \neq 7$, then $\sqrt{16+9} \neq \sqrt{16} + \sqrt{9}$.

SECTION 20-2 Exercises Simplification of Radicals

Root of a Product

Write in simplest form.

1. $\sqrt{12}$
2. $\sqrt{75}$
3. $\sqrt{72}$
4. $\sqrt{128}$
5. $\sqrt[3]{24}$
6. $\sqrt[3]{32}$
7. $\sqrt[5]{64}$
8. $\sqrt[4]{162}$
9. $\sqrt{9x^4}$
10. $\sqrt{25a^6}$
11. $\sqrt{24b^5c^2}$
12. $\sqrt{40m^7n^4}$
13. $\sqrt[3]{27x^4}$
14. $\sqrt[3]{64x^8}$
15. $\sqrt[3]{16a^7}$
16. $\sqrt[3]{54b^5}$
17. $\sqrt[4]{81x^5}$
18. $\sqrt[5]{243x^7}$
19. $\sqrt[6]{64x^7}$
20. $\sqrt[4]{16x^9y}$

Root of a Quotient

Write in simplest form.

21. $\sqrt{\dfrac{5}{9}}$
22. $\sqrt[3]{\dfrac{3}{8}}$
23. $\dfrac{2}{\sqrt{3}}$
24. $\dfrac{2}{\sqrt[3]{9}}$
25. $\sqrt{\dfrac{4}{5}}$
26. $\sqrt{\dfrac{5}{6}}$
27. $\sqrt[3]{\dfrac{2}{3}}$
28. $\sqrt[3]{\dfrac{7}{9}}$
29. $\sqrt{\dfrac{7}{4x^2}}$
30. $\sqrt{\dfrac{5}{9y^4}}$
31. $\sqrt[3]{\dfrac{2y}{27x^3}}$
32. $\sqrt[3]{\dfrac{a^2}{64}}$
33. $\dfrac{-3}{\sqrt{5x}}$
34. $\dfrac{1}{\sqrt[3]{4x}}$
35. $\sqrt{\dfrac{1}{3x}}$
36. $\sqrt{\dfrac{2}{5a}}$
37. $\sqrt{\dfrac{5y}{6}}$
38. $\sqrt{\dfrac{3a}{2}}$
39. $\sqrt[3]{\dfrac{2}{25x}}$
40. $\sqrt[3]{\dfrac{3}{4a^2}}$
41. $\sqrt{\dfrac{a}{28b^3}}$
42. $\sqrt{\dfrac{7}{18x^5}}$
43. $\sqrt{\dfrac{x^3}{50y}}$
44. $\sqrt{\dfrac{a^6}{200}}$

45. $\sqrt{\dfrac{2y^4}{3x^3}}$ **46.** $\sqrt{\dfrac{9x^5}{5y^3}}$ **47.** $\sqrt{\dfrac{4a^2c^3}{3b}}$ **48.** $\sqrt{\dfrac{25x^6}{6y^5}}$

Applications

49. The radius of a circle having area A is

$$\sqrt{\dfrac{A}{\pi}}$$

Simplify this expression.

50. The radius of a sphere having volume V is

$$\sqrt[3]{\dfrac{3V}{4\pi}}$$

Simplify this expression.

51. The natural frequency of a body under simple harmonic motion (Fig. 20-2) is

$$\dfrac{1}{2\pi}\sqrt{\dfrac{kg}{W}}$$

where W is the weight of the body and g is the acceleration due to gravity, and k is the value for the spring constant. Simplify this expression.

FIGURE 20-2

52. The resonant frequency for the circuit in Fig. 20-3 is

$$\sqrt{\dfrac{1}{LC}}$$

where L is the inductance and C is the capacitance of the circuit. Simplify this expression.

FIGURE 20-3

53. Factor the radicand and simplify.
  (a) $\sqrt{x^4 + x^2}$ (b) $\sqrt{a^2 + 2ab + b^2}$

54. Simplify by changing to exponential form and reducing the index. Express the result in radical form.
  (a) $\sqrt[4]{x^2}$ (b) $\sqrt[6]{4y^2}$

20-3 ADDING AND SUBTRACTING RADICALS

In this section we learn how to add and subtract similar radical expressions.

Similar Radical Expressions

Only *similar* (or *like*) radical expressions can be added or subtracted, just as with any algebraic expressions. Two radical expressions are *similar* if they have the same radicand and the same index.

Example 34

(a) $\sqrt{3}$ and $\sqrt{2}$ are *not* similar expressions, because they have different radicands.
(b) $3\sqrt{5}$ and $-4\sqrt{5}$ are similar expressions, since they have the same radicand, 5, and the same index, 2.
(c) $\sqrt[3]{5}$ and $\sqrt{5}$ are not similar expressions, since they have different indices.
(d) $-4\sqrt[3]{a}$ and $\sqrt[3]{a}$ are similar expressions.

Combining Radical Expressions

As with algebraic expressions, we add or subtract similar radical expressions by *combining the coefficients* of the radicals.

Example 35

$$3\sqrt{2} + 4\sqrt{2} - 5\sqrt{2} = (3 + 4 - 5)\sqrt{2} = 2\sqrt{2}$$

Note that this is just an application of the distributive property; we factor out the similar radical parts. Since this is a numerical expression, we could approximate it using a calculator:

$$2\sqrt{2} \approx 2.83$$

but in general we leave the result in its radical form.

Example 36

$$6\sqrt[3]{x} - 2\sqrt[3]{x} + \sqrt[3]{x} = 6\sqrt[3]{x} - 2\sqrt[3]{x} + 1\sqrt[3]{x}$$
$$= (6 - 2 + 1)\sqrt[3]{x}$$
$$= 5\sqrt[3]{x}$$

It's very important to note that $\sqrt[3]{x} = 1\sqrt[3]{x}$.

Not all radical expressions can be combined algebraically.

Example 37

Combine, if possible, and approximate to three significant digits:

$$\sqrt{3} + \sqrt{2}$$

Solution: These radicals are not similar, and cannot be combined algebraically. Remember that

$$\sqrt{3} + \sqrt{2} \neq \sqrt{3 + 2}$$

The algebraic sum can only be *indicated,* as given. The terms cannot be combined into a single term. However, this numerical expression can be approximated using a calculator:

$$\sqrt{3} + \sqrt{2} \approx 3.15$$

rounded to three significant digits.

> **COMMON ERROR** Remember that the sum of two roots is *not* equal to the root of the sum!
>
> $$\sqrt[n]{x} + \sqrt[n]{y} \neq \sqrt[n]{x + y}$$

Example 38

Combine $2\sqrt{3x} - 5\sqrt{2x} - 7\sqrt{3x} + 8\sqrt{2x}$.

Solution: Rearranging the terms, we get

$$2\sqrt{3x} - 5\sqrt{2x} - 7\sqrt{3x} + 8\sqrt{2x} = (2\sqrt{3x} - 7\sqrt{3x}) + (-5\sqrt{2x} + 8\sqrt{2x})$$

Combining,
$$= -5\sqrt{3x} + 3\sqrt{2x}$$

Radicals that do not appear to be similar at first may turn out to be similar after simplification. So, we express each radical in simplest form, and then combine any similar terms.

Example 39
Combine algebraically.
$$2\sqrt{48} - \sqrt{80} + \sqrt{108}$$

Solution:
$$2\sqrt{48} - \sqrt{80} + \sqrt{108} = 2\sqrt{16 \cdot 3} - \sqrt{16 \cdot 5} + \sqrt{36 \cdot 3}$$
$$= 2(4)\sqrt{3} - 4\sqrt{5} + 6\sqrt{3}$$
$$= 8\sqrt{3} - 4\sqrt{5} + 6\sqrt{3}$$

Combining,
$$= 14\sqrt{3} - 4\sqrt{5}$$

Example 40
Simplify and combine:
$$4a\sqrt[3]{16a} - 2\sqrt[3]{54a^4}$$

Solution:
$$4a\sqrt[3]{16a} - 2\sqrt[3]{54a^4} = 4a\sqrt[3]{8 \cdot 2a} - 2\sqrt[3]{27 \cdot 2 \cdot a^3 \cdot a}$$
$$= 4a(2)\sqrt[3]{2a} - 2(3)(a)\sqrt[3]{2a}$$
$$= 8a\sqrt[3]{2a} - 6a\sqrt[3]{2a}$$
$$= 2a\sqrt[3]{2a}$$

To simplify radical expressions with fractions, first rationalize the denominators of each term, and then combine terms using the least common denominator.

Example 41
Combine algebraically.
$$4\sqrt{\frac{2}{3}} - 5\sqrt{\frac{1}{6}}$$

Solution: First we rationalize denominators:
$$4\sqrt{\frac{2}{3}} - 5\sqrt{\frac{1}{6}} = 4\sqrt{\frac{2}{3} \cdot \frac{3}{3}} - 5\sqrt{\frac{1}{6} \cdot \frac{6}{6}}$$
$$= \frac{4\sqrt{6}}{3} - \frac{5\sqrt{6}}{6}$$

Getting a common denominator of 6,
$$= \frac{4\sqrt{6}}{3} \cdot \frac{2}{2} - \frac{5\sqrt{6}}{6}$$

Simplifying,
$$= \frac{8\sqrt{6}}{6} - \frac{5\sqrt{6}}{6}$$
$$= \frac{3\sqrt{6}}{6}$$
$$= \frac{\sqrt{6}}{2}$$

Example 42
Combine:

$$\sqrt{\frac{5}{2x}} - \sqrt{\frac{2}{3x}} = \sqrt{\frac{5}{2x} \cdot \frac{2x}{2x}} - \sqrt{\frac{2}{3x} \cdot \frac{3x}{3x}}$$

Rationalizing each denominator,

$$= \frac{\sqrt{10x}}{2x} - \frac{\sqrt{6x}}{3x}$$

Getting the common denominator of $6x$,

$$= \frac{\sqrt{10x}}{2x} \cdot \frac{3}{3} - \frac{\sqrt{6x}}{3x} \cdot \frac{2}{2}$$

Combining,

$$= \frac{3\sqrt{10x}}{6x} - \frac{2\sqrt{6x}}{6x}$$

$$= \frac{3\sqrt{10x} - 2\sqrt{6x}}{6x}$$

In summary,

To Add or Subtract Radicals

1. Simplify each radical expression.
2. Add or subtract similar radicals by combining their coefficients.

SECTION 20-3 Exercises Adding and Subtracting Radicals

Combine algebraically. Express the answer in radical form.
1. $5\sqrt{3} - 2\sqrt{3} + \sqrt{3}$
2. $6\sqrt{5} + 3\sqrt{5} - \sqrt{5}$
3. $2\sqrt[3]{2} - \sqrt[3]{2} + 3\sqrt[3]{2}$
4. $\sqrt[4]{3} - 5\sqrt[4]{3} + 7\sqrt[4]{3}$
5. $\sqrt{8} - \sqrt{18} + \sqrt{32}$
6. $\sqrt{12} - \sqrt{48} - \sqrt{27}$
7. $\sqrt{75} + 3\sqrt{72} - 2\sqrt{27}$
8. $4\sqrt{80} - 2\sqrt{32} - 3\sqrt{125} + \sqrt{72}$
9. $5\sqrt{150} - 2\sqrt{500} - 3\sqrt{180} - \sqrt{600}$
10. $\sqrt[3]{108} - 2\sqrt[3]{32}$
11. $8 + \sqrt{8} + \sqrt[3]{8} + \sqrt[4]{8}$
12. $16 + \sqrt{16} + \sqrt[3]{16} + \sqrt[4]{16}$
13. $\sqrt{\frac{1}{2}} + \sqrt{\frac{1}{8}}$
14. $\sqrt{\frac{2}{3}} - \sqrt{\frac{2}{27}}$

Combine algebraically.
15. $5\sqrt{x} - 7\sqrt{x} + 3\sqrt{x}$
16. $-2\sqrt[3]{a} + 9\sqrt[3]{a} - 4\sqrt[3]{a}$
17. $2\sqrt{3x} - 4\sqrt{2x} - 5\sqrt{3x} + 7\sqrt{2x}$
18. $\sqrt{10a} + \sqrt{15a} - 2\sqrt{10a} + 3\sqrt{15a}$
19. $2\sqrt{50x} + 4\sqrt{45x} - \sqrt{98x}$
20. $3a\sqrt{98a} - 5\sqrt{24a^3} + 2a\sqrt{72a} - 4\sqrt{54a^3}$
21. $\sqrt{10x} + \sqrt{100x} + \sqrt{1000x} + \sqrt{10000x}$
22. $2\sqrt[3]{24x^4} - x\sqrt[3]{81x}$

23. $x\sqrt{3x^3} + x^2\sqrt{12x} - \sqrt{27x^5}$ 24. $\sqrt{2y^2} - y\sqrt{18} + \sqrt{50y}$

25. $3\sqrt{\dfrac{2x}{5}} - \sqrt{\dfrac{x}{10}} + 2\sqrt{\dfrac{5x}{2}}$ 26. $4\sqrt{\dfrac{y}{6}} - 3\sqrt{\dfrac{2y}{3}} - 2\sqrt{\dfrac{3y}{2}}$

27. $\sqrt[3]{\dfrac{1}{2y}} - \sqrt[3]{\dfrac{1}{16y}}$ 28. $\sqrt[3]{\dfrac{3}{4x}} + \sqrt[3]{\dfrac{2}{9x}}$

29. $27^{1/3} - 16^{1/4} + 25^{-1/2} + 8^{2/3}$ 30. $4^{-3/2} + 8^{1/3} - 81^{1/4} + 49^{1/2}$

20-4 MULTIPLYING RADICALS

In this section we learn how to multiply radicals with the same index. Then we multiply multinomials with radicals.

Multiplying Radicals with the Same Index

Radicals having the *same index r* can be multiplied using Eq. 29:

$$\sqrt[r]{x} \cdot \sqrt[r]{y} = \sqrt[r]{xy}$$

Example 43
(a) $\sqrt{2} \cdot \sqrt{3} = \sqrt{6}$
(b) $\sqrt[3]{5} \cdot \sqrt[3]{x} = \sqrt[3]{5x}$
(c) $\sqrt{x} \cdot \sqrt{x} = (\sqrt{x})^2 = x$

Radical expressions having the same index may be multiplied just like algebraic expressions. The coefficient of the product is the product of the coefficients. The radicands are multiplied and the resulting radical expression should be simplified if possible.

Example 44
(a) $2\sqrt{3} \cdot 3\sqrt{5} = 2(3)\sqrt{3 \cdot 5} = 6\sqrt{15}$
(b) $5\sqrt{3} \cdot 2\sqrt{3} = 5(2)(\sqrt{3})^2 = 5(2)3 = 30$

Example 45
(a) $3\sqrt{a} \cdot 5\sqrt{b} = 3(5)\sqrt{a \cdot b} = 15\sqrt{ab}$
(b) $2x\sqrt{5y} \cdot 3x\sqrt{2y} = 2x \cdot 3x \cdot \sqrt{5y \cdot 2y}$
$= 6x^2\sqrt{10y^2}$
$= 6x^2\sqrt{10}\sqrt{y^2}$
$= 6x^2y\sqrt{10}$

It is very important to note that in the examples above, we are multiplying two radicals with the *same index*. Eq. 29 holds only for radicals with the *same index*.

Multiplying Multinomials Containing Radicals

Recall from Sec. 8-2 that a *multinomial* is an algebraic expression having more than one term. Multinomials containing radicals are multiplied like other algebraic expressions, using the distributive property. So, we follow the same general procedure that we used in multiplying polynomials with variables.

Example 46

$$\sqrt{2}(4 - 5\sqrt{3}) = \sqrt{2}(4) - \sqrt{2}(5\sqrt{3})$$
$$= 4\sqrt{2} - 5\sqrt{6}$$

Example 47

$$2\sqrt{3x}(5\sqrt{6} - 4\sqrt{x}) = 2\sqrt{3x}(5\sqrt{6}) - 2\sqrt{3x}(4\sqrt{x})$$
$$= 10\sqrt{18x} - 8\sqrt{3x^2}$$
$$= 10\sqrt{9 \cdot 2x} - 8\sqrt{x^2 \cdot 3}$$
$$= 10 \cdot 3\sqrt{2x} - 8x\sqrt{3}$$
$$= 30\sqrt{2x} - 8x\sqrt{3}$$

The product of two binomials may be found by multiplying vertically or horizontally, as illustrated below.

Example 48

Multiply $(\sqrt{x} + 5\sqrt{y})(4\sqrt{x} - \sqrt{y})$.

Solution: Multiplying vertically:

$$\begin{array}{r} \sqrt{x} + 5\sqrt{y} \\ 4\sqrt{x} - \sqrt{y} \\ \hline -\sqrt{xy} - 5 \cdot y \\ 4 \cdot x + 20\sqrt{xy} \\ \hline 4x + 19\sqrt{xy} - 5y \end{array}$$

Alternate Solution: Multiplying horizontally:

$$(\sqrt{x} + 5\sqrt{y})(4\sqrt{x} - \sqrt{y}) = \sqrt{x} \cdot 4\sqrt{x} - \sqrt{x} \cdot \sqrt{y} + 5\sqrt{y} \cdot 4\sqrt{x} - 5\sqrt{y} \cdot \sqrt{y}$$
$$= 4 \cdot x - \sqrt{xy} + 20\sqrt{xy} - 5 \cdot y$$
$$= 4x + 19\sqrt{xy} - 5y$$

Recall from Sec. 8-3 that this method is also called the FOIL method.

Example 49

Multiply $(3 + \sqrt{a})(3 - \sqrt{a})$.

Solution: Using FOIL,

$$(3 + \sqrt{a})(3 - \sqrt{a}) = 9 - 3\sqrt{a} + 3\sqrt{a} - a$$
$$= 9 - a$$

Conjugates

In Example 49, the product of two binomials with radicals was an expression without radicals. This happened because $(3 + \sqrt{a})$ and $(3 - \sqrt{a})$ are alike except for the sign of one term. Expressions such as these are called *conjugates* of each other. A conjugate of a binomial is a binomial that differs from the original binomial only in the sign of one term, usually the second term.

Example 50
The conjugate of $a + b$ is $a - b$,

The conjugate of $x - y$ is $x + y$, and

The conjugate of $\sqrt{a} + \sqrt{b}$ is $\sqrt{a} - \sqrt{b}$.

When we multiply two conjugates, the middle terms drop out, and the product has no radical in it!

Example 51
$$(\sqrt{a} + \sqrt{b})(\sqrt{a} - \sqrt{b}) = (\sqrt{a})^2 - \sqrt{a}\sqrt{b} + \sqrt{b}\sqrt{a} - (\sqrt{b})^2$$
$$= a - b$$

Example 52
Multiply $\sqrt{3} - \sqrt{x}$ by its conjugate.

Solution: The conjugate of $\sqrt{3} - \sqrt{x}$ is $\sqrt{3} + \sqrt{x}$. Multiplying, we have
$$(\sqrt{3} - \sqrt{x})(\sqrt{3} + \sqrt{x}) = \sqrt{3}^2 - \sqrt{x}^2$$
$$= 3 - x$$

SECTION 20-4 Exercises Multiplying Radicals

Multiplying Radicals with the Same Index

Multiply and simplify.

1. $\sqrt{3} \cdot \sqrt{5}$
2. $\sqrt{7} \cdot \sqrt{2}$
3. $(\sqrt{x})(-\sqrt{2})$
4. $\sqrt{5} \cdot \sqrt{2y}$
5. $\sqrt[3]{2} \cdot \sqrt[3]{3x^2}$
6. $(\sqrt[4]{3x})(\sqrt[4]{5})$
7. $\sqrt{3y} \cdot \sqrt{3y}$
8. $\sqrt{7x} \cdot \sqrt{7x}$
9. $3\sqrt{2} \cdot 4\sqrt{3}$
10. $5\sqrt{5x} \cdot 4\sqrt{2}$
11. $2\sqrt{3x} \cdot 5\sqrt{3x}$
12. $3\sqrt{2y} \cdot 4\sqrt{2y}$
13. $\sqrt[3]{6} \cdot \sqrt[3]{4}$
14. $2\sqrt[3]{18} \cdot \sqrt[3]{3}$
15. $(3\sqrt{5a})^2$
16. $(2\sqrt[3]{4y})^2$
17. $2\sqrt{10a} \cdot \sqrt{5a}$
18. $\sqrt{3x} \cdot 4\sqrt{6}$
19. $2\sqrt{15x} \cdot 3\sqrt{6}$
20. $5\sqrt{14x} \cdot 4\sqrt{10x}$
21. $3x\sqrt{2y} \cdot 2x\sqrt{5y}$
22. $4\sqrt{3x} \cdot 3y\sqrt{2x}$
23. $4\sqrt{6x} \cdot 3\sqrt{2x}$
24. $5x\sqrt{6x} \cdot 2\sqrt{15x}$
25. $(2\sqrt{5})(-4\sqrt{3})(3\sqrt{5})$
26. $(-3\sqrt{6})(2\sqrt{2})(5\sqrt{3})$

Multiplying Multinomials Containing Radicals

27. $\sqrt{3}(5 - 2\sqrt{2})$
28. $\sqrt{7}(\sqrt{2} + 3)$
29. $3\sqrt{2x}(4\sqrt{2} + 6\sqrt{x})$
30. $5\sqrt{3x}(2\sqrt{6} - 3\sqrt{2x})$
31. $2\sqrt{3}(4 - 2\sqrt{3} + \sqrt{6})$
32. $3\sqrt{5}(\sqrt{10} - 3 - 2\sqrt{5})$
33. $\sqrt{a}(\sqrt{2b} - \sqrt{ab})$
34. $\sqrt{3x}(\sqrt{6x} - 2\sqrt{xy})$
35. $(2x + \sqrt{3})(3x - \sqrt{3})$
36. $(\sqrt{2} - 5)(\sqrt{2} - 3)$
37. $(\sqrt{5} - \sqrt{2})(2\sqrt{5} + \sqrt{2})$
38. $x\sqrt{3} + 2\sqrt{5})(4x\sqrt{3} + \sqrt{5})$
39. $(\sqrt{x} - \sqrt{3})(2\sqrt{x} - \sqrt{3})$
40. $(\sqrt{a} + \sqrt{b})(\sqrt{a} - 2\sqrt{b})$
41. $(\sqrt{5} + 2a)(\sqrt{5} - 2a)$
42. $(2\sqrt{3} - \sqrt{x})(2\sqrt{3} + \sqrt{x})$

CHAPTER 20 / Exponents and Radicals

43. $(\sqrt{x} - \sqrt{y})(\sqrt{x} + \sqrt{y})$
44. $(3\sqrt{2} + 4\sqrt{5})(3\sqrt{2} - 4\sqrt{5})$
45. $(\sqrt{2y} + \sqrt{3})^2$
46. $(\sqrt{5x} - \sqrt{7})^2$
47. $(\sqrt{3} - 2\sqrt{5})(\sqrt{2} + 3)$
48. $(3\sqrt{2} - 4\sqrt{3})(2\sqrt{5} - \sqrt{7})$
49. Multiply these monomials with different indices and different bases. (Hint: Change the following factors to exponential form, express the exponents with their LCD, use Eq. 21 to "factor" out the common denominator, and change back to radical form.)

   (a) $(\sqrt{6})(\sqrt[3]{5})$ (b) $(\sqrt[3]{5})(\sqrt[4]{2})$

20-5 DIVIDING RADICALS

In this section we learn how to divide radicals that have the same index. We also learn to divide multinomials with radicals by monomials and binomials with radicals.

Dividing Radicals with the Same Index

Radicals having the same index can be divided using Eq. 30.

$$\sqrt[n]{x} \div \sqrt[n]{y} = \frac{\sqrt[n]{x}}{\sqrt[n]{y}} = \sqrt[n]{\frac{x}{y}}$$

Just divide the radicands and keep the common index.

Example 53
Divide and simplify.

(a) $\sqrt{6} \div \sqrt{3} = \dfrac{\sqrt{6}}{\sqrt{3}} = \sqrt{\dfrac{6}{3}} = \sqrt{2}$

(b) $\dfrac{\sqrt{15xy}}{\sqrt{3x}} = \sqrt{\dfrac{15xy}{3x}} = \sqrt{5y}$

Note that division problems are often expressed as fractions, as in Example 53(b) above. It is common practice to rationalize the denominator in a division problem.

Example 54

$$\sqrt{14x} \div \sqrt{7y} = \frac{\sqrt{14x}}{\sqrt{7y}} = \sqrt{\frac{14x}{7y}}$$
$$= \sqrt{\frac{2x}{y}} = \sqrt{\frac{2x}{y} \cdot \frac{y}{y}}$$
$$= \frac{\sqrt{2xy}}{y}$$

Dividing a Multinomial by a Monomial Radical

If a multinomial is divided by a monomial, the division is done *term by term* (using the distributive property). Then each term is simplified.

Example 55

Divide and simplify:

$$\frac{\sqrt{2} + 4\sqrt{x}}{\sqrt{2}}$$

Solution:

$$\frac{\sqrt{2} + 4\sqrt{x}}{\sqrt{2}} = \frac{\sqrt{2}}{\sqrt{2}} + \frac{4\sqrt{x}}{\sqrt{2}}$$

$$= 1 + \frac{4}{1}\sqrt{\frac{x}{2}} = 1 + \frac{4}{1}\sqrt{\frac{x \cdot 2}{2 \cdot 2}}$$

$$= 1 + \frac{4}{2}\sqrt{2x} = 1 + 2\sqrt{2x}$$

Dividing by a Binomial Containing a Radical

An example of an expression divided by a binomial containing a radical is

$$4 \div (3 + \sqrt{2}) \quad \text{or} \quad \frac{4}{3 + \sqrt{2}}$$

We simplify the fraction by rationalizing the denominator. To rationalize a binomial denominator containing square roots, we multiply the numerator and denominator of the fraction by the *conjugate* of the denominator. This will remove all square roots from the denominator.

Example 56

Divide and simplify:

$$\frac{4}{3 + \sqrt{2}}$$

Solution: The conjugate of $3 + \sqrt{2}$ is $3 - \sqrt{2}$. So we multiply the numerator and denominator by $3 - \sqrt{2}$:

$$\frac{4}{3 + \sqrt{2}} = \frac{(4)(3 - \sqrt{2})}{(3 + \sqrt{2})(3 - \sqrt{2})} = \frac{12 - 4\sqrt{2}}{9 - 2} = \frac{12 - 4\sqrt{2}}{7}$$

Example 57

Divide and simplify:

$$\frac{3\sqrt{x} + \sqrt{2}}{\sqrt{x} - \sqrt{2}}$$

Solution: The conjugate of $\sqrt{x} - \sqrt{2}$ is $\sqrt{x} + \sqrt{2}$, so we multiply the numerator and denominator by $\sqrt{x} + \sqrt{2}$:

$$\frac{3\sqrt{x} + \sqrt{2}}{\sqrt{x} - \sqrt{2}} = \frac{(3\sqrt{x} + \sqrt{2})(\sqrt{x} + \sqrt{2})}{(\sqrt{x} - \sqrt{2})(\sqrt{x} + \sqrt{2})} = \frac{3x + 4\sqrt{2x} + 2}{x - 2}$$

Radical expressions should be reduced to lowest terms.

SECTION 20-5 Exercises Dividing Radicals

Dividing Radicals with the Same Index

Divide and simplify. Express answers in radical form.

1. $\sqrt{6} \div \sqrt{2}$
2. $\sqrt{15} \div \sqrt{3}$
3. $\sqrt{10x} \div \sqrt{5}$
4. $\sqrt{21a} \div \sqrt{7}$
5. $\sqrt{21ab} \div \sqrt{3a}$
6. $\sqrt{55xy} \div \sqrt{5x}$
7. $\dfrac{5}{\sqrt{3}}$
8. $\dfrac{7}{\sqrt{5}}$
9. $\dfrac{x\sqrt{2}}{2\sqrt{7}}$
10. $\dfrac{y\sqrt{5}}{3\sqrt{3}}$
11. $\sqrt[3]{14xy} \div \sqrt[3]{2x}$
12. $\sqrt[4]{6ab} \div \sqrt[4]{3b}$

Dividing a Multinomial by a Monomial Radical

13. $(\sqrt{3} + 6\sqrt{2}) \div \sqrt{3}$
14. $(4\sqrt{5} - \sqrt{2}) \div \sqrt{2}$
15. $\dfrac{2 - \sqrt{x}}{\sqrt{x}}$
16. $\dfrac{2\sqrt{3} - 3}{\sqrt{3}}$
17. $\dfrac{2 + 4\sqrt{2}}{6\sqrt{2}}$
18. $\dfrac{1 - 6\sqrt{x}}{9\sqrt{x}}$

Dividing by a Binomial Containing a Radical

19. $3 \div (2 - \sqrt{3})$
20. $7 \div (\sqrt{2} + 1)$
21. $\dfrac{\sqrt{3}}{4\sqrt{2} + \sqrt{3}}$
22. $\dfrac{\sqrt{5}}{2\sqrt{3} - \sqrt{7}}$
23. $\dfrac{3 + \sqrt{x}}{1 - \sqrt{x}}$
24. $\dfrac{6 - \sqrt{y}}{3 + \sqrt{y}}$
25. $\dfrac{\sqrt{7} - \sqrt{5}}{\sqrt{7} + \sqrt{5}}$
26. $\dfrac{\sqrt{2} + \sqrt{3}}{\sqrt{2} - \sqrt{3}}$
27. $\dfrac{2\sqrt{a} - \sqrt{2}}{\sqrt{a} - 3\sqrt{2}}$
28. $\dfrac{\sqrt{m} + 2\sqrt{3}}{3\sqrt{m} + \sqrt{3}}$
29. $\dfrac{1 - \sqrt{2}}{\sqrt{3} - \sqrt{2}}$
30. $\dfrac{5 + \sqrt{3}}{\sqrt{2} - \sqrt{3}}$
31. $\dfrac{3 - \sqrt{x}}{2 + \sqrt{x}}$
32. $\dfrac{\sqrt{x} + 1}{\sqrt{x} - 1}$
33. $\dfrac{1 - \sqrt{5}}{3 + \sqrt{2} - \sqrt{5}}$
34. $\dfrac{3 - \sqrt{2}}{1 - \sqrt{2} - \sqrt{3}}$

Applications

35. A certain motion problem involves the expression

$$\dfrac{\sqrt{v}}{\sqrt{m} + \sqrt{v}}$$

Simplify this expression.

36. When studying the expansion of gases, we find the expression

$$\dfrac{\sqrt{V_2}}{\sqrt{V_2} - \sqrt{V_1}}$$

Simplify this expression.

37. When studying the theory of waves in wires, we find the expression

$$\dfrac{\sqrt{d_1} - \sqrt{d_2}}{\sqrt{d_1} + \sqrt{d_2}}$$

Simplify this expression.

38. Divide. (First rationalize the denominator, then change to exponential form, express the exponents with their LCD, use Eq. 21 to factor out the common denominator, and change back to radical form.)

(a) $\dfrac{\sqrt[3]{2}}{\sqrt{5}}$ (b) $\dfrac{\sqrt[4]{5}}{\sqrt[3]{4}}$

20-6 RADICAL EQUATIONS

In this section we define a radical equation, learn how to solve a radical equation, and learn how to recognize when a solution is extraneous.

Definition

A *radical equation* is an equation in which the variable is under a radical sign.

Example 58

$$\sqrt{x-2} = 3$$

is a radical equation, but the equation

$$x - \sqrt{2} = 3$$

is not a radical equation, because the variable is not under a radical sign.

Solution of a Radical Equation

In Chapter 10, we learned to solve an equation by performing the same operation on both sides of the equation. That is, we could add, subtract, multiply, or divide both sides of an equation by the same (nonzero) number in order to isolate the variable on one side of the equation. But, if the equation has a radical in it, like

$$\sqrt{x} - 3 = 0$$

none of these operations eliminate the radical sign.

To solve a radical equation it is necessary to transpose the radical term so it is alone on one side of the equal sign and then to raise both sides to a power which will eliminate the radical. That power is indicated by the index of the radical. So, to solve the radical equation,

$$\sqrt[r]{x} = a$$

raise both sides of the equation to the *r*th power.

Example 59

Solve and check:

$$\sqrt{x} - 3 = 0$$

Solution: First we isolate the radical part by adding 3 to both sides:

$$\sqrt{x} = 3$$

This is a *square* root, so we *square* both sides of the equation.

$$(\sqrt{x})^2 = 3^2$$

What does the squaring do? In squaring both sides of this equation, we multiply the left side by $\sqrt{x}$ and the right side by 3. Since $\sqrt{x} = 3$, we are multiplying both sides by the same number (if there is a number *x* such that $\sqrt{x}$ is equal to 3). Squaring both sides, we get

$$x = 9$$

Check:
We check in the original equation:
$$\sqrt{9} - 3 \stackrel{?}{=} 0$$
$$3 - 3 \stackrel{?}{=} 0$$
$$0 = 0 \quad \text{checks}$$
So the solution of the equation $\sqrt{x} - 3 = 0$ is $x = 9$.

This method of solving an equation by squaring both sides produces a new equation. This new equation will have all of the roots of the original equation. However, in some cases, it will also have roots that are *not* roots of the original equation. Hence we must *always check* the roots obtained from the squaring process by substituting them into the *original* equation.

Example 60

Solve and check:
$$\sqrt{x} + 3 = 0$$

Solution: We isolate the radical part:
$$\sqrt{x} = -3$$

Squaring both sides,
$$(\sqrt{x})^2 = (-3)^2$$
$$x = 9$$

Check:
Substituting into the original equation,
$$\sqrt{9} + 3 \stackrel{?}{=} 0$$
$$3 + 3 \stackrel{?}{=} 0$$
$$6 \neq 0$$

The solution does *not* check. The solution $x = 9$ is an *extraneous* solution. Recall from Sec. 18-1 that an extraneous solution is a "false" solution—one that does *not* satisfy the original equation and therefore is not a solution. This extraneous solution was introduced at the squaring step, since squaring meant multiplying the left by $\sqrt{x}$ and the right by -3. We may do this only if $\sqrt{x} = -3$, and of course, a square root may never be negative. So there is *no real solution*.

> **COMMON ERROR** The squaring process may introduce *extraneous* solutions. These are discovered only in the check. They must be discarded because they do not satisfy the original equation.

Example 61

Solve $\sqrt[3]{2x} = 4$ and check.

Solution: Cubing both sides:
$$(\sqrt[3]{2x})^3 = 4^3$$
$$2x = 64$$
$$x = 32$$

Check:

$$\sqrt[3]{2 \cdot 32} \stackrel{?}{=} 4$$
$$\sqrt[3]{64} \stackrel{?}{=} 4$$
$$4 = 4 \quad \text{checks}$$

Steps for Solving a Radical Equation with One Radical

1. Transpose the radical term until it is alone on one side of the equation.
2. Raise both sides of the resulting equation to the power indicated by the index of the radical.
3. Simplify the resulting equation and solve for the variable.
4. Check the solution in the original equation to see if it is extraneous.

Example 62

Solve and check:

$$4 + \sqrt{2x - 5} = 7$$

Solution: Transposing, we have

$$\sqrt{2x - 5} = 3$$

Squaring both sides,

$$2x - 5 = 9$$

Solving for x,

$$2x = 14$$
$$x = 7$$

Check:

$$4 + \sqrt{2 \cdot 7 - 5} \stackrel{?}{=} 7$$
$$4 + \sqrt{9} \stackrel{?}{=} 7$$
$$7 = 7 \quad \text{checks}$$

Equations with Fractional Exponents

To solve an equation with fractional exponents, we change it to radical form and then follow the steps above.

Example 63

Solve $1 + x^{1/3} = 5$.

Solution: Transposing, we have

$$x^{1/3} = 4$$

Changing to radical form,

$$\sqrt[3]{x} = 4$$

Cubing both sides,

$$(\sqrt[3]{x})^3 = 4^3$$
$$x = 64$$

Check:
Substituting into the original equation:
$$1 + 64^{1/3} \stackrel{?}{=} 5$$
$$1 + 4 \stackrel{?}{=} 5$$
$$5 = 5 \quad \text{checks}$$

Alternate Solution: We can also solve this equation in exponential form:

After transposing,
$$x^{1/3} = 4$$
we cube both sides,
$$(x^{1/3})^3 = (4)^3$$
so that we get x^1
$$x^1 = 64$$
or
$$x = 64$$

Equations with More Than One Square Root

If an equation has more than one square root radical, isolate one at a time and square both sides. Generally it's better to isolate the most complicated radical first.

Example 64

Solve $\sqrt{x} + \sqrt{x+2} = 2$.

Solution: We isolate the most complicated radical:
$$\sqrt{x+2} = 2 - \sqrt{x}$$

Squaring both sides,
$$(\sqrt{x+2})^2 = (2 - \sqrt{x})^2$$
$$x + 2 = 4 - 4\sqrt{x} + x$$

Combining terms,
$$-2 = -4\sqrt{x}$$
$$\sqrt{x} = \frac{1}{2}$$

Squaring both sides,
$$(\sqrt{x})^2 = \left(\frac{1}{2}\right)^2$$
$$x = \frac{1}{4}$$

Check:
Substituting into the original equation:
$$\sqrt{\frac{1}{4}} + \sqrt{\frac{1}{4} + 2} \stackrel{?}{=} 2$$
$$\frac{1}{2} + \frac{3}{2} \stackrel{?}{=} 2$$
$$2 = 2 \quad \text{checks}$$

20-6 Radical Equations

Graphical Solution of Radical Equations

We learned how to solve equations graphically in Sec. 12-6. Recall that to solve an equation graphically, we bring all terms to one side of the equal sign, set those terms equal to y, and graph the function. The x intercepts are the solutions of the given equation. We can use a graphing calculator to graph the function, and use the "root" or "zero" feature to find the solutions of the equation.

Example 65

Solve the equation in Example 64 graphically:
$$\sqrt{x} + \sqrt{x+2} = 2$$

Solution: We bring all terms to one side of the equal sign:
$$\sqrt{x} + \sqrt{x+2} - 2 = 0$$

set the left side equal to y and graph
$$y = \sqrt{x} + \sqrt{x+2} - 2$$

Using the "root" or "zero," we find that the solution is $x = 0.25$ (Fig. 20-4), which agrees with our answer of $\frac{1}{4}$ found in Example 64.

FIGURE 20-4

SECTION 20-6 Exercises Radical Equations

Solve and check.

1. $\sqrt{x} = 2$
2. $\sqrt{x} = 5$
3. $\sqrt{x} - 4 = 0$
4. $\sqrt{x} + 1 = 0$
5. $\sqrt{x} + 6 = 0$
6. $\sqrt{x} - 7 = 0$
7. $\sqrt[3]{x} = 3$
8. $2 - \sqrt[3]{x} = 0$
9. $\sqrt[3]{4x} + 1 = -1$
10. $\sqrt[3]{2x} = 2$
11. $\sqrt{x+1} = 4$
12. $2 - \sqrt{x} = 1$
13. $3 - \sqrt{x} = 7$
14. $2 + \sqrt{x} = 1$
15. $3 + \sqrt{x} = 7$
16. $\sqrt{x+4} = 2$
17. $\sqrt{2x+3} = 5$
18. $4 = \sqrt{3x-2}$
19. $2 - \sqrt{1+5x} = 8$
20. $5 - \sqrt{4x-3} = 2$
21. $\sqrt{2x-1} = \sqrt{x+3}$
22. $\sqrt{3x-1} = \sqrt{2x+1}$
23. $\sqrt[3]{x-1} = 2$
24. $\sqrt[3]{x+3} = 1$
25. $\sqrt[3]{2x-2} = 4$
26. $\sqrt[3]{2x-5} = 5$
27. $\sqrt{x-2} = \dfrac{1}{\sqrt{x-2}}$
28. $\dfrac{2}{\sqrt{x-1}} = \sqrt{x-1}$
29. $\sqrt{x^2-3} = x+3$
30. $x - 4 = \sqrt{x^2 - 8}$
31. $x^{1/3} = -2$
32. $x^{1/3} = 3$
33. $x^{3/2} - 8 = 0$
34. $2 - x^{3/2} = 1$
35. $\sqrt{12+x} = 2 + \sqrt{x}$
36. $\sqrt{x+7} = 7 - \sqrt{x}$
37. $\sqrt{3+2x} = \sqrt{2x} - 3$
38. $\sqrt{3x-1} = \sqrt{3x-5}$
39. $\sqrt{8+x} = 3 + \sqrt{x-1}$
40. $\sqrt{x-3} = 2 - \sqrt{x+1}$

Solve the equations graphically. Approximate the solution(s) to the nearest tenth.

41. $\sqrt{2x-1} - \sqrt{2x+6} = -1$
42. $\sqrt{3x+1} - \sqrt{3x-8} = 1$
43. $\sqrt{x^2 - 6.75} = 9.25 - x$
44. $\sqrt{x} + 2.25 = \sqrt{x+8.75}$

■ FIGURE 20-5

■ FIGURE 20-6

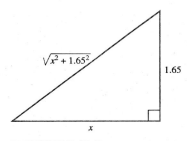

■ FIGURE 20-7

Applications

45. In simple harmonic motion, the undamped angular velocity (Fig. 20-5) is given by the equation

$$\omega = \sqrt{\frac{kg}{W}}$$

Solve this equation for k.

46. In a certain circuit (Fig. 20-6), the resonant frequency is given by the equation

$$f = \frac{1}{2\pi\sqrt{LC}}$$

Solve this equation for C.

47. The right triangle in Fig. 20-7 has one leg equal to 1.65 cm. The other leg is x and an expression for the hypotenuse is found using the Pythagorean theorem. If the perimeter is 6.65 cm, then

$$x + 1.65 + \sqrt{x^2 + 1.65^2} = 6.65$$

Solve this equation for x.

Chapter 20 REVIEW EXERCISES

Change to radical form.

1. $x^{1/5}$
2. $5^{2/3}$
3. $b^{-1/3}$
4. $-7^{3/4}$

Change to exponential form.

5. $\sqrt{3y}$
6. $\sqrt[4]{x^3}$
7. $\dfrac{a}{\sqrt[3]{5}}$
8. $\sqrt{x-y}$

Evaluate without using a calculator.

9. $8^{1/3}$
10. $-9^{3/2}$
11. $49^{-1/2}$
12. $27^{-2/3}$

Evaluate using a calculator. Round to the nearest hundredth if necessary.

13. $64^{2/3}$
14. $-9^{-3/2}$
15. $6^{1/4}$
16. $5^{-2/3}$

Simplify using the laws of exponents. Assume all variable bases are positive, and leave your answers in exponential form, with positive exponents.

17. $(x^{1/2})^{1/4}$
18. $a^{-1/3}$
19. $(x^{2/3})^{9/10}$
20. $(8x^3)^{2/3}$
21. $\dfrac{a^{3/4}}{a^{1/3}}$
22. $\left(\dfrac{x^2}{y}\right)^{1/3}$

23. A pharmaceutical company, growing an organism to be used in a vaccine, finds that the organism grows at a rate so that it quadruples every day. If the company starts with 300 units one day, it will have n units in d days according to the equation

$$n = (300)(4)^d$$

Evaluate the number of units after 7.00 hours $\left(\dfrac{7}{24} \text{ of a day}\right)$.

Write in simplest form.

24. $\sqrt{98x}$
25. $\sqrt{800y}$
26. $\sqrt[3]{54}$
27. $\sqrt[4]{32}$
28. $\sqrt{10000y^9}$
29. $\sqrt{18xy^5}$
30. $\sqrt[3]{48x^7}$
31. $\sqrt[4]{256x^9}$
32. $\sqrt{\dfrac{8}{9}}$
33. $\sqrt{\dfrac{11}{25a^6}}$
34. $\dfrac{5}{\sqrt{6x}}$
35. $\sqrt{\dfrac{3}{2}}$
36. $\sqrt[3]{\dfrac{3}{4}}$
37. $\sqrt{\dfrac{5}{6x}}$
38. $\sqrt[3]{\dfrac{5}{16x}}$
39. $\sqrt{\dfrac{y^3}{96x}}$
40. $\sqrt{\dfrac{15x^5y^8}{8z^3}}$

Combine algebraically and simplify. Leave the answer in radical form.

41. $5\sqrt{2} - 8\sqrt{2} - \sqrt{2}$
42. $\sqrt{45} - \sqrt{80} + \sqrt{20}$
43. $\sqrt[3]{250} - 2\sqrt[3]{54}$
44. $\sqrt[3]{\dfrac{5}{9}} - \sqrt[3]{\dfrac{3}{25}}$
45. $81 + \sqrt{81} + \sqrt[3]{81} + \sqrt[4]{81}$
46. $81^{1/3} - 27^{-1/2} - 12^{3/2}$
47. $3\sqrt{3}(2\sqrt{3})$
48. $3\sqrt{6}(2\sqrt{3})$
49. $\sqrt{2}(4\sqrt{3} - 5)$
50. $(\sqrt{7} - \sqrt{3})(2\sqrt{7} - \sqrt{3})$
51. $(3\sqrt{2} - 2\sqrt{3})^2$
52. $(\sqrt{5} - 4)(3\sqrt{2} + \sqrt{3})$
53. $\sqrt{15} \div \sqrt{5}$
54. $\dfrac{\sqrt{3}}{2\sqrt{5}}$
55. $\dfrac{6\sqrt{2} - \sqrt{5}}{\sqrt{5}}$
56. $\dfrac{\sqrt{3}}{\sqrt{2} - \sqrt{3}}$
57. $\dfrac{\sqrt{5} - 3\sqrt{2}}{2\sqrt{5} + \sqrt{2}}$
58. $\dfrac{2 - \sqrt{3}}{\sqrt{2} + \sqrt{5}}$

Combine algebraically and simplify.

59. $4x\sqrt[3]{2} - 3y\sqrt[3]{2}$
60. $-3\sqrt{y} + 7\sqrt{y} - 2\sqrt{y}$
61. $5\sqrt{7x} - 2\sqrt{6x} - 3\sqrt{7x} + \sqrt{6x}$
62. $5y^2\sqrt{28y} - 2y\sqrt{54y^3} + 3\sqrt{63y^5}$
63. $6\sqrt{28x} - 4\sqrt{12x} - 5\sqrt{63x} + 3\sqrt{75x}$
64. $\sqrt{5a^3} - a\sqrt{45a} + a\sqrt{20a^2}$
65. $\sqrt{\dfrac{3x}{5}} - \sqrt{\dfrac{3x}{20}}$
66. $3a\sqrt{\dfrac{3}{5}} - 2a\sqrt{\dfrac{1}{15}} + a\sqrt{\dfrac{5}{3}}$
67. $\sqrt[3]{3} \cdot \sqrt[3]{4x^2}$
68. $3\sqrt{2x} \cdot 6\sqrt{3}$
69. $(4x\sqrt{3y})(2x\sqrt{5y})$
70. $(2\sqrt[3]{a})(3\sqrt[3]{a})$
71. $4\sqrt{5x}(2\sqrt{x} - 3\sqrt{5})$
72. $(\sqrt{3} - 2x)(\sqrt{3} + x)$
73. $(\sqrt{x} + 2\sqrt{y})(3\sqrt{x} + \sqrt{y})$
74. $(4\sqrt{5a} - 2\sqrt{3})(4\sqrt{5a} + 2\sqrt{3})$
75. $\sqrt{6xy} \div \sqrt{3x}$
76. $3x \div \sqrt{7}$
77. $\dfrac{\sqrt[3]{22xy}}{\sqrt[3]{2x}}$
78. $\dfrac{2 - 6\sqrt{x}}{3\sqrt{x}}$
79. $\dfrac{3 + \sqrt{a}}{2 - \sqrt{a}}$
80. $\dfrac{3\sqrt{x} - 2\sqrt{y}}{\sqrt{x} - \sqrt{y}}$

Solve and check.
81. $\sqrt{x} - 2 = 0$
82. $\sqrt[3]{x} = -3$
83. $\sqrt{x+5} + 4 = 0$
84. $\sqrt[3]{3-x} = 5$
85. $\sqrt{3x-2} = 7$
86. $2 - \sqrt{2x-3} = 5$
87. $\sqrt{3x-5} = \sqrt{x+1}$
88. $x - 2 = \sqrt{x^2 - 4}$
89. $x^{1/3} - 1 = 3$
90. $\sqrt{x+3} = 4 - \sqrt{x-5}$

91. The diameter of a circle having area A is

$$\sqrt{\frac{4A}{\pi}}$$

Simplify this expression.

92. When studying the resistance in an electric circuit, we find the equation

$$\frac{\sqrt{R_1 R_2}}{\sqrt{R_1} + \sqrt{R_2}}$$

Simplify this expression.

93. The natural frequency f of a body under simple harmonic motion (Fig. 20-8) is given by the equation

$$f = \frac{1}{2\pi}\sqrt{\frac{kg}{W}}$$

Solve this equation for W.

■ FIGURE 20-8

WRITING

94. Explain how exponents and radicals are used to write two different forms of the same expression. Show examples of each form and explain why we need both forms.

TEAM PROJECT

95. If we know the perimeter and one leg of a right triangle, then we can find the other two sides. Suppose we know that the perimeter of a right triangle is 98.0 cm and that one leg is 35.0 cm. Cut a piece of string 98.0 cm long and mark a 35.0 cm length. Make the right triangle with these dimensions, being careful to create a precise right angle. Use this model to estimate the lengths of the other two sides, to three significant digits. Use a protractor to estimate the measures of the two acute angles to the nearest degree.
    Verify your work algebraically. Use the formula for perimeter and the Pythagorean theorem to write a radical equation that models this triangle. Solve the equation. Use right triangle trigonometry to find the acute angles. Calculate the percent error in your experimental results.

Review Exercises

21 Quadratic Equations

Outcomes

When you have completed this chapter, you will be able to

1. Identify a quadratic equation.
2. Solve a pure quadratic equation.
3. Identify quadratic equations that do not have "real" solutions.
4. Solve a quadratic equation by factoring.
5. Solve a quadratic equation by completing the square.
6. Solve a quadratic equation using the quadratic formula.
7. Use quadratic equation procedures in applications involving rates.
8. Graph a quadratic function.

In this chapter we solve quadratic equations by factoring and by using the quadratic formula. We then solve applications that involve quadratic equations. Finally, we graph the quadratic function.

21-1 INTRODUCTION TO QUADRATIC EQUATIONS

In this section we define a quadratic equation and its terminology. We learn about the different types of solutions and learn to solve a pure quadratic equation. Finally, we apply some formulas which are pure quadratic equations.

Quadratic Equation in One Variable

A *quadratic equation* (in x) is a polynomial equation of second degree. It must contain a second-degree term (x^2), but no higher degree terms. It may (or may not) have a first-degree (x) term and a zero-degree term (constant). It cannot have a variable under a radical or in the denominator of a fraction.

Example 1

The equations
$$x^2 - 2x + 3 = 0, \quad 3x^2 - x = 5, \quad \text{and} \quad x^2 = 6$$
are all quadratic equations. The equations
$$x^3 - x = 2, \quad x - \sqrt{x} + 1 = 0, \quad \text{and} \quad \frac{1}{x^2} - x - 2 = 0$$
are *not* quadratic equations.

General Form of a Quadratic Equation

A quadratic equation is said to be in *general form* when it is written in the form

$$ax^2 + bx + c = 0$$

where a, b, and c are constants, with $a \neq 0$.

GENERAL FORM OF A QUADRATIC EQUATION	$ax^2 + bx + c = 0$	58

Quadratic equations in general form are usually written *without fractions* and with the first term *positive*. It is important to be able to identify a, b, and c for later use with the quadratic formula.

--- **Example 2** ---

The equation $2x^2 + 3x - 1 = 0$ is a quadratic equation in general form, with $a = 2$, $b = 3$, and $c = -1$.

--- **Example 3** ---

Write the equation $3x - x^2 = 5$ in general form and give the values of a, b, and c.

Solution: Transposing the 5, we have

$$-x^2 + 3x - 5 = 0$$

To make the coefficient of x^2 positive, we multiply both sides of the equation by -1:

$$x^2 - 3x + 5 = 0$$

This is the general form of the equation, with $a = 1$, $b = -3$, and $c = 5$.

Solution of a Quadratic Equation

Recall from Chapter 10 that a *solution* (or *root*) of an equation is a value of the variable that satisfies the equation; that is, it makes the equation true. A quadratic equation always has *two* solutions: They may be two *different real* solutions, or two *equal real* solutions, or two *non-real* solutions. We check the solutions of a quadratic equation by substituting them back into the original equation.

--- **Example 4** ---

Determine which, if any, of these values

$$x = 2, 1, \frac{1}{3}, \text{ or } -2$$

are solutions to the equation

$$3x^2 + 5x - 2 = 0$$

Solution: Substituting $x = 2$ into the equation:

$$3(2)^2 + 5(2) - 2 \stackrel{?}{=} 0$$
$$20 \neq 0$$

So, $x = 2$ is *not* a solution to the given equation.

Substituting $x = 1$,
$$3(1)^2 + 5(1) - 2 \stackrel{?}{=} 0$$
$$6 \neq 0$$

So $x = 1$ is *not* a solution to the given equation.

Substituting $x = \frac{1}{3}$:
$$3\left(\frac{1}{3}\right)^2 + 5\left(\frac{1}{3}\right) - 2 \stackrel{?}{=} 0$$
$$\frac{1}{3} + \frac{5}{3} - 2 \stackrel{?}{=} 0$$
$$0 = 0 \quad \text{checks}$$

So $x = \frac{1}{3}$ is a solution.

Substituting $x = -2$,
$$3(-2)^2 + 5(-2) - 2 \stackrel{?}{=} 0$$
$$12 - 10 - 2 \stackrel{?}{=} 0$$
$$0 = 0 \quad \text{checks}$$

So $x = -2$ is a solution.

We have now found the *two* solutions of the quadratic equation. Thus, the two solutions are
$$\frac{1}{3} \quad \text{and} \quad -2$$

Pure Quadratic Equation

A quadratic equation in which there is no first-degree (x) term is called a *pure quadratic* equation. In a pure quadratic equation, the coefficient of the x term, b, is always 0. Its general form is
$$ax^2 + c = 0$$

Example 5

The equation
$$3x^2 - 12 = 0$$
is a pure quadratic equation, since there is no x term.

Solution of a Pure Quadratic Equation

One way to solve a pure quadratic equation is by first solving for x^2, and then taking the square root of both sides of the equation. We use this principle:

If
$$x^2 = a$$
then
$$x = \sqrt{a} \quad \text{or} \quad x = -\sqrt{a}$$
or
$$x = \pm\sqrt{a}$$

If x^2 is nonnegative, the solutions are real. If it is negative, the solutions are not real. The solutions should be simplified, if possible. As always, we check each solution by substituting back into the original equation.

Example 6

Solve and check:
$$3x^2 - 12 = 0$$

Solution: Solving for x^2,
$$3x^2 = 12$$
$$x^2 = 4$$

Taking square roots,
$$x = 2 \quad x = -2$$

(We may also write $x = \pm 2$.)
So the solutions are 2 and -2.

Check: If $x = 2$,   If $x = -2$,
$$3(2)^2 - 12 \stackrel{?}{=} 0 \quad\quad 3(-2)^2 - 12 \stackrel{?}{=} 0$$
$$0 = 0 \text{ checks} \quad\quad 0 = 0 \text{ checks}$$

> **COMMON ERROR** When taking the square root of both sides of an equation, be sure to write the positive and negative signs ($\pm$) on one side of the equation. There will always be *two* solutions to a quadratic equation.

In the example above, the solutions are two *different real (and rational)* numbers. But often the solutions to a quadratic equation are *irrational*.

Example 7

Solve $2x^2 - 3 = 0$.

Solution: Solving for x^2,
$$2x^2 = 3$$
$$x^2 = \frac{3}{2}$$

Taking square roots,
$$x = \pm\sqrt{\frac{3}{2}}$$

Rationalizing the denominator,
$$x = \pm\sqrt{\frac{3 \cdot 2}{2 \cdot 2}} = \pm\frac{\sqrt{6}}{2}$$

So the solutions are $\frac{\sqrt{6}}{2}$ and $-\frac{\sqrt{6}}{2}$.

For this equation, the solutions are two *different real (and irrational)* numbers.

Steps for the Solution of a Pure Quadratic Equation

1. Solve for x^2.
2. Take the square root of both sides of the equation.
3. Simplify the result, if necessary.
4. Check both solutions by substituting them into the original equation.

Quadratic Equations with No Real Solution

All quadratic equations have two solutions, but they may not be *real* numbers. The square root of a negative number is not a real number; we call it a *complex* number, and we will study complex numbers in Chapter 29. For now, it will suffice to know that the square root of a negative number is not a real number.

Example 8

Solve $2x^2 + 3 = 0$.

Solution: Solving for x^2,

$$x^2 = \frac{-3}{2} = -\frac{3}{2}$$

Taking the square root of both sides,

$$x = \pm\sqrt{-\frac{3}{2}}$$

We may stop at this point, since we must take the square root of a negative number, and simply state that there is *no real solution*.

Formulas

Many formulas in applications are quadratic equations. It will be helpful to be able to solve them.

Example 9

The formula for kinetic energy K of a moving body is given by

$$K = \frac{1}{2}mv^2$$

where m is the mass of a body and v is its velocity. Solve this formula for its velocity v.

Solution: Transposing, we have

$$\frac{1}{2}mv^2 = K$$

To solve for v^2, we multiply both sides by $\frac{2}{m}$,

$$\frac{2}{m}\left(\frac{1}{2}mv^2\right) = \frac{2}{m} \cdot K$$

$$v^2 = \frac{2K}{m}$$

Taking square roots,

$$v = \pm\sqrt{\frac{2K}{m}}$$

Rationalizing the denominator,

$$v = \pm\sqrt{\frac{2K \cdot m}{m \cdot m}} = \pm\frac{\sqrt{2Km}}{m}$$

Since v is a positive velocity, we choose only the *positive* root,

$$v = \frac{\sqrt{2Km}}{m}$$

---- **Example 10** ----

If current I is flowing through a resistance R, then the power dissipated is given by

$$P = I^2R$$

What current flows in a wire to a home electrical oven if the resistance is 10.0 ohms and the power is 5.30 kW?

Solution: Before substituting into the formula, we change 5.30 kW to watts:

$$5.30 \text{ kW} = 5.30 \text{ kW}\left(\frac{1000 \text{ W}}{1 \text{ kW}}\right) = 5300 \text{ W}$$

Substituting into the given equation,

$$5300 = I^2(10.0)$$
$$I^2 = 530$$
$$I = \pm\sqrt{530}$$

Taking only the positive current and approximating $\sqrt{530}$,

$$I = 23.0 \text{ A}$$

SECTION 21-1 Exercises Introduction to Quadratic Equations

Quadratic Equation in One Variable

1. Which of the following are quadratic equations?

   (a) $x^2 - x + 1 = 0$ (b) $3 - \frac{1}{x^2} = 2x$

   (c) $4 = 3x^2$ (d) $5x^2 - \sqrt{x} = 3$

   (e) $3x - 2 = 5$ (f) $\frac{1}{3}x^2 = 7x$

2. Which of the following are quadratic equations?

   (a) $x^3 + x = 0$ (b) $3x^2 = 4x$

   (c) $6x = 5$ (d) $x = \frac{3}{x^2} + 4$

   (e) $\frac{3}{4}x^2 - 2x = \frac{5}{8}$ (f) $2\sqrt{x} - x^2 + 1 = 0$

General Form of a Quadratic Equation

Write in general form, with a positive leading coefficient and no fractions.

3. $2x^2 - 3x = 4$
4. $4x^2 - 5 = 2x$
5. $7x = 4x^2$
6. $5 = 2x^2$
7. $\frac{1}{3}x - 3 = \frac{5}{6}x^2$
8. $4 + \frac{1}{5}x^2 = \frac{3}{8}x$
9. $(3x + 1)(x - 2) = 1$
10. $(3x - 2)x = 4$

Solution of a Quadratic Equation

Determine which, if any, of the given values are solutions to the given equation.

11. $x^2 - 5x + 6 = 0$; $x = -1, 1, 2, 3$
12. $2x^2 - 5x = 3$; $x = -2, \frac{1}{2}, -3, 1$
13. $3x^2 - 2x - 1 = 0$; $x = -\frac{1}{3}, -1, 1, 2$
14. $(x - 2)(x - 1) = 6$; $x = 2, 4, 1, -1$

Pure Quadratic Equation

Solve and check. If a solution is irrational, write it in its simplified radical form. If there is no real number solution, write *no real solution*.

15. $x^2 - 25 = 0$
16. $49 - x^2 = 0$
17. $12 - x^2 = 0$
18. $x^2 - 20 = 0$
19. $5x^2 + 2 = 2$
20. $4x^2 - 9 = 0$
21. $2x^2 - 5 = 0$
22. $7 - 3x^2 = 7$
23. $x^2 + 3 = 0$
24. $2x^2 = -1$

Formulas

Solve these formulas from science for the variable in parentheses. Write the solution in simplest radical form.

25. $e = mc^2$ (c)
26. $V = \pi r^2 h$ (r)
27. $V = \frac{1}{3}\pi r^2 h$ (r)
28. $P = \frac{V^2}{R}$ (V)
29. $I = \frac{k}{d^2}$ (d)
30. $W = \frac{1}{2}kx^2$ (x)
31. $F = K\frac{M_1 M_2}{d^2}$ (d)
32. $F = \frac{Mv^2}{r}$ (v)

Applications

Solve and round to three significant digits.

33. The volume V of a cylinder with radius r and height h is given by

$$V = \pi r^2 h$$

Find the radius of the cylinder in Fig. 21-1 if its volume is 7.95 cm³.

34. The volume of a cone with radius r and height h is given by

$$V = \frac{1}{3}\pi r^2 h$$

Find the diameter of a conical sand pile in Fig. 21-2 if it holds 8.50 m³ of sand.

35. The power P (in watts) dissipated in an electrical load is given by

$$P = \frac{V^2}{R}$$

where V is the voltage (in volts) across the load and R is the resistance (in ohms) of the load. What is the voltage across a $10\overline{0}$ watt light bulb whose resistance is 132 ohms?

■ FIGURE 21-1

$h = 2.56$ cm

■ FIGURE 21-2   ■ FIGURE 21-3

36. The power P (in watts) dissipated in a resistance R (in ohms) through which a current I (in amperes) is flowing is given by

$$P = I^2R$$

What current flows in a wire of a home electrical circuit whose resistance is 150.0 milliohms and is dissipating 62.0 watts of power?

37. The centripetal force F of a body of mass m revolving in a circle of radius r with a velocity v (Fig. 21-3) is

$$F = \frac{mv^2}{r}$$

Find the velocity of a body with a mass of 0.275 kg revolving around a circle with a radius of 0.215 m and a centripetal force of 6.25 N. $\left(\text{Note: N} = \frac{\text{kg} \cdot \text{m}}{\text{s}^2} \right)$

21-2   SOLUTION BY FACTORING

In this section we see how certain quadratic equations can be solved by factoring. In Chapter 16, we studied factoring expressions using integral coefficients. Here we also consider only integral coefficients. Not all quadratic equations can be solved by factoring using only integral coefficients, but many can.

Zero-Product Principle

When we can solve an equation by factoring, we use the *zero-product principle* to find the solutions:
If

$$ab = 0$$

then

either $a = 0$ or $b = 0$ (or both)

This principle says that if a product of two factors is equal to *zero*, then one (or both) of the factors must equal *zero*.

Solution of a Pure Quadratic Equation

Pure quadratic equations ($ax^2 + c = 0$) can often be solved by factoring the nonzero side of the equation.

---- **Example 11** ----

Solve and check:

$$9x^2 - 4 = 0$$

Solution: The right side of the equation is already zero, so we just factor the left:

$$(3x + 2)(3x - 2) = 0$$

Using the zero-product principle

$$3x + 2 = 0 \qquad 3x - 2 = 0$$

Solving for x.

$$x = -\frac{2}{3} \qquad x = \frac{2}{3}$$

So the solutions are $-\frac{2}{3}$ and $\frac{2}{3}$.

Check:

If $x = -\frac{2}{3}$, \qquad If $x = \frac{2}{3}$,

$$9\left(-\frac{2}{3}\right)^2 - 4 \stackrel{?}{=} 0 \qquad 9\left(\frac{2}{3}\right)^2 - 4 \stackrel{?}{=} 0$$

$$0 = 0 \quad \text{checks} \qquad 0 = 0 \quad \text{checks}$$

Quadratic Equation with No Constant Term

A quadratic equation with no constant term has the form

$$ax^2 + bx = 0$$

To solve this equation, we write the equation in general form and factor the nonzero side of the equation. We factor out a common x and perhaps a coefficient, and then use the zero-product principle to set each factor equal to zero, and solve each resulting equation for x. Of course, it's wise to check each solution by substituting back into the original equation.

---- **Example 12** ----

Solve and check:

$$2x^2 = 8x$$

Solution: We write the equation in general form:

$$2x^2 - 8x = 0$$

Factoring, we get

$$2x(x - 4) = 0$$

Setting each factor equal to zero,

$$2x = 0 \qquad x - 4 = 0$$

Solving,

$$x = 0 \qquad x = 4$$

So the solutions are 0 and 4.

Check:

If $x = 0$,	If $x = 4$,
$2(0)^2 \stackrel{?}{=} 8(0)$	$2(4)^2 \stackrel{?}{=} 8(4)$
$0 = 0$ checks	$32 = 32$ checks

In the last example, it would have been tempting to divide both sides of the given equation by $2x$. Let's look at what happens if we *do* divide by $2x$:

$$2x^2 = 8x$$
$$\frac{2x^2}{2x} = \frac{8x}{2x}$$
$$x = 4$$

We get *one* of the roots, $x = 4$, but we lose the other root, $x = 0$. (This error occurs because we divided both sides of the equation by zero, since $2x = 0$.)

COMMON ERROR In solving an equation, do *not* divide both sides by the variable or a variable expression. This improper step may cause a root to be lost.

Solution of the General Quadratic Equation

We can solve certain general quadratic equations by factoring. Just bring all terms to one side of the equal sign, and factor (if the expression is factorable using integral coefficients). Set each factor equal to zero, and solve.

Example 13

Solve $x^2 + x - 6 = 0$.

Solution: The right side is already zero, so we can begin with factoring the left side:

$$(x + 3)(x - 2) = 0$$

Using the zero-product principle,

$$x + 3 = 0 \qquad x - 2 = 0$$

Solving,

$$x = -3 \qquad x = 2$$

The solutions are -3 and 2.

Example 14

Solve $4x^2 + 6x - 4 = 0$.

Solution: If, when the equation is in general form, there is a greatest common *numerical* factor greater than 1, then we should first divide both sides of the equation by that factor so as to make the equation easier to factor. Dividing both sides of this equation by 2, we get:

$$2x^2 + 3x - 2 = 0$$

Factoring,

$$(2x - 1)(x + 2) = 0$$

Setting each factor equal to 0,

$$2x - 1 = 0 \qquad x + 2 = 0$$
$$x = \frac{1}{2} \qquad x = -2$$

The solutions are $\frac{1}{2}$ and -2.

If the given equation is not in general form, we must rewrite it in general form, with zero on one side, before we factor.

Example 15

Solve $12x = -32 + 9x^2$.

Solution: Rewriting in general form:

$$9x^2 - 12x - 32 = 0$$

Factoring,

$$(3x + 4)(3x - 8) = 0$$
$$3x + 4 = 0 \qquad 3x - 8 = 0$$
$$x = -\frac{4}{3} \qquad x = \frac{8}{3}$$

The solutions are $-\frac{4}{3}$ and $\frac{8}{3}$.

COMMON ERROR The zero-product principle can be applied only to an equation in which *one side is zero*. Always write the equation in general form before you factor and use the zero-product principle.

The procedure for solving quadratic equations by factoring is as follows:

Steps for the Solution of a Quadratic Equation by Factoring

1. Write the equation in general form.
2. Factor the nonzero side of the equation, if possible.
3. Set each factor equal to zero.
4. Solve for the variable in each new equation.
5. Check both solutions by substituting them into the original equation.

Example 16

Solve $2x^2 + 2x = 3$.

Solution: We write the equation in general form: $2x^2 + 2x - 3 = 0$.
We try to factor the left side:

$$(2x + 1)(x - 3)$$
$$(2x - 1)(x + 3)$$
$$(2x + 3)(x - 1)$$
$$(2x - 3)(x + 1)$$

but we never get a middle term of $2x$. None of these work! So, the equation cannot be solved by factoring. Alternately, we could check the left side for factorability using the test procedure in Chapter 16.

Alternate Solution: If we use $b^2 - 4ac \stackrel{?}{=}$ perfect square, we get

$$2^2 - 4(2)(-3) = 28$$

This not a perfect square so once again it cannot be solved by factoring. We will see how we can solve it in the next section.

Formulas

A formula or equation for a certain application may often be solved by factoring. Since we frequently deal with physical problems, any root that does not make sense in the problem is rejected as unsuitable.

■ FIGURE 21-4

Example 17

The total surface area A of a cone with radius r and slant height s is given by Eq. 94,

$$A = \pi r^2 + \pi rs$$

Find the radius of the cone in Fig. 21-4 if its total surface area is 21π cm^2.

Solution: We substitute $s = 4$ and $A = 21\pi$ into the formula:

$$21\pi = \pi r^2 + \pi r(4)$$

Dividing both sides by π,

$$21 = r^2 + 4r$$

Putting in general form,

$$r^2 + 4r - 21 = 0$$

Factoring,

$$(r + 7)(r - 3) = 0$$
$$r + 7 = 0 \qquad r - 3 = 0$$
$$r = -7 \qquad r = 3$$

We reject the solution $r = -7$, since the radius cannot be negative.

Check:
We check the solution $r = 3$:

$$21\pi \stackrel{?}{=} \pi(3)^2 + \pi(3)(4)$$
$$\stackrel{?}{=} 9\pi + 12\pi$$
$$= 21\pi \quad \text{checks}$$

So the radius is 3 cm.

SECTION 21-2 Exercises Solution by Factoring

Solution of a Pure Quadratic Equation

Solve by factoring and check.
1. $x^2 - 9 = 0$
2. $4x^2 - 25 = 0$
3. $2x^2 - 8 = 0$
4. $3x^2 - 48 = 0$
5. $9x^2 = 64$
6. $25x^2 = 16$

Quadratic Equation with No Constant Term

7. $x(x - 2) = 0$
8. $3x(x + 1) = 0$
9. $x^2 + 2x = 0$
10. $3x^2 - 12x = 0$
11. $2x^2 - 6x = 0$
12. $5x^2 = 15x$

Solution of the General Quadratic Equation

13. $(x + 2)(x - 3) = 0$
14. $(2x - 1)(x + 3) = 0$
15. $x^2 + 2x - 3 = 0$
16. $x^2 - 6x = 7$
17. $x^2 + 2x + 1 = 0$
18. $x^2 + 25 = 10x$
19. $2x^2 + 9x + 10 = 0$
20. $10x^2 + 3 = 11x$
21. $2x^2 - 4x - 30 = 0$
22. $6x^2 - 3x = 18$
23. $x^2 + 9 = 6x$
24. $10x^2 = 18 - 3x$

Formulas

Solve by factoring.

25. The total surface area of a right circular cylinder is given by the formula

$$A = 2\pi r^2 + 2\pi rh$$

where r is the radius and h the height of the cylinder. Find the radius of the cylinder in Fig. 21-5 if the surface area is 8π m².

26. A variable electric current I (in amperes) is given by the formula

$$I = t^2 - 9t + 23$$

where t is time in seconds. Find the time when the current is equal to
  (a) 3 A
  (b) 5 A

27. A variable voltage V (in volts) is given by the formula

$$V = t^2 - 12t + 35$$

where t is time in seconds. Find the time when the voltage is equal to
  (a) 3 volts
  (b) 24 volts

28. The cost c for manufacturing n items is approximated using the formula

$$c = -4n^2 + 5n + 625$$

Find the number of items made if the manufacturing cost is $550.

FIGURE 21-5 ($h = 3$ m)

21-3 QUADRATIC FORMULA

In the last section we saw that only certain quadratic equations could be solved by factoring. *All* quadratic equations can be solved using the quadratic formula. The quadratic formula is derived using a process called *completing the square*. We will demonstrate this process with an example, and then derive the quadratic formula. The quadratic formula is the most useful of all methods for solving quadratic equations.

Completing the Square

In the process of completing the square, we manipulate the given equation so that one side of the equation is a perfect square trinomial. Recall from Sec. 16-5 that in a perfect square trinomial, the first and last terms are perfect squares, and the middle term is twice the product of the square roots of the outer terms.

Example 18

Solve by completing the square:

$$2x^2 - 3x - 1 = 0$$

Solution: We add 1 to both sides,

$$2x^2 - 3x = 1$$

and divide both sides by the coefficient of x^2,

$$\frac{2x^2 - 3x}{2} = \frac{1}{2}$$

Simplifying,

$$x^2 - \frac{3}{2}x = \frac{1}{2}$$

We complete the square on the left by adding the square of half the coefficient of the x term to both sides. The coefficient of the x term is $\frac{-3}{2}$. We take half of $\frac{-3}{2}$,

$$\frac{1}{2} \cdot \frac{-3}{2} = \frac{-3}{4}$$

and square it, getting $\left(\frac{-3}{4}\right)^2$. Adding that to both sides, we have

$$x^2 - \frac{3}{2}x + \left(\frac{-3}{4}\right)^2 = \frac{1}{2} + \left(\frac{-3}{4}\right)^2$$

Factoring and simplifying,

$$\left(x - \frac{3}{4}\right)^2 = \frac{1}{2} + \frac{9}{16} = \frac{17}{16}$$

Taking the square root of both sides,

$$x - \frac{3}{4} = \pm\sqrt{\frac{17}{16}}$$

$$x - \frac{3}{4} = \pm\frac{\sqrt{17}}{4}$$

$$x = \frac{3}{4} \pm \frac{\sqrt{17}}{4} = \frac{3 \pm \sqrt{17}}{4}$$

So the solutions are $\frac{3 + \sqrt{17}}{4}$ and $\frac{3 - \sqrt{17}}{4}$.

Derivation of the Quadratic Formula

As you see from the example above, the method of completing the square is too cumbersome to be a practical tool for solving quadratic equations. We illustrated its use in Example 18 so we could use it to derive the quadratic formula.

We wish to find the roots of Eq. 58

$$ax^2 + bx + c = 0$$

by completing the square. We start by subtracting c from both sides,

$$ax^2 + bx = -c$$

We divide both sides by a,

$$x^2 + \frac{b}{a}x = -\frac{c}{a}$$

Completing the square, we add the square of half the coefficient of the x term, $\left(\frac{b}{2a}\right)^2$ to both sides,

$$x^2 + \frac{b}{a}x + \left(\frac{b}{2a}\right)^2 = -\frac{c}{a} + \left(\frac{b}{2a}\right)^2$$

Simplifying the right side,

$$x^2 + \frac{b}{a}x + \left(\frac{b}{2a}\right)^2 = \frac{b^2}{4a^2} - \frac{c}{a}$$

Factoring the left and simplifying the right,

$$\left(x + \frac{b}{2a}\right)^2 = \frac{b^2 - 4ac}{4a^2}$$

Taking the square root of both sides,

$$x + \frac{b}{2a} = \pm\sqrt{\frac{b^2 - 4ac}{4a^2}} = \pm\frac{\sqrt{b^2 - 4ac}}{2a}$$

Subtracting $\frac{b}{2a}$ from both sides,

$$x = -\frac{b}{2a} \pm \frac{\sqrt{b^2 - 4ac}}{2a}$$

Putting the right side over the common denominator, we have the quadratic formula:

QUADRATIC FORMULA	If $ax^2 + bx + c = 0$, then $x = \dfrac{-b \pm \sqrt{b^2 - 4ac}}{2a}$	59

Solution of a Quadratic Equation by the Quadratic Formula

To use the quadratic formula to solve a quadratic equation, write the equation in general form, determine the coefficients a, b, and c and substitute them into the formula.

Example 19

Solve the equation of Example 18 by the quadratic formula:

$$2x^2 - 3x - 1 = 0$$

Solution: The equation is in general form, with $a = 2$, $b = -3$, and $c = -1$. Substituting these values in the quadratic formula, we have

$$x = \frac{-(-3) \pm \sqrt{(-3)^2 - 4(2)(-1)}}{2(2)}$$

$$x = \frac{3 \pm \sqrt{9 + 8}}{4} = \frac{3 \pm \sqrt{17}}{4}$$

So, we found the same solutions as before,

$$\frac{3 + \sqrt{17}}{4} \quad \text{and} \quad \frac{3 - \sqrt{17}}{4}$$

TIP

1. Always rewrite the quadratic equation in general form before trying to use the quadratic formula.
2. Use only the coefficients and not the variables to substitute into the quadratic formula.
3. Be careful with negative signs.

Example 20

Solve by the quadratic formula. Express the solutions in simplest radical form and also approximated to three significant digits.

$$x^2 - 8x - 4 = 0$$

Solution: The equation is in general form, with $a = 1$, $b = -8$, and $c = -4$. Substituting into the quadratic formula, we have

$$x = \frac{8 \pm \sqrt{64 - 4(1)(-4)}}{2(1)}$$

$$= \frac{8 \pm \sqrt{80}}{2}$$

Simplifying the radical, $\quad = \dfrac{8 \pm 4\sqrt{5}}{2}$

$$= 4 \pm 2\sqrt{5}$$

Approximating to three significant digits, the solutions are

$$-0.472 \quad \text{and} \quad 8.47$$

The quadratic formula can, of course, be used to solve quadratics with rational roots. If the roots are rational, then they can also be solved by factoring.

21-3 Quadratic Formula

Example 21

Solve by the quadratic formula:

$$2x^2 - x = 6$$

Solution: First, we write the equation in general form:

$$2x^2 - x - 6 = 0$$

so that $a = 2$, $b = -1$, and $c = -6$.

Substituting into the quadratic formula,

$$x = \frac{-(-1) \pm \sqrt{(-1)^2 - 4(2)(-6)}}{2(2)}$$

$$x = \frac{1 \pm \sqrt{1 + 48}}{4}$$

$$x = \frac{1 \pm 7}{4}$$

$$x = \frac{1 + 7}{4} \quad \text{or} \quad x = \frac{1 - 7}{4}$$

$$x = \frac{8}{4} \quad\quad\quad x = \frac{-6}{4}$$

$$x = 2 \quad\quad\quad x = -\frac{3}{2}$$

So the solutions are 2 and $-\frac{3}{2}$.

Alternate Solution: We could have solved this equation by factoring,

$$2x^2 - x - 6 = 0$$
$$(2x + 3)(x - 2) = 0$$
$$2x + 3 = 0 \quad\quad x - 2 = 0$$
$$x = -\frac{3}{2} \quad\quad x = 2$$

These factors yield the identical solutions, using a much simpler method. So, if a quadratic is easily factorable, it is usually wise to use the method of factoring, rather than the quadratic formula. In the exercises, however, we offer practice using the formula, even when the solutions are rational and easy to find by factoring.

In applications, we often must solve quadratics with approximate decimal coefficients.

Example 22

Solve and round to three significant digits:

$$6.23x^2 - 5.82 = 4.93x$$

Solution: Writing the equation in general form:

$$6.23x^2 - 4.93x - 5.82 = 0$$

Substituting in the quadratic formula, we have

$$x = \frac{4.93 \pm \sqrt{4.93^2 - 4(6.23)(-5.82)}}{2(6.23)}$$

We evaluate x by calculator, and round to three significant digits,

$$x \approx 1.44 \quad \text{and} \quad -0.649$$

Note: Some calculators have a built-in program for solving quadratic equations. Other calculators are programmable. Since quadratic equations are common, a program for solving them is very useful! Check your calculator manual for specific instructions.

Some quadratic equations do not have *real* solutions. When you use the quadratic formula, the radicand may be *negative*. A negative radicand indicates that the solutions will be complex numbers. Complex numbers will be studied in more detail in Chapter 29. For now, if the radicand is negative, then we say there is *no real solution*.

Example 23

Solve by the quadratic formula:

$$3x^2 - 3x + 5 = 0$$

Solution: The equation is in general form with $a = 3$, $b = -3$ and $c = 5$. Substituting into the quadratic formula,

$$x = \frac{-(-3) \pm \sqrt{(-3)^2 - 4(3)(5)}}{2(3)}$$

$$x = \frac{3 \pm \sqrt{9 - 60}}{6}$$

$$x = \frac{3 \pm \sqrt{-51}}{6}$$

So the radicand is negative and there is *no real solution*.

Example 24

Use the quadratic formula to solve for I in this formula for voltage in an electric circuit.

$$E = IR + \frac{P}{I}$$

Solution: To put the equation in general form, we first eliminate fractions by multiplying both sides by I,

$$EI = I^2R + P$$

Then,

$$RI^2 - EI + P = 0$$

Substituting into the quadratic formula,

$$I = \frac{E \pm \sqrt{E^2 - 4RP}}{2R}$$

As we have seen, the quadratic formula may be used to solve *any* quadratic equation. It is a very important and useful method and the formula should be *memorized*.

SECTION 21-3 Exercises Quadratic Formula

Solve these quadratic equations. Give irrational solutions in simplest radical form and approximate them to three significant digits. If there are no real number solutions, write *no real solution*.

Completing the Square

Solve by completing the square.

1. $x^2 + 6x - 2 = 0$
2. $x^2 - 8x - 3 = 0$
3. $3x^2 - 4x - 8 = 0$
4. $5x^2 - 2x - 1 = 0$

Solution of a Quadratic Equation by the Quadratic Formula

Solve by using the quadratic formula.

5. $x^2 + 5x - 1 = 0$
6. $x^2 - 4x - 2 = 0$
7. $x^2 - 3x + 4 = 0$
8. $x^2 + 2x + 6 = 0$
9. $2x^2 - 7x + 4 = 0$
10. $4x^2 - 2x - 1 = 0$
11. $2x^2 + 3x + 2 = 0$
12. $3x^2 - 5x + 2 = 0$
13. $x^2 + 2x = 1$
14. $x^2 - 7 = -2x$
15. $x^2 = 3$
16. $x^2 = 5x$
17. $x^2 + 5 = 2x$
18. $2x^2 = x + 6$
19. $3x^2 + 1 = 6x$
20. $5x^2 + 8x = -2$

Solve by factoring or by the quadratic formula.

21. $x^2 - 3x = 10$
22. $x^2 = 6x + 7$
23. $x^2 + 8 = -8x$
24. $x^2 + 2 = 6x$
25. $2x^2 + 6x = 3$
26. $5x^2 + 10x = -1$
27. $3x^2 + 5x = 2$
28. $2x^2 + x = 3$
29. $x^2 - \dfrac{4}{3}x = 1$
30. $\dfrac{1}{3}x^2 + \dfrac{3}{2}x = 3$
31. $\dfrac{2}{x} + 3x = 8$
32. $-\dfrac{1}{x+1} = \dfrac{3}{x} + 2$

Solve to three significant digits.

33. $5.21x^2 + 6.13x - 9.85 = 0$
34. $0.956x^2 + 0.852x = 12.1$
35. $3.67x^2 + 12.2x = 35.6$
36. $28.3x^2 + 29.0 - 7.52 = 0$

Applications

Solve these applications. Round to three significant digits.

37. The work w done in a circuit varies with time t (in milliseconds) according to the formula

$$w = 6t^2 - 8t + 14$$

Find the time(s) when the work is 12.5.

38. A circular highway curve of radius R and roadway width w has a visual obstacle beside the roadway. The sight distance C is given by the formula

$$C^2 = 4Rw + w^2$$

How wide would a roadway have to be to have a sight distance of 160 m for a curve with a 625 m radius?

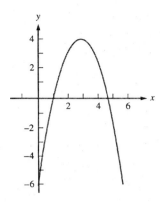

■ FIGURE 21-6   ■ FIGURE 21-7

39. The path of an object following a parabolic curve (Fig. 21-6) is described by the following equation:

$$y = -1.12x^2 + 6.52x - 5.07$$

Find the x coordinate(s) when $y = 3.00$.

40. The total surface area A of a cone with radius r and slant height s is given by

$$A = \pi r^2 + \pi r s$$

Find the radius of the cone in Fig. 21-7 if the total surface area is 112 cm².

Write each equation in general form, and use the quadratic formula to solve for the variable in parentheses.

41. The surface area of a cylinder:

$$A = 2\pi r^2 + 2\pi r h \qquad (r)$$

42. Displacement:

$$s = vt + \frac{1}{2}gt^2 \qquad (t)$$

43. The sum of the first n positive integers:

$$S = \frac{1}{2}n(n+1) \qquad (n)$$

44. The bending moment of a simply supported beam:

$$M = \frac{1}{2}wlx - \frac{1}{2}wx^2 \qquad (x)$$

21-4 APPLICATIONS OF QUADRATIC EQUATIONS

We now solve applications that require us to solve quadratic equations. You may wish to review the problem-solving techniques from Chapter 11. You should set these problems up as you did then. In this section, we solve geometric problems involving area and motion problems involving constant acceleration.

When we solve a quadratic equation, we have found that there are always two roots. We discard any roots that do not make sense in the physical problem (like a width of −3.62 metres). But do not be too hasty in rejecting solutions. Sometimes a second root will give an unexpected but equally good answer.

Area Problems

Quadratic equations are often derived from geometric problems involving areas.

Example 25

A rectangular plate of steel is needed in which the length is 3.06 m greater than twice its width and its area is 9.68 m². Find its dimensions.

■ FIGURE 21-8

Solution: First we sketch and label the rectangle (Fig. 21-8). We let w = the width of the rectangle. The length l will be

$$l = 2w + 3.06$$

We use Eq. 69 for the area of a rectangle:

$$A = lw$$

Substituting,

$$9.68 = (2w + 3.06)(w)$$
$$9.68 = 2w^2 + 3.06w$$
$$2w^2 + 3.06w - 9.68 = 0$$

We solve the equation by the quadratic formula:

$$w = \frac{-3.06 \pm \sqrt{3.06^2 - 4(2)(-9.68)}}{2(2)}$$

By calculator, we get the two solutions,

$$w = 1.564 \quad \text{and} \quad w = -3.094$$

rounded to three significant digits. Since we cannot have a negative width, we reject the second solution, $w = -3.094$.
Substituting $w = 1.564$ into the equation

$$l = 2w + 3.06 = 2(1.564) + 3.06$$

we find that

$$l = 6.19$$

So the dimensions of the rectangle are

$$1.56 \text{ m by } 6.19 \text{ m}$$

Uniformly Accelerated Motion

Uniformly accelerated motion refers to the motion of an object which has a uniform (constant) acceleration. However, its velocity is constantly changing. If an object, with an initial velocity v_0, moves with uniform acceleration a, then the distance s after time t is given by

$$s = v_0 t + \frac{1}{2} a t^2$$

An object that is dropped or thrown is an example of an object with uniformly accelerated motion—the uniform (constant) acceleration is the acceleration due to gravity. The acceleration due to gravity is called g, where

$$g = 9.81 \text{ m/s}^2 \quad \text{or} \quad g = 32.2 \text{ ft/s}^2$$

depending on whether measurements are given in SI or English units.

Example 26

If a diver falls from a diving board 10.0 m above the water, how much later does he strike the water?

Solution: Since the diver falls, the initial velocity is 0 and the acceleration is due to gravity ($g = 9.81$ m/s^2).
Using the equation

$$s = v_0 t + \frac{1}{2} a t^2$$

with $v_0 = 0$ and $a = g$, we get

$$s = \frac{1}{2} g t^2$$

Substituting $s = 10.0$ and $g = 9.81$ into this equation, we have

$$10.0 = \frac{1}{2}(9.81)t^2$$

$$t = \pm \sqrt{\frac{20}{9.81}}$$

$$t = \pm 1.43 \text{ seconds}$$

Since a negative time makes no sense in this problem, we reject the negative root.

Check: We check the positive root, $t = 1.43$:

$$10.0 \stackrel{?}{=} \frac{1}{2}(9.81)(1.43)^2$$

$$10.0 = 10.0 \quad \text{checks}$$

So it takes about 1.43 seconds for the diver to fall 10.0 metres off the diving board.

Example 27

From the top of a building, which is 62.5 metres above the ground, a ball is thrown towards the ground with an initial velocity of 13.0 metres per second. How many seconds will it take to hit the ground?

Solution: Use the equation

$$s = v_0 t + \frac{1}{2} g t^2$$

where s is the distance from the ground, v_0 is the initial velocity, g is the acceleration due to gravity and t is the time.

Substituting $s = 62.5$, $v_0 = 13.0$, and $g = 9.81$, we get

$$62.5 = 13.0 t + \frac{1}{2}(9.81) t^2$$

Rewriting the equation into the *general form* in order to use the quadratic formula and simplifying,

$$4.905 t^2 + 13.0 t - 62.5 = 0$$

Using the quadratic formula to solve for t,

$$t = \frac{-(13.0) \pm \sqrt{(13.0)^2 - 4(4.905)(-62.5)}}{2(4.905)}$$

Evaluating by calculator, we obtain two approximate solutions,

$$t = 2.48, \quad t = -5.13$$

21-4 Applications of Quadratic Equations

Again, we reject the negative solution, $t = -5.13$ seconds, since it doesn't make sense in this problem. Therefore, the ball hits the ground approximately 2.48 seconds after being thrown.

With uniform motion problems, we must pay particular attention to the directions (up or down) in the problem. In Example 27, all directions were *down*—the ground was *below* the top of the building, the ball was thrown *down,* and gravity acts *down* (toward the earth). So we actually chose *down* as the positive direction. In the next example, some directions are up and some are down, so we must choose an appropriate positive direction.

Example 28

A ball is thrown upwards at 34.1 m/s. In how many seconds will the ball be 38.0 m above the ground?

Solution: The ball is thrown upwards, and the distance of 38.0 m is also above the launching point, so we choose the up direction as positive. The force of gravity pulls the ball downwards, therefore the value of g must be negative ($g = 9.81$ m/s^2). Use the equation

$$s = v_0 t + \frac{1}{2} g t^2$$

where s is the distance from the ground, v_0 is the initial velocity, g is the acceleration due to gravity and t is the time.

Substituting $s = 38.0$, $v_0 = 34.1$, and $g = -9.81$, we get

$$38.0 = 34.1 t + \frac{1}{2}(-9.81) t^2$$

Rewriting the equation into the *general form* in order to use the quadratic formula and simplifying, we get

$$4.905 t^2 - 34.1 t + 38.0 = 0$$

Using the quadratic formula to solve for t,

$$t = \frac{-(-34.1) \pm \sqrt{(-34.1)^2 - 4(4.905)(38.0)}}{2(4.905)}$$

Evaluating by calculator, we obtain two approximate solutions,

$$t = 1.39, \quad t = 5.56$$

Both solutions are positive, indicating the ball passes 38.0 metres above the ground on its way up at approximately 1.39 seconds and again on its way down at approximately 5.56 seconds.

SECTION 21-4 Exercises Applications of Quadratic Equations

First write the equation or equations for each problem. Then solve each problem, using either factoring or the quadratic formula. Reject solutions that do not meet the conditions of the problem. Round answers appropriately.

Area Problems

1. A reception hall is three times as long as it is wide. Find its dimensions if the area is 135 m^2.

■ FIGURE 21-9

■ FIGURE 21-10

2. A small rectangular garden is to be 3.50 m longer than it is wide and to have an area of 28.0 m². Find its dimensions.

3. A sheet of tin is to be cut into 154 squares, 1.00 cm on a side. If the sheet is 3.00 cm longer than it is wide, find the dimensions of the sheet.

4. A technician must make an open pan by cutting squares from each corner of a 40.0 cm by 65.0 cm rectangular sheet of aluminum and folding up the sides (Fig. 21-9). If the base of the pan is to have an area of 1250 cm², what should be the length of the side of each square?

5. A bricklayer has bricks to lay 100.0 m² of walkway around a square garden (Fig. 21-10). The area of the garden and the area of the walkway are to be the same. How wide should he make the walkway?

6. A portrait 45.0 cm by 60.0 cm is to be set in a frame of uniform width so that the area of the frame is equal to that of the portrait (Fig. 21-11). Find the uniform width of the frame.

Uniformly Accelerated Motion

7. An object is dropped off the top of a 32 metre building. Determine how long it takes for the object to hit the ground.

8. How long will it take a drop of water to reach the bottom of a 46.6-metre-high waterfall?

9. If supplies are dropped from an airplane 258 m above the ground, how much later do they hit the ground? Give time in seconds.

■ FIGURE 21-11

10. Gravity on the moon is about $\frac{1}{6}$ the gravity on earth. About how many seconds does it take an object to fall 4.00 m from the door of the lunar module to the surface of the moon?

11. An object is thrown downward with a velocity of 8.00 m/s from the top of a 185 m building. In how many seconds will it hit the ground?

12. From an airplane flying at an altitude of 782 metres, an object is thrown downward at a rate of 29.2 metres per second. In how many seconds will it strike the ground?

13. An object is projected upward at 98.6 m/s. In how many seconds will the object first be 275 m above the ground?

14. A missile is shot upward at 147 m/s. In how many seconds will the missile hit the ground?

21-4 Applications of Quadratic Equations

15. An object is projected upward at 196.2 m/s.
    (a) In how many seconds will the object be 1962 m above the ground?
    (b) In how many seconds will the object be 1250 m above the ground? Why are there two answers?
    (c) In how many seconds will the object hit the ground?

16. Find the length and width of a rectangular field containing 8.50 acres if the length is twice the width. (1 acre = 43 560 ft²)

17. A farmer makes a rectangular enclosure using a stone wall for one side and 200.0 yd of fence for the other three sides (Fig. 21-12). Find the dimensions of the enclosure if the area enclosed is 1.00 acre. (1 acre = 4840 yd²)

18. A projectile is shot upward at 92.0 ft/s. After how many seconds will the projectile first be 125 feet above ground? ($g = 32.2$ ft/s²)

19. A ball is thrown directly upward at 128 ft/s. How long will it take the ball to return to ground level? ($g = 32.2$ ft/s²)

20. A ball is thrown upward at 96.6 feet/s.
    (a) In how many seconds will the object be 144.9 feet above the ground?
    (b) In how many seconds will the ball be 65.5 feet above the ground? Why are there two answers?
    (c) In how many seconds will the ball hit the ground?

■ FIGURE 21-12

21-5 APPLICATIONS INVOLVING RATES

Many applications involving rates lead to quadratic equations. These include uniform motion problems, rates of pay, and rates of work.

Uniform Motion

Recall that if an object is traveling with a uniform or average velocity, the distance d the object travels is given as

$$d = vt \quad \text{or} \quad d = \bar{v}t$$

where v is the uniform velocity or $\bar{v}$ is the average velocity and t is the elapsed time.

— **Example 29** —

A woman makes a regular trip driving her car a total distance of 585 kilometres. One day, because of bad weather, her average speed was reduced by 15.3 km/h. As a result, her trip took her an additional 2.00 hours. What was her usual average speed?

Solution: We make a table.

	$\bar{v}$	×	t	=	d
Regular trip	$\bar{v}$		t		585
Bad weather trip	$\bar{v} - 15.3$		$t + 2.00$		585

From this table we get two equations:

$$\bar{v}t = 585$$

and

$$(\bar{v} - 15.3)(t + 2.00) = 585$$

We want to solve for $\bar{v}$, so we solve the first equation for t resulting in

$$t = \frac{585}{\bar{v}}$$

We use this equation to substitute for t in the second equation,

$$(\bar{v} - 15.3)\left(\frac{585}{\bar{v}} + 2.00\right) = 585$$

Multiplying both sides by $\bar{v}$,

$$(\bar{v} - 15.3)(585 + 2.00\bar{v}) = 585\bar{v}$$

Multiplying out the left hand side, we get

$$2.00\bar{v}^2 + 554.4\bar{v} - 8950.5 = 585\bar{v}$$

Grouping the equation on the left hand side in *general form*

$$2.00\bar{v}^2 - 30.6\bar{v} - 8950.5 = 0$$

We solve this equation using the quadratic formula.

We find the solutions are $\bar{v} = -59.7$ and $\bar{v} = 75.0$

We reject the negative answer, since it makes no sense in this problem. The only acceptable answer is $\bar{v} = 75.0$ km/h.

Check: We calculate how long it would take the woman to travel 585 km at 75.0 km/h:

$$585 \text{ km} \div \frac{75.0 \text{ km}}{\text{h}} = 585 \text{ km} \times \frac{\text{h}}{75.0 \text{ km}} = 7.80 \text{ h}$$

In bad weather, she went 15.3 km/h slower,

$$75.0 \text{ km/h} - 15.3 \text{ km/h} = 59.7 \text{ km/h}$$

and it took her 2.00 hours longer

$$7.80 \text{ h} + 2.00 \text{ h} = 9.80 \text{ h}$$

The distance traveled then was

$$59.7 \frac{\text{km}}{\text{h}} \times 9.80 \text{ h} = 585 \text{ km}$$

Therefore the average speed during a regular trip is 75.0 km/h.

Other Rate Problems

Other types of problems involving rates may also lead to quadratic equations. These problems may concern rates of pay, rates of cost, or rates of work.

Example 30

Each year a computer company divides $7200.00 in Christmas bonuses equally among its employees. This year the company employed 6 more people than last year, and therefore each employee received $60 less. How much did each employee receive last year?

Solution: First we note that an important equation in this problem is

$$(\text{bonus rate}) \cdot (\text{number of employees}) = \text{Total bonus}$$

Using this equation, we make a table,

	Bonus rate	×	No. of employees	=	Total bonus
Last year	r		n		7200
This year	$r - 60$		$n + 6$		7200

From the table we get the two equations,

$$r \cdot n = 7200$$

and

$$(r - 60)(n + 6) = 7200$$

We wish to find r, so we eliminate n. Solving for n in the first equation, we have

$$n = \frac{7200}{r}$$

Substituting into the second equation,

$$(r - 60)\left(\frac{7200}{r} + 6\right) = 7200$$
$$(r - 60)(7200 + 6r) = 7200r$$
$$(r - 60)(1200 + r) = 1200r$$
$$1200r + r^2 - 72000 - 60r = 1200r$$
$$r^2 - 60r - 72000 = 0$$

Using the quadratic formula, we find the solutions are

$$r = 300 \quad \text{and} \quad r = -240$$

(We could have factored this quadratic, but most quadratics from applications are not factorable.) Rejecting the negative root as inappropriate for this problem, the only possible solution is $r = \$300$.

Check: If the company paid $300 in bonuses last year, for a total of $7200, there must have been 24 employees ($7200 \div 300 = 24$). This year there are 6 more, or 30 employees, and the bonus rate went down $60 to $240 per employee. Does this amount to $7200?

$$30 \cdot \$240 = \$7200 \quad \text{checks}$$

So, each employee received $300 as a Christmas bonus last year.

SECTION 21-5 Exercises Applications Involving Rates

First write the equation or equations for each problem. Then solve each problem. Reject solutions that do not meet the conditions of the problem. Round answers appropriately.

Uniform Motion

Solve by factoring or the quadratic formula.

1. A plane, making a flight of 4800 km, encountered a head wind of 160 km/h for the entire trip, making it one hour late in arriving. What was its groundspeed?

2. A man made a trip of 796 km. On his return trip, the weather was bad, and his average speed was reduced by 16 km/h. His total driving time was 20 hours. Find his average speed each way.

3. A motorboat takes 1 hour less to go 57.6 km down a river than to return. If the river flows at 4.8 km/h, find the rate at which the boat travels in still water.

4. A motorboat heads south from a dock at the same time that a sailboat heads west. The sailboat travels 11 km/h slower than the motorboat. The boats are 21 kilometres apart at the end of one hour. Find the rate of each boat.

Solve by quadratic formula (or calculator program), and round to three significant digits.

5. A jet plane, making a routine cross-country flight of 5230 km, encountered a head wind of 180 km/h for the entire trip, making it 1.00 hour late in arriving. What was its groundspeed?

6. A canoeist paddles 1670 metres across a lake. If he had paddled 15.0 m/min faster, he would have arrived at the other side 10.5 minutes earlier. How fast was he paddling?

7. A woman made a trip of 925 km. On her return trip, it was snowing, and her average speed was reduced by 16.7 km/h. As a result, her return trip took 2.25 hours longer. Find her average speed each way.

8. A train goes 335 kilometres to city A. Then, with better track, it continues on to city B, 517 kilometres from city A, at a speed of 24.7 km/h greater than on the first leg of the trip. The total travel time was 6.25 hours. Find the speed at which the train traveled to city A.

9. A man made a trip of 796 kilometres. On his return trip it was icy, and his average speed was reduced by 15.7 km/h. His total driving time was 19.5 hours. Find his average speed each way.

10. The average speed of a plane in still air is 684 km/h. With a constant wind blowing, the plane flies 1458 kilometres with a tail wind and returns with a head wind. If the round trip takes 4.25 hours of flying time, find the wind speed.

Other Rate Problems

Solve by factoring or quadratic formula.

11. The members of a technical organization decided to donate $2000 for a scholarship, and to share the cost of the contribution equally. If the organization had 20 more members, each would have had to contribute $5 less. How many members now belong to the organization?

12. A woman earned $1100 in a certain number of days. If her daily wage had been $10 more, she would have taken one less day to earn the same amount. How many days did she work at the lower rate?

13. A stonemason built 54 m of wall. If he had built 3 m less each day, it would have taken him 3 days longer. How many metres did he build each day, on the average?

14. A technician put together 48 computer boards. If she had put together 2 fewer boards each day, then it would have taken her 2 days longer to build those 48 boards. How many days did it take her to build the 48 boards?

15. Two resistors wired in series will give a total resistance R_T, where $R_T = R_1 + R_2$. Two resistors wired in parallel will give a total resistance R_T, where

$$\frac{1}{R_T} = \frac{1}{R_1} + \frac{1}{R_2}$$

What two resistors will give a total resistance of 500 ohms when wired in series and 100 ohms when wired in parallel? (Round to the nearest ohm.)

21-6 QUADRATIC FUNCTIONS

Recall from Sec. 12-5 that a quadratic function has the form

$$y = ax^2 + bx + c$$

However, quadratic functions from applications are not usually expressed with the variables x and y. Self-descriptive variables, such as t for time and r for radius, are generally used.

Example 31

These quadratic functions are examples from math, science, and technology.

(a) The area A of a circle is a quadratic function of the radius r:

$$A = \pi r^2$$

(b) The kinetic energy K of a moving body is a quadratic function of the velocity v:

$$K = \frac{1}{2} mv^2$$

(c) The displacement s of a freely falling body is a quadratic function of the time t:

$$s = vt + \frac{1}{2} gt^2$$

(d) The power P in a resistor is a quadratic function of the current I:

$$P = EI - RI^2$$

Graph of a Quadratic Function

Recall that the graph of a quadratic function is called a *parabola*. The parabola will open up or it will open down. A parabola is curved at all points (never straight). Every parabola has a maximum or a minimum point, called the *vertex*. The function may be graphed by making a table of values and plotting points, or by using a graphing calculator.

Applications

There are many examples of quadratic functions in math, science, and technology. We graph these functions, placing the independent variable along the horizontal axis and the dependent variable along the vertical axis.

Example 32

The kinetic energy K (in joules) of a moving object with mass m (in kg) and velocity v (in m/s) is given by

$$K = \frac{1}{2} mv^2$$

(a) On graph paper, graph the kinetic energy of a 2 gram rifle bullet as a function of its velocity, up to $v = 500$ m/s.
(b) Graph again, using a graphing calculator.
(c) Use the calculator to find the kinetic energy of the bullet at 355 m/s.

Solution:

(a) The formula above works for mass in kilograms and velocity in m/s. The velocity is in m/s, and we must convert the mass in grams to kilograms,

$$2 \text{ g} = 0.002 \text{ kg}$$

Substituting $m = 0.002$ into the equation for K we have

$$K = \frac{1}{2}(0.002)v^2$$

$$K = (0.001)v^2$$

The graph of this equation will be a parabola. In this problem, only positive velocities are meaningful, so our graph will be a partial parabola. We make a table:

v (m/s)	0	100	200	300	400	500
K (J)	0	10	40	90	160	250

We choose appropriate scales for each axis, plot the points, and connect them with a smooth curve (Fig. 21-13).

FIGURE 21-13

(b) Using a graphing calculator, we let the independent variable v be x and the dependent variable K be y. We graph

$$y = 0.001x^2$$

with a window of $0 \leq x \leq 500$ and $0 \leq y \leq 250$. The graph is identical to Fig. 21-13.

(c) To find the kinetic energy when $v = 355$ m/s, we use the "value" feature of the calculator, letting $x = 355$, to find that

$$y = 126$$

So the kinetic energy of the bullet is about 126 joules at 355 m/s.

──── **Example 33** ────

A ball is thrust upward at 34.1 m/s. Express the distance s of the ball above the ground as a function of time t, and **(a)** graph the function by hand, **(b)** graph the function by graphing calculator, and **(c)** use the calculator to find how long it takes for the ball to reach its highest point, to three significant digits.

Solution: We use the equation

$$s = vt - \frac{1}{2}gt^2$$

with $v = 34.1$ m/s and $g = 9.81$ m/s². Making these substitutions

$$s = (34.1)t - \frac{1}{2}(9.81)t^2$$

or

$$s = -4.905t^2 + 34.1t$$

21-6 Quadratic Functions

■ FIGURE 21-14

■ FIGURE 21-15

(a) We make a table, choosing integral values of t until we get a negative value for s. A negative value indicates the ball has hit the ground, and the equation no longer holds.

t (s)	0	1	2	3	4	5	6	7
s (m)	0.0	29.2	48.6	58.2	57.9	47.9	28.0	−1.6

We plot these points and connect them with a smooth curve, stopping between $t = 6$ and $t = 7$ where the graph goes below the t axis (Fig. 21-14).

(b) Use a graphing calculator to graph

$$y = -4.905x^2 + 34.1x$$

where $x = t$ and $y = s$, using a range of $0 \leq x \leq 7$ and $0 \leq y \leq 70$. The graph (Fig. 21-15) is identical to Fig. 21-14.

(c) We need to find the time (x) at which the ball reaches the vertex. (The vertex is the maximum point of the parabola). On the graphing calculator, we use the "maximum" feature (Fig. 21-15) to find that

$$x = 3.48 \text{ at the maximum point}$$

Therefore it takes the ball approximately 3.48 seconds to reach the highest point. As a matter of note, the y value is the maximum height reached, or 59.3 metres.

Graphical Solution of a Quadratic Equation

We can solve a quadratic equation graphically by graphing the quadratic function, and finding the x intercepts.

Example 34

Solve graphically, rounding to three significant digits.

$$x^2 + 2x = 1$$

Solution: First we bring all terms to one side of the equal sign:

$$x^2 + 2x - 1 = 0$$

Now we graph the function

$$y = x^2 + 2x - 1$$

and use the "root" or "zero" feature to find that the x intercepts are −2.41 and 0.414 (Fig. 21-16).

■ FIGURE 21-16

SECTION 21-6   Exercises   Quadratic Functions

Graph the function, placing the independent variable along the horizontal axis and the dependent variable along the vertical axis, using appropriate scales for both axes. If using the graphing calculator, determine the appropriate x and y ranges, graph the function, and sketch the graph on graph paper. Round answers to three significant digits.

FIGURE 21-17

1. The area of a circle with radius r is given by
$$A = \pi r^2$$
   (a) Graph A as a function of r, for radii from 0 to 5 cm.
   (b) Graphically find the area when the radius is 2.50 cm.

2. The power P (in watts) dissipated in an electrical load is given by
$$P = \frac{V^2}{R}$$
   where V is the voltage (in volts) across the load and R is the resistance (in ohms) of the load.
   (a) If the resistance is 100 ohms, graph P as a function of V for $0 \leq V \leq 30$ volts.
   (b) Graphically find the power dissipated if the voltage across the load is 18.0 volts.

3. The equation for the height h (in m) of a certain parabolic road arch at a distance d (in m) from the centre of the road (Fig. 21-17) is given by
$$h = 30 - \frac{1}{15}d^2$$
   (a) Graph h as a function of d for $-21 \leq d \leq 21$.
   (b) Find the height of the arch 6.70 m from the centre of the road.

4. The cost c (in dollars) for manufacturing n items is approximated using the formula
$$c = -4n^2 + 5n + 625$$
   (a) Graph c as a function of n for up to 10 items.
   (b) Find the cost for manufacturing 6 items.

5. A variable electric current I (in amperes) is given by the formula
$$I = t^2 - 9t + 23$$
   where t is the time in seconds.
   (a) Graph I as a function of t for $0 \leq t \leq 10$ s.
   (b) For what times is the current 12.5 amperes?

6. The volume V of a cylinder with radius r and height h is given by

$$V = \pi r^2 h$$

   (a) Graph V as a function of r for cylinders with a constant height $h = 10$ cm, for radii up to 8 cm.

   (b) Find the radius when the volume is 1250 cm³.

7. In a certain circuit, the power P (in watts) dissipated in the load resistor R (in ohms) with current I (in amperes) is

$$P = EI - I^2 R$$

   (a) If the voltage E is 120 volts and $R = 4$ ohms, graph P as a function of I, for I between 0 and 30 amperes.

   (b) Find the current when the power dissipated is 575 watts.

8. A projectile is shot upward at 96.6 ft/s.

   (a) Graph its distance s above the ground as a function of time t, for t between 0 and 6 seconds.

   (b) After how many seconds does it reach its highest point?

   (c) What is its maximum height?

9. The stopping distance d (in feet) for a car traveling at velocity v (in mi/h) is given by

$$d = 0.044v^2 + 1.1v$$

   (a) Graph the stopping distance as a function of the velocity, for velocities from 0 to 100 mi/h.

   (b) If it took a car 165 ft to stop, find the car's speed when the brakes were first applied.

Solve each equation graphically. Round to three significant digits.

10. $2x^2 + 10 = 9x$
11. $10x^2 + 3 = 11x$
12. $3x^2 + 2 = 8x$
13. $3x^2 - 4x = 3$

Chapter 21 REVIEW EXERCISES

1. Which of the following are quadratic equations?

   (a) $x^3 - 3x^2 - x = 8$ (b) $5x - x^2 = 3$

   (c) $\dfrac{1}{2} + \dfrac{2}{x} = \dfrac{3}{x^2}$ (d) $\dfrac{1}{4}x^2 - 3x = 1$

   (e) $3x^2 - 2 = 0$ (f) $5 = 2x$

Write in general form, with no fractions and $a > 0$.

2. $3x - x^2 = 4$ 3. $\dfrac{1}{2}x^2 = \dfrac{3}{4}$

4. $(2x - 3)(x + 2) = 1$

Determine which, if any, of the given values are solutions to the given equation.

5. $5x^2 + 9x - 2 = 0$; $x = -1, 1, -2, 2$

6. $(x - 1)(x + 2) = -4$; $x = 2, -2, 1, -3$

Solve and check.

7. $64 - x^2 = 0$
8. $3x^2 - 75 = 0$
9. $2x^2 + 3 = 3$
10. $49x^2 - 4 = 0$
11. $x^2 - 5 = 0$
12. $5.25x^2 - 3.75 = 0$
13. $3.46x^2 + 5.78 = 3.64$

Solve these formulas for the variable in parentheses. Write in simplest radical form.

14. $A = \pi r^2$ (r)
15. $s = \dfrac{1}{2}gt^2$ (t)

16. The power P (in watts) dissipated in an electrical load is given by

$$P = \dfrac{V^2}{R}$$

where V is the voltage (in volts) across the load and R is the resistance (in ohms) of the load. What is the voltage across a load whose power is 250 mW and resistance is 324 ohms?

Solve by factoring, and check.

17. $5x(x + 2) = 0$
18. $2x^2 = 10x$
19. $(2x + 1)(x - 1) = 0$
20. $x^2 + 15 = 8x$
21. $x^2 + 4x + 4 = 0$
22. $4x^2 + 18x + 20 = 0$
23. $6x^2 + x = 2$
24. $3x^2 - 3 = 6x$
25. $16x^2 - 9 = 0$
26. $8x^2 = 18$
27. $6x^2 - 6 = 5x$
28. $8x^2 + 3 = 14x$
29. $12 = x^2 - 11x$

30. A charge q (in coulombs) flows in a given circuit according to the formula

$$q = 2t^2 - 6t + 6$$

where t is time in microseconds. Find the time when the charge is equal to

(a) 2.0 coulombs

(b) 6.5 coulombs

Solve by completing the square. Give all irrational answers in simplest radical form and also approximate to three significant digits.

31. $x^2 - 7x + 2 = 0$
32. $2x^2 + 12x - 7 = 0$

Solve by using the quadratic formula. Give all irrational answers in simplest radical form and also approximate to three significant digits.

33. $x^2 - 4x + 1 = 0$
34. $2x^2 - 3x - 9 = 0$
35. $5x^2 + 6x + 4 = 0$
36. $2x^2 + 4x = 3$
37. $10 = 17x - 3x^2$
38. $2x^2 = 4x - 1$
39. $\dfrac{1}{6}x^2 + x = \dfrac{8}{3}$
40. $9x + \dfrac{25}{x} = 42$

41. $5.75x - 2.25x^2 = 1.25$
42. $3.01x^2 - 2.24 = 0.983x$

43. Write the equation in general form and use the quadratic formula to solve for r in this formula for the surface area A of a cone with slant height s:

$$A = \pi r^2 + \pi r s$$

44. In a certain circuit, the power P (in watts) dissipated in the load resistor R (in ohms) is

$$P = EI - I^2R$$

where E is the voltage (in volts) and I is the current (in amps). If the voltage E is 125 volts and $R = 103$ ohms, find the current I needed to produce a power of

(a) 20.0 watts

(b) 30.0 watts

45. The total surface area A of a cylinder with radius r and height h is given by

$$A = 2\pi r^2 + 2\pi rh$$

Find the radius of the cylindrical soup can in Fig. 21-18, if its surface area is 275 cm².

■ FIGURE 21-18

46. A rectangular field is fenced in with 725 m of fence. Find its dimensions if the area is 32 500 m².

47. If a flowerpot falls from a window and lands 78.4 m below, how many seconds does it fall?

48. An object is projected upward at 39.24 m/s.

(a) In how many seconds will the object be 30.0 m above the ground?

(b) In how many seconds will the object hit the ground?

49. A technician bought some transistors for $16.20. If she had received 3 more transistors for the same total price, then each transistor would have been 6 cents less. How much was each transistor?

Graph the function, placing the independent variable along the horizontal axis and the dependent variable along the vertical axis, using appropriate scales for both axes.

50. The volume V of a sphere with radius r is given by

$$V = \frac{4}{3}\pi r^3$$

(a) Graph V as a function of r for spheres for radii up to 8.00 cm.

(b) Find the volume for a sphere of radius 4.50 cm.

(c) Find the radius of a sphere of volume 125 cm³.

 51. An object is thrown downward from the top of a 285 ft building at a rate of 24.3 ft/s. In how many seconds does it hit the ground?

52. A jet flies a regular run of 7250 miles. With a tail wind of 103 mi/h, the trip takes 1.50 hours less than if there is no wind. How long does the regular (windless) run take?

WRITING

53. Describe four different methods of solving a general quadratic equation. What are the advantages and/or disadvantages of each method? Make up a quadratic equation and use it to illustrate all four methods of solution. Which do you prefer, and why?

TEAM PROJECT

54. A baseball player hits a pop fly directly above home plate. The ball is 1.00 m above ground when the player hits the ball. The ball goes upward at an initial velocity of 40.0 m/s.

(a) What is the height of the ball after 3.00 seconds?

(b) When is the ball 30.0 m above the ground?

(c) When does the ball hit the ground, if the catcher misses it?

(d) How high is the ball at its highest point?

(e) On the moon, the force of gravity is about one-sixth that on earth. How high would the ball have gone if it had been hit on the moon?

Oblique Triangles

Outcomes

When you have completed this chapter, you will be able to

1. Define and use the trigonometric functions at any angle using an angle θ in standard position.
2. Establish the sign of the trigonometric functions in all four quadrants.
3. Define and use the ASTC or CAST rule.
4. Define and use the primary quadrantal angles.
5. Find the trigonometric function of any angle using a calculator.
6. Find the correct angle from trigonometric functions knowing in which quadrant the angle occurs.
7. Use the reference angle to find the correct angle from the value of trigonometric functions.
8. Define and use the law of sines.
9. Define and use the law of cosines.
10. Solve oblique triangles using the law of sines and the law of cosines.

So far we have worked with only acute angles and right triangles. In this chapter we begin dealing with angles of any size and triangles that are not right triangles, *oblique triangles*. We extend the trigonometric functions of acute angles to angles of any size. We then derive the law of sines and the law of cosines, formulas that enable us to solve oblique triangles. Finally, we solve applications involving oblique triangles.

22-1 TRIGONOMETRIC FUNCTIONS OF ANY ANGLE

In this section, we extend our use of the six trigonometric functions to angles of any size. We examine the signs of the trigonometric functions in each quadrant and consider some special angles called *quadrantal* angles. Finally, we determine the trigonometric functions of any given angle, using the calculator.

The Trigonometric Functions of Any Angle

In Chapter 15, we defined the six trigonometric functions of an angle θ, in standard position (Fig. 22-1), with a point (x, y) on its terminal side, as follows:

■ FIGURE 22-1 Standard position of an angle.

TRIGONOMETRIC FUNCTIONS	$\sin \theta = \dfrac{y}{r}$	113
	$\cos \theta = \dfrac{x}{r}$	114
	$\tan \theta = \dfrac{y}{x}$	115
	$\csc \theta = \dfrac{r}{y}$	116
	$\sec \theta = \dfrac{r}{x}$	117
	$\cot \theta = \dfrac{x}{y}$	118

where $r = \sqrt{x^2 + y^2}$. We used these functions for acute angles in quadrant I only but stated that the definitions were true for angles in any quadrant. Now we consider angles of any size and in any quadrant.

Example 1

The point $(-2, -3)$ is on the terminal side of a positive angle θ in standard position. Sketch the angle and write the six trigonometric functions of θ, rounded to four significant digits.

Solution: We plot the point $(-2, -3)$ and draw the terminal side of the angle from the origin through the point (Fig. 22-2), noting that this is a quadrant III angle. Since the angle is positive, we draw the *arc* showing a counterclockwise rotation from the positive x axis to the terminal side, and label the angle θ. Drawing the perpendicular to the x axis, we find the distance r

$$r = \sqrt{x^2 + y^2}$$
$$= \sqrt{(-2)^2 + (-3)^2}$$
$$= \sqrt{13}$$

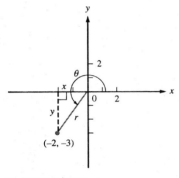

FIGURE 22-2

We leave r in radical form, as $\sqrt{13}$, rather than using its decimal approximation, so that we don't introduce rounding error. So, by Eq. 113–118, with $x = -2$, $y = -3$, and $r = \sqrt{13}$,

$$\sin \theta = \frac{y}{r} = \frac{-3}{\sqrt{13}} = -0.8321$$

$$\cos \theta = \frac{x}{r} = \frac{-2}{\sqrt{13}} = -0.5547$$

$$\tan \theta = \frac{y}{x} = \frac{-2}{-3} = -0.6667$$

$$\csc \theta = \frac{r}{y} = \frac{\sqrt{13}}{-3} = -1.202$$

$$\sec \theta = \frac{r}{x} = \frac{\sqrt{13}}{-2} = -1.803$$

$$\cot \theta = \frac{x}{y} = \frac{-3}{-2} = 1.500$$

22-1 Trigonometric Functions of Any Angle

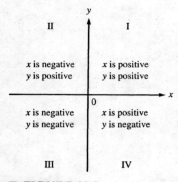

FIGURE 22-3
Signs of x and y.

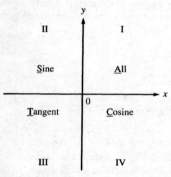

FIGURE 22-4
Positive trigonometric functions.

Algebraic Signs of the Trigonometric Functions

In Chapter 15, we saw that the six trigonometric functions of acute (quadrant I) angles were always positive. From the preceding example, it is clear that some of the trigonometric functions of angles in quadrants II, III, and IV are negative, because x or y (or both) may be negative. (Remember that r is always positive.)

The trigonometric functions are defined in terms of x, y, and r. Since r is always positive, the signs of the functions depend on the signs of x and y. Figure 22-3 shows the signs of x and y in each of the four quadrants.

Now let's determine the quadrants in which the sine, cosine, and tangent are positive:

1. Since $\sin \theta = \dfrac{y}{r}$, the sine will be positive when y is positive: quadrants I and II.

2. Since $\cos \theta = \dfrac{x}{r}$, the cosine will be positive when x is positive: quadrants I and IV.

3. Since $\tan \theta = \dfrac{y}{x}$, the tangent will be positive when x and y have the same sign: quadrants I and III.

This is summarized in Fig. 22-4.

Some students learn this as the "ASTC" graph (using the first letter of the function in each quadrant). It may help to remember the words "All Students Take Calculus" for "ASTC." Or you may find it easier just to sketch the angle and note whether x or y is positive or negative. From this you can determine whether the function you want is positive or negative.

An alternate memory trick that many others use, is to create a word starting in the fourth (IV) quadrant which results in the word "CAST." Either memory trick works quite well.

Example 2

Sketch the angle in standard position for each of the following and determine the quadrant of the terminal side. Then give the algebraic sign of each:

(a) sin 175° **(b)** tan 310° **(c)** cos 100° **(d)** cot (−120°)

Solution:

(a) We sketch a 175° angle (Fig. 22-5) and see that it is a quadrant II angle. Since

$$\sin \theta = \frac{y}{r}$$

and y is positive in Quadrant II, then sin 175° is positive.

(b) We sketch a 310° angle (Fig. 22-6) and see that it is a quadrant IV angle. Since

$$\tan \theta = \frac{y}{x}$$

and x and y have opposite signs in quadrant IV, then tan 310° is negative.

(c) We sketch a 100° angle (Fig. 22-7) and see that it is a quadrant II angle. Since

$$\cos \theta = \frac{x}{r}$$

and x is negative in quadrant II, then cos 100° is negative.

FIGURE 22-5

FIGURE 22-6

FIGURE 22-7

FIGURE 22-8

(d) Since $-120°$ is negative, we sketch the angle with a clockwise rotation (Fig. 22-8), and see that it is a quadrant III angle. Since

$$\cot \theta = \frac{x}{y}$$

and x and y have the same sign in quadrant III, then $\cot(-120°)$ is positive.

Quadrantal Angles

If the terminal side of an angle lies along one of the coordinate axes, the angle is called a *quadrantal angle*. The angles $0°$, $90°$, $180°$, and $270°$ are the *primary* quadrantal angles. Examples of other quadrantal angles are $360°$, $-180°$, and $540°$.

On the x axis, $y = 0$

and
$$r = \sqrt{x^2 + y^2} = \sqrt{x^2 + 0^2} = \sqrt{x^2}$$
$$= |x| \qquad = x \text{ on the positive } x \text{ axis}$$
$$\qquad = -x \text{ on the negative } x \text{ axis}$$

On the y axis, $x = 0$

and
$$r = \sqrt{x^2 + y^2} = \sqrt{0^2 + y^2} = \sqrt{y^2}$$
$$= |y| \qquad = y \text{ on the positive } x \text{ axis}$$
$$\qquad = -y \text{ on the negative } x \text{ axis}$$

Since the trigonometric functions are all quotients of x, y, and r, then the sine, cosine, and tangent of quadrantal angles are 0, 1, -1, or undefined. We illustrate this in the next example.

FIGURE 22-9

Example 3

Sketch each angle, and evaluate the trigonometric function without using a calculator:

(a) $\tan 0°$ **(b)** $\sin 90°$ **(c)** $\cos 180°$ **(d)** $\sec 270°$

Solution:

(a) We sketch the $0°$ angle (Fig. 22-9) and see that

$$y = 0$$

Using Eq. 115,
$$\tan 0° = \frac{y}{x}$$

Substituting $y = 0$,
$$\tan 0° = \frac{0}{x} = 0$$

(b) We sketch the 90° angle (Fig. 22-10) and see that
$$x = 0$$
and
$$r = |y| = y \qquad \text{(since } y \text{ is positive)}$$

Using Eq. 113,
$$\sin 90° = \frac{y}{r}$$

Substituting $r = y$,
$$\sin 90° = \frac{y}{y} = 1$$

■ FIGURE 22-10

■ FIGURE 22-11

(c) We sketch the 180° angle (Fig. 22-11) and see that
$$y = 0$$
and
$$r = |x| = -x \qquad \text{(since } x \text{ is negative)}$$

Using Eq. 114,
$$\cos 180° = \frac{x}{r}$$

Substituting $r = -x$,
$$\cos 180° = \frac{x}{-x} = -1$$

(d) We sketch the 270° angle (Fig. 22-12) and see that
$$x = 0$$

Using Eq. 117,
$$\sec 270° = \frac{r}{x}$$

Substituting $x = 0$,
$$\sec 270° = \frac{r}{0} \text{ undefined!}$$

■ FIGURE 22-12

Trigonometric Functions of Any Angle by Calculator

Given an angle, any of the trigonometric functions can be found using a calculator. The calculator must be in the proper angle unit mode: "degree" if the angle is in degrees, or "radian" if the angle is in radians. Refer to your calculator manual to determine how to change modes.

Example 4

Find cos 205° to four significant digits.

Solution: First, we put the calculator in the degree mode.

We find that

$$\cos 205° = -0.9063$$

rounded to four significant digits.

Example 5

Find tan 5.62 to four significant digits.

Solution: Since the angle 5.62 has no written units, we assume it is in radians and put the calculator in the radian mode.

We find that

$$\tan 5.62 = -0.7812$$

rounded to four significant digits.

> **COMMON ERROR** It is easy to forget to set the angle mode on your calculator. Be sure to set the appropriate mode (degrees or radians) before starting a problem in trigonometry!

Example 6

Find csc (−121°) to four significant digits.

Solution: We put the calculator in the degree mode. As noted in Sec. 15-5, there is no cosecant key on the calculator, so we must evaluate the cosecant as the reciprocal of the sine.

Since

$$\csc(-121°) = \frac{1}{\sin(-121°)}$$

We find that

$$\frac{1}{\sin(-121°)} = -1.167$$

So,

$$\csc(-121°) = -1.167$$

rounded to four significant digits.

22-1 Trigonometric Functions of Any Angle

SECTION 22-1 Exercises   Trigonometric Functions of Any Angle

The Trigonometric Functions of Any Angle

The given point is on the terminal side of a positive angle θ. Sketch the angle θ in standard position, showing the rotation. Then write the six trigonometric functions of θ, rounded to four significant digits.

1. $(3, -4)$
2. $(-8, 15)$
3. $(-6, 9)$
4. $(-4, -8)$
5. $(0, -4)$
6. $(6, 0)$
7. $(7, 0)$
8. $(0, -7)$

The given point is on the terminal side of a positive angle θ in standard position. Determine the quadrant of the terminal side and write the six trigonometric functions of θ in decimal form rounded to four significant digits.

9. $(-2.43, -0.750)$
10. $(1.24, -0.575)$
11. $(7.17, -3.74)$
12. $(-9.59, 8.76)$
13. $(-2.50, -\sqrt{5.75})$
14. $(\sqrt{7.25}, -3.75)$

Algebraic Signs of the Trigonometric Functions

For each of the following, an angle θ is in standard position and has its terminal side in the given quadrant. Determine the algebraic sign of the sine, cosine, and tangent of θ.

15. quadrant II
16. quadrant III
17. quadrant IV
18. quadrant I

State in what quadrant or quadrants the terminal side of θ can lie if:

19. $\sin \theta$ is positive
20. $\cos \theta$ is positive
21. $\cos \theta$ is negative
22. $\sin \theta$ is negative
23. $\tan \theta$ is negative
24. $\tan \theta$ is positive
25. $\sin \theta$ is negative and $\cos \theta$ is positive
26. $\cos \theta$ is negative and $\tan \theta$ is negative
27. both $\sin \theta$ and $\cos \theta$ are negative
28. $\sin \theta$ is positive and $\cos \theta$ is negative

Sketch the given angle and determine the algebraic signs of the sine, cosine, and tangent of the following angles.

29. 130°
30. 225°
31. 290°
32. −350°
33. 255°
34. 310°
35. −150°
36. 175°

Quadrantal Angles

Without using a calculator, find the following:

37. cos 90°
38. tan 180°
39. tan 270°
40. sin 360°
41. sin 0°
42. cos 270°
43. cos (−180°)
44. sin (−90°)
45. sin 270°
46. tan 90°
47. cot 180°
48. sec 0°

Trigonometric Functions of Any Angle by Calculator

Using a calculator, find the trigonometric function to four significant digits.

49. sin 245°
50. cos 167°
51. tan 95.7°
52. sin 331°
53. cos (−85.3°)
54. tan (−198°)
55. sec 145°
56. csc 287°
57. cot 325°
58. sec 103°
59. sin (−175°)
60. cos (−305°)
61. cos 46°50′
62. sin 25°3′
63. cos 11.5
64. sin 8.75
65. sin 3.42
66. tan 6.17

Write the sine, cosine, and tangent of θ in fraction form given the following.

67. $\sin \theta = \dfrac{-4}{5}$ and $\tan \theta$ is negative

68. $\cos \theta = \dfrac{12}{13}$ and $\sin \theta$ is negative

69. $\sec \theta = \dfrac{-17}{8}$ and $\cot \theta$ is positive

70. $\csc \theta = \dfrac{41}{9}$ and $\tan \theta$ is negative

22-2 FINDING ANGLES FROM TRIGONOMETRIC FUNCTIONS

In this section, we learn what a reference angle is. Then we use reference angles to determine the angle(s) when the value of a trigonometric function is given.

Angles with the Same Terminal Side

If we are given an angle θ in standard position, we can find another angle with the same terminal side by adding 360° (or a multiple of 360°) to θ (Fig. 22-13). We can find another angle which has the same terminal side by subtracting 360° (or a multiple of 360°) from θ (Fig. 22-14). Since these angles have the same terminal side, then the trigonometric functions of these angles have the same value.

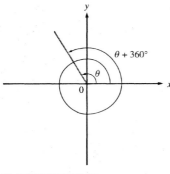

■ FIGURE 22-13
Angles with the same terminal side.

Example 7

Find a positive angle and a negative angle that have the same terminal side as the angle 145°, and show that the sine, cosine, and tangent of the three angles are the same.

Solution: To find another positive angle, just add 360° to the given angle: 145° + 360° = 505°.

To find a negative angle, just subtract 360° from the given angle:

$$145° - 360° = -215°$$

Now we'll compare the trigonometric functions. Using the calculator,

$$\sin 145° = 0.5736$$
$$\sin 505° = 0.5736$$
$$\sin (-215°) = 0.5736$$

■ FIGURE 22-14
Angles with the same terminal side.

Similarly,

$$\cos 145° = -0.8192 = \cos 505° = \cos (-215°)$$

and,

$$\tan 145° = -0.7002 = \tan 505° = \tan (-215°)$$

So the trigonometric functions of 145°, 505°, and −215° are the same.

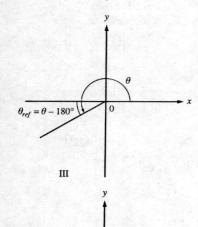

Reference Angles

Finding the value of a trigonometric function of an angle, such as sin 35°, is no problem if we use a calculator—the calculator does all the work! This is not the case, however, when we are given the value of a trigonometric function, such as $\sin \theta = 0.5$, and asked to find the angle (or angles) that have that trigonometric function value. For this, we need to make use of the *reference angle*.

The *reference angle* of an angle θ in standard position is the acute angle that the terminal side of θ makes with the x axis. We call the reference angle θ_{ref}. It is an unsigned angle (not a rotation). We sketch the reference angle by drawing the angle between the terminal side of the given angle and the x axis—either on the negative side or the positive side, whichever is closer (Fig. 22-15). This is summarized in the table:

θ	Quadrant	Reference Angle θ_{ref}
$0° < \theta < 90°$	I	θ
$90° < \theta < 180°$	II	$180° - \theta$
$180° < \theta < 270°$	III	$\theta - 180°$
$270° < \theta < 360°$	IV	$360° - \theta$

Example 8
Find the reference angle for the following:

(a) 50° **(b)** 135°

Solution: First we sketch the given angle, then sketch the reference angle.

(a) We sketch the angle $\theta = 50°$ in standard position (Fig. 22-16) and note that θ is a quadrant I angle. The reference angle, θ_{ref}, for a quadrant I angle θ is θ itself. Thus,

$$\theta_{ref} = \theta = 50°$$

FIGURE 22-16

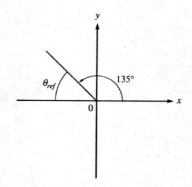

FIGURE 22-17

FIGURE 22-15 Reference angles, θ_{ref}.

(b) We sketch the angle $\theta = 135°$ and note that θ is a quadrant II angle (Fig. 22-17). The reference angle, θ_{ref}, for a quadrant II angle θ is

$$\theta_{ref} = 180° - \theta$$
$$\theta_{ref} = 180° - 135° = 45°$$

Example 9

Find the reference angle for the following:

(a) 230° (b) 345° (c) −150°

Solution:

(a) $\theta = 230°$ is a quadrant III angle (Fig. 22-18). The reference angle is

$$\theta_{ref} = \theta - 180°$$
$$\theta_{ref} = 230° - 180° = 50°$$

(b) $\theta = 345°$ is a quadrant IV angle (Fig. 22-19). The reference angle is

$$\theta_{ref} = 360° - \theta$$
$$\theta_{ref} = 360° - 345° = 15°$$

(c) $\theta = -150°$ is a negative angle whose terminal side is in the third quadrant (Fig. 22-20). From the sketch we see that

$$\theta_{ref} = 180° - 150° = 30°$$

■ FIGURE 22-18

■ FIGURE 22-19

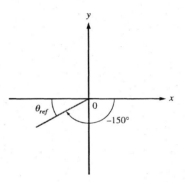
■ FIGURE 22-20

COMMON ERROR The reference angle θ_{ref} of an angle θ is always the acute angle between the terminal side of θ and the x axis, never the y axis.

■ FIGURE 22-21

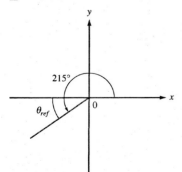

The reference angle is useful because the values of the trigonometric functions of an angle θ and of its reference angle θ_{ref} may be different only in the algebraic signs, while the absolute values are always the same.

Example 10

Find the reference angle θ_{ref} for an angle of $\theta = 215°$, and compare the sines, cosines, and tangents of θ and θ_{ref}.

Solution: First we sketch the angle θ and find its reference angle θ_{ref} (Fig. 22-21). Since θ is a quadrant III angle,

$$\theta_{ref} = 215° - 180° = 35°$$

So its reference angle is 35°. Comparing trigonometric functions, we find

$$\sin 215° = -0.5736 \quad \text{and} \quad \sin 35° = 0.5736$$
$$\text{so } \sin 215° = -\sin 35°$$
$$\cos 215° = -0.8192 \quad \text{and} \quad \cos 35° = 0.8192$$
$$\text{so } \cos 215° = -\cos 35°$$
$$\tan 215° = 0.7002 \quad \text{and} \quad \tan 35° = 0.7002$$
$$\text{so } \tan 215° = \tan 35°$$

We see that the sines and cosines of θ and of its reference angle θ_{ref} differ only in algebraic sign, and that the tangents are the same.

Finding the Angle When the Value of a Trigonometric Function Is Given

In determining an angle when the value of a trigonometric function is given, we usually need only consider nonnegative angles less than 360°. These angles are called the *primary angles*. An angle θ is a primary angle if

$$0° \le \theta < 360°$$

Recall from Sec. 15-2 that to find an angle that has a given trigonometric function value, we use an inverse trigonometric function key on the calculator. But the calculator gives us only *one* angle with the given trigonometric function value. In the next example, you will see that *there may be more than one angle having the same trigonometric function value,* and we show you how to find them. We limit our answers to *primary angles*.

■ FIGURE 22-22

■ FIGURE 22-23

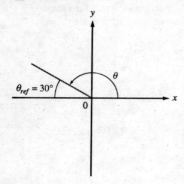

Example 11

Find all the primary angles θ for which $\sin \theta = 0.5$.

Solution: If $\sin \theta = 0.5$, then $\theta = \sin^{-1} 0.5$.

Using the calculator, we find one angle:

$$\theta = \sin^{-1} 0.5 = 30°$$

Are there any other primary angles for which the sine is 0.5? We must consider the algebraic sign of the sine function:

Since $\sin \theta = 0.5$, then the sine is positive.

Since $\sin \theta = \dfrac{y}{r}$, then $\dfrac{y}{r}$ must be positive.

Since r is always positive, then y must be positive.

And y is positive only in quadrants I and II (Fig. 22-22). Hence the sine is positive in only quadrants I and II.

We already found the quadrant I angle ($\theta = 30°$). The quadrant II angle is found using the reference angle, which is the angle itself; so $\theta_{ref} = 30°$. We sketch the reference angle θ_{ref} in quadrant II (Fig. 22-23) and sketch the positive angle θ, in standard position. We see

$$\theta = 180° - \theta_{ref}$$
$$\theta = 180° - 30° = 150°$$

So, $\theta = 150°$ has a sine of 0.5. Checking on the calculator, we see that

$$\sin 150° = 0.5$$

The two primary angles which have a sine of 0.5 are 30° and 150°.

If the reference angle is known, we can find angle θ in a specified quadrant using this table:

Quadrant	θ
I	θ_{ref}
II	$180° - \theta_{ref}$
III	$180° + \theta_{ref}$
IV	$360° - \theta_{ref}$

In general, to find all the primary angles which have a given trigonometric function:

1. Use the sign of the function to determine which are the quadrants of the primary angles.
2. Find the reference angle by calculator.
3. Determine the primary angles from the quadrants and the reference angle.

Example 12

Find the primary angles, to the nearest tenth of a degree, that have a cosine of -0.3480.

Solution: We want to find θ, so that

$$\cos \theta = -0.3480$$

Since $\cos \theta = \dfrac{x}{r}$, and r is always positive, the cosine will be negative only if x is negative. And x is negative only in quadrants II and III (Fig. 22-24). Using a calculator, we find

$$\theta = \cos^{-1}(-0.3480) = 110.4°$$

This θ is a quadrant II angle. To find the quadrant III angle, we need the reference angle. We draw the 110.4° angle (Fig. 22-25) and see that the reference angle θ_{ref} for a quadrant II angle is

$$\theta_{ref} = 180° - \theta$$
$$\theta_{ref} = 180° - 110.4° = 69.6°$$

■ FIGURE 22-24   ■ FIGURE 22-25

FIGURE 22-26

Using a reference angle of 69.6° in quadrant III (Fig. 22-26) we calculate the other primary angle,

$$\theta = 180° + \theta_{ref}$$
$$\theta = 180° + 69.6° = 249.6°$$

Checking on the calculator, we find that

$$\cos 249.6° = -0.3480$$

Thus, the primary angles that have a cosine of -0.3480 are 110.4° and 249.6°.

When the cotangent, secant, or cosecant is given, we make use of the reciprocal relationships, Eq. 119.

Example 13

Find all the primary angles, to the nearest tenth of a degree, that have a tangent of -2.5.

Solution: Since $\tan \theta = \dfrac{y}{x}$, then the tangent will be negative only if the signs of x and y are *different*. This occurs in quadrants II and IV (Fig. 22-27).

If

$$\tan \theta = -2.5$$

then

$$\theta = \tan^{-1}(-2.5)$$

By calculator, rounding to the nearest tenth of a degree,

$$\tan^{-1}(-2.5) = -68.2°$$

Sketching the angle to determine the reference angle (Fig. 22-28), we find the reference angle

$$\theta_{ref} = 68.2°$$

So, in quadrant IV

$$\theta = 360° - 68.2° = 291.8°$$

and in quadrant II (Fig. 22-29)

$$\theta = 180° - 68.2° = 111.8°$$

So the primary angles that have a tangent of -2.5 are 111.8° and 291.8°.

FIGURE 22-27

FIGURE 22-28

FIGURE 22-29

SECTION 22-2 Exercises Finding Angles from Trigonometric Functions

Find a positive angle and a negative angle which have the same terminal side as the given primary angle, and show that the sine, cosine, and tangent of the three angles are the same.

1. 164°   2. 261°   3. 328°   4. 82°

Reference Angles

Find the reference angle for each of the following angles.

5. 113°   6. 317°   7. −75°   8. −128°

Find the reference angle for each of the following, and compare the sines, cosines, and tangents of both angles.

9. 219°   10. 121°   11. 307°   12. 353°

Finding the Angle When the Value of a Trigonometric Function Is Given

Find all the primary angles, to the nearest tenth of a degree, whose functions are given below.

13. $\cos \theta = 0.5$
14. $\sin \theta = 0.7$
15. $\sin \theta = 0.1234$
16. $\cos \theta = 0.5678$
17. $\tan \theta = 1$
18. $\tan \theta = 1.5$
19. $\sin \theta = -0.5$
20. $\cos \theta = -0.5$
21. $\tan \theta = -2.348$
22. $\tan \theta = -1$
23. $\cos \theta = 0.375$
24. $\sin \theta = 0.7525$
25. $\cot \theta = 1.75$
26. $\sec \theta = -2.75$
27. $\csc \theta = -3.25$
28. $\sin \theta = -0.75$
29. $\cos \theta = 0$
30. $\sin \theta = 0$
31. $\tan \theta = 0$
32. $\cot \theta = 0$

22-3 LAW OF SINES

In this section, we derive a new formula, the law of sines, to enable us to solve certain oblique triangles quickly. Then we use the law of sines to solve some oblique triangles and applications.

Oblique Triangles

An *oblique triangle* is a triangle that has no right angle. When we sketch an oblique triangle, if we label the three angles as A, B, and C, then we label the sides *opposite* these angles as a, b, and c, respectively (Fig. 22-30).

Recall from Sec. 15-3 that to *solve* a triangle means to find all the missing sides and angles. Any triangle can be solved if we know three parts of the triangle, at least one of which is a side. If the triangle is a right triangle, we can use the Pythagorean theorem and the definitions of the six trigonometric functions. If the triangle is oblique (not a right triangle), we cannot use the Pythagorean theorem or the definitions of the six trigonometric functions.

■ FIGURE 22-30
Oblique triangle.

> **COMMON ERROR** The Pythagorean theorem and the definitions of the six trigonometric functions are used to solve only *right* triangles, and they *cannot* be used directly to solve *oblique* triangles.

Solving Oblique Triangles

Many oblique triangles can be solved using the *law of sines: In any triangle, the sides are proportional to the sines of the opposite angles.*

Mathematically, the law of sines can be stated:

| LAW OF SINES | $\dfrac{a}{\sin A} = \dfrac{b}{\sin B} = \dfrac{c}{\sin C}$ | 104 |

Derivation of the Law of Sines

We will derive the law of sines for an oblique triangle in which all three angles are acute. (The derivation is similar if one of the angles is obtuse.)

We start by separating the given triangle into two right triangles by drawing altitude h to side AB (Fig. 22-31). In right triangle ACD,

$$\sin A = \frac{h}{b} \quad \text{or} \quad h = b \sin A$$

■ FIGURE 22-31

In right triangle BCD,

$$\sin B = \frac{h}{a} \quad \text{or} \quad h = a \sin B$$

So

$$b \sin A = a \sin B$$

Dividing by $(\sin A)(\sin B)$:

$$\frac{b \sin A}{(\sin A)(\sin B)} = \frac{a \sin B}{(\sin A)(\sin B)}$$

Or,

$$\frac{b}{\sin B} = \frac{a}{\sin A}$$

Similarly, by drawing the altitude to side AC, we find

$$\frac{a}{\sin A} = \frac{c}{\sin C}$$

Combining these two equations we obtain

$$\frac{a}{\sin A} = \frac{b}{\sin B} = \frac{c}{\sin C}$$

This is the law of sines.

Classification of Triangles

Triangles can be classified according to which three parts are known. In the following, *A* refers to an angle and *S* refers to a side.

- AAS: Two angles and a side *opposite* one of the angles
- ASA: Two angles and the *included* side
- SSA: Two sides and an angle *opposite* one of the sides
- SAS: Two sides and the *included* angle
- SSS: Three sides

The notations ASA, SSS, and the like remind us of the congruence theorems and postulates of plane geometry, which establish the *uniqueness* of the triangles in all but the SSA case. In learning to solve oblique triangles, we will consider each of these cases (AAS and ASA in this section, and SSA, SAS, and SSS in the next two sections).

Solving a Triangle with Two Angles and an *Opposite* Side Known (AAS)

To use the law of sines to solve an oblique triangle, we must have a known side *opposite* a known angle (AAS). The law of sines actually gives us three equations:

$$\frac{a}{\sin A} = \frac{b}{\sin B}$$

$$\frac{a}{\sin A} = \frac{c}{\sin C}$$

$$\frac{b}{\sin B} = \frac{c}{\sin C}$$

To use the law of sines, choose whichever equation has the three known parts and the one unknown part.

■ FIGURE 22-32

─── **Example 14** ───

Solve the triangle in which $A = 65.1°$, $B = 35.2°$, and $b = 12.3$ cm (Fig. 22-32).

Solution: We can use the law of sines, because we are given a side b and the opposite angle B. Referring to Fig. 22-32, and using the law of sines, Eq. 104:

$$\frac{a}{\sin 65.1°} = \frac{12.3}{\sin 35.2°}$$

Solving for a, we get

$$a = \frac{(\sin 65.1°)(12.3)}{\sin 35.2°} = 19.4$$

In this triangle, we are given two of the three angles, so we can determine C easily:

$$C = 180° - (65.1° + 35.2°) = 79.7°$$

Again using the law of sines, Eq. 104:

$$\frac{c}{\sin 79.7°} = \frac{12.3}{\sin 35.2°}$$

Solving for c, we get

$$c = \frac{(\sin 79.7°)(12.3)}{\sin 35.2°} = 21.0$$

So, the three angles of this triangle are

$$A = 65.1°, B = 35.2°, \text{ and } C = 79.7°$$

The three sides are

$$a = 19.4 \text{ cm}, b = 12.3 \text{ cm}, \text{ and } c = 21.0 \text{ cm}.$$

22-3 Law of Sines

If you make a sketch more or less to scale, it can serve as an approximate check of your work. As another rough check, determine if the longest side is opposite the largest angle and if the shortest side is opposite the smallest angle.

Solving a Triangle with Two Angles and the *Included* Side Known (ASA)

In the preceding example, two angles and a side *opposite* one of the angles were given. We now consider an example in which two angles and the *included* side are given (ASA).

Example 15

Solve triangle ABC if $B = 123.6°$, $C = 38.3°$, and $a = 9.75$ cm.

Solution: First we sketch the triangle, more or less to scale (Fig. 22-33). In order to use the law of sines, we need a side and the angle opposite that side. The only side we have is a, and we don't have angle A. But we can find A easily:

$$A = 180° - (123.6° + 38.3°) = 18.1°$$

So now we can use the law of sines (Eq. 104):

$$\frac{9.75}{\sin 18.1°} = \frac{b}{\sin 123.6°}$$

Solving for b:

$$b = \frac{(\sin 123.6°)(9.75)}{\sin 18.1°} = 26.1$$

Again using the law of sines (Eq. 104):

$$\frac{9.75}{\sin 18.1°} = \frac{c}{\sin 38.3°}$$

Solving for c:

$$c = \frac{(\sin 38.3°)(9.75)}{\sin 18.1°} = 19.5$$

So, the missing angle and sides are

$$C = 18.1°, b = 26.1 \text{ cm, and } c = 19.5 \text{ cm}$$

■ FIGURE 22-33

Applications

As with right triangles, oblique triangles have many applications. You should follow the same procedures for setting up these problems as for other word problems. If the resulting triangle has two angles and a side given, then you can solve the triangle using the law of sines.

Example 16

A vertical antenna 4.63 m high stands on the top of a building on level ground. From a person on the street, whose eye level is 1.52 m above ground level, the angle of elevation of the top of the antenna is 36.5° and the angle of elevation of the top of the building is 33.4°. How tall is the building?

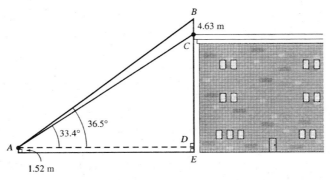

FIGURE 22-34

Solution: First we make a sketch, drawn more or less to scale (Fig. 22-34). We need to find the distance CE. If we knew AC, we could use right triangle trigonometry in triangle ACD. We use the law of sines in triangle ABC to solve for AC. We have a side ($BC = 4.63$ m) and need two angles. We can find angle BAC by subtraction:

$$\angle BAC = 36.5° - 33.4° = 3.1°$$

We can also find angle B, since angle B and angle BAD are complementary:

$$B = 90° - 36.5° = 53.5°$$

So we now have two angles and an opposite side in oblique triangle ABC. Using the law of sines (Eq. 104), we find

$$\frac{AC}{\sin 53.5°} = \frac{4.63}{\sin 3.1°}$$

Solving for AC:

$$AC = \frac{(\sin 53.5°)(4.63)}{\sin 3.1°} = 68.82$$

Now we use AC in right triangle ACD:

$$\sin 33.4° = \frac{CD}{AC}$$

Solving for CD:

$$CD = (AC)(\sin 33.4°)$$
$$CD = (68.82)(\sin 33.4°) = 37.88$$

Adding 1.52 m for the eye to ground distance,

$$37.88 + 1.52 = 39.40$$

Rounding to three significant digits, we find that the building is about 39.4 m tall.

SECTION 22-3 Exercises Law of Sines

Round sides to three significant digits and angles to the nearest tenth of a degree.

Solving a Triangle with Two Angles and an Opposite Side Known (AAS)

Solve for the missing part of these triangles which is in parentheses.
1. $A = 37.0°$, $B = 87.0°$, $a = 54.0$ cm (b)
2. $B = 68.0°$, $C = 47.0°$, $b = 72.5$ cm (c)
3. $A = 61.0°$, $C = 104.0°$, $c = 13.5$ cm (a)
4. $A = 114.0°$, $B = 51.0°$, $a = 26.7$ cm (b)
5. $B = 38.9°$, $C = 62.7°$, $b = 75.3$ cm (c)
6. $A = 78.5°$, $B = 54.1°$, $b = 6.75$ cm (a)
7. $A = 100°30'$, $C = 43°54'$, $c = 33.2$ cm (a)
8. $A = 0.900$ rad, $B = 1.50$ rad, $a = 56.5$ cm (b)

Solving a Triangle with Two Angles and the Included Side Known (ASA)

Solve the triangles for all the missing parts.
9. $A = 73.0°$, $B = 83.0°$, $c = 4.50$ cm
10. $B = 48.0°$, $C = 105.0°$, $a = 58.6$ cm
11. $A = 35.7°$, $C = 115.4°$, $b = 78.4$ cm
12. $B = 59.5°$, $C = 68.5°$, $a = 529$ cm

Applications

Solve and round to three significant digits.
13. Two stakes, A and B, are 68.0 m apart. A third stake C is placed so that angle ABC is 75.0° and angle BAC is 82.0°. Find the distance from C to each of the other stakes.
14. A surveyor needs to find the distance AB across a marsh (Fig. 22-35). A stake C is placed 155 m from A, so that angle CAB is 56.0° and angle ABC is 64.0°. Find the distance AB across the marsh.
15. Find the length of the shorter support of the shelf in Fig. 22-36, if the longer support is 37.5 cm long.
16. Two observers 1.45 km apart spot an airplane simultaneously. The angle of elevation of the airplane from observer A is 28.5°. The angle of elevation of the airplane from observer B is 47.9°. What is the height of the airplane?

■ FIGURE 22-35

■ FIGURE 22-36

522   CHAPTER 22   /   Oblique Triangles

Navigation circle
Azimuth 000° to 360°

Standard circle

■ FIGURE 22-37

17. The pilot flying a plane can see the beginning and the end of a 351-metre-long runway. The angle of depression of the beginning of the runway is 28.6°, and the angle of depression of the end of the runway is 22.5°. How far is the plane from the beginning of the runway?

18. A forest ranger in an observation tower at A spots a fire in the direction 37°15′ west of north. Another forest ranger in an observation tower 4.00 km due west of A spots the fire in the direction 43°36′ east of north. How far is the fire from each tower?

19. A lighthouse with a height of 38.1 m is on the edge of a sea cliff. From a ship, the angle of elevation of the bottom of the lighthouse is 36.5°, and the angle of elevation of the top of the lighthouse is 48.0°. How high is the cliff?

Note: In navigation, directions are usually given in terms of the size of the angle measured clockwise from the true north. This angle is usually referred to as a true azimuth angle and is quoted to three digits. For example, an aircraft flying to the northeast is said to be on a heading of 045°. (Fig. 22-37) In the following two exercises, use the navigation circle.

20. From a plane flying on a heading of due north, the azimuth of a radio station is 039.2°. After flying a distance of 4.50 km, the azimuth of the radio station from the aircraft is 046.0°. Find the distance between the aircraft and the radio station at the second azimuth.

21. In order to travel from town B to town A, we must travel 169 km at an azimuth of 198.5°. If town C is situated at an azimuth of 324.3° from town B, and also situated at an azimuth of 334.1° from town A, what is the distance between town C and town A?

22-4 AMBIGUOUS CASE

In the last section, we learned to solve oblique triangles in which two angles and one side were known (AAS and ASA). We solved these triangles using the law of sines. We now consider triangles in which two sides and an angle *opposite* one of them are known (SSA). This case is called the *ambiguous* case, because sometimes there is one *triangle* and sometimes there are *two triangles*.

Solving a Triangle with Two Sides and an *Opposite* Angle Known (SSA)

The law of sines can be used to solve triangles in which two sides and an *opposite* angle are known.

---- **Example 17** ----

Solve triangle ABC where $A = 121.0°$, $a = 24.5$ cm, and $b = 15.2$ cm.

Solution: First we will make an accurate sketch of the triangle; an accurate sketch (drawn to scale) will usually indicate whether we have one or two solutions. For consistency, we will always begin by drawing a horizontal ray with its endpoint on the left and construct the given angle ($A = 121.0°$) at the left end of the segment (Fig. 22-38). Two sides are given, one of which is opposite the given angle ($a = 24.5$ cm), and the other side is adjacent to the given angle ($b = 15.2$ cm). We use the adjacent side b to place C on the upper ray b units from A:

$$AC = b = 15.2$$

■ FIGURE 22-38

■ FIGURE 22-39

We construct side a (opposite angle A), beginning at point C. We open the compass to radius 24.5 (drawn to scale) and draw an arc and label point B at the intersection with the horizontal ray (Fig. 22-39). Since the arc intersects the horizontal ray in exactly one point, *there is exactly one triangle* with the given parts, triangle ABC.

We have an angle and the side opposite that angle, so we can use the law of sines:

$$\frac{24.5}{\sin 121.0°} = \frac{15.2}{\sin B}$$

Solving for $\sin B$:

$$\sin B = \frac{(\sin 121.0°)(15.2)}{24.5} = 0.5318$$

Taking the inverse sine,

$$B = \sin^{-1} 0.5318 = 32.13°$$

We find C by subtracting:

$$C = 180° - (121.0° + 32.13°) = 26.87°$$

To find side c, we again use the law of sines:

$$\frac{24.5}{\sin 121.0°} = \frac{c}{\sin 26.87°}$$

$$c = \frac{(\sin 26.87°)(24.5)}{\sin 121.0°} = 12.9$$

So, the missing angles and side are

$$B = 32.1°, C = 26.9°, \text{ and } c = 12.9 \text{ cm}$$

In the example above, the given angle was obtuse, and there was only one triangle that satisfied the given conditions. We can see from Fig. 22-39 that there can be *at most* one triangle if the given angle is obtuse. However, if the given angle is acute, then we must take special care when solving the triangle. We sometimes get a *second* solution (triangle) that may be just as important as the first.

Example 18

Solve triangle ABC in which $A = 31.5°$, $a = 49.3$ cm, and $b = 67.8$ cm.

Solution: As before, we sketch the given parts, starting with the given angle and the adjacent side (Fig. 22-40). With the compass point at C, we draw an arc with radius about 49.3 (to scale). We see that the arc intersects the horizontal ray in *two points, B and B'* (Fig. 22-41). So, there are *two different triangles* that have the same given parts! These triangles are overlapping in Fig. 22-41; we separate

■ FIGURE 22-40   ■ FIGURE 22-41

FIGURE 22-42
Two triangles!

them in Fig. 22-42 so we can see them better. In the triangle above, angle B' (angle $AB'C$) is obtuse. In the lower triangle, angle B (angle ABC) is acute. We can solve both triangles using the law of sines. We start with the acute triangle.

Upper Triangle: Using the law of sines, we first find the acute angle B:

$$\frac{49.3}{\sin 31.5°} = \frac{67.8}{\sin B}$$

$$\sin B = \frac{(\sin 31.5°)(67.8)}{49.3} = 0.7186$$

By calculator,

$$B = \sin^{-1} 0.7186 = 45.94°$$

We find angle C by subtracting:

$$C = 180° - (31.5° + 45.94°) = 102.56°$$

We find side c using the law of sines again:

$$\frac{49.3}{\sin 31.5°} = \frac{c}{\sin 102.56°}$$

$$c = \frac{(49.3)(\sin 102.56°)}{\sin 31.5°} = 92.1$$

So, the missing angles and side of the *upper* triangle are

$$B = 45.9°, C = 102.6°, \text{ and } c = 92.1 \text{ cm}$$

Lower Triangle: To find the lower triangle, we return to the equation above in which

$$\sin B = 0.7186$$

There are *two* angles which have a sine of 0.7186. One is in quadrant I and the other is in quadrant II. (Note that this is one place where we need that second angle!) The calculator gave us the first quadrant angle, $B = 45.94°$, and we used that angle to calculate the other parts of the first triangle. The quadrant II angle, B', is found using 45.94° as the *reference angle*.

In quadrant II,

$$B' = 180° - B$$
$$B' = 180° - 45.94° = 134.06°$$

When $B' = 134.1°$,

$$C = 180° - (31.5° + 134.06°) = 14.44°$$

Using the law of sines to find the new side c:

$$\frac{49.3}{\sin 31.5°} = \frac{c}{\sin 14.44°}$$

$$c = \frac{(49.3)(\sin 14.44°)}{\sin 31.5°} = 23.5$$

So, the missing angles and side of the *lower* triangle are

$$B' = 134.1°, C = 14.4°, \text{ and } c = 23.5 \text{ cm}$$

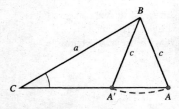

■ FIGURE 22-43

COMMON ERROR In a problem like the preceding one, it is easy to forget the possibility of a second solution, especially since a calculator gives only one angle when taking the inverse sine.

Whenever we are given two sides and an opposite angle, there are always three possibilities for the number of triangles that satisfy the given conditions: 0, 1, or 2. So how can we determine how many triangles to solve for? Well, there is a complex mathematical way of determining the number of solutions, but perhaps the *easiest* way is just to sketch the triangle fairly accurately, and see if it looks like there may be 2, 1, or 0 triangle solutions.

In all the examples above, the given parts were A, a, and b. Regardless of which parts are given, we suggest that you always start with sketching the angle on the left, labeling the adjacent side on the upper ray and then drawing the arc for the opposite side. For example, if C, a, and c were given, your sketch would be as in Fig. 22-43. You solve the triangle as shown above, using the new letters.

SECTION 22-4 Exercises Ambiguous Case

Sketch the triangle(s) with the given parts, and solve the triangle(s). Round sides to three significant digits and angles to the nearest tenth of a degree.

Solving a Triangle with Two Sides and an Opposite Angle Known (SSA)

1. $A = 126.0°$, $a = 25.2$ cm, $b = 20.8$ cm
2. $A = 135.0°$, $a = 30.5$ cm, $b = 25.9$ cm
3. $B = 36.0°$, $b = 127$ in., $c = 169$ in.
4. $C = 143.0°$, $b = 13.5$ in., $c = 18.3$ in.
5. $C = 152.0°$, $a = 74.8$ cm, $c = 89.2$ cm
6. $B = 95.0°$, $a = 53.8$ in., $b = 75.7$ in.
7. $B = 75.0°$, $a = 53.0$ cm, $b = 85.0$ cm
8. $C = 35.0°$, $b = 78.0$ in., $c = 49.0$ in.
9. $A = 16.7°$, $a = 10.3$ in., $c = 23.3$ in.
10. $B = 28.2°$, $b = 98.5$ cm, $c = 72.7$ cm
11. $C = 57.8°$, $b = 4.65$ m, $c = 4.25$ m
12. $A = 85.7°$, $a = 10.7$ m, $c = 9.25$ m
13. $C = 65.5°$, $b = 8.25$ in., $c = 9.75$ in.
14. $B = 30.7°$, $b = 44.7$ cm, $c = 87.4$ cm

Applications

15. A power pole on level ground is supported by two wires that run from the top of the pole to the ground. One wire is 12.5 m long and makes an angle of 48.6° with the ground, and the other wire is 11.0 m long. Find the angle the second wire makes with the ground.
16. A pole standing on level ground makes an angle of 84.3° with the horizontal. The pole is supported by a prop 7.77 m long whose base is 4.51 m from the base of the pole. Find the angle made by the prop with the horizontal.
17. A ship leaves a port and travels due west. At a certain point the ship turns 62.5° toward the north and continues in that direction for 45.5 km to a point 84.0 km from the port. How far did the ship travel west before turning?

18. A man owns a triangular lot. Markers at corners A and B are found but the third marker at C cannot be located. The deed states that $AB = 152$ ft, BC 115 ft, and $A = 53.5°$. Is the deed correct?

22-5 LAW OF COSINES

Not all triangles can be solved using the law of sines. In this section, we derive another formula, the law of cosines, with which we can solve the remaining two types of triangles (SAS and SSS). We solve triangles of both types and do some applications.

Law of Cosines

■ FIGURE 22-44

Can you use the law of sines to solve the triangle in Fig. 22-44? Since the side opposite angle A is not known, *you cannot use the law of sines*. We need to use a new law, the *law of cosines*, to solve this triangle. The law of cosines states that *the square of any side of a triangle is equal to the sum of the squares of the other two sides minus twice the product of these two sides and the cosine of their included angle.* This sounds a bit complicated, but it's really easy to use. Mathematically, the law of cosines tells us that in any triangle ABC (Fig. 22-45),

$$a^2 = b^2 + c^2 - 2bc \cos A$$

The same relationships hold for the other two sides.

■ FIGURE 22-45

Derivation of the Law of Cosines

Consider the oblique triangle ABC in Fig. 22-46. We start by dividing the triangle into two right triangles by drawing an altitude h from vertex C to side AB. In right triangle BCD, we have

$$a^2 = h^2 + (c - m)^2$$

In right triangle ACD, we have

$$b^2 = h^2 + m^2$$

■ FIGURE 22-46

Expanding the first equation $\quad a^2 = h^2 + c^2 - 2cm + m^2$
Rewriting the second equation $\quad \underline{b^2 = h^2 \qquad\qquad + m^2}$
Subtracting the second equation $\quad a^2 - b^2 = \qquad c^2 - 2cm$

Solving for a^2:

$$a^2 = b^2 + c^2 - 2cm$$

But, in triangle ACD,

$$\cos A = \frac{m}{b}$$

Or,

$$m = b \cos A$$

Substituting for m:

$$a^2 = b^2 + c^2 - 2c(b \cos A)$$

or,

$$a^2 = b^2 + c^2 - 2bc \cos A$$

This is one form of the law of cosines. We could derive the other forms of the law of cosines by drawing altitudes to side AC and to side BC using the same process. So there are three forms of the law of cosines, one for each side of a triangle:

LAW OF COSINES	$a^2 = b^2 + c^2 - 2bc \cos A$ $b^2 = a^2 + c^2 - 2ac \cos B$ $c^2 = a^2 + b^2 - 2ab \cos C$	105

Solving a Triangle with Two Sides and the Included Angle Known (SAS)

We need the law of the cosines to solve this type of triangle. We will solve the triangle in Fig. 22-44 that we could not solve at the beginning of this section.

Example 19

Solve triangle ABC in which $A = 41.6°$, $b = 3.27$ cm, and $c = 7.38$ cm.

Solution: We refer back to Fig. 22-44 and use the law of cosines:

$$a^2 = b^2 + c^2 - 2bc \cos A$$
$$a^2 = 3.27^2 + 7.38^2 - 2(3.27)(7.38)(\cos 41.6°)$$

The right side can be evaluated by calculator:

$$a^2 = 29.06$$

Taking the square root of both sides:

$$a = \pm 5.391$$

We keep only the positive value, since a is a distance. Rounding to three significant digits,

$$a = 5.39 \text{ cm}$$

We could now use the law of cosines to find angles B and C, or we could use the law of sines (since we now have an angle and the side opposite that angle). If one of the unknown angles is acute and one obtuse, choose to *solve for the acute angle first*, since you don't need to use reference angles for acute angles. We will solve for the acute angle B, using the law of sines.

$$\frac{5.391}{\sin 41.6°} = \frac{3.27}{\sin B}$$

$$\sin B = \frac{(\sin 41.6°)(3.27)}{5.391} = 0.4027$$

Rounding to the nearest tenth of a degree:

$$B = \sin^{-1} 0.4027 = 23.7°$$

Finding C:

$$C = 180° - 41.6° - 23.7° = 114.7°$$

So, the missing side and angles of the triangle are

$$a = 5.39 \text{ cm}, B = 23.7°, \text{ and } C = 114.7°$$

Example 20

Solve triangle ABC in which $B = 135.0°$, $a = 25.0$ cm, and $c = 37.0$ cm.

Solution: We make a sketch (Fig. 22-47). By the law of cosines

$$b^2 = a^2 + c^2 - 2ac \cos B$$
$$b^2 = 25.0^2 + 37.0^2 - 2(25.0)(37.0)(\cos 135.0°)$$
$$b^2 = 3302.1$$
$$b = 57.46$$

Rounding to three significant digits, $b = 57.5$ cm.

FIGURE 22-47

By the law of sines we find the angle C:

$$\frac{57.46}{\sin 135.0°} = \frac{37.0}{\sin C}$$

$$\sin C = \frac{(\sin 135.0°)(37.0)}{57.46} = 0.4553$$

$$C = 27.1°$$

We find angle A by subtraction:

$$A = 180° - 135.0° - 27.1° = 17.9°$$

So, the missing parts of the triangle are

$$b = 57.5 \text{ cm}, C = 27.1°, \text{ and } A = 17.9°$$

COMMON ERROR The cosine of an obtuse angle is negative. Be sure to use the proper algebraic sign when applying the law of cosines to an obtuse angle.

Solving a Triangle with Three Sides Known (SSS)

When three sides of an oblique triangle are known, we can use the law of cosines to solve for one of the angles. A second angle is found using the sine law, and the third angle is found by subtracting the other two from 180°.

Example 21

Solve triangle ABC, in which $a = 195$ cm, $b = 243$ cm, and $c = 137$ cm.

Solution: We sketch the triangle in Figure 22-48. Using the law of cosines we find the acute angle A:

$$a^2 = b^2 + c^2 - 2bc \cos A$$

$$195^2 = 243^2 + 137^2 - 2(243)(137)(\cos A)$$

$$\cos A = \frac{195^2 - 243^2 - 137^2}{-2(243)(137)} = 0.5977$$

$$A = 53.30°$$

FIGURE 22-48

We find the acute angle C by the law of sines:

$$\frac{195}{\sin 53.30°} = \frac{137}{\sin C}$$

$$\sin C = \frac{(\sin 53.30°)(137)}{195} = 0.5633$$

$$C = 34.28°$$

Finally,

$$B = 180° - 53.30° - 34.28° = 92.42°$$

So, the angles of the triangle are $A = 53.3°$, $B = 92.4°$, and $C = 34.3°$.

In the example above, we could have used the law of cosines to find the obtuse angle B, with no "ambiguity," since the cosine of an acute angle is positive, and the cosine of an obtuse angle is negative. Thus the law of cosines would tell us directly if the angle was acute or obtuse.

COMMON ERROR When working with the "Three Sides Known (SSS)" case, it is possible to be trapped into the ambiguous case unless you are careful. To avoid this problem, solve in either of these sequences. A) Solve for the obtuse angle first, using the law of cosines, then use the law of sines to solve for the other angles, or alternately, B) solve for the two acute angles first using the law of cosines for the first angle, the law of sines for the second angle, and find the third angle by difference as shown in Example 21.

SECTION 22-5 Exercises Law of Cosines

Sketch and solve each triangle ABC with the given parts. Round sides to three significant digits and angles to the nearest tenth of a degree.

Solving a Triangle with Two Sides and the Included Angle Known (SAS)

1. $A = 63.2°$, $b = 476$ cm, $c = 583$ cm
2. $B = 59.8°$, $a = 36.7$ cm, $c = 45.5$ cm
3. $C = 35.8°$, $a = 175$ cm, $b = 156$ cm
4. $A = 63.9°$, $b = 35.7$ cm, $c = 73.6$ cm
5. $B = 52.7°$, $a = 41.5$ cm, $c = 98.0$ cm
6. $C = 27.7°$, $a = 754$ m, $b = 123$ m
7. $C = 142°$, $a = 89.2$ mm, $b = 96.1$ mm
8. $B = 121°$, $a = 425$ cm, $c = 237$ cm

Solving a Triangle with Three Sides Known (SSS)

9. $a = 7.23$ m, $b = 7.64$ m, $c = 7.45$ m
10. $a = 546$ cm, $b = 487$ cm, $c = 528$ cm
11. $a = 329$ cm, $b = 651$ cm, $c = 442$ cm
12. $a = 98.3$ mm, $b = 158$ mm, $c = 67.4$ mm
13. $a = 309$ m, $b = 128$ m, $c = 182$ m
14. $a = 28.6$ cm, $b = 20.5$ cm, $c = 10.4$ cm

Applications

15. To measure the width BC of a pond, a transit is placed at A, and angle CAB is measured to be 27.4° (Fig. 22-49). The distances AB and AC are measured and found to be 73.5 m and 84.2 m, respectively. Find the width of the pond.
16. A triangular lot measures 768 m, 867 m, and 935 m along its sides. Find the angles between the sides.
17. Two boats are 68.5 km apart. Both are traveling toward the same port, which is 82.1 km from one of them and 96.4 km from the other. Find the angle at which their paths intersect.
18. Two straight roads intersect at an angle of 48.5°. A car on one road is 145 m from the crossing and a car on the other road is 179 m from the crossing. Find the distance between the two cars.
19. A ship is 7.75 km directly west of a port. If the ship sails northwest (45.0° west of north) for 2.14 km, how far will it be from the port?
20. From a point on level ground between two same-height telephone poles, cables are stretched to the top of each pole. One cable is 17.98 m long and the other is 21.95 m long. If the angle between the two cables is 135°, find the distance between the poles.

FIGURE 22-49

FIGURE 22-50

FIGURE 22-51

FIGURE 22-52

21. A ship leaves a port traveling due west at 21.0 km/h. Three hours later the ship turns 18.0° north of west. How far is the ship from the port 5.00 hours after it left the port?

22. A surveyor measures one angle of a triangular lot to be 68.0° and the two sides that meet at this corner to be 375 m and 256 m. Find the length of the third side.

23. Solar panels are to be placed on the southern roof of a home. The panels are 12.0 ft long and make an angle of 41.0° with the horizontal (Fig. 22-50). How long must the rafter on the north be?

24. Find the angle θ between the rafters in Fig. 22-51.

25. A vertical radio tower is built on the side of a hill, which makes an angle of 23.0° with the horizontal. Two guy wires support the tower 185 ft up and are attached to the ground 90.0 ft downhill and 90.0 ft uphill from the tower (Fig. 22-52). What are the lengths of the guy wires?

22-5 Law of Cosines

Chapter 22 REVIEW EXERCISES

1. Sketch the angle θ in standard position that has the point $(-4, -5)$ on the terminal side of the angle. Then write the six trigonometric functions of θ, rounded to four significant digits.

2. The point $(3.4, -2.6)$ is on the terminal side of an angle θ in standard position. Determine the quadrant of the terminal side, and write the six trigonometric functions of θ in decimal form with four significant digits.

3. An angle θ in standard position has its terminal ray in quadrant II. Determine the algebraic sign of the sine, cosine, and tangent of θ.

4. In what quadrant must the terminal side of θ lie if
   (a) the sine is positive and the cosine is negative?
   (b) both the cosine and the tangent are negative?

5. Without using a calculator, find the following functions of quadrantal angles:
   (a) tan 0°   (b) cos 270°   (c) sin 180°   (d) sec 90°

6. Find a positive angle and a negative angle which have the same terminal side as the primary angle $A = 154°$, and show that the sine, cosine, and tangent of the three angles are the same.

7. Find the reference angle for $A = 248°$, and compare the sines, cosines, and tangents of both angles.

8. Find the primary angles whose functions are given below. Round to the nearest degree.
   (a) $\cos \theta = -0.5878$ (b) $\sin \theta = -1$ (c) $\tan \theta = 0$ (d) $\csc \theta = -1.794$

Sketch the triangle and solve for all missing parts. Round all sides to three significant digits and angles to the nearest tenth of a degree.

9. $A = 68.3°$, $C = 75.2°$, $c = 5.37$ cm
10. $A = 21.0°$, $a = 7.20$ m, $c = 8.40$ m
11. $B = 82.0°$, $a = 47.0$ mm, $c = 75.0$ mm
12. $C = 31.0°$, $a = 14.5$ cm, $c = 43.7$ cm
13. $a = 6.30$ cm, $b = 7.50$ cm, $c = 5.90$ cm

■ FIGURE 22-53

532   CHAPTER 22 / Oblique Triangles

14. $A = 72.3°$, $B = 31.7°$, $c = 56.8$ m
15. A woman owns a triangular lot. Markers at corners A and B are found, but the third marker at C cannot be located. The deed states that $AB = 187$ m, $BC = 209$ m, and $A = 38.7°$. How far is C from A?
16. A drafting triangle has sides 10.2 cm, 24.5 cm, and 30.6 cm. Find the angles between the sides.
17. If $\tan \theta = -\dfrac{7}{24}$ and $\cos \theta$ is negative, write the sine and cosine of θ.
18. Find the length of the rafters PR and QR for the roof in Fig. 22-53.
19. A ranger leaves his station and walks due east 4.50 miles. He then turns and walks 3.50 miles in a direction 30.0° north of east. How far away from the station is he, and in what direction must he walk if he wants to go directly back to the station?

WRITING

20. Suppose you work for a company that builds bridges. You analyzed a proposed bridge truss made up of many triangles, sometimes using the law of sines and sometimes the more time-consuming law of cosines. Your client's accountant, angry over the size of your bill, has attacked your method of sometimes using the longer law of cosines when it is clear to him that the shorter law of sines is also "good for solving triangles." Write a letter to your client explaining why you sometimes had to use one law and sometimes the other.

TEAM PROJECT

21. Four mutually tangent circles are shown in Fig. 22-54. Find the radius r of the green circle.

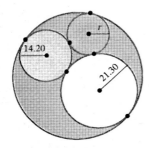

■ FIGURE 22-54

23

Vectors

Outcomes

When you have completed this chapter, you will be able to

1. Define vectors and their components.
2. Represent vectors algebraically and graphically.
3. Resolve a vector into its horizontal and vertical components.
4. Apply vectors to real applications.
5. Add vectors graphically.
6. Add vectors by their components.

In this chapter, we illustrate vectors graphically and use right triangle trigonometry to resolve vectors into components. We add vector quantities using two methods: addition by components and addition using the law of cosines. We look at many applications of vectors, including displacement, velocity, and force.

23-1 VECTORS AND COMPONENTS

In this section, we define a vector quantity and show the notation used for vectors. We show how to graphically represent a vector and the rectangular components of a vector.

Definitions

Many quantities can be described completely by giving their *magnitude* (a number indicating size). Such quantities are called *scalar* quantities.

Example 1

(a) Length is a scalar quantity, since it can be completely described by a number indicating its size: A certain board is 1.37 m long.
(b) Temperature is a scalar quantity: The temperature of a steel forging is 815°C.

Other quantities need to be described by giving not only the magnitude of the quantity but also its *direction*. Such quantities are called *vector quantities* (or vectors for short).

Example 2

(a) A plane takes off from Philadelphia and flies 79 miles. This is a distance, a scalar quantity. If we knew that the plane flew 79 miles from Philadelphia to New York City in the direction of 45° north of east, we would be specifying both magnitude and direction, or a vector quantity. This vector is called the *displacement* of an object. The displacement vector includes the direction of the displacement as well as the magnitude (the distance from the starting point).

(b) A speed of 90 km/h is a scalar quantity. If we say we are driving due south at 90 km/h, we are specifying a vector, called *velocity*. The velocity vector has both a magnitude (speed) and a direction (of motion).

(c) A *force* of 45 newtons is necessary to push a block *up* a 23° inclined plane. A force vector includes not only the magnitude of the force (45 newtons) but also the direction (23° from the horizontal).

Algebraic Representation of a Vector

Vectors are generally represented by boldface type, scalars by nonboldface type. Thus **V** (in boldface) represents a vector quantity, and V (in nonboldface) represents a scalar quantity. We use V to denote the *magnitude only* of the vector **V**. When writing by hand, it is customary to place an arrow over a vector quantity. So the vector **V** may be handwritten as $\vec{V}$.

Graphical Representation of a Vector

FIGURE 23-1
Vector **V**.

A vector is represented graphically by a line segment with an arrowhead at one end (a *directed line segment*). The length of the line segment represents the magnitude of the vector. The arrowhead is used to indicate the direction of the vector. The endpoint with the arrowhead is called the *tip* of the vector and the other endpoint is called the *tail* (Fig. 23-1).

When we draw a vector **V** on the coordinate axes, we generally start at the origin and draw the directed line segment at an angle θ specified by the direction of the vector (Fig. 23-2). The length of the directed line segment is drawn to scale, according to the magnitude of the vector. We can specify the vector **V** using the notation

$$V = \text{"magnitude"}, \quad \theta_V = \text{"direction"}$$

FIGURE 23-2

Which means that vector **V** has a magnitude V and is in the direction indicated by angle θ_V. This is called the polar form of the vector. There is no standard notation for representing the polar form of a vector, but this form of notation is least likely to be confused with a complex number. This notation will be used in this text.

Example 3

The force vector in Example 2c is drawn on the coordinate plane as a directed line segment 11 units long (drawn to scale) at an angle of 23° from the positive x axis (Fig. 23-3). The vector **F** is written in polar form as

$$F = 11, \quad \theta_F = 23°$$

FIGURE 23-3

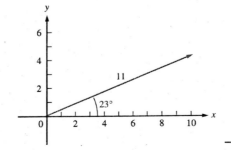

23-1 Vectors and Components

Rectangular Components of a Vector

Any vector can be replaced by two (or more) vectors which, acting together, exactly duplicate the effect of the original vector. These vectors are called the *components* of the original vector. We shall be concerned with components that are perpendicular to each other, called *rectangular components*. We sometimes use the word *component* to refer to only the magnitude of a component vector.

Resolving a Vector Graphically

To *resolve* a vector means to find its components. To resolve a vector graphically (to find its rectangular components), we sketch the vector and then draw a rectangle that has the given vector as one diagonal. From this rectangle, we can approximate the components of the vector.

FIGURE 23-4
Velocity vectors.

---- **Example 4** ----

Figure 23-4 shows a velocity vector **V** of an airplane flying in a direction 75° north of east. From due east (the horizontal axis) we draw an angle of 75° toward the north. We resolve the vector **V** to find the east component V_E and the north component V_N, by sketching a rectangle with a horizontal base that has **V** as its diagonal.

FIGURE 23-5
Force vector on an inclined plane.

---- **Example 5** ----

The weight **W** (a force vector) of a block on an inclined plane (Fig. 23-5) is shown resolved into two perpendicular components, **N** and **T**. **N** is called the *normal* force, acting perpendicular to the plane (holding the block on the plane), and **T** is called the *tangential* force, acting parallel, or "tangent," to the plane (causing the block to slide down the plane).

Often we need to resolve a vector into its horizontal and vertical components. If **V** is a vector, then the horizontal component is called the *x component* V_x and the vertical component is called the *y component* V_y (Fig. 23-6).

FIGURE 23-6
Components of a vector.

---- **Example 6** ----

Graphically resolve the vector **V** where

$$V = 4.8, \ \theta_V = 32°$$

into its vertical and horizontal components. Use the sketch to approximate V_x and V_y.

Solution: Using a protractor, we sketch a 32° angle in standard position. Next we draw the length of the vector to scale; since $V = 4.8$, we use 4.8 cm as the length of the directed line segment (Fig. 23-7). We draw a rectangle and use a metric scale to measure the length of $\mathbf{V}_x$ and $\mathbf{V}_y$ to be:

$$V_x \approx 4.1$$

and

$$V_y \approx 2.5$$

FIGURE 23-7

FIGURE 23-8
Resolving a vector.

Resolving a Vector Algebraically

Any vector **V** can be resolved algebraically into a horizontal x component $\mathbf{V}_x$ and a vertical y component $\mathbf{V}_y$. In Fig. 23-6, the vector **V** is the diagonal of a rectangle. Since opposite sides of a rectangle are equal, the right side of this rectangle is equal to the vector $\mathbf{V}_y$ (Fig. 23-8). The vector **V** is the hypotenuse of a right triangle in which its components $\mathbf{V}_x$ and $\mathbf{V}_y$ are the two legs. Using right triangle trigonometry, we have

$$\cos\theta = \frac{V_x}{V}$$

Solving for V_x:

$$V_x = V \cos\theta$$

And,

$$\sin\theta = \frac{V_y}{V}$$

Solving for V_y:

$$V_y = V \sin\theta$$

So, these are the algebraic equations needed to find the x and y components of a vector if the magnitude and direction of the vector are given:

x AND y COMPONENTS OF A VECTOR V	$V_x = V \cos\theta$	120
	$V_y = V \sin\theta$	121

Example 7

Algebraically find the components of the vector **V** in Example 6.

$$V = 4.8, \quad \theta_V = 32°$$

Solution: Using Eqs. 120 and 121,

$$V_x = 4.8 \cos 32° = 4.1$$
$$V_y = 4.8 \sin 32° = 2.5$$

This verifies our graphical result in Example 6 (Fig. 23-7).

If a vector **V** is given in polar form

$$V = \text{``magnitude''}, \quad \theta_V = \text{``direction''}$$

we use Eqs. 120 and 121 to find its rectangular components, V_x and V_y. The components are the coordinates of the tip of the vector **V**. We can specify the vector **V** by giving the coordinates of the tip of the vector:

$$\mathbf{V} = (V_x, V_y)$$

This is called the *rectangular form* of the vector.

FIGURE 23-9

Example 8

Sketch and express the vector **V** in rectangular form.

$$V = 45.7, \theta_V = 256°$$

Solution: We sketch the vector in Fig. 23-9. Using Eqs. 120 and 121 above, with $V = 45.7$ and $\theta_V = 256°$, we have

$$V_x = 45.7 \cos 256° = -11.1$$

and

$$V_y = 45.7 \sin 256° = -44.3$$

Thus, the rectangular form of vector **V** is $(-11.1, -44.3)$.

SECTION 23-1 Exercises Vectors and Components

Determine whether each of the following describes a scalar quantity or a vector quantity.

1. The area of a page.
2. The ship traveled 8 km north.
3. The airplane flew 15° west of south at 360 km/h.
4. We averaged 79.8 km/h on our trip.
5. The block has a mass of 75 kg.
6. The newspaper boy had a route 12.4 km long.
7. The woman pushed the block up the 5% incline with a force of 198 N.
8. A circuit has an impedance of 975 Ω and a phase angle of 28°.
9. A free-falling object accelerates at approximately 9.8 m/s².
10. The pipe held 45.7 litres of water.
11. A temperature of 58°C is the world's highest recorded temperature.
12. The sled went down the 34° hill at 15.6 m/s.

Resolving a Vector Graphically

Sketch these displacement vectors, and graphically approximate the magnitudes of the rectangular components to two significant digits. (use Fig. 22-37)

13. 25 km at azimuth 58°
14. 5.6 km at azimuth 305°

Sketch these velocity vectors, and graphically approximate the magnitudes of the rectangular components. (Use Fig. 22-37)

15. 6.8 km/h at azimuth 227°
16. 83 mi/h at azimuth 142°

Sketch these force vectors on a plane inclined at the given angle, and graphically approximate the magnitudes of the normal and tangential components.

17. 54 N at 15°
18. 15 N at 28°

Resolving a Vector Algebraically

Given the magnitude V of each vector and the angle θ that it makes with the positive x axis, find the x and y components, rounded to three significant digits.

19. $V = 785$, $\theta_V = 35.0°$
20. $V = 346$, $\theta_V = 67.2°$
21. $V = 135$, $\theta_V = 225°$
22. $V = 758$, $\theta_V = 145°$

Write the vector **V** in rectangular form, rounded to three significant digits.

23. $V = 4.77$, $\theta_V = 115°$
24. $V = 6.85$, $\theta_V = 315°$
25. $V = 0.125$, $\theta_V = 285°$
26. $V = 0.375$, $\theta_V = 255°$

23-2 APPLICATIONS OF VECTOR COMPONENTS

In this section, we consider applications of vectors, such as displacement, velocity, and force vectors. When solving vector applications, it's always helpful to draw a sketch. Often a sketch on the coordinate plane is useful.

Example 9

A pilot is flying a plane headed 21.5° north of east with a speed of 440 km/h. What are the east and north components of her velocity?

Solution: First we sketch the velocity vector **V** (Fig. 23-10), noting that the angle of 21.5° is drawn from the *east* direction, toward the north. Then we sketch the east and north components, V_E and V_N. It is clear from the sketch that we are asked to find the rectangular components of the velocity vector **V**. Using Eqs. 120 and 121 with $V = 440$ and $\theta = 21.5°$,

$$V_E = 440 \cos 21.5° = 409$$
$$V_N = 440 \sin 21.5° = 161$$

So, the east component of the velocity is 409 km/h, and the north component is 161 km/h.

FIGURE 23-10

The sketch for Example 9 (Fig. 23-10) was made on the coordinate axes. That was convenient, since the angle of 21.5° was in standard position. In some applications, the given angle may not be in standard position, and we have to calculate the angle in standard position.

Example 10

A ship pilot heads the ship due west at a velocity of 21.0 km/h. However, he finds that the ship is pushed off course by the water current, so that his actual direction of travel is 12.0° south of west. If the current is heading due south, find the ship's actual velocity in the direction of travel.

Solution: We make a sketch (Fig. 23-11), letting

V = actual velocity
V_x = west component of the actual velocity = -21.0 km/h
V_y = velocity of the current

FIGURE 23-11

The angle θ, in standard position, is

$$\theta = 180° + 12.0° = 192.0°$$

Using Eq. 120

$$V_x = V \cos \theta$$
$$-21.0 = V \cos 192.0°$$

Solving for V,

$$V = \frac{-21.0}{\cos 192.0°}$$
$$= 21.5 \text{ km/h}$$

So the ship's actual velocity is 21.5 km/h in the direction of 12.0° south of west.

Alternate Solution: It's not always necessary to use Eqs. 120 and 121 in which the angle is in standard position. Sometimes it's more reasonable to sketch a right triangle and use right triangle trigonometry.

We sketch the vectors (Fig. 23-12), letting

■ **FIGURE 23-12**

V = actual velocity
V_W = west component of the actual velocity
V_C = velocity of the current

We are given V_W,

$$V_W = 21.0 \text{ km/h}$$

and we need to find V. Using right triangle trigonometry,

$$\cos 12.0° = \frac{V_W}{V}$$

Or,

$$V = \frac{V_W}{\cos 12.0°} = \frac{21.0}{\cos 12.0°} = 21.5 \text{ km/h}$$

This confirms our first solution.

In some applications, the direction of the given vector has nothing to do with the positive x axis, and we just sketch a right triangle and apply simple right triangle trigonometry to the problem.

Example 11

A force of 69.5 N is being applied in the direction of the handle to push a lawn mower. The handle is at a 36.0° angle to the horizontal force. Find the vertical component which is pushing the mower into the ground.

■ FIGURE 23-13

Solution: First we make a sketch (Fig. 23-13), letting F be the 69.5 N force exerted on the handle of the lawn mower. We want to find F_y, the vertical component of the force. We're given that $F = 69.5$ N and $\theta = 36.0°$. From Eq. 121 we get

$$F_y = F \sin \theta$$

So,

$$F_y = 69.5 \text{ N} \sin 36.0°$$
$$F_y = 40.9 \text{ N}$$

So, the vertical component (the force pushing the mower into the ground) is 40.9 newtons.

In some applications, the components asked for are not x and y components, and we must use right triangle trigonometry to solve them.

Example 12

A block with a mass of 225 kg is placed on the inclined plane in Fig. 23-14. Find the two forces acting on the block, the normal force vector holding the block on the plane and the tangential force vector causing the block to slide down the plane. Round results to three significant digits.

Solution: First we find the force W of the vector **W** using the acceleration due to gravity

$$W = 225 \text{ kg} \times 9.81 \frac{\text{m}}{\text{s}^2} = 2207 \text{ N}$$

It can be shown (using geometry) that the angle θ between the force vector **W** and the normal vector **N** is the same as the angle of the inclined plane. Therefore,

$$\theta = 13.0°$$

■ FIGURE 23-14

23-2 Applications of Vector Components

Using right triangle trigonometry,

$$N = W \sin \theta = (2207 \text{ N}) \cos 13.0° = 2150 \text{ N}$$

and

$$T = W \sin \theta = (2207 \text{ N}) \sin 13.0° = 496 \text{ N}$$

Therefore, the normal force is 2150 N and the tangential force is 496 N.

SECTION 23-2 Exercises Applications of Vector Components

FIGURE 23-15

FIGURE 23-16

Solve these applications. Round results to three significant digits.

1. A ship travels 425 kilometres in a direction 36.3° north of east. What are the north and east components of the displacement?
2. An airplane is 126 km in a direction 64.4° south of west of an airport. What are the components of the plane's displacement from the airport?
3. A helicopter pilot, flying at an altitude of 751 m, sights a landing position at an angle of depression of 21.2°. What is the horizontal component of her displacement from the landing, and what is her distance from the landing?
4. An escalator has an angle of inclination of 32.7° between two floors that are 4.69 m apart. When going from the bottom of the escalator to the top, what is the horizontal component of the displacement, and how long is the escalator?
5. A plane is flying in a direction 35.5° west of north at 523 km/h. Find the north and west components of its velocity.
6. The wind is blowing at 91.1 km/h from the southeast (toward 45.0° west of north). Find its velocity components directed toward the north and the west.
7. Find the normal and tangential force vectors of this 125 kg block on the inclined plane (Fig. 23-15).
8. A cable exerts a force of 456 N at an angle of 48.7° with the horizontal (Fig. 23-16). Resolve this force into vertical and horizontal components. What is the maximum weight (in Newtons) that can be lifted by this cable?
9. At what speed should an airplane head east in order to follow a course 75.3° east of north if a 18.5 km/h wind is blowing from due south?
10. What force, neglecting friction, must be exerted to push a 75.0 lb weight up a slope inclined 14.7° from the horizontal?
11. A woman is pulling a sled with a rope with a force of 40.7 lb. If the rope makes an angle of 35.8° with the ground, find the component of the pulling force that pulls the sled along the ground and the component which is lifting the sled off the ground.
12. At what speed with respect to the water should a ship head south in order to follow a course 12.3° west of south if a current is flowing west at the rate of 8.75 mi/h?

23-3 GRAPHICAL ADDITION OF VECTORS

In Sec. 23-1 we saw that any vector can be replaced by its component vectors which, acting together, exactly duplicate the effect of the original vector. Conversely, any two vectors can be combined into a single vector, which has the same effect as the original two vectors. This vector is called the *resultant* of the two vectors. In this

FIGURE 23-17
Two vectors with an angle θ between them.

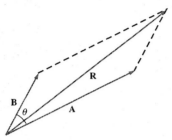

FIGURE 23-18
Vector addition: parallelogram method.

section, we estimate graphically the magnitude and direction of the resultant of two vectors.

Addition of Vectors

The process of combining vectors into a resultant is called *vector addition*. The resultant is also called the *vector sum*. If the vector sum of vectors **A** and **B** is the resultant **R**, we can write

$$\mathbf{A} + \mathbf{B} = \mathbf{R}$$

Note that this is *not* ordinary addition, but *vector* addition.

In general, if the vectors **A** and **B** are two vectors, with an angle θ between them (Fig. 23-17), we can find the resultant graphically using either of two methods: the *parallelogram* method or the *triangle* method.

Parallelogram Method

The method of adding two vectors *tail to tail* is called the *parallelogram* method. First we draw the vectors with their *tails* at a common endpoint and with the given angle between them. Then we complete a parallelogram by drawing line segments from their tips parallel to the given vectors (Fig. 23-18). The resultant **R** is a vector represented by the diagonal of the parallelogram, drawn from the common endpoint of the original vectors. Since the resultant is a vector, it has both magnitude and direction. If the sketch is drawn to scale, both the magnitude and direction (the angle it makes with one of the vectors) can be estimated using a ruler and protractor.

FIGURE 23-19

FIGURE 23-20

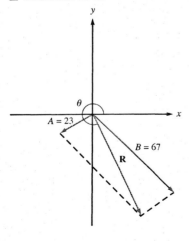

Example 13

Two vectors, **A** and **B** make an angle of 110° with each other and have magnitudes $A = 4.8$ and $B = 2.7$. Sketch the vectors and show their resultant **R**. Estimate the magnitude of **R** and the angle that it makes with vector **A**.

Solution: The vectors **A** and **B** can be drawn in any direction, as long as we maintain the 110° angle between them. Using a ruler, we draw the vectors using a convenient scale, say 1 unit = 1 cm. So we make vector **A** 4.8 cm long and vector **B** 2.7 cm long. We sketch the other two sides of the parallelogram, draw the resultant (Fig. 23-19), and measure the length of the resultant. We find it is about 4.6 cm long, so

$$R \approx 4.6$$

Using a protractor, we measure the angle θ that **R** makes with **A**, and find

$$\theta_R \approx 33°$$

Example 14

Find the resultant (magnitude and direction) of two force vectors, **A** and **B**, with $A = 23$ N at 210°, and $B = 67$ N at 315°.

Solution: We sketch these vectors, to scale, in standard position with the given angles: $\theta_A = 210°$ and $\theta_B = 315°$ (Fig. 23-20). Using ruler and protractor we find,

$$R \approx 65 \text{ N} \quad \text{and} \quad \theta_R \approx 296°$$

FIGURE 23-21
Vector sum.

FIGURE 23-22

If the vectors to be added are perpendicular to each other, the parallelogram is a rectangle, and the process is the same as above. A good example of perpendicular vectors are horizontal and vertical components (Fig. 23-21). *The vector sum of $\mathbf{R}_x$ and $\mathbf{R}_y$ is the resultant vector $\mathbf{R}$:*

$$\mathbf{R}_x + \mathbf{R}_y = \mathbf{R}$$

Example 15

Vector $\mathbf{R}$ has horizontal and vertical components

$$R_x = 41 \quad \text{and} \quad R_y = 23$$

Estimate the magnitude R and the angle θ that $\mathbf{R}$ makes with $\mathbf{R}_x$.

Solution: We sketch vectors $\mathbf{R}_x$ and $\mathbf{R}_y$ (Fig. 23-22), and sketch the rectangle and the diagonal from the common tail of the two vectors. From the sketch we estimate that

$$R \approx 47 \quad \text{and} \quad \theta_R \approx 29°$$

Triangle Method

The method of adding two vectors *tip to tail* is called the vector *triangle* method. We will again find the magnitude and direction of the resultant $\mathbf{R}$ of the two vectors in Fig. 23-17. First we draw one vector, say $\mathbf{A}$ (it doesn't matter which vector you start with). Then we draw vector $\mathbf{B}$ by connecting the tail of $\mathbf{B}$ to the tip (arrowhead) of $\mathbf{A}$, keeping the angle θ between vector $\mathbf{A}$ and vector $\mathbf{B}$ (Fig. 23-23). Note that we need to extend the line in the direction of vector $\mathbf{A}$ before we sketch angle θ. The resultant $\mathbf{R}$ is drawn by connecting the tail of $\mathbf{A}$ to the tip of $\mathbf{B}$. To find the resultant $\mathbf{R}$, we need to measure the magnitude (length) of $\mathbf{R}$ and the angle α that $\mathbf{R}$ makes with $\mathbf{A}$. Notice that this triangle is identical to the lower triangle of Fig. 23-18. Thus the two methods give the same results.

FIGURE 23-23
Vector addition: vector triangle method.

Example 16

Two vectors $\mathbf{A}$ and $\mathbf{B}$ make an angle of 65° with each other. If their magnitudes are

$$A = 39 \quad \text{and} \quad B = 24$$

use the vector triangle method to estimate the magnitude of $\mathbf{R}$ and the angle that $\mathbf{R}$ makes with $\mathbf{B}$.

Solution: Since we need to find the angle that $\mathbf{R}$ makes with $\mathbf{B}$, we begin by drawing vector $\mathbf{B}$. We chose to draw $\mathbf{B}$ horizontal and to the right, since we're accustomed to sketching angles in standard position. We choose a convenient scale, say 1 unit = 0.1 cm, and draw $\mathbf{B}$ to scale. Vector $\mathbf{A}$ is drawn to scale from the arrowhead (tip) of $\mathbf{B}$ at a 65° angle with the extension of $\mathbf{B}$. Then the resultant $\mathbf{R}$ is drawn from the tail of $\mathbf{B}$ to the tip of $\mathbf{A}$ (Fig. 23-24). With a ruler, we find that

$$R \approx 54$$

and with a protractor, we find that the angle α that $\mathbf{R}$ makes with $\mathbf{B}$ is

$$\alpha \approx 41°$$

FIGURE 23-24

Addition of Three or More Vectors

The vector triangle method (tip to tail) is especially useful when three or more vectors are to be added. However, with three or more vectors, the sketch is not a triangle, but a *polygon*.

FIGURE 23-25

--- Example 17 ---

A pilot flies due west for 345 km, turns and flies southwest (45° west of south) for 275 km, turns again and flies south-southeast (30° east of south) for 125 km. What is his displacement (distance and direction) from the starting point?

Solution: Using ruler and protractor, we make a sketch using the tip to tail method (Fig. 23-25). We measure the resultant vector and estimate that

$$R \approx 565 \text{ km}$$

Using a protractor, we estimate that

$$\theta_R \approx 58°$$

So the pilot is about 565 km 58° west of south of his starting point.

SECTION 23-3 Exercises Graphical Addition of Vectors

Parallelogram Method

In Exercises 1–8, use the parallelogram method to sketch the resultant of the vectors.

The angle θ between the vectors is given. Estimate the magnitude of the resultant **R** and the angle α that **R** makes with **A**.

1. $A = 73$, $B = 89$, $\theta = 56°$
2. $A = 8.3$, $B = 6.1$, $\theta = 152°$

The angle θ between the vectors is given. Estimate the magnitude of the resultant **R** and the angle α that **R** makes with **B**.

3. $A = 0.75$, $B = 0.25$, $\theta = 130°$
4. $A = 38$, $B = 78$, $\theta = 35°$

Estimate the magnitude and direction of the resultant of these vectors.

5. $A = 4.5$, $\theta_A = 80°$, $B = 3.3$, $\theta_B = 160°$
6. $A = 60$, $\theta_A = 340°$, $B = 85$, $\theta_B = 200°$

The rectangular components of a vector **R** are given. Estimate the magnitude of **R** and the positive angle that **R** makes with $\mathbf{R}_x$.

7. $R_x = 85$, $R_y = -51$
8. $R_x = -1.2$, $R_y = 2.1$

Triangle Method

In Exercises 9–17, use the vector triangle method to sketch the resultant of the vectors.

The angle θ between the vectors is given. Estimate the magnitude of the resultant **R** and the angle α that **R** makes with **A**.

9. $A = 83$, $B = 98$, $\theta = 100°$
10. $A = 4.2$, $B = 8.6$, $\theta = 42°$

The angle θ between the vectors is given. Estimate the magnitude of the resultant **R** and the angle α that **R** makes with **B**.

11. $A = 80$, $B = 50$, $\theta = 23°$
12. $A = 350$, $B = 410$, $\theta = 160°$

Addition of Three or More Vectors

Estimate the magnitude and direction of the resultant of the three given vectors.

13. $A = 6$, $\theta_A = 50°$, $B = 9$, $\theta_B = 130°$, $C = 4.8$, $\theta_C = 245°$
14. $A = 23$, $\theta_A = 315°$, $B = 42$, $\theta_B = 195°$, $C = 65$, $\theta_C = 75°$

■ FIGURE 23-26

Sketch and estimate.

Applications

15. A boat sails 57° west of north at 15 km/h. There is a current from the south at 3.5 km/h. Find the boat's actual velocity (speed and direction).

16. A plane flies 130 kilometres due north, turns and flies 190 kilometres northeast (45° east of north), turns again and flies 250 kilometres southeast-east (60° east of south). Find the displacement from the starting point (magnitude and direction).

 17. Three forces are acting on a body as shown (Fig. 23-26). If $F_1 = 95$ lb, $F_2 = 65$ lb, and $F_3 = 50$ lb, find the magnitude and direction of the resultant force.

23-4 ADDITION OF PERPENDICULAR VECTORS

In the last section we saw how the resultant of two or more vectors could be represented graphically and the magnitude and direction estimated from the graph. In this section we show how to calculate the magnitude and direction of resultants using trigonometry. We begin with perpendicular vectors and then consider applications. Non-perpendicular vectors are treated in the next two sections.

In general if **A** and **B** are two perpendicular vectors (Fig. 23-27), we may use the vector triangle method (or the parallelogram method) to sketch the resultant. If we are asked to find the angle between **R** and **A**, we choose to draw **A** first, and then connect the tail of **B** to the tip of **A** (Fig. 23-28).

Since the triangle is a right triangle, we use the Pythagorean theorem to determine the magnitude of **R**:

$$R = \sqrt{A^2 + B^2}$$

■ FIGURE 23-27
Perpendicular vectors.

■ FIGURE 23-28
Addition of perpendicular vectors.

The direction of **R** can be described by giving the angle α that **R** makes with one of the vectors, say **A**.

Using right triangle trigonometry,

$$\tan \alpha = \frac{B}{A}$$

or,

$$\alpha = \tan^{-1} \frac{B}{A}$$

where α is the angle between the resultant **R** and the vector **A**.

RESULTANT OF TWO PERPENDICULAR VECTORS $\mathbf{R} = \mathbf{A} + \mathbf{B}$	Magnitude	$R = \sqrt{A^2 + B^2}$	122
	Angle between $\mathbf{R}$ and $\mathbf{A}$	$\alpha = \tan^{-1}\dfrac{B}{A}$	123

FIGURE 23-29

Example 18

Vectors **A** and **B** are perpendicular vectors, with

$$A = 481 \quad \text{and} \quad B = 762$$

Find the magnitude of the resultant and the angle that the resultant makes with vector **A**.

Solution: Using the vector triangle method, we sketch the vectors (Fig. 23-29). From Eqs. 122 and 123, we have

$$R = \sqrt{481^2 + 762^2} = 901$$

and

$$\alpha = \tan^{-1}\frac{762}{481} = 57.7°$$

where $R = 901$ and $\alpha = 57.7°$ is the angle between **R** and **A**.

FIGURE 23-30

Take special note that in Eq. 123, α is the angle between the resultant and vector **A**. If we need to find the angle between **R** and vector **B** (Fig. 23-30), then using right triangle trigonometry, we see that

$$\tan \alpha = \frac{A}{B}$$

or,

$$\alpha = \tan^{-1}\frac{A}{B}$$

So, if α is the angle between **R** and **B**, then B must be in the *denominator*.

Example 19

Vectors **A** and **B** are perpendicular vectors, with

$$A = 16.2 \quad \text{and} \quad B = 28.5$$

Find the magnitude of the resultant and the angle that the resultant makes with the longer vector.

Solution: Using the vector triangle method, we sketch the vectors, with the vertex at the origin (Fig. 23-31). From Eq. 122 we have

$$R = \sqrt{16.2^2 + 28.5^2} = 32.8$$

Since **B** is the longer vector, we use the equation above to find the angle α, taking special care that $B = 28.5$ is in the denominator,

$$\alpha = \tan^{-1}\frac{16.2}{28.5} = 29.6°$$

FIGURE 23-31

23-4 Addition of Perpendicular Vectors

Applications

Many vector problems involve finding the resultant of two perpendicular vectors.

Example 20

A river flows with a current, at a velocity of 5.75 km/h. A rower, who can row 8.25 km/h in still water, heads directly across the current but is pushed downstream by the current. Find the actual velocity (rate and direction of travel) of the boat.

Solution: Let

V_c = the current's velocity vector,
V_r = the rower's velocity vector (in still water), and
V_b = the boat's actual velocity vector with respect to the shore.

Now we sketch the vectors (Fig. 23-32). We can see that V_b is the resultant vector of V_c and V_r.

Applying Eqs. 122 and 123, with $V_c = 5.75$ and $V_r = 8.25$, we have

$$V_b = \sqrt{V_c^2 + V_r^2}$$
$$= \sqrt{5.75^2 + 8.25^2}$$
$$= 10.1$$

And,

$$\alpha = \tan^{-1} \frac{V_r}{V_c}$$
$$= \tan^{-1} \frac{8.25}{5.75} = 55.1°$$

where α is the angle between the boat and the current.

So, the boat's actual rate with respect to the shore is 10.1 km/h, and the direction of travel is 55.1° from the shoreline (the current's direction).

■ FIGURE 23-32

SECTION 23-4 Exercises   Addition of Perpendicular Vectors

The magnitudes of two perpendicular vectors are given. Sketch the vectors and the resultant. Find the magnitude of the resultant **R** and the positive angle that **R** makes with vector **A**, rounded to three significant digits.

1. $A = 585$, $B = 358$
2. $A = 846$, $B = 297$
3. $A = 7.53$, $B = 8.96$
4. $A = 239$, $B = 382$

The magnitude of two perpendicular vectors are given. Sketch the vectors and the resultant. Find the magnitude of the resultant **R** and the positive angle that **R** makes with vector **B**.

5. $A = 16.5$, $B = 50.1$
6. $A = 396$, $B = 677$
7. $A = 5.13$, $B = 8.95$
8. $A = 9.46$, $B = 4.57$

Vector **V** has the given x and y components. Find the magnitude of **V** and the positive angle it makes with the positive x axis.

9. $V_x = 12.4$, $V_y = 15.6$
10. $V_x = -6.35$, $V_y = 4.77$
11. $V_x = -7.42$, $V_y = -5.16$
12. $V_x = 876$, $V_y = -124$

Applications

Sketch, solve, and round to three significant digits.

13. A cyclist bikes 14.0 kilometres north, makes a left turn, and bikes 23.4 kilometres west. Find her displacement from her starting point (both distance and direction).

14. An airplane flies due east for 185 km, makes a 90.0° turn and flies due south for 275 km. Find its displacement from its starting point.

15. Given two displacements with magnitudes of 12.0 km and 15.0 km.
    (a) Find the magnitude of the maximum resultant displacement possible.
    (b) Find the magnitude of the minimum resultant displacement possible.
    (c) Find the magnitude of the resultant if the given displacements are perpendicular.

16. A driver goes 27.8 kilometres north, turns and drives 18.2 kilometres east, turns and drives 32.1 kilometres south. Find the total distance traveled and the displacement (both magnitude and direction) from his starting point.

17. A plane is headed due west with an airspeed of 235 km/h. It is driven from its course by a wind *from* due south blowing at 35.5 km/h. Find the groundspeed of the plane and the actual direction of travel.

18. A canoeist is paddling straight across a river with a current of 6.50 km/h. Find the rate and the direction of travel, with respect to the shore, of the canoe, if the canoeist can paddle 7.45 km/h in still water.

19. A plane is traveling 325 km/h in still air.
    (a) Find its groundspeed if it had an 85.0 km/h head wind.
    (b) Find its groundspeed if it had an 85.0 km/h tail wind.
    (c) Find its groundspeed if it were flying perpendicular to an 85.0 km/h wind. Find the angle the flight path makes with the direction of the wind.

20. A boater wants to go straight across a river, whose current rate is 9.25 km/h. If the boat travels at 35.7 km/h in still water, in what direction, with respect to the shore, should the boater head to go straight across the river?

21. A pilot wants to fly due north at a groundspeed of 475 km/h. The wind is 45.0 km/h *from* the east. At what heading (direction) and speed must the pilot fly the plane to compensate for the wind?

■ FIGURE 23-33

22. Two forces of 67.6 N and 78.9 N act on a body at right angles to each other. Find the magnitude of the resultant force and the angle it makes with the 78.9 N force.

23. Two people are pushing a rectangular box on adjacent sides. One pushes with a force of 55.3 lb and the other with a force of 75.8 lb. What is the magnitude and direction, with respect to the larger force, of the resultant force?

24. The lift of an airplane wing is 852 lb force and the drag is 325 lb force. Find the magnitude F and direction α of the resultant force (Fig. 23-33).

25. A sign is being acted on by a horizontal force of 25.6 lb acting to the left and a vertical force of 15.7 lb acting down. Find the magnitude and direction of the resultant force.

23-5   VECTOR ADDITION BY COMPONENTS

In the last section we added perpendicular vectors using the Pythagorean theorem and right triangle trigonometry. In this section we add non-perpendicular vectors algebraically, by components. We resolve the vectors into horizontal and vertical components and combine like components. Then we use the Pythagorean theorem and right triangle trigonometry to find the resultant.

Vector Addition by Components

1. Sketch the vectors and their resultant **R** on the coordinate plane and determine the quadrant of the resultant.
2. Find the x and y components, algebraically, of each given vector.
3. Add the x components to get the x component R_x.
4. Add the y components to get the y component R_y.
5. Find the magnitude R using the Pythagorean theorem:

$$R = \sqrt{R_x^2 + R_y^2}$$

6. Find the reference angle of the resultant:

$$\theta_{ref} = \tan^{-1}\left|\frac{R_y}{R_x}\right|$$

7. Find the angle θ that **R** makes with the positive x axis, from the quadrant of the resultant.

The absolute value sign in Step 6 above is necessary because a reference angle is never negative. If the vectors being added are in quadrants II, III, or IV, some components will be negative. By taking the absolute value, we disregard the signs of R_x and R_y, so that the reference angle is not negative.

---- **Example 21** ----

Find the resultant of vector **A** = 3.2 at 27° and vector **B** = 6.8 at 81°.

Solution: We sketch the vectors and their resultant (Fig. 23-34) and note that the resultant is in quadrant I. Next we resolve vectors **A** and **B** into components and place the results in a table. We carry a third significant digit in the calculations, so that we can round to two significant digits at the end.

■ FIGURE 23-34

Vector	x component	y component
A	3.2 cos 27° = 2.85	3.2 sin 27° = 1.45
B	6.8 cos 81° = 1.06	6.8 sin 81° = 6.72
Sums	R_x = 3.91	R_y = 8.17

To find R:

$$R = \sqrt{R_x^2 + R_y^2}$$
$$= \sqrt{3.91^2 + 8.17^2}$$
$$= 9.1$$

Finding the reference angle θ_{ref}:

$$\theta_{ref} = \tan^{-1}\left|\frac{R_y}{R_x}\right|$$

$$\theta_{ref} = \tan^{-1}\left|\frac{8.17}{3.91}\right|$$

$$\theta_{ref} = 64.4°$$

Since θ_R is a quadrant I angle,

$$\theta_R = \theta_{ref} = 64.4°$$

So the resultant vector **R** in polar form is

$$R = 9.1, \theta_R = 64.4°$$

We do a visual check by estimating the magnitude and direction of the resultant in Fig. 23-34. Indeed, our sketched resultant agrees with our numerical results.

Example 22

Find the resultant of vector **A** = 85.1 at 15.5° and vector **B** = 39.2 at 254°.

Solution: We sketch the vectors and their resultant (Fig. 23-35) and note that the resultant is in quadrant IV. Next we resolve vectors **A** and **B** into components and place the results in a table:

Vector	x component		y component	
A	85.1 cos 15.5° =	82.00	85.1 sin 15.5° =	22.74
B	39.2 cos 254° =	−10.80	39.2 sin 254° =	−37.68
Sums	R_x =	71.20	R_y =	−14.94

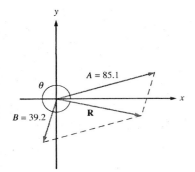

FIGURE 23-35

To find R:

$$R = \sqrt{R_x^2 + R_y^2}$$
$$= \sqrt{71.20^2 + (-14.94)^2}$$
$$= 72.8 \text{ (rounding to three significant digits)}$$

Finding the reference angle θ_{ref}:

$$\theta_{ref} = \tan^{-1}\left|\frac{R_y}{R_x}\right|$$

$$\theta_{ref} = \tan^{-1}\left|\frac{-14.94}{71.20}\right|$$

$$\theta_{ref} = 11.9°$$

Since the resultant **R** is a quadrant IV angle, θ_R is

$$\theta_R = 360° - 11.9° = 348.1°$$

So the resultant vector **R** in polar form is

$$R = 72.8, \theta_R = 348.1°$$

We do a visual check by estimating the magnitude and direction of the resultant in Fig. 23-35. Indeed, our sketched resultant agrees with our numerical results.

FIGURE 23-36

Example 23

Two vectors of magnitudes $A = 25.2$ and $B = 72.8$ have an angle of 126° between them. Find the magnitude of the resultant **R** and the angle between **R** and **A**.

Solution: Since the directions of **A** and **B** are not specifically given, we choose to draw **A** along the positive x axis for simplicity (Fig. 23-36). Vector **A** is then represented by

$$A = 25.2, \theta_A = 0°$$

We represent vector **B** as

$$B = 72.8, \theta_B = 126°$$

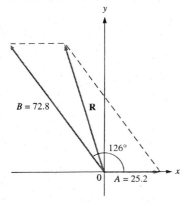

We sketch the resultant and see that it is in quadrant II. We calculate the resultant as we did in Example 21.

Vector	x component	y component
A	25.2 cos 0° = 25.2	25.2 sin 0° = 0
B	72.8 cos 126° = −42.79	72.8 sin 126° = 58.90
Sums	R_x = −17.59	R_y = 58.90

To find R:

$$R = \sqrt{R_x^2 + R_y^2}$$
$$= \sqrt{(-17.59)^2 + 58.90^2}$$
$$= 61.5$$

Finding the reference angle θ_{ref}:

$$\theta_{ref} = \tan^{-1}\left|\frac{R_y}{R_x}\right|$$

$$\theta_{ref} = \tan^{-1}\left|\frac{58.90}{-17.59}\right|$$

$$\theta_{ref} = 73.4°$$

Since the resultant **R** is a quadrant II angle, θ_R is

$$\theta_R = 180° - 73.4° = 106.6°$$

So, the resultant **R** is a vector whose magnitude is 61.5 and makes an angle of 107° with vector **A**.

We do a visual check by estimating the magnitude and direction of the resultant in Fig. 23-36. Indeed, our sketched resultant agrees with our numerical results.

Resultant of Three or More Vectors

The component method is especially useful in finding the resultant of three or more vectors.

Example 24

Find the resultant of the three vectors in Fig. 23-37. Round results to three significant digits.

Solution: First we determine the angles the given vectors make with the positive x axis (Fig. 23-38) and find that

Vector **A** (in polar form): $A = 3$, $\theta_A = 130°$
Vector **B** (in polar form): $B = 9$, $\theta_B = 270°$
and Vector **C** (in polar form): $C = 8$, $\theta_C = 30°$

Next we sketch the vector sum, on the coordinate plane with the vertex at the origin, using the vector triangle (tip to tail) method (Fig. 23-39), and see that the resultant **R** is in quadrant IV.

FIGURE 23-37

FIGURE 23-38

FIGURE 23-39

To find the magnitude and direction of **R**, we make the table of components:

Vector	x component	y component
A	$3 \cos 130° = -1.928$	$3 \sin 130° = 2.298$
B	$9 \cos 270° = 0$	$9 \sin 270° = -9$
C	$8 \cos 30° = 6.928$	$8 \sin 30° = 4$
Sums	$R_x = 5.000$	$R_y = -2.702$

Next we find R:

$$R = \sqrt{(5.000)^2 + (-2.702)^2} = 5.68$$

Finding the reference angle θ_{ref} in order to find θ_R:

$$\theta_{ref} = \tan^{-1}\left|\frac{-2.702}{5.000}\right| = 28.4°$$

Since the resultant **R** is a quadrant IV angle, θ_R is

$$\theta_R = 360° - 28.4° = 331.6°$$

Therefore the resultant vector **R** in polar form is

$$R = 5.68, \; \theta_R = 331.6°$$

We do a visual check by estimating the magnitude and direction of the resultant in Fig. 23-39. Indeed, our sketched resultant agrees with our numerical results.

SECTION 23-5 Exercises Vector Addition by Components

Sketch the given vectors **A** and **B** and their resultant. Find the resultant of the given vectors by components. Round to three significant digits.

1. $A = 57, \theta_A = 20°, B = 87, \theta_B = 75°$
2. $A = 3.5, \theta_A = 36°, B = 2.6, \theta_B = 80°$
3. $A = 243, \theta_A = 301°, B = 726, \theta_B = 155°$
4. $A = 1.37, \theta_A = 248°, B = 1.73, \theta_B = 65.8°$

The magnitudes of vectors **A** and **B** and the angle between them are given. Sketch the vectors and their resultant. Use components to find the magnitude of the resultant **R** and the angle that **R** makes with **A**.

5. $A = 6.88, B = 5.25, \theta = 112°$
6. $A = 737, B = 811, \theta = 85.0°$

The magnitudes of vectors **A** and **B** and the angle between them are given. Sketch the vectors and their resultant. Use components to find the magnitude of the resultant **R** and the angle that **R** makes with **B**.

7. $A = 0.151, B = 0.752, \theta = 165°$
8. $A = 739, B = 837, \theta = 65.4°$

■ FIGURE 23-40
Resultant of Three or More Vectors

■ FIGURE 23-41

■ FIGURE 23-42

Solve, and round to three significant digits.

9. Find the resultant of the three force vectors in Fig. 23-40, given that **B** is a horizontal force.
10. Find the resultant of the three force vectors in Fig. 23-41.

Solve, and round to three significant digits.

11. A ship leaves port and sails 345 km due west, then turns and sails 275 km southwest (45° south of west). Find its displacement from the port (both distance and direction).
12. A pilot is heading his plane due south with an airspeed of 523 km/h. The wind is from the southeast (45° east of south) at 94.5 km/h. Find the actual direction of travel and the plane's groundspeed.
13. A tractor and a truck are trying to pull a car out of a snowbank, with respective forces $F_1 = 8160$ N and $F_2 = 6200$ N (Fig. 23-42). Find the magnitude and direction (from the front of the car) of the resultant force.
14. An airplane takes off from an airport, flies 78.8 km 35.7° east of north, turns and flies 95.0 km 58.0° east of north, and again turns and flies 46.5 km 75.8° east of north. Find the plane's displacement from the airport (both distance and direction).

23-6 VECTOR ADDITION USING THE LAWS OF SINES AND COSINES

■ FIGURE 23-43
Vector triangle method.

In the last section we added non-perpendicular vectors by the component method. In this section we add vectors using the law of sines and the law of cosines.

Given two non-perpendicular vectors and the angle between them, we can sketch the resultant using the vector triangle method (Fig. 23-43) on the coordinate plane. The resultant is one side of an *oblique* triangle. We are given two sides of the oblique triangle and can easily calculate the included angle, since it is supplementary to the angle between the vectors. To find the length of the resultant and the angle it makes with one of the original vectors, we simply solve the oblique triangle. To find the magnitude R of the resultant, we use the law of cosines. To find the angle the resultant makes with one of the vectors, we use the law of sines.

Vector Addition Using the Laws of Sines and Cosines

1. Use the vector triangle method to sketch the two vectors and their resultant.
2. Find the angle between the two given sides of the triangle.
3. Use the law of cosines to calculate R.
4. Use the law of sines to calculate the angle (direction) of **R**.

Example 25

Two vectors, **A** and **B**, make an angle of 41.5° with each other. If their magnitudes are $A = 56.7$ and $B = 87.6$, find the magnitude of the resultant **R** and the angle it makes with vector **B**.

■ FIGURE 23-44

Solution: First we sketch the vectors and the resultant, using the vector triangle method. We start with vector **B** (Fig. 23-44), since we need to find the angle that **R** makes with **B**. To use the law of cosines to find R, we need the included angle β between the given sides A and B. Since β is supplementary to the 41.5° angle,

$$\beta = 180° - 41.5° = 138.5°$$

Using the law of cosines,

$$R^2 = 87.6^2 + 56.7^2 - 2(87.6)(56.7)(\cos 138.5°) = 18329$$
$$R = 135.4$$

Now we use the law of sines to find the angle α between **R** and **B**:

$$\frac{135.4}{\sin 138.5°} = \frac{56.7}{\sin \alpha}$$

$$\sin \alpha = \frac{(\sin 138.5°)(56.7)}{135.4} = 0.2775$$

$$\alpha = 16.1°$$

As usual, we round our answers to the same number of significant digits as in the given data, even though we have kept more digits for intermediate steps. So, the resultant **R** has a magnitude of 135 and makes a 16.1° angle with vector **B**.

Example 26

Find the resultant of vectors **A** and **B** where:

$$A = 3.25, \theta_A = 76.5°, \quad \text{and} \quad B = 6.75, \theta_B = 11.2°$$

■ FIGURE 23-45

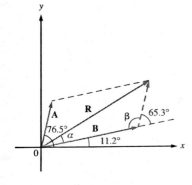

Solution: First we sketch the vectors (Fig. 23-45). To use the vector triangle method, we need to find the angle between the vectors:

$$76.5° - 11.2° = 65.3°$$

In the oblique triangle, the angle β between the two given sides is

$$\beta = 180° - 65.3° = 114.7°$$

Now we use the law of cosines to find R:

$$R^2 = 6.75^2 + 3.25^2 - 2(6.75)(3.25)(\cos 114.7°) = 75.46$$

Thus,

$$R = 8.629$$

We use the law of sines to find the angle α between **R** and **B**.

$$\frac{8.629}{\sin 114.7°} = \frac{3.25}{\sin \alpha}$$

$$\sin \alpha = 0.3421$$

$$\alpha = 20.0°$$

So the resultant is at an angle of

$$11.2° + 20.0° = 31.2°$$

or,

$$R = 8.63, \theta_R = 31.2°$$

SECTION 23-6 Exercises Vector Addition Using the Laws of Sines and Cosines

In Exercises 1–8, the magnitudes of vectors **A** and **B** and the angle θ between them are given. Round to three significant digits.

Find the magnitude of the resultant and the angle it makes with vector **A**.

1. $A = 345, B = 652, \theta = 75.2°$
2. $A = 5.32, B = 1.61, \theta = 158°$
3. $A = 8.22, B = 3.45, \theta = 164°$
4. $A = 735, B = 858, \theta = 63.4°$

Find the magnitude of the resultant and the angle it makes with vector **B**.

5. $A = 378, B = 294, \theta = 102°$
6. $A = 2.78, B = 1.05, \theta = 89.0°$
7. $A = 74.5, B = 27.7, \theta = 15.6°$
8. $A = 83.7, B = 27.4, \theta = 178°$

Find the resultant of vectors **A** and **B**, rounded to three significant digits.

9. $A = 343, \theta_A = 29.4°, B = 734, \theta_B = 85.2°$
10. $A = 5.61, \theta_A = 175°, B = 2.26, \theta_B = 319°$
11. $A = 6.39, \theta_A = 152°, B = 4.23, \theta_B = 295°$
12. $A = 29.3, \theta_A = 258°, B = 25.4, \theta_B = 48.3°$

Applications

Solve, and round to three significant digits.

13. A motorist drives 75.0° west of north for 35.2 kilometres and then turns and drives 60.0° west of south for 25.3 kilometres. Find the distance and direction of her displacement.

14. A surveyor locates a marker 56.3 m southwest (45.0° west of south) of him. He knows that the marker is 32.3 m south of a telephone pole. Find the displacement (distance and direction) of the telephone pole from the surveyor.

■ FIGURE 23-46

■ FIGURE 23-47

15. Two tugboats are pulling a barge toward the west as shown (Fig. 23-46), with pulling forces

$$F_1 = 8140 \text{ newtons} \quad \text{and} \quad F_2 = 5544 \text{ newtons}$$

Find the magnitude and direction of the resultant force.

16. A ship needs to sail directly to a port at 35.0° west of south. The current is flowing due north at 18.0 km/h. If the ship's speed in still water is 38.5 km/h, find the heading the ship must take and its actual velocity with respect to land.

17. Two tugboats are trying to pull a barge due north. One tugboat, with a pulling force of $F_1 = 2070$ pounds, pulls at a 65.0° angle as shown (Fig. 23-47). If the smaller tugboat pulls with a force of $F_2 = 1580$ pounds, what must the angle θ of the smaller tugboat be?

18. A pilot wishes to fly 30.0° east of north with a groundspeed of 285 mi/h. A wind is blowing from the west at 65.0 mi/h. What should the pilot's heading and airspeed be?

■ Chapter 23 REVIEW EXERCISES

1. Sketch the vector **V** whose magnitude is 75.0 and makes an angle of 125° with the positive x axis. From the sketch, estimate the horizontal and vertical components, V_x and V_y. Then calculate the components.

2. Find the x and y components of the vector **V** where

$$V = 4.96, \quad \theta_V = 265°$$

3. A ship travels 275 km in a direction 64.7° south of west. What are the south and west components of the displacement?

4. An airplane pilot, flying at an altitude of 203 m, sights a runway at an angle of depression of 18.0°. What is the horizontal component of his displacement from the runway, and what is his distance from the runway?

5. An escalator travels at a rate of 15.4 m/min and its angle of inclination is 34.6°. What is the vertical component of the velocity? How long will it take a passenger to travel 8.05 m vertically?

6. A projectile is launched at an angle of 49.0° to the horizontal with a speed of 6750 m/min. Find the vertical and horizontal components of this velocity.

7. A truck weighing 17.0 tonnes stands on a hill inclined 16.0° from the horizontal. How large must the braking force be to prevent the truck from rolling downhill?

8. Use the vector triangle method to sketch the resultant **R** of two perpendicular vectors that have magnitudes of 278 and 847. Estimate the magnitude R and the angle **R** makes with the larger vector.

9. Use the parallelogram method to sketch the resultant **R** of the two vectors **A** and **B**. Estimate the magnitude R and direction of θ.

$$A = 6.85, \theta_A = 48.8°, B = 5.65, \theta_B = 197.5°$$

10. Sketch the resultant **R** of the three vectors **A**, **B**, and **C**. Estimate the magnitude and direction of **R**. Round to three significant digits.

$$A = 5, \theta_A = 150°, B = 9, \theta_B = 65°, C = 7, \theta_C = 245°$$

11. Two perpendicular vectors have magnitudes of 278 and 847. Calculate the magnitude of the resultant **R** and the angle **R** makes with the larger vector. Check your results with the answer in Exercise 8.

12. The components of vector **V** are $V_x = -0.376$ and $V_y = 0.644$. Find the magnitude and direction of **V**.

13. A jogger jogs 13.0 km north, then turns and jogs 11.0 km west. Find her displacement (both distance and direction) from her starting point.

14. A swimmer wants to swim straight across a river, whose current rate is 1.00 km/h. If her swimming rate is 2.00 km/h in still water, in what direction should she head to go straight across the river? How long will it take her (in minutes) to cross the river, if it is 0.40 km wide?

15. Two forces of 45.5 N and 73.8 N act on a body at right angles to each other. Find the magnitude of the resultant force and the angle it makes with the smaller force.

16. Use the method of components to find the resultant of the vectors **A** and **B**

$$A = 6.85, \theta_A = 48.8°, B = 5.65, \theta_B = 197.5°$$

Check your results with the answer in Exercise 9.

17. Calculate the resultant of the vectors **A**, **B**, and **C**.

$$A = 5.00, \theta_A = 150.0°, B = 9.00, \theta_B = 65.0°, C = 7.00, \theta_C = 245°$$

Check your results with the answer in Exercise 10.

18. Two vectors of magnitude 76.0 and 32.0 have a 84.0° angle between them. Using the laws of cosines and sines, find the magnitude of the resultant and the angle it makes with the smaller vector.

19. Use the laws of cosines and sines to find the magnitude and direction of the resultant of the two vectors **A** and **B**:

$$A = 3.57, \theta_A = 175°, B = 1.46, \theta_B = 287°$$

 20. A driver, traveling at 65.5 mi/h, throws an apple core out his window perpendcular to his direction at a speed of 3.00 mi/h. What is the velocity (magnitude and direction) of the apple core as it leaves his hand?

21. Two forces of 754 pounds and 986 pounds act on a point of a bridge girder at an angle of 84.8° to each other. Find the resultant force and the angle it makes with the larger force.

22. As an airplane heads due north with an airspeed of 375 mi/h, a wind with a speed of 45.0 mi/h causes the plane to travel slightly west of north with a groundspeed of 355 mi/h. What is the wind direction and in what direction does the plane actually fly?

WRITING

23. We have learned three methods of adding vectors. Describe each method, including its advantages and disadvantages. Which do you prefer and why? Make up two non-perpendicular vectors and use these to illustrate all three methods.

TEAM PROJECT

24. A displacement vector is a distance (magnitude) traveled in a specific direction. In this problem your team will create displacement vectors and find their vector sum.

Begin in an open area of your campus. Mark a beginning point A. Measure 12 m along a compass heading of due east; mark that point B. This is your first displacement vector **AB**. Now measure 19 m along the heading 45° north of east, and mark that point C, to create vector **BC**. From point C, walk 15 m in a direction 35° west of north, and mark that point D to create vector **CD**.

The vector from A to D is the vector sum of the three displacement vectors you have just walked. With a tape or other measuring device, determine the distance (magnitude) from A to D. Use your compass to determine the direction from A to D.

Draw a scaled vector diagram of your displacement vectors. Verify the vector sum using one of the analytical methods learned in this chapter. What is the percentage error? How do you account for this error?

24 Applications of Radian Measure

Outcomes

When you have completed this chapter, you will be able to

1. Define the radian.
2. Convert between radians and degrees.
3. Calculate trigonometric functions using radians.
4. Compute arc length of a circle and the area of a sector of a circle.
5. Calculate the angular velocity of an object in uniform circular motion.

We introduced radian measure in Chapter 13 and learned that angles are measured in *radians* for some applications. In this chapter we review radian measure and consider applications of radian measure. Although we often work with angles in degrees, radians are very useful for topics such as arc length, area of a sector, and angular velocity.

24-1 RADIAN MEASURE

In this section, we learn to express radian measure in terms of π. Recall that one radian (1 rad) is the measure of a central angle that intercepts an arc whose length is equal to the radius (Fig. 24-1). (A central angle is an angle with its vertex at the centre of a circle and has radii for sides.) In a circle, one revolution (360°) is equal to 2π radians.

| CONVERSIONS | 1 rev = 360° = 2π radians | 79 |

We also learned to convert an angle from degrees to radians by multiplying the degree measurement by the conversion factor $\left(\dfrac{2\pi \text{ rad}}{360°}\right)$. This gives us the radian measure of an angle *in decimal form*.

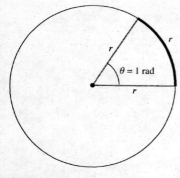

FIGURE 24-1
Angle of 1 radian.

Example 1
Convert 65.8° to radians.

Solution:

$$65.8° = (65.8°)\left(\frac{2\pi \text{ rad}}{360°}\right) = 1.15 \text{ rad}$$

Radian Measure in Terms of π

It is quite common to express the radian measure of an angle in terms of π. Eq. 79 provides the basic conversion:

$$360° = 2\pi \text{ rad}$$

Dividing, we get

$$180° = \pi \text{ rad}$$
$$90° = \frac{\pi}{2} \text{ rad}$$
$$60° = \frac{\pi}{3} \text{ rad}$$

and so forth. Generally, radian measure in terms of π is expressed *in fraction form*, rather than in decimal form. This table shows the common conversions of angles in degrees to radians in terms of π.

Angle measure	
Degrees	Radians
360°	2π
180°	π
90°	$\frac{\pi}{2}$
60°	$\frac{\pi}{3}$
45°	$\frac{\pi}{4}$
30°	$\frac{\pi}{6}$

In Example 1, we converted the angle from degrees to radians when we multiplied by the conversion factor $\frac{2\pi \text{ rad}}{360°}$, but the conversion factor may be reduced:

$$\frac{2\pi \text{ rad}}{360°} = \frac{\pi \text{ rad}}{180°}$$

To convert angles in degrees to radian measure in terms of π, we multiply the angle in degrees by $\frac{\pi \text{ rad}}{180°}$ *in fraction form*, and then reduce the fraction to lowest terms.

Example 2

Convert 120° to radians in terms of π.

Solution:

$$120° = 120° \left(\frac{\pi \text{ rad}}{180°} \right)$$

Reducing,

$$120° = \frac{2\pi}{3} \text{ rad}$$

To convert an angle in radians to degrees, we multiply the angle in degrees by the conversion factor $\dfrac{180°}{\pi \text{ rad}}$ and express the result in decimal form.

Example 3
Convert 1 rad to degrees, rounded to tenths.

Solution:
$$1 \text{ rad} = (1 \text{ rad})\left(\dfrac{180°}{\pi \text{ rad}}\right) = 57.3°$$

So, an angle of one radian is approximately 57.3° (Fig. 24-1).
$$1 \text{ rad} \approx 57.3°$$

Example 4
Convert $\dfrac{5\pi}{6}$ rad to degrees.

Solution:
$$\dfrac{5\pi}{6} \text{ rad} = \left(\dfrac{5\pi \text{ rad}}{6}\right)\left(\dfrac{180°}{\pi \text{ rad}}\right)$$
$$= 150°$$

In Chapter 13, we noted that when the unit of the angle measure is not marked, it is assumed that the angle is in radians.

Example 5
An angle of $\dfrac{5\pi}{6}$ rad can be expressed as just $\dfrac{5\pi}{6}$ (without the "rad"). So
$$\dfrac{5\pi}{6} \text{ rad} = \dfrac{5\pi}{6}$$

COMMON ERROR Students sometimes confuse the decimal value of π with the degree equivalent of π radians.

What is π? 180 or 3.1416?

A decimal approximation of the number π is
$$\pi \approx 3.1416$$
But an angle measurement of π is π radians,
$$\pi = \pi \text{ rad} = 180°$$

Trigonometric Functions of Angles in Radians

In Chapters 15 and 22, we found the trigonometric functions of angles in radians in decimal form, but not in terms of π. The process is the same when the angle is given in terms of π. When using a calculator, put the calculator in radian mode, and be careful to use parentheses properly when you enter the function.

Example 6

Use a calculator to evaluate to four significant digits:

$$\cos \frac{5\pi}{6}$$

Solution: Since the angle $\frac{5\pi}{6}$ has no written units, it is assumed to be in radians. With the calculator in radian mode, we enter

$$\cos\left(\frac{5\pi}{6}\right)$$

Note the use of parentheses when we enter the function into the calculator! We find that

$$\cos \frac{5\pi}{6} = -0.8660$$

COMMON ERROR Remember, if an angle measure has no written units, it is assumed to be in *radians*!

SECTION 24-1 Exercises Radian Measure

Radian Measure in Terms of π

Convert each angle from degrees to radian measure in terms of π (in fraction form).

1. 360°
2. 180°
3. 90°
4. 45°
5. 60°
6. 30°
7. 15°
8. 75°
9. 135°
10. 225°
11. 72°
12. 40°
13. 340°
14. 36°
15. 84°
16. 210°
17. 270°
18. 144°
19. 200°
20. 324°

Convert each angle from radians to degrees.

21. $\frac{\pi}{6}$
22. $\frac{\pi}{3}$
23. $\frac{\pi}{4}$
24. $\frac{\pi}{2}$
25. $\frac{2\pi}{3}$
26. $\frac{3\pi}{4}$
27. $\frac{4\pi}{5}$
28. $\frac{4\pi}{3}$
29. $\frac{17\pi}{10}$
30. $\frac{23\pi}{20}$
31. $\frac{11\pi}{15}$
32. $\frac{13\pi}{12}$

Trigonometric Functions of Angles in Radians

Use a calculator to evaluate to four significant digits.

33. $\sin \frac{\pi}{3}$
34. $\cos \frac{\pi}{6}$
35. $\tan \frac{\pi}{6}$
36. $\sin \frac{\pi}{4}$
37. $\cos \frac{5\pi}{8}$
38. $\tan \frac{4\pi}{3}$
39. $\sin \frac{6\pi}{5}$
40. $\cos \frac{7\pi}{9}$
41. $\tan\left(-\frac{7\pi}{4}\right)$
42. $\sin \frac{3\pi}{10}$
43. $\cos \frac{19\pi}{12}$
44. $\tan\left(-\frac{7\pi}{18}\right)$

24-2 ARC LENGTH AND AREA OF A SECTOR

In this section, we discuss arc length and the area of a sector. These are useful for topics such as latitude and longitude, magnetic tapes and disks, circular highway curves, and pendulums.

Radian Measure and Arc Length

In a circle of radius r, a central angle of 1 radian intercepts an arc of length r (see Fig. 24-1). The measure of any central angle θ, in radians, is the number of "radius units" contained in the arc.

Example 7

In Fig. 24-2, it looks like the arc length s is about 2 "radius units," so the angle θ is about 2 radians.

FIGURE 24-2
Radian measure of an angle.

We define the radian measure of an angle as the length of the intercepted arc s divided by the radius r (Fig. 24-2).

| RADIAN MEASURE OF AN ANGLE | $\theta = \dfrac{s}{r}$ | 77 |

Example 8

Find the angle that intercepts an arc of 8.52 cm in a circle of radius 7.00 cm (Fig. 24-3).

Solution: We use Eq. 77 with $s = 8.52$ cm and $r = 7.00$ cm.

$$\theta = \frac{s}{r} = \frac{8.52 \text{ cm}}{7.00 \text{ cm}} = 1.22$$

Notice that the units (cm) cancel out and the angle measure is a pure number (*dimensionless*). This feature makes it more useful than degrees for some calculations. We call this measure "radians" as an indication that it is an angle measurement. Thus, the angle measurement is 1.22 radians, and we write

$$\theta = 1.22 \text{ rad} \quad \text{or} \quad \theta = 1.22$$

FIGURE 24-3

Arc Length

FIGURE 24-4

To find an expression for arc length s, we solve Eq. 77 for s.

$$s = \theta r, \text{ where } \theta \text{ is in radian measure}$$

Example 9

Find the length of the arc intercepted by a central angle of $\dfrac{\pi}{4}$ in a circle of radius 5.00 cm (Fig. 24-4).

Solution: We use the equation for arc length with $\theta = \dfrac{\pi}{4}$ and $r = 5.00$ cm

$$s = \left(\frac{\pi}{4}\right)(5.00) = 3.93 \text{ cm}$$

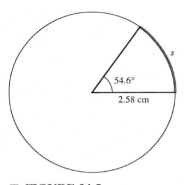

■ FIGURE 24-5

Example 10

Find the length of the arc intercepted by a central angle of 54.6° in a circle of radius 2.58 cm (Fig. 24-5).

Solution: Before we can use the equation for arc length, we must convert 54.6° to radians:
$$54.6° = 0.9529 \text{ rad}$$
Now, substituting $\theta = 0.9529$ and $r = 2.58$ into
$$s = \theta r$$
$$s = (0.9529)(2.58) = 2.46 \text{ cm}$$
Note that the "rad" may be put in or taken out as needed.

Example 11

Find the radius of a circle in which a central angle of 75.8° intercepts an arc of length 8.75 cm (Fig. 24-6).

Solution: We solve Eq. 77 for r:
$$r = \frac{s}{\theta}$$
First we convert $\theta = 75.8°$ to radians:
$$\theta = 75.8° = 1.323 \text{ rad}$$
Substituting $s = 8.75$ and $\theta = 1.323$ into the equation for r, we have
$$r = \frac{s}{\theta} = \frac{8.75}{1.323} = 6.61 \text{ cm}$$

■ FIGURE 24-6

> **COMMON ERROR** Remember, when using Eq. 77 that the angle θ must be in radians.

Area of a Sector

■ FIGURE 24-7
Sector of a circle.

Recall from Sec. 13-5 that a *sector* of a circle is a pie-shaped region bounded by two radii and an arc (Fig. 24-7).

The area of a sector having a radius r and a central angle θ (in radians) is found using ratios that compare a sector of a circle to the whole circle. The ratio of the areas is equal to the ratio of the central angles:

$$\frac{\text{area } A \text{ of the sector}}{\text{area of the whole circle}} = \frac{\text{central angle } \theta \text{ of the sector}}{\text{central angle of the whole circle}}$$

Or,

$$\frac{A}{\pi r^2} = \frac{\theta}{2\pi}$$

where A is the area of the sector and θ is the central angle of the sector, in radians. Solving for A, we have

$$A = \left(\frac{\theta}{2\pi}\right) \pi r^2$$

or,

$$A = \frac{r^2 \theta}{2}$$

24-2 Arc Length and Area of a Sector

| AREA OF A SECTOR | $A = \dfrac{r^2\theta}{2}$, where θ is in radians | 78 |

FIGURE 24-8

Example 12

Find the area of a sector having a radius 2.50 m and a central angle of $\dfrac{3\pi}{4}$ (Fig. 24-8).

Solution: Using Eq. 78 with $r = 2.50$ and $\theta = \dfrac{3\pi}{4}$, we have

$$A = \frac{r^2\theta}{2} = \frac{(2.50)^2\left(\dfrac{3\pi}{4}\right)}{2} = 7.36 \text{ m}^2$$

Example 13

Find the area of a sector having a radius 5.60 cm and a central angle of 36.0° (Fig. 24-9).

Solution: To use Eq. 77, we need to convert $\theta = 36.0°$ to radians:
$$\theta = 36.0° = 0.6283 \text{ rad}$$

Using Eq. 78, we have

$$A = \frac{r^2\theta}{2} = \frac{(5.60)^2(0.6283)}{2} = 9.85 \text{ cm}^2$$

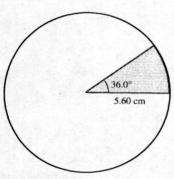

FIGURE 24-9

Applications

Eqs. 77 and 78 can be used to solve many applications involving arc length and sectors.

FIGURE 24-10

Example 14

A community is building a new baseball field in the shape of a sector (Fig. 24-10). The central angle is 100.0° and the radius is 96.30 m.

(a) How much fence is needed for the backfield?
(b) What is the area of the entire field?

Solution: To use Eq. 77, we change 100.0° to radians.
$$\theta = 100.0° = 1.7453 \text{ rad}$$

Using the equation derived from Equation 77 to find the arc length, the length of the fence, with $r = 96.30$ m:

$$s = \theta r = (1.7453)(96.30)$$
$$s = 168.1 \text{ m}$$

We use Eq. 78 to find the area of the sector, with $\theta = 1.7453$ rad and $r = 96.30$ m:

$$A = \frac{r^2\theta}{2} = \frac{(96.30)^2(1.7453)}{2}$$
$$A = 8093 \text{ m}^2$$

So, (a) the fence needs to be 168.1 m long, and (b) the area of the field is 8093 m².

SECTION 24-2 Exercises Arc Length and Area of a Sector

Arc Length

Find the angle (in radians) that intercepts an arc s in a circle of radius r.
1. $s = 6.23$ cm, $r = 4.50$ cm
2. $s = 3.54$ in., $r = 1.27$ in.

Find the angle (in degrees) that intercepts an arc s in a circle of radius r.
3. $s = 15.0$ ft, $r = 3.43$ ft
4. $s = 67.4$ mm, $r = 78.2$ mm

Find the length of the arc intercepted by a central angle θ in a circle of radius r.
5. $\theta = \dfrac{\pi}{3}$ rad, $r = 4.83$ m
6. $\theta = \dfrac{5\pi}{6}$ rad, $r = 97.4$ in.
7. $\theta = 105°$, $r = 538$ mm
8. $\theta = 235°$, $r = 12.5$ cm

Find the radius of a circle in which a central angle θ intercepts an arc of length s.
9. $\theta = 1.56$ rad, $s = 9.56$ cm
10. $\theta = 3.17$ rad, $s = 56.7$ m
11. $\theta = 315°$, $s = 14.5$ in.
12. $\theta = 335°$, $s = 9.61$ ft

Area of a Sector

Find the area of a sector having a radius r and a central angle θ.
13. $r = 3.80$ cm, $\theta = \dfrac{2\pi}{3}$ rad
14. $r = 17.1$ ft, $\theta = 4.00$ rad
15. $r = 12.4$ mm, $\theta = 125°$
16. $r = 145$ m, $\theta = 80.5°$

Applications

17. Latitude is the angle, measured at the earth's centre, between a point on the earth and the equator. Halifax, Nova Scotia, is at a latitude of 44.55° N. Find the distance in kilometres from Halifax to the equator (Fig. 24-11). The radius of the earth is approximately 6378 km.

18. One "track" on a magnetic disk used for computer data storage is located at a radius of 121 mm from the centre of the disk. If 10 "bits" of data can be stored on 1.00 mm of track, how many bits can be stored in the length of track intercepted by an arc of 15.3°?

19. A circular highway curve has a radius of 107 m and a central angle of 15.0° measured to the centre line of the road. Find the length of the curve.

20. Tape passes over a circular magnetic tape head, forming a central angle of 125° (Fig. 24-12). If the radius of the tape head is 2.50 cm, how long is the section of tape that is in contact with the tape head?

21. A pendulum swings through an angle of 12.4°. If the pendulum is 72.5 cm long, find the area swept by the pendulum.

22. Use Eqs. 77 and 78 to find another formula for the area of a sector in terms of r and s. Use the new formula to find the area of a sector with radius 5.63 cm and arc length 14.5 cm.

■ FIGURE 24-11
Latitude.

■ FIGURE 24-12

23. A cone is to be made from a circle of sheet metal by removing a 135° sector from the circle and bending the remaining sector to form the cone. If the radius of the circle is 18.0 in., find the lateral area of the cone.
24. Which has the greater area: a piece of a 10 in. pizza which has been cut into 6 pieces, or a piece of a 12 in. pizza which has been cut into 8 pieces?
25. The area of a sector-shaped piece of land is 1.75 acres. Find the radius if the central angle is 142°. (1 acre = 43 560 ft²)

24-3 UNIFORM CIRCULAR MOTION

In this section we discuss uniform circular motion, including angular velocity, angular displacement, and linear speed. Many applications involve rotations such as rotating disks, propellers, flywheels, satellites, and turbines.

Uniform Circular Motion

If an object is rotating in a circle about a point O, the *angular velocity* ω is the rate at which the object is rotating. If the angular velocity is constant, the motion is called *uniform circular motion*.

$\omega = 66$ rev/min

■ FIGURE 24-13

Example 15

The fan in Fig. 24-13 is rotating about the centre point with a uniform circular motion. The fan rotates at an angular velocity ω of 66 revolutions per minute, or 66 rev/min. An angular velocity ω can also be described in terms of radians per second.

Angular Displacement and Angular Velocity

Velocity v is described as displacement D per unit time t. Or

$$v = \frac{D}{t}$$

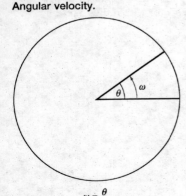

■ FIGURE 24-14
Angular velocity.

When we consider a rotating object the displacement D at any point, such as at the end of the rotating fan, is dependent on the radius of rotation r and the *angular displacement* θ through which it turns. The total length of displacement D is equal to the length of arc s as described in Equation 77, or $D = s = r\theta$.

$$D = r\theta$$

Substituting for D in the velocity formula, we get

$$v = \frac{r\theta}{t}$$

Therefore, the velocity of a rotating object at a distance r from the centre of rotation is dependent on the angular displacement θ per unit time t. This is called *angular velocity*. (Fig. 24-14)

$$v = r\left(\frac{\theta}{t}\right)$$

The standard symbol for angular velocity is ω giving us $\omega = \frac{\theta}{t}$ and

$$v = r\omega$$

568 CHAPTER 24 / Applications of Radian Measure

Angular velocity ω is generally specified in radians per second (SI). In practical applications, the angular velocity is commonly expressed in revolutions per unit time such as revolutions per minute or revolutions per second.

Example 16

Convert the angular velocity ω = 66 rev/min to rad/s (the SI unit of angular velocity).

Solution: We convert using the conversion procedure described in Chapter 4.

$$\omega = 66 \text{ rev/min} = \left(\frac{66 \text{ rev}}{\text{min}}\right)\left(\frac{2\pi \text{ rad}}{1 \text{ rev}}\right)\left(\frac{1 \text{ min}}{60 \text{ s}}\right) = \frac{11\pi}{5}\frac{\text{rad}}{\text{s}} = 6.91 \text{ rad/s}$$

Since the radian is a dimensionless unit, the exact angular velocity can also be stated as $\omega = \frac{11\pi}{5 \text{ s}}$ or in approximate form $\omega = 6.91 \text{ s}^{-1}$.

Example 17

A propeller on a wind generator rotates 90.0° in 1.45 s. Find the angular velocity in radians per second, and in revolutions per minute.

Solution: First we convert the angle to radians

$$90.0° = 90°\left(\frac{\pi}{180°}\right) \text{rad} = \frac{\pi}{2} \text{ rad} = 1.57 \text{ rad}$$

then using $\omega = \frac{\theta}{t}$, we get the angular velocity

$$\omega = \frac{1.57 \text{ rad}}{1.45 \text{ s}} = 1.08 \text{ rad/s}$$

Next we convert to revolutions per minute.

$$\omega = 1.08\frac{\text{rad}}{\text{s}}\left(\frac{1 \text{ rev}}{2\pi \text{ rad}}\right)\left(\frac{60 \text{ s}}{1 \text{ min}}\right)$$

$$\omega = 10.3 \text{ rev/min}$$

The angular velocity ω is 10.3 rev/min.

Example 18

A flywheel makes 721 revolutions in a minute. What is the angular displacement expressed in radians during 0.500 second?

Solution: We need a common unit of time so we convert the angular velocity ω from revolutions per minute to revolutions per second.

$$\omega = 721\frac{\text{rev}}{\text{min}}\left(\frac{1 \text{ min}}{60 \text{ s}}\right) = 12.01\overline{6}\frac{\text{rev}}{\text{s}}$$

Next we find the angular velocity in radians per second

$$\omega = 12.01\overline{6}\frac{\text{rev}}{\text{s}}\left(\frac{2\pi \text{ rad}}{1 \text{ rev}}\right) = 75.5\frac{\text{rad}}{\text{s}}$$

Finally, we find the angular displacement θ after 0.500 seconds. Using $\theta = \omega t$ we get

$$\theta = \omega t$$

$$\theta = \left(75.50\,\frac{\text{rad}}{\text{s}}\right)(0.500\,\text{s})$$

$$\theta = 37.75\,\text{rad}$$

Therefore, the angular displacement θ is 37.8 rad (rounded) after 0.500 s.

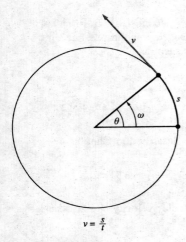

$v = \frac{s}{t}$

■ **FIGURE 24-15**
Linear velocity.

Linear Velocity

For any point on a rotating body, the linear displacement per unit time along the circular path is called the *linear velocity* v (Fig. 24-15). The linear velocity is zero for a point at the centre of rotation, and it increases as the distance from the centre increases.

Example 19

Two bugs are riding on a ceiling fan (Fig. 24-16). They are both rotating at the same angular velocity ω. But, during any specific period of time the bug closer to the end of the fan blade moves a greater distance than the bug closer to the centre. The bug on the outer portion of the fan blade also travels faster because the distance traveled is greater in the same amount of time. The linear velocity of the outer bug is greater than the linear velocity of the bug closest to the centre of rotation.

■ **FIGURE 24-16**

To derive an expression for linear velocity v, we begin with the equation

$$s = \theta r, \text{ where } \theta \text{ is in radians}$$

Dividing both sides by t, we have

$$\frac{s}{t} = \frac{\theta}{t} r$$

Since $\frac{s}{t} = v$ and $\frac{\theta}{t} = \omega$, we have

$$v = \omega r$$

where ω is in *radians* per unit time.

Example 20

A flywheel is rotating at 33.3 rev/min. Find the linear velocity of a bug 15.2 cm from the centre.

Solution: First we change 33.3 rev/min to rad/min.

$$\omega = \left(\frac{33.3\,\text{rev}}{\text{min}}\right)\left(\frac{2\pi\,\text{rad}}{1\,\text{rev}}\right) = 209.2\,\text{rad/min}$$

Substituting into

$$v = \omega r$$
$$v = (209.2\,\text{rad/min})(15.2\,\text{cm})$$

we have

$$v = 1380\,\text{cm/min}$$

SECTION 24-3 Exercises Uniform Circular Motion

Angular Velocity

Find the missing values. Round to three significant digits.

	rev/min	rad/s	deg/s
1.	1520		
2.		4.50	
3.		$4\pi/5$	
4.			135

All angular velocities are to be calculated in rad/s unless otherwise specified.

5. A propeller on a wind generator rotates 54.5° in 0.750 s. Find the angular velocity of the propeller in rad/min.

6. A satellite circles the earth 6.25 times each 24.0 hours. Find its angular velocity in rad/s. Give results in scientific notation.

7. The flywheel of an engine is rotating at 452 rpm. Find the angular displacement of the flywheel in 5.00 s.

8. A blade on a water turbine turns 175° in 1.25 s. Find its angular velocity, and the angular displacement of the blade in 10.0 s.

Linear Velocity

9. A flywheel rotates at an angular velocity of 3.00 rad/s. If the radius of the flywheel is 37.6 cm, find the linear velocity of a point on the rim, in m/s.

10. A car wheel 61.0 cm in diameter is rotating at 775 rev/min. Find the linear velocity of the car (in km/h).

11. The earth rotates on its axis at 1.00 revolution every 24.0 hours. Assuming that the earth is a sphere 6378 km in radius, find the linear velocity (in km/h) of a point on the earth's equator.

12. A car is traveling at a rate of 88.0 km/h and has tires that have a diameter of 63.2 cm. Find the angular velocity of the wheels of the car.

♚ 13. The earth rotates around the sun at 1 revolution every 365 days. Assuming that the earth's orbit around the sun is a circle with a radius of 92 960 000 miles, find the linear velocity of the earth around the sun, in miles per hour.

14. A race car is driven around a circular track at an average velocity of 85.0 mi/h. Find the angular velocity of the car (in rad/min) if the radius of the track is 842 ft.

Chapter 24 REVIEW EXERCISES

Convert each angle from degrees to radian measure in terms of π (in fraction form).

1. 12°   2. 80°   3. 216°   4. 300°

Convert each angle from radians to degrees.

5. $\dfrac{\pi}{5}$ 6. $\dfrac{5\pi}{4}$ 7. $\dfrac{17\pi}{12}$ 8. $\dfrac{14\pi}{9}$

9. Find the angle (in radians) that intercepts an arc of 4.65 cm in a circle of radius 3.15 cm.
10. Find the angle (in degrees) that intercepts an arc of 12.5 cm in a circle of radius 8.50 cm.
11. Find the length of the arc intercepted by a central angle of 175° in a circle of radius 75.0 mm.
12. Find the radius of a circle in which a central angle of 105° intercepts an arc of length 7.25 m.
13. Find the area of a sector having a radius of 1.53 m and a central angle of 75.8°.
14. Find the area of a sector-shaped window with a radius of 87.4 cm and a central angle of 152°.
15. Change 315°/s to rev/min.
16. Find the angular velocity (in rad/s) of a wheel that rotates 37.5 rev in 1.75 min.

 17. A pendulum of a clock swings through an angle of 8.50°. If the pendulum is 30.0 inches long, find the length of the arc determined by the end of the pendulum.
18. The propeller of an airplane is 6.00 ft in diameter and rotates at 2250 rpm. Find the linear speed of the tip of the propeller (in ft/s).

WRITING

19. We were pretty happy measuring angles in degrees. Why do we need to study angles measured in radians? What applications use angles measured in radians, but not in degrees? Make up a problem for which you need to use radian measure. Solve your problem and revise it if necessary.

TEAM PROJECT

20. The kinetic energy of a rotating body is given by

$$E = \frac{1}{2}mk^2\omega^2 \text{ Joules}$$

where m is the mass in kilograms, ω is the angular velocity in radians/second, and k is the radius of rotation in metres. For a sphere of radius r,

$$k = \frac{\sqrt{2}}{5}r$$

Weigh and measure the radius of five balls of various sizes (such as a golf ball, a billiard ball, a baseball, a softball, and a basketball). Then compute the kinetic energy of rotation for each ball, assuming that it is rotating at 3.00 revolutions per second. What conclusions can you draw about mass, radius, and kinetic energy?

Graphs of Trigonometric Functions

Outcomes

When you have completed this chapter, you will be able to

1. Graph the trigonometric functions.
2. Graph $y = a \sin x$.
3. Graph $y = a \sin bx$.
4. Graph $y = a \sin (bx + c)$.
5. Graph $y = a \cos (bx + c)$.
6. Graph $y = a \tan (bx + c)$.
7. Determine the period, amplitude, and phase shift for a given sine or cosine function.

In this chapter, we apply the graphing techniques of Chapter 12 to graph some *periodic* functions—the sine, cosine, and tangent—and we'll see how the constants in the equation affect the shape and position of the curves. We define some terms that will be applicable to the study of periodic functions in general, such as *cycle, period,* and *amplitude*. We solve applications with topics such as current, voltage, and simple harmonic motion.

25-1 GRAPH OF $y = a \sin x$

In this section we graph the functions $y = \sin x$ and $y = a \sin x$. We discuss the concepts of period and amplitude of periodic functions and use our knowledge of period and amplitude to make a "quick sketch" of the graph.

FIGURE 25-1
Point P on a rotating unit circle.

Graph of $y = \sin \theta$

We can visualize the graph of the sine function

$$y = \sin \theta$$

Consider a point $P(x, y)$ on a rotating circle of radius 1 (a unit circle), with its centre at the origin and θ the central angle as shown in Fig. 25-1. As the radius OP revolves counterclockwise through one revolution, beginning with P at $(1, 0)$, θ goes from $0°$ to $360°$. At any instant, the y coordinate of the point P is a function of θ. In fact, since $r = 1$,

$$\sin \theta = \frac{y}{r} = \frac{y}{1} = y$$

So

$$y = \sin \theta$$

■ FIGURE 25-2
Vertical displacement *y* on the unit circle.

■ FIGURE 25-3
Vertical displacement *y* as a function of θ.

Thus, in the unit circle, $\sin \theta$ is the *vertical displacement* of the point P from the x axis (Fig. 25-2). As θ changes from 0° to 90°, the vertical displacement ($\sin \theta$) increases from 0 to 1, changing rapidly at first, and more slowly as it nears 1. As θ changes from 90° to 180°, the vertical displacement ($\sin \theta$) decreases from 1 to 0, slowly at first and more rapidly as it nears 0. As θ changes from 180° to 270°, the vertical displacement ($\sin \theta$) decreases from 0 to -1, rapidly at first and then more slowly as it nears -1. And as θ changes from 270° to 360°, the vertical displacement ($\sin \theta$) increases from -1 to 0, slowly at first and more rapidly as it nears 0 (Fig. 25-3). From this model we can see that the curve pattern *repeats itself* with each revolution, or every 360°. A curve that repeats itself is called *periodic*.

Now that we have a general idea of how the sine function behaves, let's graph the function $y = \sin \theta$ in the same way as we graphed other equations in Chapter 12. We select values for the independent variable θ and compute values for the dependent variable y, producing a table of values for θ and y. We then plot these points and connect them with a smooth curve.

Example 1

Graph $y = \sin \theta$ for θ from 0° to 360°.

Solution: We make a table of values for θ and y, choosing intervals for θ of 30° each. We use the calculator to find the y values, making sure that the calculator is in the degree mode. If y values are not exact, we round to tenths. (When you make a graph "by hand," you can't plot points with much more accuracy than tenths.)

Note: Many calculators have a "table" feature you can use to make a table. See your manual for specific instructions.

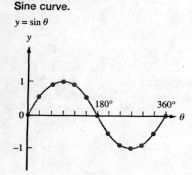

■ FIGURE 25-4
Sine curve.
$y = \sin \theta$

θ	0°	30°	60°	90°	120°	150°	180°	210°	240°	270°	300°	330°	360°
y	0	0.5	0.9	1	0.9	0.5	0	-0.5	-0.9	-1	-0.9	-0.5	0

These points are plotted and connected with a smooth curve in Fig. 25-4.

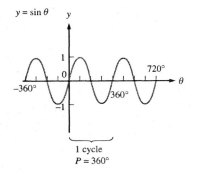

■ **FIGURE 25-5**
Three cycles of the sine curve.

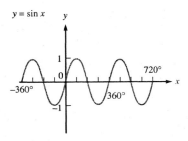

■ **FIGURE 25-6**
Sine as a function of x.

If we had plotted points for θ from 360° to 720°, we would have gotten another curve pattern, identical to the one just sketched. The curve will *repeat* to the right of $\theta = 360°$ and also to the left of $\theta = 0°$ (Fig. 25-5).

Period

Since the sine curve repeats, it is periodic. The smallest repeating portion of the graph is called a *cycle* and the length of one cycle is called the *period*. A cycle of the curve in Fig. 25-5 repeats every 360°, so the period P of $y = \sin \theta$ is 360°:

$$P = 360° \quad \text{in degrees}$$

Graph of $y = \sin x$

In the model above, we used θ to represent the angle, and we expressed the sine function as a function of the angle θ. However, the sine is often expressed as a function of the variable x, where x is an angle, expressed either in degrees or in radians. From this point on, we write the sine function as a function of x; that is, the independent variable will be x, not θ. The graph of $y = \sin x$ is similar to Fig. 25-5, except that the horizontal axis is the x axis, not the θ axis. Fig. 25-6 shows several cycles of the curve $y = \sin x$.

Graphing Calculator

A graphing calculator makes it easy to graph the trigonometric functions and to explore how the constants in the equation affect the graph. First we must set the calculator to the appropriate angle mode, either degree or radian. We define the window variables, which are determined by the domain and the range of the function. Then we enter the function and graph it. Check your calculator manual for specific instructions.

─── **Example 2** ───

Use a graphing calculator to graph $y = \sin x$ for x from $-180°$ to $450°$.

Solution: First we set the calculator angle mode to degree. We are given the domain (x is between $-180°$ and $450°$), and we have noted that y values for $y = \sin x$ are always between -1 and 1. Thus, we set the window so that

■ **FIGURE 25-7**
Sine curve on a graphing calculator.

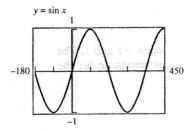

We enter the function $y = \sin x$, and graph (Fig. 25-7). The graphing calculator does not show labels on the axes; you must check the window variables to determine what each tick mark represents on the graph. In this book, when a figure contains a graphing calculator screen, we show the values of Xmin, Xmax, Ymin, and Ymax, although they do not appear on the graphing calculator screen. Experiment with the different predefined windows on your calculator. Many calculators have a special window for graphing trigonometric functions.

25-1 Graph of $y = a \sin x$

Many applications of the sine function use angles measured in radians, so we must also be able to graph the sine function in radians.

Example 3

Use a graphing calculator to graph $y = \sin x$, for x from -2π to 2π radians.

Solution: First we set the calculator angle mode to radians. Then we set the window.

Note that when we enter -2π for Xmin, the calculator immediately evaluates this as a decimal and displays it as $-6.2831853\ldots$. Similarly, it evaluates 2π and $\pi/2$ as decimals. We enter the function

$$y = \sin x$$

and graph (Fig. 25-8). The graph shows two complete cycles of the sine curve. In radians, the period P of the function is 2π, so

$$P = 2\pi \quad \text{in radians}$$

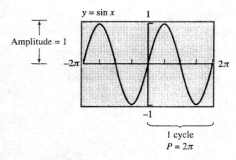

■ **FIGURE 25-8**

Amplitude

We see from Fig. 25-8 that if $y = \sin \theta$, the y values vary between -1 and 1. The maximum variation of the curve $y = \sin \theta$ from zero is called its *amplitude*. So, the amplitude of the curve $y = \sin \theta$ is 1.

Graph of $y = a \sin x$

Now we'll look at the effect of a constant coefficient a on the graph of the equation.

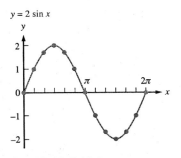

FIGURE 25-9
A change in amplitude.

Example 4

Graph $y = 2 \sin x$, for x from 0 to 2π radians.

Solution: We make a table of values, taking x intervals of $\pi/6$. We use a calculator in radian mode to find the y values and round to tenths.

x	0	$\frac{\pi}{6}$	$\frac{\pi}{3}$	$\frac{\pi}{2}$	$\frac{2\pi}{3}$	$\frac{5\pi}{6}$	π	$\frac{7\pi}{6}$	$\frac{4\pi}{3}$	$\frac{3\pi}{2}$	$\frac{5\pi}{3}$	$\frac{11\pi}{6}$	2π
y	0	1	1.7	2	1.7	1	0	-1	-1.7	-2	-1.7	-1	0

We plot the points and draw a smooth curve (Fig. 25-9). The graph shows one complete cycle. The curve repeats every 2π units, and therefore the period P is:

$$P = 2\pi$$

We also note that for $y = 2 \sin x$, the y values vary between -2 and 2, so

$$\text{amplitude} = 2$$

The amplitude represents the maximum y value of the curve. The graph of the function $y = a \sin x$ will rise up to $y = |a|$ and down to $y = -|a|$. We use absolute values here, because a may be negative. In general, if $y = a \sin x$, the amplitude will be the *absolute value* of the coefficient of sin x. So

$$\text{amplitude} = |a|$$

The period P of the curve $y = a \sin x$ is always 2π or 360°, depending on whether x is in radians or degrees.

	amplitude = $|a|$	146
$y = a \sin x$		
	Period = 2π or 360°	147

Example 5

Given the function $y = -3 \sin x$,

(a) Find the amplitude and the period, in degrees.
(b) Graph the function for x from 0° to 360°.

Solution:

(a) We use Eq. 146 to find the amplitude. Since $a = -3$:

$$\text{amplitude} = |-3| = 3$$

This means that y varies between -3 and 3. Since a is negative ($a = -3$), the curve is *inverted* (turned upside down). That is, if we start at $x = 0$, the curve goes *down* to -3, instead of *up* to 3.
We use Eq. 147 to find the period P of $y = -3 \sin x$. Since x is in degrees:

$$P = 360°$$

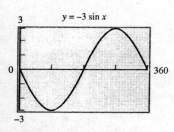

■ FIGURE 25-10

(b) To graph the function, we use a graphing calculator in degree mode, setting the window so that

We enter the function and graph (Fig. 25-10).

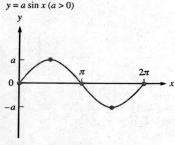

■ FIGURE 25-11

If $y = a \sin x$, the coefficient a acts as a vertical curve *stretcher* if $|a| > 1$ or as a vertical curve *compressor* if $|a| < 1$. The curves of $y = \sin x$ and $y = a \sin x$ have the same general shape; the second curve is just stretched or compressed vertically depending on the value of a.

All the curves $y = a \sin x$ have the same general shape and the same period. They vary only in their amplitude. To make a "complete" graph of a sine curve, we graph *at least one* complete cycle. This includes three x intercepts and two turning points—a maximum point and a minimum point for each cycle. We plot these five key points and connect them with a smooth curve (Fig. 25-11).

COMMON ERROR Remember that the sine curve is *never straight* and *never pointy*.

— **Example 6** —
Given the function

$$y = \frac{1}{2} \sin x$$

where x is in radians,

(a) Find the period and the amplitude.
(b) Graph the function.

■ FIGURE 25-12

Solution:

(a) Since the equation is of the form $y = a \sin x$, the period P is 2π (in radians). The amplitude is

$$\text{amplitude} = \left|\frac{1}{2}\right| = \frac{1}{2}$$

(b) Using $P = 2\pi$ and $a = \frac{1}{2}$, we sketch the graph in Fig. 25-12.

Example 7

 Using a graphing calculator in degree mode, graph these three functions on the same screen, for x from $0°$ to $360°$:

(a) $y = \sin x$
(b) $y = 3 \sin x$
(c) $y = -2 \sin x$

Solution: We set the window so that

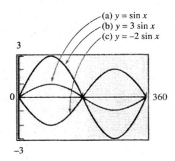

■ FIGURE 25-13
Three sine curves with different amplitudes.

We graph the functions, as in Fig. 25-13. Now we can really see the effect of the coefficient a on the curve. The coefficient 3 in equation (b) stretches the curve vertically, to three times the original amplitude. The coefficient -2 in equation (c) inverts the curve and stretches it vertically, to two times the original amplitude.

SECTION 25-1 Exercises Graph of $y = a \sin x$

Graph of $y = \sin x$

1. Complete the table for the equation

$$y = \sin x$$

rounding to tenths, and graph the equation for x from $-\pi$ to 3π.

x	$-\pi$	$-\frac{3\pi}{4}$	$-\frac{\pi}{2}$	$-\frac{\pi}{4}$	0	$\frac{\pi}{4}$	$\frac{\pi}{2}$	$\frac{3\pi}{4}$	π	$\frac{5\pi}{4}$	$\frac{3\pi}{2}$	$\frac{7\pi}{4}$	2π	$\frac{9\pi}{4}$	$\frac{5\pi}{2}$	$\frac{11\pi}{4}$	3π
y																	

2. Complete the table for the equation

$$y = 3 \sin x$$

rounding to tenths, and graph the equation for x from $-90°$ to $360°$.

x	$-90°$	$-60°$	$-30°$	$0°$	$30°$	$60°$	$90°$	$120°$	$150°$	$180°$	$210°$	$240°$	$270°$	$300°$	$330°$	$360°$
y																

Graph of y = a sin x

Write the equations of the sine curves.

3.

FIGURE 25-14

4.

FIGURE 25-15

5.

FIGURE 25-16

6.

FIGURE 25-17

7. Write an equation of the sine curve that has an amplitude of 2 and a period of 360°.
8. Write an equation of the sine curve that has an amplitude of 5 and a period of 2π.

Find the amplitude and the period, in degrees. Graph each equation for x from 0° to 360°.

9. $y = 5 \sin x$
10. $y = 4 \sin x$
11. $y = -\sin x$
12. $y = -2 \sin x$
13. $y = \frac{1}{3} \sin x$
14. $y = -\frac{3}{2} \sin x$

Find the amplitude and the period, in radians. Graph each equation for x from 0 to 2π.

15. $y = -2.5 \sin x$ **16.** $y = 7.5 \sin x$
17. $y = 100 \sin x$ **18.** $y = -60 \sin x$

Graph each equation for integral values of x from 0 to 7 radians.

19. $y = \sin x$ **20.** $y = -\dfrac{1}{2} \sin x$

25-2 GRAPH OF $y = a \sin bx$

In this section we graph the function

$$y = a \sin bx$$

and see how the period of the curve is affected by the constant b.

Graph of $y = \sin bx$

First we graph the function

$$y = \sin bx$$

to see the effect of b, the coefficient of x.

Example 8

Graph the function $y = \sin 2x$, for x from 0 to 2π.

Solution: We make a table of values, taking intervals of $\pi/8$ and computing the values for y using a calculator, in radian mode, rounding to tenths.

x	0	$\frac{\pi}{8}$	$\frac{\pi}{4}$	$\frac{3\pi}{8}$	$\frac{\pi}{2}$	$\frac{5\pi}{8}$	$\frac{3\pi}{4}$	$\frac{7\pi}{8}$	π	$\frac{9\pi}{8}$	$\frac{5\pi}{4}$	$\frac{11\pi}{8}$	$\frac{3\pi}{2}$	$\frac{13\pi}{8}$	$\frac{7\pi}{4}$	$\frac{15\pi}{8}$	2π
y	0	0.7	1	0.7	0	−0.7	−1	−0.7	0	0.7	1	0.7	0	−0.7	−1	−0.7	0

These points are plotted and connected with a smooth curve in Fig. 25-18. We see from the graph that the amplitude is still 1. But the curve repeats itself every π units, instead of repeating every 2π. So the period of the graph is *not* 2π, but π. Now we can see that the effect of the 2 in the equation $y = \sin 2x$ is to divide the period (of the basic curve $y = \sin x$) in half.

FIGURE 25-18
A change in period.

Example 9

Graph $y = \sin x$ and $y = \sin 2x$ on the same axes, for x from 0 to 2π.

Solution: We graph these functions on a graphing calculator in radian mode, setting the window so that

FIGURE 25-19
Two sine curves with different periods.

We enter the functions and graph (Fig. 25-19). Notice how the constant 2 compresses the curve horizontally, indicating a change in the period.

In general, the graph of the function

$$y = \sin bx$$

has a period P equal to 2π divided by b (in radians), or $360°$ divided by b (in degrees):

$$P = \frac{2\pi}{b} \quad \text{or} \quad \frac{360°}{b}$$

If $|b| > 1$, then the curve is horizontally *compressed*. If $|b| < 1$, then the curve is horizontally *stretched*.

Example 10

Find the period of the function $y = \sin 2x$, where x is in radians.

Solution: Since x is in radians, we use

$$P = \frac{2\pi}{b}$$

We see that $b = 2$, so

$$P = \frac{2\pi}{2} = \pi$$

This period is verified by the graph in Fig. 25-18, since the curve repeats itself every π units.

Graph of $y = a \sin bx$

The constant coefficient a of the sine function

$$y = a \sin bx$$

determines the amplitude of the curve:

$$\text{amplitude} = |a|$$

The constant a does not affect the period of the curve $y = a \sin bx$, so we still divide 2π (or $360°$) by b to find the period.

$y = a \sin bx$	amplitude $= \|a\|$	146
	Period $= \dfrac{2\pi}{b}$ or $\dfrac{360°}{b}$	147

─── **Example 11** ───────

Find the amplitude and the period, in degrees, of the function

$$y = \sin 3x$$

Solution: Since

$$y = 1 \sin 3x$$

then $a = 1$ and by Eq. 146

$$\text{amplitude} = |1| = 1$$

We use Eq. 147 to find the period P, in degrees, with $b = 3$:

$$P = \frac{360°}{b} = \frac{360°}{3} = 120°$$

Each cycle of the graph of $y = \sin 3x$ repeats every $120°$.

─── **Example 12** ───────

Find the amplitude and the period, in degrees, of the function

$$y = 2 \sin \frac{x}{4}$$

Solution: Using Eq. 146, with $a = 2$

$$\text{amplitude} = |2| = 2$$

To find the period, we note that

$$\sin \frac{x}{4} = \sin \frac{1}{4}x, \quad \text{so} \quad b = \frac{1}{4}$$

Using Eq. 147, with $b = \frac{1}{4}$, the period P is

$$P = \frac{360°}{\frac{1}{4}} = 360° \div \frac{1}{4} = 360° \left(\frac{4}{1}\right) = 1440°$$

Each cycle of the graph repeats every $1440°$.

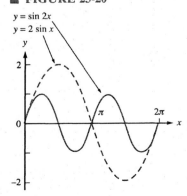

■ FIGURE 25-20
$y = \sin 2x$
$y = 2 \sin x$

─── **Example 13** ───────

On the same axes, graph the functions

$$y = \sin 2x \quad \text{and} \quad y = 2 \sin x$$

for x from 0 to 2π. Compare the amplitudes, periods, and graphs of the two functions.

Solution: For the function

$$y = \sin 2x$$

the 2 is the coefficient of x. The constant 2 affects the *period* of the graph of $y = \sin 2x$. The graph of the function $y = \sin 2x$ has a period of π and an amplitude of 1 (Fig. 25-20).

For the function

$$y = 2 \sin x$$

the 2 is the coefficient of *sin x*. Here, the 2 affects the *amplitude*, but not the period, of the graph of $y = 2 \sin x$. The graph of the function $y = 2 \sin x$ has a period of 2π and an amplitude of 2.

So we see in Fig. 25-20 that the graphs of these two functions have different periods and different amplitudes.

Quick Sketch of the Sine Function

A fast way to graph one complete cycle of a sine function is to make a "quick sketch." All the curves $y = a \sin bx$ have the same general shape; they vary only in their period and amplitude.

Quick Sketch of the Sine Curve

1. Find the period P and amplitude $|a|$ of the curve.
2. Draw two horizontal lines $|a|$ units above and below the x axis, and label $|a|$ on the y axis.
3. Draw two vertical lines, one on the y axis and another P units to the right of the y axis. Label P on the x axis. (This completes a rectangle, bounded by the two horizontal lines drawn in Step 2 and the two vertical lines drawn in Step 3.)
4. On the x axis, make three marks to subdivide the period P into four equal parts; draw vertical lines through the marks.
5. Use the lines as guides to sketch one complete cycle of the sine curve.

One complete cycle of the sine curve includes three x intercepts, one maximum point, and one minimum point. Plot these five key points and sketch the sine curve in the rectangle. The quick sketch is illustrated in Fig. 25-21, for $a > 0$.

FIGURE 25-21
Quick sketch of the sine curve.

> **COMMON ERROR** Remember that the sine curve is *never straight* and *never pointy.*

---- **Example 14** ----

Given $y = 2 \sin \frac{x}{3}$, where x is in radians,

(a) Find the amplitude and the period.
(b) Make a quick sketch of the graph.

Solution:

(a) We use Eq. 146 to find the amplitude:

$$\text{amplitude} = |a| = |2| = 2$$

We use Eq. 147 to find the period, with $b = \frac{1}{3}$

$$P = \frac{2\pi}{b} = \frac{2\pi}{\frac{1}{3}} = 2\pi \left(\frac{3}{1}\right) = 6\pi$$

So one cycle of the curve repeats every 6π radians.

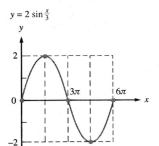

FIGURE 25-22

(b) We make a quick sketch of the graph, using $a = 2$ and $P = 6\pi$. After drawing and labeling the x and y axes, we draw two horizontal lines, one at $y = 2$ and the other at $y = -2$. Then we draw two vertical lines at $x = 0$ (y axis) and $x = 6\pi$, forming a rectangle of width 6π and height 2. We subdivide the period 6π into four equal parts and draw the three inner vertical lines. We mark the five key points, and connect them with a smooth curve (Fig. 25-22).

TIP When graphing a trigonometric function in which the domain is not specified, graph at least one complete cycle of the curve.

Example 15

Given the function $y = \sin \pi x$, where x is in radians,

(a) Find the amplitude and the period.
(b) Graph the function.

Solution:

(a) Since $y = \sin \pi x = 1 \sin \pi x$, the amplitude is 1. Since $b = \pi$, the period is

$$P = \frac{2\pi}{\pi} = 2$$

(b) To graph the function, we either make a quick stretch or use a graphing calculator (Fig. 25-23).

FIGURE 25-23

SECTION 25-2 Exercises Graph of $y = a \sin bx$

Graph of $y = \sin bx$

1. Complete the table for the equation

$$y = \sin 3x$$

rounding to tenths, and graph the equation for x from 0 to 2π. Find the amplitude and the period. On the same axes, graph $y = \sin x$ with a dotted line.

x	0	$\frac{\pi}{12}$	$\frac{\pi}{6}$	$\frac{\pi}{4}$	$\frac{\pi}{3}$	$\frac{5\pi}{12}$	$\frac{\pi}{2}$	$\frac{7\pi}{12}$	$\frac{2\pi}{3}$	$\frac{5\pi}{6}$	π	$\frac{7\pi}{6}$	$\frac{4\pi}{3}$	$\frac{3\pi}{2}$	$\frac{5\pi}{3}$	$\frac{11\pi}{6}$	2π
y																	

2. Complete the table for the equation

$$y = \sin 5x$$

rounding to tenths, and graph the equation for x from 0° to 180°. Find the amplitude and the period. On the same axes, graph $y = \sin x$ with a dotted line.

x	0°	9°	18°	27°	36°	45°	54°	63°	72°	90°	108°	126°	144°	162°	180°
y															

Write the equation of the curve.

3.

■ FIGURE 25-24

4.
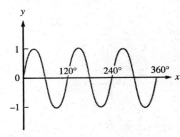
■ FIGURE 25-25

5. Write an equation of the sine curve that has an amplitude of 1 and a period of π.
6. Write an equation of the sine curve that has an amplitude of 1 and a period of 60°.

Find the amplitude and the period, in degrees. Then graph the equation.

7. $y = \sin \dfrac{x}{2}$ 8. $y = \sin 8x$

Find the amplitude and the period, in radians. Then graph the equation.

9. $y = \sin 6x$ 10. $y = \sin \dfrac{x}{4}$

Graph of $y = a \sin bx$

Write the equation of the sine curve.

11.

■ FIGURE 25-26

12.

■ FIGURE 25-27

13. Write an equation of the sine curve that has an amplitude of 5 and a period of 90°.
14. Write an equation of the sine curve that has an amplitude of 4 and a period of $\dfrac{\pi}{3}$.

■ FIGURE 25-28

Find the amplitude and the period, in degrees. Then graph the equation.

15. $y = 2 \sin 3x$ 16. $y = -3 \sin 2x$
17. $y = -3 \sin \dfrac{x}{3}$ 18. $y = 5 \sin \dfrac{x}{5}$

Find the amplitude and the period, in radians. Then graph the equation.

19. $y = 3 \sin 2x$ 20. $y = 2 \sin 3x$
21. $y = -\dfrac{\sin \pi x}{2}$ 22. $y = -\sin \dfrac{\pi x}{2}$

23. The weight in Fig. 25-28 is pulled down 6.0 inches and released, causing it to oscillate with simple harmonic motion. If it takes 0.5 seconds to return to its lowest position, then the equation of motion is

$$y = 6 \sin 4\pi t$$

where y is the vertical displacement from its rest position (in inches) and t is the elapsed time (in seconds). Find the amplitude and the period of the oscillation, and graph the equation for t from 0 to 3.0 seconds.

25-3 GRAPH OF $y = a \sin(bx + c)$

In the previous two sections we considered examples in which the amplitude and the period of the sine curve were changed. In this section we consider how a certain change in the equation causes a *shift* of the sine curve to the right or left.

Phase Shift

If a constant c is added to the x term of a sine function, it causes the graph to be shifted in the x direction (to the right or to the left). The amount of shift is called the *phase shift*.

---- **Example 16** ----

Graph the function $y = \sin(x - 60°)$, for x from $0°$ to $420°$.

Solution: We make a table of values, taking intervals of $30°$. We use a calculator, in degree mode, to find the values for y, rounded to tenths.

x	$0°$	$30°$	$60°$	$90°$	$120°$	$150°$	$180°$	$210°$	$240°$	$270°$	$300°$	$330°$	$360°$	$390°$	$420°$
y	-0.9	-0.5	0	0.5	0.9	1	0.9	0.5	0	-0.5	-0.9	-1	-0.9	-0.5	0

These points are plotted and connected with a smooth curve in Fig. 25-29.

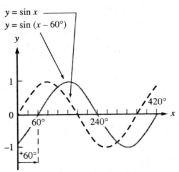

FIGURE 25-29
Sine curve with phase shift.

We see from the graph that the curve does not pass through the origin, but has been shifted in the x direction by an amount $60°$ to the *right*. For example, $\sin x = 0$ at $x = 60°$ instead of at $x = 0$. We call this a phase shift of $+60°$. The phase shift is the amount of horizontal shift in the curve.

To compare the graphs of $y = \sin x$ and $y = \sin(x - 60°)$, we graph both functions on the same axes (Fig. 25-30), using a dotted line for $y = \sin x$ and a solid line for $y = \sin(x - 60°)$. Note that the amplitude of both curves is 1 and that the period of both curves is $360°$. The $60°$ causes the graph (of the basic curve $y = \sin x$) to be shifted $60°$ to the right.

Graph of $y = \sin(bx + c)$

Now we'll see what happens if the coefficient of x is not 1.

FIGURE 25-30

FIGURE 25-31

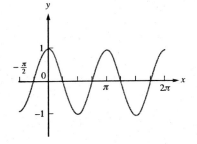

---- **Example 17** ----

Graph the equation $y = \sin\left(2x + \dfrac{\pi}{2}\right)$, for x from $-\dfrac{\pi}{2}$ to 2π.

Solution: Once again, we graph this equation by plotting points. We take intervals of $\dfrac{\pi}{4}$ and compute the values for y, using the calculator, in radian mode.

x	$-\dfrac{\pi}{2}$	$-\dfrac{\pi}{4}$	0	$\dfrac{\pi}{4}$	$\dfrac{\pi}{2}$	$\dfrac{3\pi}{4}$	π	$\dfrac{5\pi}{4}$	$\dfrac{3\pi}{2}$	$\dfrac{7\pi}{4}$	2π
y	-1	0	1	0	-1	0	1	0	-1	0	1

We plot these points and connect them with a smooth curve in Fig. 25-31.

FIGURE 25-32

To help us visualize the phase shift we graph both $y = \sin 2x$ and $y = \sin\left(2x + \frac{\pi}{2}\right)$ on the same axes (Fig. 25-32). We see from the graph that the curve $y = \sin\left(2x + \frac{\pi}{2}\right)$ is shifted $\frac{\pi}{4}$ to the left of the curve $y = \sin 2x$. We call this a phase shift of $-\frac{\pi}{4}$. You will note that the amplitude of both curves is 1 and the period of both curves is π. In the equation $y = \sin\left(2x + \frac{\pi}{2}\right)$, the only effect of the $\frac{\pi}{2}$ is to cause a phase shift of $-\frac{\pi}{4}$ (to the left). This phase shift is found by dividing $-\frac{\pi}{2}$ by the coefficient 2.

Graph of $y = a \sin (bx + c)$

In general, if

$$y = a \sin (bx + c)$$

then the constant c affects the phase shift, and the phase shift is found by dividing $-c$ by b, so

$$\text{Phase shift} = -\frac{c}{b}$$

When the phase shift is *positive*, the curve will be shifted to the *right*. When the phase shift is *negative*, the curve will be shifted to the *left*. The constant c does not affect the amplitude or the period.

The general sine curve

$$y = a \sin (bx + c)$$

contains three constants: a, b, and c. Their effects are summarized as follows:

	amplitude = $|a|$	146
$y = a \sin (bx + c)$	Period = $\frac{2\pi}{b}$ or $\frac{360°}{b}$	147
	Phase shift = $-\frac{c}{b}$	148

Quick Sketch of $y = a \sin (bx + c)$

We can use the quick sketch method of Sec. 25-2 to graph a sine function with a phase shift. First we make a *dotted* quick sketch of the graph of $y = a \sin bx$ (without the phase shift). Then calculate the phase shift and shift the curve by the phase shift.

Example 18

Given $y = 2 \sin \left(3x + \dfrac{\pi}{2}\right)$,

(a) Find the amplitude, the period, and the phase shift.
(b) Graph the function.

Solution:

(a) For this function, $a = 2$, $b = 3$, and $c = \dfrac{\pi}{2}$. Thus,

$$\text{amplitude} = 2$$

$$\text{Period} = \dfrac{2\pi}{3}$$

$$\text{Phase shift} = \dfrac{-\dfrac{\pi}{2}}{3} = -\dfrac{\pi}{6}$$

(b) First we make a quick sketch of $y = 2 \sin 3x$, shown with a *dotted* line. Since the phase shift is negative, we shift the curve $\dfrac{\pi}{6}$ units to the left as shown in Fig. 25-33.

Check: You can verify that this graph is correct by graphing the function on a graphing calculator.

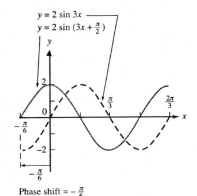

■ FIGURE 25-33

SECTION 25-3 Exercises Graph of $y = a \sin (bx + c)$

Phase Shift

Complete the table for the equation for x from $-90°$ to $360°$, rounding to tenths. Sketch the graph and give the period, amplitude, and phase shift. On the same axes, graph $y = a \sin bx$, for the given a and b, with a dotted line, and indicate the phase shift on the graph.

x	$-90°$	$-60°$	$-30°$	$0°$	$30°$	$60°$	$90°$	$120°$	$150°$	$180°$	$210°$	$240°$	$270°$	$300°$	$330°$	$360°$
y																

1. $y = \sin (x - 90°)$
2. $y = \sin (3x + 180°)$

Graph of $y = a \sin (bx + c)$

Write the equation of the curve.

3.

■ FIGURE 25-34

4.

■ FIGURE 25-35

5.

■ FIGURE 25-36

6.

■ FIGURE 25-37

7. Write an equation of the sine curve that has an amplitude of 2, a period of π, and a phase shift of $\frac{\pi}{2}$.

8. Write an equation of the sine curve that has an amplitude of 3, a period of 90°, and a phase shift of 45°.

For each function, find the amplitude, period, and phase shift. Then sketch at least one complete cycle of the graph.

9. $y = \sin(x + 45°)$
10. $y = \sin\left(x - \frac{\pi}{3}\right)$

11. $y = \sin\left(2x - \frac{\pi}{3}\right)$
12. $y = \sin(2x + 180°)$

13. $y = -2\sin\left(x - \frac{\pi}{6}\right)$
14. $y = 3\sin(x + 90°)$

15. $y = 0.5\sin(3x + 180°)$
16. $y = -2\sin(3x - 2\pi)$

17. $y = \sin(x + 1)$
18. $y = -2\sin(3x - 1)$

19. In a certain AC circuit, the current i (in amps) is given by the equation

$$i = 60\sin\left(120\pi t + \frac{\pi}{2}\right)$$

where t is the time (in seconds). Find the amplitude, period, and phase shift, and graph two complete cycles of the curve.

25-4   GRAPH OF $y = a\cos(bx + c)$

In this section we see that the graph of the cosine function is just a sine curve with a phase shift. We find the period, amplitude, and phase shift of the graph of $y = a\cos(bx + c)$ and sketch its graph.

Graph of $y = \cos x$

Now we graph the function

$$y = \cos x$$

by plotting points from a table of values.

FIGURE 25-38
Cosine curve.

Example 19

Graph $y = \cos x$ for x from 0 to 2π.

Solution: We make a table of values using intervals of $\frac{\pi}{6}$ each, rounding y values to tenths.

x	0	$\frac{\pi}{6}$	$\frac{\pi}{3}$	$\frac{\pi}{2}$	$\frac{2\pi}{3}$	$\frac{5\pi}{6}$	π	$\frac{7\pi}{6}$	$\frac{4\pi}{3}$	$\frac{3\pi}{2}$	$\frac{5\pi}{3}$	$\frac{11\pi}{6}$	2π
y	1	0.9	0.5	0	-0.5	-0.9	-1	-0.9	-0.5	0	0.5	0.9	1

These points are plotted and connected with a smooth curve in Fig. 25-38. It looks like the *shape* of the cosine curve is identical to that of the sine curve.

Example 20

Graph $y = \cos x$ and $y = \sin x$ on the same axes, for x from $-90°$ to $360°$.

Solution: We graph the functions using a graphing calculator, in degree mode, and the window set so that

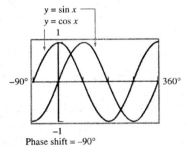

FIGURE 25-39
Sine and cosine curves.

and sketch the curve in Fig. 25-39. We see that the *shapes* of the curves are identical (the period of $y = \cos x$ is 2π and the amplitude is 1), and that the cosine curve is just a sine curve with a phase shift of $\frac{\pi}{2}$ to the left. Thus,

$$\cos x = \sin\left(x + \frac{\pi}{2}\right)$$

Graph of $y = a \cos (bx + c)$

For the function

$$y = a \cos (bx + c)$$

the constants a, b, and c have the same effect as for the sine function.

	amplitude = $|a|$	150
$y = a \cos (bx + c)$	Period = $\frac{2\pi}{b}$ or $\frac{360°}{b}$	151
	Phase shift = $-\frac{c}{b}$	152

25-4 Graph of $y = a \cos (bx + c)$

Quick Sketch of $y = a \cos(bx + c)$

We can use the quick sketch method to graph a cosine function. We first graph the unshifted curve

$$y = a \cos bx$$

as we did for the sine function. Then we shift the curve as before.

Example 21

Given $y = -3 \cos(2x - 180°)$,

(a) Find the amplitude, period, and phase shift.
(b) Graph the function.

Solution:

(a) For this equation, $a = -3$, $b = 2$, and $c = -180°$. Thus, from Eqs. 150–152, we have

$$\text{amplitude} = |-3| = 3$$

$$\text{Period} = \frac{360°}{2} = 180°$$

$$\text{Phase shift} = -\frac{-180°}{2} = +90°$$

(b) First we make a dotted quick sketch of $y = -3 \cos 2x$ (Fig. 25-40). Since the phase shift is positive, we shift the curve 90° units to the right, as shown in Fig. 25-41.

Check: We verify that the graph is correct by graphing the function on the graphing calculator, setting the window so that

```
Xmin = 0
Xmax = 270
Xscl = 90
Ymin = -3
Ymax = 3
Yscl = 1
```

■ FIGURE 25-40

■ FIGURE 25-41

Example 22

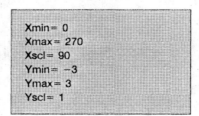

Use a graphing calculator to graph one complete cycle of the function

$$y = 2.75 \cos\left(4.85x + \frac{\pi}{3}\right)$$

Solution: For this equation, $a = 2.75$, $b = 4.85$, and $c = \frac{\pi}{3}$. Thus, from Eqs. 150–152, we have

$$\text{amplitude} = |2.75| = 2.75$$

$$\text{Period} = \frac{2\pi}{4.85} = 1.30$$

$$\text{Phase shift} = -\frac{\frac{\pi}{3}}{4.85} = -0.216$$

$y = 2.75 \cos (4.85x + \frac{\pi}{3})$

■ FIGURE 25-42

On a graphing calculator, in radian mode, we set the window so that

```
Xmin= 0
Xmax= 1.30
Xscl= 1
Ymin= -2.75
Ymax= 2.75
Yscl= 1
```

We enter the function and graph (Fig. 25-42).

SECTION 25-4 Exercises Graph of $y = a \cos (bx + c)$

Graph of $y = \cos x$

Complete the table for the given equation and plot the curve for x from 0° to 360°, rounded to tenths.

x	0°	30°	60°	90°	120°	150°	180°	210°	240°	270°	300°	330°	360°
y													

1. $y = \cos x$

2. $y = \cos \dfrac{x}{2}$

Graph of $y = a \cos (bx + c)$

Write the equation of the cosine curve.

3.

■ FIGURE 25-43

4.

■ FIGURE 25-44

5.

■ FIGURE 25-45

6.
■ FIGURE 25-46

7. Write an equation of a cosine curve that has an amplitude of 2 and a period of $\dfrac{2\pi}{3}$.

8. Write an equation of a cosine curve that has an amplitude of 5 and a period of 90°.

9. Write an equation of a cosine curve that has an amplitude of 3, a period of 720°, and a phase shift of π.

10. Write an equation of a cosine curve that has an amplitude of 10, a period of 6π, and a phase shift of -3π.

For each function, find the amplitude and the period (in degrees). Then graph the function for x from 0° to 360°.

11. $y = 3 \cos x$
12. $y = -2 \cos x$
13. $y = \cos 3x$
14. $y = \cos 4x$

For each function, find the amplitude, the period (in radians), and the phase shift. Then graph at least one complete cycle of the function.

15. $y = -5 \cos 4x$
16. $y = 3 \cos 2x$
17. $y = \cos (x + \pi)$
18. $y = \cos \left(x - \dfrac{\pi}{2}\right)$

For each function, find the amplitude, period, and phase shift. Then graph at least one complete cycle of the function.

19. $y = \cos (2x - 180°)$
20. $y = \cos (3x + 360°)$
21. $y = 2 \cos (3x + \pi)$
22. $y = -4 \cos \left(2x + \dfrac{\pi}{2}\right)$

Use a graphing calculator to graph one complete cycle of the function.

23. $y = 4.25 \cos (2.56x - 328°)$
24. $y = -2.48 \cos (3.87x + 225°)$

Applications

25. The equation for the current i (in amps) in a certain alternating current circuit with a frequency of 60 hertz (cycles/s) is

$$i = 1.5 \cos 120\pi t$$

where t is time in seconds. Find the amplitude and the period, and graph the current i as a function of t for two complete cycles.

26. The equation for the voltage V in a certain alternating current circuit is given by

$$V = 120 \cos \left(120\pi t + \dfrac{\pi}{2}\right)$$

where t is time in seconds. Find the amplitude, period, and phase shift, and graph the voltage V as a function of t for two complete cycles.

25-5   GRAPH OF $y = a \tan (bx + c)$

In this section we graph the tangent function $y = \tan x$ and find its period. Then we examine the graph of $y = a \tan (bx + c)$ and see how the coefficients a, b, and c affect the graph. The function $y = \tan x$ will be graphed by plotting points from a table of values.

---- **Example 23** ----
Graph the function $y = \tan x$ for x from $-90°$ to $180°$.

Solution: With the aid of a calculator, in degree mode, we make a table, rounding to tenths.

x	$-90°$	$-60°$	$-45°$	$-30°$	$0°$	$30°$	$45°$	$60°$	$90°$	$120°$	$135°$	$150°$	$180°$
y	none	-1.7	-1	-0.6	0	0.6	1	1.7	none	-1.7	-1	-0.6	0

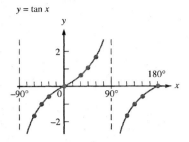

FIGURE 25-47
Tangent curve.

Note that the calculator gives no value for tan −90° and tan 90°; it gives an *ERROR* message. In Sec. 22-1, we saw that some trigonometric functions are *undefined* for some angles. Indeed, the tangent function $y = \tan x$ is undefined for $x = -90°$ and $x = 90°$. On the graph (Fig. 25-47), at the points where the tangent is undefined, we draw dotted vertical lines. These lines are used as guidelines in sketching the curve. We plot the remaining points and connect them with a smooth curve, taking care *not* to cross these vertical lines.

Asymptotes

Now we look at the behaviour of the function near the vertical line at $x = 90°$. For angles slightly *smaller* than 90° (just to the left of 90°), say $x = 89°$, the tangent is large and positive:

tan 89°   = 57.3
tan 89.9° = 573.0
tan 89.99° = 5729.6

As the angle x approaches 90° from the left, tan x gets larger and larger.

For angles slightly *larger* than 90° (just to the right of 90°), say $x = 91°$, the tangent is negative:

tan 91°   = −57.3
tan 90.1° = −573.0
tan 90.01° = −5729.6

As the angle x approaches 90° from the right, tan x gets more and more negative. We see that the vertical line at $x = 90°$ is a line which the curve *approaches but never reaches*. Such a line is called an *asymptote*. We complete the graph of the tangent function by showing the curve approaching the asymptote but not reaching it (Fig. 25-47).

─── **Example 24** ───

Use a graphing calculator to graph $y = \tan x$ for x between $-\dfrac{3\pi}{2}$ and $\dfrac{3\pi}{2}$. Find the period of the curve.

Solution: On a graphing calculator, in radian mode, we set the window so that

```
Xmin = -3π/2
Xmax = 3π/2
Xscl = π/2
Ymin = -3
Ymax = 3
Yscl = 1
```

FIGURE 25-48
$y = \tan x$

We enter the function and graph (Fig. 25-48). The tangent function is a series of "wavy curves" between asymptotes at $x = \ldots -\dfrac{\pi}{2}, \dfrac{\pi}{2}, \dfrac{3\pi}{2}, \ldots$. The period of the curve $y = \tan x$ is the distance between two consecutive asymptotes, say $x = -\dfrac{\pi}{2}$ and $x = \dfrac{\pi}{2}$. So, the period P of the tangent function is

$$P = \dfrac{\pi}{2} - \left(-\dfrac{\pi}{2}\right) = \pi$$

> **TIP** If your graphing calculator is in "connected" graph mode, then it may show vertical lines near the position of the asymptotes. The calculator has really just connected a point high on one wave with a point low on the next wave. You can avoid this "error" by graphing in the "dot" mode, instead of "connected."

As with sine and cosine functions, the period of function $y = \tan x$ depends on whether x is in degrees or radians. Example 24 shows that, if x is in radians, the period P is

$$P = \pi \quad \text{in radians}$$

If x is in degrees, as in Example 23 (Fig. 25-47), then the period P is

$$P = 90° - (-90°)$$

or

$$P = 180° \quad \text{in degrees}$$

Graph of $y = a \tan x$

The constant a has the effect of stretching (or compressing) the curve vertically.

---- **Example 25** ----

Graph $y = \tan x$ and $y = 2 \tan x$ on the same axes, for x from $-\dfrac{\pi}{2}$ to $\dfrac{3\pi}{2}$.

Solution: We use a graphing calculator in radian mode, setting the window so that

FIGURE 25-49

We enter the two functions and graph (Fig. 25-49) in the "connected" mode. Note that the constant 2 has the effect of stretching the curve vertically.

Graph of $y = a \tan bx$

As with the sine and cosine curves, the coefficient b in $y = a \tan bx$ affects the period of the curve.

---- **Example 26** ----

Graph $y = \tan 2x$ for x from $-90°$ to $90°$. Find the period of the function.

Solution: We use a graphing calculator in degree mode, and set the window so that

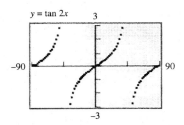

FIGURE 25-50
Calculator graph in dot mode.

We enter the function and graph the function in the "dot" graph mode (Fig. 25-50). Note that there are asymptotes at $x = -45°$ and at $x = 45°$, but they are not shown on the graph.

> **TIP** When graphing a tangent function "by hand," it is customary and helpful to sketch the asymptotes as guidelines, with dotted vertical lines.

The period of this curve is

$$P = 45° - (-45°) = 90°$$

This period is exactly half the period of the curve $y = \tan x$.

In general, the period P of the function $y = a \tan bx$ is

$$P = \frac{\pi}{b} \quad \text{or} \quad \frac{180°}{b}$$

depending on whether x is in radians or degrees.

The graph of $y = a \tan bx$ always passes through the origin $(0, 0)$. One asymptote is at $x = P/2$ and another asymptote is at $x = -P/2$. One complete cycle of the function is between this pair of asymptotes.

 Example 27

Find the period of the function $y = 2 \tan 3x$ (x in radians), and graph two complete cycles of the curve.

Solution: We use the equation for the period of the tangent function to calculate the period:

$$P = \frac{\pi}{b} = \frac{\pi}{3}$$

To find a pair of asymptotes, we take half of $\frac{\pi}{3}$

$$\frac{1}{2}\left(\frac{\pi}{3}\right) = \frac{\pi}{6}$$

So there will be asymptotes at $x = -\frac{\pi}{6}$ and $x = \frac{\pi}{6}$. To get two complete cycles, we set

$$\text{Xmin} = -\frac{\pi}{6} \quad \text{and} \quad \text{Xmax} = \frac{\pi}{6} + \frac{\pi}{3} = \frac{\pi}{2}$$

25-5 Graph of $y = a \tan (bx + c)$

■ FIGURE 25-51

Using a graphing calculator in radian mode, we set the window to

```
Xmin= -π/6
Xmax= π/2
Xscl= π/6
Ymin= -3
Ymax= 3
Yscl= 1
```

We enter the function and graph in connected mode (Fig. 25-51).

Graph of $y = a \tan (bx + c)$

Similarly, the constant c effects a phase shift of $-c/b$ units, just as with the sine and cosine curves. In general,

$y = a \tan (bx + c)$	Period $= \dfrac{\pi}{b}$ or $\dfrac{180°}{b}$	154
	Phase shift $= -\dfrac{c}{b}$	155

─── **Example 28** ───

Given $y = \tan \left(x + \dfrac{\pi}{3} \right)$,

(a) Find the period and the phase shift.
(b) Graph the function.

Solution:

(a) In this equation $a = 1$, $b = 1$, and $c = \dfrac{\pi}{3}$. Using Eq. 154, with $b = 1$, the period is

$$P = \pi$$

$$\text{Phase shift} = -\dfrac{\pi}{3}$$

which indicates a shift of $\dfrac{\pi}{3}$ units to the left.

(b) We graph the function $y = \tan \left(x + \dfrac{\pi}{3} \right)$ for x between $-\dfrac{5\pi}{6}$ and $\dfrac{7\pi}{6}$ (Fig. 25-52).

■ FIGURE 25-52
Tangent curve with phase shift.
$y = \tan (x + \dfrac{\pi}{3})$
Phase shift $= -\dfrac{\pi}{3}$

SECTION 25-5 Exercises Graph of $y = a \tan(bx + c)$

Graph of $y = a \tan bx$

Complete the table for the given equation, and graph the equation for x from $-90°$ to $360°$, rounding to tenths.

x	$-90°$	$-60°$	$-45°$	$-30°$	$0°$	$30°$	$45°$	$60°$	$90°$	$135°$	$180°$	$225°$	$270°$	$315°$	$360°$
y															

1. $y = \tan x$
2. $y = -3 \tan x$

Find the period, in degrees, and graph two complete cycles of each equation, indicating asymptotes by dotted lines.

3. $y = 3 \tan x$
4. $y = 4 \tan x$
5. $y = \tan 3x$
6. $y = \tan 4x$

Find the period, in radians, and graph two complete cycles of each equation, indicating asymptotes by dotted lines.

7. $y = -2 \tan x$
8. $y = -5 \tan x$
9. $y = \tan \dfrac{x}{2}$
10. $y = \tan \dfrac{x}{3}$

Graph of $y = a \tan(bx + c)$

For each equation, (a) find the period and the phase shift; and (b) graph two complete cycles of the equation, indicating asymptotes by dotted lines.

11. $y = \tan\left(x - \dfrac{\pi}{2}\right)$
12. $y = \tan(x - 45°)$
13. $y = \tan(x + 45°)$
14. $y = \tan(x + \pi)$
15. $y = 2 \tan(2x + \pi)$
16. $y = -3 \tan(4x - 360°)$

CHAPTER 25 REVIEW EXERCISES

For each equation, find the amplitude, the period (in degrees), and the phase shift. Graph at least one complete cycle of the function.

1. $y = -4 \cos x$
2. $y = 3 \sin 2x$
3. $y = \sin(x - 45°)$
4. $y = -\dfrac{2}{3} \cos x$
5. $y = \cos 2x$
6. $y = 0.5 \sin(2x + 60°)$
7. $y = \cos(2x - 90°)$
8. $y = \cos(x + 180°)$
9. $y = 5 \sin 3x$
10. $y = 75 \sin x$

For each equation, find the period (in radians) and the phase shift. Graph two complete cycles of the function, indicating asymptotes by dotted lines.

11. $y = \tan \dfrac{x}{4}$
12. $y = \tan\left(x + \dfrac{3\pi}{2}\right)$

For each function, find the amplitude, the period (in radians), and the phase shift. Graph at least one complete cycle of the function.

13. $y = \sin 3x$
14. $y = \sin \dfrac{x}{4}$

15. $y = 5 \cos x$ 16. $y = -2 \sin \dfrac{\pi x}{3}$

17. $y = -3 \sin \left(x + \dfrac{\pi}{8}\right)$ 18. $y = \sin(3x - \pi)$

19. $y = 2 \cos 3x$ 20. $y = -3 \cos(x + \pi)$

21. Write an equation of a sine curve that has an amplitude of 2, a period of 180°, and a phase shift of 90°.

22. Write an equation of a cosine curve that has an amplitude of 3, a period of 4π, and a phase shift of -2π.

For each function, find the period (in radians) and the phase shift. Graph two complete cycles of the function, indicating asymptotes by dotted lines.

23. $y = \tan 5x$ 24. $y = 2 \tan \left(x - \dfrac{\pi}{2}\right)$

WRITING

25. Given the general sine function, $y = a \sin(bx + c)$, explain in your own words how a change in each of the constants (a, b, or c) will affect the graph of the function. Show three examples (one for each constant) which illustrate the effect of each constant.

TEAM PROJECT

FIGURE 25-53

26. When a sound is made, it disturbs nearby air molecules, creating regions of higher-than-normal pressure and regions of lower-than-normal pressure. If we graph pressure y versus time t, we get a sine wave which represents the sound. Important characteristics of the sound wave are its period, frequency, and amplitude. Using a tuning fork and a microphone, pressure readings were taken every 0.0002 second for about 0.025 second, and plotted on a pressure versus time graph to get the graphing calculator graph in Fig. 25-53.

Note that the y axis is the left border of the graph. The coordinates of the first and last maximum points are

(0.003 1030, 0.022 389) and (0.022 2112, 0.019 033)

Use this information to estimate the amplitude a of the sound wave. Round all calculations to three significant digits. How many cycles of the sine curve are there between the first and last maximum points? Estimate the period T of the sound wave, in seconds per cycle. The frequency f is found using

$$f = \dfrac{1}{T}$$

Frequency is measured in cycles per second, or hertz (Hz). This tuning fork was labeled 256 Hz. Compare this frequency to your estimated frequency and calculate the percentage error. The equation for a sound wave is

$$y = a \sin 2\pi f t$$

Write the equation of a sound wave using your estimated values for *a* and *f*. Graph the equation. How does your graph compare to the graph in Fig. 25-53? The sine curve in Fig. 25-53 has been shifted to the right or left, so you must find the phase shift. (Note that the phase shift is not a property of the sound, but that the phase shift depends on when the first data point was taken.) Estimate the phase shift *d* for this data. The equation for this sound wave is

$$y = a \sin (2\pi f(t + d))$$

Write the equation of this sound wave using your estimated values for *a*, *f*, and *d*. Graph the equation. Your graph should match the graph in Fig. 25-53 pretty well.

26 Ratio, Proportion, Variation

Outcomes

When you have completed this chapter, you will be able to

1. Define the concept of a ratio.
2. Define a proportion.
3. Find missing terms of a proportion.
4. Apply the principles of proportions to rates and similar figures.
5. Define a direct variation.
6. Determine a constant of proportionality.
7. Apply joint, inverse, and combined variation.

In this chapter, we deal with ratio, proportion, the power function, and four types of variation: direct, joint, inverse, and combined. Many applications use the concepts of ratio, proportion, and variation, such as maps and scale drawings.

26-1 RATIO

The ratio concept is an extremely important one for math students. Maps are printed to scale where the scale is the ratio of a map distance to an actual distance. Auto mechanics are concerned with compression ratios, machinists with gear ratios, and carpenters with the pitch ratio of a roof. In this section, we define a ratio, consider two forms of a ratio, learn to simplify a ratio, and do calculations with similar figures and rates.

We are often concerned with comparing one quantity to another. For example, we may need to compare the weights of two objects, the first 6 kg and the second 3 kg. We could say that the first weighs 3 kg *more* than the second; this is a comparison by *subtraction* (the *difference*). Or we could say that the first weighs *twice as much as* the second; this is a comparison by *division*. The comparison of two quantities by division is called a *ratio*.

Notation

The ratio of quantity a to quantity b may be written with a colon or as a fraction.

| RATIO OF a TO b | $a:b$ or $\dfrac{a}{b}$ | 45 |

The quantities a and b are called the *terms* of the ratio.

Example 1

The ratio of 3 to 2 may be written with a colon as

$$3:2$$

or as a fraction

$$\frac{3}{2}$$

Simplification of Ratios

To *simplify* a ratio, we write the ratio as a fraction, and reduce that fraction.

Example 2

Simplify:

(a) the ratio of 12 cm to 9 cm
(b) the ratio of 6 kilograms to 3 kilograms

Solution:

(a) The ratio of 12 cm to 9 cm may be written as a fraction:

$$\frac{12 \text{ cm}}{9 \text{ cm}}$$

We simplify the ratio by reducing the fraction:

$$\frac{12 \text{ cm}}{9 \text{ cm}} = \frac{4}{3}$$

Thus the ratio of 12 cm to 9 cm is the same as the ratio of 4 to 3, or 4:3. Notice that since the quantities were expressed in the same units, the ratio is *dimensionless*.

(b) The ratio of 6 kilograms to 3 kilograms may be written in fractional form and reduced:

$$\frac{6 \text{ kg}}{3 \text{ kg}} = \frac{2}{1}$$

Notice that the denominator 1 is retained. Since a ratio is a comparison of *two* quantities, one of them cannot disappear by being canceled. Thus the ratio of 6 kilograms to 3 kilograms is the same as the ratio of 2 to 1, or 2:1.

Dimensionless Ratios

If the quantities being compared can be expressed in the same units, it is customary to do so, so that the units will cancel and leave the ratio dimensionless.

FIGURE 26-1

Example 3

A laboratory is 1176 cm wide and 15.12 m long (Fig. 26-1). Find the ratio of the length to width and simplify.

Solution: It is important to note that we need to find the ratio of length to width. Since length is the first term of the ratio, the *length* must be the *numerator* of the fraction. We change 15.12 m to 1512 cm and proceed to simplify the ratio.

$$\frac{15.12 \text{ m}}{1176 \text{ cm}} = \frac{1512 \text{ cm}}{1176 \text{ cm}} = \frac{9}{7}$$

So the ratio of length to width is 9 to 7, or 9:7.

Dimensionless ratios are often used in mathematics.

Example 4

(a) The number *pi* (π) is the ratio of circumference C to diameter D.

$$\pi = \frac{C}{D}$$

(b) The radian measure θ of an angle is the ratio of arc length s to radius r.

$$\theta = \frac{s}{r}$$

(c) Trigonometric ratios are ratios of two sides of a right triangle.

$$\tan \theta = \frac{\text{opposite leg}}{\text{adjacent leg}}$$

Other examples of ratios in technical fields are gear ratio, turn ratio, load ratio, fuel-air ratio, efficiency, Mach number, mechanical advantage, and the pitch of a roof. The word *specific* is often used to denote a ratio, such as in specific heat (the ratio of two thermal capacities) and specific gravity (the ratio of two densities).

Example 5

A flywheel has 75 teeth and a starter drive gear has 15 teeth (Fig. 26-2). Find the ratio of flywheel teeth to starter drive teeth.

Solution:

$$\frac{\text{flywheel teeth}}{\text{starter drive teeth}} = \frac{75}{15} = \frac{5}{1}$$

So the ratio is 5 to 1 or 5:1. Note that we do *not* change $\frac{5}{1}$ to 5, since this is a *ratio*, not a fraction.

FIGURE 26-2

Scale Drawings

An important application of ratio is in the use of scale drawings, such as maps, engineering drawings, surveying layouts, and so on. The ratio of a distance on the scale drawing to the corresponding distance on the actual object is called the *scale* of the drawing. A scale is often given using a colon.

Scale = distance on the drawing : actual distance

A scale on a drawing is often given as a ratio of two distances with different units (dimensions). To express the scale as a dimensionless ratio, convert the distances to the same unit. This may be done before writing the ratio, or afterward, as in the following example.

---- **Example 6** ----

On a certain map, 1 inch represents 10 miles. Express the scale of the map as a dimensionless ratio.

Solution: We set up a ratio for the scale and express it in fraction form:

$$\text{Scale} = 1 \text{ in.} : 10 \text{ mi} = \frac{1 \text{ in.}}{10 \text{ mi}}$$

Convert 10 mi to in.:

$$10 \text{ mi} = (10 \text{ mi}) \left(\frac{5280 \text{ ft}}{1 \text{ mi}}\right)\left(\frac{12 \text{ in.}}{1 \text{ ft}}\right) = 633\,600 \text{ in.}$$

Expressing the scale as a dimensionless ratio, we have

$$\frac{1 \text{ in.}}{10 \text{ mi}} = \frac{1 \text{ in.}}{633\,600 \text{ in.}} = \frac{1}{633\,600}$$

So the scale of the map, as a dimensionless ratio, is $1 : 633\,600$.

The advantage of a dimensionless ratio on a map is in the application. The person using the map is free to use dimensions of their own choosing. Beside inches, this ratio could be read

$$\frac{1 \text{ cm}}{633\,600 \text{ cm}} \quad \text{or} \quad \frac{1 \text{ m}}{633\,600 \text{ m}} \quad \text{or} \quad \frac{1 \text{ ft}}{633\,600 \text{ ft}}$$

SECTION 26-1 Exercises Ratio

Notation

Write the ratio in both colon and fractional form.
1. 5 to 4
2. 3 to 5

Simplification of Ratios

Simplify the ratio in fraction form.
3. 6 cm to 15 cm
4. 4 lb to 6 lb
5. 8 ft^2 to 4 ft^2
6. 15 kg to 5 kg
7. 7.1 mL to 28.4 mL
8. 9.3 in. to 18.6 in.

Dimensionless Ratios

Simplify the ratio in fraction form.
9. 4 inches to 1 foot
10. 4 feet to 18 inches
11. 1 m to 25 cm
12. 400 g to 2 kg
13. 3 quarts to 1 gallon
14. 1 pound to 4 ounces
15. 2 mL to 2 L
16. 45 minutes to 1 hour
17. 75 cents to 2 dollars
18. 1500 m to 1 km

Applications

19. A corridor is 9 ft wide and 15 yd long. Find the ratio of length to width.
20. A bronze alloy is made by mixing 1.40 kg of tin with 11.2 kg of copper. Find the ratio of copper to tin.
21. Estimate the value of π if a circle with a diameter of 2.50 cm has a circumference of 7.85 cm.

■ FIGURE 26-3

22. Find the radian measure of an angle θ that intercepts an arc of 45.4 mm in a circle of radius 22.5 mm (Fig. 26-3).
23. The *mechanical advantage* of a simple machine is the ratio of load force to effort force. Find the mechanical advantage of a pulley system in which a force of 85.6 N (effort) is needed to lift a 3470 N load. Express the ratio as a decimal.
24. A mixture of gasohol contains 8.60 gallons of alcohol and 90.3 gallons of gasoline. Find the ratio of gasoline to alcohol.
25. The power output of any machine or device is always less than the power input, because of inevitable power losses within the device. The *efficiency* of the device is a measure of those losses, and is defined as the ratio of output to input. Find the efficiency of an electric motor which consumes 795 W and has an output of 1 horsepower. Express the efficiency as a percent. (1 hp = 746 W)
26. Carpenters define the pitch of a roof as the ratio of the rise to the run of the rafter (Fig. 26-4). What is the pitch of a roof with a rise of 9.0 feet and a run of 12.0 feet?
27. The *Mach number* of a moving object is defined as the ratio of its velocity to the velocity of sound (340.2 m/s). A Mach number is generally expressed as a decimal. Find the Mach number of a jet traveling at 275 m/s.

 28. The *specific gravity* (also called *relative density*) of a substance is the ratio of its density to the density of water. Specific gravity is generally expressed as a decimal. Taking the density of water as 62.4 lb/ft^3, find the specific gravity of copper if its density is 554.7 lb/ft^3.

■ FIGURE 26-4

26-2 PROPORTION

In this section, we define a proportion and discuss the terms, means, and extremes of a proportion. We learn how to find a missing term in a proportion, state the fundamental principle of proportions, and solve some application problems.

The Fraction Form of a Proportion

A *proportion* is a statement that two ratios are equal. A proportion is an equation, which in fraction form looks like this:

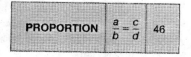

CHAPTER 26 / Ratio, Proportion, Variation

The proportion in Eq. 46 is read

"the ratio of a to b equals the ratio of c to d"

or

"a is to b as c is to d"

Example 7

Write as a proportion in fraction form.

(a) The ratio of x to y equals the ratio of 9 to 12.
(b) x is to 5 as y is to 15.

Solution: (a) $\dfrac{x}{y} = \dfrac{9}{2}$ (b) $\dfrac{x}{5} = \dfrac{y}{15}$

In the proportion in Eq. 46, the quantities $a, b, c,$ and d are called the *terms* of the proportion. The first and fourth terms, a and d, are called the *extremes* and the second and third terms, b and c, are called the *means* of the proportion.

$$\underset{\text{means}}{\overset{\text{extremes}}{\dfrac{a}{b} = \dfrac{c}{d}}}$$

Fundamental Principle of Proportions

A very important principle of proportions is found by multiplying both sides of the general proportion

$$\dfrac{a}{b} = \dfrac{c}{d}$$

by the LCD, bd:

$$\dfrac{a}{b}(bd) = \dfrac{c}{d}(bd)$$

$$ad = bc$$

Thus we have the *fundamental principle of proportions*:

FUNDAMENTAL PRINCIPLE OF PROPORTIONS	In a proportion, the product of the extremes equals the product of the means. If $\quad \dfrac{a}{b} = \dfrac{c}{d}$ then $\quad ad = bc$	47

Note: This procedure is often referred to as "cross-multiplication," and ad and bc are called cross-products.

Example 8

Use the fundamental principle of proportions to show that this is a true proportion:

7 is to 4 as 63 is to 36

Solution: First we write the proportion in fraction form:

$$\frac{7}{4} = \frac{63}{36}$$

Calculating cross-products, we have

$$(7)(36) = 252 \quad \text{and} \quad (4)(63) = 252$$

Since the cross-products are equal, the proportion is true.

Finding a Missing Term in a Proportion

To find a missing term in a proportion, cross-multiply and solve the resulting equation.

Example 9

Find x in the proportion:

$$\frac{3}{5} = \frac{2}{x}$$

Solution: Cross-multiplying, we have

$$3x = 10$$

Solving for x:

$$x = \frac{10}{3}$$

Applications

Many applications use proportions. Set these up as you set up the word problems in Chapter 11. Be careful not to reverse the terms in the proportion. Match the first item in the verbal statement with the first number in the ratio, and the second with the second.

Example 10

Separate $180 into two parts so that one part is to the other part as 7 is to 5.

Solution: We follow the process of Chapter 11, starting with defining the variables.

Let x = one part
and $180 - x$ = other part

Writing the proportion, we have

$$\frac{\text{one part}}{\text{other part}}: \quad \frac{x}{180 - x} = \frac{7}{5}$$

Cross-multiplying,
$$5x = 7(180 - x)$$

Solving for x,
$$5x = 1260 - 7x$$
$$12x = 1260$$
$$x = 105$$

Finding the other part:
$$180 - x = 180 - 105 = 75$$

So the two parts of $180 are $105 and $75.

Check: Verify that these two parts have a ratio of 7 to 5:
$$\frac{105}{75} \stackrel{?}{=} \frac{7}{5}$$

Calculating cross-products:
$$(105)(5) = 525 \text{ and } (75)(7) = 525 \quad \text{checks}$$

Rate Problems

Many applications involving rates may be solved using the method of proportions.

Example 11

If a car can go 210 kilometres using 23.9 litres of gasoline, how far can it travel on 54.5 litres of gasoline?

Solution: Set up a proportion with dimensionless ratios, using like quantities in each ratio (that is, kilometres with kilometres and litres with litres).

$$\underset{\text{kilometres}}{\frac{210 \text{ km}}{x}} = \underset{\text{litres}}{\frac{23.9 \text{ L}}{54.5 \text{ L}}}$$

We cancel the units on the right:
$$\frac{210 \text{ km}}{x} = \frac{23.9}{54.5}$$

Cross-multiply and solve:
$$23.9x = (210 \text{ km})(54.5)$$
$$x = 479 \text{ km}$$

The car will go 479 kilometres using 54.5 litres of gasoline.

Similar Figures

Two geometric figures are said to be similar if they have the *same shape*, but not necessarily the same size (Fig. 26-5). This means that the angles of the two figures have the same measure, but the sides are not necessarily equal. Figure 26-5 shows two similar triangles. There are three pairs of equal angles:

$$A = D \quad B = E \quad C = F$$

Each pair of equal angles are called *corresponding* angles. So, we say, angle A and angle D are corresponding angles, and so forth.

FIGURE 26-5
Similar triangles.

Sides opposite corresponding angles are called *corresponding sides*. In Fig. 26-5, corresponding pairs of sides are a and d, b and e, and c and f. In similar figures, the ratios of these corresponding sides are equal, so

$$\frac{a}{d} = \frac{b}{e} = \frac{c}{f}$$

We can use this to set up a proportion to solve similar triangles.

Example 12

A right triangle has sides 3.00 cm, 4.00 cm, and 5.00 cm. The hypotenuse of a similar triangle is 12.7 cm.

(a) What is the ratio of corresponding sides?
(b) Find the other two sides of the similar triangle.

Solution: We sketch the two similar triangles in Fig. 26-6.

(a) Since the triangles are similar, the ratio of corresponding sides must be the same for all pairs of sides. The hypotenuse of the first triangle (5.00 cm) corresponds to the hypotenuse of the similar triangle (12.7 cm). We form the ratio and note that we can cancel the units:

$$\frac{5.00 \text{ cm}}{12.7 \text{ cm}} = \frac{5.00}{12.7}$$

(b) To find the other two sides, x and y, we need to identify their corresponding sides in the first triangle. Side x is the shorter leg in the second triangle and corresponds to the shorter leg in the first triangle (the 3.00 cm side). We set up a proportion:

$$\frac{3.00}{x} = \frac{5.00}{12.7}$$

Cross-multiply and solve:

$$5.00x = (3.00)(12.7)$$
$$x = 7.62 \text{ cm}$$

Side y is the longer leg in the second triangle and corresponds to the longer leg in the first triangle (the 4.00 cm side).

We set up a proportion:

$$\frac{4.00}{x} = \frac{5.00}{12.7}$$

Cross-multiply and solve:

$$5.00x = (4.00)(12.7)$$
$$x = 10.2 \text{ cm}$$

So the other two sides of the similar triangle are 7.62 cm and 10.2 cm long.

■ FIGURE 26-6

SECTION 26-2 Exercises Proportion

The Fraction Form of a Proportion

Write as a proportion in fraction form.
1. The ratio of x to y equals the ratio of 6 to 9.
2. The ratio of 15 to 18 equals the ratio of x to y.

3. p is to 20 as q is to 5.
4. q is to 25 as p is to 100.
5. The ratio of 12 cm to 30 cm equals the ratio of y to 5.
6. 6 kg is to 33 kg as x is to 11.

Fundamental Principle of Proportions

Use the fundamental principle of proportions to show that these proportions are true proportions.

7. The ratio of 2 to 3 equals the ratio of 6 to 9.
8. The ratio of 15 to 18 equals the ratio of 5 to 6.
9. 12 is to 20 as 3 is to 5.
10. 13 is to 25 as 52 is to 100.
11. The ratio of 12 cm to 30 cm equals the ratio of 2 to 5.
12. 6 kg is to 33 kg as 2 is to 11.

Finding a Missing Term in a Proportion

Solve the proportion.

13. $\dfrac{x}{2} = \dfrac{15}{6}$
14. $\dfrac{5}{x} = \dfrac{25}{30}$
15. $\dfrac{12}{6} = \dfrac{6}{x}$
16. $\dfrac{3}{7} = \dfrac{x}{3}$
17. $\dfrac{4}{x} = \dfrac{x}{9}$
18. $\dfrac{x}{3} = \dfrac{12}{x}$
19. $\dfrac{x}{5} = \dfrac{6}{7}$
20. $\dfrac{7}{4} = \dfrac{x}{9}$
21. $\dfrac{x}{8} = \dfrac{18}{x}$
22. $\dfrac{8}{x} = \dfrac{x}{11}$
23. $\dfrac{x}{4} = \dfrac{x+3}{5}$
24. $\dfrac{x-2}{3} = \dfrac{x}{5}$
25. $\dfrac{x}{30-x} = \dfrac{7}{8}$
26. $\dfrac{75-x}{x} = \dfrac{8}{7}$

Applications

27. A board 488 cm long must be cut into two pieces so that the lengths of the pieces are in a ratio of 3 to 5. Find the length of the two pieces.
28. An electrician and an apprentice worked together on a job and received $585 in wages. Separate the wages so that the parts are in a ratio of 8 to 5.
29. Separate $650 into two parts that are in a ratio of 4 to 9.
30. Two women jointly bought some stock. The first put in $600 more than the second, and the stock of the first was to that of the second as 8 is to 7. How much money did each invest?
31. A certain automobile cooling system contains a coolant that is a mixture of antifreeze and water in a ratio of 2 to 11. How much antifreeze is contained in 9.00 litres of coolant?
32. A certain solder contains tin and lead in a ratio of 2 to 3. How much tin is contained in 65.5 kg of the solder?

In the following problems, assume that the rates remain constant.

33. If a bag of sand with a weight of 5.00 lb has a metric equivalent of 2.27 kg, what would the equivalent number of kilograms be for a bag weighing 8 lb?
34. The property taxes on a property valued at $95 500 are $1710. What are the property taxes on a similar property valued at $115 000?

FIGURE 26-7

35. A certain map uses a scale of 1:2 278 000 (Fig. 26-7). At opposite ends of Lake Superior are Duluth, Minnesota and Sault Ste. Marie, Ontario. The length measured on the map is 26.0 cm. What is the distance between these two cities?
36. A map of Canada uses a scale of 1 cm = 391 km. If Vancouver, British Columbia is 12.8 cm from St. John's, Newfoundland, on the map, what is the actual distance between these two cities?
37. A carpenter finds that a standard two-by-four with a length 244 cm has a mass of 4.55 kg. What is the mass of a 275 cm length of a similar board?
38. A technician can assemble 3.5 instruments in a 7.5-hour day. How many hours would it take her to assemble 25 instruments? How many days?
39. An airplane travels 585 km in 2.75 hours. At the same rate, how long would it take the airplane to travel 825 km?
40. If a box of 250 screws costs $3.49, how much would a box of 400 screws cost?
41. A truck can carry 3.86 m³ of sand with a mass of 5.96 t. A larger truck can carry 6.31 m³ of sand. What is the mass of the sand?
42. A right triangle has sides 5.00 cm, 12.0 cm, and 13.0 cm. The hypotenuse of a similar triangle is 18.5 cm. Find the other two sides of the similar triangle.
43. A contractor can build an 1850 ft² house for $81 900. How much would it cost to build a similar house of 2250 ft²?
44. A triangle has sides 24.0 in., 28.0 in., and 32.0 in. The shortest side of a similar triangle is 15.5 in. Find the other two sides of the similar triangle.
45. A tree casts a shadow 36.0 ft long at the same time a boy 4.00 feet tall casts a shadow 6.25 ft long (Fig. 26-8). How tall is the tree?
46. Use similar triangles to find the width of the river in Fig. 26-9.

FIGURE 26-8

■ FIGURE 26-9

26-3   DIRECT VARIATION

We saw in Chapter 12 that two quantities, x and y, may be related in such a way, that if the value of one quantity changes, then the value of the other quantity also changes. If a change in x causes a specific and predictable change in y, then we say that *y varies as x varies.*

In this section we will look at one type of variation: direct variation. We will see that direct variation is closely associated with the concepts of ratio and proportion. We consider many applications of direct variation and solve problems involving direct variation.

Direct Variation

If two variables, x and y, are related in such a way that the ratio $\dfrac{y}{x}$ is a constant, then we say *y varies directly as x*, and we may write

$$y \propto x$$

The symbol $\propto$ means *is directly proportional to.* We may also say that *y is proportional to x.*

Example 13

It is obvious that the circumference C of a circle changes as its diameter D changes. Consider this table of pairs of values of D (in cm) and C (in cm):

D	1.50	2.00	2.50	3.00	3.50
C	4.71	6.28	7.85	9.42	11.0

If we calculate the ratio $\dfrac{C}{D}$ for each pair of values, we find that the ratio $\dfrac{C}{D}$, rounded to three significant digits, is a constant:

$$\frac{4.71}{1.50} = \frac{6.28}{2.00} = \frac{7.85}{2.50} = \frac{9.42}{3.00} = \frac{11.0}{3.50} = 3.14$$

Thus, we say that the circumference of a circle *varies directly as* the diameter, or the circumference *is directly proportional to* the diameter, and we write $C \propto D$. (In this case, the ratios are equal to the familiar mathematical constant π.)

One important property of direct variation is that if one variable *increases,* then the other *increases,* and if one variable *decreases,* then the other also *decreases.* In Example 13, we see that the circumference of a circle varies directly as the diameter. If the diameter *increases,* then the circumference *increases.* Similarly, if the diameter *decreases,* then the circumference *decreases.*

Constant of Proportionality

If the ratio of two variables, y and x, is equal to a constant k, then we express the relationship by the equation

$$\frac{y}{x} = k$$

where k is called the *constant of proportionality.* If we multiply both sides of the equation by x, we have a familiar form of the equation for direct variation.

DIRECT VARIATION	$y = kx$ where k is the constant of proportionality	48

FIGURE 26-10
Direct variation.

Example 14

Write an equation for the force needed to stretch a spring, if the force F varies directly as the distance x the spring is stretched (Fig. 26-10).

Solution: We use Eq. 48 to write the equation. Substituting the force F for y, we get

$$F = kx$$

(This is called Hooke's Law and k is called the spring constant.)

Example 15

Write an equation for the electrical resistance R of a wire, if the resistance is proportional to the length L of the wire.

Solution: We use Eq. 48 to write the equation. Substituting the resistance R for y and the length L for x, we get

$$R = kL$$

where k is related to the coefficient of resistivity of the wire.

Example 16

Write an equation for the force F due to gravity on an object, if the force is directly proportional to its mass m.

Solution: Again we use Eq. 48 to write the equation.

$$F = km$$

In this equation, k is the constant rate of acceleration due to gravity g.

In most direct variation problems involving physical relationships, the constants of proportionality have *units* associated with them. In Example 16, if the force F is in newtons and the mass is in kilograms, then the constant k is the rate of acceleration due to gravity,

so that

$$k = 9.81 \text{ m/s}^2$$

The units are an *essential* part of the constant of proportionality.

Solution of Variation Problems

To solve a variation problem in which a value is unknown, at least one pair of corresponding values of the related variables must be known.

Example 17

If s varies directly as t, and $s = 118$ km when $t = 1.25$ hours, find s when $t = 2.75$ hours.

Solution: Since s varies directly as t, we use Eq. 48 to write the equation

$$s = kt$$

To find the constant of proportionality, k, we substitute the given pair of values for s and t into the equation.

$$118 \text{ km} = k(1.25 \text{ h})$$

Solving for k:

$$k = \frac{118 \text{ km}}{1.25 \text{ h}} = 94.4 \text{ km/h}$$

Substituting $k = 94.4$ (km/h) into the equation $s = kt$, we have

$$s = 94.4t$$

where s is in kilometres and t is in hours.

Although the units of the constant of proportionality are of the utmost importance, we usually don't write the units of k in the equation. We do, however, state the units used for the other variables. The constant is 94.4 (km/h) *only if* s is in kilometres and t is in hours.

To find the missing s, we substitute $t = 2.75$ into the equation $s = 94.4t$ to get

$$s = (94.4)(2.75) = 260 \text{ km}$$

The example above suggests a method for solving variation problems in which a value is missing.

The Solution of Variation Problems

1. Using Eq. 48, write an equation that describes the expressed relationship.
2. Substitute the two given values in the equation and solve for the constant of proportionality, k.
3. Rewrite the equation by substituting the value for the constant into the equation from Step 1.
4. To find the missing value, substitute the corresponding given value into the equation and solve.

Sometimes we find that one quantity varies directly as a *power* of another quantity.

Example 18

The area A of a circle varies directly as the square of its radius r (Eq. 76):

$$A = \pi r^2$$

where π is the constant of proportionality.

If y varies as a power of x, then we write $y = kx^n$, where k is the constant of proportionality.

Example 19

(a) If *y varies directly as the square of x,* then

$$y = kx^2$$

(b) The period T of a pendulum *is proportional to the square root of* its length L, so

$$T = k\sqrt{L}$$

Example 20

The volume V of a sphere *is directly proportional to the cube of* the radius r, so

$$V = kr^3$$

This reminds us of the familiar formula for the volume of a sphere, Eq. 98:

$$V = \frac{4}{3}\pi r^3$$

We see that the constant of proportionality is

$$k = \frac{4}{3}\pi$$

Applications

Many practical problems can be solved using the idea of direct variation.

Example 21

The electric current i through a resistor is directly proportional to the voltage drop E across it. If the current through a resistor is 18.6 amperes when the voltage drop across it is 45.6 volts, find the current when the voltage drop is 60.0 volts.

Solution: Use Eq. 48 to find the equation of proportionality:

$$I = kE$$

We substitute the given values ($I = 18.6$ A when $E = 45.6$ V) and solve for k:

$$18.6 \text{ A} = k(45.6 \text{ V})$$

$$k = \frac{18.6 \text{ A}}{45.6 \text{ V}} = 0.4079 \frac{\text{A}}{\text{V}}$$

(Note that for accuracy, we're carrying one extra digit for k.)
Substituting $k = 0.4079$ into the equation:

$$I = 0.4079E \quad \text{where } I \text{ is in amperes and } E \text{ is in volts}$$

Substitute $E = 60.0$ to find the current I:

$$I = (0.4079)(60.0) = 24.5 \text{ A}$$

Example 22

The flow rate Q through a pipe varies directly as the square of the diameter D of the pipe. If water flows through a pipe with a diameter of 9.53 cm at a rate of 737 L/min, what is the rate of flow, under similar conditions, for a pipe with a diameter of 5.72 cm?

Solution: We write an equation for Q:

$$Q = kD^2$$

We substitute $Q = 737$ L/min and $D = 9.53$ cm, into the equation.

$$736 \text{ L/min} = k(9.53 \text{ cm})^2$$

We solve for k:

$$k = \frac{736 \text{ L/min}}{(9.53 \text{ cm})^2} = \frac{736 \text{ L/min}}{(9.53)^2 \text{ cm}^2} = 8.104 \frac{\text{L/min}}{\text{cm}^2}$$

Then we substitute the value k into the equation $Q = kD^2$ to get

$$Q = \left(8.104 \frac{\text{L/min}}{\text{cm}^2}\right) D^2 \text{ where } D \text{ is in centimetres, and } Q \text{ will be in L/min.}$$

To find the rate of flow through a 5.72 cm pipe we substitute $D = 5.72$ cm into the equation.

$$Q = \left(8.104 \frac{\text{L/min}}{\text{cm}^2}\right)(5.72 \text{ cm})^2 = 265 \text{ L/min}$$

SECTION 26-3 Exercises Direct Variation

Show that the following tables of values represent direct variation by calculating the ratio $\frac{y}{x}$ for each pair of values.

1.
x	1	2	3	4	5
y	3	6	9	12	15

2.
x	1	3	5	7	9
y	4	12	20	28	36

3. Given the following experimental data, show that the velocity v of a freely falling object varies directly as the time of the fall.

t (in s)	1.50	2.00	2.50	3.00	3.50
v (in m/s)	14.7	19.6	24.5	29.4	34.3

4. Given the following experimental data, show that the force F applied to a spring is proportional to the distance x the spring is stretched.

x (in cm)	1.25	2.50	3.75	5.00	6.25
F (in N)	0.50	1.00	1.50	2.00	2.50

Constant of Proportionality

Write an equation that describes the relationship.

5. y varies directly as x.
6. p is directly proportional to q.
7. The mass m of steel balls is proportional to the number n of balls.
8. The number of parts n that a machine can make varies directly as the time t of operation.
9. The power P generated by a hydroelectric plant is directly proportional to the flow rate r through the turbines.
10. The volume V of a cylinder varies directly as its height h.
11. y varies as the cube of x.
12. p varies as the square root of r.
13. The distance s varies as the square of the time t.
14. The volume V of a cylinder is proportional to the square of the radius.

Solution of Variation Problems

Find the constant of proportionality and write the equation that describes the relationship.

15. y varies directly as x, and $y = 20$ when $x = 5$.
16. w is proportional to z, and $w = 6.4$ when $z = 4$.
17. F is directly proportional to L, and $F = 12.0$ N when $L = 15$ cm.
18. E varies directly as I, and $E = 12$ volts when $I = 5$ amperes.

Write the equation and find the missing value.

19. If y is proportional to x, and $y = 50.4$ when $x = 3.20$, find y when $x = 2.50$.
20. If p varies directly as q, and $p = 5.7$ when $q = 15$, find p when $q = 35$.

Find the constant of proportionality and write the equation.

Applications

21. The area A of a circle is directly proportional to the square of the radius, and $A = 63.6$ cm^2 when $r = 4.50$ cm.
22. The power P dissipated in a resistor varies directly as the square of the current I in the resistor, and the power dissipated in a resistor is 502 W when the current is 2.80 A.

Find the missing value.

23. According to Charles' Law, the volume V of a gas at constant pressure is directly proportional to its absolute temperature T (in Kelvin units). If the volume of a gas is 1220 cm^3 at 295 K, what is the volume of the gas, at the same pressure, at 273 K?
24. The amount of heat Q necessary to melt ice is directly proportional to the mass m of ice melted. If it takes 717 calories of heat to melt 9.00 grams of ice, how much heat is required to melt 5.00 grams of ice?

25. A certain automobile engine has a displacement (the total volume swept out by the pistons) of 3.85 litres and delivers 50.6 kW. If the power is proportional to the displacement, what power would you expect from a similar engine that has a displacement of 3.00 litres?

26. The exposure time t for a photograph is directly proportional to the square of the f stop. A certain photograph will be correctly exposed at a shutter speed of $\frac{1}{50}$ s with a lens opening of $f8 (f = 8)$. What shutter speed is required if the lens opening is changed to $f5.6$? (Hint: Round the shutter speed, as a decimal, to one significant digit and express as a fraction.)

27. The surface area A of a sphere varies directly as the square of its radius r. If the surface area of a sphere of radius 1.50 cm is 28.3 cm², find the surface area of a sphere of radius 2.50 cm.

28. The period T of a pendulum is proportional to the square root of its length L. If the period of a 1.50 m long pendulum is 1.75 s, what is the period of a 1.25 m long pendulum?

29. The rate of flow of liquid from a hole in the bottom of a tank is proportional to the square root of the liquid depth. If the flow rate is 175 L/min when the depth is 2.75 m, find the flow rate when the depth is 2.00 m.

Find the constant of proportionality and write the equation.

30. The power P of an engine varies directly with the square of the radius r of its piston, and a 90.0-hp engine has a piston of radius 3.00 inches.

31. The kinetic energy E_K of a moving object varies directly as the square of its velocity v and the kinetic energy is 9360 ft·lb at 10.0 ft/s.

Find the missing value.

32. The thermal resistance R of a wall (the R value) is proportional to the thickness T of the wall. If the R value of an 8.00 in. wall is 2.40, find the R value of a 9.00 in. wall.

26-4 JOINT, INVERSE, AND COMBINED VARIATION

In this section we study quantities that vary directly as *two or more* quantities. We also study quantities that vary *inversely* as other quantities, and quantities that vary directly as some quantities and inversely as other quantities. In mathematics, science, and technology there are many such quantities.

Joint Variation

In the last section, we studied examples of a quantity that varied directly as *one* other quantity. That is, if y varied directly as x, then y was a function of the *single* variable x. But in the real world, many quantities are functions of *two or more* variables. For example, the area A of a triangle varies directly as both its base b and its height h (Fig. 26-11). We call this *joint variation*.

In general, if y varies directly as x, and also if y varies directly as w, we say that

y varies *jointly* as x and w.

Or, we may say that

y is proportional to x and w.

■ FIGURE 26-11
Joint variation.

$A = k\,bh$

We may write

$$y \propto xw$$

This means that y varies directly as the product of x times w. Therefore, the three variables (y, x, and w) and the constant of proportionality, k, can be related by the equation:

| JOINT VARIATION | $y = k\,xw$ where k is the constant of proportionality | 49 |

─── **Example 23** ───

(a) If p varies jointly as q and r, then

$$p = k\,qr$$

(b) The area A of a triangle varies jointly as the base b and the height h (Fig. 26-11), so

$$A = k\,bh$$

■ **FIGURE 26-12**
Joint variation.

$V = k\,r^2 h$

─── **Example 24** ───

The volume V of a conical pendulum weight is jointly proportional to the height h and the square of the radius r (Fig. 26-12), so

$$V = k\,r^2 h$$

This reminds us of the familiar formula for the volume of a cone, Eq. 92:

$$V = \frac{1}{3}\pi r^2 h$$

We see that the constant of proportionality is

$$k = \frac{1}{3}\pi$$

Inverse Variation

We recall that when two quantities vary directly, and one quantity increases, then the other increases. Sometimes, however, quantities are so related that if one *increases*, the other *decreases*.

─── **Example 25** ───

A woman regularly drives between two towns, 100 miles apart. The rate of travel (speed) and the time of travel are related. If we *increase* the rate, then the travel time *decreases*. Similarly, if we decrease the rate, then the travel time increases. We say that, for a car traveling a constant distance with uniform motion, the travel time t varies inversely as the rate r.

In general, if two variables, x and y, are related in such a way that $y = \dfrac{k}{x}$, then we say that *y varies inversely as x*, or *y is inversely proportional to x*. We may write

$$y \propto \frac{1}{x}$$

| INVERSE VARIATION | $y = \dfrac{k}{x}$ where k is the constant of proportionality | 50 |

Example 26

(a) If s varies inversely as w, then

$$s = \frac{k}{w}$$

(b) The current I in a resistor is inversely proportional to the resistance R, so

$$I = \frac{k}{R}$$

(c) The force F of attraction between two bodies varies inversely as the square of the distance d between them (Fig. 26-13).

$$F = \frac{k}{d^2}$$

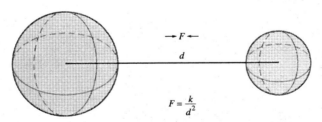

FIGURE 26-13
Inverse variation.

If y varies inversely as x, then $y = \dfrac{k}{x}$, where k is a constant. If we multiply both sides of this equation by x, we find that

$$k = xy$$

or that k is the product of x times y. In other words, in inverse variation the *product* of the quantities is a constant.

26-4 Joint, Inverse, and Combined Variation

Example 27

Show from the following data that the current I in a resistor is inversely proportional to the resistance R.

R (in ohms)	10	25	50	100	250
I (in amperes)	25	10	5	2.5	1

Solution: To show that I is inversely proportional to R, we calculate the product RI for each pair of numbers. If the product is a constant, then I varies inversely as R.

$$10(25) = 25(10) = 50(5) = 100(2.5) = 250(1) = 250$$

Since

$$RI = 250$$

for all R and I, then

$$k = 250$$

and the current I varies inversely as the resistance R, so that

$$I = \frac{250}{R}$$

Combined Variation

If one quantity varies with two or more quantities in a way that combines direct and inverse variation, we have what is called *combined variation*. In general, if y varies directly as x and inversely as w, we may write

$$y \propto \frac{x}{w}.$$

We write an equation by placing the variables which vary directly in the numerator and the variables which vary inversely in the denominator. Of course, we must include the constant of proportionality in the numerator.

COMBINED VARIATION	If y varies directly as x and inversely as w then $y = \dfrac{kx}{w}$	51

Example 28

(a) If p varies inversely as q and directly as r, then

$$p = \frac{kr}{q}$$

(b) The resistance R in a wire is directly proportional to the length L and inversely proportional to the square of the diameter d.

$$R = \frac{kL}{d^2}$$

FIGURE 26-14

Solution of Variation Problems

To solve a variation problem for an unknown quantity, we use the method described in the last section.

--- **Example 29** ---

The volume V of a cone varies as the square of the radius r of the base, and directly as the height h. When the radius is 1.50 cm and the height is 4.00 cm, the volume is 9.42 cm³ (Fig. 26-14). Find the constant of proportionality and write the equation for V.

Solution: We use Eq. 49 for joint variation to write the equation:

$$V = k\, r^2 h$$

We substitute the given values ($V = 9.42$ cm³, $r = 1.50$ cm, and $h = 4.00$ cm) into the equation above:

$$9.42 \text{ cm}^3 = k\,(1.50 \text{ cm})^2 (4.00 \text{ cm})$$

Solving for k:

$$k = 1.047$$

Substituting k into the equation $V = k\, r^2 h$, we get

$$V = 1.047\, r^2 h$$

Applications

Many applications involve variation.

--- **Example 30** ---

Boyle's Law states that for a confined gas at a constant temperature, the pressure P is inversely proportional to the volume V. If the air in a cylinder is at a pressure of 101.35 kPa and occupies a volume of 0.3687 m³, find the pressure when it is compressed to 0.2048 m³.

Solution: We use Eq. 50 for inverse variation to write the equation:

$$P = \frac{k}{V}$$

where k is the constant of proportionality. To find k we substitute the given values ($P = 101.35$ kPa and $V = 0.3687$ m³) and solve for k.

$$101.35 \text{ kPa} = \frac{k}{0.3687 \text{ m}^3}$$

$$k = (101.35 \text{ kPa})(0.3687 \text{ m}^3) = 37.367\,745 \text{ kPa} \cdot \text{m}^3$$

To find the equation, we substitute the value k in the equation $P = \dfrac{k}{V}$

And

$$P = \frac{37.367\,745 \text{ kPa} \cdot \text{m}^3}{V}$$

where V is in m³.

26-4 Joint, Inverse, and Combined Variation

To find the unknown pressure, we substitute the volume associated with it ($V = 0.2048$ m³) into the equation and solve.

$$P = \frac{37.367\,745 \text{ kPa} \cdot \text{m}^3}{0.2048 \text{ m}^3} = 182.5 \text{ kPa}$$

SECTION 26-4 Exercises Joint, Inverse, and Combined Variation

Joint Variation

Show that the following tables of values represent that z varies jointly as x and y by calculating the ratio $\dfrac{z}{xy}$ for each set of values.

1.
x	1	2	3	4	5
y	2	4	6	8	10
z	4	16	36	64	100

2.
x	2	4	6	8	10
y	1	3	5	7	9
z	6	36	90	168	270

3. Given the following data, show that the interest i earned on a savings account varies jointly as the interest rate r and the time t.

r (in %)	6.28	6.54	6.86	6.98
t (in yr)	0.25	0.50	1.25	1.50
i (in \$)	3.14	6.54	17.15	20.94

4. Given the following experimental data, show that the expansion E of a metal rod varies jointly as its original length L and the temperature change T.

L (in cm)	25	50	50	75
T (in °C)	10	10	20	20
E (in cm)	0.005	0.010	0.020	0.030

Write an equation.

5. p varies jointly as q and r.
6. x is jointly proportional to y and to z.
7. The expansion E of a metal rod varies jointly as its original length L and the temperature change T.
8. The kinetic energy E_K of a moving object is jointly proportional to its mass m and the square of its velocity v.

Inverse Variation

Show that the following tables of values represent that y varies inversely as x by calculating the product xy for each pair of values.

9.
x	1	2	3	4	5
y	60	30	20	15	12

10.
x	2	4	6	8	10
y	30	15	10	7.5	6.0

Write an equation.

11. z varies inversely as y.
12. p is inversely proportional to q.
13. y is inversely proportional to the square of z.
14. a varies inversely as the cube of b.
15. The capacitive reactance X_c of a circuit varies inversely as the capacitance C of the circuit.
16. The amount of light I that falls on an object is inversely proportional to the square of the distance d from the source of light.

Combined Variation

Write an equation.

17. y is inversely proportional to x and directly proportional to z.
18. p varies directly as q and inversely as r.
19. Electric current I flowing in a circuit varies directly as the voltage V and inversely as the resistance R.
20. The intensity of illumination I at a given point is directly proportional to the intensity of the light source L and inversely proportional to the square of the distance d from the light source.

Solution of Variation Problems

Write the equation and find the unknown value.

21. x varies jointly as y and z, and $x = 36$ when $y = 3$ and $z = 4$. Find x when $y = 4$ and $z = 5$.
22. a is directly proportional to b^2 and to c, and $a = 4.5$ when $b = 1.5$ and $c = 4$. Find a when $b = 2.5$ and $c = 6$.
23. If y varies directly as x and inversely as z, and $y = 12$ when $x = 3$ and $z = 4$, find y when $x = 5$ and $z = 8$.
24. a is inversely proportional to the square of b and directly proportional to c. If $a = 7.5$ when $b = 12$ and $c = 10$, find a when $b = 10$ and $c = 12$.

Solve.

Applications

25. The amount A paid to a work crew varies jointly as the number n of persons working and the length of time t worked. If a crew of 4 workers earns $150.00 in 5.00 hours, how much will a crew of 7 workers earn in 6.00 hours?
26. The amount W of work done varies jointly as the number n of persons working and the length of time t worked. If 6 stonemasons can build 20.0 m of a stone wall in 4.00 hours, how long a wall can 8 stonemasons build in 3.00 hours?
27. The time t needed to empty a vertical cylindrical tank varies as the square root of the height h of the tank and the square of the radius r. If it takes 81.0 minutes to empty a tank whose height is 2.00 m and whose radius is 0.75 m, how long would it take to empty a tank 1.50 m high with radius 0.48 m?
28. The mass m of a solid cylinder varies jointly as the height h and the square of the radius r. If a solid cylinder of height 3.50 cm and radius 1.20 cm has a mass of 0.179 kg, find the mass of a solid cylinder of height 2.50 cm and a radius of 1.10 cm.
29. When an object travels a constant distance in uniform motion, the travel time t is inversely proportional to the velocity v. If a driver usually makes a trip in 1.5 hours at an average speed of 88 km/h, find the traveling time if the speed were reduced to 66 km/h.
30. The wavelength w of a radio wave varies inversely as its frequency f. If the wavelength is 70.0 m when the frequency is 4250 kHz, find the wavelength when the frequency is 4560 kHz.
31. Newton's law of gravitation states that any two bodies attract each other with a force F that is inversely proportional to the square of the distance d between them. The force of attraction between two certain spheres is 2.25×10^{-10} N when the spheres are 0.240 m apart. What is the force of attraction when the spheres are 0.150 m apart?
32. The intensity I of sound is inversely proportional to the square of the distance d from the source. If the intensity is 3.6×10^{-7} W/m² at a distance of 6.0 m from a speaker, find the intensity at 9.0 m.
33. When an electric current flows through a wire, the resistance R varies directly as the length L and inversely as the cross-sectional area A of the wire. If the length of the wire is tripled and the cross-sectional area is doubled, by what factor will the resistance change?

34. The volume V of a given weight of gas varies directly as the absolute temperature T and inversely as the pressure p. If the volume is 3.76 m³ when the pressure is 275 kPa and the absolute temperature is 315 K, find the volume when the pressure is 285 kPa and the absolute temperature is 295 K.

35. The number n of vibrations per second made when a stretched wire is plucked varies directly as the square root of the tension T in the wire, and inversely as the length L. If a 1.50 m wire vibrates 226 times a second when the tension is 125 N, find the number of vibrations per second if the wire is shortened to 1.25 m and the tension increased to 175 N.

36. Newton's law of gravitation states that every body in the universe attracts every other body with a force that varies directly as the product of their masses and inversely as the square of the distance between them. By what factor will the force change when the distance is tripled and each mass is doubled?

Chapter 26 REVIEW EXERCISES

Solve the proportion.

1. $\dfrac{3}{7} = \dfrac{x}{42}$ 2. $\dfrac{x}{21} = \dfrac{5}{6}$ 3. $\dfrac{9}{x} = \dfrac{x}{16}$ 4. $\dfrac{8}{9} = \dfrac{x}{136 - x}$

Simplify the ratio in fraction form.

5. 9 kg to 15 kg   6. 35 m to 7 m   7. 600 g to 3 kg

Write an equation.

8. y varies jointly as x and z.

9. p varies inversely as the square of r.

10. t varies directly as the square root of s and inversely as r.

Write an equation that describes the relationship.

11. The thermal resistance R of a wall is proportional to the thickness T of the wall.

12. According to Newton's second law, the acceleration a of a body is directly proportional to the force F applied to it.

Solve.

13. A beam 16 metres long must be separated into two parts so that the parts are in a ratio of 5 to 3. How long must the parts be?

14. A settling tank contains 23 700 litres of water and can be emptied in 16.0 hours. How much water was in the tank if it was emptied in 9.75 hours? (Assume the volume is directly proportional to time.)

15. Given the following experimental data, show that the resistance R of a wire is directly proportional to the length L of the wire.

L (m)	0.5000	1.500	2.500	3.500	4.500
R (ohms)	57.50	172.5	287.5	402.5	517.5

16. Steel is made by mixing nickel and iron. If 10 tonnes of nickel are mixed with 95 tonnes of iron, find the ratio of iron to nickel.

FIGURE 26-15
Transformer.

17. The turn ratio of a transformer (Fig. 26-15) is

$$a = \frac{\text{number of turns in primary winding}}{\text{number of turns in secondary winding}} = \frac{N_1}{N_2}$$

Find the turn ratio of a transformer that has 250 turns in the secondary winding and 10 000 turns in the primary winding.

18. Write the following as a proportion in fraction form and then use the fundamental principle of proportions to show that the proportion is true.

$$63 \text{ is to } 81 \text{ as } 7 \text{ is to } 9$$

Give the general equation and find the missing value.

19. The interest i earned on a savings account varies jointly as the interest rate r and the time t. A certain savings account earned \$176.40 interest at a rate of 6.72% for $\frac{3}{4}$ of a year. How much interest would the account earn at 6.68% for $1\frac{1}{4}$ year?

20. When an electric current flows through a wire, the resistance R varies directly as the length and inversely as the square of the diameter of the wire. If 958 m of wire 4.00 mm in diameter has a resistance of 19.7 ohms, find the resistance in a wire 825 m long with a diameter of 4.50 mm.

Write the equation and find the missing value.

21. If E varies directly as I, and $E = 24.8$ volts when $I = 15.2$ amperes, find E when $I = 20.5$ amperes.

22. The amount of sulfuric acid in a solution varies directly as the concentration. If 3.65 litres of sulfuric acid are used to make a quantity of a 12.0% solution, how many litres are needed to make the same quantity of a 15.0% solution?

23. The horizontal distance s traveled by a projectile is directly proportional to the square of its initial velocity v. If a projectile travels 728 m horizontally when its initial velocity is 157 m/s, how far will it travel when its initial velocity is 218 m/s?

24. Find the constant of proportionality and give the general equation of variation that describes this relationship: y varies directly as x, and $y = 9.6$ when $x = 1.6$.

25. The rate r of heat loss through insulation is inversely proportional to the thickness t of the insulation. If 2200 Btu/h are lost through 3.5 inches of insulation, what would the rate of heat loss be through 5.0 inches of insulation?

26. The time required to fill a tank varies inversely as the square of the diameter of the pipe used. If it takes 3.5 hours to fill a tank using a pipe with a diameter of 2.5 inches, how long would it take to fill the tank using a pipe with a diameter of 4.5 inches?

27. If w is proportional to r^3, and $w = 2.13$ lb when $r = 1.25$ in., find w when $r = 1.50$ in.

28. The velocity v of a ship varies directly as the cube root of the power P used to propel the ship. If 6570 hp can propel the ship at 15.0 knots, what will be the velocity under 4950 hp?

WRITING

29. Suppose your company makes plastic trays and is planning new ones with dimensions double those now being made. Your company president is convinced that the new trays will need only twice as much plastic as the older version. "Twice the size, twice the plastic," he proclaims, and no one is willing to challenge him. Write a presentation to the president in which you tactfully point out that he is wrong and you explain that the new trays will require eight times as much plastic.

TEAM PROJECT

30. We saw that the volumes (and hence the weights) of solids are proportional to the cube of corresponding dimensions. Applying that idea, suppose that a sporting goods company has designed a new type of ski based on the statement:

The weights of people of similar build are proportional to the cube of their heights.

The goal of this project is to prove or disprove the given statement. Gather data from students on your campus to reach a conclusion. In addition to a person's height and weight, it may be useful to note a person's build, such as thin, average, heavy. Should you note gender and/or age? Make a conclusion and prove it.

Exponential Functions

Outcomes

When you have completed this chapter, you will be able to

1. Define exponential functions.
2. Apply the laws of exponents.
3. Graph exponential functions.
4. Apply exponential functions to growth.
5. Apply exponential functions to decay of radioactive materials and to other examples.

In this chapter we define the exponential function, examine its graph, and review the laws of exponents. We also study two very important and special examples of exponential functions: exponential growth and exponential decay. We look at applications of the exponential function, such as biological growth, compound interest, population growth, radioactive decay, electromotive force, heat loss, inflation, electric current, light intensity, atmospheric pressure, and capacitance.

27-1 INTRODUCTION TO EXPONENTIAL FUNCTIONS

In this section we define the exponential function and see how the laws of exponents are applied with exponential functions.

Definitions

An *exponential function* is a function in which the independent variable is in the exponent. The quantity that is raised to the power is called the *base*.

Example 1

Some exponential functions are

$$y = 4^x, \; y = 2 \cdot 3^x, \; y = b^x, \text{ and } y = 3e^{2x}$$

where b and e represent positive constants, not equal to 1.

Note that the exponential function is different from the *power function* of Chapter 12. In the power function $y = x^n$, the exponent n is a constant and the base x is a variable. In the exponential function $y = b^x$, the exponent x is a variable and the base b is a constant.

Example 2

The function $y = 3^x$ is an *exponential* function, since the variable is in the exponent. However, the function $y = x^3$ is a *power* function, since the variable is in the base.

Laws of Exponents

All of the laws of exponents previously studied in Chapter 20 are also true here. We rewrite the laws below for expressions in which the base is a positive constant and the exponent is a variable.

LAWS OF EXPONENTS			
	Product	$b^x \cdot b^y = b^{x+y}$	19
	Quotient	$\dfrac{b^x}{b^y} = b^{x-y}$	20
	Power	$(b^x)^y = b^{xy}$	21
	Product Raised to a Power	$(ab)^x = a^x b^x$	22
	Quotient Raised to a Power	$\left(\dfrac{a}{b}\right)^x = \dfrac{a^x}{b^x}$	23
	Negative Exponent	$b^{-x} = \dfrac{1}{b^x}$	25

The following examples illustrate the laws of exponents for exponents that are variables.

Example 3

$$2^x 2^y = 2^{x+y}$$

Example 4

$$\frac{3^z}{3^y} = 3^{z-y}$$

Example 5

$$(4^x)^y = 4^{xy}$$

Example 6

$$(5a)^x = 5^x a^x$$

Example 7

$$\left(\frac{2}{3}\right)^x = \frac{2^x}{3^x}$$

Example 8

$$3^{-x} = \left(\frac{1}{3^x}\right)$$

The next two examples follow from the laws of exponents above and the fact that

$$\frac{1}{2} = 2^{-1}$$

Example 9

$$\left(\frac{1}{2}\right)^x = (2^{-1})^x = 2^{-x}$$

Example 10

$$\left(\frac{1}{2}\right)^{-x} = (2^{-1})^{-x} = 2^x$$

SECTION 27-1 Exercises Introduction to Exponential Functions

Definitions

1. Define an exponential function.
2. What is the difference between an exponential function and a power function? Give an example of each.

Laws of Exponents

Use the laws of exponents to complete these equations.

3. $5^x 5^y = ?$
4. $\dfrac{6^y}{6^z} = ?$
5. $(8^y)^z = ?$
6. $(9m)^x = ?$
7. $\left(\dfrac{5}{3}\right)^x = ?$
8. $\left(\dfrac{1}{3}\right)^x = ?$
9. $2^{-x} = ?$
10. $8^{-x} = ?$
11. $\left(\dfrac{1}{4}\right)^{-x} = ?$
12. $\left(\dfrac{1}{5}\right)^{-x} = ?$

27-2 GRAPH OF THE EXPONENTIAL FUNCTION

In this section we graph exponential functions and solve some applications involving exponential functions.

We make a graph of an exponential function in the same way we graphed other functions in Chapter 12. We choose values of the independent variable x and compute the corresponding values of the dependent variable y. We then plot the resulting ordered pairs and connect the points with a smooth curve.

--- **Example 11** ---

Graph the exponential function $y = 3^x$ for x from -2 to $+2$.

Solution: We compute the values of y and obtain the following table of values:

x	−2	−1	0	1	2
y	$\frac{1}{9}$	$\frac{1}{3}$	1	3	9

These pairs are plotted and connected with a smooth curve in Fig. 27-1. Note that the y intercept is 1 and there is no x intercept. In fact, the x axis is an asymptote. (Recall that an asymptote is a line that a curve approaches but never reaches.) Also as *x increases, y increases,* and the curve rises and gets steeper. We say that y is an *increasing* function.

■ FIGURE 27-1
An exponential function.

--- **Example 12** ---

Use the graphing calculator to graph $y = 3^x$ for x from -2 to $+2$.

Solution: We are given the domain of the function (x is between -2 and 2). In the table of values above, y is between $\frac{1}{9}$ and 9. We set the window so that

```
Xmin=−2
Xmax=2
Xscl=1
Ymin=0
Ymax=9
Yscl=1
```

We enter the function $y = 3^x$ and graph (Fig. 27-2).

■ FIGURE 27-2

--- **Example 13** ---

Graph $y = \left(\frac{1}{2}\right)^x$ for x from -3 to 3.

Solution: We compute the values of y and make the table of values:

x	−3	−2	−1	0	1	2	3
y	8	4	2	1	$\frac{1}{2}$	$\frac{1}{4}$	$\frac{1}{8}$

We plot these points and connect them with a smooth curve in Fig. 27-3. Notice that the y intercept is 1 again, there again is no x intercept, and the x axis is an asymptote. Also, as *x increases, y decreases* and the curve "falls" and gets "flatter." We say that y is a *decreasing* function.

■ FIGURE 27-3

Example 14

Use a graphing calculator to graph $y = 2^{-x}$ for x from -3 to 3.

Solution: We set the window so that

■ FIGURE 27-4

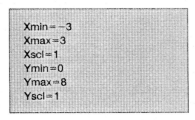

We enter the function and graph (Fig. 27-4). Note that this graph of $y = 2^{-x}$ is identical to the graph of $y = \left(\dfrac{1}{2}\right)^x$ in Fig. 27-3. This is because the functions are the same function, since

$$\left(\frac{1}{2}\right)^x = \frac{1^x}{2^x} = \frac{1}{2^x} = 2^{-x}$$

■ FIGURE 27-5

Example 15

Graph $y = 4(1.5)^{-2x}$ for x from -1 to 3.

Solution: We make a table of values, using a calculator to compute the y values, and rounding to the nearest tenth.

x	-1	0	1	2	3
y	9	4	1.8	0.8	0.4

The points are plotted and connected with a smooth curve in Fig. 27-5.

Applications

Applications of exponential functions include biological growth of organisms such as bacteria and viruses.

Example 16

A biologist has a culture consisting of about 1500 bacteria, which doubles its number every hour.

(a) Write an equation for the number of bacteria n after t hours.
(b) Find the number of bacteria in the culture after 8.0 hours.
(c) Graph the equation.
(d) Graphically find the number of bacteria after 6.0 hours.

FIGURE 27-6

FIGURE 27-7

FIGURE 27-8

Solution:

(a) To find an equation for the number of bacteria n as a function of time t, we make a table to see if there is a pattern:

t	0.0	1.0	2.0	3.0	...	t
n	1500	1500(2)	1500(2)(2)	1500(2)(2)(2)	...	$1500(2^t)$

Indeed there is a pattern and we see that an equation for n as a function of t is

$$n = 1500(2^t)$$

(b) To find how many bacteria there are in the culture after 8.0 hours, we substitute $t = 8.0$ into the equation above:

$$n = 1500(2^{8.0}) = 380\,000 \text{ bacteria, after 8.0 hours}$$

(c) We graph this function on the graphing calculator. Using y as a function of x instead of n as a function of t, we enter the function

$$y = 1500(2^x)$$

and set the window so that

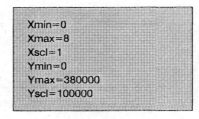

and graph (Fig. 27-6). To graph the function

$$n = 1500(2^t)$$

on paper, we label the axes with n and t, and label the scales (Fig. 27-7).

(d) We find how many bacteria there are after 6.0 hours graphically. We find the point where $x = 6.0$ and read the y value. Many graphing calculators will find this value for you (Fig. 27-8). So, after 6.0 hours, there are about 96 000 bacteria in the culture.

SECTION 27-2 Exercises Graph of the Exponential Function

Graph each exponential function for x from -3 to 3.

1. $y = 2^x$
2. $y = 4^x$
3. $y = -3^x$
4. $y = -2^x$
5. $y = 3(2^x)$
6. $y = 2(4^x)$
7. $y = 3^{-x}$
8. $y = 4^{-x}$
9. $y = 2^{3x}$
10. $y = 2^{-2x}$
11. $y = \left(\dfrac{2}{3}\right)^x$
12. $y = \left(\dfrac{3}{2}\right)^{-x}$
13. $y = (2.3)^{-x}$
14. $y = (1.4)^x$
15. $y = 4(1.5)^{3x}$
16. $y = 0.3(2.1)^{-2x}$

634 CHAPTER 27 / Exponential Functions

Applications

Solve. If answers are not exact, round to two significant digits.

17. A biologist has a culture of 2500 bacteria, which triples its number every hour.
    (a) Write an equation for the number of bacteria n after t hours.
    (b) Find the number of bacteria in the culture after 5.0 hours.
    (c) Graph the equation for t from 0 to 5 hours.
    (d) Graphically find the number of bacteria after 3.5 hours.

18. A biologist has a culture of 200 bacteria, which quadruples its number every day.
    (a) Write an equation for the number of bacteria n after t days.
    (b) Find the number of bacteria in the culture after 4.0 days.
    (c) Graph the equation for t from 0 to 4 days.
    (d) Graphically find the number of bacteria after 2.5 days.

19. The radioactive isotope technetium-99m is used to identify brain tumors. Half of it disintegrates in 6.0 hours. If a physician injects 4.8 mg, then the amount present y (in mg) after t hours is found using the equation

    $$y = 4.8 \left(\frac{1}{2}\right)^{t/6}$$

    (a) How much technetium-99m will be present after 12.0 hours?
    (b) Graph y as a function of t, for t from 0 to 12 hours.
    (c) Graphically find how many hours must elapse for the amount present to be 2.0 mg.

20. A census in Ethiopia found that the population was about 55 million people. It was expected to double every 22 years, so that the expected population P, in millions of people, was found using

    $$P = 55(2^{t/22})$$

    where t is the number of years after the census.
    (a) Find the expected population in 50.0 years.
    (b) Graph the expected population as a function of time, for t from 0 to 50 years.
    (c) After how many years was the population expected to reach 100 million people?

21. A yeast manufacturer finds that the yeast will double in weight every hour.
    (a) Find the mass of yeast the manufacturer must start with to obtain 650 kilograms in 4.0 hours.
    (b) Using the result in part (a), write an equation for the mass m of yeast as a function of time t.
    (c) What is the mass of yeast after 2.5 hours?
    (d) Graph the equation for t from 0 to 4 hours.
    (e) Graphically find how many hours it takes to produce 500 kg of yeast.

22. A pharmaceutical company is growing an organism to be used in a vaccine. The organism grows at a rate so it triples every day.
    (a) Find the number of organisms the company must start with to have 2800 units after 6.0 days.
    (b) Using the result in part (a), write an equation for the number n of organisms as a function of time t.
    (c) How many organisms will there be after 10 days?
    (d) Graph n as a function of t, for t from 0 to 10 days.
    (e) Graphically find how many days it takes to grow 100 000 organisms.

27-3 EXPONENTIAL GROWTH

In the last section, we looked at growth that occurred in *steps* (by doubling or tripling periodically). In this section we look at a type of growth that is *continuous;* it is called *exponential growth.* The irrational number *e* is introduced, as an important constant in the equation for exponential growth. We study the graphs of exponential functions which use *e* as the base. We also look at applications, including compound interest and inflation.

Compound Interest

When an amount of money (the *principal*) is invested for a period of time, it will earn *interest* at a specified rate. If this interest is then *added* to the principal and both continue to earn interest, we have what is called *compound interest,* the type of interest commonly given on bank accounts. The bank determines how often the interest is compounded; it may be compounded every year, month, day, hour, minute, second, or even continuously. If the interest is compounded *m* times per year at an annual interest rate *r*, then the original amount *a* will accumulate to an amount *y* in *t* years, according to the formula

$$y = a\left(1 + \frac{r}{m}\right)^{tm}$$

The interest rate *r* is generally given as a percent, but it must be expressed as a decimal in the equation. Also, the units for time *t* and rate *r* must agree; for example, if the rate is an annual rate, then the time must be in years.

---- **Example 17** ----

Find the amount to which $700 will accumulate in 4 years at an annual interest rate of 6.5%, if the interest is compounded annually. Round to the nearest cent.

Solution: Converting the rate *r* to a decimal,

$$r = 6.5\% = 0.065$$

Interest is compounded annually (once a year), so $m = 1$. Using the formula for compound interest with $a = 700$, $r = 0.065$, $m = 1$, and $t = 4$,

$$y = 700\left(1 + \frac{0.065}{1}\right)^4$$
$$= 900.53$$

So if the interest is compounded annually, $700 will accumulate to $900.53 after 4 years.

---- **Example 18** ----

Find the amount to which $700 will accumulate in 4 years at an annual interest rate of 6.5%, if the interest is compounded monthly. Round to the nearest cent.

Solution: Interest is compounded monthly, so it is compounded 12 times each year, making $m = 12$. Using the formula for compound interest with $a = 700$, $r = 0.065$, $m = 12$, and $t = 4$,

$$y = 700\left(1 + \frac{0.065}{12}\right)^{4(12)}$$
$$= 907.21$$

So, if the interest is compounded monthly, $700 will accumulate to $907.21 after 4 years. Compare that to the $900.53 we found in Example 17 with annual compounding!

But interest can even be compounded *continuously!* When interest is compounded continuously, then the accumulated amount y grows *exponentially*.

Exponential Growth

If a quantity grows continuously in this manner, we use the equation for *exponential growth:*

| EXPONENTIAL GROWTH | $y = ae^{rt}$ | 142 |

where a is the original amount of a quantity which is growing continuously at a rate r and y is the amount present after time t. The base e is a constant, described below. We say that y *increases* (or *grows*) *exponentially* with respect to t.

The number e

So what is this number e, which is the base in the exponential equation for continuous growth? The number e is an irrational number and has been calculated to millions of digits. The first 16 digits of e are

$$e = 2.718281828459045$$

Evaluating e^x

With a calculator, we can evaluate e or any power of e.

Example 19

Use a calculator to find the value of the number e.

Solution: Since $e = e^1$, we use the $\boxed{e^x}$ key on a calculator to find

$$e = e^1 = 2.718281828 \quad \text{to 10 significant digits}$$

Example 20

Use a calculator to approximate $e^{2.75}$.

Solution: Using the $\boxed{e^x}$ key and rounding to three significant digits,

$$e^{2.75} = 15.6$$

Example 21

Find the amount to which $700 will accumulate in 4 years at an annual interest rate of 6.5%, if the interest is compounded continuously. Round to the nearest cent.

Solution: Since the interest is being compounded continuously, the accumulated amount grows exponentially and is given by Eq. 142, with $a = 700$, $r = 0.065$, and $t = 4$:

$$y = 700e^{0.065(4)} = 907.85$$

So, $700 will accumulate to $907.85 in 4 years. Compare that to the $907.21 we found with monthly compounding. We earn *a little more* with continuous compounding!

Graph of Exponential Growth

We saw in Sec. 27-2 that an exponential function $y = b^x$ is an increasing function if $b > 1$. The function $y = e^x$ is a special exponential function in which the base is the positive number e. Since $e > 1$, then $y = e^x$ is an increasing function. The graph of $y = e^x$ rises to the right and gets steep, and the x axis is an asymptote.

---- **Example 22** ----

Graph the function $y = e^x$ for x from -3 to 3.

Solution: We make a table of values and use a calculator to find the y values, rounding to tenths.

x	-3	-2	-1	0	1	2	3
y	0.0	0.1	0.4	1	2.7	7.4	20.1

FIGURE 27-9
Exponential growth.

We plot the points and connect them with a smooth curve (Fig. 27-9). Note that the function is increasing, the y intercept is 1, the x axis is an asymptote, and the values of y are all positive.

---- **Example 23** ----

Graph $y = 3e^{2x}$ for x from -2 to 2.

Solution: We use a graphing calculator to graph this function. But first we determine the range of values for y. The values of y will be positive, and the largest y value will be when $x = 2$, since this is an increasing function. So, we calculate the y value when $x = 2$:

$$3e^{2(2)} = 164$$

We set the window so that

```
Xmin=-2
Xmax=2
Xscl=1
Ymin=0
Ymax=164
Yscl=100
```

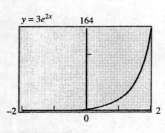

FIGURE 27-10

and graph (Fig. 27-10).

---- **Example 24** ----

The prices of some goods rise exponentially due to inflation. The price of a certain new radio is $125. Assume that the rate of inflation remains constant at 4.50% per year.

(a) Find the price of the radio 5.00 years from now.
(b) Graph the price p as a function of time t from 0 to 25 years.
(c) Graphically, find the price of the radio 20.0 years from now.

Solution:

(a) Since price is increasing exponentially, we use Eq. 142, with $y = p$, $a = 125$, $r = 4.50\% = 0.0450$, and $t = 5.00$

$$p = 125e^{0.0450(5.00)}$$
$$= \$157$$

(b) We graph the function on a graphing calculator. The price p as a function of time t is

$$p = 125e^{0.0450t}$$

We enter the function as

$$y = 125e^{0.0450x}$$

To find the range of the function, we evaluate y for $x = 25$ to find the maximum y value:

$$y = 125e^{0.0450(25)}$$
$$= \$385$$

So we set the window so that

■ FIGURE 27-11

```
Xmin=0
Xmax=25
Xscl=10
Ymin=0
Ymax=385
Yscl=100
```

and graph (Fig. 27-11).

(c) To find the price of the radio 20.0 years from now, we find the point on the graph where x is 20.0 (Fig. 27-12), and read the y value. The price of this radio will be $307 in 20.0 years.

■ FIGURE 27-12

TIP Many graphing calculators will evaluate a function at a specified value. Check your calculator manual for instructions.

SECTION 27-3 Exercises Exponential Growth

Evaluating e^x

Use a calculator to evaluate these expressions, and round to three significant digits.
1. e^2
2. e^3
3. $e^{1.36}$
4. $e^{2.75}$
5. $e^{-0.234}$
6. $e^{-1.75}$
7. $3e^5$
8. $7e^{-2}$
9. $4.56e^{0.765}$
10. $78.3e^{-8.56}$

Compound Interest

Round to the nearest cent.

11. Find the amount to which $650 will accumulate in 4 years at an annual interest rate of 6.5%, if the interest is compounded (a) annually, (b) monthly, and (c) continuously.

12. Find the amount to which $2800 will accumulate in 5 years at an annual interest rate of 7.5%, if the interest is compounded (a) annually, (b) daily, and (c) continuously.

13. Find the amount to which $1.00 will accumulate in 10 years at an annual interest rate of 10%, if the interest is compounded (a) annually, (b) monthly, (c) daily, and (d) continuously.

14. Find the amount to which $1.00 will accumulate in 10 years at an annual interest rate of 12%, if the interest is compounded (a) annually, (b) monthly, (c) daily, and (d) continuously.

15. A savings bank is offering an annual interest rate of 9.5%, compounded monthly. How much do you need to invest to have an accumulated amount of $5250 at the end of 5 years?

16. A savings bank is offering an annual interest rate of 8.5%, compounded daily. How much do you need to invest to have an accumulated amount of $1550 at the end of 3 years?

Graph of Exponential Growth

17. Graph on the same set of axes, for x from -2 to 2:
    (a) $y = e^x$ (b) $y = e^{2x}$ (c) $y = 2e^x$

18. Graph on the same set of axes, for x from -2 to 2:
    (a) $y = e^x$ (b) $y = 3e^x$ (c) $y = e^{3x}$

19. Graph $y = 2e^{3x}$, for x from -2 to 2.

20. Graph $y = 4e^{2x}$, for x from -2 to 2.

Round to three significant digits.

Applications

21. The population of Canada on July 1, 1997 was estimated at 30 286 600 people. The growth rate for the previous four years averaged 1.14% per year. If the population grows exponentially, what will the population be on July 1, 2000?

22. If the rate of inflation is 0.83% per year, how much should the minimum wage of $7.20 per hour be increased to five years from now, assuming that the rate of inflation remains constant and the rate increases exponentially?

23. A yeast manufacturer finds that a certain kind of yeast will grow exponentially at a rate of 13.0% per hour.
    (a) How many kilograms of yeast will the manufacturer have at the end of 10.0 hours, if there are 45.0 kg originally?
    (b) Graph the mass of yeast as a function of t, for t from 0 to 10 hours.
    (c) Graphically find the mass of yeast after 7.25 hours.
    (d) Graphically find how many hours it takes to grow 100 kg of yeast.

24. A utility company claims that the demand for electricity in its region is growing exponentially at an annual rate of 11.5%. At present, the company's customers use 275 billion kW·h of electricity.
    (a) Assuming that the customers make no effort to conserve electricity, how much electrical power will be needed in 10 years?
    (b) Graph the demand for electricity as a function of time, for t from 0 to 10 years.
    (c) Graphically find the demand for electricity after 7.50 years.
    (d) Graphically find how many years it would take for the demand to be 500 billion kW·h.

27-4 EXPONENTIAL DECAY

There are many quantities in nature that *decrease* continuously; we say that the quantity *decays exponentially,* usually with respect to time. In this section we solve applications involving quantities including radioactive materials and temperature.

Exponential Decay

The equation for exponential decay is the same as for exponential growth, except that the rate is expressed as $-r$ (a *negative* rate, since the quantity is *decreasing*):

| EXPONENTIAL DECAY | $y = ae^{-rt}$ | 143 |

where a is the original amount of a quantity that is decaying continuously at a rate r and y is the amount present after time t. We say that *y decreases* (or *decays*) *exponentially* with respect to t.

Example 25

Radioactive materials disintegrate (decay) exponentially. Strontium-90, a material found in nuclear power plants, decays at the rate of 3.05% per year. A certain power plant had 125 kg of strontium-90.

(a) Write the equation for the amount of strontium-90 remaining t years later.
(b) Find the amount remaining 1.00 year later.

Solution:

(a) First we change the rate to a decimal,

$$r = 3.05\% = 0.0305$$

We use Eq. 143, with $a = 125$ and $r = 0.0305$,

$$y = 125e^{-0.0305t}$$

(b) To find the amount remaining one year later, we let $t = 1.00$ in the equation above:

$$y = 125e^{-0.0305(1)} = 121 \text{ kg}$$

Graph of Exponential Decay

We saw in Sec. 27-1 that an exponential function $y = b^{-x}$ is a decreasing function if $b > 1$. So $y = e^{-x}$ is a decreasing function, since $e > 1$.

Example 26

Graph $y = e^{-x}$ for x from -3 to 3.

Solution: We make a table of values, rounding to the nearest tenth.

x	-3	-2	-1	0	1	2	3
y	20.1	7.4	2.7	1	0.4	0.1	0.0

We sketch the graph in Fig. 27-13. We see that the function is decreasing and that the graph "falls" to the right and gets "flatter." The x axis is an asymptote.

FIGURE 27-13
Exponential decay.

Example 27

In Example 25, the amount y of strontium-90 remaining after t years was found to be

$$y = 125e^{-0.0305t}$$

(a) Graph this function for t from 0 to 100 years.
(b) Graphically find the amount of strontium-90 remaining after 50.0 years.

Solution:

(a) We use a graphing calculator to graph the function, letting $t = x$:

$$y = 125e^{-0.0305x}$$

The domain is given (x is between 0 and 100) and we must find the range of y values. Since the amount of strontium-90 is decaying, the maximum amount present is 125 kg (at $x = 0$). Certainly the amount will never be negative. So y must be between 0 and 125. We set the window so that

■ FIGURE 27-14

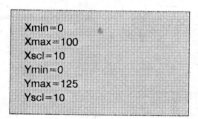

```
Xmin=0
Xmax=100
Xscl=10
Ymin=0
Ymax=125
Yscl=10
```

We enter the function and graph (Fig. 27-14). This curve is a typical exponential decay curve. Note that the x axis is an asymptote, indicating that the mass remaining *approaches* (but never reaches) zero. In other words, there will *always* be *some* radioactive material left.

(b) To graphically find how much remains after 50.0 years, we find the point on the graph where $x = 50.0$ (Fig. 27-15), and read that its y value is 27.2. So, after 50.0 years, there will be 27.2 kg of strontium-90 remaining.

■ FIGURE 27-15

Other quantities that decay exponentially include temperature, light intensity, and current through a capacitor. We examine temperature in the next example.

Example 28

If an object is placed in a freezer at 0°C, the temperature of the object decreases exponentially according to the formula

$$T = T_0 e^{-rt}$$

where T_0 is the original temperature of the object, r is the rate of temperature change, and T is the temperature of the object t time units later.

A man puts a hot cup of coffee at 100.0°C in a freezer at 0°C. If it cools at the rate of 5.15% per minute, what is the temperature of the coffee after 1.00 hour in the freezer?

Solution: First we note that the given units of the rate and the time do not agree (yet).

$$r = 5.15\% \text{ per minute} \quad \text{and} \quad t = 1.00 \text{ hour}$$

We change the time to minutes, to agree with the rate units:

$$t = 1.00 \text{ hour} = 60.0 \text{ minutes}$$

Changing the rate to a decimal,

$$r = 5.15\% = 0.0515$$

Using Eq. 143 with $a = 100$, $r = 0.0515$, and $t = 60.0$,

$$T = 100e^{-0.0515(60.0)}$$
$$= 4.55°C$$

COMMON ERROR When using Eq. 143 for exponential decay, the time units of rate r and time t must agree.

SECTION 27-4 Exercises Exponential Decay

Graph of Exponential Decay

1. Graph $y = 2e^{-x}$ for x from -3 to 3.
2. Graph $y = 3e^{-0.5x}$ for x from -3 to 3.
3. Graph $y = 0.8e^{-0.3t}$ for t from 0 to 5.
4. Graph $y = 0.5e^{-2t}$ for t from 0 to 3.

Solve. Round to three significant digits.

Applications

5. Tritium, a radioactive isotope of hydrogen, decays at the rate of 6.02% per year. A power plant has 175 kg of tritium.
   (a) Write the decay equation.
   (b) Graph the equation for 0 to 25 years.
   (c) Find the amount remaining after 7.50 years.

6. In a certain chemical reaction, the amount of salt in a solution decreases exponentially at the rate of 35.3% per hour. A solution has 7.85 grams of salt when the chemical reaction begins.
   (a) Write the equation for the amount of salt after time t.
   (b) Find the amount of salt in the solution 2.00 hours later.
   (c) Graph the amount of salt as a function of time for 0 to 5 hours.

7. A flywheel is rotating at 2400 rev/min. When the power is disconnected, the angular velocity decreases exponentially at the rate of 28.2% per minute. Find the angular velocity 25.0 seconds after the power has been disconnected.

8. A steel forging is 2500°F above room temperature. If it cools exponentially at the rate of 2.38% per minute, how many degrees above room temperature will it be in 2.00 hours?

9. A certain isotope of uranium decays at the rate of 2.85% per day. What percent of the original amount of radioactive uranium will remain after 31 days?

10. In a certain fabric mill, cloth is removed from a dye bath and is then observed to dry exponentially at the rate of 22.5% per hour. What percent of the original moisture will still be present after 40.0 minutes?

11. When a capacitor with capacitance C (in farads), charged to a voltage v_0 (in volts), is discharged through a resistor, the current i (in amperes) will decay exponentially with respect to time (in seconds) according to the formula

$$i = \frac{v_0}{R} e^{-t/(RC)} \quad \text{amperes}$$

Find the current, in milliamps, after 37.0 milliseconds in a circuit where $v_0 = 220$ volts, $C = 125$ microfarads, and $R = 2500$ ohms.

12. Graph $y = 3(1 - e^{-t})$ for t from 0 to 8.

13. When an object is placed in surroundings that are at a higher temperature T_0, it will increase in temperature exponentially according to the equation

$$T = T_0(1 - e^{-n})$$

A steel casting initially at 0°C is placed in a furnace at 1000°C.

 (a) If it increases in temperature at the rate of 5.00% per minute, find its temperature after 120 minutes.

 (b) Graph T as a function of t, for t from 0 to 120 minutes.

Chapter 27 REVIEW EXERCISES

Graph each function for x between -3 and 3.

1. $y = 4^x$
2. $y = 2 \cdot 3^x$
3. $y = 4e^x$
4. $y = e^{4x}$
5. $y = 3e^{-x}$
6. $y = 2e^{-2x}$
7. $y = 5^{-x}$
8. $y = 3(2.5)^{2x}$

Use the laws of exponents to complete these equations.

9. $\left(\frac{1}{4}\right)^x = ?$
10. $3^0 = ?$
11. $2^{-x} = ?$
12. $\left(\frac{1}{8}\right)^{-x} = ?$

13. Find the amount to which $500 will accumulate in 5.00 years at an annual interest rate of 7.25%, if the interest is compounded (a) annually, (b) monthly, (c) daily, (d) continuously. Round to the nearest cent.

14. A certain virus will increase according to the equation

$$N = N_0 2^{0.2t}$$

where N_0 is the original number of viruses and N is the number after time t (in hours). How many viruses will there be after 12.0 hours if there were 1250 to begin with?

15. The demand y for a certain mineral grows exponentially at the rate of 15.7% per year.

 (a) If the demand was for 12.3 tonnes in 1997, what will the demand be in 2007?

 (b) Graph the demand as a function of time t for 0 to 10 years.

16. The atmospheric pressure P (in kPa) at an altitude h (in kilometres) is given approximately by

$$P = 101e^{-0.143h}$$

assuming constant air temperature. Find the atmospheric pressure at the summit of British Columbia's Mount Robson, which is approximately 3.90 km high.

17. A savings bank is offering an annual interest rate of 7.5%, compounded continuously. How much do you need to invest to have an accumulated amount of $3000 at the end of 4 years? Round to the nearest cent.

18. Radioactive carbon-14 (C-14) is present at a constant level in all living things, but decays exponentially after death at the rate of $r = 1.24 \times 10^{-4}$ per year. Scientists use the formula

$$m = m_0 e^{-rt}$$

where m_0 is the original amount of C-14 in the material and m is the amount of C-14 present after t years, to determine the age of organic material.

(a) What percent of the original C-14 (m_0) would remain in the body of a person who died 100 years ago?

(b) Graph the percent of m as a function of time t, for 0 to 10 000 years.

19. A certain yeast will triple every 6.00 hours.

(a) If you start with 15.0 pounds, how many pounds of yeast will there be after 18.0 hours?

(b) How many pounds do you need to start with to have 1500 pounds after 1.00 day?

WRITING

20. You are applying for a job as the business manager of a new biotech company. On the application you are asked how exponential functions are used in business and in biotechnology. Write a response.

TEAM PROJECT

21. A kettle of boiling water is taken off the stove, and the temperature of the water is measured every minute, as the water cools from 100.0°C at $t = 0$. The room temperature is 21.0°C. Data is recorded as follows:

| 100.0 | 91.4 | 86.9 | 79.8 | 71.8 | 68.5 | 63.6 | 60.8 | 54.4 | 53.1 | 48.9 |
| 46.8 | 42.8 | 40.3 | 39.5 | 39.2 | 38.8 | 35.5 | 33.9 | 32.8 | 31.7 | 30.6 |

Make a graph of temperature versus time. Plot the data points and write an exponential equation for the temperature as a function of time, with all constants evaluated.

Logarithms

Outcomes

When you have completed this chapter, you will be able to

1. Define logarithms and logarithmic functions.
2. Convert between exponential and logarithmic forms of equations.
3. Solve logarithmic equations.
4. Apply common and natural logarithms.
5. Use the properties of logarithms to solve equations.
6. Solve exponential equations using logarithms.

In this chapter we define a logarithm, evaluate common and natural logarithms, graph the logarithmic function, learn the properties of logarithms, and solve exponential and logarithmic equations.

Logarithms have many applications, including the measurement of sound intensity in decibels, pH levels in chemistry, earthquake intensity by the Richter scale, and applications in electronics and mechanics.

28-1 INTRODUCTION TO LOGARITHMS

We introduce logarithms in this section. We look at the exponential and logarithmic forms of an equation and solve some simple logarithmic equations.

Recall from Sec. 27-2 the 1500 bacteria that double their number every hour. How long would it take to have 900 000? The equation for the number of bacteria after t hours was

$$n = 1500 \cdot 2^t$$

So, letting $n = 900\,000$,

$$900\,000 = 1500 \cdot 2^t$$

Dividing both sides by 1500, we have

$$600 = 2^t$$

We need to solve this equation. Try it. What is the value of t?

We *can't* solve it with the operations of adding, subtracting, multiplying, or dividing both sides of the equation. We need a new operation. The base, 2, is being raised to the tth power. We need a new operation—we need to be able to "find the exponent"; that is,

$$\text{if } 2^t = 600, \quad \text{then } t = ?$$

We call the exponent t a *logarithm* and the new operation is called "taking the logarithm." We use the symbol *log* for logarithm, and we write

$$\text{if } 2^t = 600, \quad \text{then } t = \log_2 600$$

This is read "t is the logarithm of 600 to the base 2."

Definition of a Logarithm

A logarithm is defined from the exponential function.

$$\text{If } y = b^x \quad \text{then } x = \log_b y$$

where x is called the *logarithm* of y to the base b.

The two equations above are equivalent equations. That is, they are two different forms of the same equation.

EXPONENTIAL AND LOGARITHMIC FORMS	$y = b^x$ is called the *exponential form* $x = \log_b y$ is called the *logarithmic form*	124

So, we see that a logarithm is really an *exponent*. The logarithm $\log_b y$ is the exponent (the power) that you raise the base b to, to get y.

Example 1

Consider the equation $4^3 = 64$, in which 4 is the base and 3 the exponent. The exponent 3 is a logarithm, and we can write

$$3 = \log_4 64$$

We read this equation "3 is the logarithm of 64 to the base 4."

In working with logarithmic and exponential expressions, we sometimes need to change an equation from exponential form to logarithmic form, or vice versa.

Example 2

Change $2^{-3} = \frac{1}{8}$ to logarithmic form.

Solution: Using Eq. 124, we get

$$-3 = \log_2 \left(\frac{1}{8}\right)$$

Example 3

Change $7^{1.50} = 18.5$ to logarithmic form.

Solution: Using Eq. 124, we get

$$1.50 = \log_7 18.5$$

Example 4
Change $\log_5 125 = 3$ to exponential form.

Solution: Using Eq. 124, we get

$$5^3 = 125$$

The constant e is often used in logarithmic and exponential equations. Recall from Sec. 27-3, that e is an irrational number, approximately equal to 2.718.

Example 5
Find x, rounded to three significant digits, and write the equation in logarithmic form.

$$x = e^{1.62}$$

Solution: Using the calculator with the $\boxed{e^x}$ key, we find that

$$x = e^{1.62} = 5.05$$

In logarithmic form, $e^{1.62} = 5.05$ becomes

$$1.62 = \log_e 5.05$$

Logarithmic Equation

A *logarithmic equation* is an equation that contains a logarithm of a variable.

Example 6
The equation

$$\log_{10} x = 3$$

is a logarithmic equation, but the equation

$$x \log_{10} 2 = 5$$

would not usually be called a logarithmic equation, since it does not contain a logarithm of a variable.

Solution of Simple Logarithmic Equations

We know that the two equations

$$y = b^x \quad \text{and} \quad x = \log_b y$$

are equivalent equations; that is, they represent the same relationship between x and y. To solve a simple logarithmic equation, write the equation in logarithmic form, change it to the exponential form, and solve.

Example 7
Solve for y: $\log_7 y = 2$.

Solution: Changing to exponential form, we have

$$7^2 = y$$

or

$$y = 49$$

Example 8

Solve for b: $\log_b 8 = 3$.

Solution: Changing to exponential form, we have
$$b^3 = 8$$

To solve for b, we take the cube root of both sides,
$$\sqrt[3]{b^3} = \sqrt[3]{8}$$
$$b = 2$$

Example 9

Solve for x: $\log_2 32 = x$.

Solution: Changing to exponential form, we have
$$2^x = 32$$

Now we express 32 as a power of 2,
$$32 = 2^5$$

So,
$$2^x = 2^5$$

This can only be true if
$$x = 5$$

Therefore,
$$\log_2 32 = 5$$

SECTION 28-1 Exercises Introduction to Logarithms

Change to logarithmic form.

1. $2^3 = 8$
2. $4^2 = 16$
3. $625 = 5^4$
4. $16 = 2^4$
5. $10^6 = 1\,000\,000$
6. $e^2 = 7.4$
7. $\dfrac{1}{81} = \left(\dfrac{1}{3}\right)^4$
8. $\dfrac{1}{9} = 3^{-2}$
9. $1 = 78^0$
10. $2 = 8^{1/3}$
11. $6^{1.27} = 9.73$
12. $10^{2.48} = 200$

Without using a calculator, find the indicated power and write the equation in logarithmic form.

13. $4^3 = ?$
14. $2^5 = ?$
15. $10^5 = ?$
16. $10^{-3} = ?$
17. $9^0 = ?$
18. $\left(\dfrac{1}{4}\right)^2 = ?$
19. $3^{-3} = ?$
20. $16^{1/4} = ?$

Use a calculator to find the indicated power, rounded to three significant digits. Then write the equation in logarithmic form.

21. $4^{2.37} = ?$
22. $1.5^{5.9} = ?$
23. $2.4^{-3.29} = ?$
24. $e^{3.21} = ?$

Express each of the following in logarithmic form.

25. $5^x = 120$
26. $y^8 = 65$
27. $a = 9^5$
28. $10 = 7^p$
29. $4^{-3} = q$
30. $10^x = 4.5$
31. $e^x = 45$
32. $a^5 = b$

Change to exponential form.

33. $\log_7 343 = 3$
34. $\log_2 64 = 6$
35. $3 = \log_{10} 1000$
36. $\log_e 5 = 1.609$
37. $\log_9 3 = \dfrac{1}{2}$
38. $\log_5 1 = 0$
39. $1 = \log_7 7$
40. $-3 = \log_{1/2} 8$
41. $\log_5 \dfrac{1}{25} = -2$
42. $\log_x 7 = 34$
43. $6.59 = \log_4 y$
44. $a = \log_7 9$
45. $\log_b x = y$
46. $\log_c xy = p$
47. $\log_e 5 = q$
48. $\log_{10} 0.01 = -2$

Solution of Simple Logarithmic Equations

Change to exponential form and solve for x. (Do *not* use your calculator.)

49. $x = \log_3 27$
50. $x = \log_2 16$
51. $\log_5 125 = x$
52. $\log_{10} 1000 = x$
53. $x = \log_2 1$
54. $\log_6 1 = x$
55. $\log_e e^4 = x$
56. $x = \log_e 1$
57. $\log_x 64 = 3$
58. $\log_x 81 = 4$
59. $\log_3 x = 0$
60. $\log_x 4 = \dfrac{1}{2}$
61. $2 = \log_x 100$
62. $1 = \log_x e$
63. $\log_5 x = 3$
64. $\log_{10} x = 4$
65. $-3 = \log_6 x$
66. $-4 = \log_{1/2} x$
67. $\dfrac{2}{3} = \log_8 x$
68. $-\dfrac{2}{3} = \log_8 x$
69. $\log_x 3 = \dfrac{1}{3}$
70. $\log_4 x = -2$
71. $\log_x e^2 = 2$
72. $-3 = \log_x 0.001$

28-2 COMMON AND NATURAL LOGARITHMS

In the last section we worked with logarithms to many different bases: b, 2, e, 3, 10, and so on. In practice however, we usually use only two bases, the base 10 and the base e. In this section we examine common and natural logarithms, and use a calculator to find values of logarithms.

Common Logarithms

A logarithm to the base 10 is called a *common* logarithm. If a logarithm is written without naming the base, then it is understood to be a common logarithm—that is, the base is 10.

$$\log N = \log_{10} N$$

Example 10

$$\log 12.5 = \log_{10} 12.5$$

Exponential and Logarithmic Forms

If x is a common logarithm, we have

$$\log N = x \quad \text{the } \textit{logarithmic form}$$
$$10^x = N \quad \text{the } \textit{exponential form}$$

Example 11

Change $10^3 = 1000$ to logarithmic form.

Solution: The logarithmic form is

$$\log 1000 = 3$$

Example 12

Complete the equation and then write in logarithmic form:

$$10^{-2} = ?$$

Solution: Completing the equation (without using a calculator),

$$10^{-2} = \frac{1}{10^2} = \frac{1}{100}$$

Or, as a decimal,

$$10^{-2} = 0.01$$

The logarithmic form is

$$\log 0.01 = -2$$

Example 13

Write the exponential form of the equation

$$\log 1.74 = 0.241$$

Solution: The exponential form is

$$10^{0.241} = 1.74$$

Common Logarithms by Calculator

Common logarithms may be found using the $\boxed{\log}$ key on a calculator. Express the result using the same number of significant digits as in the given number. From now on, the term *logarithm* will refer to the *common* logarithm.

Example 14
Find log 25.9 by calculator. Then write the equation in exponential form.

Solution: Using the $\boxed{\log}$ key, we get

$$\log 25.9 = 1.41$$

The exponential form is

$$10^{1.41} = 25.9$$

To find a number whose logarithm is given, use the $\boxed{10^x}$ key on your calculator. (On some calculators, the 10^x may be called the "inverse log" key.)

Example 15
Find the number whose logarithm is 1.37.

Solution: Let x be the number whose logarithm is 1.37. Then

$$\log x = 1.37$$

Changing to exponential form, we have

$$10^{1.37} = x$$

Using the $\boxed{10^x}$ key to evaluate $10^{1.37}$, we get

$$10^{1.37} = 23.4$$

So, the number whose logarithm is 1.37 is about 23.4.

Example 16
Solve $10^x = 7.34$.

Solution: Changing from exponential form to logarithmic form, we get

$$x = \log 7.34$$
$$x = 0.866$$

Example 17
Solve $1.45(10^x) = 38.2$.

Solution: We divide both sides of the equation by 1.45 to write the equation in exponential form,

$$\frac{1.45(10^x)}{1.45} = \frac{38.2}{1.45}$$
$$10^x = 26.34$$

Changing to logarithmic form,

$$x = \log 26.34$$
$$x = 1.42$$

Natural Logarithms

A *natural* logarithm is a logarithm to the base e. Recall that we discussed the irrational number e in Sec. 27-3, and found that it is approximately equal to 2.718, rounded to four significant digits. The natural logarithm of N is written $\ln N$

$$\ln N = \log_e N$$

If x is a natural logarithm, we have

$$\ln N = x \quad \text{the } \textit{logarithmic form}$$
$$e^x = N \quad \text{the } \textit{exponential form}$$

Example 18

Change $e^2 = 7.39$ to logarithmic form.

Solution: The logarithmic form is

$$\ln 7.39 = 2$$

Natural Logarithms by Calculator

To find the natural logarithm of a number by calculator, use the $\boxed{\ln}$ key.

Example 19

Find $\ln 78.5$ and write the exponential form.

Solution: Use the $\boxed{\ln}$ key to get

$$\ln 78.5 = 4.36$$

Changing to exponential form, we have

$$e^{4.36} = 78.5$$

To find a number whose natural logarithm is given, use the $\boxed{e^x}$ key on a calculator. (On some calculators, the e^x key may be called the "inverse ln" key.)

Example 20

Use a calculator to find the number whose natural logarithm is 5.43.

Solution: Let x be the number whose natural logarithm is 5.43. Then we need to solve the equation

$$\ln x = 5.43$$

Changing to exponential form, we have

$$e^{5.43} = x$$

Use the $\boxed{e^x}$ key to find

$$e^{5.43} = 228$$

So the number whose natural logarithm is 5.43 is about 228.

Graph of the Logarithmic Function

The function $y = \log_b x$ is called the *logarithmic function*. The base b is a positive number other than 1.

FIGURE 28-1
Common logarithmic function.

Example 21

Make a table of values for the common logarithmic function

$$y = \log x$$

for x from $\frac{1}{10}$ to 10. Graph $y = \log x$ for x from $\frac{1}{10}$ to 10.

Solution: We make a table of values for x from $\frac{1}{10}$ to 10, and use a calculator to determine the y values, rounding to tenths.

x	$\frac{1}{10}$	$\frac{1}{5}$	$\frac{1}{2}$	1	2	5	10
y	-1	-0.7	-0.3	0	0.3	0.7	1

We plot these points and connect them with a smooth curve in Fig. 28-1. We see that the graph of $y = \log x$ is asymptotic to the y axis and that the x intercept is 1. The equation is defined for only positive values of x. The curve is not asymptotic to any horizontal line, and indeed, the y values increase without bound. This will be true of all graphs of the equation $y = \log_b x$, where $b > 0$ and $b \neq 1$. The domain of logarithmic function $y = \log_b x$ is all positive numbers (x), and the range is all real numbers (y).

FIGURE 28-2
Natural logarithmic function.

Example 22

Use a graphing calculator to graph the natural logarithmic function

$$y = \ln x$$

for x from 0.1 to 10.

Solution: To determine the range of the function, we find its smallest and its largest values. The logarithmic function is an increasing function, so its smallest value (in the given domain) is

$$y = \ln 0.1 = -2.30$$

and its largest value is

$$y = \ln 10 = 2.30$$

We set the window so that $0 \leq x \leq 10$ and $-3 \leq y \leq 3$. We enter the function $y = \ln x$ and graph (Fig. 28-2). Again we see that the graph of $y = \ln x$ is asymptotic to the y axis and that the x intercept is 1. The equation is defined for only positive values of x. The curve is not asymptotic to any horizontal line, and indeed, the y values increase without bound.

SECTION 28-2 Exercises Common and Natural Logarithms

Common Logarithms

Write the logarithmic form of the equation.

1. $10^2 = 100$
2. $10^4 = 10\,000$
3. $10^{-3} = 0.001$
4. $10^{-1} = 0.1$

Without using your calculator, complete the equation, and then write the logarithmic form.

5. $10^5 = ?$
6. $10^0 = ?$
7. $10^{-4} = ?$
8. $10^1 = ?$

Write the equation in exponential form.

9. $\log 15.5 = 1.19$ **10.** $\log 155 = 2.19$
11. $\log 1.55 = 0.19$ **12.** $\log 1550 = 3.19$

Common Logarithms by Calculator

Use a calculator to find the value of the following logarithms, rounding to three significant digits. Then write the equation in exponential form.

13. $\log 5.37$ **14.** $\log 34.5$ **15.** $\log 74.5$
16. $\log 6.97$ **17.** $\log 0.456$ **18.** $\log 0.624$

Write the equation in exponential form and evaluate x to three significant digits.

19. $\log x = 3$ **20.** $\log x = 5$ **21.** $\log x = 1.56$
22. $\log x = 2.48$ **23.** $\log x = -1.62$ **24.** $\log x = -0.521$

Natural Logarithms

Write the logarithmic form of the equation.

25. $e^3 = 20.1$ **26.** $e^{-1} = 0.368$

Complete the equation, and write the logarithmic form.

27. $e^{2.42} = ?$ **28.** $e^{1.45} = ?$ **29.** $e^0 = ?$ **30.** $e^{-2} = ?$

Find the value of the natural logarithm and then write the equation in exponential form.

31. $\ln 56.7 = ?$ **32.** $\ln 1.05 = ?$ **33.** $\ln 0.0238 = ?$ **34.** $\ln e = ?$

Write the equation in exponential form and evaluate x to three significant digits.

35. $\ln x = 3.54$ **36.** $\ln x = 8.64$
37. $\ln x = -0.00950$ **38.** $\ln x = -2.75$

Graph of the Logarithmic Function

39. Complete this table of values for the logarithmic function $y = \ln x$, rounding to tenths.

x	$\frac{1}{10}$	$\frac{1}{5}$	$\frac{1}{2}$	1	2	5	10
y							

Graph the function $y = \ln x$ by plotting these points and connecting them with a smooth curve.

40. Use a graphing calculator to graph the function $y = \log x$.

28-3 PROPERTIES OF LOGARITHMS

In this section we study the properties of logarithms, including the logarithm of a product, quotient, power, and root. We will use these properties in the next section to solve both logarithmic and exponential equations using logarithms.

Before discussing any specific properties of logarithms, we need to consider an important fact concerning powers of numbers: *Any positive number can be expressed as a power of any other positive number, other than 1.*

Example 23

We know that 8 can be expressed as a power of 2. Indeed,

$$8 = 2^3 \quad \text{and in logarithmic form,} \quad \log_2 8 = 3$$

But 8 can also be written as a power of any other positive number, other than 1:

$$8 = 10^{0.9031} \quad \text{and in logarithmic form,} \quad \log 8 = 0.9031$$

and

$$8 = 7^{1.0686} \quad \text{and in logarithmic form,} \quad \log_7 8 = 1.0686$$

In general, if M is any positive number, then M can be expressed as a power of a base b, where $b \neq 1$:

$$M = b^c \quad \text{and in logarithmic form,} \quad \log_b M = c$$

Similarly, if N is a positive number, it also can be expressed as

$$N = b^d \quad \text{and in logarithmic form,} \quad \log_b N = d$$

We use these equations to derive the property for the logarithm of a product.

Logarithm of a Product

We consider the product of two positive numbers M and N. Substituting $M = b^c$ and $N = b^d$ in the product MN, we have

$$MN = b^c \cdot b^d$$

By the laws of exponents,

$$MN = b^{c+d}$$

Changing to logarithmic form,

$$\log_b MN = c + d$$

Substituting $c = \log_b M$ and $d = \log_b N$, we have

LOGARITHM OF A PRODUCT	$\log_b MN = \log_b M + \log_b N$	125

In other words, *the logarithm of a product equals the sum of the logarithms of the factors.* This is reasonable, since logarithms are *exponents,* and when two quantities with the same base are multiplied, we add the exponents!

Example 24

Using Eq. 125,

$$\log_b 2x = \log_b 2 + \log_b x$$

Example 25

Express $\log_4 3 + \log_4 y$ as a single logarithm.

Solution:

$$\log_4 3 + \log_4 y = \log_4 3y$$

COMMON ERRORS The logarithm of a product is *not* equal to the product of the logarithms.

$$\log_b MN \neq (\log_b M)(\log_b N)$$

The logarithm of a sum is *not* equal to the sum of the logarithms.

$$\log_b (M + N) \neq \log_b M + \log_b N$$

The product of two logs is *not* equal to the sum of the logarithms.

$$(\log_b M)(\log_b N) \neq \log_b M + \log_b N$$

Logarithm of a Quotient

Similarly, the property for the logarithm of the quotient of two numbers is

LOGARITHM OF A QUOTIENT	$\log_b \dfrac{M}{N} = \log_b M - \log_b N$	126

In other words, *the logarithm of a quotient equals the logarithm of the numerator minus the logarithm of the denominator*. This is reasonable, since logarithms are exponents, and when two quantities with the same base are divided, we subtract the exponents!

---- **Example 26** ----

Using Eq. 126,

$$\log_5 \frac{x}{3} = \log_5 x - \log_5 3$$

---- **Example 27** ----

Using Eq. 126,

$$\log \frac{2}{3x} = \log 2 - \log 3x$$
$$= \log 2 - (\log 3 + \log x)$$
$$= \log 2 - \log 3 - \log x$$

Similar errors are made with quotients as with products.

COMMON ERRORS

$$\log_b \frac{M}{N} \neq \frac{\log_b M}{\log_b N}$$

$$\log_b (M - N) \neq \log_b M - \log_b N$$

$$\frac{\log_b M}{\log_b N} \neq \log_b M - \log_b N$$

Logarithm of a Power

Now we write an expression for the logarithm of a power, say M^p, where M is a positive number. Again we let $M = b^c$, so that $c = \log_b M$. Substituting $M = b^c$, we get

$$M^p = (b^c)^p$$

By the laws of exponents,

$$M^p = b^{cp}$$

Changing to logarithmic form,

$$\log_b M^p = cp$$
$$\log_b M^p = pc$$

Substituting $c = \log_b M$, we get

| LOGARITHM OF A POWER | $\log_b M^p = p \cdot \log_b M$ | 127 |

In other words, *the logarithm of a power of a number equals the exponent times the logarithm of the number.*

Example 28
Using Eq. 127,
$$\log_b x^2 = 2 \cdot \log_b x$$

Example 29
Express as a single logarithm, with a coefficient of 1.
$$3 \ln 2 + 2 \ln x - \ln 3$$

Solution:
$$3 \ln 2 + 2 \ln x - \ln 3 = \ln 2^3 + \ln x^2 - \ln 3$$
$$= \ln\left(\frac{2^3 \cdot x^2}{3}\right)$$
$$= \ln\left(\frac{8x^2}{3}\right)$$

Logarithm of a Root

Similarly, the property for the logarithm of the root of a number is

| LOGARITHM OF A ROOT | $\log_b \sqrt[r]{M} = \frac{1}{r} \log_b M$ | 128 |

In other words, *the logarithm of the rth root of a number equals $\frac{1}{r}$ times the logarithm of the number.*

Example 30
Using Eq. 128,
$$\log_2 \sqrt[3]{x} = \frac{1}{3} \log_2 x$$

Example 31
Using Eq. 128,
$$\log_b \sqrt{y} = \log_b \sqrt[2]{y} = \frac{1}{2} \log_b y$$

---- **Example 32** ----

Using Eq. 128,

$$\log_3 \sqrt{x^5} = \frac{1}{2}\log_3 x^5 = \frac{5}{2}\log_3 x$$

Let's consider some special logarithms now.

Logarithm of 1

First we look at $\log_b 1$. We know that

$$b^0 = 1$$

Changing $b^0 = 1$ to logarithmic form, we get

$$\log_b 1 = 0$$

| LOGARITHM OF 1 | $\log_b 1 = 0$ | 129 |

In other words, *the logarithm of 1 to any base is always 0.*

Logarithm of the Base

Now we look at $\log_b b$. We know that

$$b^1 = b$$

Changing $b^1 = b$ to logarithmic form, we get

$$\log_b b = 1$$

| LOGARITHM OF THE BASE | $\log_b b = 1$ | 130 |

Logarithm of the Base Raised to a Power

The logarithm of a base b raised to a power n is $\log_b b^n$. By Eq. 130,

$$\log_b b^n = n \cdot \log_b b$$

Substituting $\log_b b = 1$, we get

$$\log_b b^n = n \cdot 1$$

| LOGARITHM OF THE BASE RAISED TO A POWER | $\log_b b^n = n$ | 131 |

---- **Example 33** ----

By Eq. 131, $\log_2 2^3 = 3$.

The properties in Eqs. 130 and 131 are especially important for common and natural logarithms:

$$\log 10 = 1 \qquad \log 10^n = n$$
$$\ln e = 1 \qquad \ln e^n = n$$

Example 34
By Eq. 131, $\ln e^3 = 3$.

Example 35
By Eq. 131, $\log 0.01 = \log(10^{-2}) = -2$.

SECTION 28-3 Exercises Properties of Logarithms

Logarithms of Products and Quotients

Use the properties of logarithms to expand the expression.

1. $\log_3 7x$
2. $\ln 3y$
3. $\log_5 \dfrac{x}{4}$
4. $\log_2 \dfrac{5}{y}$
5. $\ln 3xy$
6. $\log abc$
7. $\log_2 \dfrac{5x}{3}$
8. $\log_3 \dfrac{xy}{7}$
9. $\log_b \dfrac{3}{4a}$
10. $\log_b \dfrac{2}{5x}$

Express as a single logarithm.

11. $\log_4 5 + \log_4 a$
12. $\log_7 8 - \log_7 c$
13. $\log_b 2 - \log_b y$
14. $\log_3 z + \log_3 4$
15. $\log_2 x + \log_2 y + \log_2 z$
16. $\log_b x - \log_b y + \log_b 3$
17. $\log_3 a - \log_3 5 - \log_3 c$
18. $\log_4 3 + \log_4 a - \log_4 7$

Logarithms of Powers and Roots

Use the properties of logarithms to expand the expression.

19. $\log_b x^3$
20. $\log a^x$
21. $\log_3 2^x$
22. $\ln x^{1.76}$
23. $\log_b \sqrt{x}$
24. $\ln \sqrt[5]{2}$
25. $\log_2 \sqrt[3]{x}$
26. $\log \sqrt[4]{x}$
27. $\log_b x^2 y^3$
28. $\log_2 5\sqrt{3}$

Express as a single logarithm with a coefficient of 1.

29. $5 \log_2 x$
30. $x \log_2 5$
31. $3 \ln x + 4 \ln y$
32. $2 \log a - 3 \log c - 4 \log d$
33. $\dfrac{1}{2} \log_b x$
34. $\dfrac{1}{3} \log_2 7$
35. $\dfrac{\ln 2}{4}$
36. $\dfrac{\log 6}{3}$
37. $\dfrac{1}{2} \log_2 x + \dfrac{1}{3} \log_2 y$
38. $\dfrac{1}{3} \ln 4 - 2 \ln a$
39. $\dfrac{2}{3} \log_5 x$
40. $\dfrac{3}{4} \log_2 y$

Simplify.

41. $\log_2 1$
42. $\log_7 1$
43. $\log_3 3$
44. $\log_b b$
45. $\log_5 5^9$
46. $\log_2 2^7$
47. $\ln e^{5.3}$
48. $\log_b b^2$
49. $\ln \sqrt[5]{e}$
50. $\log_b \sqrt[3]{b^4}$

28-4 SOLUTION OF EXPONENTIAL EQUATIONS USING LOGARITHMS

In this section we use logarithms to solve exponential equations. We solve applications, including the growth of bacteria, compound interest, population growth, and radioactive decay.

To solve an exponential equation we generally just *take the logarithm of both sides of the equation*. We can use *either* common logarithms *or* natural logarithms. If the exponential equation uses the base e, then we take the natural logarithm of both sides. If the exponential equation uses the base 10, then we take the common logarithm of both sides. If it uses a base other than 10 or e, we may use either natural or common logarithms.

Example 36

Solve to three significant digits:

$$10^x = 7.34$$

Solution: In this equation, the base is 10, so take the common log of both sides:

$$\log 10^x = \log 7.34$$

Using Eq. 131,

$$x = \log 7.34$$

Evaluating log 7.34,

$$x = 0.866$$

Example 37

Solve $75.3 = 91.2e^{-0.417t}$ to three significant digits.

Solution: First we divide both sides by the constant 91.2:

$$\frac{75.3}{91.2} = \frac{91.2e^{-0.417t}}{91.2}$$

Carrying one extra digit, we have

$$0.8257 = e^{-0.417t}$$

Taking the natural log of both sides,

$$\ln 0.8257 = \ln e^{-0.417t}$$

By Eq. 131

$$\ln 0.8257 = -0.417t$$

Dividing both sides by -0.417,

$$t = \frac{\ln 0.8257}{-0.417}$$

$$t = 0.459$$

Check: Substitute $t = 0.459$ into the original equation,

$$75.3 \stackrel{?}{=} 91.2e^{-0.417(0.459)}$$

Evaluate the right side by calculator,

$$91.2e^{-0.417(0.459)} = 75.3 \quad \text{checks}$$

Example 38

Solve $588 = 2^t$ to three significant digits.

Solution: In this equation the base is 2 (neither 10 nor e), so we may use either common or natural logarithms. Let us take the common logarithm of both sides:

$$\log 588 = \log 2^t$$

By Eq. 127,

$$\log 588 = t \cdot \log 2$$

Dividing both sides by log 2,

$$t = \frac{\log 588}{\log 2}$$

Evaluating the right side,

$$t = 9.20$$

Check: We substitute $t = 9.20$ into the right side of the equation and evaluate:

$$588 \stackrel{?}{=} 2^{9.20}$$
$$588 = 588 \quad \text{checks}$$

Applications

In many applications we find it necessary to solve exponential equations. A number of these applications were introduced in Chapter 27 on exponential functions. These include biological growth, compound interest, and exponential growth and decay.

Example 39

How long would it take for *any* amount of money, compounded continuously at 5.75% per year, to triple?

Solution: Use Eq. 142 for continuous compounding,

$$y = ae^{rt}$$

where a is the original amount invested.
 Since the money is being tripled, the amount a accumulates to $3a$. So,

$$y = 3a \quad \text{and} \quad r = 5.75\% = 0.0575$$

Substituting, we get

$$3a = ae^{0.0575t}$$

Dividing both sides by a,

$$3 = e^{0.0575t}$$

Taking the natural log of both sides,

$$\ln 3 = \ln e^{0.0575t}$$

By Eq. 131,

$$\ln 3 = 0.0575t$$

Dividing by 0.0575,

$$t = \frac{\ln 3}{0.0575}$$

$$t = 19.1$$

So, it would take about 19.1 years for any amount of money to triple, at 5.75% per year, compounded continuously.

Example 40

Radium D decays exponentially at the rate of 3.22% per year. Find the time necessary for 5.00 kg of Radium D to disintegrate to 4.00 kg.

Solution: Since radioactive decay is an example of exponential decay, we use Eq. 143:

$$y = ae^{-nt}$$

Substituting,

$$4.00 = 5.00e^{-0.0322t}$$

Solving,

$$0.800 = e^{-0.0322t}$$

$$\ln 0.800 = \ln e^{-0.0322t}$$

$$\ln 0.800 = -0.0322t$$

$$t = \frac{\ln 0.800}{-0.0322}$$

$$t = 6.93$$

So, 5.00 kg of Radium D will disintegrate to 4.00 kg in about 6.93 years.

SECTION 28-4 Exercises Solution of Exponential Equations Using Logarithms

Solve each exponential equation, and round to three significant digits. Check each solution.

1. $10^x = 25.4$
2. $8.97 = 10^x$
3. $e^x = 0.789$
4. $33.8 = e^x$
5. $4.89 = 1.06^x$
6. $1.76^x = 525$
7. $0.476 = 10^{-x}$
8. $10^{-x} = 87.9$
9. $e^{-x} = 1.25$
10. $0.376 = e^{-x}$
11. $1.29 = 2.30^{-x}$
12. $5.45^{-x} = 4.98$
13. $6.45 = 4.64(10^x)$
14. $5.25 \cdot 10^x = 9.69$
15. $2e^x = 3$
16. $7 = 3e^x$
17. $12 = 24e^{-x}$
18. $15e^{-x} = 5$
19. $700 = 400(1.01)^t$
20. $225(1.025)^{3t} = 500$
21. $6.79 = 10^{2x}$
22. $1200 = 300e^{0.04x}$
23. $8.5e^{55t} = 145$
24. $0.023 = 0.01e^{5x}$
25. $7.6 = 9.2e^{-75x}$
26. $3 = 6.25e^{-150x}$
27. $6 = 18e^{-0.4t}$
28. $25 = 100e^{-0.075t}$

Solve for *t*.

29. $N = 100 \cdot 2^t$ **30.** $S = 500(1.12^t)$
31. $1000 = P \cdot e^{0.12t}$ **32.** $m = m_0 e^{-kt}$

Applications

Solve. If answers are not exact, round to three significant digits.

33. A certain bacterium triples its number every hour. If we begin with 1000 bacteria, how long would it take to have a million?

34. A yeast manufacturer finds that the yeast will quadruple every hour. If the manufacturer begins with 5 kilograms of yeast, how long will it take to grow 100 kilograms of yeast?

35. How long will it take for $100 to grow to $450, if it is compounded continuously at 11.5% per year?

36. How long would it take for an amount of money to quadruple, if it were compounded continuously at 12.5% per year?

37. The population of a city is growing exponentially at the rate of 1.85% per year. How many years will it take for the population to double?

38. If the rate of inflation is 3.45% per year, how long will it take for a $10.00 bag of groceries to cost $15.00? Assume the cost of goods rises exponentially.

39. A certain radioactive isotope of Uranium, Uranium Z, decays at the rate of 10.3% per hour. Find the time necessary for 75.0 grams of Uranium Z to decay to 30.0 grams.

40. A radioactive isotope called Radium A decays at the rate of 22.7% per minute. Find the time necessary for 20.0 kg of Radium A to decay to 10.0 kg.

41. The current *i* in a certain circuit is given by

$$i = 4.75e^{-125t} \text{ amperes}$$

where *t* is the time in seconds. At what time will the current be 2.00 amperes?

42. The discharge current *i* of a capacitor is given by

$$i = 0.0338e^{-50.2t}$$

where *t* is the time in seconds. At what time will the discharge current be 0.00500 amperes?

43. A cup of coffee will cool exponentially in a certain freezer at the rate of 5.06% per minute. How long will it take for a cup of coffee at 99.5°C to cool to 37.0°C?

44. A flywheel is rotating at 2450 rev/min. When the power is disconnected, the angular velocity decreases exponentially at the rate of 31.4% per minute. How long would it take the flywheel to slow to an angular velocity of 1.00 rev/min (virtually stop)?

45. In 1996, the world population was 5.8 billion people and expected to double in 45 years.

  (a) Assuming that the growth is exponential, find the rate of growth (as a percent).

  (b) In what year would we expect there to be only 1 square metre of land per person? (The earth contains approximately 1.40×10^{14} square metres of land.)

46. If a leukemic cell is injected into a healthy mouse, that cell will divide into two cells in half a day. The mouse will die when the number of leukemic cells reaches one billion.

  (a) Write an equation for the number of leukemic cells *n* present after *t* days.

  (b) After how many days will the mouse die?

47. When an object at 0°C is placed in surroundings that are at a higher temperature c, its temperature T will increase exponentially according to the equation

$$T = c(1 - e^{-rt})$$

where r is the rate of increase and t is the time. A steel casting initially at 0°C is placed in a furnace at 1000°C. If it increases in temperature at the rate of 5.16% per minute, how long will it take to reach 700°C? Round to three significant digits.

48. Solve for t:

$$T = c(1 - e^{-rt})$$

28-5 APPLICATIONS OF LOGARITHMIC EQUATIONS

In this section we solve applications with logarithmic equations. Logarithmic equations are used to define many important quantities such as the loudness of sound, the pH value of a solution, and the magnitude of an earthquake.

Example 41

The loudness of sound β (in decibels) is given by the formula

$$\beta = 10 \log \frac{I}{I_0}$$

where I is the intensity of the sound and I_0 is the faintest sound detectable, about 1×10^{-12} W/m². Normal hearing ranges from 0 to 120 decibels (dB).

The intensity of sound for street traffic is about 3.25×10^{-5} W/m². How loud is the street traffic in decibels?

Solution: Substituting $I = 3.25 \times 10^{-5}$ and $I_0 = 10^{-12}$ into the given equation:

$$\beta = 10 \log \frac{3.25 \times 10^{-5}}{10^{-12}}$$

$$\beta = 10 \log (3.25 \times 10^{7})$$

$$\beta = 75.1 \text{ dB}$$

Example 42

The pH value of a solution having a concentration C of hydrogen ions (in moles per litre) is

$$pH = - \log C$$

Acids have a pH value less than 7.00 and bases have a pH value greater than 7.00.

Vinegar has a concentration of 0.006 moles per litre. Find its pH value to three significant digits.

Solution: Using the given equation for the pH value, we have

$$pH = - \log 0.006$$

$$pH = 2.22$$

Example 43

The difference in elevation h (in metres) between two locations having barometer readings of B_1 and B_2 kilopascals is given by the equation

$$h = 18\,421 \log\left(\frac{B_2}{B_1}\right)$$

Find the difference in elevation between two stations having barometer readings of 96.2 kPa at the lower station and 87.4 kPa at the upper station, where B_1 is the pressure at the upper station.

Solution: Using the equation above for the difference in elevation, we have

$$h = 18\,421 \log\left(\frac{96.2}{87.4}\right)$$

$$h = 767 \text{ m} \quad \text{(Rounding to three significant digits)}$$

SECTION 28-5 Exercises Applications of Logarithmic Equations

Solve the applications, and round the answer to three significant digits.

1. The intensity of sound for a whisper is about 10^{-10} W/m². How many decibels is a whisper?

2. The intensity of sound for normal conversation is about 10^{-6} W/m². What is the loudness in decibels?

3. The loudness of sound for the threshold of pain is 120 dB. What is the intensity of the sound?

4. The loudness of sound for soft music is 30.0 dB. What is the intensity of the sound?

5. Wine has a hydrogen ion concentration of 0.0004 moles per litre. Find its pH value.

6. Pure water has a hydrogen ion concentration of 10^{-7} moles per litre. Find its pH value.

7. The pH of ammonia is 10.8. Find its hydrogen ion concentration.

8. The pH of beer is 6.34. Find its hydrogen ion concentration.

9. Find the difference in elevation between two stations having barometer readings of 100.2 kPa at the lower station and 83.0 kPa at the upper station.

10. Find the barometer reading 283 m above a station having a barometer reading of 93.1 kPa.

11. If interest is compounded continuously, the time necessary for an account to double is given by

$$t = \frac{\ln 2}{r}$$

where r is the rate of interest. How long will it take for an account to double if the interest is 6.25% per year compounded continuously?

12. What interest rate would be necessary for an account to double in 4 years, if the interest were compounded continuously?

13. The magnitude R of an earthquake of intensity I is defined on the Richter scale, as

$$R = \log\left(\frac{I}{I_0}\right)$$

where I_0 is the minimum level for comparison. Since the Richter scale is a logarithmic base 10 scale, a magnitude 7 quake is 10 times more intense than a magnitude 6 quake; a magnitude 7 quake is 100 times more intense than a magnitude 5 quake; and a magnitude 7 quake is 1000 times more intense than a magnitude 4 quake. How many times more intense was the 1964 Alaska earthquake whose magnitude was 9.2 on the Richter scale than the 1906 San Francisco earthquake whose magnitude was 7.9?

Chapter 28 REVIEW EXERCISES

If answers are not exact, round to three significant digits.
Express each of the following in logarithmic form.

1. $4^3 = 64$
2. $\frac{1}{8} = 2^{-3}$
3. $4 = 10^x$
4. $y^5 = 2.45$

Use the properties of logarithms to expand the expression.

5. $\log_2 ab$
6. $\log_3 \frac{6}{7}$
7. $\log_7 \frac{5a}{b}$
8. $\log_5 y^3$
9. $\ln \sqrt[3]{x}$
10. $\log \sqrt{x^3}$
11. $\log_3 a^2 b^3$
12. $\ln \frac{\sqrt{3}}{x}$

Complete the equation, and write the logarithmic form.

13. $10^4 = ?$
14. $10^{-5} = ?$
15. $e^{-1.23} = ?$
16. $e^{0.6} = ?$

Change to exponential form and find x, without using a calculator.

17. $x = \log_2 8$
18. $\log_5 1 = x$
19. $\ln e^3 = x$
20. $\log 0.01 = x$
21. $\log_x 9 = \frac{1}{2}$
22. $\log_8 x = 2$
23. $\log 100\,000 = x$
24. $\log_x \sqrt[3]{b} = \frac{1}{3}$
25. $x = \log_4 \frac{1}{16}$
26. $x = \log_7 \sqrt{7}$
27. $x = \log_b 1$
28. $x = \log_8 8^7$

Using a calculator, determine the following logarithms. Then write the equations in exponential form.

29. $\log 1.70$
30. $\log 35.7$
31. $\ln 4.50$
32. $\ln 0.625$

Solve each equation. Check your solution.

33. $2.50^x = 7.20$ **34.** $e^x = 6.15$
35. $0.090 = e^{-x}$ **36.** $8.34 = 3(1.23)^x$
37. $4e^x = 7$ **38.** $8.32 = 3e^{-x}$
39. $5.25 = 6.50(1.02)^{2x}$ **40.** $92.6 = 6.18e^{-0.3x}$

Use a calculator to find the indicated power, rounded to three significant digits. Then write the equation in logarithmic form.

41. $6^{2.30} = ?$ **42.** $e^{-1.52} = ?$
43. $10^{-3.65} = ?$ **44.** $2.5^{3.18} = ?$

Solve for x and evaluate.

45. $\log x = 7$ **46.** $\log x = -2.35$
47. $\ln x = -1.52$ **48.** $\ln x = 0.327$

Without using a calculator, find the indicated power and write the equation in logarithmic form.

49. $5^3 = ?$ **50.** $10^{-2} = ?$
51. $\left(\dfrac{1}{3}\right)^{-1} = ?$ **52.** $8^{1/3} = ?$

Express as a single logarithm with a coefficient of 1.

53. $\log_3 5 + \log_3 x$ **54.** $\log_2 5 + \log_2 x - \log_2 3 - \log_2 y$
55. $4 \log 3x$ **56.** $x \log_b 5$
57. $\dfrac{1}{2} \log_4 a$ **58.** $\dfrac{\log_5 x}{3}$
59. $3 \ln x - 2 \ln y$ **60.** $\dfrac{1}{4} \ln 3 + \dfrac{2}{3} \ln y$

Change to exponential form.

61. $\log_3 81 = 4$ **62.** $2 = \log 100$
63. $\ln 7 = 1.95$ **64.** $\log_4 \left(\dfrac{1}{16}\right) = -2$

Evaluate.

65. $\log_3 3^7$ **66.** $\log_2 4^3$
67. $\ln \sqrt[3]{e}$ **68.** $\log_b \sqrt{b^9}$

Solve.

69. A certain bacterium quadruples its number every day. If we begin with 1000 bacteria, how long would it take to have a million?

70. How long would it take $250 to grow to $800, if it were invested at 6.25% per year, compounded continuously?

71. A radioactive isotope called Radium A decays at the rate of 22.7% per minute. Find the time necessary for 16.0 kg of Radium A to decay to 12.0 kg.

72. The time necessary for half the quantity of radioactive tritium to decay is 11.6 years (its half-life). Find the rate of decay of tritium.

73. The discharge current i of a capacitor is given by

$$i = 0.050 e^{-65.0t}$$

where t is the time in seconds. At what time will the discharge current be 0.001 amperes?

74. The intensity of sound for loud music is about 10^{-2} W/m². Use the equation

$$\beta = 10 \log \frac{I}{I_0}$$

where $I_0 = 10^{-12}$ W/m², to find the loudness in decibels of loud music.

75. Use the equation

$$pH = -\log C$$

to find the hydrogen ion concentration C (in moles/litre) for eggs, which have a pH of 8.

76. Find the time necessary for a casting to cool from 1520°C to 825°C, if it cools at 25.1% per minute.

WRITING

77. What is a logarithm? Explain the relationship between the logarithmic form and the exponential form of an equation. Give three examples of how logarithms are used in everyday life.

TEAM PROJECT

78. The students in Ms. Smith's math class were asked to record the number of hours spent studying for their math test. For each student, Ms. Smith wrote an ordered pair (x, y). The value of x represented the number of hours of study and the value of y was the student's grade on the test. Jenni's ordered pair was (3.0, 82) because she spent 3.0 hours studying and received a grade of 82 on her math test. The set of ordered pairs which Ms. Smith recorded are shown here:

(3.0, 82), (5.5, 78), (1.0, 60), (4.9, 93), (5.1, 86), (2.5, 71), (4.2, 90), (0.5, 40), (3.5, 88), (7.0, 96), (1.5, 73), (2.4, 82), (2.0, 53), (6.2, 87), (8.4, 100), (2.6, 75), (3.7, 85), (5.4, 70), (9.3, 89), (7.6, 87)

Make a graph of test grades versus the number of hours studied. Is there is a relationship between the number of hours of study and the test grade? Find an equation of a line which seems to "fit" this data. Graph the line. Might a logarithmic curve fit this data? Find a logarithmic equation that fits the data and graph the curve. Which curve do you think yields the better fit? Use that curve to estimate the grade Becky would achieve with 5.2 hours of study.

29 Complex Numbers

Outcomes

When you have completed this chapter, you will be able to

1. Define complex numbers.
2. Plot complex numbers.
3. Define an *imaginary* number.
4. Perform additions and subtractions of complex numbers.
5. Multiply and divide complex numbers.
6. Convert complex numbers between polar form, rectangular form, and trigonometric form.

Thus far we have dealt entirely with real numbers; in this chapter we introduce another set of numbers that are not real; they are called "imaginary" numbers. We also learn what complex numbers are, how to express them in several different forms, and how to perform the basic operations on them.

So why do we need to study imaginary and complex numbers? There are many reasons. The solutions of many quadratic and higher degree equations are imaginary or complex numbers; complex numbers are used in the study of electricity; and many manipulations of vectors are simplified if the vectors are represented by complex numbers.

29-1 IMAGINARY NUMBERS

In this section, we define the basic imaginary unit and imaginary number, and learn to simplify, add, subtract, multiply, and divide imaginary numbers.

Imaginary Unit

In Chapter 6 we learned that, in the real number system, we cannot take the square root of a negative number. For example, $\sqrt{-1}$ is *not* a *real* number. But $\sqrt{-1}$ *is* a number; we call it an *imaginary* number. The choice of the words *real* and *imaginary* is unfortunate. *Imaginary* was chosen as the opposite of *real*. But *imaginary* numbers are actual solutions of equations and are very *real* to people using them.

We define the *imaginary unit* as the square root of -1, and represent it by the symbol j. Some textbooks use i as the imaginary unit. Since i frequently represents electrical current, it has become common to represent the imaginary unit by j instead of i. This symbol is sometimes called the *j-operator*.

| IMAGINARY UNIT | $j = \sqrt{-1}$ | 156 |

Imaginary Numbers

An *imaginary number* is the product of the imaginary unit j times a real number b; it is generally expressed as

$$jb$$

where $j = \sqrt{-1}$ and b is a real number.

Example 1

An example of an imaginary number is $j2$. It is the product of the imaginary unit j and the real number 2.

Example 2

The imaginary number $j(-3)$ is usually written as $-j3$:

$$j(-3) = -j3$$

because if the real coefficient b is negative, the negative sign is usually written in *front* of j.

Imaginary Numbers in *jb* Form

The square root of a negative number is an imaginary number. It can be expressed in *jb* form by factoring the -1 from the radicand, and expressing $\sqrt{-1}$ as j.

Example 3

Express the imaginary number $\sqrt{-7}$ in *jb* form.

Solution: First we factor -1 from the radicand -7:

$$\sqrt{-7} = \sqrt{(-1)(7)}$$

and take the square root of each factor

$$= \sqrt{-1} \cdot \sqrt{7}$$

Expressing $\sqrt{-1}$ as j, we get

$$= j\sqrt{7}$$

In Example 3, the square root of the positive factor $\sqrt{7}$ was already simplified. Often, however, the square root of the positive factor can be simplified further using the rules of radicals from Chapter 20.

Example 4

Express the imaginary numbers in *jb* form.

(a) $\sqrt{-4}$ **(b)** $\sqrt{-45}$

Solution:

(a) Following the same process as in Example 3, we get

$$\sqrt{-4} = \sqrt{(-1)(4)}$$
$$= \sqrt{-1} \cdot \sqrt{4}$$
$$= j\sqrt{4}$$

Now we simplify $\sqrt{4}$ to get

$$= j2$$

(b) Similarly,

$$\sqrt{-45} = \sqrt{-1 \cdot 45}$$
$$= \sqrt{-1} \cdot \sqrt{45}$$

Simplifying $\sqrt{45}$,

$$= j \cdot \sqrt{9 \cdot 5}$$
$$= j \cdot \sqrt{9} \cdot \sqrt{5}$$
$$= j3\sqrt{5}$$

Example 5

Express $-2\sqrt{-18}$ in jb form.

Solution: First we express $\sqrt{-18}$ in jb form, and then multiply by the factor -2.

$$-2\sqrt{-18} = -2 \cdot \sqrt{(-1)(18)}$$
$$= -2 \cdot \sqrt{-1} \cdot \sqrt{18}$$
$$= -2 \cdot j\sqrt{9 \cdot 2}$$
$$= -2 \cdot j3\sqrt{2}$$

Now multiply by the -2 to get

$$= -j6\sqrt{2}$$

Adding and Subtracting Imaginary Numbers

Imaginary numbers may be added and subtracted in their jb form. They are added or subtracted by combining the coefficients of j (as we combine other like algebraic terms).

Example 6

Combine $j2 + j5 - j3$.

Solution:

$$j2 + j5 - j3 = j(2 + 5 - 3)$$
$$= j4$$

Example 7

Combine $\sqrt{-25} - \sqrt{-16} + 2\sqrt{-9}$.

Solution:

$$\sqrt{-25} - \sqrt{-16} + 2\sqrt{-9} = j5 - j4 + 2j3$$
$$= j5 - j4 + j6$$
$$= j7$$

Multiplying Imaginary Numbers

Imaginary numbers may be multiplied by real numbers or by other imaginary numbers. First we look at the product of an imaginary number and a real number. Just treat j as a variable, multiply the real factors together, and simplify.

Example 8

Multiply and simplify.

(a) $5 \cdot j3$ **(b)** $-\sqrt{-49} \cdot -3$

Solution:

(a) Multiplying the real factors, we get
$$5 \cdot j3 = j15$$

(b) Writing in jb form, we get
$$-\sqrt{-49} \cdot -3 = -j7 \cdot -3$$
$$= j21$$

In Example 8, we see that the product of an imaginary number and a real number is an *imaginary* number.

Now we learn to multiply two imaginary numbers. But first we must define j^2, using the fact that $j = \sqrt{-1}$:
$$j^2 = -1$$

Example 9

Multiply and simplify:
$$j2 \cdot -j\sqrt{6}$$

Solution: First, multiplying the j's and the real coefficients,
$$j2 \cdot -j\sqrt{6} = j^2 \cdot -2\sqrt{6}$$

Changing j^2 to -1,
$$= (-1)(-2\sqrt{6})$$

So we get,
$$= 2\sqrt{6}$$

Recall from the product rule in Chapter 20 that
$$\sqrt{a} \cdot \sqrt{b} = \sqrt{ab}$$
is true for all *real* numbers. This product rule is true only if $\sqrt{a}$ and $\sqrt{b}$ are real numbers. Since the square root of a negative number is not real, then it must be true that
$$a \geq 0 \text{ and } b \geq 0$$

We can use the product rule when we multiply imaginary numbers, but only after expressing the imaginary numbers in jb form.

To multiply two imaginary numbers, convert them to jb form, multiply the j's, multiply the real coefficients, and then simplify.

Example 10
Multiply and simplify:

$$\sqrt{-4} \cdot \sqrt{-9}$$

Solution: Changing to *jb* form,

$$\sqrt{-4} \cdot \sqrt{-9} = j2 \cdot j3$$
$$= j^2 \cdot 6$$
$$= -1 \cdot 6$$
$$= -6$$

Now watch what happens if we don't change to *jb* form first. We are tempted to use the product rule and multiply the radicands:

$$\sqrt{-4} \cdot \sqrt{-9} \stackrel{?}{=} \sqrt{-4 \cdot -9}$$
$$\stackrel{?}{=} \sqrt{+36}$$
$$\stackrel{?}{=} 6$$

We get the wrong answer when we use the product rule with numbers that are not real; that is, when the radicand is negative, and we don't change to *jb* form first!

COMMON ERROR Never use the product rule with negative radicands. Always change square roots of negative numbers to *jb* form before operating on them.

So we see in Examples 9 and 10 that the product of two imaginary numbers is a *real* number.

Dividing Imaginary Numbers

To divide with imaginary numbers, express the division problem as a fraction and divide by canceling like factors. Then express the result in *jb* form.

Example 11
Divide and simplify.

(a) $j6 \div j3$ (b) $-j12 \div 4$

Solution:

(a) $j6 \div j3 = \dfrac{\cancel{j6}^{2}}{\cancel{j3}_{1}} = 2$

(b) $-j12 \div 4 = \dfrac{-j\cancel{12}^{3}}{\cancel{4}_{1}} = -j3$

If a factor of *j* remains in the denominator after you have canceled, multiply the fraction by 1 in the form of $\dfrac{j}{j}$.

Example 12

Divide 8 by $j2$ and simplify.

Solution:

$$8 \div j2 = \frac{\overset{4}{\cancel{8}}}{\underset{1}{\cancel{j2}}} = \frac{4}{j}$$

Since j is in the denominator, we multiply by $\frac{j}{j}$

$$= \frac{4 \cdot j}{j \cdot j} = \frac{j4}{-1}$$
$$= -j4$$

SECTION 29-1 Exercises Imaginary Numbers

Imaginary Numbers in jb Form

Express the imaginary number in jb form and simplify.

1. $\sqrt{-5}$
2. $\sqrt{-2}$
3. $\sqrt{-900}$
4. $\sqrt{-16}$
5. $\sqrt{-8}$
6. $\sqrt{-27}$
7. $-\sqrt{-12}$
8. $-\sqrt{-200}$
9. $2\sqrt{-7}$
10. $-3\sqrt{-3}$
11. $-5\sqrt{-50}$
12. $7\sqrt{-48}$
13. $\sqrt{-\frac{1}{9}}$
14. $\sqrt{-\frac{4}{25}}$
15. $-\sqrt{-0.16}$
16. $\sqrt{-0.0001}$

Adding and Subtracting Imaginary Numbers

Combine and simplify.

17. $j3 - j7 + j5$
18. $-j6 + j4 - j2$
19. $\sqrt{-25} + \sqrt{-16} - \sqrt{-100}$
20. $\sqrt{-36} - \sqrt{-64} + \sqrt{-81}$
21. $2\sqrt{-49} - 3\sqrt{-900} + 5\sqrt{-121}$
22. $4\sqrt{-400} + 2\sqrt{-144} - 3\sqrt{-169}$

Multiplying Imaginary Numbers

Multiply and simplify.

23. $6 \cdot j3$
24. $-3 \cdot j8$
25. $-j7 \cdot 4$
26. $j9 \cdot 5$
27. $j\sqrt{2} \cdot 5$
28. $-j3 \cdot \sqrt{7}$
29. $-\sqrt{5} \cdot j\sqrt{6}$
30. $\sqrt{6} \cdot j\sqrt{3}$
31. $5 \cdot \sqrt{-64}$
32. $-7 \cdot \sqrt{-81}$
33. $-\sqrt{-36} \cdot -5$
34. $\sqrt{-49} \cdot -6$
35. $j8 \cdot j2$
36. $j5 \cdot j4$
37. $j\sqrt{2} \cdot -j\sqrt{7}$
38. $j\sqrt{3} \cdot j\sqrt{5}$
39. $j\sqrt{3} \cdot j\sqrt{3}$
40. $j\sqrt{5} \cdot j\sqrt{5}$
41. $j3\sqrt{2} \cdot -j2\sqrt{3}$
42. $-j5\sqrt{2} \cdot j3\sqrt{2}$
43. $\sqrt{-4} \cdot \sqrt{-9}$
44. $-\sqrt{-25} \cdot \sqrt{-36}$
45. $\sqrt{-7} \cdot \sqrt{-7}$
46. $\sqrt{-3} \cdot \sqrt{-3}$
47. $-\sqrt{-3} \cdot \sqrt{-12}$
48. $\sqrt{-5} \cdot \sqrt{-10}$
49. $-4\sqrt{-2} \cdot -6\sqrt{-2}$
50. $-5\sqrt{-7} \cdot -3\sqrt{-7}$

Dividing Imaginary Numbers

Divide and simplify.

51. $j14 \div j2$
52. $j15 \div j3$
53. $j12 \div -j8$
54. $-j20 \div j12$
55. $-j16 \div 4$
56. $j18 \div -6$
57. $-j20 \div -8$
58. $j16 \div -6$
59. $5 \div j10$
60. $-12 \div j4$
61. $-16 \div j20$
62. $-15 \div -j25$

29-2 COMPLEX NUMBERS

In this section, we define a complex number and its conjugate and learn that the solutions of certain quadratic equations are complex numbers.

Complex Number

A *complex number* is a number that can be expressed in the form

$$a + jb$$

where a and b are real numbers, and j is the imaginary unit ($j = \sqrt{-1}$).

Example 13

The number $2 + j3$ is a complex number, in which $a = 2$, $b = 3$, and $j = \sqrt{-1}$.

Rectangular Form of a Complex Number

A complex number

$$a + jb$$

is said to be in *rectangular form*. The number a is called the *real part* and the number jb is called the *imaginary part*.

| RECTANGULAR FORM OF A COMPLEX NUMBER | $a + jb$
 real part   imaginary part | 157 |

If the coefficient b is negative, the negative sign is usually written in front of the j.

Example 14

The complex number $2 + j(-3)$ is usually written as $2 - j3$.

Example 15

Express the complex number in rectangular form:

$$-3 - \sqrt{-8}$$

Solution: Factoring $\sqrt{-1}$ from $\sqrt{-8}$, we have

$$-3 - \sqrt{-8} = -3 - \sqrt{-1}\sqrt{8}$$
$$= -3 - j\sqrt{8}$$

Simplifying $\sqrt{8}$, we have

$$= -3 - j2\sqrt{2}$$

If the real part a of a complex number is zero ($a = 0$), then the number, in rectangular form, is

$$a + jb = 0 + jb = jb$$

The number jb is called a *pure imaginary number*. Every pure imaginary number is also a complex number, since it can be written as $a + jb$.

If the coefficient b of a complex number is zero ($b = 0$), then the number, in rectangular form, is

$$a + jb = a + j0 = a$$

The number a is a *real* number, but it is also a *complex* number, since it can be written as $a + jb$. Every real number is a complex number.

Example 16

Express each number as a complex number in rectangular form.

(a) 5   (b) $\sqrt{-50}$

Solution:

(a) $5 = 5 + j0$

So the number 5 is both a real number and a complex number.

(b) $\sqrt{-50} = j\sqrt{50} = 0 + j5\sqrt{2}$

So the number $\sqrt{-50}$ is both a pure imaginary number and a complex number.

Conjugate of a Complex Number

The *conjugate* of a complex number $a + jb$ is the complex number $a - jb$. A complex number and its conjugate differ only in the *sign* of the imaginary part. Conjugates are used in division with complex numbers.

Example 17

Write the conjugate of the complex numbers:

(a) $-4 + j3$ (b) $5 - j2$

Solution:

(a) The conjugate of $-4 + j3$ is $-4 - j3$.
(b) The conjugate of $5 - j2$ is $5 + j2$.

Since real numbers and pure imaginary numbers are complex numbers, they also have conjugates.

Example 18

Write the conjugate of each number.

(a) -8 (b) $j7$

Solution:

(a) First we express the real number -8 as a complex number in rectangular form:

$$-8 = -8 + j0$$

The conjugate of $-8 + j0$ is $-8 - j0$. But
$$-8 - j0 = -8$$
So the conjugate of -8 is -8. In fact, *the conjugate of any real number is the real number itself.*

(b) We write the pure imaginary number $j7$ as a complex number in rectangular form:
$$j7 = 0 + j7$$
The conjugate of $0 + j7$ is $0 - j7$. But
$$0 - j7 = -j7$$
So the conjugate of $j7$ is $-j7$. In fact, *the conjugate of any pure imaginary number is its opposite.*

Complex Solutions of a Quadratic Equation

We saw in Chapter 21 that every quadratic equation has two solutions. In that chapter, we solved quadratic equations with solutions that were real numbers. Now we can solve quadratic equations with solutions that are not real numbers—they are non-real complex numbers.

Example 19

Solve the quadratic equation:
$$x^2 + 27 = 0$$

Solution: We use the method of Sec. 21-1 to solve the pure quadratic $x^2 + 27 = 0$.
$$x^2 = -27$$
$$x = \pm\sqrt{-27}$$
$$x = \pm j3\sqrt{3}$$

The solutions of this quadratic equation are two non-real pure imaginary complex numbers.

Solution by Quadratic Formula

To solve the quadratic equation $ax^2 + bx + c = 0$, we may use the quadratic formula. If the discriminant, $b^2 - 4ac$, is negative, then the two solutions of the quadratic equation are non-real complex numbers.

Example 20

Solve the quadratic equation $x^2 - 6x + 13 = 0$.

Solution: Applying the quadratic formula to the equation $x^2 - 6x + 13 = 0$, we have
$$x = \frac{-(-6) \pm \sqrt{(-6)^2 - 4(1)(13)}}{2(1)} = \frac{6 \pm \sqrt{-16}}{2}$$

The discriminant is negative, so we know the solutions are complex. Simplifying $\sqrt{-16}$, we get
$$x = \frac{6 \pm j4}{2}$$

Simplifying the fraction, we get

$$x = \frac{2(3 \pm j2)}{2} = 3 \pm j2$$

So, the two solutions of the equation $x^2 - 6x + 13 = 0$ are $3 + j2$ and $3 - j2$. The two solutions of this quadratic equation are conjugates of each other. In fact, *the complex solutions of a quadratic equation are always conjugates.*

SECTION 29-2 Exercises Complex Numbers

Rectangular Form of a Complex Number

Express the complex number in rectangular form.
1. $3 + \sqrt{-2}$ 2. $-2 - \sqrt{-5}$ 3. $\sqrt{-12} - 5$ 4. $-\sqrt{-18} - 4$
5. -7 6. 9 7. $\sqrt{-24}$ 8. $-\sqrt{-27}$

Conjugate of a Complex Number

Write the conjugate of the complex number.
9. $3 + j8$ 10. $-4 - j7$ 11. $-5 - j\sqrt{2}$ 12. $6 + j\sqrt{5}$
13. -2 14. 5 15. $j4$ 16. $-j5$

Complex Solutions of a Quadratic Equation

Solve the quadratic equation. Write the complex solutions in rectangular form.
17. $x^2 + 25 = 0$ 18. $x^2 + 49 = 0$
19. $x^2 + 48 = 0$ 20. $x^2 + 75 = 0$
21. $x^2 + 4x + 13 = 0$ 22. $x^2 - 10x + 34 = 0$
23. $x^2 - 8x + 20 = 0$ 24. $x^2 + 2x + 2 = 0$
25. $2x^2 + 2x + 3 = 0$ 26. $3x^2 - 4x + 16 = 0$

29-3 OPERATIONS WITH COMPLEX NUMBERS

In this section we learn to perform the basic operations of addition, subtraction, multiplication, and division with complex numbers. Before performing any operations, the complex numbers should be simplified and expressed in rectangular form.

Adding Complex Numbers

The addition of complex numbers is similar to the addition of binomial expressions. To add complex numbers, first we remove parentheses, and then combine the real parts and the imaginary parts separately. We simplify if possible and express the result in rectangular form.

Example 21

Add $(2 - j5) + (-4 + j8)$.

Solution: Removing parentheses:

$$(2 - j5) + (-4 + j8) = 2 - j5 - 4 + j8$$

Combining parts separately,

$$= (2 - 4) + (-j5 + j8)$$

Simplifying,

$$= -2 + j3$$

Alternate Solution: Complex numbers can also be added vertically.

$$\text{Add:} \quad \begin{array}{r} 2 - j5 \\ -4 + j8 \\ \hline -2 + j3 \end{array}$$

Subtracting Complex Numbers

To subtract complex numbers, we first remove the parentheses by changing signs, and then combine the real parts and the imaginary parts separately.

Example 22

Subtract $(-3 + j7) - (5 - j2)$.

Solution: Removing parentheses:

$$(-3 + j7) - (5 - j2) = -3 + j7 - 5 + j2$$

Combining parts separately,

$$= (-3 - 5) + (j7 + j2)$$

Simplifying,

$$= -8 + j9$$

Alternate Solution: To subtract vertically, we change the signs of the second (lower) number and add.

$$\text{To subtract:} \quad \begin{array}{r} -3 + j7 \\ 5 - j2 \end{array} \quad \text{add:} \quad \begin{array}{r} -3 + j7 \\ -5 + j2 \\ \hline -8 + j9 \end{array}$$

Multiplying Complex Numbers

Use the distributive property to multiply complex numbers just as you multiplied algebraic expressions. Whenever j^2 occurs, replace j^2 with -1. Express the result in rectangular form.

Example 23

Multiply:

(a) $3(-2 + j5)$ **(b)** $j4(3 - j7)$

Solution: We use the familiar distributive property to multiply.

(a)
$$3(-2 + j5) = 3 \cdot -2 + 3 \cdot j5$$
$$= -6 + j15$$

(b)
$$j4(3 - j7) = j4 \cdot 3 - j4 \cdot j7$$
$$= j12 - j^2 \cdot 28$$

Replacing j^2 with -1,

$$= j12 - (-1)(28)$$
$$= j12 + 28$$

Writing in rectangular form,

$$= 28 + j12$$

Example 24
Multiply $(2 - j3)(-1 + j4)$.

Solution: We multiply as with binomials, using FOIL:

$$(2 - j3)(-1 + j4) = 2(-1) + 2(j4) + (-j3)(-1) + (-j3)(j4)$$
$$= -2 + j8 + j3 + j^2(-12)$$
$$= -2 + j8 + j3 + (-1)(-12)$$
$$= -2 + j8 + j3 + 12$$
$$= 10 + j11$$

Example 25
Multiply $(3 + j2)(3 - j2)$.

Solution: Again we multiply as with binomials:

$$(3 + j2)(3 - j2) = 3(3) + 3(-j2) + (j2)(3) + (j2)(-j2)$$
$$= 9 - j6 + j6 - j^2(4)$$
$$= 9 - j6 + j6 - (-4)$$
$$= 13$$

Note that the product in Example 25 is a real number, and that we were multiplying conjugates. In fact, *the product of complex conjugates is always a real number*. These factors have the same form as the difference of two squares, and the middle terms (the j terms) combine to make zero, leaving only the real part.

Dividing Complex Numbers

In the division of complex numbers, we generally want to express the quotient in rectangular form. Division by a single term, real or complex, is done term by term in fraction form.

Example 26
Divide $(6 - j4) \div 2$.

Solution: First we change the division problem to fraction form,

$$(6 - j4) \div 2 = \frac{6 - j4}{2}$$

Divide term by term and cancel,

$$= \frac{\overset{3}{\cancel{6}}}{\underset{1}{\cancel{2}}} - \frac{j\overset{2}{\cancel{4}}}{\underset{1}{\cancel{2}}}$$

and write in rectangular form,

$$= 3 - j2$$

We want to express a quotient in rectangular form, so whenever j appears in the denominator of a fraction, it is common practice to eliminate it. The j can be eliminated from the denominator of a fraction by multiplying the fraction by 1, in the form of $\frac{j}{j}$.

Example 27
Divide $(-12 + j8) \div j4$.

Solution:

$$(-12 + j8) \div j4 = \frac{-12 + j8}{j4}$$

$$= \frac{-\overset{3}{\cancel{12}}}{\underset{1}{\cancel{j4}}} + \frac{\overset{2}{\cancel{j8}}}{\underset{1}{\cancel{j4}}}$$

$$= \frac{-3}{j} + 2$$

In this case, j appears in the denominator of the fraction $\frac{-3}{j}$, and we eliminate it from the denominator by multiplying the fraction by $\frac{j}{j}$, so

$$(-12 + j8) \div j4 = \frac{-3}{j} \cdot \frac{j}{j} + 2$$

$$= \frac{-j3}{-1} + 2$$

$$= j3 + 2$$

$$= 2 + j3$$

If the divisor in a division problem is a complex number, $a + jb$, where neither a nor b is 0, then eliminating the j from the denominator is more complicated. Recall from Example 25 that multiplying a complex number by its conjugate gives you a product that is a real number. So we can eliminate the j from the denominator by multiplying both the numerator and the denominator by the conjugate of the denominator. Then we simplify and express the result in rectangular form. We observe that this process is similar to *rationalizing the denominator* of a fraction containing radicals.

Example 28
Divide $(1 + j4) \div (3 + j2)$.

Solution: We write the division problem in fraction form:

$$(1 + j4) \div (3 + j2) = \frac{1 + j4}{3 + j2}$$

Multiplying the numerator and denominator by the conjugate of the denominator $(3 - j2)$,

$$(1 + j4) \div (3 + j2) = \frac{(1 + j4)}{(3 + j2)} \cdot \frac{(3 - j2)}{(3 - j2)}$$

Simplifying,

$$= \frac{3 - j2 + j12 - j^2 \cdot 8}{9 - j^2 \cdot 4}$$

$$= \frac{3 + j10 + 8}{9 + 4}$$

$$= \frac{11 + j10}{13}$$

Expressing in rectangular form:

$$= \frac{11}{13} + j\frac{10}{13}$$

SECTION 29-3 Exercises Operations with Complex Numbers

Adding and Subtracting Complex Numbers

Combine and simplify.
1. $2 - j4 + j7 - 5$
2. $j5 - 7 + j2 - 4$
3. $(-3 + j8) + (2 - j5)$
4. $(5 - j3) + (-3 - j4)$
5. $(4 - j3) + (4 + j3)$
6. $(-5 + j4) + (-5 - j4)$
7. $(-7 + j7) - (3 - j4)$
8. $(5 + j4) - (8 - j3)$
9. $(-1 + j2) - (-1 - j2)$
10. $(6 - j) - (6 + j)$

Add.

11. $-8 + j6$
    $\underline{5 - j3}$

12. $9 - j8$
    $\underline{-7 + j7}$

13. $6 - j5$
    $-8 + j$
    $-5 - j4$
    $\underline{4 - j2}$

14. $-1 + j6$
    $7 - j$
    $-5 - j5$
    $\underline{8 - j4}$

Subtract.

15. $9 - j7$
    $\underline{5 + j4}$

16. $-4 + j3$
    $\underline{-5 - j8}$

17. $-5 - j$
    $\underline{-6 - j3}$

18. $7 + j4$
    $\underline{-3 + j}$

Multiplying Complex Numbers

Multiply and simplify.
19. $5(-3 + j2)$
20. $-6(4 - j5)$
21. $-7(-4 - j6)$
22. $8(-5 + j)$
23. $j(2 - j3)$
24. $-j3(-3 + j4)$
25. $-j5(-6 + j4)$
26. $j7(8 - j)$
27. $(3 - j)(5 - j2)$
28. $(5 + j4)(-3 + j)$
29. $(-4 + j3)(1 - j5)$
30. $(-7 - j4)(-6 - j5)$
31. $(9 + j5)(-4 + j8)$
32. $(-8 - j3)(7 + j7)$
33. $(7 + j4)(7 - j4)$
34. $(-5 - j3)(-5 + j3)$
35. $(-6 - j5)(-6 + j5)$
36. $(8 + j3)(8 - j3)$
37. $(4 - j3)^2$
38. $(5 + j6)^2$
39. $(-7 + j2)^2$
40. $(-3 - j5)^2$

Dividing Complex Numbers

Divide and express in rectangular form.
41. $(4 - j8) \div 2$
42. $(6 - j9) \div 3$
43. $(-9 + j12) \div -6$
44. $(-10 - j25) \div -15$
45. $(15 - j20) \div j5$
46. $(-12 + j15) \div -j3$
47. $(-18 - j24) \div -j12$
48. $(8 - j6) \div j4$
49. $3 \div (1 - j)$
50. $-5 \div (2 + j3)$
51. $-j4 \div (3 + j2)$
52. $j2 \div (2 + j)$
53. $(2 + j5) \div (1 + j)$
54. $(1 + j) \div (-4 - j)$
55. $(-3 - j4) \div (2 - j)$
56. $(-3 + j) \div (-2 - j)$
57. $(6 + j5) \div (-2 - j3)$
58. $(-1 - j) \div (3 + j5)$

29-4 GRAPH OF A COMPLEX NUMBER

In this section we define the complex plane and show that every complex number can be represented by a point on the complex plane. We show that a vector can be represented as a complex number and vice versa, and that vector addition is greatly simplified using complex number representation.

FIGURE 29-1
The complex plane.

Complex Plane

A complex number, $a + jb$, consists of two parts: a real part a and an imaginary part jb. To graph a complex number we use two axes, one for the real part and one for the imaginary part. We use a form of the rectangular coordinate system (Fig. 29-1). The horizontal axis is called the *real axis*. The vertical axis is called the *imaginary axis*. Such a coordinate system is called the *complex plane*.

Graph of a Complex Number

Every complex number can be represented by a *point* on the complex plane. To plot a complex number $a + jb$ on the complex plane, locate a point with a horizontal coordinate a and a vertical coordinate b. (Start at the origin. Go a units to the right or left of the origin, and b units up or down from the horizontal axis.)

--- **Example 29** ---

Graph the complex number $2 + j3$.

Solution: Draw the two perpendicular axes, label the axes *real* and *imaginary*, and label the units on each axis. To graph the complex number $2 + j3$, plot the point which is 2 units to the right of the origin on the real (horizontal) axis and 3 units up (Fig. 29-2). This point on the complex plane is labeled by writing the number $2 + j3$, and do not enclose the complex number with parentheses.

FIGURE 29-2

This process is similar to plotting the point (a, b) on the standard rectangular coordinate system. However, a point on the complex plane represents a single complex number, *not* an ordered pair of real numbers.

--- **Example 30** ---

Graph the complex number $1 - j2$.

Solution: To graph the complex number $1 - j2$, plot the point that is 1 unit to the right of the origin and two units *down*, since the imaginery coefficient b is -2 (Fig. 29-3).

FIGURE 29-3

Graphs of Real and Pure Imaginary Numbers

Real and pure imaginary numbers can also be graphed on the complex plane. Real numbers are graphed on the real axis and pure imaginary numbers are graphed on the imaginary axis.

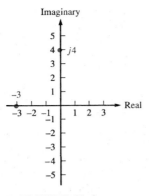

■ FIGURE 29-4

Example 31

Graph the numbers on the same complex plane:
(a) -3 (b) $j4$

Solution:
(a) First we write -3 as a complex number in rectangular form:

$$-3 = -3 + j0$$

To graph the complex number $-3 + j0$, we plot the point which is 3 units to the left of the origin and 0 units up (Fig. 29-4). Note that the point is *on* the real axis. This is reasonable, since -3 is a real number.

(b) First we write $j4$ as a complex number in rectangular form:

$$j4 = 0 + j4$$

To graph the number $0 + j4$ on the complex plane, plot the point which is 0 units to the right of the origin and 4 units up. You will note that the point is *on* the imaginary axis. This is reasonable, since $j4$ is a pure imaginary number (Fig. 29-4).

■ FIGURE 29-5
Vector representation of a complex number.

Vectors and Complex Numbers

One of the most important applications of complex numbers is that they can represent vectors, and vice versa.

Example 32

Consider the complex number $3 + j2$, which is plotted as a point in Fig. 29-5. If we draw a line segment from the origin to the point, then we have a *vector*, which we call **R**. The vector **R** has a horizontal component of 3 units and a vertical component of 2 units.

Vector Addition

The sum of two complex numbers may be found graphically. In Sec. 23-3 we learned the parallelogram method of adding two vectors. When we consider complex numbers as vectors, we may use the parallelogram method to add them graphically.

Example 33

Add $(3 + j2) + (1 + j4)$ graphically on the complex plane.

Solution: We plot the two points on the complex plane and draw the vectors associated with those points. We complete the other two sides of the parallelogram and draw the diagonal (Fig. 29-6). The fourth vertex is the point which represents the sum of the two complex numbers, $4 + j6$. Of course, if we had added the complex numbers algebraically we would have found the same result:

$$(3 + j2) + (1 + j4) = 4 + j6$$

So we have a very important result here: *The resultant of two vectors representing complex numbers is a vector that represents the sum of the complex numbers.*

■ FIGURE 29-6
Sum of two complex numbers.

29-4 Graph of a Complex Number

SECTION 29-4 Exercises Graph of a Complex Number

Complex Plane

Graph each complex number.
1. $1 + j2$
2. $3 - j4$
3. $4 - j5$
4. $5 + j2$
5. $-3 + j3$
6. $-4 - j$
7. $-2 - j$
8. $-1 + j2$
9. 4
10. -2
11. $j2$
12. $-j3$
13. -5
14. 1
15. $-j4$
16. j

17. Write the complex numbers represented by the points in Fig. 29-7.
18. Write the complex numbers represented by the points in Fig. 29-8.

■ FIGURE 29-7

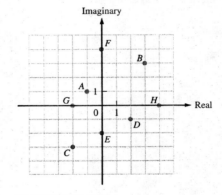

■ FIGURE 29-8

Vector Addition

Add the complex numbers graphically, and check the result algebraically.

19. $(2 + j) + (3 + j4)$
20. $(4 + j3) + (2 - j)$
21. $(3 - j2) + (-6 - j3)$
22. $(-5 - j4) + (3 - j2)$

23. $-1 + j2$
    $-1 - j2$

24. $3 - j5$
    $3 + j5$

25. $2 + j$
    $-3 - j4$
    $5 - j2$

26. $-4 - j3$
    $-5 + j$
    $7 - j2$

29-5 POLAR AND TRIGONOMETRIC FORMS

Thus far we have expressed complex numbers in only one form, the rectangular form, $a + jb$. In this section we learn to express complex numbers in two other forms, the polar form and the trigonometric form.

■ FIGURE 29-9
Polar form of a complex number.

Polar Form of a Complex Number

Figure 29-9 shows a complex number $a + jb$ and its vector representation. We let the length of the vector be r. The vector is at an angle of θ with the horizontal axis (in standard position), where $0 \leq \theta < 360°$. The radius r is called the *absolute value* of the complex number and can be found using the Pythagorean theorem:

| ABSOLUTE VALUE r OF THE COMPLEX NUMBER $a + jb$ | $r = \sqrt{a^2 + b^2}$ | 161 |

686 CHAPTER 29 / Complex Numbers

The angle θ is called the *argument* of the complex number. Note that in solving for θ, we get

| ARGUMENT θ OF THE COMPLEX NUMBER $a + jb$ | $\tan \theta = \dfrac{b}{a}$ | 162 |

where $0 \leq \theta < 360°$.

Any complex number can be expressed in terms of r and θ, by writing the absolute value r, the angle symbol $\angle$, and then the argument θ. This is called the *polar form* of the complex number.

| POLAR FORM OF A COMPLEX NUMBER $a + jb$ | $a + jb = r\angle\theta$ | 163 |

■ FIGURE 29-10

Example 34

Given the complex number $3 + j2$.
(a) Graph the number as a vector.
(b) Find the absolute value r and the argument θ, in degrees, rounding to three significant digits.
(c) Express the number in polar form.

Solution:
(a) We graph the number $3 + j2$ in Fig. 29-10.
(b) Using Eq. 161, we find the absolute value r:

$$r = \sqrt{3^2 + 2^2} = \sqrt{13} = 3.61$$

Using Eq. 162,

$$\tan \theta = \frac{2}{3}$$

Solving for θ, in quadrant I,

$$\theta = \tan^{-1} \frac{2}{3} = 33.7°$$

(c) Using Eq. 163, $3 + j2$ can be expressed in polar form, as

$$3 + j2 = 3.61 \angle 33.7°$$

If the graph of a complex number is in quadrant II, III, or IV, then we find the reference angle of the argument, θ_{ref}, using

$$\theta_{ref} = \tan^{-1} \left| \frac{b}{a} \right|$$

Absolute value is used, since the reference angle is always positive. We use the quadrant and the reference angle θ_{ref} to find θ, as we did in Chapter 22.

FIGURE 29-11

Example 35

Given the complex number $-3 + j5$,
(a) Graph the number as a vector,
(b) Express it in polar form.

Solution:

(a) We graph the number $-3 + j5$ in Fig. 29-11, and see that its graph is in quadrant II.

(b) Using Eq. 161 to find r,

$$r = \sqrt{(-3)^2 + 5^2} = \sqrt{34} = 5.83$$

To find θ_{ref}, we use the equation

$$\theta_{ref} = \tan^{-1}\left|\frac{b}{a}\right|$$

or,

$$\theta_{ref} = \tan^{-1}\left|\frac{5}{-3}\right| = 59.0°$$

In quadrant II,

$$\theta = 180° - \theta_{ref}$$

Substituting,

$$\theta = 180° - 59.0°$$

Or,

$$\theta = 121°$$

So,

$$-3 + j5 = 5.83 \,\underline{/121°} \quad \text{in polar form}$$

Suppose we know the polar form of a complex number, $r\,\underline{/\theta}$, and we wish to express it in rectangular form, $a + jb$. That is, we know the absolute value r and the argument θ of the complex number, and we want to find a and b to express it in rectangular form. Referring to Fig. 29-9, we see that both a and b can be found by trigonometry:

$$\cos\theta = \frac{a}{r} \quad \text{and} \quad \sin\theta = \frac{b}{r}$$

Solving for a and b, respectively, we get

POLAR TO RECTANGULAR FORM CONVERSIONS	$a = r\cos\theta$	159
	$b = r\sin\theta$	160

Example 36

The polar form of a complex number is $9.00\,\underline{/105°}$. Graph the number as a vector and express it in rectangular form.

Solution: We graph the number in Fig. 29-12 and use Eqs. 159 and 160 with $r = 9.00$ and $\theta = 105°$ to get

$$a = 9.00 \cos 105° = -2.33$$
$$b = 9.00 \sin 105° = 8.69$$

So, rounding to three significant digits, and expressing the number in rectangular form, we have

$$9.00 \;\underline{/105°} = -2.33 + j8.69$$

FIGURE 29-12

Trigonometric Form of a Complex Number

If we use Eqs. 159 and 160 and substitute the values for a and b into the rectangular form, we have

$$a + jb = r \cos \theta + j \cdot r \sin \theta$$

Factoring out the r from the right side, we have the *trigonometric form* of a complex number.

TRIGONOMETRIC FORM OF A COMPLEX NUMBER, $a + jb$	$a + jb = r(\cos \theta + j \sin \theta)$	158

Example 37

Write the complex number $3 + j2$ in trigonometric form, accurate to three significant digits.

Solution: In Example 34, we expressed $3 + j2$ in polar form, finding that

$$r = 3.61 \quad \text{and} \quad \theta = 33.7°$$

Using Eq. 158, we have

$$3 + j2 = 3.61(\cos 33.7° + j \sin 33.7°)$$

Example 38

Verify that the right side of the answer in Example 37 equals the left side.

Solution: We distribute the 3.61 on the right side and use the calculator to find the products:

$$3.61(\cos 33.7° + j \sin 33.7°) = 3.61 \cos 33.7° + j(3.61 \sin 33.7°)$$
$$= 3.00 + j2.00 \quad \text{(accurate to three significant digits)}$$

Example 39

Write the complex number

$$5.28(\cos 68.1° + j \sin 68.1°)$$

(a) in polar form, and **(b)** in rectangular form, rounded to three significant digits.

Solution:

(a) We are given the trigonometric form, and can see from Eq. 158 that

$$r = 5.28 \quad \text{and} \quad \theta = 68.1°$$

Using Eq. 163, we write the complex number in polar form:

$$5.28\underline{/68.1°}$$

(b) To find the rectangular form we use the trigonometric form, distribute the 5.28, and use a calculator to find the products:

$$5.28(\cos 68.1° + j \sin 68.1°) = 5.28 \cos 68.1° + j(5.28 \sin 68.1°)$$
$$= 1.97 + j4.90$$

SECTION 29-5 Exercises Polar and Trigonometric Forms

Polar Form of a Complex Number

Graph each complex number and express it in polar form, rounding to three significant digits.

1. $2 + j3$
2. $4 - j$
3. $-4 - j2$
4. $-3 + j4$

Graph each complex number and express it in rectangular form, rounding to three significant digits.

5. $8.00\underline{/56.5°}$
6. $43.7\underline{/235°}$
7. $549\underline{/235°}$
8. $0.702\underline{/345°}$

Trigonometric Form of a Complex Number

Graph each complex number and express it in trigonometric form.

9. $1 + j$
10. $3 + j4$
11. $3 - j2$
12. $-6 - j$
13. $-4 + j7$
14. $-5 + j3$
15. $-6 - j3$
16. $1 - j7$

Express each complex number in both polar and rectangular forms (rounded to three significant digits).

17. $2(\cos 30° + j \sin 30°)$
18. $4(\cos 45° + j \sin 45°)$
19. $3.56(\cos 135° + j \sin 135°)$
20. $5.74(\cos 121° + j \sin 121°)$
21. $8.45(\cos 297° + j \sin 297°)$
22. $78.9(\cos 247° + j \sin 247°)$

23. Verify the following rule for the product of two complex numbers in polar form

$$(r_1\underline{/\theta_1})(r_2\underline{/\theta_2}) = r_1 r_2 \underline{/\theta_1 + \theta_2}$$

by showing that

$$(2\underline{/45°})(5\underline{/30°}) = 10\underline{/75°}.$$

(Hint: Change to rectangular form, multiply, and change back to polar form.)

24. Verify the following rule for the quotient of two complex numbers in polar form

$$\frac{r_1\underline{/\theta_1}}{r_2\underline{/\theta_2}} = \frac{r_1}{r_2}\underline{/\theta_1 - \theta_2}$$

by showing that

$$\frac{2\underline{/45°}}{5\underline{/30°}} = \frac{2}{5}\underline{/15°}$$

Chapter 29 REVIEW EXERCISES

Graph and express each complex number in polar form, rounding to three significant digits.

1. $-2 + j5$ 2. $5 - j7$ 3. $7 + j$ 4. $-4 - j4$

Perform the indicated operations and express in rectangular form.

5. $2\sqrt{-64} - 5\sqrt{-25} + 4\sqrt{-36}$ 6. $-j\sqrt{3} \cdot -\sqrt{6}$
7. $(4 - j3)(1 - j2)$ 8. $(7 - j5)(7 + j5)$
9. $2\sqrt{-3} \cdot -3\sqrt{-3}$ 10. $-j18 \div j8$
11. $(-6 + j8) \div 4$ 12. $(12 - j9) \div -j6$
13. $(3 - j5) + (-2 - j7)$ 14. $(-3 - j4) - (-6 - j5)$
15. $-4(6 - j3)$ 16. $-j3(2 - j4)$
17. $(-3 + j4)^2$ 18. $(-2 - j)^2$
19. $-5 \div (-1 + j3)$ 20. $j4 \div (2 + j3)$
21. $-2 \cdot \sqrt{-8}$ 22. $-j7 \cdot -j6$
23. $j15 \div -3$ 24. $-26 \div -j8$
25. $j\sqrt{5} \cdot j\sqrt{5}$ 26. $\sqrt{-16} \cdot \sqrt{-64}$
27. $(5 - j3) \div (7 + j2)$ 28. $(3 - j4) \div (3 + j4)$

Add graphically on the complex plane and check the result algebraically.

29. $(3 - j5) + (-4 - j)$
30. $-2 - j5$
    $\underline{4 + j3}$

31. $-2 + j5$
    $5 - j$
    $\underline{-5 - j4}$

32. $-4 - j2$
    $\underline{-3 + j5}$

Graph and express each complex number in trigonometric form, rounding to three significant digits.

33. $2 + j2$ 34. $-4 + j6$ 35. $-5 - j3$ 36. $5 - j6$

Solve the quadratic equations. Write the complex solutions in rectangular form.

37. $x^2 + 72 = 0$ 38. $x^2 + 6x + 10 = 0$

Express each complex number in both polar and rectangular forms, rounding to three significant digits.

39. $4.25(\cos 123° + j \sin 123°)$ 40. $35.6(\cos 347° + j \sin 347°)$

WRITING

41. Some complex numbers are real, some are pure imaginary, some are neither. Explain the relationship among real, pure imaginary, and complex numbers. Give an example of each type and graph each number on the complex plane. Explain how the graph of the number relates to the type of number.

TEAM PROJECT

42. We know that $\sqrt[3]{1} = 1$. So, in the real number system, the solution of the equation

$$x^3 = 1$$

is

$$x = 1$$

But, in the complex number system, the equation

$$x^3 = 1$$

has three solutions. These three solutions can be found by factoring the equation

$$x^3 - 1 = 0$$

and setting the factors equal to zero. Find the three complex cube roots of 1. Check, by cubing each number.

Summary of Facts and Formulas

	No.				Page
ALGEBRAIC LAWS	1	Commutative Law	Addition	$a + b = b + a$	8
	2		Multiplication	$ab = ba$	13, 149
	3	Associative Law	Addition	$a + (b + c) = (a + b) + c = (a + c) + b$	8
	4		Multiplication	$a(bc) = (ab)c = (ac)b = abc$	13
	5	Distributive Law		$a(b + c) = ab + ac$	13, 151, 333
RULES OF SIGNS	6	Addition and Subtraction		$a + (-b) = a - (+b) = a - b$	111
	7			$a + (+b) = a - (-b) = a + b$	112
	8			$-(a + b) = -a - b \qquad -(a - b) = b - a$	362
	9	Multiplication		$(+a)(+b) = (-a)(-b) = +ab$	114, 149
	10			$(+a)(-b) = (-a)(+b) = -(+a)(+b) = -ab$	114, 149
	11	Division		$\dfrac{+a}{+b} = \dfrac{-a}{-b} = -\dfrac{-a}{+b} = -\dfrac{+a}{-b} = \dfrac{a}{b}$	117, 163, 361
	12			$\dfrac{+a}{-b} = \dfrac{-a}{+b} = -\dfrac{-a}{-b} = -\dfrac{a}{b}$	117, 163, 361
PERCENTAGE	13			Amount = base × rate, $\quad A = BP$	97
	14			Percent change = $\dfrac{\text{new value} - \text{original value}}{\text{original value}} \times 100$	101
	15			Percent error = $\dfrac{\text{measured value} - \text{known value}}{\text{known value}} \times 100$	104
	16			Percent concentration of ingredient $A = \dfrac{\text{amount of } A}{\text{amount of mixture}} \times 100$	104, 199
	17			Percent efficiency = $\dfrac{\text{output}}{\text{input}} \times 100$	105

Source: Adapted from Paul Calter, *Technical Mathematics, Third Edition,* Prentice-Hall, 1995.

	No.				Page
EXPONENTS	18		Definition	$x^n = \underbrace{x \cdot x \cdot x \cdot \ldots \cdot x}_{n \text{ factors}}$	141
	19	Laws of Exponents	Product	$x^m \cdot x^n = x^{m+n}$	142, 439, 630
	20		Quotient	$\dfrac{x^m}{x^n} = x^{m-n} \quad (x \neq 0)$	142, 164, 439, 630
	21		Power	$(x^m)^n = x^{mn} = (x^n)^m$	143, 439, 630
	22		Product Raised to a Power	$(xy)^n = x^n \cdot y^n$	143, 439, 630
	23		Quotient Raised to a Power	$\left(\dfrac{x}{y}\right)^n = \dfrac{x^n}{y^n} \quad (y \neq 0)$	144, 439, 630
	24		Zero Exponent	$x^0 = 1 \quad (x \neq 0)$	144, 440
	25		Negative Exponent	$x^{-n} = \dfrac{1}{x^n} \quad (x \neq 0)$	145, 440, 630
	26	Fractional Exponents		$\sqrt[r]{x} = x^{\frac{1}{r}}$	440
	27			$x^{\frac{p}{r}} = (\sqrt[r]{x})^p = \sqrt[r]{x^p}$	441
RADICALS	28	Rules of Radicals		$\sqrt[r]{x^r} = x \text{ and } (\sqrt[r]{x})^r = x$	441
	29		Root of a Product	$\sqrt[r]{xy} = \sqrt[r]{x}\sqrt[r]{y}$	445
	30		Root of a Quotient	$\sqrt[r]{\dfrac{x}{y}} = \dfrac{\sqrt[r]{x}}{\sqrt[r]{y}}$	447
SPECIAL PRODUCTS AND FACTORING	31	Binomials	Difference of Two Squares	$x^2 - y^2 = (x - y)(x + y)$	153, 339
	32		Sum of Two Cubes	$x^3 + y^3 = (x + y)(x^2 - xy + y^2)$	356
	33		Difference of Two Cubes	$x^3 - y^3 = (x - y)(x^2 + xy + y^2)$	356
	34	Trinomials	Test for Factorability	$ax^2 + bx + c$ is factorable if $b^2 - 4ac$ is a perfect square	350
	35		Leading Coefficient = 1	$x^2 + (a + b)x + ab = (x + a)(x + b)$	154, 343
	36		General Quadratic Trinomial	$acx^2 + (ad + bc)x + bd = (ax + b)(cx + d)$	154, 348
	37		Perfect Square Trinomials	$x^2 + 2xy + y^2 = (x + y)^2$	157, 354
	38			$x^2 - 2xy + y^2 = (x - y)^2$	157, 354

	No.						Page
FRACTIONS	39	Simplifying		$\dfrac{ac}{bc} = \dfrac{a}{b}$			363, 364
	40			$\dfrac{a \div c}{b \div c} = \dfrac{a}{b}$			363
	41	Multiplication		$\dfrac{a}{b} \cdot \dfrac{c}{d} = \dfrac{ac}{bd}$			368
	42	Division		$\dfrac{a}{b} \div \dfrac{c}{d} = \dfrac{a}{b} \cdot \dfrac{d}{c} = \dfrac{ad}{bc}$			369
	43	Addition and Subtraction	Same Denominators	$\dfrac{a}{b} \pm \dfrac{c}{b} = \dfrac{a \pm c}{b}$			372
	44		Different Denominators	$\dfrac{a}{b} \pm \dfrac{c}{d} = \dfrac{ad}{bd} \pm \dfrac{bc}{bd} = \dfrac{ad \pm bc}{bd}$			376
RATIO AND PROPORTION	45	Ratio of a to b		$a : b$ or $\dfrac{a}{b}$			602
	46	Proportion		$\dfrac{a}{b} = \dfrac{c}{d}$			606
	47	Fundamental Principle of Proportions $\dfrac{a}{b} = \dfrac{c}{d}$		The product of the extremes equals the product of the means		$ad = bc$	607
VARIATION	48	k = Constant of Proportionality	Direct	$y \propto x$ or $y = kx$			614
	49		Joint	$y \propto xw$ or $y = kxw$			620
	50		Inverse	$y \propto \dfrac{1}{x}$ or $y = \dfrac{k}{x}$			621
	51		Combined	$y \propto \dfrac{x}{w}$ or $y = \dfrac{kx}{w}$			622
SYSTEMS OF LINEAR EQUATIONS	52	Algebraic Solution	$\begin{aligned} a_1 x + b_1 y &= c_1 \\ a_2 x + b_2 y &= c_2 \end{aligned}$	$x = \dfrac{b_2 c_1 - b_1 c_2}{a_1 b_2 - a_2 b_1}$, and $y = \dfrac{a_1 c_2 - a_2 c_1}{a_1 b_2 - a_2 b_1}$		where $a_1 b_2 - a_2 b_1 \ne 0$	427
	53		$\begin{aligned} a_1 x + b_1 y + c_1 z &= k_1 \\ a_2 x + b_2 y + c_2 z &= k_2 \\ a_3 x + b_3 y + c_3 z &= k_3 \end{aligned}$	$x = \dfrac{b_2 c_3 k_1 + b_1 c_2 k_3 + b_3 c_1 k_2 - b_2 c_1 k_3 - b_3 c_2 k_1 - b_1 c_3 k_2}{a_1 b_2 c_3 + a_3 b_1 c_2 + a_2 b_3 c_1 - a_3 b_2 c_1 - a_1 b_3 c_2 - a_2 b_1 c_3}$ $y = \dfrac{a_1 c_3 k_2 + a_3 c_2 k_1 + a_2 c_1 k_3 - a_3 c_1 k_2 - a_1 c_2 k_3 - a_2 c_3 k_1}{a_1 b_2 c_3 + a_3 b_1 c_2 + a_2 b_3 c_1 - a_3 b_2 c_1 - a_1 b_3 c_2 - a_2 b_1 c_3}$ $z = \dfrac{a_1 b_2 k_3 + a_3 b_1 k_2 + a_2 b_3 k_1 - a_3 b_2 k_1 - a_1 b_3 k_2 - a_2 b_1 k_3}{a_1 b_2 c_3 + a_3 b_1 c_2 + a_2 b_3 c_1 - a_3 b_2 c_1 - a_1 b_3 c_2 - a_2 b_1 c_3}$			432
	54	Value of a Determinant	Second Order	$\begin{vmatrix} a_1 & b_1 \\ a_2 & b_2 \end{vmatrix} = a_1 b_2 - a_2 b_1$			428
	55		Third Order	$\begin{vmatrix} a_1 & b_1 & c_1 \\ a_2 & b_2 & c_2 \\ a_3 & b_3 & c_3 \end{vmatrix} = a_1 b_2 c_3 + a_3 b_1 c_2 + a_2 b_3 c_1 - (a_3 b_2 c_1 + a_1 b_3 c_2 + a_2 b_1 c_3)$			433
	56	Determinants / Cramer's Rule	Two Equations	$x = \dfrac{\begin{vmatrix} c_1 & b_1 \\ c_2 & b_2 \end{vmatrix}}{\begin{vmatrix} a_1 & b_1 \\ a_2 & b_2 \end{vmatrix}}$ and $y = \dfrac{\begin{vmatrix} a_1 & c_1 \\ a_2 & c_2 \end{vmatrix}}{\begin{vmatrix} a_1 & b_1 \\ a_2 & b_2 \end{vmatrix}}$	where $\begin{vmatrix} a_1 & b_1 \\ a_2 & b_2 \end{vmatrix} \ne 0$		429
	57		Three Equations	$x = \dfrac{\begin{vmatrix} k_1 & b_1 & c_1 \\ k_2 & b_2 & c_2 \\ k_3 & b_3 & c_3 \end{vmatrix}}{\begin{vmatrix} a_1 & b_1 & c_1 \\ a_2 & b_2 & c_2 \\ a_3 & b_3 & c_3 \end{vmatrix}}$, $y = \dfrac{\begin{vmatrix} a_1 & k_1 & c_1 \\ a_2 & k_2 & c_2 \\ a_3 & k_3 & c_3 \end{vmatrix}}{\begin{vmatrix} a_1 & b_1 & c_1 \\ a_2 & b_2 & c_2 \\ a_3 & b_3 & c_3 \end{vmatrix}}$, $z = \dfrac{\begin{vmatrix} a_1 & b_1 & k_1 \\ a_2 & b_2 & k_2 \\ a_3 & b_3 & k_3 \end{vmatrix}}{\begin{vmatrix} a_1 & b_1 & c_1 \\ a_2 & b_2 & c_2 \\ a_3 & b_3 & c_3 \end{vmatrix}}$ where $\begin{vmatrix} a_1 & b_1 & c_1 \\ a_2 & b_2 & c_2 \\ a_3 & b_3 & c_3 \end{vmatrix} \ne 0$			434

	No.				Page
QUADRATICS	58	General Form		$ax^2 + bx + c = 0$	469
	59	Quadratic Formula		$x = \dfrac{-b \pm \sqrt{b^2 - 4ac}}{2a}$	482
INTERSECTING LINES	60			Vertical angles are equal	249
	61			When two lines intersect, the adjacent angles are supplementary, $A + B = 180°$	250
	62			If two parallel lines are intersected by a transversal, corresponding angles are equal and alternate interior angles are equal	251
QUADRILATERALS	63		Quadrilateral	$A + B + C + D = 360°$	267
	64			Perimeter = $a + b + c + d$	268
	65		Trapezoid	Area = $\dfrac{(a + b)h}{2}$	268
	66		Parallelogram	Perimeter = $2(a + b)$	269
	67			Area = bh	269
	68		Rectangle	Perimeter = $2(a + b)$	269
	69			Area = ab	269
	70		Rhombus	Perimeter = $4a$	270
	71			Area = ah	270
	72		Square	Perimeter = $4a$	271
	73			Area = a^2	271
CIRCLES	74			Diameter = $d = 2r$	274
	75			Circumference = $2\pi r = \pi d$	275
	76			Area = πr^2	275
	77			$\theta = \dfrac{s}{r}$ (in radians)	564
	78			Area of sector = $\dfrac{r^2 \theta}{2}$	566
	79			1 revolution = 2π radians = $360°$ $1° = 60'$ $1' = 60''$	246, 560

No.				Page
80		Prism	Volume = (area of base)(altitude) = Bh	281
81			Lateral area = (perimeter of base)(altitude) = ph	282
82			Surface area = 2(area of base) + lateral area = $2B + L$	283
83		Rectangular Solid	Volume = lwh	283
84			Surface area = $2(lw + hw + lh)$	283
85		Cube	Volume = a^3	284
86			Surface area = $6a^2$	284
87		Cylinder	Volume = $\pi r^2 h$	287
88			Surface area = $2\pi r^2 + 2\pi rh$	288
89		Pyramid	Volume = $\frac{1}{3}$ (area of base)(altitude) = $\frac{1}{3} Bh$	291
90			Lateral area = $\frac{1}{2}$ (perimeter of base) × (slant height) = $\frac{1}{2} ps$	292
91			Surface area = area of base + lateral area = $B + L$	292
92		Cone	Volume = $\frac{1}{3} \pi r^2 h$	293
93			Lateral area = πrs	294
94			Surface area = $\pi r^2 + \pi rs$	294
95		Frustum Cone or Pyramid	Volume = $\frac{1}{3} h (A_1 + A_2 + \sqrt{A_1 + A_2})$	295
96			$V_{Cone} = \frac{1}{3} \pi h (R^2 + r^2 + Rr)$	295
97			Lateral Area = $\frac{s}{2}(P_1 + P_2)$	297
98		Sphere	Volume = $\frac{4}{3} \pi r^3$	299
99			Surface area = $4\pi r^2$	300

SOLIDS

Appendix A / Summary of Facts and Formulas

	No.				Page
ANY TRIANGLE	100		Perimeter	$p = a + b + c$	257
	101		Area	$\text{Area} = \frac{1}{2} bh$	257
	102		Hero's Formula	$\text{Area} = \sqrt{s(s-a)(s-b)(s-c)}$ where $s = \frac{1}{2}(a+b+c)$	257
	103		Sum of the Angles	$A + B + C = 180°$	255
	104		Law of Sines	$\dfrac{a}{\sin A} = \dfrac{b}{\sin B} = \dfrac{c}{\sin C}$	518
	105		Law of Cosines	$a^2 = b^2 + c^2 - 2bc \cos A$ $b^2 = a^2 + c^2 - 2ac \cos B$ $c^2 = a^2 + b^2 - 2ab \cos C$	527
RIGHT TRIANGLES	106			The acute angles of a right triangle are complementary, $A + B = 90°$	261, 312
	107		Pythagorean Theorem	$c^2 = a^2 + b^2$	262, 312
	108		Hypotenuse of a Right Triangle	$c = \sqrt{a^2 + b^2}$	262, 312
	109		Leg of a Right Triangle	$a = \sqrt{c^2 - b^2}$	262, 312
	110			30-60-90 Right Triangle Relationships	263
	111			Isosceles Right Triangle Relationships	265
	112	Trigonometric Ratios	Hypotenuse	$r = \sqrt{x^2 + y^2}$	305
	113		Sine	$\sin \theta = \dfrac{y}{r} = \dfrac{\text{opposite leg}}{\text{hypotenuse}}$	306, 312, 505
	114		Cosine	$\cos \theta = \dfrac{x}{r} = \dfrac{\text{adjacent leg}}{\text{hypotenuse}}$	306, 312, 505
	115		Tangent	$\tan \theta = \dfrac{y}{x} = \dfrac{\text{opposite leg}}{\text{adjacent leg}}$	306, 312, 505
	116		Cosecant	$\csc \theta = \dfrac{r}{y} = \dfrac{\text{hypotenuse}}{\text{opposite leg}}$	325, 505
	117		Secant	$\sec \theta = \dfrac{r}{x} = \dfrac{\text{hypotenuse}}{\text{adjacent leg}}$	325, 505
	118		Cotangent	$\cot \theta = \dfrac{x}{y} = \dfrac{\text{adjacent leg}}{\text{opposite leg}}$	325, 505
	119	Reciprocal Relations	$\csc \theta = \dfrac{1}{\sin \theta}$	$\sec \theta = \dfrac{1}{\cos \theta}$ $\quad$ $\cot \theta = \dfrac{1}{\tan \theta}$	327
VECTORS	120		Components of a Vector **V**	$V_x = V \cos \theta$	537
	121			$V_y = V \sin \theta$	537
	122		Resultant of Two Perpendicular Vectors	$R = \sqrt{A^2 + B^2}$	547
	123			$\tan \theta = \dfrac{B}{A}$	547

Appendix A / Summary of Facts and Formulas

	No.				Page
LOGARITHMS	124	Exponential and Logarithmic Forms		If $b^x = y$ then $x = \log_b y$ $(y > 0, b > 0, b \neq 1)$	647
	125	Laws of Logarithms	Product	$\log_b MN = \log_b M + \log_b N$	656
	126		Quotient	$\log_b \dfrac{M}{N} = \log_b M - \log_b N$	657
	127		Power	$\log_b M^p = p \log_b M$	658
	128		Root	$\log_b \sqrt[r]{M} = \dfrac{1}{r} \log_b M$	658
	129		Log of 1	$\log_b 1 = 0$	659
	130		Log of the Base	$\log_b b = 1$	659
	131		Log of the Base Raised to a Power	$\log_b b^n = n$	659

No.					Page
132			Slope	$m = \dfrac{y_2 - y_1}{x_2 - x_1}$	221
133			$m = \tan$ (angle of inclination) $= \tan \theta$ $0 \leq \theta < 180°$		221
134	Straight Line		Linear Equation	$Ax + By = C$	402
135			Horizontal Line	$y = b$	223
136			Vertical Line	$x = a$	224
137			Slope-Intercept Form	$y = mx + b$	221
138	Quadratic Function			$y = ax^2 + bx + c$	226
139				Vertex V at $x = \dfrac{-b}{2a}$	227
140	Power Function			$y = ax^n$	616
141	Exponential Function			$y = a \cdot b^x$ where $b > 0$ and $b \neq 1$	629
142	Exponential Growth			$y = ae^{rt}$	637
143	Exponential Decay			$y = ae^{-rt}$	641
144	Logarithmic Function			$y = \log_b x$ $(x > 0, b > 0, b \neq 1)$	653

SOME USEFUL FUNCTIONS

	No.					Page
TRIGONOMETRIC FUNCTIONS	145	Sine Function	(graph)		$y = a \sin(bx + c)$	588
	146				amplitude $= |a|$	577, 582, 588
	147				Period $= \dfrac{2\pi}{b} = \dfrac{360°}{b}$	577, 582, 588
	148				Phase shift $= -\dfrac{c}{b}$	588
	149	Cosine Function	(graph)		$y = a \cos(bx + c)$	591
	150				amplitude $= |a|$	591
	151				Period $= \dfrac{2\pi}{b} = \dfrac{360°}{b}$	591
	152				Phase shift $= -\dfrac{c}{b}$	591
	153	Tangent Function	(graph)		$y = a \tan(bx + c)$	598
	154				Period $= \dfrac{\pi}{b} = \dfrac{180°}{b}$	598
	155				Phase shift $= -\dfrac{c}{b}$	598
COMPLEX NUMBERS	156	Imaginary Unit			$j = \sqrt{-1}$	670
	157		Rectangular Form		$a + jb$	676
	158		Trigonometric Form		$a + jb = r(\cos\theta + j\sin\theta)$	689
	159	(diagram: Imaginary, $a+jb$, r, b, θ, a, Real)		where	$a = r\cos\theta$	688
	160				$b = r\sin\theta$	688
	161				$r = \sqrt{a^2 + b^2}$	686
	162				$\tan\theta = \dfrac{b}{a}$	687
	163		Polar Form		$a + jb = r\angle\theta$	687

Appendix A / Summary of Facts and Formulas

Conversion Factors

Conversion factors marked with an asterisk (*) are exact. All other conversion factors are rounded to four significant digits. The first conversion factor listed for each unit is the SI factor for that unit.

Unit		Equals	
Length			
1 angstrom	Å	1×10^{-10} metre*	m
1 chain	ch	20.1168 metres*	m
		66 feet*	ft.
		100 links*	
1 foot	ft	0.3048 metre*	m
		12 inches*	in.
1 furlong		201.2 metres	m
		660 feet*	
1 inch	in.	0.0254 metre*	m
1 light year	l.y.	9.461×10^{12} kilometres	km
		5.879×10^{12} statute miles	mi
1 nautical mile (international)		1852 metres*	m
		6076 feet	
		1.151 statute miles	
1 rod	rd	5.0292 metres*	m
		1 pole*	
		1 linear perch*	
		16½ feet*	ft
1 statute mile	mi	1609.344 metres*	m
		1760 yards*	yd
		5280 feet*	ft
		63 360 inches*	in.
1 yard	yd	0.9144 metre*	m
		3 feet*	ft
		36 inches*	in.
Angles			
1 degree	°	$\pi/180$ radians*	
		0.01745 radians	
		60 minutes of arc*	
		3600 seconds of arc*	
		1/360 revolution*	
		10/9 grad*	
1 minute of arc	'	$\pi/10\,800$ radians*	
		60 seconds of arc*	
1 second of arc	"	$\pi/648\,000$ radians*	
1 radian	rad	$180/\pi$ degrees*	
1 revolution	rev, r	2π radians*	
		360 degrees*	
1 grad		$\pi/200$ radian*	
		1/400 revolution*	
		9/10 degree*	

Unit		Equals	
Area			
1 acre		4047 square metres	m²
		0.4047 hectare	ha
		43 560 square feet*	ft²
		10 square chains*	
1 hectare	ha	1 square hectometre*	hm²
		10 000 square metres*	m²
		2.471 acres	
		107 600 square feet	ft²
1 square foot	ft²	0.092 90 square metre	m²
		144 square inches*	in.²
1 square inch	in.²	6.4516 × 10⁻⁴ square metre*	m²
		6.4516 square centimetres*	cm²
1 square mile	mi²	2.589 988 110 square kilometres*	km²
		640 acres*	
		1 section*	
1 square yard	yd²	0.8361 square metres	m²
		9 square feet*	ft²
		1296 square inches*	in.²
Volume			
1 acre-foot	acre·ft	1233 cubic metres	m³
		43 560 cubic feet*	ft³
1 board-foot	board·ft	0.002 360 cubic metres	m³
		2360 cubic centimetres	ft³
		144 cubic inches*	in.³
1 bushel (Imperial)	bu (Imp.)	0.036 37 cubic metre	m³
		1.284 cubic feet	ft³
		36.37 litres	L
1 bushel (U.S.)	bu (U.S.)	0.035 24 cubic metre	m³
		1.244 cubic feet	ft³
		35.24 litres	L
1 cord		3.625 cubic metres	m³
		128 cubic feet*	ft³
1 cubic foot	ft³	0.028 32 cubic metre	m³
		6.226 gallons (Imperial)	gal (Imp.)
		7.481 gallons (U.S.)	gal (U.S.)
		1728 cubic inches*	in.³
1 cubic inch	in.³	16.387 064 cubic centimetres*	cm³
1 cubic yard	yd³	0.7646 cubic metres	m³
		27 cubic feet*	ft³
1 gallon (Imperial)	gal (Imp.)	4.456 090 cubic decimetres*	dm³
		0.004 546 cubic metres	m³
		4 quarts (Imperial)*	qt (Imp.)
		160 ounces (Imperial)*	oz (Imp.)
1 gallon (U.S.)	gal (U.S.)	3.785 411 784 cubic decimetres*	dm³
		0.003 785 cubic metres	m³
		4 quarts (U.S.)*	qt (U.S.)
		128 ounces (U.S.)*	oz (U.S.)
1 litre	L	1 cubic decimetre*	dm³
1 ounce (Imperial)	fl oz (Imp.)	28.41 cubic centimetres	cm³
1 ounce (U.S. fluid)	fl oz (U.S.)	29.57 cubic centimetres	cm³
1 quart (Imperial)	qt (Imp.)	0.001 136 cubic metre	m³
		1.136 litres	L
		40 ounces (Imperial)*	oz (Imp.)
1 quart (U.S. liquid)	liq qt (U.S.)	0.000 964 4 cubic metre	m³
		0.9464 litres	L
		32 ounces (U.S.)*	oz (U.S.)

Unit		Equals	
Mass			
1 carat		0.22 gram*	g
1 hundredweight (long)		50.80 kilograms	kg
		112 pounds*	lb
1 hundredweight (short)		45.36 kilograms	kg
		100 pounds*	lb
1 grain	gr	0.06480 gram	g
1 ounce (avoirdupois)	oz	28.35 grams	g
1 ounce (troy)	oz	31.10 grams	g
1 pound (avoirdupois)	lb	0.453 592 37 kilogram*	kg
		16 ounces (avoirdupois)*	oz
		7000 grains*	gr
1 pound (troy)	lb	0.3732 kilogram	kg
1 slug		14.59 kilograms	kg
		32.17 pounds (avoirdupois)	
1 ton (long)		1016 kilograms	kg
		2240 pounds*	lb
1 ton (short)		907.2 kilograms	kg
		2000 pounds*	lb
1 tonne (metric)	t	1000 kilograms*	kg
		1 megagram*	Mg
Force			
1 dyne	dyn	1×10^{-5} newton*	N
1 kip		4448 newtons	N
1 ounce-force		0.2780 newton	N
1 pound-force	lb, lbf	4.448 newtons	N
1 poundal		0.1383 newton	N
Velocity			
1 foot per minute	ft/min	5.08×10^{-3} metre per second*	m/s
		0.3048 metre per minute*	m/min
		18.288 metres per hour*	m/h
1 foot per second	ft/s	0.3048 metre per second*	m/s
		18.288 metres per minute*	m/min
		1097 metres per hour	m/h
1 kilometre per hour	km/h	0.2778 metre per second	m/s
		16.67 metres per minute	m/min
1 knot (international)		0.5144 metre per second	m/s
		30.87 metres per minute	m/min
		1.852 kilometres per hour*	km/h
		1.151 miles per hour	mi/h
		1 nautical mile per hour*	
1 mile per hour	mi/h	0.447 04 metre per second*	m/s
		1.609 344 kilometres per hour*	km/h
Energy, Work			
1 British thermal unit (IT)	Btu (IT)	1055 joules	J
1 calorie (thermochemical)	cal (th)	4.184 joules*	J
1 foot-pound-force	ft · lbf	1.356 joules	J
1 joule	J	1 watt second*	W · s
		1 newton metre*	N · m
		1 volt ampere second*	V · A · s
1 kilowatt hour	kW · h	3.6 megajoules*	MJ

Unit		Equals	
Power			
1 British thermal unit (IT) per hour	Btu(IT)/h	0.2931 watt	W
1 British thermal unit (IT) per second	Btu(IT)/s	1055 watts	W
1 calorie (thermochemical) per second	cal(th)/s	4.184 watts*	W
1 foot-pound-force per second	ft·lbf/s	1.356 watts	W
1 horsepower (electric)	hp(electric)	746 watts*	W
		0.7071 Btu(IT) per second	Btu(IT)/s
1 horsepower	hp	745.7 watts	W
		550 foot-pound-force per second*	ft·lbf/s
		0.7068 Btu(IT) per second	Btu(IT)/s
1 watt	W	1 newton metre per second*	N·m/s
		1 joule per second*	J/s
		1 volt ampere*	V·A
Pressure			
1 atmosphere (standard)	atm	101.325 kilopascals*	kPa
		14.70 pounds-force per square inch	lbf/in.^{2}
		29.92 inches of mercury (32°F)	
		760 millimetre of mercury (0°C)	
1 bar		100 kilopascals*	kPa
1 dyne per square centimetre	dyn/cm^2	0.1 pascal*	Pa
1 foot of water (39.2°F)		2.989 kilopascals	kPa
1 inch of mercury (32°F)		3.386 kilopascals	kPa
1 inch of mercury (60°F)		3.377 kilopascals	kPa
1 millimetre of mercury (0°C)		0.1333 kilopascal	kPa
1 pascal	Pa	1 newton per square metre*	N/m^2
1 pound-force per square foot	lbf/ft^2	0.04788 kilopascal	kPa
1 pound-force per square inch	lbf/in.^{2}, psi	6.895 kilopascals	kPa

APPENDIX B / Conversion Factors

Answers to Selected Odd-Numbered Exercises

Exercises expressed in imperial units are indicated by italicizing the exercise number.

CHAPTER 1 Arithmetic with Whole Numbers

EXERCISE 1-1
The Real Numbers

1. integer, whole, natural, a digit, real, rational, positive
3. real, rational, positive
5. real, decimal, negative
7. real, rational, negative
9. real, decimal, irrational, negative
11. ten thousands, tens
13. exact
15. approximate
17. to 27.

$$\leftarrow \!\!\! \begin{array}{c} \downarrow \quad \downarrow \quad \quad \downarrow \quad \quad \downarrow \quad \downarrow \\ -6 \;\; -4 \;\; -2 \;\; 0 \;\; 2 \;\; 4 \;\; \underset{23.}{\overset{\uparrow}{6}} \;\; 8 \end{array} \!\!\! \rightarrow$$

29. $8 < 11$
31. $-2 < 6$
33. $\dfrac{1}{4} = 0.25$

EXERCISE 1-2
Adding and Subtracting Whole Numbers

1. 1521   3. 1731   5. 16 786   7. 16 057
9. 164 303   11. 87 934   13. 2466   15. 1649
17. 340   19. 131   21. 463   23. 489
25. 3836   27. 62   29. 62   31. $42 129
33. 99

EXERCISE 1-3
Multiplying Whole Numbers

1. 2 081 937   3. 24 064   5. 410 688   7. 36 204
9. 7056   11. 21 868   13. 5200   15. 5915 grams
17. $299

EXERCISE 1-4
Dividing Whole Numbers

1. 3   3. 12   5. 5   7. 0
9. undefined   11. 3 m³   13. 810 km   *15.* 4 A

EXERCISE 1-5
Powers and Roots of Whole Numbers

1. 9   3. 4   5. 121   7. 529   9. 64
11. 125   13. 625   15. 128   17. 256   19. 512
21. 4   23. 45   25. 3   27. 2   29. 2
31. 2645 W   *33.* 14 cm

EXERCISE 1-6
Scientific Notation

1. 100 000   3. 10   5. 10 000   7. 10 000 000
9. 10^3 11. 10^4 13. 3×10^4 15. 2×10^6
17. 7×10^3 19. 3000   21. 200   23. 7 000 000

EXERCISE 1-7 Combined Operations	**1.** 3340 **3.** 5940 **5.** 5 **7.** 3 **9.** 121 **11.** 27 **13.** 27 **15.** 24 **17.** 12 **19.** 2 **21.** 30
CHAPTER 1 Review Exercises	**1.** 441 **3.** 451 **5.** 63 **7.** 7 **9.** 0 **11.** 812 **13.** 32 **15.** 625 **17.** 7 **19.** 5 **21.** 559 **23.** 20 988 **25.** 11 **27.** undefined **29.** 343 **31.** 12 **33.** 21 **35.** 4 **37.** 10 000 **39.** The 5 is in the hundreds place, the 6 is in the ten-thousands place. **41.** 400 000 **43.** and **45.** **47.** 3×10^5 **49.** $2 < 3$ **51.** $\frac{6}{3} = 2$ **53.** 10^5 **55.** 3 m^3 **57.** 300 km **59.** 3 A **61.** 44 **63.** $752 **65.** 3960° **67.** 6 ft

CHAPTER 2 Common Fractions

EXERCISE 2-1 Terminology	**1.** common proper **3.** common improper **5.** algebraic **7.** common improper **9.** algebraic complex **11.** mixed number **13.** similar **15.** neither
EXERCISE 2-2 Changing the Form of a Fraction	**1.** $\frac{1}{3}$ **3.** 2 **5.** $\frac{4}{9}$ **7.** $\frac{4}{3}$ **9.** $\frac{9}{12}$ **11.** $\frac{56}{21}$ **13.** $2\frac{2}{5}$ **15.** $2\frac{1}{2}$ **17.** $11\frac{5}{7}$
EXERCISE 2-3 Multiplying Fractions	**1.** $\frac{2}{7}$ **3.** $\frac{5}{12}$ **5.** $\frac{3}{10}$ **7.** $\frac{7}{27}$ **9.** $\frac{1}{8}$ **11.** $\frac{1}{10}$ **13.** $\frac{7}{2}$ **15.** $\frac{34}{27}$ **17.** $\frac{15}{2}$ **19.** $\frac{1813}{135}$ **21.** $\frac{17}{8}$ **23.** $\frac{11}{24}$ **25.** $\frac{77}{24}$ **27.** $\frac{1225}{192}$ **29.** $9\frac{17}{40}$ rev **31.** $152\frac{17}{30}$ loads
EXERCISE 2-4 Dividing Fractions	**1.** $\frac{1}{2}$ **3.** $\frac{3}{2}$ **5.** $\frac{13}{12}$ **7.** $\frac{14}{15}$ **9.** $\frac{9}{16}$ **11.** $\frac{4}{15}$ **13.** $\frac{3}{25}$ **15.** 49 **17.** $\frac{27}{2}$ **19.** $\frac{1}{60}$ **21.** $\frac{2}{315}$ **23.** $\frac{20}{9}$ **25.** $\frac{21}{5}$ **27.** $\frac{4}{9}$ **29.** $\frac{237}{140}$ **31.** $\frac{49}{6}$

33. $\frac{3}{2}$ 35. $\frac{17}{80}$ 37. $\frac{13}{90}$ 39. $\frac{416}{81}$

41. $\frac{147}{295}$ 43. $\frac{17}{32}$ %   45. $21\frac{2}{3}$ V

EXERCISE 2-5
Adding and Subtracting Fractions

1. $\frac{5}{7}$ 3. $\frac{3}{4}$ 5. $\frac{3}{13}$ 7. $\frac{1}{21}$

9. $\frac{4}{5}$ 11. $\frac{5}{13}$ 13. 16   15. 12

17. 35   19. $\frac{5}{6}$ 21. $\frac{7}{16}$ 23. $\frac{13}{22}$

25. $\frac{5}{24}$ 27. $\frac{17}{30}$ 29. $\frac{53}{156}$ 31. $\frac{20}{3}$

33. $\frac{99}{16}$ 35. $\frac{19}{5}$ 37. $\frac{7}{4}$ 39. $5\frac{1}{2}$

41. $\frac{121}{16}$ 43. $\frac{60}{11}$ 45. $\frac{35}{24}$ 47. $\frac{139}{30}$

49. $\frac{799}{156}$ 51. $33\frac{7}{32}$ %   53. $5\frac{43}{60}$ minutes   55. $84\frac{19}{20}$ ft

EXERCISE 2-6
Complex Fractions

1. $\frac{1}{20}$ 3. $\frac{24}{5}$ 5. $\frac{15}{14}$ 7. $\frac{10}{3}$ 9. $\frac{1}{2}$

11. $\frac{7}{13}$ 13. $\frac{1}{2}$

CHAPTER 2
Review Exercises

1. $\frac{8}{15}$ 3. $\frac{17}{16}$ 5. $\frac{5}{48}$ 7. $\frac{104}{135}$

9. $\frac{7}{16}$ 11. $\frac{71}{16}$ 13. $\frac{11}{48}$ 15. $\frac{47}{6}$

17. $\frac{9}{35}$ 19. $\frac{3}{110}$ 21. $\frac{27}{44}$ 23. $\frac{1}{6}$

25. $\frac{111}{8}$ 27. $\frac{50}{9}$ 29. $\frac{61}{105}$ 31. $\frac{7}{24}$

33. 1   35. $\frac{1}{27}$ 37. $\frac{38}{5}$ 39. $\frac{15}{8}$

41. $\frac{1}{7}$ 43. $\frac{16}{3}$ 45. $3\frac{1}{7}$ 47. $17\frac{4}{5}$ hours

49. $230\frac{2}{5}$ rotations   51. $6\frac{13}{24}$ %   53. $89\frac{41}{60}$ lb   55. $45\frac{13}{15}$ lbf/in.

CHAPTER 3 Arithmetic with Decimal Numbers

EXERCISE 3-1
Decimal Numbers

1. 1, 3   3. 3, 4   5. 0, 1   7. 5, 5   9. 58.31
11. 86.04   13. 689.05   15. 9876.76   17. 46.0   19. 9.8
21. 3984.7   23. 4358.4   25. 58 000   27. 8 475 300   29. 7350
31. 0.0476   33. 231 000   35. 43.6

EXERCISE 3-2
Adding and Subtracting Decimal Numbers

1. 1361.6   3. 5.02   5. 37.3   7. 7334   9. 22.14
11. 0.001 09   13. 35.92   15. $20 258   17. $70 504   19. 11.125 ft

EXERCISE 3-3 Multiplying Decimal Numbers

1. 4.32
3. 1881
5. 997.78
7. 62.8
9. 21.22
11. 0.220
13. 81.7
15. 2673.0
17. 88.5
19. 252
21. 65.6
23. 866
25. 42.2
27. 4433 m, $217
29. 293 W
31. $430.75
33. 14.8 mi

EXERCISE 3-4 Dividing Decimal Numbers

1. 5.07
3. 1.808
5. 77.81
7. 0.301
9. 0.2168
11. 1.82
13. 17.35
15. 80.34
17. 2183
19. 0.010 66
21. 0.001 066
23. 0.104
25. 0.001 78
27. 0.105
29. 6618 km
31. 16.1 days
33. 108 cm
35. 298 Ω
37. 138 ft^3

EXERCISE 3-5 Powers and Roots of Decimal Numbers

1. 4.54
3. 346
5. 0.0348
7. 1.018
9. 9460
11. 572
13. 5.559
15. 4.28
17. 22.2
19. 4.08
21. 4.56
23. 21.00
25. 9.18
27. 3.76
29. 91.79
31. 3.85
33. 4.44
35. 559 cm^3
37. 135 m
39. 728 Ω
41. 2918 W

EXERCISE 3-6 Scientific Notation

1. 2.373×10^3
3. 7.26×10^3
5. 1.92×10^6
7. 5.228×10^3
9. 7.01×10^3
11. 902.3
13. 4010
15. 952 000
17. 8.74×10^4
19. 5.49×10^3
21. 1.36×10^3
23. 10^7
25. 10^9
27. 3.10×10^{12}
29. 4.94×10^9
31. 2.51×10^9
33. 10^3
35. 7.81×10^4
37. 2×10^4
39. 30
41. 5.65×10^7
43. 2.09 A
45. 1.69×10^7 N/m^2

EXERCISE 3-7 Combined Operations

1. 105
3. 97 800
5. 2.28
7. 0.160
9. 189
11. 3704
13. 3.59
15. 7.17
17. 3.23
19. 0.871
21. 7.93

CHAPTER 3 Review Exercises

1. 0.003 05
3. 0.4241
5. 0.0188
7. 2.3×10^3
9. 8.39×10^5
11. 3.02×10^3
13. 302.3
15. 8010
17. 41.0
19. 25.4
21. 2270
23. 1.878
25. 29.9
27. 41.0
29. 89.09
31. 14.83
33. 20.31
35. 28.9
37. 30.9
39. 0.212
41. 3.52
43. 3.32
45. 1.36
47. 34.6
49. 1.5×10^6
51. 3.27×10^{11}
53. 5.07
55. 0.153
57. 4.52×10^{15}
59. 6.44×10^4
61. 3.41×10^4
63. 8.89×10^7 bits
65. 4.87 h

CHAPTER 4 Units of Measurement

EXERCISE 4-1 The SI (Metric) System of Units

1. 22 300 m
3. 47.835 mm
5. 0.0452 Mm
7. 3476 J
9. 3450 nmol
11. 3.01 m
13. 6.09 m
15. 13 040 m

EXERCISE 4-2 British/American Customary System of Weights and Measure

1. 12.7 ft
3. 9144 in.
5. 58 000 lb
7. 44.8 short tons
9. 1.09 nautical miles
11. 0.217 cubic ft.
13. 45.0 board feet
15. 665 cubic inches
17. 0.314 gal (Imp.)
19. 0.2028 acres
21. 2.154 acres
23. 117 gallons

EXERCISE 4-3
Converting Between British/American and Metric Units

1. 228 miles
3. 39.7 L
5. 207 mph
7. 14.5 m³
9. 88 ft/s
11. 1.19×10^6 ft
13. 3.337×10^5 kg
15. 17.6 litres
17. 3.60 m
19. 125 g
21. 21 100 kJ
23. 150 kL/s

EXERCISE 4-4
Arithmetic with Numbers Having Units of Measure

1. 151.4 cm
3. 136 acres
5. 317 N·m
7. 30.6 cm³
9. 29.3 km/h
11. 1.65 m/s
13. 25.2
15. 1.66
17. 27.4 cm²
19. 86 900 mm³
21. 105 m³
23. 5.08 mm
25. 6.34 cm
27. 5250 m²
29. 14.7 in.³

EXERCISE 4-5
Converting Areas and Volumes

1. 4327 m²
3. 2.30 m²
5. 0.863 in.²
7. 243 000 acres
9. 982.8 yd²
11. 1.27×10^7 in.³
13. 56.4 m³
15. 0.0163 in.³
17. 29.6 gal
19. 97.8 yd³
21. 6160 m²
23. 1.44×10^7 in.³

EXERCISE 4-6
Converting Rates

1. 4.00 mi/h
3. 107 km/h
5. 18.3 births/week
7. 4860 ha/year
9. 19.8¢/m²
11. 236¢/lb
13. 46.8 L/ha

CHAPTER 4
Review Exercises

1. 1.068 kJ/s
3. 2.28×10^3 megohms
5. 7.52×10^{-3} μF
7. 969.2 in.²
9. 487 ha/year
11. 15.6 slug
13. 0.0832 km
15. 283.2 kg
17. 24.6¢/m²
19. 2820 N
21. 19.2 mi/h
23. 89.3 km/h
25. 21.3 m²
27. 268 m³

CHAPTER 5 Percentage

EXERCISE 5-1
Converting to and from Percent

1. 629%
3. 0.46%
5. 65.3%
7. 395.9%
9. 60%
11. 90%
13. 28.6%
15. 4%
17. 33.3%
19. 46.7%
21. 0.45
23. 3.655
25. 0.074
27. 0.52
29. $\frac{1}{8}$
31. $2\frac{1}{4}$
33. $\frac{5}{8}$
35. $\frac{9}{20}$

EXERCISE 5-2
Solving Percentage Problems

1. 75 tonnes
3. 980 kg
5. 1.5 m³
7. 126 km
9. 528 ha
11. 5000
13. 300
15. 72
17. $2\frac{4}{7}$
19. 50%
21. 20%
23. 30%
25. 10%
27. 6.67%
29. 2118 Ω
31. $1050
33. 2.31 cm
35. 119 tonnes
37. 107 km
39. $3668
41. $3000
43. 18.2 million barrels
45. 77.8%
47. 56.3%
49. 64.4%
51. $182 000
53. $520
55. $5567.02

EXERCISE 5-3
Percent Change

1. +97.7%
3. +37.2%
5. $28\overline{0}$
7. 8306
9. 360
11. $40\overline{0}$
13. $2\overline{0}$
15. 775
17. 0.68% increase
19. $33 000 and $33 990
21. 3.72%
23. $89.88
25. 22.9% increase
27. 1.33%
29. 38.3 million barrels

EXERCISE 5-4
Percent Error, Percent Concentration and Percent Efficiency

1. 1.66% low   3. 120V and 225V   5. 0.392% high
7. 16 L   9. 86%   11. 62.7%
13. 74.2%   15. 24.2%

CHAPTER 5
Review Exercises

1. (a) 61.5% increase   (b) 18.2% decrease
3. (a) 841   (b) 22.8
5. (a) 309   (b) 646
7. (a) $\frac{21}{40}$ (b) $\frac{19}{40}$
   (c) $3\frac{1}{4}$ (d) $\frac{9}{100}$
9. (a) 14.3%   (b) 46.7%
   (c) 12.0%   (d) 71.4%
11. (a) 824%   (b) 62.7%
    (c) 0.27%   (d) 35.8%
13. (a) 171   (b) 73.6
15. (a) 1358 kg   (b) 95.3 kJ
    (c) 331 km   (d) 134 litres
17. (a) 7030   (b) 388
    (c) 368
19. (a) 0.15   (b) 0.0764
    (c) 0.655   (d) 0.283
21. (a) 35.7%   (b) 38.1%
    (c) 52%   (d) 24.3%

CHAPTER 6 Arithmetic with Signed Numbers

EXERCISE 6-1
Signed Numbers

1. and 3.

5. $5 > 1$ 7. $-5 < 1$ 9. $-12.2 > -12.3$
11. 6   13. 3   15. 1

EXERCISE 6-2
Adding and Subtracting Signed Numbers

1. 2   3. 53   5. 2   7. -12 9. -89
11. -7.56 13. -5.56 15. 59.5   17. -41.16 19. 10
21. 126   23. -71 25. -6 27. 42   29. 26.30
31. 12.10   33. -127.9 35. 86.48   37. $-\$5218.60$ 39. -415 m

EXERCISE 6-3
Multiplying Signed Numbers

1. -32 3. -77 5. -27 7. 15   9. 46
11. -1250 13. $-10\overline{0}0$ 15. 20.8   17. 3250   19. 42
21. 240   23. 144   25. -9.20 27. 22.6   29. 2140
31. -1.36 m/s   33. 13 950 N·m

EXERCISE 6-4
Dividing Signed Numbers

1. -4 3. -0.3333 5. -2 7. 1.6000
9. 0.3478   11. -0.493 13. -61.1 15. 0.816
17. 0.0531   19. -11.5V/min

Answers to Selected Odd-Numbered Exercises   711

EXERCISE 6-5 Converting Within SI	1. 23 670 000 MW	3. 84 500 pF	5. 0.055 23 Mg
	7. 4.4598 hm²	9. 4800 dm³	11. 60.032 ha, $1 416 755
	13. 1.3 km³	15. 6250 m²	

EXERCISE 6-6 Powers and Roots of Signed Numbers

1. -32 3. 25 5. 36 7. 64 9. -2.93
11. 5.52 13. 5.57 15. 88.2 17. 1/81 19. 1/8
21. 1/49 23. 1/729 25. 0.003 28 27. 0.0325 29. 0.0195
31. 1.74×10^{-5} 33. -2 35. -2 37. -3 39. -2.03
41. -4.01 43. -2.85 45. -2.12 47. -2.21 49. -1.39

EXERCISE 6-7 Scientific and Engineering Notation

1. $3.73 \times 10^{-1}, 373 \times 10^{-3}$ 3. $2.60 \times 10^{-2}, 26.0 \times 10^{-3}$
5. $2.0000 \times 10^{-3}, 2.0000 \times 10^{-3}$ 7. $2.28 \times 10^{-2}, 22.8 \times 10^{-3}$
9. 0.023 23   11. 0.007 01
13. 0.000 033 70   15. 2.02×10^{-2}
17. 0.436   19. 2.24×10^{-10}
21. 5640   23. 2.82×10^{5}
25. 3.40×10^{-5} 27. 3.45 A
29. 1.57×10^{8} N/m², 157×10^{6} N/m²   31. 5.15×10^{8} lb/in.²

EXERCISE 6-8 Combined Operations

1. -10.4 3. -1.06 5. 0.0651 7. -1.34 9. 2.55×10^{-5}
11. 0.214 13. 1.77 15. 0.629 17. 8.48

CHAPTER 6 Review Exercises

1. $-3 < -1$ 3. $-13.2 > -18.6$ 5. $8.54 \times 10^{-3}, 8.54 \times 10^{-3}$
7. 3 9. 8 11. 0.054704
13. 0.197 15. 2.91×10^{-4} 17. -22.5
19. $-91\overline{0}$ 21. -1230 23. -3520
25. 8.96×10^{-5} 27. 0.0234 29. -2.41
31. -2.70 33. -2.23 35. -2.18
37. 2840 39. -13.5 41. -1.84
43. -0.281 45. -13.0 47. -18.2
49. -0.0213 51. 0.00707 53. 15.79
55. 12.10 57. -2.26 59. 30.5
61. 25.8 63. 2480 65. 1.10×10^{-8}
67. $85.45/share 69. $-$7218.64 71. 21.5°C
73. 5.08 A 75. -655 ft

CHAPTER 7 Introduction to Algebra

EXERCISE 7-1 Algebraic Expressions

1. yes 3. no 5. no 7. yes 9. 2 11. 1
13. $3, a, x$ 15. $7, x, y$ 17. 6 19. -1 21. $\dfrac{a}{2}$ 23. $3(c - 2)$
25. 1st 27. 2nd 29. 1st 31. 2nd

EXERCISE 7-2 Adding and Subtracting Polynomials

1. $10y$ 3. $21.0a$
5. $-6ac$ 7. $2.71y$
9. $-2x$ 11. $52x + 91y + 22z$
13. $2.83a - 7.32x - 9.59z$ 15. $5z - 3$
17. $-2c - 11$ 19. $10.48z + 0.97$

21. $0.553c - 8.13$
23. $7a^4 - 8a^3 - 2$
25. $2bc - 2c + a$
27. $7.74a - 7.56b + 10.58c$
29. $8ab + 3x^2 - 7z + 2y^2$
31. $y - 2y^2 + 6b - 5$
33. $a + 14b - 3c + 3d$
35. $-5ab - 4cd$
37. $8b - 17p - 14r + 9s$
39. $6x^3 - 11x^2 + 5x - 2a^3 - 5a^2 + 2a - 15$
41. $2b$
43. $12b - 2a - 16c - 8d$
45. $5by^5 - 72bx^5 - 8bx^4 + 23by^4$
47. $26a^2 + 5bd - 14c$
49. $5(b + c)$

EXERCISE 7-3 Laws of Exponents

1. 27
3. 16
5. 1×10^{-9}
7. x^6
9. a^9
11. 10^8
13. a^2
15. y^3
17. 10^{b-a+2}
19. b
21. a^8
23. z^{ac}
25. a^{3x-3}
27. $9a^2$
29. $32x^5y^5z^5$
31. $-8/125$
33. $\dfrac{16x^4}{9y^4}$
35. 1
37. 82
39. 1
41. $\dfrac{1}{x}$
43. $\dfrac{x^4}{16}$
45. $\dfrac{27y^6}{64x^9}$
47. $\dfrac{4}{w^4} - \dfrac{3}{z^2}$
49. a^{-1}
51. c^2d^{-3}
53. x^4y^2

CHAPTER 7 Review Exercises

1. 81
3. -64
5. $35w^2 - 2y + 4xz$
7. $\dfrac{1}{b}$
9. x
11. 1
13. b^3
15. b^{x+3}
17. 10^{14}
19. $256x^{12}y^8$
21. $\dfrac{1}{27}$
23. $\dfrac{27a^3x^9}{64y^6z^3}$
25. $-2a$
27. $2w + 3y - 2z$
29. w^{-1}
31. $x^2y^2z^{-4}$
33. $108w^3$
35. 7
37. 2
39. $\dfrac{1}{a}$
41. $\dfrac{w}{x^5y^2}$
43. $\dfrac{27b^6}{64a^9}$
45. $5z - 3$
47. $11b^2 - 15ab + w$
49. $5w - 5z - 2y - 4x$
51. $2w + 12a - w^3 - 9a^2$

CHAPTER 8 Multiplication of Algebraic Expressions

EXERCISE 8-1 Product of Two or More Monomials

1. $-xy$
3. $-abc$
5. $9.19xy$
7. $16x^2y$
9. a^7
11. $-15.8a^4x^3$
13. $-30x^5y^5z^3$
15. $-40x^4y^5z^2$
17. $9.14a^{m+3}x^2$
19. $3x^{m+3}y^{n+2}$

EXERCISE 8-2 Product of a Multinomial and a Monomial

1. $-6 - 3x$
3. $2a + 6b$
5. $3.83b^3 + 4.86b$
7. $-30x^2 - 21x$
9. $2a^3b + a^2b^2 - a^3b^2$
11. $18a^3b + 12a^2b^2 - 6ab^3$
13. $-22.0x^3y^3 - 6.35x^3y^2 + 30.1x^2y^3 + 15.2x^2y^2$

EXERCISE 8-3 Product of Two Binomials

1. $x^2 + xy + xz + yz$
3. $8m^3 + 2m^2n - 4mn - n^2$
5. $2x^2 + xy - y^2$
7. $12x^2y^4 + 7a^3bxy^2 - 12a^6b^2$
9. $2a^2 - 21x^2 - 11ax$
11. $a^2x^2 - 25b^2$
13. $2.93x^2 + 1.82xy - 1.11y^2$
15. $13.1y^4 + 6.28a^3by^2 - 18.5a^6b^2$
17. $LW - 3L + 2W - 6$

EXERCISE 8-4
Product of Two Multinomials

1. $x^2 - xy + x + 3y - 12$
3. $4w^3 + 2w^2 - 22w + 10$
5. $b^8 + 2.93b^7 - 2.82b^6 - 8.26b^5 + 4.27b^4 + 12.5b^3$
7. $4c^4 + 7c^3 + c^2 + 7c - 3$
9. $x^3 + b^3$
11. $b^3 - x^3$
13. $a^3 - xyz - axy + a^2z + 5az + 5a^2$
15. $x^5 - bx^4 - x^3y^2 + abx^3y - x^2y^3 - ax^2y^2 + bx^2y^2 - abxy^3 + y^5 + ay^4$
17. $25dz^2 - 40dwz + 16dw^2$
19. $d^3 - 6.78d^2 - 0.168d + 45.8$
21. $c^3 + c^2m + c^2n - c^2p + cmn - cmp - cnp - mnp$
23. $y^5 - 2y^4 - 16y + 32$
25. $25x^5 - 10x^4 + 10x^3y^2 - 10x^3 - 4x^2y^2 + 4x^2$
27. $x^2 - 2xz - y^2 + z^2$
29. $m^5 - 3.37m^4 + 34.7m^3 - 114m^2 + 89.5m - 10.9$

EXERCISE 8-5
Powers of Multinomials

1. $x^2 + 2xy + y^2$
3. $a^2 - 2ad + d^2$
5. $B^2 + 2BD + D^2$
7. $24.2y^2 + 30.7yz + 9.73z^2$
9. $36x^2 + 60nx + 25n^2$
11. $w^2 - 2w + 1$
13. $b^6 - 26b^3 + 169$
15. $x^2 + 2xy + 2xz + y^2 + 2yz + z^2$
17. $a^2 + 2ab - 2a + b^2 - 2b + 1$
19. $c^4 - 2c^3d + 3c^2d^2 - 2cd^3 + d^4$
21. $x^3 - 3x^2y + 3xy^2 - y^3$
23. $27.5m^3 + 59.1m^2n + 42.3mn^2 + 10.1n^3$
25. $c^3 + 3c^2d + 3cd^2 + d^3$
27. $x^2 + 4x + 4$
29. $4.19r^3 - 25.1r^2 + 50.3r - 33.5$

EXERCISE 8-6
Removing Symbols of Grouping

1. $2a + b$
3. $2x - y$
5. $9.03x + 34.0y$
7. $-5x - 11$
9. $14p + 2q + 2$

CHAPTER 8
Review Exercises

1. $-12a^4y^3$
3. $a^2 - ab + a + 3b - 12$
5. $x^2 - 3.45xy + 5.83x + 2.45y^2 - 5.83y$
7. $x^4 + 2x^2y - 15y^2$
9. $8a^3 + 4a^2b - 4ab - 2b^2$
11. $c^3 + c^2 - 21c + 45$
13. $x^3 + p^3$
15. $25a^5 - 10a^4 + 10a^3b^2 - 10a^3 - 4a^2b^2 + 4a^2$
17. $p^3 + p^2q - p^2r - pq^2 - 2pqr - pr^2 - q^3 - q^2r + qr^2 + r^3$
19. $3m^5 - 14m^4 + 2m^3 + 32m^2 - 35m + 12$
21. $a^4 - a^3b + ab^3 - b^4$
23. $b^2m^2 - b^2y^2 - bnyz + bmnz$
25. $-720x^3y^6z^5$
27. $4p^2 + 12pq + 9q^2$
29. $x^6 + 2x^5 + x^4$
31. $x^4 - 2x^3y + 3x^2y^2 - 2xy^3 + y^4$
33. $7x + 4y$
35. $6x + 2y + 2$
37. $64x^3 + 144x^2y + 108xy^2 + 27y^3$

CHAPTER 9 Division of Algebraic Expressions

EXERCISE 9-1
Quotient of Two Monomials
1. x^3 3. $5z$ 5. $-2a$ 7. $31b$ 9. $-33d$
11. $6p^2q^3r$ 13. $-8mn$ 15. $-4a^3bc$ 17. $-6n^3$ 19. $5a^2y$
21. $-4ac$ 23. $-3by$ 25. $8xy$ 27. y^6 29. a^{x-y}
31. $-5a^2z$ 33. $\dfrac{19}{ab^2}$

EXERCISE 9-2
Dividing a Polynomial by a Monomial
1. $15x^2 + 3x$ 3. $12d^4 - 2d$
5. $4c^2 + 3c^3$ 7. $-3 - 4p$
9. $-5x^2 - 3x$ 11. $bmn + 3b$
13. $2a^2 + ab$ 15. $y^2z - xy$
17. $mn^2 - mn - m^2n$ 19. $x^3 - x^2y^2 - y^3$
21. $\dfrac{c^2}{d^2} - 4c + \dfrac{d}{c}$ 23. $\dfrac{a^2}{b^2} + 2 - \dfrac{b^2}{a^2}$
25. $\dfrac{3z^3}{x} - 4x^2 - 2z$ 27. $r^3 - \dfrac{pq^3}{r} - \dfrac{q^2r^2}{p}$
29. $d^3 - \dfrac{c^3}{d} - 3cd$ 31. $12c - \dfrac{8c^2}{b} - \dfrac{4}{bc}$

EXERCISE 9-3
Quotient of Two Polynomials
1. $a + 8$ 3. $a + 8$ 5. $x + 5$ 7. $a - 8$
9. $-3x - 2$ 11. $a^4 - 2$ 13. $a^2 + 3a + 1$

CHAPTER 9 Review Exercises
1. $q^2r - pq$ 3. $m - 8$ 5. $p - q - r$
7. $(u - v)^3$ 9. $x^3 - xy - y^3$ 11. $-\dfrac{11b}{a}$
13. $p + 2q$ 15. $x - y$ 17. $a^2 + 2a - 1$
19. $z^4 + \dfrac{1}{z^4} + 1$ 21. $y - x^2y^2 - x^2y$

CHAPTER 10 Simple Equations

EXERCISE 10-1
Substituting into Equations and Formulas
1. 15 3. -23 5. 28
7. 19.78 9. 515 11. 10.7
13. $3050 15. $40.6°C$ 17. 17 N·m
19. $9756 21. 6.1 ft (U.S. gal), 8.8 ft (imp gal)

EXERCISE 10-2
Solving Simple Equations
1. 7 3. 2 5. $\dfrac{9}{4}$ 7. -4 9. 2
11. $\dfrac{9}{4}$ 13. 11 15. 2 17. 4 19. $-\dfrac{1}{2}$
21. 5 23. $\dfrac{1}{21}$ 25. 3 27. 18 29. 7
31. $-\dfrac{66}{5}$ 33. 13 35. $\dfrac{1}{7}$ 37. 16 39. 6
41. 0 43. 5 45. $\dfrac{2}{3}$ 47. 22 49. 5

Answers to Selected Odd-Numbered Exercises

51. $\dfrac{59}{3}$		**53.** -6		**55.** $-\dfrac{12}{7}$		**57.** 2		**59.** 26	
61. 56		**63.** 4		**65.** $\dfrac{15}{11}$					

EXERCISE 10-3
Simple Equations with Approximate Coefficients and Literals

1. 0.583 **3.** 28.5 **5.** -6.19 **7.** -7.16 **9.** -10.3
11. 3.89 **13.** $\dfrac{3}{d}$ **15.** $\dfrac{(c-b+d)}{a}$ **17.** $-\dfrac{a+2b}{a-b}$

CHAPTER 10 Review Exercises

1. 12 **3.** $\dfrac{5}{14}$ **5.** 1 **7.** 49
9. 4 **11.** $\dfrac{6}{5}$ **13.** 63 **15.** -8
17. 11 **19.** $-5/11$ **21.** -4 **23.** 25/33
25. $\dfrac{24}{23}$ **27.** 84 **29.** 54 **31.** -142
33. 14.7 **35.** 105 **37.** -191 m **39.** 3.9 N·m
41. 2.91×10^{11} N/m² **43.** 1020 Ω **45.** -0.361 **47.** 31.4
49. -0.504 **51.** $\dfrac{(d-5)}{c}$ **53.** $\dfrac{a}{m}$

CHAPTER 11 Solving Word Problems

EXERCISE 11-1
Solving Word Problems

1. x and $3x + 10$ **3.** x and $x + 42$ or x and $x - 42$
5. $\dfrac{x}{(6x+4)}$ **7.** $0.11x$ L **9.** (a) $32 - x$ (b) $\dfrac{320}{x}$
11. 8 **13.** 9 **15.** 5

EXERCISE 11-2
Uniform Motion Problems

1. 3.38 h **3.** $13\overline{0}$ km, 1.52 h **5.** 68.1 km/h
7. 51.8 mi **9.** 254 in./min

EXERCISE 11-3
Financial Problems

1. \$63 644 **3.** 6 masons
5. \$34 027 at 6.75% **7.** \$398 for computer
\$139 897 at 8.24% \$597 for printer

EXERCISE 11-4
Mixture Problems

1. 11.0 L **3.** 89.1% **5.** 5.71 tonnes of 1.15% chromium steel
2.29 tonnes of 1.50% chromium steel
7. 0.180 tonnes **9.** 27.8 lb

CHAPTER 11 Review Exercises

1. \$165 420 at 5.94% **3.** 6.44 litres **5.** \$68 493 **7.** 7 technicians
\$58 392 at 8.56%
9. 10 **11.** 5.40 gal **13.** 1130 lb **15.** 3060 lb

CHAPTER 12 Functions and Graphs

EXERCISE 12-1
Functions

1. 1 **3.** -64
5. explicit, x-independent, y-dependent
7. implicit **9.** $y = 5x^2$

11. $y = x + \dfrac{1}{x}$ **13.** $A = 4\pi r^2$
15. $v = at$ **17.** $d = 90t$
19. $C = 0.75 + 0.53(t - 1)$ **21.** $A = 5h$
23. $A = \dfrac{p^2}{16}$ **25.** $A = \dfrac{C^2}{4\pi}$
27. $(-2, -5), (-1, -2), (0, 1), (1, 4), (2, 7)$

EXERCISE 12-2
Functional Notation

1. $f(x) = 2x - 1$ **3.** $f(C) = \dfrac{9}{5}C + 32$
5. domain: all real numbers x; range: all real numbers y
7. domain: $x \neq 0$; range: $y \neq 0$ **9.** domain: $d \geq 0$; range: $W \geq 0$
11. $1, -8, 3a - 2, 14$ **13.** $\dfrac{7}{4}, 1, 3c^2 + 1, -9$
15. 1890 V²/Ω, 184 V²/Ω, 2.77 V²/Ω

EXERCISE 12-3
Rectangular Coordinates

1. $(4, 2)$ **3.** $(-4, 3)$ **5.** $(0, -4)$ **7.** $(-2.5, -2.5)$
9. to **25.**

27. IV **29.** II, III **31.** 0 **33.** II, IV **35.** parallelogram
37.

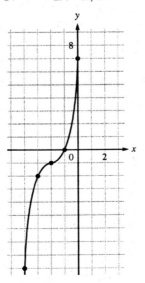

EXERCISE 12-4
Linear Functions

1. (a) −2, −1, 0, 1, 2   (b)
   (c) $x = 0, y = 0$ (d) $m = 1$
3. (a) −1, 1, 3, 5, 7   (c) $x = -\dfrac{3}{2}, y = 3$
   (d) $m = 2$
5. (a) 7, 4, 1, −2, −5   (b)
   (c) $x = \dfrac{1}{3}, y = 1$ (d) $m = -3$
7. (a) −2, −1.5, −1, −0.5, 0   (c) $x = 2, y = -1$
   (d) $m = \dfrac{1}{2}$
9. (a) 3, 3, 3, 3, 3   (b)

(c) no x intercept, $y = 3$ (d) $m = 0$
11. (b) $x = -1$, no y intercept   (c) m is undefined
13.   15. $y = -\dfrac{1}{5}x - \dfrac{6}{5}$

$y = 2x$

718 Answers to Selected Odd-Numbered Exercises

17.

19. $V = 4.5I$

21. $W = 7.87 V$

23. $C = 100x + 250000$

25.

EXERCISE 12-5
Quadratic and Other Functions

1. $6, 1, -2, -3, -2, 1, 6$

3. $V(0, 0)$

5. $V(-1, 2)$

7. $V(0, 0)$

9. $-8, -1, -\dfrac{1}{8}, 0, \dfrac{1}{8}, 1, 8$

11.
x	−3	−2	−1	0	1	2	3
y	9	2.7	0.3	0	−0.3	−2.7	−9

13.
x	−3	−2	−1	0	1	2	3
y	0	4	2	0	4	20	54

15.
x	−3	−2	−1	0	1	2	3
y	0.3	0.5	1	undef	−1	−0.5	−0.3

17.
x	−3	−2	−1	0	1	2	3
y	−0.5	−1	undef	1	0.5	0.3	0.2

19.
x	0	1	4	9
y	0	−1	−2	−3

21.
x	−8	−3	0	1
y	3	2	1	0

25.

EXERCISE 12-6
Reading Graphs

1. (a) $s = 50t - 4.9t^2$
   (c) 99 m
   (b)

3. (a) $A = 12\pi h + 72\pi$
   (d) 380 cm²
   (c) 7.3 cm

5. **(a)** $v = 5.0 + 2.0t$ **(b)**

(c) 16 m/s **(d)** 7.5 s

7. **(a)**   **(b)** 2.5 s

9. **(a)** l (cm)   **(b)** 14 N

11. **(a)**   **(b)** 96 ft/s **(c)** 3.0 s

13. **(a)**   **(b)** $140

15. $-0.6, 4.6$ 17. $-5.2, -0.8$ 19. 2.4

EXERCISE 12-7
Graphing Calculators

1.   5.

9.

11. $x = -2.4, 0.4$; $y = 1$; max$(-1, 2)$

13.

$x = -1, 2$
$y = -4$
min $(0, -4)$
max $(2, 0)$

15. $x = -1.45, 3.45$; $y = 5$; max$(1, 6)$

17.

$x = -0.62, 1, 1.62$
$y = -1$
min $(-1, 0)$
max $(1.33, 0.19)$

19. $y = 1.32x^2 + 1.95x - 4.02$; $x = -2.63, 1.16$

21. $y = 2.56x^3 - 6.75 - 7.25x^2$

$x = 3.11$

CHAPTER 12 Review Exercises

1. $y = 3x^2$

3. domain: $x \geq 3$; range: $y \geq 0$

5. $0, 6, b - b^3, \dfrac{3}{4}$

7. $-8, -5, -2, 1, 4$

9. implicit

11.

x	-3	-2	-1	0	1	2	3
y	0	-4	-6	-6	-4	0	6

13.

$x = -3, 0.5$
$y = 3.00$
max $(-1.25, 6.12)$

15. $C = 6.65 + 0.16(x - 200)$

17. $y = 300(1 + r)^2$
$340, $347, $364

21. Work to lift to 3.8 ft is 285 ft·lbf

23. $x = 3.81$; $y = -30$; max $(0.33, -29.85)$; min $(1.00, -30)$

CHAPTER 13 Plane Geometry

EXERCISE 13-1
Lines and Angles

1. R •——3 cm——• S

3. θ, $\angle Y$, $\angle XYZ$, $\angle ZYX$

5. 40°, 150°, −120°

7. A radian is the measure of a central angle that cuts off an arc of length r (the radius of the circle). A radian is about 57.3°.

9. 32.30°   **11.** 93.7914°
13. 235.3750°   **15.** 56°15′
17. 12°20′42″   **19.** 263°52′30″
21. 15.7 rad   **23.** 1.05 rad
25. 2.1962 rad   **27.** 0.557 rev
29. 0.836 rev   **31.** 0.209487 rev
33. 1620°   **35.** 293°33′29″

37.

39.

41. 225°

43. 150°

45.

47. 64.7°, 127.4°, 123.7°

49. $A = C = D = F = 117.9°$, $B = E = G = 62.1°$

51. $A = C = 145.2°$, $B = 34.8°$ **53.** due north

EXERCISE 13-2 Triangles

1. 62.1°, isosceles, acute, oblique   **3.** 61.7°, scalene, right
5. 27.9°, isosceles, obtuse, oblique   **7.** sum = 180°
9. 41.4°   **11.** 65°5′
13. yes   **15.** yes
17. yes, yes, yes   **19.** 4.19 in., 0.33 in.²
21. 11.44 cm, 2.98 cm²   **23.** 78.6°
25. 73.9°   **27.** $728
29. $22 980   **31.** $184

EXERCISE 13-3 Right Triangles

1. The hypotenuse is the longest side of a *right* triangle.
3. (a) 9.96 cm   (b) 6.18 cm
5. (a) $x = 7.79$ in., $y = 4.50$ in.   (b) $x = 4.60$ cm, $y = 3.25$ cm
7. 4.11 m, 9.77 m²   **9.** 79.2 cm
11. 51.62 cm   **13.** 188.2 mm, 2160 mm²
15. 57.0 m   **17.** 13.1 m
19. 10.2 m   **21.** 22.4 ft
23. 0.866 in.

EXERCISE 13-4 Quadrilaterals

1. 360°

3. 4 properties of a parallelogram: opposite sides are parallel, opposite sides are equal, opposite angles are equal, diagonals bisect each other

5. 6 properties of a rhombus: opposite sides are parallel, all sides are equal, opposite angles are equal, diagonals bisect each other, diagonals are perpendicular to each other, diagonals bisect the angles

7. (a) 27.50 cm, 41.6 cm²   (b) 13.50 in., 7.88 in.²   (c) 209.6 mm, 2750 mm²
9. 50.0 cm, 110 cm²   **11.** $1926   **13.** 1944 tiles
15. $315 000   **17.** 0.258 acre   **19.** $557

EXERCISE 13-5 Circles

1. C, $\overline{AC}$ or $\overline{BC}$ or $\overline{CD}$, $\overline{BD}$, $\overline{AD}$ or $\overline{BD}$, $\widehat{AB}$ or $\widehat{AD}$ or $\widehat{BD}$, $\widehat{BD}$
3. 22.6 cm, 40.7 cm²   **5.** 12.9 cm²
7. The radius is perpendicular to the tangent.
9. 78.5 m   **11.** 942 000 000 km   **13.** 71.3 cm²   **15.** 20.3 cm
17. 440 m   **19.** 37.7 mi   **21.** 6.61 in.   **23.** 1.03 in.²

CHAPTER 13 Review Exercises

1. 68.4°, 56.9°   **3.** 80.8°   **5.** 51.3°
7. 28°   **9.** 5.66 cm, 18.5 cm²   **11.** 36.3 cm, 50.9 cm²

13. 4 properties of a parallelogram: opposite sides are parallel, opposite sides are equal, opposite angles are equal, diagonals bisect each other

6 properties of a rectangle: opposite sides are parallel, opposite sides are equal, opposite angles are equal, all angles are right angles, diagonals bisect each other, diagonals are equal

6 properties of a rhombus: opposite sides are parallel, all sides are equal, opposite angles are equal, diagonals bisect each other, diagonals are perpendicular to each other, diagonals bisect the angles

7 properties of a square: opposite sides are parallel, all sides are equal, all angles are right angles, diagonals bisect each other, diagonals are equal, diagonals are perpendicular to each other, diagonals bisect the angles

15. 193.8 mm, 1170 mm² **17.** 21.8 acres **19.** 3610 ft, 235 acres
21. 14.0 in. **23.** 96 ft

CHAPTER 14 Solid Geometry

EXERCISE 14-1
Prisms
1. 268 cm³, 330 cm² **3.** 0.928 cm³, 7.37 cm² **5.** 66.7 cm³, 105 cm²
7. 61.6 cm³, 98.4 cm² **9.** 43.6 m³, 74.3 m² **11.** 0.165 m³, 1.81 m²
13. 1070 cm³ **15.** 15.3 cm³ **17.** 290̄ m²
19. 50.2 m³, $4390 **21.** 3.0 ft³ **23.** 16.3 loads
25. 561 in.²

EXERCISE 14-2
Cylinders
1. 358 m³ **3.** 205 cm² **5.** 22 600 L **7.** 37 200 cm³
9. 152 cm³ **11.** 0.495 m³ **13.** 348 m² **15.** 346 cm³
17. 4750 000 L **19.** 1.70 ft³ **21.** 449 000 gal (Imp) **23.** 2.65 in.

EXERCISE 14-3
Pyramids and Cones
1. 94.6 cm³ **3.** 114 cm³ **5.** 110 m² **7.** 377 cm²
9. 471 cm² **11.** 1.83 kg **13.** 1.20 kg **15.** 173 000 kg
17. 17.6 kg **19.** 101 m³ **21.** 14.1 t **23.** 10.6 m²
25. 186 in.², 3.47 lb **27.** 21.7 ft **29.** 3.93 gal (Imp)

EXERCISE 14-4
Spheres
1. 153 cm³, 139 cm² **3.** 1.07×10^{12} km³, 5.07×10^8 km²
5. 8.18 m³ **7.** 3.76 L
9. 630 L **11.** $63 700
13. 57.0 kg **15.** 18.6 lb
17. 1.24 ft, 4.84 ft² **19.** 0.0904 ft³

CHAPTER 14
Review Exercises
1. 179 cm³, 194 cm² **3.** 88.5 cm³, 190 cm²
5. 32.3 L **7.** 10.5 m²
9. 47.4 kL **11.** 3.77 cm³
13. 2.554×10^7 m³, 1.374×10^5 m² **15.** 160 m³
17. 8.21 lb (mass) = 9.98 lb(troy) = 120 oz(troy), $46 800
19. 215 lb

CHAPTER 15 Right Triangle Trigonometry

EXERCISE 15-1
Sine, Cosine, and
Tangent Functions
1. **3.**

Answers to Selected Odd-Numbered Exercises

5. (graph with 30° angle) **7.** (graph with 10° angle)

9. sin θ = 0.923, cos θ = 0.385, tan θ = 2.4
11. sin θ = 0.471, cos θ = 0.882, tan θ = 0.533
13. sin θ = 0.832, cos θ = 0.555, tan θ = 1.5
15. sin θ = 0.5, cos θ = 0.866, tan θ = 0.577
17. sin θ = 0.916, cos θ = 0.401, tan θ = 2.29
19. sin 60° = 0.866, cos 60° = 0.5, tan 60° = 1.73
21. sin 45° = 0.707, cos 45° = 0.707, tan 45° = 1
23. tan θ = 0.75   **25.** tan θ = 6
27. sin 45° = 0.707, cos 45° = 0.707, tan 45° = 1
29. 0.789   **31.** 0.222   **33.** 0.801   **35.** 0.980   **37.** 2.38
39. 0.723   **41.** 32.5   **43.** 1.00

EXERCISE 15-2
Inverse Trigonometric Functions

1. θ, angle, sine, 0.636
3. α, angle, tangent, 3.45
5. 30°, angle, cosine, a
7. $\frac{\pi}{2}$, angle, tangent, x
9. B, angle, sine, a
11. 1, angle, sine, 0.8415
13. 66.5°, angle, cosine, 0.399
15. $\alpha = \sin^{-1} 0.634$, α = arcsin 0.634
17. $\theta = \tan^{-1} 4.67$, θ = arctan 4.67
19. $b = \cos 60°$, 60° = arccos b
21. $\sin \alpha = 0.268$, $\alpha = \sin^{-1} 0.268$
23. $x = \cos 0.785$, $\cos^{-1} x = 0.785$
25. 89.0° = $\tan^{-1}$ 57.3, 89.0° = arctan 57.3
27. 45.0°   **29.** 78.4°   **31.** 30°   **33.** 45°   **35.** 30.0°
37. 0.39   **39.** 1.55   **41.** 80.8°   **43.** 45.0°   **45.** 60.0°
47. 1.47   **49.** 0.52   **51.** $\sin^{-1} 0.738 = 0.830$, $(\sin 0.738)^{-1} = 1.49$

EXERCISE 15-3
Solution of Right Triangles

1. $c = 25.0$, tan $A = 3.43$, $A = 73.7°$
3. $a = 40.0$, tan $A = 43.6$, $A = 12.7°$
5. $b = 24.8$, tan $A = 0.524$, $A = 27.7°$
7. $c = 11.7$, tan $A = 0.854$, $A = 40.5°$
9. $b = 1.25$, tan $A = 0.248$, $A = 13.9°$
11. $A = 42.0°$, $a = 23.4$ cm, $b = 26.0$ cm
13. $B = 21.0°$, $b = 1.66$ m, $c = 4.63$ m
15. $A = 64.8°$, $b = 26.7$ cm, $c = 62.8$ cm
17. $B = 47°45'$, $b = 1.82$ m, $c = 2.45$ m
19. $A = 29.7°$, $B = 60.3°$, $c = 8.06$ cm
21. $A = 41.8°$, $B = 48.2°$, $b = 5.14$ cm
23. $A = 30.2°$, $B = 59.8°$, $a = 20.8$ m

EXERCISE 15-4
Applications

1. 155 cm   **3.** 69.1°   **5.** 53.8°
7. 38.5 m   **9.** 10 300 m   **11.** 18.4 m
13. 9.90 m   **15.** 63.4 m   **17.** 39$\overline{0}$ m
19. 29.4 m   **21.** 23.8 km   **23.** 19$\overline{0}$ km north, 292 km west
25. 39.4°   **27.** 21.8°, 8.1 m   **29.** 58.8°
31. 2.95 cm, 53.6°   **33.** 7.066 mm   **35.** 4.85 cm

EXERCISE 15-5
Cosecant, Secant, and Cotangent Functions

1. $\csc\theta = 1.25$, $\sec\theta = 1.67$, $\cot\theta = 0.75$
3. $\csc\theta = 3.57$, $\sec\theta = 1.04$, $\cot\theta = 3.43$
5. $\csc\theta = 1.05$, $\sec\theta = 3.16$, $\cot\theta = 0.333$
7. $\csc\theta = 2.35$, $\sec\theta = 1.11$, $\cot\theta = 2.12$
9. $\csc\theta = 1.05$, $\sec\theta = 3.22$, $\cot\theta = 0.326$
11. $\csc A = 1.12$, $\sec A = 2.12$, $\cot A = 0.533$, $A = 61.9°$
13. $\csc A = 1.02$, $\sec A = 4.56$, $\cot A = 0.225$, $A = 77.3°$
15. $\csc A = 1.33$, $\sec A = 1.51$, $\cot A = 0.882$, $A = 48.6°$
17. 0.503 19. 1.13 21. 24.0 23. 0.760 25. 3.17
27. 29.6° 29. 4.5° 31. 0.224 33. 55.2° 35. 0.190

CHAPTER 15 Review Exercises

1. $\sin\theta = 0.980$, $\cos\theta = 0.198$, $\tan\theta = 4.95$, $\csc\theta = 1.02$, $\sec\theta = 5.05$, $\cot\theta = 0.202$, $\theta = 78.6°$
3. 0.974 5. 0.332 7. 0.224 9. 30.0°
11. 78.1° 13. 0.603 15. 34.1°
17. $\sin A = 0.421$, $\cos A = 0.907$, $\tan A = 0.464$, $\csc A = 2.37$, $\sec A = 1.10$, $\cot A = 2.15$, $A = 24.9°$
19. $B = 52.4°$, $a = 40.1$ cm, $b = 52.1$ cm
21. $A = 0.746$, $b = 3.52$ cm, $c = 4.79$ cm
23. 68.5 cm 25. 1660 cm² 27. 3.88 cm 29. 8210 ft
31. 19.5 ft, 77.0°

CHAPTER 16 Factoring

EXERCISE 16-1 Common Factors

1. $x(y + z)$
3. $3(a - b)$
5. $x(x - y)$
7. $6(x + 3)$
9. $3(8x^2 - 1)$
11. $a(c + 1)$
13. $x^2(4x - 3)$
15. $xy(24x^2 + 15xy - 40)$
17. $2x(2xy + 3x^2 - 1)$
19. $2a^2b^3(3 + 2a - 5b)$
21. $(2 + x)(x + y)$
23. $(p - 1)(q - p)$
25. $(c - d)(d + 1)$
27. $(x - y)(x - y + 1)$
29. $(3 + a)^2(2 + a)$
31. $(x + 3)(y + z)$
33. $(x + 3)(y - 2)$
35. $(x - 2)(x + y)$
37. $(m^2 + 3)(m - 5)$
39. $2\pi r(r + h)$
41. $\pi r(r + s)$
43. $i(R_1 + R_2 + R_3)$
45. $16(4 + 2t + t^2)$

EXERCISE 16-2 Difference of Two Squares

1. $(m + n)(m - n)$
3. $(x + 4)(x - 4)$
5. $(5 + p)(5 - p)$
7. $(3y + 1)(3y - 1)$
9. $(2a + 3b)(2a - 3b)$
11. $(p^2 + 2)(p^2 - 2)$
13. $3(x + 2y)(x - 2y)$
15. $4(ab + c)(ab - c)$
17. $(a^2 + b^2)(a + b)(a - b)$
19. $(p^3 + q^4)(p^3 - q^4)$
21. $(10x^2 + 9)(10x^2 - 9)$
23. $(x^4 + 1)(x^2 + 1)(x + 1)(x - 1)$
25. $(a + b + c)(a + b - c)$
27. $(p + q + r)(p - q - r)$
29. $(1 + a - b)(1 - a + b)$
31. $(s + d)(s - d)$
33. $\pi(R + r)(R - r)$
35. $\pi LD(R + r)(R - r)$

EXERCISE 16-3
Simple Trinomials

1. $(x + 7)(x + 1)$
3. $(a - 3)(a + 2)$
5. $(u + 9)(u - 2)$
7. $(y - 6)(y - 1)$
9. $(7 + k)(2 + k)$
11. $(x - 5y)(x - 3y)$
13. $(6 + x)(3 - x)$
15. $(a - 4b)(a + 2b)$
17. $(w + 4)(w - 15)$
19. $(x - 162)(x - 200)$
21. $(x - 10)(x - 1)$
23. $(x + y - 4)(x + y - 1)$
25. $(t + 1)^2(t - 1)^2$

EXERCISE 16-4
General Trinomials

1. $(2x + 1)(x + 1)$
3. $(5y - 2)(y - 1)$
5. $(3x + 1)(2x - 5)$
7. $(3a - 2)(a + 3)$
9. $(3 + 4x)(5 - 3x)$
11. $(3t + 2s)(2t + 7s)$
13. $3(b - 6)(b + 1)$
15. $2ab(b - 4)(b - 3)$
17. $3(5k - 7)(k + 1)$
19. $5(3 - t)(2 + t)$ or $-5(t - 3)(t + 2)$
21. $6(x - 15)(x + 6)$
23. $(2y - 5)(3y - 4)$
25. $(3 - x + y)(5 - x + y)$
27. $(7m + 5n)(2m)$

EXERCISE 16-5
Perfect Square Trinomials

1. $(x - 2)^2$
3. $(y + 3)^2$
5. $(6 - s)^2$
7. $(3a + 2b)^2$
9. $(1 - 11y)^2$
11. $(10 + 3x)^2$
13. $(9y + 4)^2$
15. $(4t - 3uv)^2$
17. $7y(y + 1)^2$
19. $5x(2y - 3)^2$
21. $(x + y + 1)^2$
23. $(2 - a + b)^2$
25. $(9m - n)^2$

EXERCISE 16-6
Sum or Difference of Two Cubes

1. $(a - 1)(a^2 + a + 1)$
3. $(x + 3)(x^2 - 3x + 9)$
5. $(b - 4)(b^2 + 4b + 16)$
7. $(1 - 2a)(1 + 2a + 4a^2)$
9. $(10 + 3x)(100 - 30x + 9x^2)$
11. $(4a - 5b)(16a^2 + 20ab + 25b^2)$
13. $3(y - 2)(y^2 + 2y + 4)$
15. $2p(3p + 1)(9p^2 - 3p + 1)$
17. $(a + b - x)[(a + b)^2 + (a + b)x + x^2]$
19. $(2x - a + b)[4x^2 + 2x(a - b) + (a - b)^2]$
21. $(a - b + x + y)[(a - b)^2 - (a - b)(x + y) + (x + y)^2]$
23. $(x^2 + 1)(x^4 - x^2 + 1)$
25. $(a - 1)(a^2 + a + 1)(a^6 + a^3 + 1)$
27. $\frac{4}{3}\pi(R - r)(R^2 + Rr + r^2)$
29. (a) $(a + b)(a - b)(a^4 + a^2b^2 + b^4)$
    (b) $(a + b)(a^2 - ab + b^2)(a - b)(a^2 + ab + b^2)$
    difference of two squares

CHAPTER 16
Review Exercises

1. $3y(5y - 9)$
3. $3a(5a - 7b + c)$
5. $(3 + y)(x - y)$
7. $(2x + 5)(2x - 5)$
9. $3(3 + 2y)(3 - 2y)$
11. $(m + n - 1)(m - n + 1)$
13. $4\pi(R + r)(R - r)$
15. $(m + 10)(m - 9)$
17. $(b - 7)(b - 4)$
19. $(b - 18)(b - 2)$
21. $(8y - 5)(3y + 4)$
23. $(5x - 8)(3x + 5)$
25. $(y + 8)^2$
27. $(4x + 9)^2$
29. $(1 - y)(1 + y + y^2)$
31. $(3x + 2y)(9x^2 - 6xy + 4y^2)$
33. $(a - b + 2)[a^2 + a(b - 2) + (b - 2)^2]$

CHAPTER 17 Algebraic Fractions

EXERCISE 17-1
Equivalent Fractions

1. $\dfrac{r}{s}$
3. $\dfrac{x}{2}$
5. $\dfrac{b-a}{1-b}$
7. $\dfrac{s-r}{r-1}$
9. $\dfrac{12x^2}{20xy}$
11. $\dfrac{a^2-b^2}{a^2-2ab+b^2}$
13. $\dfrac{3x}{2y}$
15. $\dfrac{6x^2}{5y^2}$
17. $\dfrac{9p^5}{5q^4}$
19. $\dfrac{4}{x+2}$
21. $\dfrac{4(x-2)}{3(x+2)}$
23. $\dfrac{5}{3}$
25. $\dfrac{x+4}{x-3}$
27. $\dfrac{x+7}{x-7}$
29. $\dfrac{x-2}{x+2}$
31. $\dfrac{1+2x}{3x-2}$
33. $\dfrac{x^2-xy+y^2}{x-y}$
35. $\dfrac{(x-y)^2}{x^2+xy+y^2}$
37. -1
39. 1
41. $b-a$
43. $\dfrac{1-2x}{2+x}$
45. $\dfrac{x-2}{x+2}$
47. $-\dfrac{3x}{3+x}$
49. $-\dfrac{y}{y+2x}$
51. $-\dfrac{x^2+2x+4}{2+x}$
53. $-\dfrac{3x^2}{x^2+2xy+4y^2}$
55. $\dfrac{m+M}{m}$
57. $\dfrac{R^2+Rr+r^2}{R+r}$

EXERCISE 17-2
Multiplying and Dividing Algebraic Fractions

1. $\dfrac{2ax}{3by}$
3. $\dfrac{20}{x}$
5. $\dfrac{2a}{c}$
7. $\dfrac{ab}{a^2-b^2}$
9. $\dfrac{2}{a}$
11. $\dfrac{3a+2b}{2(2a+3b)}$
13. $\dfrac{x+2}{x-2}$
15. $\dfrac{35ay}{18bx}$
17. $\dfrac{3x^2}{2y}$
19. $2x^2$
21. $\dfrac{(x+2)(x-1)}{6}$
23. $\dfrac{a+5}{3(6-a)}$
25. $\dfrac{4}{3}$
27. $\dfrac{3(x+y)}{x-y}$
29. $\dfrac{2c}{b}$
31. $\dfrac{2}{3}$
33. $\dfrac{2(2x-1)}{x^2-2x+4}$
35. $\dfrac{(x+1)^2}{(x-2)^2}$
37. $\dfrac{3x+1}{4(x-3)}$
39. $\dfrac{5x^2y^2}{a^4}$
41. $\dfrac{5xy}{12}$
43. $\dfrac{3x-1}{x-1}$
45. $\dfrac{2(x-1)(x-2)}{x^2+x+1}$
47. $\dfrac{2(x+1)}{2x+1}$
49. $\dfrac{3t+1}{2t^3}$
51. $\dfrac{3m}{\pi r^2 h}$

EXERCISE 17-3
Adding and Subtracting Algebraic Fractions

1. $\dfrac{7}{y}$
3. $\dfrac{7}{ab}$
5. $\dfrac{8x-5y}{a^2b^3}$
7. $\dfrac{4a}{x-y}$
9. $\dfrac{3}{a-2}$
11. $c+1$
13. $x-1$
15. $\dfrac{1}{n-m}$
17. $\dfrac{2(a+b)}{a-b}$

Answers to Selected Odd-Numbered Exercises

19. $\dfrac{x+y}{xy}$ 21. $\dfrac{7a-5}{a^2}$ 23. $\dfrac{7a}{6x}$

25. $\dfrac{3x-18y+8z}{12}$ 27. $\dfrac{yz-xz+xy}{xyz}$ 29. $\dfrac{12a^2b^2-20a+9b}{24a^2b^2}$

31. $-\dfrac{x+5y}{(x+y)(x-y)}$ 33. $\dfrac{2p^2+7}{(p+2)(p-3)}$ 35. $-\dfrac{13b+19}{(b+3)(b-2)(b+1)}$

37. $\dfrac{1}{x-1}$ 39. $\dfrac{6+5x}{2x(x-1)}$ 41. $\dfrac{a^2+1}{a}$

43. $\dfrac{c^2-2c+3}{c-2}$ 45. $\dfrac{-3x^3+2x^2+1}{x^3}$ 47. $\dfrac{-x^2-9x+6}{x(3x-2)}$

49. $\dfrac{3\pi\ell^3 - 12c\ell^2 + \pi c^3}{3\pi\ell^3}$ 51. $\dfrac{R_1R_2 + R_1R_3 + R_2R_3}{R_2 + R_3}$

EXERCISE 17-4
Complex Fractions

1. $\dfrac{2a}{b}$ 3. $\dfrac{xz}{y} \neq \dfrac{x}{yz}$ 5. $\dfrac{ad}{bc}$

7. $\dfrac{x}{2}$ 9. $\dfrac{b}{4}$ 11. $\dfrac{1}{3y-2}$

13. $\dfrac{3a}{4(3a+2)}$ 15. $\dfrac{b}{b-a}$ 17. $\dfrac{2a+1}{2a(a+4)}$

19. $\dfrac{x+2y}{x-2y}$ 21. $\dfrac{R_1R_2}{R_1+R_2}$ 23. $\dfrac{C_1C_2C_3}{C_2C_3 + C_1C_3 + C_1C_2}$

CHAPTER 17
Review Exercises

1. $\dfrac{x+6}{x-6}$ 3. $-a-2b$ 5. $\dfrac{y^2+xy+x^2}{y-x}$

7. $\dfrac{6(x+5)}{x(x+3)}$ 9. $-\dfrac{4}{n^2+2n+4}$ 11. $1+a$

13. $\dfrac{2y}{y-x}$ 15. $\dfrac{17x}{24}$ 17. $\dfrac{17-x^2}{(x+4)(x-4)}$

19. $\dfrac{a-15}{5}$ 21. $\dfrac{-6p^2+17p+15}{p(p-3)}$ 23. $-\dfrac{x}{3}$

25. $\dfrac{x}{x^2+x+1}$ 27. $\dfrac{t\lambda - xT}{T\lambda}$

CHAPTER 18 Fractional Equations

EXERCISE 18-1
Solution of Fractional Equations

1. 18 3. 5 5. 3 7. 60
9. 18 11. $-\dfrac{29}{8}$ 13. -1 15. 9
17. 2 19. 14 21. 8 23. 3
25. 2 27. -4 29. no solution 31. no solution
33. $\dfrac{5}{2}$ 35. no solution 37. no solution 39. 5.55
41. 0.386

EXERCISE 18-2
Literal Equations and Formulas

1. $x = ac - ab$ 3. $x = \dfrac{bc}{a}$ 5. $x = \dfrac{ab}{bc+ac}$

7. $x = ac - b$ 9. $x = \dfrac{bc}{a-b}$ 11. $a = \dfrac{2A}{b}$

13. $g = \dfrac{4\pi^2 L}{T^2}$ 15. $e = IR - E$ 17. $R_1 = \dfrac{RR_2R_3}{R_2R_3 - RR_3 - RR_2}$

19. $r = \dfrac{eR}{E - e}$ 21. $T_1 = \dfrac{P_1V_1T_2}{P_2V_2}$ 23. $g = \dfrac{2vt - 2s}{t^2}$

25. $B = \dfrac{2A - bh}{h}$ 27. 33 mg   29. 17.5 ohms

31. 599 mm   33. 7.58 μF

EXERCISE 18-3 Applications

1. 93.0 cm × 124 cm   3. 96.0 m, 48.0 m, 40.0 m   5. $324
7. 2.49 h   9. 3.39 days   11. 6.95 h
13. 23.7 km/h   15. 73.5 km/h, 4.05 h   17. no, all or no hits

CHAPTER 18 Review Exercises

1. 4.5   3. −12   5. 10
7. no solution   9. $M_1 = \dfrac{Fd^2}{kM_2}$ 11. $C = \dfrac{5F - 160}{9}$
13. 18 cm × 27 cm   15. 11.1 days   17. 3.36 h
19. $T = 2, W = 3, O = 7, S = 5, E = 6, V = 1, N = 9$

CHAPTER 19 Systems of Linear Equations

EXERCISE 19-1 Graphical Method

1. yes   3. no   5. yes
7. (−4, 4)   9. (−1, −2)   11. (9, 3)
13. (−2, 1)   15. (−1, 3)   17. (−5.3, 5.7)
19. all (x, y) such that $x - 2y = 6$ 21. (12, 7)
23. (0.3, 0.2)   25. no solution   27. (0.8, −0.5)
29. (3, −1)   31. all (x, y) such that $2x - 5y = -2$
33. (2.1, −0.2)   35. no solution   37. (2.5, 1.3)
39. (0, 1)   41. (3, 2)   43. no solution
45. (1.75, 0.39)   47. (−0.12, 1.31)

EXERCISE 19-2 Addition Method

1. (5, 2)   3. $\left(2, -\dfrac{1}{3}\right)$ 5. (−1, 3)
7. $\left(\dfrac{5}{2}, \dfrac{1}{2}\right)$ 9. (−2, 3)   11. (0.90, −0.38)
13. no solution   15. all (x, y) such that $2x - y = 3$
17. (2, 3)   19. $\left(2, -\dfrac{1}{2}\right)$ 21. no solution
23. (5.04, 2.17)   25. (0.427, 1.45)   27. (2, 3)
29. $\left(\dfrac{7}{5a}, \dfrac{1}{5b}\right)$ 31. $\left(\dfrac{ce - bf}{ae - bd}, \dfrac{af - cd}{ae - bd}\right)$

EXERCISE 19-3 Substitution Method

1. (5, 4)   3. (3, −1)   5. (4, 1)
7. $\left(\dfrac{2}{11}, -\dfrac{7}{11}\right)$ 9. (0, 3)   11. $\left(-\dfrac{14}{5}, -\dfrac{16}{5}\right)$
13. (2, −1)   15. all (x, y) such that $3x + 2y = -1$
17. no solution   19. $\left(\dfrac{2}{3}, \dfrac{7}{3}\right)$ 21. $\left(\dfrac{6a + 2b}{3}, \dfrac{3a + 4b}{3}\right)$

23. $\left(\dfrac{1}{3}, \dfrac{1}{4}\right)$ 25. (12, 8) 27. (1, 1)
29. (−2, 1) 31. (3, −2) 33. (4, 3)

EXERCISE 19-4
Applications of Systems of Two Equations

1. Gear (a) = 45 teeth, gear (b) = 28 teeth
3. Electrician = $18.00/h, assistant = $13.00/h
5. River = 0.772 km/h
7. Airplane = 1350 km/h, wind = 150 km/h
9. $v_0 = 4.47$ m/s, $a = 0.448$ m/s^2 11. 200 kg of 63% Cu, 800 kg of 78% Cu
13. 40.0 mL of 75% acid, 110 mL of water
15. 180 L   17. $3600 at 6.0%, $4400 at 10.5%
19. $2400 at 9.0%, $1800 at 12.0%   21. Carpenter 18 days, helper 36 days
23. Faster: 4 parts/h, slower: 3 parts/h   25. 9.00 ft, 12.0 ft
27. Train 7 h for 490 mi, bus 3 h for 165 mi

EXERCISE 19-5
Systems of Three Equations

1. (4, 2, 0)   3. (3, 2, 1)   5. no solution   7. (2, 2, −1)
9. (1, 0, −1)   11. (2, −1, 3)   13. $\left(\dfrac{1}{2}, \dfrac{1}{2}, \dfrac{1}{4}\right)$
15. 70 mL of 20%, 10 mL of 30%, 20 mL of 40%
17. 105 V, 65 V, 170 V
19. $1000 at 8%, $2000 at 9%, $5000 at 12%
21. 75 g of A, 150 g of B, 25 g of C

EXERCISE 19-6
Solution of Systems of Two Linear Equations by Determinants

1. 5   3. −1   5. 23.21
7. −8.1162   9. (4, −1)   11. $\left(\dfrac{1}{14}, -\dfrac{31}{14}\right)$
13. $\left(\dfrac{16}{11}, \dfrac{67}{11}\right)$ 15. (4, 0)   17. (4, −1)
19. (4.83, 1.09)   21. (0.824, 1.22)   23. 34 ohms, 53 ohms
25. 12 kg at $1.90, 4 kg at $1.50   27. 4.5 servings of A, 5 servings of B

EXERCISE 19-7
Solution of Systems of Three Linear Equations by Determinants

1. 19   3. −3   5. −29.568
7. (5, 6, 7)   9. (10, 15, 20)   11. (4, −3, 3)
13. (3, 2, −1)   15. (−5.4, −4.6, −3.0)   17. 1440 ohms, 340 ohms, 820 ohms

CHAPTER 19
Review Exercises

1. no   3. (−1, 0)   5. no solution   7. (3, −1)
9. (34, 25)   11. all (x, y) such that $y = 5 - 2x$
13. (4, 5)   15. no solution   17. (0, 2)   19. 14
21. (−2, 2)   23. all (x, y) such that $5x - 2y = 4$
25. no solution   27. (6, 0)   29. 50 V, 36 V
31. 15 ha at $1100, 25 ha at $1300   33. 40 kg of 30% tin, 80 kg of 70% tin
35. 4 kg at $1.40, 12 kg at $3.00   37. (3, −4, 2)
39. 33.951   41. 150 mL of 10%, 450 mL of 30%, 400 mL of 60%
43. 86.0 lb of zinc, 0.512 lb of tin, 4.62 lb of lead, 2.25 lb of nickel, 9.05 lb of manganese

CHAPTER 20 Exponents and Radicals

EXERCISE 20-1
Fractional Exponents

1. $\sqrt{x}$
3. $\sqrt[3]{y}$
5. $\dfrac{1}{\sqrt{a}}$
7. $\sqrt{x-y}$
9. $\left(\sqrt[3]{-c}\right)^2 = \sqrt[3]{(-c)^2}$
11. $-\left(\sqrt[3]{c}\right)^2 = -\sqrt[3]{c^2}$
13. $x^{1/2}$
15. $y^{1/4}$
17. $b^{2/3}$
19. $y^{-1/2}$
21. $(x+y)^{1/2}$
23. 4
25. 4
27. -3
29. 1
31. 0.1
33. 125
35. $\dfrac{1}{9}$
37. $\dfrac{1}{16}$
39. 16
41. $-\dfrac{1}{27}$
43. 27
45. 4
47. 0.111
49. -64
51. 2.24
53. -1.43
55. 4.33
57. 0.397
59. undefined
61. $a^{7/12}$
63. $b^{3/7}$
65. $2a$
67. $x^{1/2}$
69. $\dfrac{x^{3/2}}{y^3}$
71. $\dfrac{1}{a^{1/3}}$
73. $x^{3/2}$
75. $a^{2/3}$
77. $a^{10/3}$
79. $\sqrt[3]{a^2}$
81. 660 bacteria

EXERCISE 20-2
Simplification of Radicals

1. $2\sqrt{3}$
3. $6\sqrt{2}$
5. $2\sqrt[3]{3}$
7. $2\sqrt[5]{2}$
9. $3x^2$
11. $2b^2c\sqrt{6b}$
13. $3x\sqrt[3]{x}$
15. $2a^2\sqrt[3]{2a}$
17. $3x\sqrt[4]{x}$
19. $2x\sqrt[6]{x}$
21. $\dfrac{\sqrt{5}}{3}$
23. $\dfrac{2\sqrt{3}}{3}$
25. $\dfrac{2\sqrt{5}}{5}$
27. $\dfrac{\sqrt[3]{18}}{3}$
29. $\dfrac{\sqrt{7}}{2x}$
31. $\dfrac{\sqrt[3]{2y}}{3x}$
33. $\dfrac{-3\sqrt{5x}}{5x}$
35. $\dfrac{\sqrt{3x}}{3x}$
37. $\dfrac{\sqrt{30y}}{6}$
39. $\dfrac{\sqrt[3]{10x^2}}{5x}$
41. $\dfrac{\sqrt{7ab}}{14b^2}$
43. $\dfrac{x\sqrt{2xy}}{10y}$
45. $\dfrac{y^2\sqrt{6x}}{3x^2}$
47. $\dfrac{2ac\sqrt{3bc}}{3b}$
49. $\dfrac{\sqrt{\pi A}}{\pi}$
51. $\dfrac{\sqrt{kgW}}{2\pi W}$
53. $x\sqrt{x^2+1}$

EXERCISE 20-3
Adding and Subtracting Radicals

1. $4\sqrt{3}$
3. $4\sqrt[3]{2}$
5. $3\sqrt{2}$
7. $18\sqrt{2} - \sqrt{3}$
9. $15\sqrt{6} - 38\sqrt{5}$
11. $10 + 2\sqrt{2} + \sqrt[4]{8}$
13. $\dfrac{3\sqrt{2}}{4}$
15. $\sqrt{x}$
17. $3\sqrt{2x} - 3\sqrt{3x}$
19. $3\sqrt{2x} + 12\sqrt{5x}$
21. $11\sqrt{10x} + 110\sqrt{x}$
23. 0
25. $\dfrac{3\sqrt{10x}}{2}$
27. $\dfrac{\sqrt[3]{4y^2}}{4y}$
29. $\dfrac{26}{5}$

EXERCISE 20-4
Multiplying Radicals

1. $\sqrt{15}$
3. $-\sqrt{2x}$
5. $\sqrt[3]{6x^2}$
7. $3y$
9. $12\sqrt{6}$
11. $30x$
13. $2\sqrt[3]{3}$
15. $45a$
17. $10a\sqrt{2}$
19. $18\sqrt{10x}$
21. $6x^2y\sqrt{10}$
23. $24x\sqrt{3}$
25. $-120\sqrt{3}$
27. $5\sqrt{3} - 2\sqrt{6}$
29. $24\sqrt{x} + 18x\sqrt{2}$

31. $8\sqrt{3} - 12 + 6\sqrt{2}$ 33. $\sqrt{2ab} - a\sqrt{b}$ 35. $6x^2 + x\sqrt{3} - 3$
37. $8 - \sqrt{10}$ 39. $2x - 3\sqrt{3x} + 3$ 41. $5 - 4a^2$
43. $x - y$ 45. $2y + 2\sqrt{6y} + 3$
47. $\sqrt{6} + 3\sqrt{3} - 2\sqrt{10} - 6\sqrt{5}$
49. (a) $\sqrt[6]{5400}$ (b) $\sqrt[12]{5000}$

EXERCISE 20-5 Dividing Radicals

1. $\sqrt{3}$ 3. $\sqrt{2x}$ 5. $\sqrt{7b}$
7. $\dfrac{5\sqrt{3}}{3}$ 9. $\dfrac{x\sqrt{14}}{14}$ 11. $\sqrt[3]{7y}$
13. $1 + 2\sqrt{6}$ 15. $\dfrac{2\sqrt{x}}{x} - 1$ 17. $\dfrac{\sqrt{2}}{6} + \dfrac{2}{3}$
19. $6 + 3\sqrt{3}$ 21. $\dfrac{4\sqrt{6} - 3}{29}$ 23. $\dfrac{3 + 4\sqrt{x} + x}{1 - x}$
25. $6 - \sqrt{35}$ 27. $\dfrac{2a + 5\sqrt{2a} - 6}{a - 18}$ 29. $\sqrt{3} - \sqrt{6} + \sqrt{2} - 2$
31. $\dfrac{6 - 5\sqrt{x} + x}{4 - x}$ 33. $\dfrac{4 - 3\sqrt{2} - \sqrt{10}}{6}$ 35. $\dfrac{\sqrt{mv} - v}{m - v}$
37. $\dfrac{(\sqrt{d_1} - \sqrt{d_2})^2}{d_1 - d_2}$

EXERCISE 20-6 Radical Equations

1. 4   3. 16   5. no real solution   7. 27
9. -2 11. 9   13. no real solution   15. 16
17. 11   19. no real solution   21. 4   23. 9
25. 33   27. 3   29. -2 31. -8
33. 4   35. 4   37. no real solution   39. 1
41. 5   43. 5.0   45. $k = \dfrac{\omega^2 W}{g}$ 47. 2.23 cm

CHAPTER 20 Review Exercises

1. $\sqrt[5]{x}$ 3. $\dfrac{1}{\sqrt[3]{b}}$ 5. $(3y)^{1/2}$
7. $5^{-1/3}a$ 9. -2 11. $\dfrac{1}{7}$
13. 16   15. 1.57   17. $x^{1/8}$
19. $x^{3/5}$ 21. $a^{5/12}$ 23. 449 units
25. $20\sqrt{2y}$ 27. $2\sqrt[4]{2}$ 29. $3y^2\sqrt{2xy}$
31. $4x^2\sqrt[4]{x}$ 33. $\dfrac{\sqrt{11}}{5a^3}$ 35. $\dfrac{\sqrt{6}}{2}$
37. $\dfrac{\sqrt{30x}}{6x}$ 39. $\dfrac{y\sqrt{6xy}}{24x}$ 41. $-4\sqrt{2}$
43. $-\sqrt[3]{2}$ 45. $93 + 3\sqrt[3]{3}$ 47. 18
49. $4\sqrt{6} - 5\sqrt{2}$ 51. $30 - 12\sqrt{6}$ 53. $\sqrt{3}$
55. $\dfrac{6\sqrt{10} - 5}{5}$ 57. $\dfrac{16 - 7\sqrt{10}}{18}$ 59. $(4x - 3y)\sqrt[3]{2}$
61. $2\sqrt{7x} - \sqrt{6x}$ 63. $7\sqrt{3x} - 3\sqrt{7x}$ 65. $\dfrac{\sqrt{15x}}{10}$
67. $\sqrt[3]{12x^2}$ 69. $8x^2y\sqrt{15}$ 71. $8x\sqrt{5} - 60\sqrt{x}$
73. $3x + 7\sqrt{xy} + 2y$ 75. $\sqrt{2y}$ 77. $\sqrt[3]{11y}$
79. $\dfrac{6 + 5\sqrt{a} + a}{4 - a}$ 81. 4   83. no real solution

85. 17   **87.** 3   **89.** 64

91. $\dfrac{\sqrt{4\pi A}}{\pi}$ **93.** $W = \dfrac{kg}{4\pi^2 f^2}$ **95.** 21.8 cm, 41.2 cm, 32°, 58°

CHAPTER 21   Quadratic Equations

EXERCISE 21-1
Introduction to Quadratic Equations

1. a, c, f **3.** $2x^2 - 3x - 4 = 0$ **5.** $4x^2 - 7x = 0$
7. $5x^2 - 2x + 18 = 0$ **9.** $3x^2 - 5x - 3 = 0$ **11.** 2, 3
13. $-\dfrac{1}{3}, 1$ **15.** ± 5 **17.** $\pm 2\sqrt{3}$
19. 0   **21.** $\pm\dfrac{\sqrt{10}}{2}$ **23.** no real solution
25. $\dfrac{\sqrt{em}}{e}$ **27.** $\dfrac{\sqrt{3\pi hV}}{\pi h}$ **29.** $\dfrac{\sqrt{kI}}{I}$
31. $\dfrac{\sqrt{KFM_1M_2}}{F}$ **33.** 0.994 cm   **35.** 115 V
37. 2.21 m/s

EXERCISE 21-2
Solution by Factoring

1. $-3, 3$ **3.** $-2, 2$ **5.** $-\dfrac{8}{3}, \dfrac{8}{3}$ **7.** 0, 2
9. $-2, 0$ **11.** 0, 3   **13.** $-2, 3$ **15.** $-3, 1$
17. -1 **19.** $-\dfrac{5}{2}, -2$ **21.** $-3, 5$ **23.** 3
25. 1 m   **27. (a)** 4 s, 8 s   **(b)** 1 s, 11 s

EXERCISE 21-3
Quadratic Factoring

1. $-3 \pm \sqrt{11} = -6.32, 0.317$ **3.** $\dfrac{2 \pm 2\sqrt{7}}{3} = -1.10, 2.43$
5. $\dfrac{-5 \pm \sqrt{29}}{2} = -5.19, 0.193$ **7.** no real solution
9. $\dfrac{7 \pm \sqrt{17}}{4} = 0.719, 2.78$ **11.** no real solution
13. $-1 \pm \sqrt{2} = -2.41, 0.414$ **15.** $\pm\sqrt{3} = -1.73, 1.73$
17. no real solution   **19.** $\dfrac{3 \pm \sqrt{6}}{3} = 0.184, 1.82$
21. $-2, 5$ **23.** $-4 \pm 2\sqrt{2} = -6.83, -1.17$
25. $\dfrac{-3 \pm \sqrt{15}}{2} = -3.44, 0.436$ **27.** $-2, \dfrac{1}{3}$
29. $\dfrac{2 \pm \sqrt{13}}{3} = -0.535, 1.87$ **31.** $\dfrac{4 \pm \sqrt{10}}{3} = 0.279, 2.39$
33. $-2.08, 0.907$ **35.** $-5.19, 1.87$
37. 0.226 ms, 1.11 ms   **39.** 1.79, 4.04
41. $\dfrac{-\pi h \pm \sqrt{\pi^2 h^2 + 2\pi A}}{2\pi}$ **43.** $\dfrac{-1 \pm \sqrt{1 + 8S}}{2}$

EXERCISE 21-4
Applications of Quadratic Equations

1. 6.71 m, 20.1 m   **3.** 11 cm, 14 cm   **5.** 4.14 m
7. 2.07 m   **9.** 7.26 s   **11.** 5.38 s
13. 3.35 s   **15. (a)** 20.0 s   **(b)** 7.95 s, 32.0 s   **(c)** 40.0 s

17. Two possible answers
   (a) 41.06 yd wide, 117.9 yd long
   (b) 58.94 yd wide, 82.1 yd long

19. 3.98 s

EXERCISE 21-5 Applications Involving Rates

1. Airspeed = 960 km/h
Ground speed = 800 km/h

3. Speed = 24 km/h

5. Airspeed = 884 km/h
Ground speed = 552 km/h

7. To = 91.6 km/h
Return = 74.9 km/h

9. 56.1 mi/h, 46.4 mi/h

11. 80 members

13. 9 m

15. 362 ohms, 138 ohms

EXERCISE 21-6 Quadratic Functions

1. (a)   **(b)** 19.6 cm²

3. (a)   **(b)** 27.0 m

5. (a)   **(b)** 1.38 s, and 7.62 s

7. (a)   **(b)** 5.99 A, and 24.0 A

9. (a)   **(b)** 50 mi/h

11. 0.5, 0.6   **13.** −0.535, 1.87

CHAPTER 21 Review Exercises	1. (b), (d), (e)	3. $2x^2 - 3 = 0$
	5. -2	7. $x = \pm 8$
	9. 0	11. $\pm\sqrt{5}$
	13. no real solution	15. $\pm\dfrac{\sqrt{2gs}}{g}$
	17. $-2, 0$	19. $-\dfrac{1}{2}, 1$
	21. -2	23. $-\dfrac{2}{3}, \dfrac{1}{2}$
	25. $\pm\dfrac{3}{4}$	27. $-\dfrac{2}{3}, \dfrac{3}{2}$
	29. $-1, 12$	31. $\dfrac{7 \pm \sqrt{41}}{2} = 0.298, 6.70$
	33. $2 \pm \sqrt{3} = 0.268, 3.73$	35. no real solution
	37. $\dfrac{2}{3}, 5$	39. $-8, 2$
	41. $0.240, 2.32$	43. $\pi r^2 + \pi s r - A = 0, r = \dfrac{-\pi s \pm \sqrt{\pi^2 s^2 + 4\pi A}}{2\pi}$
	45. 3.29 cm	47. 4.00 s
	49. $60¢$	51. 3.52 s

CHAPTER 22 Oblique Triangles

**EXERCISE 22-1
Trigonometric Functions of Any Angle**

1. $\sin\theta = -0.8$, $\cos\theta = 0.6$, $\tan\theta = -1.333$, $\csc\theta = -1.25$, $\sec\theta = 1.667$, $\cot\theta = -0.75$
3. $\sin\theta = 0.8321$, $\cos\theta = -0.5547$, $\tan\theta = -1.5$, $\csc\theta = 1.202$, $\sec\theta = -1.803$, $\cot\theta = -0.6667$
5. $\sin\theta = -1$, $\cos\theta = 0$, $\tan\theta = $ undefined, $\csc\theta = -1$, $\sec\theta = $ undefined, $\cot\theta = 0$
7. $\sin\theta = 0$, $\cos\theta = 1$, $\tan\theta = 0$, $\csc\theta = $ undefined, $\sec\theta = 1$, $\cot\theta = $ undefined
9. III, $\sin\theta = -0.2949$, $\cos\theta = -0.9555$, $\tan\theta = 0.3086$, $\csc\theta = -3.391$, $\sec\theta = -1.047$, $\cot\theta = 3.24$
11. IV, $\sin\theta = -0.4625$, $\cos\theta = 0.8866$, $\tan\theta = -0.5216$, $\csc\theta = -2.162$, $\sec\theta = 1.128$, $\cot\theta = -1.917$
13. III, $\sin\theta = -0.6922$, $\cos\theta = -0.7217$, $\tan\theta = 0.9592$, $\csc\theta = -1.445$, $\sec\theta = -1.386$, $\cot\theta = 1.043$
15. sine positive, cosine negative, tangent negative
17. sine negative, cosine positive, tangent negative
19. I, II   21. II, III   23. II, IV   25. IV   27. III
29. sine positive, cosine negative, tangent negative
31. sine negative, cosine positive, tangent negative
33. sine negative, cosine negative, tangent positive
35. sine positive, cosine negative, tangent negative
37. 0 39. undefined   41. 0 43. -1
45. -1 47. undefined   49. -0.9063 51. -10.02

53. 0.08194 **55.** −1.221 **57.** −1.428 **59.** −0.08716
61. 0.6841 **63.** 0.4833 **65.** −0.2748
67. $\sin \theta = -\frac{4}{5}$, $\cos \theta = \frac{3}{5}$, $\tan \theta = -\frac{4}{3}$
69. $\sin \theta = -\frac{15}{17}$, $\cos \theta = -\frac{8}{17}$, $\tan \theta = \frac{15}{8}$

EXERCISE 22-2
Finding Angles from Trigonometric Functions

1. 524°, −196° **3.** 688°, −32° **5.** 67° **7.** 75°
9. 39°, sin 219° = −sin 39°, cos 219° = −cos 39°, tan 219° = tan 39°
11. 53°, sin 307° = −sin 53°, cos 307° = cos 53°, tan 307° = −tan 53°
13. 60°, 300° **15.** 7.1°, 172.9° **17.** 45°, 225°
19. 210°, 330° **21.** 113.1°, 293.1° **23.** 68.0°, 292.0°
25. 29.7°, 209.7° **27.** 197.9°, 342.1° **29.** 90°, 270°
31. 0°, 180°

EXERCISE 22-3
Law of Sines

1. 89.6 cm **3.** 12.2 cm **5.** 107 cm **7.** 47.1 cm
9. $a = 10.6$ cm, $b = 11.0$ cm, $C = 24.0°$
11. $a = 94.7$ cm, $c = 147$ cm, $B = 28.9°$
13. 172 m to B, 168 m to A
15. 25.9 cm **17.** 1260 m **19.** 76.1 m **21.** 805 km

EXERCISE 22-4
Ambiguous Case

1. $c = 6.53$ cm, $B = 41.9°$, $C = 12.1°$
3. $a = 216$ in., $A = 92.5°$, $C = 51.5°$; or $a = 57.6$ in., $A = 15.5°$, $C = 128.5°$
5. $b = 16.0$ cm, $A = 23.2°$, $B = 4.8°$
7. $c = 81.6$ cm, $A = 37.0°$, $C = 68.0°$
9. $b = 30.1$ in., $B = 122.8°$, $C = 40.5°$; or $b = 14.5$ in., $B = 23.8°$, $C = 139.5°$
11. $a = 4.08$ m, $A = 54.4°$, $B = 67.8°$; or $a = 0.871$ m, $A = 10.0°$, $B = 112.2°$
13. $a = 9.64$ in., $A = 64.1°$, $B = 50.4°$
15. 58.5° **17.** 52.7 km

EXERCISE 22-5
Law of Cosines

1. $a = 562$ cm, $B = 49.1°$, $C = 67.7°$
3. $c = 103$ cm, $A = 82.2°$, $B = 62.0°$
5. $b = 80.0$ cm, $A = 24.4°$, $C = 102.9°$
7. $c = 175$ mm, $A = 18.3°$, $B = 19.7°$
9. $A = 57.2°$, $B = 62.7°$, $C = 60.1°$
11. $A = 27.4°$, $B = 114.4°$, $C = 38.2°$
13. $A = 170.7°$, $B = 3.9°$, $C = 5.5°$
15. 38.8 m **17.** 44.2° **19.** 9.39 km
21. 104 km **23.** 22.4 ft **25.** 171 ft, 235 ft

CHAPTER 22
Review Exercises

1. $\sin \theta = -0.7809$, $\cos \theta = -0.6247$, $\tan \theta = 1.25$, $\csc \theta = -1.281$, $\sec \theta = -1.601$, $\cot \theta = 0.8$
3. sine positive, cosine negative, tangent negative
5. (a) 0 (b) 0 (c) 0 (d) undefined
7. $\theta_R = 68°$, sin 248° = −sin 68°, cos 248° = −cos 68°, tan 248° = tan 68°
9. $a = 5.16$ cm, $b = 3.30$ cm, $B = 36.5°$
11. $b = 82.8$ mm, $A = 34.2°$, $C = 63.8°$
13. $A = 54.5°$, $B = 75.8°$, $C = 49.7°$

15. 319 m

17. $\sin \theta = \dfrac{7}{25}$, $\cos \theta = -\dfrac{24}{25}$

19. 7.73 mi, 46.9° west of south

21. 11.21

CHAPTER 23 Vectors

**EXERCISE 23-1
Vectors and Components**

1. scalar 3. vector 5. scalar
7. vector 9. vector 11. scalar
13. 21 km east, 13 km north 15. 5.0 km/h west, 4.6 km/h south
17. $N = 52$ N, $T = 14$ N 19. $V_x = 643$, $V_y = 450$
21. $V_x = -95.5$, $V_y = -95.5$ 23. $\mathbf{V} = (-2.02, 4.32)$
25. $\mathbf{V} = (0.0324, -0.121)$

**EXERCISE 23-2
Applications of Vector Components**

1. 343 km east, 252 km north
3. 1940 m horizontal, distance = 2080 m
5. 426 km/h north, 304 km/h west
7. $N = 119$ N, $T = 39.7$ N
9. The East component = 70.5 km/h
11. 33.0 lbf pull, 23.8 lbf lift

**EXERCISE 23-3
Graphical Addition of Vectors**

1. $R = 143$, $\theta_R = 31°$ 3. $R = 0.62$, $\theta_R = 112°$
5. $R = 6.0$, $\theta_R = 113°$ 7. $R = 99$, $\theta_R = 329°$
9. $R = 117$, $\theta_R = 56°$ 11. $R = 128$, $\theta_R = 14°$
13. $R = 8.7$, $\theta_R = 115°$ 15. 17.2 km, 47° west of north
17. 40 lbf, 306°

**EXERCISE 23-4
Addition of Perpendicular Vectors**

1. $R = 686$, $\theta_R = 31.5°$ 3. $R = 11.7$, $\theta_R = 50.0°$
5. $R = 52.7$, $\theta_R = 18.2°$ 7. $R = 10.3$, $\theta_R = 29.8°$
9. $R = 19.9$, $\theta_R = 51.5°$ 11. $R = 9.04$, $\theta_R = 215°$
13. 27.3 km, 59.1° west of north
15. (a) 27.0 km (b) 3.0 km (c) 19.2 km
17. 238 km/h, 81.4° west of north
19. (a) 240 km/h (b) 410 km/h (c) 336 km/h, 75.3°
21. 5.41° east of north 23. $F = 93.8$ lb(f), $\theta_F = 36.1°$
25. 30.0 lbf, 31.5° down from left

**EXERCISE 23-5
Vector Addition by Components**

1. $R = 128$, $\theta_R = 53.7°$ 3. $R = 542$, $\theta_R = 170°$
5. $R = 6.92$, $\theta_R = 44.7°$ 7. $R = 0.607$, $\theta_R = 161°$
9. $R = 4.06$, $\theta_R = 9.49°$ 11. 573 km, 70.2° west of south
13. 14 200 N, 53.5°

**EXERCISE 23-6
Vector Addition Using the Law of Sines and Cosines**

1. $R = 812$, $\theta_R = 53.7°$ 3. $R = 5.00$, $\theta_R = 11.0°$ 5. $R = 428$, $\theta_R = 59.8°$
7. $R = 101$, $\theta_R = 11.4°$ 9. $R = 969$, $\theta_R = 68.2°$ 11. $R = 3.94$, $\theta_R = 192°$
13. 56.0 km, 176° west of south
15. 11 300 N, 77.4° west of north
17. $R = 3190$ lbf, $\theta = 56.4°$

Answers to Selected Odd-Numbered Exercises

CHAPTER 23 Review Exercises

1. $V_x = -43.0$, $V_y = 61.4$
3. 118 km west, 249 km south
5. 8.74 m/min, 0.921 min
7. 4.69 tonnes or 46.0 kN
9. $R = 3.56$, $\theta_R = 104.2°$
11. $R = 981$, $\theta_R = 18.2°$
13. 17.0 km, 40.2° west of north
15. $R = 86.7$, $\theta_R = 58.3°$
17. $R = 5.54$, $\theta_R = 129°$
19. $R = 3.31$, $\theta_R = 199°$
21. 1290 lbf, $\theta_R = 35.5°$

CHAPTER 24 Applications of Radian Measure

EXERCISE 24-1 Radian Measure

1. 2π
3. $\dfrac{\pi}{2}$
5. $\dfrac{\pi}{3}$
7. $\dfrac{\pi}{12}$
9. $\dfrac{3\pi}{4}$
11. $\dfrac{2\pi}{5}$
13. $\dfrac{17\pi}{9}$
15. $\dfrac{7\pi}{15}$
17. $\dfrac{3\pi}{2}$
19. $\dfrac{10\pi}{9}$
21. 30°
23. 45°
25. 120°
27. 144°
29. 306°
31. 132°
33. 0.8660
35. 0.5774
37. -0.3827
39. -0.5878
41. 1
43. 0.2588

EXERCISE 24-2 Arc Length and Area of a Sector

1. 1.38 rad
3. 251°
5. 5.06 m
7. 986 mm
9. 6.13 cm
11. 2.64 in.
13. 15.1 cm²
15. 168 mm²
17. 4959 km
19. 28.0 m
21. 569 cm²
23. 382 in.²
25. 248 ft

EXERCISE 24-3 Uniform Circular Motion

1. 159 rad/s, 9120°/s
3. 24.0 rev/min, 144°/s
5. 76.1 rad/min
7. 237 rad
9. 1.13 m/s
11. 1660 km/h
13. 66 700 mi/h

CHAPTER 24 Review Exercises

1. $\dfrac{\pi}{15}$
3. $\dfrac{6\pi}{5}$
5. 36°
7. 255°
9. 1.48 rad
11. 229 mm
13. 1.55 m²
15. 52.5 rev/min
17. 4.45 in.

CHAPTER 25 Graphs of Trigonometric Functions

EXERCISE 25-1 Graph of $y = a \sin x$

1. 0, −0.7, −1, −0.7, 0, 0.7, 1, 0.7, 0, −0.7, −1, −0.7, 0, 0.7, 1, 0.7, 0

3. $y = 4 \sin x$
5. $y = -\sin x$
7. $y = 2 \sin x$

9. $|a| = 5$, $P = 360°$

11. $|a| = 1$, $P = 360°$

13. $|a| = \dfrac{1}{3}$, $P = 360°$

15. $|a| = 2.5$, $P = 2\pi$

17. $|a| = 100$, $P = 2\pi$

19.

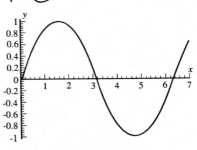

EXERCISE 25-2
Graph of $y = a \sin bx$

1. $0, 0.7, 1, 0.7, 0, -0.7, -1, -0.7, 0, 1, 0, -1, 0, 1, 0, -1, 0$

$|a| = 1$, $P = \dfrac{2\pi}{3}$

3. $y = \sin 4x$ **5.** $y = \sin 2x$ **7.** $|a| = 1$, $P = 720°$

Answers to Selected Odd-Numbered Exercises   **741**

9. $|a| = 1, P = \dfrac{\pi}{3}$

11. $y = 3 \sin 3x$ 13. $y = 5 \sin 4x$ 15. $|a| = 2, P = 120°$

17. $|a| = 3, P = 1080°$

19. $|a| = 3, P = \pi$

21. $|a| = \dfrac{1}{2}, P = 2$

23. $|a| = 6$ in., $P = 0.5$ s

EXERCISE 25-3
Graph of
$y = a \sin (bx + c)$

1. $0, -0.5, -0.9, -1, -0.9, -0.5, 0, 0.5, 0.9, 1, 0.9, 0.5, 0, -0.5, -0.9, -1$

$P = 360°, |a| = 1, -\dfrac{c}{b} = 90°$

3. $y = \sin\left(x + \dfrac{\pi}{6}\right)$ 5. $y = 2 \sin (x - 60°)$ 7. $y = 2 \sin (2x - \pi)$

9. $|a| = 1, P = 360°, -\dfrac{c}{b} = -45°$ **11.** $|a| = 1, P = \pi, -\dfrac{c}{b} = \dfrac{\pi}{6}$

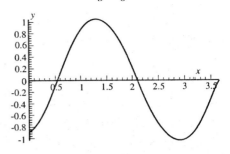

13. $|a| = 2, P = 2\pi, -\dfrac{c}{b} = \dfrac{\pi}{6}$ **15.** $|a| = 0.5, P = 120°, -\dfrac{c}{b} = -60°$

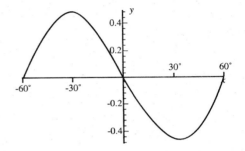

17. $|a| = 1, P = 2\pi, -\dfrac{c}{b} = -1$ **19.** $|a| = 60$ A, $P = \dfrac{1}{60}$ s, $-\dfrac{c}{b} = -\dfrac{1}{240}$ s

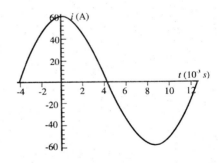

EXERCISE 25-4
Graph of
$y = a \cos (bx + c)$

1. $1, 0.9, 0.5, 0, -0.5, -0.9, -1, -0.9, -0.5, 0, 0.5, 0.9, 1$

3. $y = 2 \cos x$ **5.** $y = -\cos 2x$

7. $y = 2 \cos 3x$ **9.** $y = 3 \cos\left(\dfrac{x}{2} - \dfrac{\pi}{2}\right)$ **11.** $|a| = 3, P = 360°$

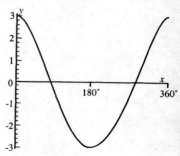

13. $|a| = 1, P = 120°$ **15.** $|a| = 5, P = \dfrac{\pi}{2}, -\dfrac{c}{b} = 0$

17. $|a| = 1, P = 2\pi, -\dfrac{c}{b} = -\pi$ **19.** $|a| = 1, P = 180°, -\dfrac{c}{b} = 90°$

21. $|a| = 2, P = \dfrac{2\pi}{3}, -\dfrac{c}{b} = -\dfrac{\pi}{3}$ **23.** $|a| = 4.25, P = 141°, -\dfrac{c}{b} = 128°$

25. $|a| = 1.5$ amp, $P = \dfrac{1}{60}$ s

EXERCISE 25-5
Graph of
$y = a \tan (bx + c)$

1. $\infty, -1.7, -1, -0.6, 0, 0.6, 1, 1.7, \infty, -1, 0, 1, \infty, -1, 0$

3. $P = 180°$

5. $P = 60°$

7. $P = \pi$

9. $P = 2\pi$

11. $P = \pi, -\dfrac{c}{b} = \dfrac{\pi}{2}$

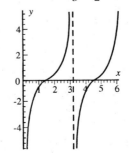

13. $P = 180°, -\dfrac{c}{b} = -45°$

15. $P = \dfrac{\pi}{2}, -\dfrac{c}{b} = -\dfrac{\pi}{2}$

CHAPTER 25
Review Exercises

1. $|a| = 4, P = 360°, -\dfrac{c}{b} = 0°$

3. $|a| = 1, P = 360°, -\dfrac{c}{b} = 45°$

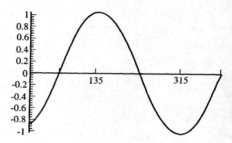

5. $|a| = 1, P = 180°, -\dfrac{c}{b} = 0°$

7. $|a| = 1, P = 180°, -\dfrac{c}{b} = 45°$

9. $|a| = 5, P = 120°, -\dfrac{c}{b} = 0°$

11. $P = 4\pi, -\dfrac{c}{b} = 0$

13. $|a| = 1, P = \dfrac{2\pi}{3}, -\dfrac{c}{b} = 0$

15. $|a| = 5, P = 2\pi, -\dfrac{c}{b} = 0$

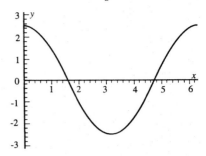

17. $|a| = 3, P = 2\pi, -\dfrac{c}{b} = -\dfrac{\pi}{8}$

19. $|a| = 2, P = \dfrac{2\pi}{3}, -\dfrac{c}{b} = 0$

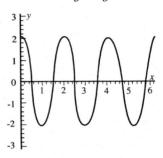

21. $y = 2 \sin (2x - 180°)$

23. $P = \dfrac{\pi}{5}$

CHAPTER 26 Ratio, Proportion, Variation

EXERCISE 26-1
Ratio

1. $5:4, \dfrac{5}{4}$ **3.** $\dfrac{2}{5}$ **5.** $\dfrac{2}{1}$ **7.** $\dfrac{1}{4}$

9. $\dfrac{1}{3}$ **11.** $\dfrac{4}{1}$ **13.** $\dfrac{3}{4}$ **15.** $\dfrac{1}{1000}$

17. $\dfrac{3}{8}$ **19.** $\dfrac{5}{1}$ **21.** 3.14 **23.** 40.5

25. 93.8% **27.** 0.808

EXERCISE 26-2
Proportion

1. $\dfrac{x}{y} = \dfrac{6}{9}$ **3.** $\dfrac{p}{20} = \dfrac{q}{5}$ **5.** $\dfrac{12}{30} = \dfrac{y}{5}$

7. $2 \cdot 9 = 3 \cdot 6$ **9.** $12 \cdot 5 = 20 \cdot 3$ **11.** $12 \cdot 5 = 30 \cdot 2$

13. 5		**15.** 3		**17.** −6, 6	
19. $\dfrac{30}{7}$		**21.** −12, 12		**23.** 12	
25. 14		**27.** 183 cm, 3.5 cm		**29.** $200, $450	
31. 1.38 gal		**33.** 3.63 kg		**35.** 579 km	
37. 5.13 kg		**39.** 3.88 h		**41.** 9.79 t	
43. $99 600		**45.** 23.0 ft			

EXERCISE 26-3
Direct Variation

1. 3   **3.** 9.80
5. $y = kx$ **7.** $m = kn$
9. $P = kr$ **11.** $y = kx^3$
13. $s = kt^2$ **15.** $k = 4, y = 4x$
17. $k = \dfrac{4}{5}$ N/m, $F = \dfrac{4}{5}L$ **19.** $y = 15.75x, y = 39.4$
21. $k = 3.14, A = 3.14r^2$ **23.** $k = 64.0\,\dfrac{\text{W}}{\text{A}^2}, P = 64.0I^2$
25. 39.4 kW   **27.** 78.3 cm^2
29. 149 L/min   **31.** $k = 93.6\,\dfrac{\text{lbf}\cdot\text{s}^2}{\text{ft}}, E_k = 93.6\,v^2$

EXERCISE 26-4
Joint, Inverse, and Combined Variation

1. 2   **3.** 2   **5.** $p = kqr$
7. $E = kLT$ **9.** 60   **11.** $z = \dfrac{k}{y}$
13. $y = \dfrac{k}{z^2}$ **15.** $X_c = \dfrac{k}{C}$ **17.** $y = \dfrac{kz}{x}$
19. $I = \dfrac{kV}{R}$ **21.** $x = 3yz, x = 60$ **23.** $y = \dfrac{16x}{z}, y = 10$
25. $315   **27.** 28.7 min   **29.** 2.0 h
31. 5.76×10^{-10} N   **33.** 1.5   **35.** 321 vibrations/s

CHAPTER 26
Review Exercises

1. 18   **3.** −12, 12
5. $\dfrac{3}{5}$ **7.** $\dfrac{1}{5}$
9. $p = \dfrac{k}{r^2}$ **11.** $R = kT$
13. 10 m, 6 m   **15.** 115
17. $\dfrac{40}{1}$ **19.** $292.25
21. $E = 1.632I, E = 33.4$ V   **23.** 1400 m
25. 1540 Btu/h   **27.** 3.68 lb

CHAPTER 27 Exponential Functions

EXERCISE 27-1
Introduction to Exponential Functions

1. An exponential function is a function in which the independent variable is in the exponent.

3. 5^{x+y} **5.** 8^{yz} **7.** $\dfrac{5^x}{3^x}$ **9.** $\dfrac{1}{2^x}$ **11.** 4^x

EXERCISE 27-2
Graph of the Exponential Function

1.

3.

5.

7.

9.

11.

13.

15.

17. (a) $n = 2500(3)^t$ (b) 610 000 bacteria
(c) n (bacteria) (d) 120 000 bacteria

Answers to Selected Odd-Numbered Exercises 749

19. (a) 1.2 mg (b)
    (c) 7.6 h

21. (a) 40.6 kg (b) $m = 40.6(2)^t$
    (c) 230 kg   (d)
    (e) 3.6 h

EXERCISE 27-3
Exponential Growth

1. 7.39 3. 3.90 5. 0.791 7. 445 9. 9.80
11. (a) $836.20 (b) $842.41 (c) $843.00
13. (a) $2.59 (b) $2.71 (c) $2.72 (d) $2.72
15. $3271.01
17.   19.
21. 31.7 million
23. (a) 165 kg   (b) $y = 45.0e^{0.130t}$

(c) 115 kg   (d) 6.14 h

EXERCISE 27-4
Exponential Decay

1.   3.

5. (a) $y = 175e^{-0.0602t}$ (b)
   (c) 111 kg

7. 2130 rev/min   9. 41.3%   11. 78.2 mA

13. (a) 997°C (b)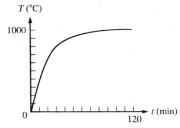

CHAPTER 27
Review Exercises

1.   3.

5.   7.

9. $\dfrac{1}{4^x}$ or 4^{-x} 11. $\dfrac{1}{2^x}$

13. (a) $709.51   (b) $717.68   (c) $718.43   (d) $718.46

15. (a) 59.1 tonnes   (b) $y = 12.3e^{0.157t}$

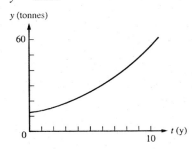

17. (a) $2222.45
19. (a) 405 lb   (b) 18.5 lb

21.

$T = 79e^{-0.1t} + 21$

CHAPTER 28 Logarithms

EXERCISE 28-1
Introduction to Logarithms

1. $\log_2 3 = 3$ **3.** $\log_5 625 = 4$
5. $\log_{10} 1\,000\,000 = 6$ **7.** $\log_{1/3}\left(\dfrac{1}{81}\right) = 4$
9. $\log_{78} 1 = 0$ **11.** $\log_6 9.73 = 1.27$
13. $64,\ \log_4 64 = 3$ **15.** $100\,000,\ \log_{10} 100\,000 = 5$
17. $1,\ \log_9 1 = 0$ **19.** $\dfrac{1}{27},\ \log_3\left(\dfrac{1}{27}\right) = -3$
21. $26.7,\ \log_4 26.7 = 2.37$ **23.** $0.0561,\ \log_{2.4} 0.0561 = -3.29$
25. $\log_5 120 = x$ **27.** $\log_9 a = 5$
29. $\log_4 q = -3$ **31.** $\log_e 45 = x$
33. $7^3 = 343$ **35.** $10^3 = 1000$ **37.** $9^{1/2} = 3$ **39.** $7^1 = 7$
41. $5^{-2} = \dfrac{1}{25}$ **43.** $4^{6.59} = y$ **45.** $b^y = x$ **47.** $e^q = 5$
49. 3 **51.** 3 **53.** 0 **55.** 4
57. 4 **59.** 1 **61.** 10 **63.** 125
65. $\dfrac{1}{216}$ **67.** 4 **69.** 27 **71.** e

EXERCISE 28-2
Common and Natural Logarithms

1. $\log 100 = 2$ **3.** $\log 0.001 = -3$
5. $100\,000,\ \log 100\,000 = 5$ **7.** $0.0001,\ \log 0.0001 = -4$
9. $10^{1.19} = 15.5$ **11.** $10^{0.19} = 1.55$
13. $0.730,\ 10^{0.730} = 5.37$ **15.** $1.87,\ 10^{1.87} = 74.5$
17. $-0.341,\ 10^{-0.341} = 0.456$ **19.** $x = 10^3 = 1000$

21. $x = 10^{1.56} = 36.3$ **23.** $x = 10^{-1.62} = 0.0240$
25. $\ln 20.1 = 3$ **27.** $11.2, \ln 11.2 = 2.42$
29. $1, \ln 1 = 0$ **31.** $4.04, e^{4.04} = 56.7$
33. $-3.74, e^{-3.74} = 0.0238$ **35.** $x = e^{3.54} = 34.5$
37. $e^{-0.00950} = 0.991$ **39.** $-2.3, -1.6, -0.7, 0, 0.7, 1.6, 2.3$

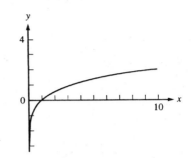

EXERCISE 28-3
Properties of Logarithms

1. $\log_3 7 + \log_3 x$ **3.** $\log_5 x - \log_5 4$
5. $\ln 3 + \ln x + \ln y$ **7.** $\log_2 5 + \log_2 x - \log_2 3$
9. $\log_b 3 - \log_b 4 - \log_b a$ **11.** $\log_4 5a$
13. $\log_b \dfrac{2}{y}$ **15.** $\log_2 xyz$ **17.** $\log_3 \dfrac{a}{5c}$
19. $3 \log_b x$ **21.** $x \log_3 2$ **23.** $\dfrac{1}{2} \log_b x$
25. $\dfrac{1}{3} \log_2 x$ **27.** $2 \log_b x + 3 \log_b y$ **29.** $\log_2 x^5$
31. $\ln x^3 y^4$ **33.** $\log_b \sqrt{x}$ **35.** $\ln \sqrt[4]{2}$
37. $\log_2 \sqrt{x}\sqrt[3]{y}$ **39.** $\log_5 \sqrt[3]{x^2}$ **41.** 0
43. 1 **45.** 9 **47.** 5.3
49. $\dfrac{1}{5}$

EXERCISE 28-4
Solution of Exponential Equations Using Logarithms

1. 1.40 **3.** -0.237 **5.** 27.2
7. -0.322 **9.** -0.223 **11.** -0.306
13. 0.143 **15.** 0.405 **17.** 0.693
19. 56.2 **21.** 0.416 **23.** 0.0516
25. 0.00255 **27.** 2.75 **29.** $t = \dfrac{\ln \dfrac{N}{100}}{\ln 2}$
31. $t = \dfrac{\ln \dfrac{1000}{P}}{0.12}$ **33.** 6.29 h **35.** 13.1 y
37. 37.5 y **39.** 8.90 h **41.** 0.00692 s
43. 19.6 min **45. (a)** 1.54% **(b)** year 2621 **47.** 23.3 min

EXERCISE 28-5
Applications of Logarithmic Equations

1. 20 dB **3.** 1 W/m² **5.** 3.40 **7.** 1.58×10^{-11} moles/L
9. 1510 m **11.** 11.1 years **13.** 20.0

CHAPTER 28
Review Exercises

1. $\log_4 64 = 3$ **3.** $\log 4 = x$
5. $\log_2 a + \log_2 b$ **7.** $\log_7 5 + \log_7 a - \log_7 b$

Answers to Selected Odd-Numbered Exercises

9. $\dfrac{1}{3} \ln x$
11. $2 \log_3 a + 3 \log_3 b$
13. $10000, \log 10000 = 4$
15. $0.292, \ln 0.292 = -1.23$
17. 3
19. 3
21. 81
23. 5
25. -2
27. 0
29. $0.230, 10^{0.230} = 1.70$
31. $1.50, e^{1.50} = 4.50$
33. 2.15
35. 2.41
37. 0.560
39. -5.39
41. $61.6, \log_6 61.6 = 2.30$
43. $0.000224, \log 0.000224 = -3.65$
45. $x = 10^7 = 10000000$
47. $x = e^{-1.52} = 0.219$
49. $125, \log_5 125 = 3$
51. $3, \log_{1/3} 3 = -1$
53. $\log_3 5x$
55. $\log (3x)^4$
57. $\log_4 \sqrt{a}$
59. $\ln \dfrac{x^3}{y^2}$
61. $3^4 = 81$
63. $e^{1.95} = 7$
65. 7
67. $\dfrac{1}{3}$
69. 4.98 days
71. 1.27 min
73. 0.0602 s
75. 10^{-8} moles/L

CHAPTER 29 Complex Numbers

EXERCISE 29-1
Imaginary Numbers

1. $j\sqrt{5}$
3. $j30$
5. $j2\sqrt{2}$
7. $-j2\sqrt{3}$
9. $j2\sqrt{7}$
11. $-j25\sqrt{2}$
13. $j\dfrac{1}{3}$
15. $-j0.4$
17. j
19. $-j$
21. $-j21$
23. $j18$
25. $-j28$
27. $j5\sqrt{2}$
29. $-j30$
31. $j40$
33. $j30$
35. -16
37. $\sqrt{14}$
39. -3
41. $6\sqrt{6}$
43. -6
45. -7
47. 6
49. -48
51. 7
53. $-\dfrac{3}{2}$
55. $-j4$
57. $j\dfrac{5}{2}$
59. $-j\dfrac{1}{2}$
61. $j\dfrac{4}{5}$

EXERCISE 29-2
Complex Numbers

1. $3 + j\sqrt{2}$
3. $-5 + j2\sqrt{3}$
5. $-7 + j0$
7. $0 + j2\sqrt{6}$
9. $3 - j8$
11. $-5 + j\sqrt{2}$
13. -2
15. $-j4$
17. $-j5, j5$
19. $-j4\sqrt{3}, j4\sqrt{3}$
21. $-2 - j3, -2 + j3$
23. $4 - j2, 4 + j2$
25. $-\dfrac{1}{2} - j\dfrac{\sqrt{5}}{2}, -\dfrac{1}{2} + j\dfrac{\sqrt{5}}{2}$

EXERCISE 29-3
Operations with Complex Numbers

1. $-3 + j3$
3. $-1 + j3$
5. 8
7. $-10 + j11$
9. $j4$
11. $-3 + j3$
13. $-3 - j10$
15. $4 - j11$
17. $1 + j2$
19. $-15 + j10$
21. $28 + j42$
23. $3 + j2$
25. $20 + j30$
27. $13 - j11$
29. $11 + j23$
31. $-76 + j52$
33. 65
35. 61
37. $7 - j24$
39. $45 - j28$
41. $2 - j4$
43. $\frac{3}{2} - j2$
45. $-4 - j3$
47. $2 - j\frac{3}{2}$
49. $\frac{3}{2} + j\frac{3}{2}$
51. $-\frac{8}{13} - j\frac{12}{13}$
53. $\frac{7}{2} + j\frac{3}{2}$
55. $-\frac{2}{5} - j\frac{11}{5}$
57. $-\frac{27}{13} + j\frac{8}{13}$

EXERCISE 29-4
Graph of a Complex Number

1. to 15.

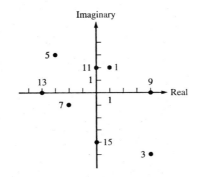

17. $A: 4 + j2$ $B: -2 + j3$ $C: -3 - j3$ $D: 3 - j4$
    $E: 2$ $F: j4$ $G: -4$ $H: -j$
19. $5 + j5$ 21. $-3 - j5$ 23. -2 25. $4 - j5$

EXERCISE 29-5
Polar and Trigonometric Forms

1. $3.61\ \underline{/56.3°}$
3. $4.47\ \underline{/207°}$
5. $4.42 + j6.67$
7. $-388 + j388$
9. $1.41\ (\cos 45° + j \sin 45°)$
11. $3.61\ (\cos 326° + j \sin 326°)$
13. $8.06\ (\cos 120° + j \sin 120°)$
15. $6.71\ (\cos 207° + j \sin 207°)$
17. $2\ \underline{/30°},\ 1.73 + j$
19. $3.56\ \underline{/135°},\ -2.52 + j2.52$
21. $8.45\ \underline{/297°},\ 3.84 - j7.53$
23. $2.588 + j9.658$

CHAPTER 29 Review Exercises

1. $5.39\ \underline{/112°}$
3. $7.07\ \underline{/8.13°}$
5. $j15$
7. $-2 - j11$
9. 18
11. $-\frac{3}{2} + j2$
13. $1 - j12$
15. $-24 + j12$
17. $-7 - j24$
19. $\frac{1}{2} + j\frac{3}{2}$
21. $-j4\sqrt{2}$
23. $-j5$
25. -5
27. $\frac{29}{53} - j\frac{31}{53}$
29. $-1 - j6$
31. -2
33. $2.83\ (\cos 45° + j \sin 45°)$
35. $5.83\ (\cos 211° + j \sin 211°)$
37. $-j6\sqrt{2},\ j6\sqrt{2}$
39. $4.25\ \underline{/123°},\ -2.31 + j3.56$

General Index

Absolute value, 29, 109
  of complex number, 686
Abscissa, 214
Acute triangle, 255
Addition and subtraction:
  of algebraic expressions, 136
  of algebraic fractions, 372
  of approximate numbers, 58
  by calculator, 7
  of complex and imaginary
      numbers, 672, 679
  of decimal numbers, 58,
  of denominate numbers, 85
  of fractions, 41,
  horizontal and vertical, 7, 138
  of integers, 7,
  method, for solution of sets of
      equations, 408
  of mixed numbers, 44,
  of radicals, 450
  in scientific notation, 68
  of signed numbers, 110
  of vectors, 542, 546, 549, 554,
      685
  of whole numbers, 7
Additive identity, 10
Adjacent angles, 249
Adjacent side, 312
Algebra, introduction to, 131-147
Algebraic expression, 131
Algebraic fraction, 29, 360-399
Alternate interior angle, 251
Altitude:
  of a cone, 293
  of a pyramid, 290
  of a triangle, 256
Ambiguous case, 523
Amplitude, 577
Angle(s):
  adjacent, 249
  alternate interior, 251
  between intersecting lines, 251
  complementary and
      supplementary, 249
  conversion of, 246
  corresponding, 250
  definition of, 244, 304
  of elevation and depression,
      318
  exterior, 250
  formed by rotation, 304
  interior, 250
  measure of, 245
  opposite or vertical, 249
  quadrantal, 504, 507
  reference, 512
  sides of, 244
  in standard position, 304
  types of, 248
  units of measure, 245

Angle sum principle, 255, 261,
      312
Angular displacement, 568
Angular velocity, 568
Approximate numbers, 4, 54
Arc, 275
  length of, 564
Arc functions, 511
Area(s):
  of circle, 275
  of cone, 293
  conversion of, 89
  of cube, 284
  of cylinder, 287
  of frustums, 297
  of prism, 281
  of pyramid, 290
  of quadrilaterals, 269-271
  of rectangular solid, 283
  of sector of a circle, 565
  of sphere, 299
  of triangle, 257
Argument of a complex number,
      687
Associative law:
  for addition, 8,
  for multiplication, 13,
A.S.T.C. "rule", 506
Asymptote, 229, 595
Axis (Axes), coordinate, 214
  abscissa, 214
  ordinate, 214
Azimuth circle, 523, 538

Base:
  of a decimal number, 3,
  of exponents, 141
  of logarithms, 646
  of a percentage, 96
  of prism, 280
  of a triangle, 255
  unit, 77,
Binomial, 151
  conjugate of, 455
Braces and brackets (see
      Grouping)
British/American system, 77, 81

Canceling, 32, 33, 35, 163
Capacity, liquid, 288
C.A.S.T. "rule", 506
Centre, 273
Central angle, 560
Chord, 274
Circle, 273-277
Circular cone, 293
Circular cylinder, 286
Circular motion, 568
Circumference, 274
Coefficient, 131

Column of a determinant, 427
Combined calculator operations,
      22, 72, 127
Combined variation, 622
Common denominator (see
      Least common
      denominator)
Common factors (see Factor)
Common fractions, 27,
Common logarithms, 650
  by calculator, 651
Commutative law:
  for addition, 8, 137
  for multiplication, 13, 149
Complementary angles, 249
Completing the square, 481
Complex conjugate, 677
Complex fractions (see Fraction)
Complex numbers, 670-692
Complex numbers, represented
      by vectors, 685
Complex plane, 684
Complex roots, 678
Components of a vector, 534
Compound fractions (see
      Complex fractions)
Concentration, percent, 104, 199
Concentric, 274
Conditional equation, 172
Cone, 293
Conjugate:
  of a binomial, 455
  complex, 677
Constant, 131
Constant of proportionality, 614
Conversion:
  of angular units, 246
  of areas and volumes, 88
  of decimals and fractions to
      percent, 95
  between exponential and
      logarithmic form, 647
  between exponential and
      radical form, 440
  between improper fractions
      and mixed numbers, 33
  of rates, 90
  between rectangular and polar
      coordinates, 214, 688
  to and from percent, 94
  to and from scientific notation,
      20, 21, 67, 124
  of temperatures, 177, 185
  of units, 78, 79, 81
Conversion factors, 81,
  SI conversion factors, 78, 79
  table of, 702
Coordinates, rectangular, 214
Corresponding dimensions, 250
Cosecant of an angle, 326

Cosine of an angle, 306, 312
Cosine curve, 590
Cosines, law of, 527
Cotangent of an angle, 326
Cramer's rule, 429, 434
Cube, 284
Cube of a binomial, 157
Cubes, sum and difference of
      two, 356
Cursor, 239
Cylinder, 286
  hollow, 288

Data, 216
Decay, exponential, 641
Decimal number, 3, 54
Decimal point, 3, 55
Degree, angular, 245
  of an expression, 134
Denominate number, 85
Denominator, 27, 162, 360
  lowest common (LCD) (see
      Least common
      denominator)
Dependent variable, 206
Derived Unit, 77
Determinant(s), 427-435
  calculator evaluation of, 430,
      433
  of the coefficients, 432
  second order, 427
  third order, 432
Diagonals:
  of a determinant, 427
  of a parallelogram, 268
  of a rhombus, 270
Diameter, 274
Difference of two cubes, 356
Difference of two squares, 153,
      339
Digit, 2
Dimensionless ratio, 603
Direction of a vector, 535
Direct variation, 613
Distributive law, 13, 151
Dividend, 15, 28, 162
Division:
  involving one, 15
  of algebraic expressions,
      162-171
  of algebraic fractions, 369
  of approximate numbers, 62
  by calculator, 15
  of complex and imaginary
      numbers, 674, 681
  of decimal numbers, 62
  of denominate numbers, 85
  of fractions, 37
  of mixed numbers, 38
  of monomials, 162

756

Division: (*continued*)
  of numbers in scientific notation, 70
  of polynomial by monomial, 166
  of powers, 142
  of radicals, 457
  of signed numbers, 117
  of two polynomials, 167
  of whole numbers, 15
  by zero, 28, 131
Divisor, 15, 28, 162
Domain, 211
Drawings, scale, 603

e (base of natural logarithms), 637
Efficiency, percent, 105
Element of a determinant, 427
Elevation, angle of, 318
Engineering notation, 68, 126
Endpoints of a line, 244
Equal sign, 5, 109,172
Equation(s):
  with approximate coefficients, 183
  conditional, 172
  definition of, 131
  exponential, 661
  first degree, 178
  fractional, 179, 385–399
  graphical solution of, 233
  linear, 178
  literal, 183, 390
  logarithmic, 648
  of a parabola, 227, 496
  quadratic, 226, 468–503, 678
  radical, 460–464
  simple, 172–186
  of a straight line, 178
  substituting into, 172
  systems of (see Systems of linear equations)
Equilateral triangle, 255
Equivalent (equal) fractions, 30, 360, 363
Error, percent, 104
Estimation, 206
Euclidean geometry (see Plane geometry)
Exact numbers, 4
Explicit form, 205
Explicit function, 205
Exponent(s):
  base of, 141
  definition of, 141
  fractional, 64, 439
  integral, integer, 140, 141
  laws of, 141, 439, 442, 629
  negative, 122, 144
  zero, 144
Exponential decay, 641
Exponential equations, 661
Exponential form, 440, 647
Exponential functions, 629–645

Factor(s), 332
  common, 13, 332, 333, 351
  greatest common, 335
  prime, 32, 332
Factorability, test for, 349
Factoring, 332–359
  solving quadratics by, 475
Factors, conversion (see Conversion factors)
First-degree equations, 178
  systems of (see Systems of linear equations)
FOIL rule, 153

Formula(s), 174, 175, 207, 224, 390, 472
  substituting into, 174, 391
  summary of (table), 684–701
Fraction(s), 15, 27–53, 162
  addition and subtraction of, 41
  algebraic, 360–399
  common, 27–53
  complex (compound), 30, 48, 379
  equivalent (equal), 30, 360, 363
  fundamental principle of, 31
  multiplication and division of, 34, 37
  proper and improper, 29
  signs of (see Rules of signs)
  similar (like), 30
  simplifying, 31
Fractional equations, 179, 385–399
Fractional exponents, 64, 441
Fraction line, 28
Frustums
  of a cone or pyramid, 295
  lateral area of, 297
  volume of, 295
Function(s), 204–212
  explicit and implicit, 205, 206
  exponential, 629
  graphs of, 219, 236
  linear, 218
  logarithmic, 653
  quadratic, 226, 496
  substituting into, 212
  trigonometric (see Trigonometric function)
Functional notation, 210
Functional variation, 613
Fundamental principle of fractions, 31, 363

General form for quadratic equation, 477
Geometry:
  plane, 244–279
  solid, 280–302
Grad, 246
Graph(s), 214–241
  complete, 237
  of complex numbers, 684
  of equations and functions, 219, 236
  of experimental data, 216
  of an exponential function, 631, 638, 641
  of linear equation, 402
  of ordered pairs, 216
  of quadratics, 666
  of trigonometric functions, 573-600
  by calculator, 576
Graphical addition of vectors, 542
Graphical representation of vectors, 535
Graphical solution:
  of equations, 233, 240
  of pairs of linear equations, 364, 403
  by calculator, 406
  of quadratics, 498
  of radical equations, 463
Graphing calculator, 236–240
Great circle, 299
Grouping symbols of, 8, 23, 28, 136, 137, 158, 179
  on calculator, 10

Grouping method for factoring, 337, 350
Growth, exponential, 636

Height (see Altitude)
Hemisphere, 299
Hero's formula, 257
Horizontal lines, 222
Hypotenuse, 261, 305

I (the imaginary unit), 670
Imaginary numbers, 3, 670
Implicit form, 206
Implicit function, 206
Improper fraction, 29, 33
Independent variable, 206
Index of a radical, 18
Inequality signs, 5, 109
Initial position, 245, 304
Integers, 1
Intercept, 220
Interior angles of intersecting lines, 250
Inverse variation, 620
Irrational numbers, 2
Isosceles triangle, 255

j (the imaginary unit), 670
Joint variation, 619
J operator, 670

Lateral area of solids, 281
Latitude, 567
Law(s):
  of cosines, 527
  of exponents (see Exponent)
  of sines, 517
Least common denominator (LCD), 41, 374
Legs of a right triangle, 261
Like terms, 137
Line, straight, 244, 402
Linear equations, 178, 401
  systems of (see Systems of linear equations), 218
Line segment, 244
Lines, intersecting, 251
Liquid capacity, 288
Literal equations, 183, 390
Literal expression, 131, 150
Logarithmic equations, 648
Logarithmic form, 670
Logarithmic function, 653
Logarithm of one (1), 659
Logarithm of the base, 659
Logarithm of the base to a power, 659
Logarithms, 646–669
  common, 650
  natural, 652
  properties of, 655
Lowest common denominator (see Least common denominator)

Magnitude of a vector, 535
Mass, 294
Mathematical expression, 131
Maximum and minimum points, 227
Measure of an angle, 245
Median of a triangle, 256
Metric system (see SI units)
Minute of arc, 246
Mixed expression, 271
Mixed number, 29, 33, 36, 38, 43
Monomial, 148, 149, 162
Multinomial, 151

Multiplication:
  dot, 12
  of algebraic expressions, 148–160
  of algebraic fractions, 368
  of approximate numbers, 61
  of binomials, 152
  by calculator, 12
  of complex and imaginary numbers, 672, 680
  of decimal numbers, 60
  of denominate numbers, 85
  of fractions, 34
  of mixed numbers, 36
  of monomials, 148
  of monomials and multinomials, 151
  of multinomials, 155
  of numbers in scientific notation, 69
  of powers, 141
  of radicals, 454
  of signed numbers, 114
  symbols for, 12
  of whole numbers, 12
  by zero, 14
Multiplicative identity, 14

Natural logarithms, 652
  by calculator, 653
Natural numbers, 1, 2
Negative exponent, 122
Negative numbers, 1, 108
  on calculator, 109
Normal component, 536
Number line, 5, 109, 214
Numbers:
  approximate, 4, 55
  complex, 670
  decimal, 3, 54
  exact, 4
  imaginary, 2, 670
  irrational, 2
  mixed, 29
  natural, 1, 2
  negative, 1, 108
  ordered pairs of, 207
  rational, 2
  real, 1, 3
  signed, 108
  whole, 1
Numerator, 27, 162, 360
Numerical coefficient, 131

Oblique triangles, 255, 504–533
Obtuse angle, 248
Obtuse triangle, 255
Opposite of a number, 109
Opposite side, 312
Order of a determinant, 427
Ordinate, 214
Ordered pairs of numbers, 207
  graphing of, 215, 216
Order of operations, 22
Origin, 214

Pairs, ordered, 207
Parabola, 227, 496
Parallel lines, 250
Parallelogram, 268
Parallelogram law for vectors, 543
Parentheses (see Grouping)
Percent:
  percentage, 94
  change, 101
  concentration, 104, 199
  efficiency, 105
  error, 104

Perfect square trinomials (see Trinomial)
Perimeter, 254, 257, 269
Period, 575
Perpendicular lines, 250
Phase shift, 586
Pi (n), 274
Place value, 3, 50
Plane, complex, 684
Plane geometry, 244–279
Plotting (see Graph)
Point, decimal, 3, 55
Polar form of a complex number, 686
Polygon, 254
Polynomial, 136, 151, 167
Power(s):
  by calculator, 17, 64, 121
  of denominate numbers, 86
  logarithm of, 657
  of multinomials, 156
  multiplication of, 141
  raised to a power, 142
  of signed numbers, 118
  of ten, 20, 124
  of whole numbers, 17
Prefixes, SI, 77
  table of, 78
Prime:
  factoring, 32
  factors, 42, 332
  numbers, 32
Principal diagonal, 427, 432
Principal root, 122
Prism, 280
Product, 12
Proper fraction, 29
Proportion, 606
Proportionality, constant of, 614
Protractor, 246
Pure quadratic, 470
Pyramid, 290
Pythagorean theorem, 261, 312

Quadrant, 214
Quadrantal angle, 504, 507
Quadratic equation(s), 468–503, 678
Quadratic formula, 480
Quadratic function, 496
Quadratic trinomial (see Trinomial)
Quadrilateral, 267
Quotient, (see also Division) 15, 28, 142, 162

Radian(s), 246, 560
Radical(s), 444–460
Radical equations, 460
Radical form, 440
Radical sign, 18
Radicand, 18
Radius, 274
Range, 211
Rate(s):
  applications of, 492, 609
  conversion of, 90
  of a percentage, 94, 97
Ratio, 15, 162, 602
  dimensionless, 603
  trigonometric (see Trigonometric function)
Rationalizing the denominator, 447
Rational numbers,
Ray, 244
Real numbers, 2
Reciprocal(s), 37, 62
  by calculator, 62

Reciprocal relationships, 327
Rectangle, 269
Rectangular components, 535
Rectangular coordinates, 214
Rectangular form of a complex number, 676
Rectangular solid, 283
Reference angle, 512
Regular polygon, 254
Repeating decimal, 15
Resolution of vectors, resultant, 535, 552
Revolution, 245
Rhombus, 270
Right angle, 248
Right cone, 293
Right cylinder, 286
Right prism, 280
Right triangle(s), 261, 304–331
  solution of 311–318
Roots
  by calculator, 18, 65, 122
  complex, 678
  of decimal numbers, 65
  of denominate numbers, 86
  of equations, 240
  logarithm of, 658
  principal, 122
  of a product, 444
  of quadratics (see Quadratic equation)
  of a quotient, 447
  of signed numbers, 122
  of whole numbers, 18
Rotation, 568
Rounding, 56
Rules of signs:
  for addition and subtraction, 111, 112
  for division, 117, 162
  for fractions, 361
  for multiplication, 114, 148

Scalar, 534
Scale drawings, 604
Scalene triangle, 255
Scientific notation, 20, 67, 124
  on the calculator, 68, 125
Secant of an angle, 326
Second of arc, 246
Secondary diagonal, 427, 432
Second order determinants, 427
Sector, 275
  area of, 565
Segment line, 244
Segment of a circle, 275
Semicircle, 275
Sets of equations (see Systems of linear equations)
Side of an equation, 172
Signed number, 108
Significant digits, 55, 82
Signs of fractions (see Rules of signs)
Signs of the trigonometric functions, 506
Signs, rules of (see Rules of signs)
Similar figures, 609
Similar fractions, 30
Similar radicals, 450
Simultaneous equations (see Systems of linear equations)
Sine of an angle, 306, 312
Sine curve, 573
Sines, law of, 517

SI:
  base units, 77
  converting areas and volumes, 119
  converting within, 78
  converting within, using exponents, 119
  converting with Brit/Am., 84
  derived units, 77
  prefixes, 77, 78
  table of prefixes, 78
  table of units allowed with, 79
Slant height of cone, pyramid, 291, 293
Slope, 221
Slope-intercept form, 221
Solid geometry, 280–302
Special products, 153, 332
Specific gravity, 606
Sphere, 299
Square, 271
Square root (see Roots)
Squares, difference of two, 153, 339
Standard position for angles, 304, 504
Substitution:
  into equations, 173
  into formulas, 174, 391
  into functions, 212
Substitution method, 413
Subtraction (see Addition and subtraction)
Sum or difference of two cubes, 356
Supplementary angles, 249
Symbols of grouping (see Grouping)
Systems of linear equations, 364–438
  by determinants, 427–435

Tables:
  of conversion factors, 702–705
  of units allowed with SI, 79
  of facts and formulas, 693–701
  of SI (metric) unit prefixes, 78
Tangent:
  of an angle, 306, 312
  to a circle, 275
Tangential component, 536
Terminal position 245, 304, 504
Term of an expression, 131
Terms, like, 137
Third-order determinants, 432
Trace, 230
Transversal, 250
Trapezoid, 268
Triangle(s), 255–267
  area of, 257
  oblique (see types of), 255
Trigonometric form of a complex number, 689
Trigonometric function(s), 305, 505
  by calculator, 307
  graphs of 573–600
Trigonometric ratio (see Trigonometric function)
Trinomial, 151, 343
  factoring of, 343–355
  perfect square, 156, 353
  quadratic, 154, 343
  test for factorability, 349

Uniform motion. 396, 492
  circular, 568
Unit, imaginary, 670
Unit circle, 573

Units of measure, 77–93
  angular, 245
  conversion of, 78
  table of, and conversion factors, 702
Unknown, 188

Value:
  absolute, 29, 109
  of a determinant, 428
Variables, 131
  dependent and independent, 206
Variation:
  combined, 622
  direct, 613
  functional, 613
  inverse, 620
  joint, 619
Vector(s), 534–559
  addition by components, 549
  addition by graphical method, 542
  addition by laws of sines and cosines, 554
  addition by tip to tail method, 544
  addition of perpendicular vectors, 546
  components of (x,y), 537
  definition of, 534
  direction of, 535
  magnitude of, 534, 535
  tail of, 535
  tip of, 535
Vector sum (see Addition and subtraction, of vectors)
Velocity
  angular, 568
  linear, 570
Vertex:
  of an angle, 244
  of a cone, 293
  of a parabola, 227, 496
  of a polygon, 254, 255
Vertical angles, 249
Vinculum, 8
Volume(s):
  conversion of, 89
  of a cone, 293
  of cube, 284
  of cylinder, 287
  of frustums, 295
  of prism, 280
  of pyramid, 290
  of rectangular solid, 283
  of sphere, 299

Whole numbers, 1
Window, calculator, 236
Word problems, solution of, 187–203
Work problems, 394

Zero: 1
  adding or subtracting, 10
  denominator, 28
  division by, 28, 162
  multiplication by, 14, 28
  numerator, 28
  of a function, 240
  overscore (overbar) on a, 56
  place holder, 55
  significant digit, 55
  value, 1
Zero exponent, 144
Zero product principle, 458, 475
Zoom, 239